# 2018 14th Conference on Ph.D. Research in Microelectronics and Electronics (PRIME 2018)

Prague, Czech Republic
2-5 July 2018

IEEE Catalog Number: CFP18622-POD
ISBN: 978-1-5386-5388-3

**Copyright © 2018 by the Institute of Electrical and Electronics Engineers, Inc.
All Rights Reserved**

*Copyright and Reprint Permissions*: Abstracting is permitted with credit to the source. Libraries are permitted to photocopy beyond the limit of U.S. copyright law for private use of patrons those articles in this volume that carry a code at the bottom of the first page, provided the per-copy fee indicated in the code is paid through Copyright Clearance Center, 222 Rosewood Drive, Danvers, MA 01923.

For other copying, reprint or republication permission, write to IEEE Copyrights Manager, IEEE Service Center, 445 Hoes Lane, Piscataway, NJ 08854. All rights reserved.

***\*\*\* This is a print representation of what appears in the IEEE Digital Library. Some format issues inherent in the e-media version may also appear in this print version.***

IEEE Catalog Number:      CFP18622-POD
ISBN (Print-On-Demand):   978-1-5386-5388-3
ISBN (Online):            978-1-5386-5387-6

**Additional Copies of This Publication Are Available From:**

Curran Associates, Inc
57 Morehouse Lane
Red Hook, NY 12571 USA
Phone:      (845) 758-0400
Fax:        (845) 758-2633
E-mail:     curran@proceedings.com
Web:        www.proceedings.com

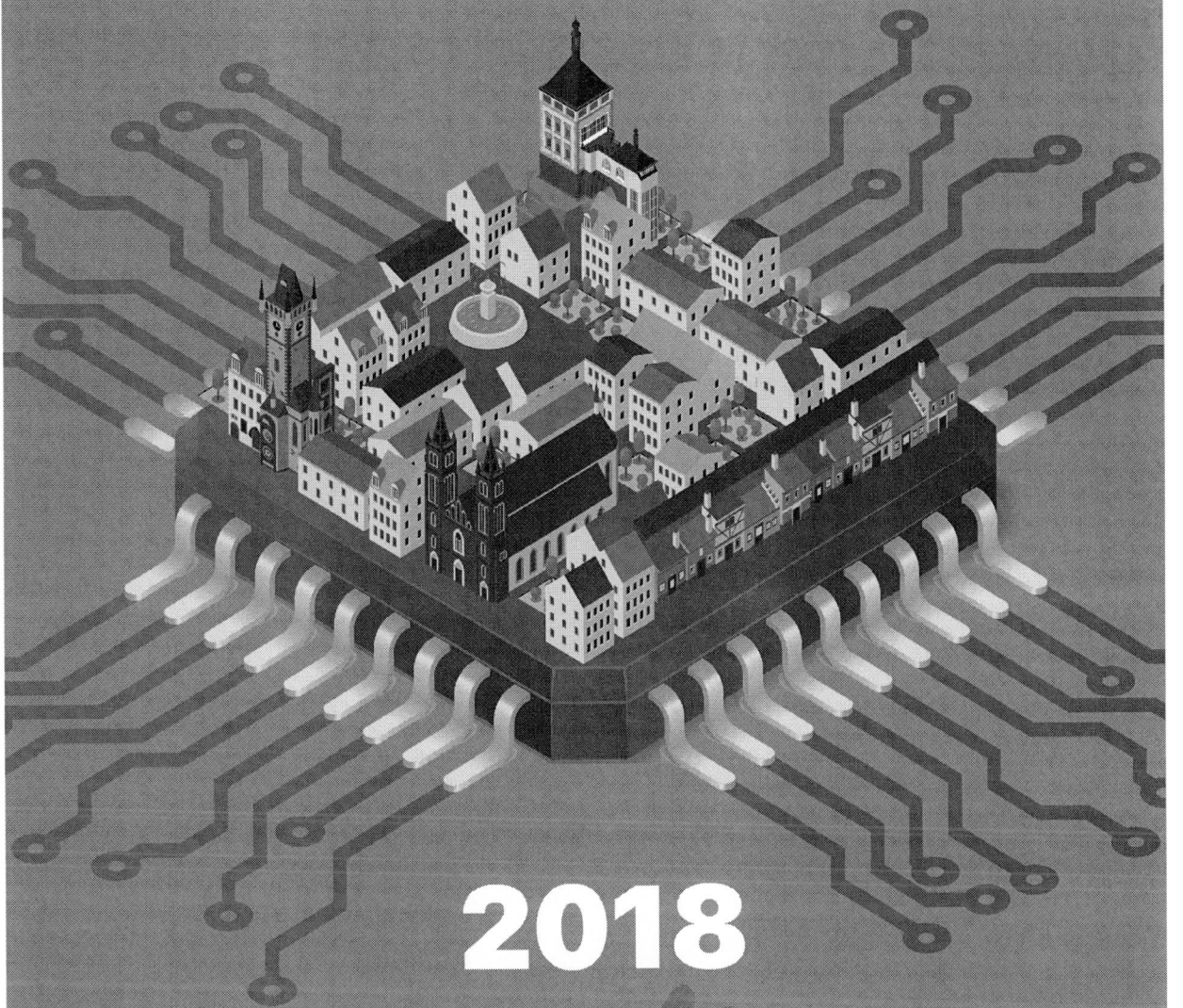

**July 2nd - July 5th, 2018**
**PRAGUE, Czech Republic**

# 2018

# PRIME

**14th Conference on PhD Research**
**in Microelectronics and Electronics**

## Gold Sponsors

## Silver Sponsors

## Bronze Sponsors

## Other Sponsors

## Technical Sponsorships

## Institutional Sponsors and Organizers

Czech Technical University in Prague

Faculty of Nuclear Sciences

*PRIME 2018, Prague, Czech Republic*

# Table of Contents

**Welcome** xiii

**Committees** xv

**Company Fair** xix

**Social Events** xxi

**Plenaries** xxiii

**Regular Papers**

**Session: Analog Circuits I**

**Improved Class-AB Output Stage for Sub-1-V Fully-Differential Operational Amplifiers**
*A. Ria, S. Del Cesta, A. Catania, M. Piotto, P. Bruschi*
University of Pisa, Italy 1

**A Fully-Differential, 200-MHz, Programmable Gain, Level-Shifting, Hybrid Amplifier/Power Combiner/Test Buffer, Using Pre-Distortion for Enhanced Linearity**
*V. Kampus[1], T. Rang[1], D. Knaller[2], C. Fleischhacker[2], M. Korak[2], J. Kiss[2]*
[1]Tallinn University of Technology, Estonia, [2]Intel, Austria 5

**New Resistor-Free Current Mode Wheatstone Bridge Topologies with Intrinsic Linearity**
*L. Safari, G. Barile, V. Stornelli, G. Ferri, A. Leoni*
University of L'Aquila, Iran 9

**A Fully Differential Charge-Sensitive Amplifier for Dust-Particle Detectors**
*S. Kelz, T. Veigel, M. Grözing, M. Berroth*
University of Stuttgart, Germany 13

**A ZTC-based 0.5-V CMOS Voltage Reference**
*Y. Wenger, B. Meinerzhagen*
TU Braunschweig, Germany 17

**Session: Data Converters**

**Generalization of Reference-Less Timing Mismatch Calibration Methods for Time-Interleaved ADCs**
*A. Uran, M. Kilic, Y. Leblebici*
EPFL, Switzerland 21

**A 6.5-Œ°W 70-dB 0.18-Œ°m CMOS Potentiostatic Delta-Sigma for Electrochemical Sensors**
*J. Aymerich, M. Dei, L. Terés, F. Serra-Graells*
IMB-CNM (CSIC), Spain 25

**11.7-Bit Time-to-Digital Converter with 0.82-ps Resolution in 130-nm CMOS Technology**
R. Granja[1], M. Santos[1], J. Guilherme[2], N. Horta[3]
[1]Instituto de Telecomunicações, Portugal, [2]Instituto Politécnico de Tomar, Portugal, [3]Instituto Superior Técnico Lisboa, Portugal
29

**A High-Resolution Δ-Modulator ADC with Oversampling and Noise-Shaping for IoT**
A. Correia[1], P. Barquinha[1], J. Marques[2], J. Goes[1]
[1]Universidade NOVA de Lisboa, Portugal, [2]S3 Semiconductors, Portugal
33

**A 12.4-fJ-FoM 4-Bit Flash ADC Based on the Strong-Arm Architecture**
A. Almansouri, A. Alturki, H. Fariborzi, K. Salama, T. Al-Attar
King Abdullah University of Science and Technology, Saudi Arabia
37

## Session: Modeling, Optimization, and Characterization

**A Low-Power Voice Activity Detector for Portable Applications**
G. Meoni, L. Pilato, L. Fanucci
University of Pisa, Italy
41

**Fractional-Order Hartley Oscillator**
A. Agambayev[1], A. Kartci[2], A. Hassan[1], N. Herencsar[2], H. Bagci[1], K. Salama[1]
[1]King Abdullah University of Science and Technology, Saudi Arabia, [2]Brno University of Technology, Czech Republic
45

**System-Level Simulation Framework for the ASICs Development of a Novel Particle Physics Detector**
A. Caratelli[1], S. Scarfi[1], D. Ceresa[2], K. Kloukinas[2], Y. Leblebici[1]
[1]EPFL, Switzerland, [2]CERN, Switzerland
49

**A Bayesian Indicator for Run-to-Run Performance Assessment in Semiconductor Manufacturing**
T. Korabi[1, 3], G. Graton[2, 3], E. El Adel[3], M. Ouladsine[3], J. Pinaton[1]
[1]STMicroelectronics, France, [2]Ecole Centrale Marseille, France, [3]University of Aix-Marseille, France
53

**UTBB FD-SOI Circuit Design Using Multi-Finger Transistors: A Circuit-Device Interaction Perspective**
A. Sharma[1], N. Alam[2], A. Bulusu[1]
[1]Indian Institute of Technology Roorkee, India, [2]Aligarh Muslim University, India
57

**Let's Make it Noisy: A Simulation Methodology for Adding Intrinsic Physical Noise to Cryptographic Designs**
K. Nawaz, L. Van Brandt, F. Standaert, D. Flandre
KU Leuven, Belgium
61

## Session: Circuits for Memories and Security

**Analysis on Sensing Yield of Voltage Latched Sense Amplifier for Low Power DRAM**
S. Kim, B. Song, T. Oh, S. Jung
Yonsei University, South Korea
65

**The Key Impact of Incorporated $Al_2O_3$ Barrier Layer in W-Based ReRAM Switching Performance**
E. Shahrabi[1], C. Giovinazzo[2], J. Sandrini[1], Y. Leblebici[1]
[1]EPFL, Switzerland, [2]Politecnico di Torino, Italy
69

**A Variability-Aware Analysis and Design Guideline for Write and Read Operations in Crosspoint STT-MRAM Arrays**
Y. Belay, A. Cabrini, G. Torelli
University of Pavia, Italy
73

**A Simulated Approach to Evaluate Side Channel Attack Countermeasures for the Advanced Encryption Standard**
L. Sarti, L. Baldanzi, L. Crocetti, B. Carnevale, L. Fanucci
University of Pisa, Italy
77

**Enabling Secure Boot Functionality by Using Physical Unclonable Functions**
K. Müller, R. Ulrich, A. Stanitzki, R. Kokozinski
Fraunhofer IMS Duisburg, Germany
81

**Encryption of Test Data: Which Cipher is Better?**
M. Da Silva, E. Valea, M. Flottes, S. Dupuis, G. Di Natale, B. Rouzeyre
LIRMM, France
85

## Session: Reliability and Resiliency

**Increasing EM Robustness of Placement and Routing Solutions Based on Layout-Driven Discretization**
S. Bigalke[1], T. Casper[2], S. Schöps[2], J. Lienig[1]
[1]TU Dresden, Germany, [2]TU Darmstadt, Germany
89

**Characterising Soft-Failures in Component-Level ESD Testing**
P. Schrey
Graz University of Technology, Austria
93

**Torus Topology Based Fault-Tolerant Network-on-Chip Design with Flexible Spare Core Placement**
P. Veda Bhanu[1], P. Kulkarni[1], J. Soumya[1], L. Cenkeramaddi[2], H. Idsøe[2]
[1]BITS-PILANI Hyderabad, India, [2]University of Agder, Norway
97

**Design and Analysis of Energy-Efficient Self-Correcting Latches Considering Metastability**
C. Indra Kumar, A. Bulusu
Indian Institute of Technology Roorkee, India
101

## Session: Power Circuits and Harvesting

**A CMOS Gate Driver with Ultra-Fast dV/dt Embedded Control Dedicated to Optimum EMI and Turn-On Losses Management for GaN Power Transistors**
P. Bau[1, 2], M. Cousineau[1], B. Cougo[2], F. Richardeau[1], D. Colin[3], N. Rouger[1]
[1]University of Toulouse, France, [2]IRT Saint Exupéry, France, [3]Association des industriels PRIMES, France
105

**Design of a SIBO DC-DC Converter for AMOLED Display Driving**
F. Boera, A. Salimath, E. Bonizzoni, F. Maloberti
University of Pavia, Italy
109

**A Human Body Powered Sensory Glove System Based on Multi-Source Energy Harvester**
A. Leoni[1], V. Stornelli[1], G. Ferri[1], V. Errico[2], M. Ricci[2], A. Pallotti[2], G. Saggio[2]
[1]University of L'Aquila, Italy, [2]University of Roma Tor Vergata, Italy
113

**HW Platform for BMS Algorithm Validation**
L. Buccolini, F. Garbuglia, M. Unterhorst, M. Conti
Università Politecnica delle Marche, Italy
117

## Session: Analog Circuits II

**Ultra Low Frequency Low Power CMOS Oscillators for MPPT and Switch Mode Power Supplies**
F. Galea, O. Casha, I. Grech, E. Gatt, J. Micallef
University of Malta, Malta
121

### Analysis of Gain and Bandwidth Limitations of Operational Amplifiers in Sigma-Delta Modulators

T. Saalfeld, A. Meyer, E. Schulte Bocholt, R. Wunderlich, S. Heinen

RWTH Aachen University, Germany

125

### Design of a Low-Power Ultrasound Transceiver for Underwater Sensor Networks

G. Berkol, P. Baltus, P. Harpe, E. Cantatore

Eindhoven University of Technology, Netherlands

129

### On the Design of a Linear Delay Element for the Triggering Module at CERN LHC

J. Gauci[1], E. Gatt[1], O. Casha[1], G. De Cataldo[2], I. Grech[1], J. Micallef[1]

[1]University of Malta, Malta, [2]INFN Bari, Italy

133

## Session: Radio Frequency Circuits and Systems I

### A 4-W 37.5-42.5-GHz Power Amplifier MMIC in GaN on Si Technology

F. Costanzo, R. Giofrè, A. Salvucci, G. Polli, E. Limiti

University of Roma Tor Vergata, Italy

137

### An Integrated Power Detector for a 5-GHz RF PA

V. Qunaj, U. Celik, P. Reynaert

KU Leuven, Belgium

141

### Design of an E-Band Doherty Power Amplifier

M. Najmussadat[1], R. Ahamed[1], D. Parveg[2], M. Varonen[2], K. Halonen[1]

[1]Aalto University, Finland, [2]VTT Technical Research Centre of Finland, Finland

145

### A Novel Multi-level CMOS Switching Mode Amplifier for Mobile Communication Signals

R. Bieg, M. Schmidt, M. Grözing, M. Berroth

University of Stuttgart, Germany

149

### Down-Converter Solutions for 77-GHz Automotive Radar Sensors in 28-nm FD-SOI CMOS Technology

C. Nocera[1], A. Cavarra[1], G. Papotto[2], E. Ragonese[1], G. Palmisano[1]

[1]University of Catania, Italy, [2]STMicroelectronics, Italy

153

## Session: Digital Circuits and Sub-Systems

### A Novel Very Low Voltage Topology to Implement MCML XOR Gates

D. Bellizia[1], G. Palumbo[2], G. Scotti[1], A. Trifiletti[1]

[1]University of Roma La Sapienza, Italy, [2]University of Catania, Italy

157

### VLSI Design of Frequent Items Counting Using Binary Decoders Applied to 8-Bit per Item Case-Study

K. Inoue[1], T. Hoang[2], X. Nguyen[2], H. Nguyen[2], C. Pham[2]

[1]Advanced Original Technologies, Japan, [2]University of Electro-Communications, Japan

161

### Multi-Stage Complex Notch Filtering for Interference Detection and Migitation to Improve the Acquisition Performance of GPS

S. Arif, A. Coskun, I. Kale

University of Westminster, United Kingdom

165

### Stall-Aware Fixed-Point Implementation of LMS Filters

D. Esposito, G. Di Meo, D. De Caro, E. Napoli, N. Petra, A. Strollo

University of Napoli Federico II, Italy

169

### A Space-Fibre Multi-Lane CODEC System-on-a-Chip: Enabling Technology for Low-Cost Satellite EGSE

P. Nannipieri[1], G. Dinelli[1], D. Davalle[2], L. Fanucci[1]

[1]University of Pisa, Italy, [2]IngeniArs, Italy

173

**Design and Implementation of a Complete Test Equipment Solution for Space-Wire Links**
A. Marino, L. Dello Sterpaio, L. Fanucci
University of Pisa, Italy
177

## Session: Radio Frequency Circuits and Systems II

**A Novel Hybrid Polar-I/Q Modulation Method Relaxing RF Phase Modulator Design Requirements**
T. Buckel[1], P. Preyler[1], E. Hager[1], T. Mayer[1], S. Tertinek[1], A. Springer[2], R. Weigel[3]
[1]Intel, Austria, [2]University of Linz, Austria, [3]University of Erlangen-Nuremberg, Austria
181

**A Sub-1V, 72-Œ°W Stacked LNA-VCO for Wireless Sensor Network Applications**
E. Kargaran, D. Manstretta, R. Castello
University of Pavia, Italy
185

**Low-Power Locking Detector for Frequency Calibration of Multi-Frequency Injection Locked Oscillators**
A. Boulmirat, C. Jany, A. Siligaris, J. Gonzalez Jimenez
CEA-TECH, France
189

**A Novel True Logarithmic Amplifer in 0.25-$\mu$m GaN on SiC Technology for Radar Applications**
A. Salvucci, M. Vittori, S. Colangeli, G. Polli, E. Limiti
University of Roma Tor Vergata, Italy
193

**Ka-/V-Band Self-Biased LNAs in 70-nm GaAs/InGaAs Technology**
G. Polli, M. Vittori, W. Ciccognani, S. Colangeli, F. Costanzo, A. Salvucci, E. Limiti
University of Roma Tor Vergata, Italy
197

**Single MMIC Receivers for C-band T/R Module in 0.25-$\mu$m GaN Technology**
A. Salvucci, G. Polli, A. De Padova, S. Colangeli, F. Costanzo, W. Ciccognani, E. Limiti
University of Roma Tor Vergata, Italy
201

## Session: Sensing and Biomedical Circuits I

**Decreasing the Actuation Voltage in Electrowetting on Dielectric with Thin and Micro-Structured Dielectric**
S. Türk[1], E. Verheyen[1], R. Viga[1], S. Allani[2], A. Jupe[2], H. Vogt[2]
[1]University of Duisburg-Essen, Germany, [2]Fraunhofer IMS Duisburg, Germany
205

**Exploiting Non-Linearities to Improve the Linear Region in an Electrostatic MEMS Demodulator**
J. Scerri, B. Portelli, I. Grech, E. Gatt, O. Casha
University of Malta, Malta
209

**Multi-Channel Electrotactile Stimulation System for Touch Substitution: A Case Study**
H. Fares[1], L. Seminara[1], H. Chible[2], S. Dosen[3], M. Valle[1]
[1]University of Genova, Italy, [2]Lebanese University, Lebanon, [3]Aalborg University, Denmark
213

**Modeling of a Capacitive Sensor Dedicated to Drug Injection**
S. Joly[1], A. Lepple-Wienhues[1], C. Dehollain[2]
[1]Valtronic Technologies, Switzerland, [2]EPFL, Switzerland
217

**Lock-In Based Differential Front-End for Raman Spectroscopy Applications**
A. Ragni, G. Ferrari, G. Sciortino, M. Sampietro, D. Polli, V. Kumar, F. Crisafi
Politecnico di Milano, Italy
221

## Session: Automotive Circuits and Systems

**Electric Vehicle Battery Management Sytem Using Power Line Communication Technique**
A. Pake[1], W. Pribyl[2], G. Hofer[1]
[1]Infineon Technologies, Austria, [2]Graz University of Technology, Austria
225

**Evaluation of Front-End Readout Circuits for High Performance Automotive MEMS Accelerometers**
A. Lanniel[1], T. Alpert[1], T. Boeser[1], M. Ortmanns[2]
[1]Bosch, Germany, [2]University of Ulm, Germany
229

**Real-Time Defect Detection of Wheel Bearing by Means of a Wirelessly Connected Microphone**
E. Raviola, F. Fiori
Politecnico di Torino, Italy
233

**Multi-Object Detection in Direct Time-of-Flight Measurements with SPADs**
J. Haase, M. Beer, J. Ruskowski, H. Vogt
Fraunhofer IMS Duisburg, Germany
237

## Session: Sensing and Biomedical Circuits II

**Super-Capacitors for Implantable Medical Devices with Wireless Power Transmission**
P. Mendoza Ponce, B. John, D. Schroeder, W. Krautschneider
TU Hamburg-Harburg, Germany
241

**Design and Modelling of a Super-Regenerative Receiver for Medical Implant Devices**
N. Pekçokgüler[1], G. Dündar[1], C. Dehollain[2]
[1]Bogazici University, Turkey, [2]EPFL, Switzerland
245

**A Fully Fail-Safe Capacitive-Based Charge Metering Method for Active Charge Balancing in Deep Brain Stimulation**
R. Ranjandish[1], O. Shoaei[2], A. Schmid[1]
[1]EPFL, Switzerland, [2]University of Tehran, Iran
249

**Current Controlled CMOS Stimulator with Programmable Pulse Pattern for a Retina Implant**
P. Raffelberg[1], R. Burkard[1], R. Viga[1], W. Mokwa[2], P. Walter[2], A. Grabmaier[3], R. Kokozinski[3]
[1]University of Duisburg-Essen, Germany, [2]RWTH Aachen University, Germany, [3]Fraunhofer IMS Duisburg, Germany
253

**A Laser Diode-Based Wireless Optogenetic Headstage**
A. Mesri[1], A. Cunha[2], Ø. Martinsen[2], M. Sampietro[1], G. Ferrari[1]
[1]Politecnico di Milano, Italy, [2]University of Oslo, Norway
257

**An Active Charge Balancing Method Based on Chopped Anodic Phase**
R. Ranjandish, A. Schmid
EPFL, Switzerland
261

## Session: Emerging and Non-CMOS Technologies

**Fabrication of Full-3D Printed Electronics RF Passive Components and Circuits**
A. Salas Barenys[1], N. Vidal[1], J. Sieiro[1], J. Lopez Villegas[1], B. Medina[2], F. Ramos[2]
[1]University of Barcelona, Spain, [2]Francisco Albero S.A.U. (FAE), Spain
265

**Modeling and Simulation of Novel GaN-based Light Emitting Transistor for Display Applications**
S. Lee, I. Yun
Yonsei University, South Korea
269

**Improving Deep Learning with a Customizable GPU-Like FPGA-Based Accelerator**
M. Gagliardi, E. Fusella, A. Cilardo
University of Napoli Federico II, Italy
273

**Analysis and Verification of Identical-Order Mixed-Matrix Fractional-Order Capacitor Networks**
*A. Kartci[1], A. Agambayev[2], N. Herencsar[1], K. Salama[2]*
[1]Brno University of Technology, Czech Republic, [2]King Abdullah University of Science and Technology, Saudi Arabia                                                                277

# Author Index                                                                281

*PRIME 2018, Prague, Czech Republic*

# Welcome

On behalf of the Organizing Committees of **SMACD 2018** and **PRIME 2018**, it is our pleasure and honor to welcome you to the **15th International Conference on Synthesis, Modeling, Analysis, and Simulation Methods and Applications to Circuit Design** and the **th Conference on Ph. D. Research in Microelectronics and Electronics**.

PRIME has been established over the recent years as the most important conference where Ph. D. students and post-docs with less than one-year post-Ph. D. experience can present their research results and network with experts from industry, academia, and research.

SMACD is an international conference focused on Design Methods and Tools for Analog, Mixed-signal, RF (AMS/RF), and multi-domain (MEMS, nanoelectronic, optoelectronic, biological, etc.) integrated circuits and systems. SMACD's mission has also taken shape through the EDA Competition. Now in its 5th edition, the Competition is the perfect setting for students to pitch their ideas on Design Automation to a Jury of renowned experts from the Industry and the Academia.

As for the past two editions, PRIME 2018 and SMACD 2018 are running in parallel, broadly covering many aspects of modern electronic systems, circuits, devices and modeling. The Venue for the Conferences, the Faculty of Nuclear Sciences and Physical Engineering (from Czech Technical University in Prague), is located amidst the best spots in Prague: The Old Town (Staré Město). There, streets full of history and architecture open up to one of the most beautiful scenes in Europe, the Prague Castle and St. Vitus Cathedral, right across the Vltava river through the quintessential, 14th-century Charles Bridge.

PRIME 2018 received a total of 101 submissions from 25 countries, 15 of which in Europe, testifying the international relevance the conference is gaining year after year. The Technical Program Committee, led by Jiri Jakovenko, Gianluca Giustolisi, and Viera Stopjakova, with the support of 137 recognized experts from academia and industry, selected 70 papers with an acceptance rate of around 70%. All manuscripts received at least 3 reviews (and 59 manuscripts more than 4 reviews).

SMACD 2018 received a total of 104 submissions from 29 different countries, from Europe, Asia, Americas, and Africa. The 53 members of the Technical Program Committee, led by the Technical Program Committee Chairs, Francisco V. Fernandez, Nuno Horta, Günhan Dündar and Giulia di Capua, with the support of 41 recognized expert reviewers, have carried out the always difficult job of selecting 61 papers for Oral presentation and 9 papers for Poster presentation. SMACD 2018 includes 3 Special Sessions, on "Modeling, Design and Control of Power Converters with Non-Linear Passive Power Components" organized by Giulia di Capua, Nicola Femia, Alberto Oliveri, and Marco Storace, on "New Solutions for Analog and Radio-Frequency Layout Synthesis" organized by Ricardo Martins and Nuno Lourenço, and on the "Latest Advances in Variability Impact on Devices and Circuits Functionality" organized by Hussam Amrouch and Rosana Rodríguez. In this year's SMACD EDA Competition, we have selected 7 contestants, who will present and Live Demo their ideas and solutions in 2 sessions on July 3rd and July 4th.

All accepted papers will be submitted for inclusion in IEEEXplore. The most outstanding contributions to PRIME will be awarded with the Gold, Silver, and Bronze Leaf certificates as typical in this conference. The most outstanding contributions to SMACD will be awarded with 3 Best Paper Certificates (Best Paper, Runner-Up Best Paper, and Honorable Mention). The winner of the EDA Competition will take home not only the Best EDA Demo Certificate, but the invaluable recognition and feedback of the Jury and the stimulating IEEE CEDA-sponsored prize of $1,000. The winners of all these awards will be selected during the Conferences and announced on the last day (July 5th), during the Closing Session. Additionally, a selection of the very best contributions will be invited to submit extended versions for a special issue of "Integration, the VLSI

Journal", from Elsevier, reserved to PRIME and SMACD.

The program will include three days of presentations and will be preceded by Tutorial half-day (July 2$^{nd}$), with two brilliant talks, from great experts in the fields, about Reliability and Power Management. Moreover, SMACD & PRIME 2018 are proud to feature 4 Plenary Talks, given by prominent experts: Mr. Marcel Urban (ams AG), Prof. Sachin S. Sapatnekar (University of Minnesota), Prof. Richard Shi (University of Washington) and Mr. Roger Panicacci (ON Semiconductor)

We are truly indebted with an exceptional team of volunteers in the Organizing Committees and in the Technical Program Committees. Our warmest thanks go to all of them for their commitment and hard work. We would also like to thank the Steering Committee of PRIME and SMACD, for giving us the opportunity to organize this event and supporting us in every step of the way. Our deepest thanks certainly must go to our friends in Prague, Varclav Vrba and Michal Marcisovsky, for making it possible to have our conference in such a wonderful Venue. Likewise, we must take our hats off to Jakub Uher, from AMCA, for nothing would have run as smoothly as it did. Last but certainly not least, we owe special thanks to the past General Chairs, Giulia di Capua and Salvatore Pennisi, for sharing their precious experience with us.

We thank our Technical sponsors, IEEE, IEEE Circuits and Systems Society, and IEEE Council on Electronic Design Automation. We also thank our Institutional sponsorships from Czech Technical University in Prague, the Faculty of Nuclear Sciences and Physical Engineering of CTU in Prague, and the Instituto de Microelectrónica de Sevilla.

Moreover, we have been tremendously lucky to have the support of our generous Sponsors: Dialog Semiconductor, Allegro MicroSystems LLC and STMicroelectronics, as Gold Sponsors, ON Semiconductor, ASI-Centrum, S3 Semiconductors, and Infineon Technologies, as Silver Sponsors, and Coilcraft, ams AG and Europractice, as Bronze Sponsors. We also want to thank River Publishers for its generous donation.

Finally, we would like to express our greatest and most sincere appreciation to all the Authors who submitted their papers to the Conferences, to all the Delegates and to the Plenary Speakers.

Our wish is that you fully enjoy the 2018 edition of PRIME and SMACD. Our aim is that you have the greatest time meeting old friends and making new ones. Our hope is that you also find the time to explore this gem of a city. Our aspiration is that, after saying goodbye, you will think back of the Conferences and Prague in the same way as Richard Wagner did: "the ancient splendor and beauty of Prague, a city like no other, left an impression on my imagination that will never fade away".

Welcome! Vítejte!


Rafael Castro
SMACD 2018 General Chair

Alberto Gola
PRIME 2018 General Chair

*PRIME 2018, Prague, Czech Republic*

# Committees

## Local SMACD and PRIME Chairman

- Varclav Vrba, Czech Technical University, Czech Republic

## General Chairman

- Alberto Gola, ams Italy srl, Italy

## Technical Program Committee Chairmen

- Jiri Jakovenko, Czech Technical University, Czech Republic
- Gianluca Giustolisi, University of Catania, Italy
- Viera Stopjakova, Slovak Technical University, Slovakia

## Publication Chairmen

- Piero Malcovati, University of Pavia, Italy
- Oleksandr Korchak, Czech Technical University, Czech Republic

## Local Arrangement Chairmen

- Michal Marcisovsky, Czech Technical University, Czech Republic
- Jakub Uher, AMCA, Czech Republic

## Steering Committee

- Franco Maloberti, University of Pavia, Italy
- Catherine Dehollain, EPFL, Switzerland
- Alberto Gola, ON-Semiconductors, Czech Republic
- Bernd Deutschmann, Graz University of Technology, Austria
- Elena Blokhina, University College Dublin, Ireland
- Gunhan Dundar, Bogazici University, Turkey
- Anton Klotz, Cadence, Germany
- Ravinder Dahiya, University of Glasgow, United Kingdom
- Nuno Horta, University of Lisbon, Portugal
- Salvatore Pennisi, University of Catania, Italy

## Technical Program Committee

- Nidhi Agrawal, Micron Technology
- Aisha Alhammadi, University of Sharjah
- Federico Alimenti, University of Perugia
- Judy Amanor-Boadu, Texas A&M University
- Salvatore Amoroso, Synopsys
- Arbet Arbet, Slovak University of Technology
- Paolo Arena, University of Catania
- Laurent Artola, ONERA
- Sergio Bampi, Federal University of Rio Grande do Sul
- Roc Berenguer, CEIT
- Andoni Beriain, CEIT and Tecnun (University of Navarra)
- Elena Blokhina, University College Dublin
- Edoardo Bonizzoni, University of Pavia
- Francesco Brandonisio, Infineon Technologies
- Alessandro Busacca, University of Palermo
- Paulo F. Butzen, Federal University of Rio Grande do Sul
- Alessandro Cabrini, University of Pavia
- Hao Cai, Telecom Paristech
- Marco Cerchi, ams Italy srl
- Ren Chen, University of Southern California
- Yuan-Ho Chen, Chang Gung University
- Massimo Conti, Università' Politecnica delle Marche
- Fernando Corinto, Politecnico di Torino
- Ravinder Dahiya, University of Glasgow
- Gian-Franco Dalla Betta, University of Trento
- Marcello De Matteis, University of Milano-Bicocca
- Carl Debono, University of Malta
- Catherine Dehollain, EPFL
- Manuel Delgado-Restituto, Institute of Microelectronics of Seville
- Richard Dorrance, Intel
- Li Du, University of California at Los Angeles
- Gunhan Dundar, Bogazici University
- Luca Fanucci, University of Pisa
- Jorge Fernandes, University of Lisbon
- Vittorio Ferrari, University of Brescia
- Giuseppe Ferri, University of L'Aquila
- Carlo Fiocchi, ams Italy srl
- Edoardo Fusella, University of Napoli
- Peter Gadfort, US Army Research Laboratory
- Dimitri Galayko, UPMC-LIP6
- Tzeno Galchev, University of Michigan
- Horst Gieser, Fraunhofer IMS
- Joao Goes, Universidade NOVA de Lisboa
- Marco Grassi, University of Pavia
- Alfio Dario Grasso, University of Catania
- Jorge Guilherme, Instituto Politecnico Tomar
- Xinfei Guo, University of Virginia
- Mohit Kumar Gupta, Krishna Institute of Engineering and Technology
- Ku He, University of Texas at Austin
- Hadi Heidari, University of Glasgow

- Stefan Heinen, RWTH Aachen University
- Robert Henderson, University of Edinburgh
- Frank Henkel, IMST
- Nuno Horta, University of Lisbon
- Jiří Hospodka, CTU
- Po-Tsang Huang, National Chiao Tung University
- Guoxian Huang, University of Connecticut
- Yanxiang Huang, Nvidia
- Mohsen Imani, University of California at San Diego
- Jiří Jakovenko, Czech Technical University
- Vladimir Janicek, Czech Technical University
- Izzet Kale, University of Westminster
- Michael Peter Kennedy, University College Dublin
- Tony Kim, Nanyang Technological University
- Ji-Hoon Kim, Seoul National University of Science and Technology
- Juha Kostamovaara, University of Oulu
- Pavel Kulha, Czech Technical University
- Akhilesh Kumar, ANSYS
- Marco Lanuzza, University of Calabria
- Long Le, University of Illinois at Urbana-Champaign
- Sun Lei, Chinese University of Hong Kong
- Xueqing Li, Tsinghua University
- Chunshu Li, Marvell
- Salvatore Lombardo, CNR-IMM
- Antonio Lopez-Martin, Public University of Navarra
- Leandro Lorenzelli, Fondazione Bruno Kessler
- Marie-Minerve Louerat, Pierre and Marie Curie University
- Nuno Lourenço, Instituto de Telecomunicações
- Chao Lu, Southern Illinois University
- Paolo Madoglio, Intel
- Torsten Maehne, Berner Fachhochschule
- Franco Maloberti, University of Pavia
- Davide Marano, INAF Catania
- Ricardo Miguel Martins, Instituto de Telecomunicações
- Rui Martins, University of Macau
- Nicola Massari, Fondazione Bruno Kessler
- Pascal Meinerzhagen, Intel
- Ivan Miro Panades, CEA-LETI
- David Moran, University of Glasgow
- Dominique Morche, CEA-LETI
- Pierluigi Nuzzo, University of Southern California
- Tom O'Dwyer, Analog Devices
- Giuseppe Palmisano, University of Catania
- Rajesh Pamula, imec
- Luis Parrilla, University of Granada
- Daniele Passeri, University of Perugia
- Roberto Passerone, University of Trento
- Salvatore Pennisi, University of Catania
- Valerio Pisati, ams Italy srl
- Jean-Michel Portal, IM2NP/AMU
- Wolfgang Pribyl, Graz University of Technology

*Committees*                                                                                          *PRIME 2018, Prague, Czech Republic*

- Sai Manoj Pudukotai Dinakarrao, George Mason University
- Ruediger Quay, Fraunhofer IMS
- Egidio Ragonese, University of Catania
- Jaime Ramirez-Angulo, New Mexico State University
- Lodovico Ratti, University of Pavia
- Valerio Re, University of Bergamo
- Patrick Reynaert, Katholieke Universiteit Leuven
- Angel Rodriguez-Vazquez, IMSE-CNM/CSIC and University of Seville
- Nicolas Rouger, University of Toulouse
- Khaled Nabil Salama, KAUST
- Sergio Saponara, University of Pisa
- Ladislav Seliga, ON Semiconductor
- Qixian Shi, imec
- Naohiko Shimizu, Tokai University School of Information and Telecommunication Engineering
- Gilles Sicard, TIMA Laboratory and University of Grenoble
- Pietro Siciliano, CNR-IMM
- Carlos Silva Cardenas, Pontificia Universidad Católica del Perù
- Bala Sivakumar, Qualcomm Technologies
- Stefan Slesazeck, NaMLab
- Hector Solar, CEIT
- Fei Song, Ubilinx Technology
- Gino Sorbello, University of Catania
- Marc Sorel, University of Glasgow
- Viera Stopjakova, Slovak University of Technology
- Roland Thewes, TU Berlin
- Guido Torelli, University of Pavia
- Alberto Tosi, Politecnico di Milano
- Alessandro Trifiletti, University of Rome La Sapienza
- Maurizio Valle, University of Genova
- Elena Ioana Vatajelu, INP-TIMA Laboratory
- Vincenzo Vinciguerra, STMicroelectronics
- Jan Voves, Czech Technical University
- Nicoleta Wacker, Infineon Technologies
- Edward Wasige, University of Glasgow
- Sai Zhang, University of Illinois at Urbana-Champaign
- Domenico Zito, Aarhus University

*PRIME 2018, Prague, Czech Republic*

# Company Fair

On July 3$^{rd}$ and 4$^{th}$ in the afternoon, specific time slots will be dedicated to presentations by selected institutions (companies and research centers). Moreover, every day from July 3$^{rd}$ to 5$^{th}$, during lunch and coffee breaks it will be possible to establish new contacts between Ph. D. students and institutions for the creation of future joint activities (like, for instance, fellowships, internships, or also employment). The following institution confirmed their presence:

- Dialog Semiconductor

- Allegro MicroSystems

- STMicroelectronics

- ON Semiconductor

- S3 Semiconductors

- Infineon Technologies

- ASICentrum

- Coilcraft

- ams AG

- Europractice

## Gold Sponsors

## Silver Sponsors

## Bronze Sponsors

# Social Events

## Welcome Reception

### Tuesday, July 3<sup>rd</sup>, 2018 at 19:00

**Velkoprevorsky Palace**

Near the Maltese Square (Maltézské Náměstí) and in front of the French Embassy (Velkopřevorské Náměstí) there is one of the top sights of Prague: the **John Lennon Wall**. The wall is the property of the Knights of Malta, and they have repainted it several times, but it soon gets covered with more Lennon images, peace messages and inconsequential tourist graffiti. Behind that wall there is an exquisite and quiet garden, part of the **Velkoprevorsky Palace**, property also of the Knights of Malta. There, on Tuesday July 3$^{rd}$, weather permitting, we will be hosting our Welcome Reception for SMACD and PRIME 2018 delegates.

## Conference Dinner

### Wednesday, July 4<sup>th</sup>, 2018 at 19:00

**Lobkowicz Palace at Prague Castle**

The SMACD and PRIME 2018 Conference Dinner will be held on Wednesday July 4$^{th}$ in a stunning, privileged location, with one of the most amazing views of the city: the **Prague Castle** (Pražský Hrad, or just Hrad). Looming above the Vltava's left bank, its serried ranks of spires, palaces, and towers dominate the city centre like a fairy-tale fortress. Within its walls lies a varied and fascinating collection of historic buildings. One of these buildings is the venue of our Conference Dinner: the 16$^{th}$-century **Lobkovický Palác**, housing a private museum known as the Princely Collections, which includes priceless paintings, furniture and musical memorabilia. Before the dinner there will be a panoramic sightseeing tour across Prague.

*PRIME 2018, Prague, Czech Republic*

# Plenaries

## Magnetic and Inductive Position Sensors for Industrial and Automotive Application: From Silicon Design to System Design

### Tuesday, July 3rd, 2018 from 09:00 to 10:00

**Marcel Urban**
*ams AG, Austria*

Intelligent magnetic and inductive position sensor solutions gain more and more market share compared to optical, resolver, and other technologies. High level CMOS integration on digital, analog, and high-voltage periphery elements support an accurate, cost-effective, and reliable sensor solution. This speech will give an overview about the implementation path of a monolithic magnetic sensor IC, together with system level interactions, like magnetics and mechanics. This presentation will cover industrial and automotive use cases of a position sensor IC and will highlight the multidisciplinary interaction in sensor systems. This talk will compare different position sensor technologies and will highlight important functions of integrated circuits. Several new market trends in robotics, automotive, e. g. autonomous driving, and electrification will further drive IC developments and further shrink process technology nodes. Time to market will further shorten development times in engineering areas in general.

   **Marcel Urban** joined ams AG as IC design engineer for magnetic position sensors in 2000. He is currently Head of Product and Application Management in the ams AG business line Position Sensors. Marcel graduated at the Carinthian Institute of Technology located in Villach and received a master degree in electrical engineering. He is leading the application support team for position sensors in Austria and is covering automotive, industrial, and consumer market. Marcel holds several patents for magnetic position sensors and published several technical articles in this domain.

## Challenges in Analog/Mixed-Signal Design Automation

### Wednesday, July 4th, 2018 from 09:00 to 10:00

**Sachin S. Sapatnekar**
*University of Minnesota, USA*

The traditional view of electronic design automation has intensively focused on the design, synthesis, and layout of digital circuits. This perspective has been reinforced by the trends from Moore's law, which have seen digital system complexities grow exponentially, prompting an acute need for efficient design tools and flows. In contrast, analog design has remained largely focused on the expert designer. This world view is now changing, for several reasons. First, several tasks in analog design are now at a point where they can be realistically automated, notably tasks related to layout automation. In advanced finFET technologies, the reduction in the degrees of freedom due to restricted design rules actually makes layout automation easier. Second, the clear distinction between analog and digital designs has blurred, with modern designs seeing a great deal of digital-like circuitry that assists in implementing analog functionalities. For these structures, established techniques from digital system design can carry over to enable design automation. Third, the complexity of the mixed-signal design space makes it difficult for designers to fully comprehend and compensate for the impact

xxiii

*Plenaries*                                                                    *PRIME 2018, Prague, Czech Republic*

of phenomena such as process variations and device aging. Especially under stringent design specifications, these complexities create openings for design automation tools that can complement the knowledge of the expert designer. Thus, analog and mixed-signal design, which has long been the bastion of the expert designer, is projected to be the new frontier in design automation. This talk will present a brief history of prior efforts and will overview the set of opportunities and challenges in this emerging field.

**Sachin Sapatnekar** received the Ph. D. degree from the University of Illinois at Urbana-Champaign in 1992, after which he joined the faculty at Iowa State University. Since 1997, he has been teaching at the University of Minnesota, where he is a Distinguished McKnight University Professor and the Henle Chair in Electrical and Computer Engineering. His research is related to developing CAD techniques for the analysis and optimization of circuit performance, currently focused on both CMOS circuits and spintronics technologies. He has served as Editor-in-Chief of the IEEE Transactions on CAD and General Chair for the ACM/IEEE Design Automation Conference (DAC). He is a recipient of several conference Best Paper Awards, ten-year retrospective Best Paper Awards, the Semiconductor Research Corporation's Technical Excellence Award, and the Semiconductor Industry Association University Research Award, and a Fulbright award. He is a Fellow of the IEEE and the ACM.

# Accelerating Mixed-Signal Design Verification: Turn a SPICE netlist into a SystemVerilog Model

## Thursday, June 5th, 2018 from 09:00 to 10:00

### Richard Shi
*University of Washington, USA*

Design verification is a bottleneck on modern SoC design. Very often, modern SoCs contain both digital blocks and analog blocks. While the metric-driven verification methodology exists for digital verification, analog and mixed-signal design verification relies heavily on SPICE simulation and manual modeling, a process known to be time consuming and error prone. In this talk, we will introduce a new technology that can turn automatically a SPICE netlist into a SystemVerilog model, and thus allows metric-driven digital verification methodologies and tools to be used for analog and mixed-signal design verification. This breakthrough is based on a new theory of signal abstraction developed under a recent DARPA sponsored research program. We will show that how signal-driven abstraction allows various circuit analysis techniques developed in the past several decades including symbolic analysis, Laplace transform, pole/zero extraction and fractional expansion, event-driven analog modeling, interval mathematics, modified nodal analysis, regression, wreal and real number modeling, language compilation, analog assertion, can all be integrated in this unified framework, and to be used in a methodology transparent to designers. Practical examples from 28-Gb/s SerDes design will be used to illustrate the methodology. In particular, this talk will show the design verification of 28-Gb/s serial transceiver link adaptation and equalization, which were not feasible previously. This research has been supported by the US DARPA IRIS program.

**Richard Shi** is currently a Professor in Electrical Engineering at the University of Washington, Seattle, WA, where he joined in 1998. He received a prestigious Doctoral Prize from the Natural Science and Engineering Research Council (NSERC) of Canada and a Governor-General's Silver Medal in 1995 for his Ph. D. Dissertation in computer science. His current research interest includes energy-efficient circuit and system design for sensing, computing, learning and communication. Since 2006, he has directed several DoD-sponsored projects in the area of ADC, PLL, SerDes and LDPC design. Previously, he worked in the area of computer-aided design of mixed-signal integrated circuits, in which for his contribution he was elevated to a Fellow of IEEE in 2005. He received several awards for his research including Donald O. Pederson Best IEEE Transactions on CAD Paper Award, Best Paper Awards from the IEEE/ACM Design Automation Conference, the IEEE VLSI Test Symposium, and the SRC Technical Conference, and an NSF CAREER Award. He has served ten years as an Associate Editor of the IEEE Transactions on Computer-Aided Design of Integrated Circuits and Systems twice as an Associate Editor, once as a Guest Editor, of the IEEE Transactions on Circuits and Systems-II: Analog and Digital Signal Processing, as well as Transactions Briefs. He is a key contributor to the IEEE

*PRIME 2018, Prague, Czech Republic*                                                        *Plenaries*

1076.1 standard on mixed-signal circuit description and simulation language. He co-founded three startups in EDA and one in chip design, which have been acquired. He is a founder of Orora Design Technologies, Inc., a pioneer in automating mixed-signal design verification.

# Innovating Vision Beyond the Human Eye: Navigating the Expanding CMOS Image Sensor Design Space

## Thursday, June 5th, 2018 from 13:00 to 14:00

### Roger Panicacci
*ON Semiconductor, USA*

Today's image sensor solutions for demanding applications like automotive driver assist image sensing, high speed machine vision, and low power battery operated cameras started from relatively modest solutions for pixels and analog readout design. These initial solution options have evolved along different paths. Now the designer must understand the overall image system requirements to decide where and what to process in the pixel domain, analog circuit domain, and digital processing domain. These decisions rely on understanding the fundamentals of imager photon capture, noise components, analog/digital signal processing topology efficiency, and now the impact of 3D wafer stacking on sensor architecture solutions. This talk will review some of the fundamentals of image sensor technology, design, how it evolved to its current state, and the trajectory for what is possibly next.

**Roger Panicacci** is Vice President of Imager Design Platform Technology at On Semiconductor where he focuses on circuits, sensor architecture, and system design for next generation image sensor products. He started work on CMOS active pixel image sensor design at the Jet Propulsion Laboratory in the 1990s during the early development of the technology and was one of founding members of Photobit Corporation that commercialized CMOS active pixel technology. He stayed with the same team as it became part of Micron Technology, then Aptina, and now On Semiconductor. He holds a B. S. EECS from the University of California, Berkeley, and M. S. EE from the University of California, Santa Barbara.

PRIME 2018, Prague, Czech Republic

Session: Analog Circuits I

# Improved class-AB output stage for Sub-1 V fully-differential operational amplifiers

A. Ria*
Dipartimento di Ingegneria
dell'Informazione
University of Pisa
Pisa, Italy
andrea.ria@ing.unipi.it

S. Del Cesta
Dipartimento di Ingegneria
dell'Informazione
University of Pisa
Pisa, Italy
simone.delcesta@ing.unipi.it

A. Catania
Dipartimento di Ingegneria
dell'Informazione
University of Pisa
Pisa, Italy
alessandro.catania@ing.unipi.it

M. Piotto
Dipartimento di Ingegneria
dell'Informazione
University of Pisa
Pisa, Italy
massimo.piotto@unipi.it

P. Bruschi
Dipartimento di Ingegneria
dell'Informazione
University of Pisa
Pisa, Italy
paolo.bruschi@unipi.it

*Abstract*—An existent architecture for low voltage class-AB output stages is analyzed finding critical issues for which effective original solutions are proposed. The approach has been applied to the design of a compact class-AB fully-differential operational amplifier, capable of operating at a supply voltage of 0.8 V, providing a maximum output current of 7.5 mA with only 156 uA of quiescent supply current. The proposed amplifier constitutes a convenient building block for switched-capacitor circuits and low-voltage sensor interfaces. The performances of the amplifier are demonstrated by means of electrical simulations performed on a prototype designed with the UMC 0.18 um CMOS process. The total estimated area of the cell is 0.023 mm².

*Keywords— Class-AB, Fully-differential, CMOS, Operational amplifier, sub-1V.*

## I. INTRODUCTION

In the last decade, the demand of non-intrusive systems capable of achieving information on a wide set of environment parameters raised the interest on complex mixed-signal systems on a chip (SoC). In very compact, low-cost units for portable or wearable devices, it is desirable to use a single power supply for the whole circuit. In these conditions, the analog cells should keep pace with the continuous scaling down of the digital supply voltage, currently crossing the 1 V barrier. In cost-constrained designs, also the area is becoming extremely critical for analog blocks. The degraded performances in terms of offset voltage and low frequency noise density can be counterbalanced by the use of correlated double sampling (CDS) techniques [1], requiring the use of switched capacitor (SC) circuits. The reduced range available for signal, following $V_{dd}$ scaling down, should be compensated by fully-differential architectures, that are also essential to reject the noise injected by the on-chip digital sub-units.

The operational amplifier (op-amp) is the most versatile building block for the synthesis of analog circuits. Frequently, the mentioned low-voltage and small-area specifications should be met while maintaining the capability of sinking and sourcing relatively high output currents in order to enable fast charging of capacitors, as in SC circuits, and stimulation of resistive sensors. This involves the design of class-AB output stages capable of operating at 1 V and below.

The most popular class-AB architectures present in the literature introduce serious limitation to the lowest applicable supply voltage. The traditional technique of splitting the output MOSFET gate voltages by means of a voltage shifter [2] requires a minimum $V_{dd}$ in excess of $3V_{GS}$. Substitution of the voltage shifter with the well-known Monticelli's head-to-tail MOSFET cell [4] lowers the $V_{dd}$ limit to $2V_{GS} + V_{DSAT}$, which, with conventional threshold voltages, remains still well above 1 V. Replacement of the voltage shifter with capacitors may be an option to increase the maximum output current during transients [4]. However, this kind of "ac" class-AB solution is completely ineffective in terms of dc performances.

To our knowledge, sub-1V class-AB fully-differential operational amplifiers, capable of providing output dc currents in the mA range have not yet been proposed in the literature. An interesting architecture for class-AB output stages that, in principle, may operate at $V_{dd}$ values down to $V_{GS} + 2V_{DSAT}$ was described in [5]. Nevertheless, such an approach does not seem to have received much interest after its early proposal, maybe for the difficulty of dealing with the internal feedback loop that sets the quiescent current. A recent application of this technique to a single-ended op-amp, designed to work at $V_{dd}$ down to 0.9 V, is presented in [6]. In this work we revise the architecture of [6] pursuing two distinct goals: (i) to design a fully-differential version of the previous op-amp, introducing an efficient output common mode voltage control; (ii) to improve the technique proposed in [3] in order to mitigate the aspects that make its application critical, and, possibly, discourage its use in modern designs. The result is an original fully-differential op-amp that uses an output stage which exploits a modified version of the topology proposed in [5] to provide a 7.5 mA output current at 0.8 V supply voltage, while the quiescent current consumption is only 156 µA. The effectiveness of the proposed topology is demonstrated by means of accurate electrical simulations performed on a prototype designed with the UMC 0.18 µm CMOS process.

978-1-5386-5388-3/18 $31.00 © 2018 IEEE

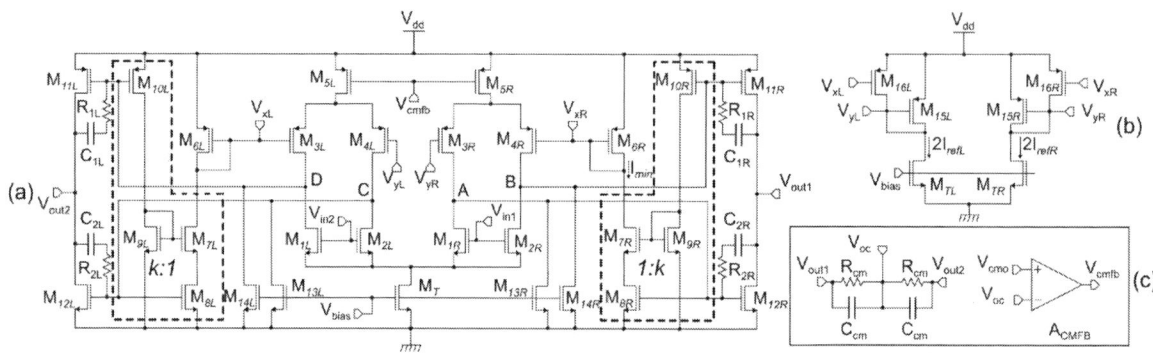

Fig. 1. Schematic view of the proposed class-AB FD operational amplifier (a-b) with a simplified view of the common mode feedback loop circuit (c).

## I. CIRCUIT TOPOLOGY AND BEHAVIOUR

The schematic view of the proposed amplifier is shown in Fig. 1. The amplifier has a symmetric structure: to distinguish MOSFETs in the right section from their analogous components in the left one, letters "R" and "L" have been added to the device names. The input stage is formed by devices $M_{1L-5L}$ and $M_{1R-5R}$, biased by current source $M_T$. This is a conventional fully-differential gain stage, where both the right and left branches are split to create independent drive voltages for the p and n MOSFETs of the output stages. Considering the right section, the gate voltages in the output devices ($M_{12R}$ and $M_{11R}$) are indicated with A and B. To understand the principle of operation of the class-AB control, derived from [5], it is convenient to indicate the input common mode and differential mode voltages of the A, B nodes pair with $V_{CA-B}$ and $V_{DA-B}$, respectively. It can be easily shown [5] that a $V_{CA-B}$ variation tends to cause concordant drain current variations in the output devices, so that the output current is strongly affected. Through this mechanism, $V_{CA-B}$ affects the output voltage. Note that, correctly, $V_{CA-B}$ is an amplified version of the input differential voltage of the amplifier. On the other hand, $V_{DA-B}$ coincides with the voltage shift between $M_{12R}$, $M_{11R}$ gates, which, as in more conventional class-AB stages, controls the bias current that flows from $V_{dd}$ to gnd through the series of the output devices. In quiescent conditions, this current has to be fixed to the desired value; in all other conditions (non-zero input signal) the current should not differ much from the quiescent value. This result is obtained thanks to the local negative feedback loop, formed by the minimum current selector [5], enclosed in the dashed box of Fig.1, and by the differential pair $M_{3R}$-$M_{4R}$. The minimum current selector senses the gate voltages of the output MOSFETs and produces a replica ($I_{min}$) of the minimum of $M_{12R}$ and $M_{11R}$ drain currents, scaled down by a factor $k$. Consequently, this current modifies the input voltage ($V_{xR}$) of the pair $M_{3R}$-$M_{4R}$, changing the $V_{DA-B}$ value, as required.

In the previous version [5], $M_{3R}$ gate voltage ($V_{yR}$) was kept at a constant reference value and due to the high gain of the feedback loop, in stationary conditions, $V_{xR}$ approaches $V_{yR}$. In this way, the minimum current $I_{min}$ was fixed. Unfortunately, this approach is not robust against output impedance reduction of the $M_{5R}$ ($M_{5L}$) tail current source. The problem is that the

differential pair $M_{3R}$-$M_{4R}$ is stimulated with a pseudo differential signal. As frequency increases, parasitic capacitances from $M_{3R}$-$M_{4R}$ sources to ground block the path from $V_{xR}$ to node A. Thus, at high frequencies, $V_{xR}$ affects practically only node B, so that the sensitivity of $V_{C-AB}$ to $V_{xR}$ is not negligible. The net result is the onset of a feedback loop that involves also the useful signal. This is clearly visible in the AC open-loop frequency response as a sharp notch occurring well below the unity gain frequency. The altered frequency response hinders the ability to find a decent phase margin for the amplifier.

Similarly, even in DC conditions, crosstalk between the quiescent current control and the main signal path can occur when, due to low power supply requirements, it is not possible to bias $M_{5R}$ in full saturation region. In this case, the effect is a reduction in the dc open loop gain. Finally, in particular conditions that can be experienced by the amplifier in normal operation, the quiescent current control circuit can turn into a positive feedback loop, resulting in hysteresis of the output current. Indeed, when $M_{2R}$ is in triode region, an $I_{min}$ increase tends to rise $M_{1R}$ source voltage through the reduced $M_{2R}$ resistance. Depolarization of $M_{1R}$ causes node A to rise, increasing $M_{12R}$ current, and then, unless $I_{12R} \gg I_{11R}$, also $I_{min}$, confirming the positive sign of feedback.

We have obtained decisive mitigation of all these adverse effects by simply driving the $M_{3R}$-$M_{4R}$ pair with a pure differential signal, using the circuit of Fig.1(b) to derive $V_{yR}$ from $V_{xR}$. Considering that $M_{15R}=M_{16R}=M_{6R}$, then variations of $V_{yR}$ and $V_{xR}$ are opposite, as required. Note that $V_{xR}=V_{yR}$ for $I_{min}=I_{ref}$. Then $I_{ref}$ sets the $I_{min}$ value.

Two stage fully differential amplifiers generally require two separated Common-Mode-Feedback systems (CMFB), one for each stage. Instead, we have opted for a single CMFB circuit for the whole amplifier. The output common mode voltage is extracted by a simple AC-compensated voltage divider, as shown in Fig.1(c). Common mode voltage is then compared with the target value by a differential amplifier, which drives the gate voltage of $M_{5R}$. This architecture presents an incorrect stable closed loop operation point, in which the output common mode voltage ($V_{OC}$) may be trapped when the amplifier is used in closed loop configuration that forces the input common mode voltage ($V_{IC}$) to coincide with $V_{OC}$. If $V_{OC}$ gets accidentally close to ground, so does $V_{IC}$, turning off the input pair $M_{1R}$-$M_{2R}$. The $A_{CMFB}$ amplifier in Fig.1(c), turns $M_{5R-L}$ off in an attempt to lower nodes A-D, raising $V_{OC}$. Due to the input pair being turned off, a pulldown path for A-D is missing. To avoid this problem, we have introduced transistors $M_{13}$, $M_{14}$ acting as weak pulldown devices. Their effect is sufficient to force $V_{OC}$ and $V_{IC}$ to increase, resuming correct operation of the CMFB.

For what concerns the minimum applicable supply voltage the following relationships hold true:

$$V_{dd} \geq V_{DSAT\_T} + V_{DSAT\_2} + \left| V_{GS\_11} \right|, \quad (1)$$

$$V_{dd} \geq V_{DSAT\_5} + V_{DSAT\_3} + \left| V_{GS\_12} \right|, \quad (2)$$

Both equations indicate a minimum $V_{dd}$ of $V_{GS} + 2V_{DSAT}$, enabling sub 1 V operation. In order to keep the minimum $V_{dd}$ as low as possible we have biased $M_{11}$ and $M_{12}$ in weak inversion obtaining also a high current efficiency. The latter has been expressed as the ratio between the maximum output current and the quiescent current. It is worth noting that the input common mode voltage should stay within $V_{GS1} + V_{DSAT\_T}$ and $V_{GS12} + V_{TH2}$. The small common mode range is generally not particularly critical in fully-differential architectures.

## II. SIMULATION RESULTS

A prototype of the proposed amplifier was designed with the UMC 0.18 μm CMOS process with an estimated area occupation of 0.023 mm². The main area occupation is due to the output and input MOSFETs. $M_{11R-L}$ and $M_{12R-L}$ are designed with W/L dimensions of 300/0.5 and 750/0.34, respectively, while $M_{1R-L}$ and $M_{2R-L}$ are designed with W/L ratio of 400/1. This prototype was optimized to work properly with supply voltages in the range of 0.8 V-1.4 V and to be stable with capacitive loads up to 20 pF. Performances of the circuit were estimated by means of accurate transistor- level simulations, with $V_{dd} = 0.8$ V unless differently specified.

The open loop behavior of the proposed amplifier was evaluated by means of AC simulations with different $V_{dd}$ and a capacitive load of 20 pF. As in Fig. 2, the prototype exhibited a DC gain greater than 83 dB, a gain-bandwidth product (GBW) of 5.7 MHz with a phase margin of 49°.

The output short circuit current of the amplifier ($I_{cc}$) was evaluated considering the current flowing through short circuit placed across the two output ports. As shown in Fig. 3, the maximum output current ($I_{cc-max}$) at 0.8 V of voltage supply is 7.48 mA. This value rises up to 79.23 mA at 1.4 V. The total quiescent current ($I_{supply-q}$) was around 156 μA at 0.8 V, corresponding to a current efficiency ($I_{cc-max} / I_{supply-q}$) of 48.

Monte Carlo runs have been performed to evaluate the robustness of the proposed circuit. As reported in Table I, the standard deviation of the input offset voltage is around 2.8 mV, a value that does not change significantly with $V_{dd}$. About the quiescent current of the output stage, Monte Carlo runs showed a mean value of 60 μA, with a standard deviation of 5 μA confirming the reliability of the internal current feedback.

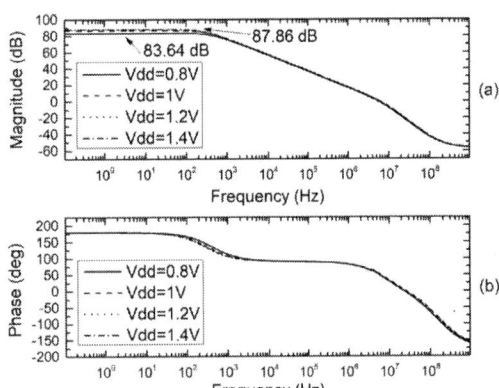

Fig. 2. AC simulated response of the amplifier at different supply voltages with a capacitive load of 20pF. (a-b);

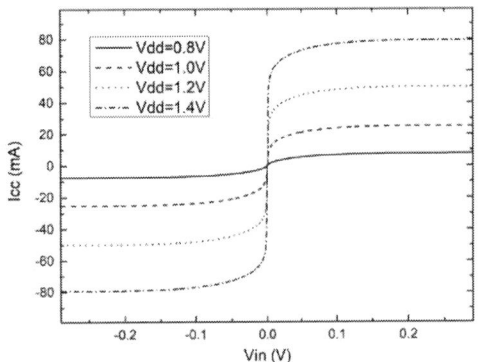

Fig. 3. Simulated operational amplifier output short circuit current at different supply voltages.

The input referred noise of the proposed prototype was estimated with AC-noise simulations at different supply voltages. As shown in Table I, the input referred noise estimated over a bandwidth from 0.1 Hz to GBW for $V_{dd}$ at 0.8 V, is 48.8 μV rms, with a flicker corner frequency around 90 kHz.

To study the step response of the op-amp, the test-bench in the inset of Fig. 4, has been used. The amplifier is closed with a resistive feedback loop (R=100 kΩ) and a 100 mV amplitude differential square waveform is applied to its inputs. The response of the amplifier, shown in Fig.4, exhibits a monotone response with a 1%-settling time of 0.29 μs.

Finally, we have built a simple test-bench aimed at investigating the suitability of the proposed amplifier for fully differential switched capacitor architectures. Fig. 5 shows the simple SC amplifier designed around the proposed prototype for a gain of 10 and stimulated with a 140 mV peak-to-peak amplitude sinusoidal signal at 10 Hz frequency. The simulation was performed at 0.8 V of supply voltage and with a clock frequency of 10 kHz. As shown in Fig. 6, the output differential signal ($V_{out1} - V_{out2}$) approaches the supply rail exhibiting a rail-to-rail output swing.

TABLE I. AMPLIFIER SIMULATED PERFORMANCES AT DIFFERENT SUPPLY VOLTAGES

| Vdd (V) | 0.8 | 1 | 1.2 | 1.4 |
|---|---|---|---|---|
| $I_{supply}$ | 156 µA | 205 µA | 205 µA | 205 µA |
| Input common mode range | 0.55 V to $V_{dd}$ | | | |
| DC Gain | 83.6 dB | 86.9 dB | 87.5 dB | 87.8 dB |
| GBW and Phase Margin @CL=20pF | 5.36 MHz 50° | 5.87 MHz 49° | 5.92 MHz 49° | 5.94 MHz 49° |
| $I_{cc}$ range (mA) | ±7.48 | ±25.1 | ±49.81 | ± 79.23 |
| Offset spread | 3.32 mV | 2.31 mV | 2.70 mV | 2.64 mV |
| $Vn_{RMS}$ | 48.80 µV | 48.16 µV | 48.16 µV | 48.16 µV |
| $f_k$ | 92 kHz | | | |
| Area | 0.023 mm² | | | |

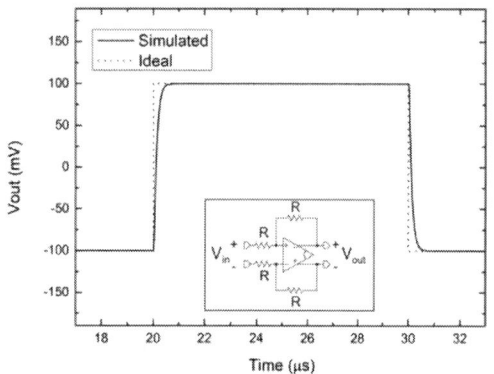

Fig. 4. Simulated step response of the operational amplifier with a resistive feedback loop. The overall amplifier is shown in the inset.

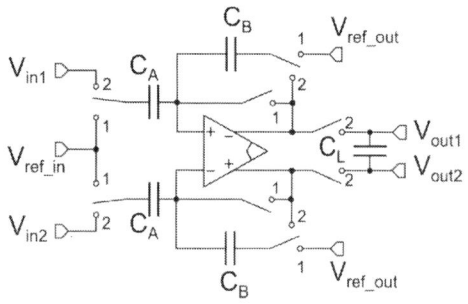

Fig. 5. SC amplifier based on the proposed prototype.

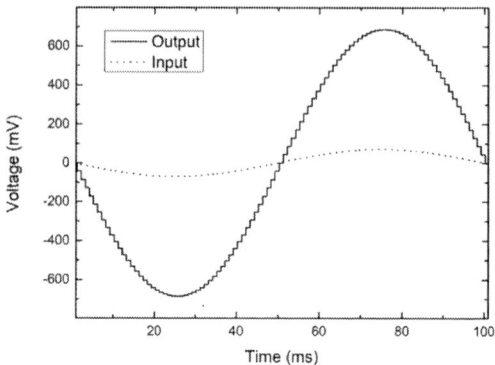

Fig. 6. Simulated behaviour of the SC amplifier based on the proposed prototype.

## III. CONCLUSIONS

In this work, a previously proposed class-AB architecture suitable for sub 1V operation has been analyzed, highlighting critical issues that could have limited its massive applications. The simple topological improvement proposed in this paper have proven very effective in solving the mentioned issues. Application to a fully-differential op-amp showed that a current efficiency ratio close to 50 can be obtained at supply voltages down to 0.8 V. Suitability for switched capacitor applications have been demonstrated with a simple test-bench.

## REFERENCES

[1] C.C. Enz, G.C. Temes, "Circuit techniques for reducing the effects of op-amp imperfections: autozeroing, correlated double sampling, and chopper stabilization," Proceedings of the IEEE, vol. 84, no. 11, pp. 1584–1614, November 1996.

[2] R. Gregorian, G.C. Temes, "Analog MOS integrated circuits for signal processing", John Wiley and Sons, New York 1986, p.214.

[3] J. Ramirez-Angulo, R.G. Carvajal, R.G. Lopez-Martin, "A free but efficient low voltage class AB two stage operational amplifier", IEEE Trans. Circ. Syst. II, 2006, 53, (7), pp. 568-571

[4] D. M. Monticelli, "A quad CMOS single-supply op amp with rail-to-rail output swing", IEEE Journal of Solid-State Circuits, vol. SC-21, no. 6 December 1986, pp.1026-1034.

[5] K. de Langen, J. H. Huijsing "Compact low voltage power effecient operational amplifier cells for VLSI", IEEE Journal of Solid-State Circuits, vol. 33, no. 10 October 1998, pp. 1482-1496.

[6] S. Del Cesta, A. Catania, P. Bruschi, M. Piotto, "A compact sub-1V class AB operational amplifier for low-voltage switched-capacitor circuits", European Conference on Circuit Theory and Design (ECCTD) 2017, September 2017.

[7] S. Rabii, B. A. Wooley, "A 1.8-V Digital-Audio Sigma–Delta Modulator in 0.8- m CMOS", IEEE Journal of Solid State Circuits, vol. 32, no. 6, June 1997.

PRIME 2018, Prague, Czech Republic

Session: Analog Circuits I

# A fully differential, 200MHz, programmable gain, level-shifting, hybrid amplifier/power combiner/test buffer, using pre-distortion for enhanced linearity

Vahur Kampus
and Toomas Rang
Department of Electronics
Tallinn University of Technology
Tallinn, Estonia 19086

Daniel Knaller,
Christian Fleischhacker,
Markus Korak,
and Jozef Kiss
Intel Austria GmbH
Villach, Austria 9528

*Abstract*—**With the continuous advancement of standards in high-end telecommunication systems, requirements for analog circuitry have become ever more demanding. The new standards not only want to support better modulation schemes, demanding less noise with higher linearity, but also require increased bandwidth from analog circuitry. This paper describes a fully differential, programmable gain, baseband power combiner with a bandwidth of 200MHz, using pre-distortion for extra linearity, that offers also level-shifting and test buffer functionalities. The pre-distortion allows the improvement of linearity in the amplifiers AB stage, what otherwise would be a limiting factor in rail-to-rail swing operations and lowering the maximum throughput.**

*Index Terms*—**CMOS, semiconductor, analog, opamp, power combiner, pre-distortion, class-AB, multi-path, transmit chain**

## I. INTRODUCTION

Standard OFDM-like signals lose about 3dB of single sub-carrier strength for every doubling of the bandwidth (BW). To benefit from all the effort of increasing the BWs and maintain the same throughput per sub-carrier, analog blocks have to be designed with 3dB better noise and linearity specs for each doubling of the BW. If new standards incorporate within them new modulation schemes, these requirements can be even more challenging [1]. Modern day high-BW transmitter chains tend to consist of digital blocks dedicated to signal processing, a digital-to-analog converter (DAC), analog amplifiers or buffers and either a line driver (for wired systems) or a power amplifier (for wireless systems) (Fig. 1). High performance line drivers and power amplifiers are generally built in different technologies than the rest of the transmit chain – meaning the signal path has to go off-chip. Those external blocks can often be the biggest contributors to the total chain noise figure and non-linearities [2]. Thus lowering the performance and gain requirements for them is both highly welcomed and beneficial. Because the DAC and the external amplifier are often designed for one optimum operating condition, the power combiner can also be the primary means to change the total chain gain. Modern day analog technologies usually offer devices with at least two flavors – one in core voltage domain and one in input/output (I/O) voltage domain. In order to have the most

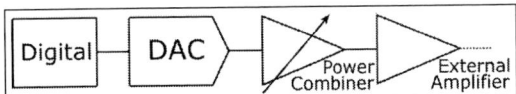

Fig. 1. Typical signal path in high-BW, high-performance transmitters.

effective level-planning and increase the signal amplitude at the input of the external amplifier, therefore requiring less gain and performance from it, but at the same time taking full advantage of the shrink what the technology node can offer, one would like to design all the digital blocks and the DAC in the core voltage domain and then amplify the signal with the power combiner using I/O devices to the needed levels while also level-shifting the common-mode before going off-chip and arriving at the externals.

## II. POWER COMBINER

The proposed power combiner is part of a VLSI system designed in a standard C40 technology with I/O devices allowing max VDD of 2.5V. It is designed to provide programmable gain from -15dB up to +15dB with a level-shifting function to a fixed output common-mode voltage of VDD/2 (Fig. 2a). It can be operated with any input common mode ranging from VSS to VDD for gain configurations less than 5dB. If the configured gain is higher, the input common mode will have a slight range limitation depending a lot on the actual gain used.

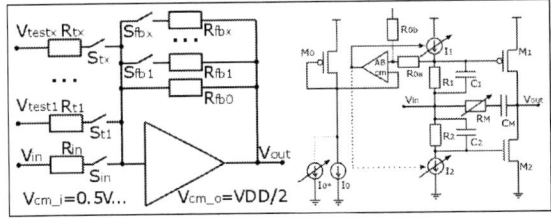

Fig. 2. a) topology of the power combiner. b) topology of a classical push-pull symmetrical high linearity AB stage.

978-1-5386-5388-3/18 $31.00 © 2018 IEEE

In the buffer mode (0dB), there are no special limitations and the input common mode can actually range from a bit below VSS to a bit above VDD. The combiner has a single optimized main input ($V_{in}$) connected to the transmitter DAC used in the signal path and several test-mux inputs ($V_{test}$) that can be used to test the performance of various other IPs across the chip (Fig. 2a). Since there will be a steady DC current flowing though the feedback network, if the common-mode voltages at the input and output differ, the other IPs in test must be able to sink or source the extra current coming from the level shifting functionality, if operating with any other different common-mode. The amount of current needed to be sinked or sourced is proportional to the difference between the two common-mode voltages and inverse proportional to the total resistance of the feedback resistor $R_{fb}$ and the input resistor $R_{in}$:

$$I_{DC} = 2 * \frac{V_{cm_o} - V_{cm_i}}{R_{fb} + R_{in}} \qquad (1)$$

The four main contributors for the non-linearities of the combiner are the finite loop gain of the amplifying circuit, the inherited non-linearity coming from the AB stage, the non-linearities of the signal path switches ($S_{fb}$, $S_{in}$, $S_t$) required for the gain and input programming and the non-linearities associated with the resistors ($R_{fb}$, $R_{in}$, $R_t$). Assuming proper care on the feedback network design – so the switches are sized correctly, that the on-resistance of the switch is non-dominant and the resistance of the resistor is constant with different potential across it – all the dominant non-linearities will come from the amplifier. This is a critical issue in modern high-speed data communications requiring full-rail swing operation for max throughput [3].

### III. STRUCTURE OF THE AMPLIFIER

The designed amplifier has a three stage feed-forward compensated structure for the signal path, with one two-stage Miller compensated common-mode loop, one three-stage Miller compensated common-mode loop and one feed-forward compensated three-stage common-mode loop (Fig. 3). The feed-forward structure is used to have a sufficiently high loop-gain in the desired operating BW. The very high DC gain of the multi stage structure is not needed or intentionally designed for, but rather comes as a side benefit, for free. The real value of the feed-forward structure comes from the fact that it allows the designer to place two or more poles in close proximity

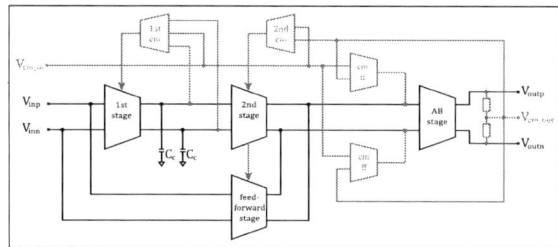

Fig. 3. Topology of the designed multiple feed-forward opamp.

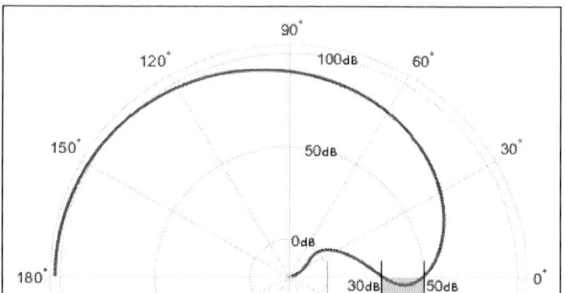

Fig. 4. Nyquist plot of the designed amplifier in buffer (0dB) mode. The amplifier has two clearly visible uncompensated poles between 30dB and 50dB and an optimum PM (˜45°) at 0dB crossing.

to have a higher-order (more than 2nd order) roll-off and compensate this with one or more zeros coming from the feed-forward branch before the unity gain frequency (UFG). This can be potentially dangerous if not used properly, because we are creating a conditionally stable system, where the phase can shift easily more 180 degrees before the UGF. Even worse, the location of the UGF changes with the gain configuration of the power combiner. In buffer-mode the amplifier for example has more than 2 clearly visible uncompensated poles between 30dB and 50dB (Fig. 4). However, since it is under the control of the system at all times and cannot be programmed by the user to just about any configuration as it would be with a generic stand-alone amplifier, we can still easily stabilize the system with different predefined biasing programming currents, what shift the singularities for different gains.

### IV. AMPLIFIER PROGRAMMING

The basic stability programming of the amplifier is done totally autonomous, with no extra effort from the system, with the help of the basic gain programming. The gain of the amplifier is determined with a binary weighted network of resistive ladders, so the opamp actually has all the information about the gain setting used and can determine itself the compensation needed for best stability. For that reason, the information provided by the programming of the 3MSBs of the resistive feedback is taken and re-used. This means, a stability compensation system, what divides the compensation information into eight sections, has been developed. The amplifier singularities (poles and zeros) are programmed to move together across the frequency range, ensuring in optimum phase margin (PM) for each gain setting (Fig. 4). The lowest gains also result in configuration to lowest open-loop BWs. Doing this, current consumption can be saved in low-gain configurations as shown in Fig. 5.

| Setting | 7 | 6 | 5-3 | 2 | 1 | 0 |
|---------|---|---|-----|---|---|---|
| gain (dB) | -15; -14 | -13; -12 | | -2; +2 | +3; +5 | >+6 |
| $I_Q$ (mA) | $9 + I_{DC}$ | $10 + I_{DC}$ | $R_M$ | $18 + I_{DC}$ | $21 + I_{DC}$ | $27 + I_{DC}$ |

Fig. 5. Eight modes of compensation for different gains. Switchover of $R_M$.

In order not to change the noise performance of the amplifier over different programming modes, the operating point of the first stage should not be programmed quite the same way as for other stages. For that reason, the BW of the first state is programmed rather with programmable capacitive loads $C_c$, what shift the singularities coming from there without altering the noise performance associated mostly with the first stage in such structures (Fig. 3).

In addition to the autonomous stability programming, the opamp also has one more manual lever to alter performance. The function of the AB stage is not only to provide linear rail-to-rail swing but also to drive the uncertain, sometimes undefined and unknown loads – for example when under mass production test. For that reason an AB boost programming option was added, when the external load is higher than expected, a programmable biasing current $I_{0*}$ only for the AB stage can be increased and the pole coming from the AB stage and output capacitance re-adjusted (Fig. 2b). This allows to change the capacitive driving capabilities of the combiner.

## V. AB STAGE

As mentioned before, if the in-band loop gain of the amplifier is sufficiently high, the dominant part of the non-linearities from a rail-to-rail swing tends to come from the AB stage. The designed AB stage itself has a three-stage Miller-compensated common-mode regulation loop (Fig. 2b). This is the only place in the entire amplifier where high DC gain is desired and designed for, because with three gain stages in the common-mode loop, the DC gain should be sufficiently large, that the operating point is set with a high degree accuracy, so that the biasing current for the AB stage would not vary. To improve the linearity of the AB stage, the amplifier has a built in AB stage pre-distort function, what modulates the signal before reaching the AB stage to enhance the gm matching between the pMOS and nMOS output devices.

Inherently one of the most linear push-pull AB stages, where the p-, and n-channel devices have roughly equal amplification and driving strengths is the so called flating battery symmetrical push-pull structure presented in Fig. 2b [4]. The current sources $I_1$ and $I_2$ are controlled by a three-stage AB common mode loop, regulated to have an accurate and fixed DC operating point if the DC gain is sufficiently large. The current is regulated to match a reference circuit current composed of $M_0$ and $I_0$. The ratio between $M_0$ and $M_1$ will set the current ratio between the reference circuit and the current in the AB stage. The circuit is fully differential, so one reference can regulate the DC current for both the n-, and p-side via a resistive common-mode divider composed of $R_{0a}$ and $R_{0b}$. The output stage biasing is then done by controlling the potential across resistors $R_1$ and $R_2$. Capacitors $C_1$ and $C_2$ are sized to add additional high order zeros after the desired BW but before the UGF to further improve the differential stability.

The AB stage itself incorporates a Miller compensation scheme composing of a Miller capacitor $C_M$ and a Miller resistor $R_M$. This allows us to control the process variables a

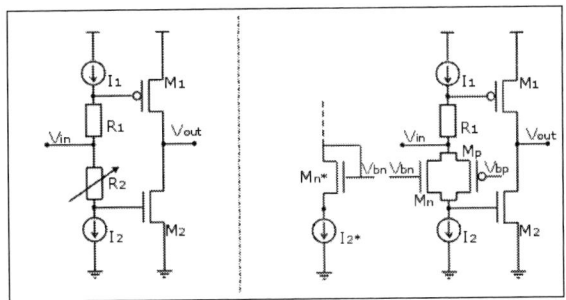

Fig. 6. AB stage pre-distortion. Basic idea and actual implementation.

lot better resulting in a more predictable and better defined UGF, what is important for a conditionally stable system. However controlling the optimum placement of the Miller zero over a large gain range is challenging, so to optimize and simplify matters, the gain range is divided into two subsections, where $R_M$ can be halved with the gain and the Miller zero can be shifted (Fig. 2b). While lowering the frequency of the zero will add extra stability to the lower gain range, it also lowers the inherited performance, so the placement of the switchover is important and should be done with care, as late as possible, for minimum loss of performance (Fig. 5).

## VI. AB-STAGE PRE-DISTORTION

In order to have a highly linear push-pull AB stage operation for a large distortion-free range driving capability, one would like to operate both the p-, and n-channel devices of the AB-stage always in similar operating conditions – meaning the gain product coming from one or the other should be similar and matched over the entire output swing. This is very difficult to achieve in full swing, because the last stage of any amplifier structure has only finite gain and its input will therefore already have a considerable amount of signal on top of it. Assuming roughly 20dB of gain from the stage, one will already have a tenth of the total opamp output swing at the input of the last stage, creating a considerable operation point swift during different stages of the swing. This is less apparent in previous gain stages, because with

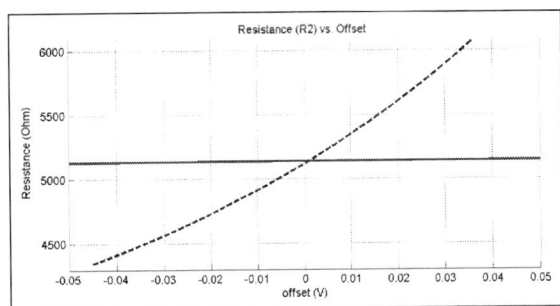

Fig. 7. Modulated resistance of R2 (Dashed line) vs original resistance.

each next stage, the signal residue on top increases. So with near rail-to-rail output swings, every AB structure will start to exhibit some non-linear tendencies, because of the input swing that changes the gms in different directions for the p-, and n-channel devices used for the AB stage. Fortunately, the distinctive and unique structures of AB stages, what are different from all the stages prior, allow us to add some wanted defects to pre-distort the signal seen by the output drivers, to help linearize the signal. For that purpose, a simple modulation scheme has been developed to pre-distort only one of the two signal paths in the AB stage, in order to compensate for the shift of gms over the full rail-to-rail operation. $R_1$ and $R_2$ together with a fixed current $I_1$ and $I_2$ are typically used in such a structure to bias the drivers $M_1$ and $M_2$. The current sources $I_2$ are used to create a voltage drop across resistors $R_1$ and $R_2$. The smaller the fixed current and resistances, the smaller the voltage drop across them. This will in return bias the output drivers $M_1$ and $M_2$ in wanted operating condition. By keeping $R_1$ constant and modulating $R_2$, the input can be pre-distorted for only the n-side in a beneficial way, so that the total output linearity is actually enhanced (Fig. 6a). The wanted $R_2$ characteristic can be built with a combination of a pure resistor and output resistances of devices $M_n$ and $M_p$. Total $R_2$ is modulated to have a signal dependent resistance by using the output resistances of $M_n$ and $M_p$ by adding another replica circuit composed of current source $I_{2*}$ and $M_{n*}$, plus a similar topology for $M_p$, to properly bias the devices according to current $I_2$ (Fig. 6b). By doing this, any needed slope for the required modulated resistance can be acquired (Fig. 7).

## VII. MEASURED RESULTS

The function of the power combiner is to level-shift, amplify and buffer the output of the DAC or to use it as a vehicle to test some other IPs across the die. It has a signal BW of about 200MHz, with a relatively smooth roll-off desired for any blocker scenarios and mask requirements. The internal programmability of the compensation keeps the performance relatively constant over the gain range seen in Fig. 8. The similarities in the slight signal peaking at band-edge over

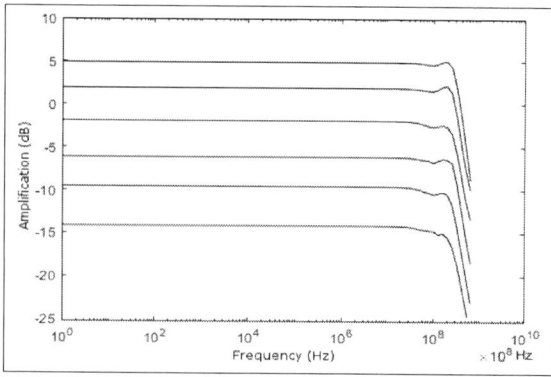

Fig. 8. Measured STF of the designed power combiner over different gains.

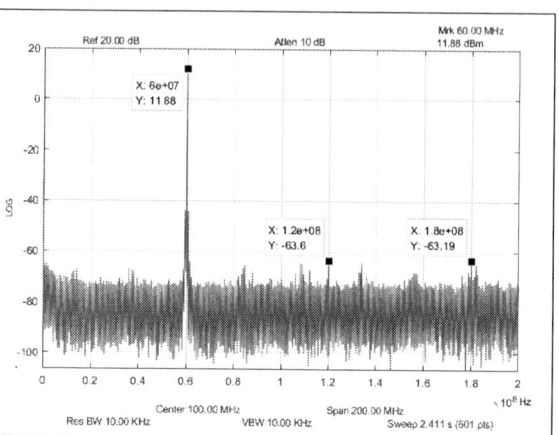

Fig. 9. Measured output spectrum of the power combiner with 5dB gain and with 12dBm output power.

different gains prove the effectiveness of the compensation programming, keeping the loop-gain constant over gain programming. It has a HD2/HD3 performance of $<$-75dBc, when driving a 10pF load, with a near full-scale output, with no extra current for the AB stage boosting (Fig. 9). The power consumption of the power combiner is variable, depending from the compensation biasing schemes used, described before, starting from 9mA (+$I_{DC}$ for level-shifting – Eq. 1) for low gains (-15dB...) and ending with 27mA (+$I_{DC}$) for the highest gain and load configurations from the a 2.5V rail. The capacitive load, the combiner is able to drive, ranges from up to 20pF when boosting capabilities for extra load driving are disabled and up to 200pF when they are enabled. Summary:

| Gain | $C_{LOAD}$ | BW | $I_Q$(mA) | PSRR@2MHz |
|---|---|---|---|---|
| $\pm15dB$ | $<$200pF | 200MHz | 9 - 27 | $<$-40dB |

## VIII. CONCLUSION

A 200MHz, programmable gain power combiner with pre-modulation for enhanced linearity has been presented. It also offers level-shifting functions to VDD/2 of the combiner, to be able to maximize the signal swing at the output. The measured results show a decent HD2/HD3 performance with even a near rail-to-rail operation, when the compensation schemes and power consumption are autonomously self-programmed.

## REFERENCES

[1] A. Mehmia, "Optimum DAC resolution for WMAN, WLAN and WPAN OFDM-based standards," ICCE, Jan. 2005.

[2] H. Pan, Y. Yao, M. Hammad, J. Tan, K. Abdelhalim, E. W. Wang, R. C. J. Hsu, D. Tam and I. Fujimori, "A Full-Duplex Line Driver for Gigabit Ethernet With Rail-to-Rail Class-AB Output Stage in 28 nm CMOS," IEEE J. Solid-State Circuits, vol. 49, no. 12, Dec. 2014.

[3] S-C. Lee, Y-D. Jeon, J-K. Kwon and J. Kim, A 10-bit 205-MS/s 1.0-mm2 90-nm CMOS Pipeline ADC for flat Panel Display Applications, IEEE J. Solid-State Circuits, vol. 42, no. 12, Dec. 2007.

[4] R. Hogervorst, J. P. Tero, R. G. H. Eschauzier and J. H. Huijsing, "A Compact Power-Efficient 3 V CMOS Rail-to-Rail Input/Output Operational Amplifier for VLSI Cell Libraries," IEEE J. Solid-State Circuits, vol. 29, no. 12, Dec. 1994

*PRIME 2018, Prague, Czech Republic*

*Session: Analog Circuits I*

# New Current Mode Wheatstone Bridge Topologies with Intrinsic Linearity

L. Safari[1], G. Barile[1], V. Stornelli[1], G. Ferri[1], A. Leoni[1]

[1]Department of Industrial and Information Engineering and Economics, University of L'Aquila, Italy
Email:Leilasafari@yahoo.com, gianluca.barile@graduate.univaq.it, vincenzo.stornelli@univq.it, giuseppe.ferri@univaq.it, alfiero.leoni@graduate.univaq.it

*Abstract*— **In this paper two new topologies for current-mode Wheatstone bridge (CMWB) are presented. The circuits, unlike other CMWB topologies, are based on two second generation voltage conveyors (VCII) as basic building blocks and two nMOS transistors operating as variable resistor. The outputs of both circuits are intrinsically linear function of ΔR. Compared to previously reported CMWB circuits, the proposed topologies offer several advantages. Firstly, they do not require any extra voltage buffer at output i.e. the produced output voltage can be directly used in practical applications. Secondly, they do not employ any passive resistor while there are multiple grounded and floating resistors in other CMWBs. Thirdly, they have the capability to electronically control the gain without a significant impact on consumed power. To confirm the proposed theory, PSpice simulation results using 0.35μm CMOS technology parameters are presented.**

*Keywords*— *Current-Mode, Current Mode Wheatsone Bridge, Resistor Sensor Applications, VCII.*

## I. INTRODUCTION

For many years, Voltage-Mode Wheatstone Bridge (VMWB) has been widely used in applications of temperature, pressure, and resistive measurements [1-3]. Generally, it is a network of four resistors excited by a reference voltage producing a proportional output voltage in response to any variation in the value of resistors. Various voltage handling circuits have been employed to process the output signal of Wheatstone bridge. Later, in [4], using the duality concept, Current Mode Wheatstone Bridge (CMWB) was introduced which offered significant improvements compared to conventional VMWB. Shown in Fig.1, it consists of only two resistors excited by a current source. Its output signals are of current kind and current mode techniques can be used to process the output signals of CMWB. Therefore, CMWB can take advantage of current mode signal processing such as higher bandwidth, larger dynamic range, simpler circuitry etc. [4]. In CMWB, the resistors are connected at one end and are forced to have equal and constant voltage ($V_c$) at the other end. Therefore, in Fig.1 the following conditions must be fulfilled by the signal conditioning circuit:

$$V_1 = V_2 = V_c \qquad (1)$$

$$I_{ref} = I_1 + I_2 \qquad (2)$$

Because the resistors behave like parallel resistors, in the CMWB of Fig.1, $I_1$ and $I_2$ can be expressed as:

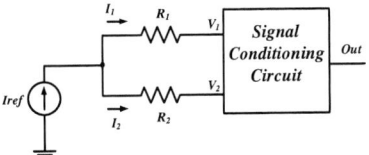

Figure 1. Current Mode Wheatstone bridge [4]

$$I_1 = \frac{R_2}{R_1+R_2} I_{ref}; \; I_2 = \frac{R_1}{R_1+R_2} I_{ref} \qquad (3)$$

By assuming $R_1=R_0\pm\Delta R$ and $R_2=R_0\mp\Delta R$, the signal conditioning circuit is used to produce an output signal proportional to the difference between $I_1$ and $I_2$:

$$I_1 - I_2 = \Delta I = \pm \frac{\Delta R}{R_0} I_{ref} \qquad (4)$$

Therefore, any variation in the value of resistors can be measured. However, in case of one sensor applications where $R_1=R_0\pm\Delta R$ and $R_2=R_0$, Eq. (4) will be a nonlinear function of $\Delta R$. Therefore, special linearization techniques must be employed to make the output signal a linear function of $\Delta R$. The other problem of the CMWB is its large common mode current which is equal to $I_{ref}/2$ and can produce large offsets at output. Many signal conditioning circuits have been reported for CMWB [4-10]. The main drawback which is common in all the previously reported CMWBs is that their output is a non-linear function of $\Delta R$ in one sensor applications. Therefore, they all need special linearization techniques. In addition, these circuits have several other drawbacks. For example, the operational floating conveyor (OFCC) based circuits of [4-5], current operational amplifier-based circuits of [6-7] and Op-Amp based circuit of [4] require large number of resistors. In addition, matching between resistors is very critical in their performance. Circuit reported in [8] employs Op-Amp in its structure which complicates the design and increases power consumption. The circuits proposed in [9-10] are the only ones providing electronic controlling feature and do not require any passive resistors. In these circuits, gain is varied by means of bias current. Therefore, higher gain is achieved at the expense of increased power consumption.

In this paper, two new structures for implementing electronically controllable CMWB are presented. The first one

978-1-5386-5388-3/18 $31.00 © 2018 IEEE

is suitable for two sensors while the second for one sensor application. The outputs of both circuits are intrinsically linear function of *ΔR*. The proposed circuits are based on second generation voltage conveyor (VCII) and there is no passive resistor in their structure. A simple active resistor made of two nMOS transistors is used to obtain variable gain. Therefore, different values of gain are achieved without a significant change in power consumption. The proposed circuits provide output signal in the form of voltage and exhibit very low output impedance. Both circuits are able to eliminate common mode signals and to produce negligible value of offset voltage at output. The proposed circuits take the advantage of interesting properties of VCII and are power and area efficient compared to previously circuits of [4-10].

The organization of this paper is as follows. In Section I, an introduction on VCII block and its implementation is given. The proposed CMWB topologies are presented in Section II. Simulation results are shown in Section III and finally Section IV concludes the paper.

## II. VCII FEATURES AND IMPLEMENTAION

By applying the duality concept on CCII, a new device is formed which is called VCII [11]. Fig.2 shows the symbolic representation of VCII and its function is characterized by Eq. (5):

$$i_x \approx \pm\beta i_y, \quad V_z = \alpha V_x, \quad V_Y = 0 \quad (5)$$

where β and α (close to unity) are current gain between Y and X terminals and voltage gain between X and Z terminals, respectively. For -β we have a VCII⁻ and for +β we have a VCII⁺. From Eq. (5), it is deduced that unlike CCII, the Y terminal is a current input terminal with ideally zero input impedance, the X terminal is a current output terminal with ideally infinite impedance and finally the Z terminal is a voltage output terminal which has ideally zero impedance. Therefore, the internal structure of a VCII consists of a current buffer between Y and X terminals and a voltage buffer between X and Z terminals. Comparing to a CCII that includes a voltage buffer between Y and X terminals and a current buffer between X and Z terminals, a VCII can be achieved if we exchange the voltage buffer and current buffer of CCII. This implies that both CCII and its dual circuit VCII have equal power consumption and similar circuit implementation. Fig. 3 shows a simple implementation of VCII which includes the current buffer reported in [12] and the voltage buffer presented in [13]. The circuit operates in class A. The VCII circuit is simulated using 0.35μm CMOS parameters and supply voltage of ±1.65V. The used transistors aspect ratios are reported in Table-1. The important parameters of the VCII are also summarized in Table-2. As it is seen, the circuit provides voltage and current gains very close to unity, low input impedances at Y and Z terminals and a high impedance at X terminal.

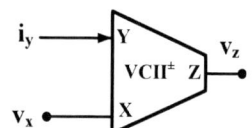

Figure 2. Symbol representation of VCII

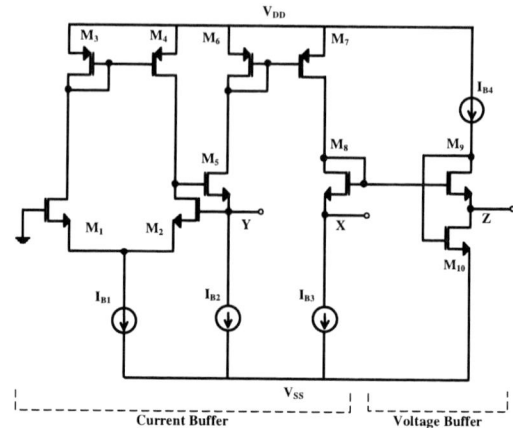

Figure 3. The VCII circuit implementation

Table-1. Transistors aspect ratios

| Transistor | Width | Length |
|---|---|---|
| $M_{1/2}$ | 21μm | 0.35μm |
| $M_{3/4}$ | 4.2μm | 0.35μm |
| $M_5$ | 7μm | 0.35μm |
| $M_{6/7}$ | 72.8μm | 4.6μm |
| $M_{8/9}$ | 28.7μm | 1.4μm |
| $M_{10}$ | 14μm | 52μm |
| **Bias Current** | **Value** | |
| $I_{B1}$ | 30μA | |
| $I_{B2/B3/B4}$ | 40μA | |

Table-2. VCII performance parameters

| Parameter | Value |
|---|---|
| Impedance at X node | 802 kΩ |
| Impedance at Y node | 49 Ω |
| Impedance at Z node | 79 Ω |
| α (DC value, BW) | 0.995, 340 MHz |
| β (DC value, BW) | 0.996, 14.6 MHz |
| Static Power Consumption | 700 μW |

## III. THE PROPOSED CMWB STRUCTURES

The first proposed CMWB, designed for two sensor applications, is shown in Fig.4. It consists of two VCIIs and one variable resistor. The X output of VCII₂ is directly connected to the Y input of VCII₁ to make a current subtraction node. Performing KCL analysis at Y input of VCII₁, it results:

$$I_{Y1} = I_1 - \beta_2 I_2 \quad (6)$$

where $I_{YI}$ is the input current to Y port of VCII₁. As a result of the current following action between Y and X terminals of VCII, $I_{YI}$ is transferred to X port of VCII₁ and converted to a proportional voltage by $R_3$ expressed by:

$$V_{X1} = \beta_1(I_1 - \beta_2 I_2)R_3 \quad (7)$$

where $\beta_1$ is the current gain between Y and X terminals of VCII$_1$. The voltage following action between X and Z terminals of VCII$_1$ (with gain of $\alpha_1$) results the output voltage as:

$$V_{out} = \beta_1(I_1 - \beta_2 I_2)\alpha_1 R_3 \qquad (8)$$

In Eq. (8), $\beta_1$, $\beta_2$ and $\alpha_1$ are defined as:

$$\beta_1 = 1 \pm \varepsilon_1; \; \beta_2 = 1 \pm \varepsilon_2; \; \alpha_1 = 1 \pm \varepsilon_3 \qquad (9)$$

where $\varepsilon_1$, $\varepsilon_2$, $\varepsilon_3 \ll 1$ are error terms. For $R_1 = R_2$, input currents are common mode currents $I_1 = I_2 = I_{ref}/2 = I_{cm}$ and using Eq. (8) the (common mode) output voltage is:

$$V_{outc} \cong \varepsilon_2 \alpha_1 R_3 I_{cm} \qquad (10)$$

According to Eq. (10), in the case that there is no change in $R_1$ and $R_2$, due to low value of $\varepsilon_2$, a negligible output voltage is produced. The input currents for a change of $\pm\Delta R$ in $R_1$ and $\mp\Delta R$ in $R_2$ are:

$$I_1 = \frac{R_0 \mp \Delta R}{R_1 + R_2} I_{ref}; \; I_2 = \frac{R_0 \pm \Delta R}{R_1 + R_2} I_{ref} \qquad (11)$$

Inserting Eq. (11) into Eq. (8) gives the output voltage as:

$$V_{outd} = \frac{\varepsilon_2 R_0 \pm 2\Delta R}{R_1 + R_2}\alpha_1 R_3 I_{ref} \cong \frac{\pm\Delta R}{R_0}\alpha_1 R_3 I_{ref} \qquad (12)$$

According to Eq. (12), if $\beta_2$ is made very close to unity, that is $\varepsilon_2 \ll 1$, it is possible to obtain a linear relationship between $\Delta R$ and $V_{out}$. This condition is always satisfied in the VCII because the parameter $\beta$ can be easily designed very close to unity. However, in case only one sensor is used i.e. $R_1 = R_0 \pm \Delta R$ and $R_2 = R_0$, Eq. (8) is a nonlinear function of $\Delta R$.

The second proposed circuit, which is suitable for one sensor applications, is shown in Fig.5 and exhibits a linear relationship between output voltage and $\Delta R$. In this circuit $R_1$ is varying sensor, $R_2$ is constant with value equal to $R_0$ and $R_3$ is variable resistor. A simple analysis gives:

$$V_{out} = \left(\frac{\beta_1 \alpha_1 R_1}{R_2} - 1\right)I_{ref}\alpha_2\beta_2 R_3 \qquad (13)$$

By inserting $R_2 = R_0$ and $R_1 = R_0 \pm \Delta R$ into Eq. (13), we have:

$$V_{out} = (\beta_1\alpha_1 - 1)\alpha_2\beta_2 R_3 I_{ref} + \frac{\pm\Delta R}{R_0}\alpha_2\beta_2 R_3 I_{ref} \qquad (14)$$

The first term in Eq. (14) is an offset voltage which is negligible because the condition $\beta_1\alpha_1 = 1$ is always met. The second term is a linear function of $\Delta R$ controllable by $R_3$.

Noticeable, the accuracy of the examined topology is affected by the matching between the two current sources shown in Fig. 5. However, being external generators, they can be precisely tuned so to match each other. The same generators can also be adjusted at slightly different values achieving the compensation of possible VCII non-idealities.

Fig.6 shows the implementation of variable resistor based on two nMOS. Its value can be changed by control voltages $V_C$ and $-V_C$. The equivalent resistance is expressed as [14]:

$$R_3 = \frac{L}{2\mu C_{ox} W(V_C - V_{TH})} \qquad (15)$$

where $W/L$ is the aspect ratio of $M_{R1}$-$M_{R2}$ transistors.

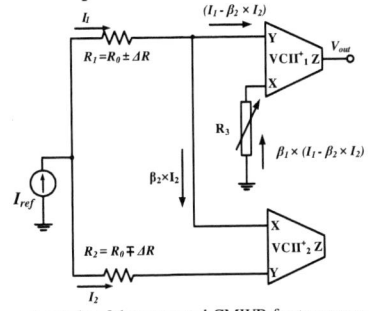

Figure 4. The schematic of the proposed CMWB for two sensor applications

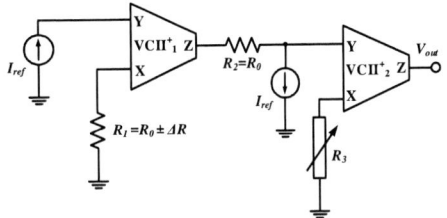

Figure 5. The schematic of the proposed CMWB for one sensor applications

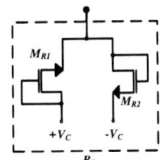

Figure 6. Implementation of variable resistor $R_3$ [14]

IV. SIMULATION RESULTS

The proposed CMWB is simulated using PSPICE and CMOS 0.35µm technology parameters with supply voltage of ±1.65V. DC characteristics of the proposed circuit of Fig.4-5 are shown in Fig.7 for three values of control voltage $V_C$=0.5V, 0.6V, 1.65V and setting $R_1$=10kΩ and different values of $R_2$. As it is seen, for $R_1 = R_2$=10kΩ, the circuit shows negligible offset at output caused by VCII $\alpha$ parameter being slightly different from unity. Moreover, it shows a linear response. Fig.8 shows differential mode performances of the proposed circuits where $R_1$=10kΩ and $R$=20kΩ and for different values of control voltages which shows variable gain with a -3dB bandwidth of 33MHz for the first topology and 6MHz for the second one. Circuits gain variation against different temperatures with $R_1$=10kΩ and $R_2$=20kΩ are shown in Fig.9. The result of Fig.9 proves robust performance against temperature variation and suitability of the proposed circuits for temperature measurements applications. The output impedances of the proposed circuits are 79Ω. The power consumption of the circuits is 1.437mW and 1.436mW for $V_C$=0.5V and $V_C$=0.9V respectively, showing a negligible variation in power consumption for different values of gain.

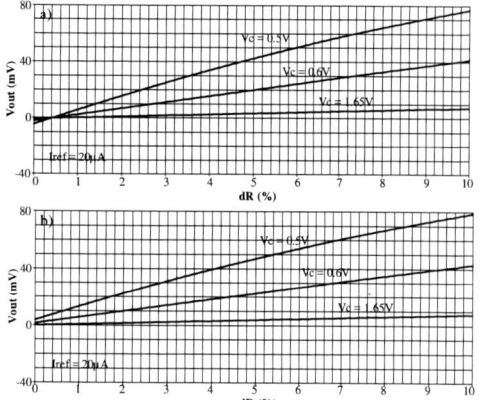

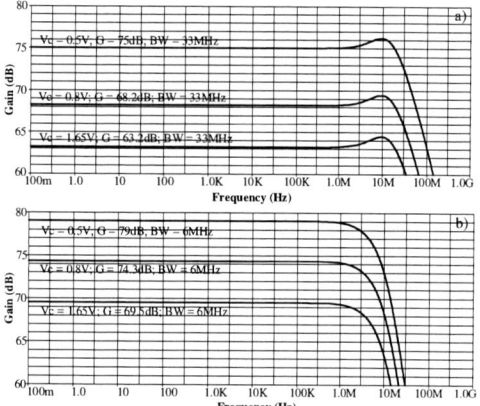

Figure 7. DC response of the proposed CMWB of a) Fig.4 b) Fig.5

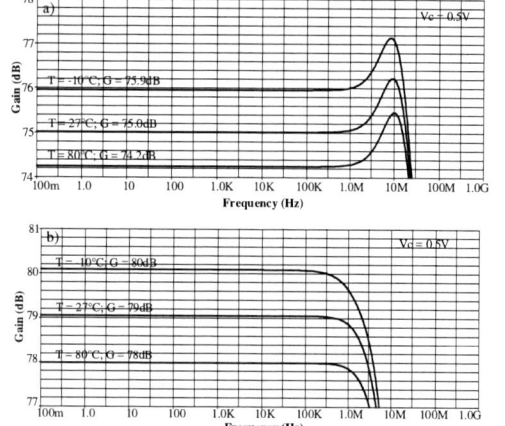

Figure 8. Differential mode frequency response of the proposed CMWB of a) Fig.4 b) Fig.5

Figure 9. Differential mode response of the proposed CMWBs of a) Fig.4 b) Fig.5 with $R_1$=10 kΩ and R2=20 kΩ at different temperatures

## V. CONCLUSION

In this paper, two new CMWB topologies based on VCII are presented. The proposed circuits employ two VCIIs as active elements and one active resistor made of two nMOS transistors. Compared to the previously reported works, the advantages of the proposed circuits are: intrinsically linear performance, resistor-free structure, electronically variable gain and suitability for practical applications due to very low output impedance.

## REFERENCES:

[1] S. Ekelof, "The genesis of the Wheatstone bridge", *Engineering Science & Education Journal*, vol. 10, no. 1, pp. 37-40, 2001.

[2] J. Fraden, *Handbook of Modern Sensors*. Cham: Springer International Publishing, 2016.

[3] G. Ong and P. Chan, "A Power-Aware Chopper-Stabilized Instrumentation Amplifier for Resistive Wheatstone Bridge Sensors", *IEEE Transactions on Instrumentation and Measurement*, vol. 63, no. 9, pp. 2253-2263, 2014.

[4] H. Kaabi and S. Azhari, "AZKA cell, the current-mode alternative of Wheatstone bridge", *IEEE Transactions on Circuits and Systems I: Fundamental Theory and Applications*, vol. 47, no. 9, pp. 1277-1284, 2000.

[5] Y. Ghallab and W. Badawy, "A new topology for a current-mode wheatstone bridge", *IEEE Transactions on Circuits and Systems II: Express Briefs*, vol. 53, no. 1, pp. 18-22, 2006.

[6] I. Mucha, "Current operational amplifiers: Basic architecture, properties, exploitation and future", *Analog Integrated Circuits and Signal Processing*, vol. 7, no. 3, pp. 243-255, 1995.

[7] L. Safari and S. Minaei, "A novel COA-based electronically adjustable current-mode instrumentation amplifier topology", *AEU - International Journal of Electronics and Communications*, vol. 82, pp. 285-293, 2017.

[8] E. Farshidi, "Simple realization of CMOS current-Mode Wheatstone bridge", in *Signals, Circuits and Systems, 2008. SCS 2008. 2nd International Conference on*, Monastir, Tunisia, 2008.

[9] C. Tanaphatsiri, W. Jaikla and M. Siripruchyanun, "A current-mode wheatstone bridge employing only single DO-CDTA", in *Circuits and Systems, 2008. APCCAS 2008. IEEE Asia Pacific Conference on*, Macao, China, 2008.

[10] W. Jaikla and M. Siripruchyanun, "New Low Temperature-sensitive and Electronically Controllable Configurations for the Measurement of Small Resistance Changes", in *Circuits, systems, computers, and communications. ITCCSCC 2006. International technical conference on*, Chiang Mai, Thailand, 2006.

[11] J. Čajka and K. Vrba, "The Voltage Conveyor May Have in Fact Found its Way into Circuit Theory", *AEU - International Journal of Electronics and Communications*, vol. 58, no. 4, pp. 244-248, 2004.

[12] L. Safari, S.J.Azhari "A Novel Low Input Impedance Low Power Fully Differential Current Buffer with ±0.65V Supply Voltage and high bandwidth of 520MHz" WSEAS Transactions on Circuits and Systems vol.11, no.8.pp.272-284, 2012.

[13] R. Carvajal, J. Ramirez-Angulo, A. Lopez-Martin, A. Torralba, J. Galan, A. Carlosena and F. Chavero, "The flipped voltage follower: a useful cell for low-voltage low-power circuit design", *IEEE Transactions on Circuits and Systems I: Regular Papers*, vol. 52, no. 7, pp. 1276-1291, 2005.

[14] Z. Wang, "Reply: 2-MOSFET transresistor with extremely low distortion for output reaching supply voltages", *Electronics Letters*, vol. 26, no. 25, p. 2127, 1990.

# A Fully Differential Charge-Sensitive Amplifier for Dust-Particle Detectors

S. Kelz, T. Veigel, M. Grözing, M. Berroth

Institute of Electrical and Optical Communications Engineering, University of Stuttgart
70569 Stuttgart, Germany

*Abstract*— **This paper presents the design and measurements of a fully differential charge sensitive amplifier operating in the frequency range from 7 Hz to 300 kHz. In comparison to the typically employed single input transistor topology, the shown differential approach greatly simplifies the suppression of common mode noise. The theory of the fully differential charge sensitive amplifier and analytical rules for the correct sizing of the feedback resistor and the input transistor are derived. Finally a double-cascode-amplifier is presented, achieving an equivalent noise charge of 114 elementary charges rms in the 7 Hz to 300 kHz frequency band at 5.4 pF differential detector capacitance. The amplifier is realized in a 0.35 μm standard CMOS technology and consumes 26.4 mW at 3.3 V supply voltage.**

*Keywords—charge-sensitive amplifier, cosmic dust, low noise amplifier, pseudo resistor*

## I. INTRODUCTION

Charge sensitive amplifiers (CSAs) are widely used with different kind of detectors. While typical semiconductor-based-detector systems assume step like charge input signals and relatively narrowband pulse-shaping filters [1], dust-particle detectors require a much larger relative bandwidth.

These dust-particle detector systems are based on electrostatic induction and consist of single or arrayed electrodes, in which charge is influenced as previously charged up particles fly by [2]. The charge up of particles is easily achieved in space by solar UV-radiation. A simple type of differential dust-particle detector consists of two shielded tubes (see fig. 1). As a dust particle passes through a tube-segment an approximately trapezoidal charge signal is first influenced on $\varphi_{det1}$ and thereafter influenced on $\varphi_{det2}$. This charge transfer can be considered as a current $i_{det}$, that is amplified by the CSA as shown in fig. 2. The slope, the amplitude and the duration of the signal can be evaluated to estimate the corresponding speed and surface of the particle. More advanced detectors can also estimate the complete trajectory of the particle by using shifted grids [3] or electrode arrays [2]. All these detectors show a direct

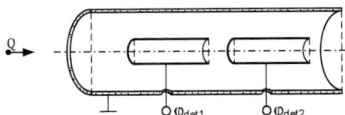

Fig. 1. Simplified concept of an electrostatic induction based dust detector consisting of two tube-electrodes (cross-section).

Partially supported by Deutsche Forschungsgemeinschaft (DFG) grant BE 2256/31-1 and the University of Stuttgart research fund

dependence between the particle speed and the observed output spectrum of the charge signal. As the particle speed varies from 1 m/s [4] to 100 km/s [5], depending on the application, special CSAs are needed that provide the lowest possible noise level over a wide bandwidth. Traditionally these amplifiers cover the 1 kHz to 10 MHz range [5, 6]. New applications on lunar and planetary surfaces require amplifier with a lower frequency corner of 7 Hz [4]. This paper addresses the challenges observed during the design of such an amplifier and presents a fully differential CSA for frequencies from 7 Hz to 300 kHz, implemented in 0.35 μm CMOS, which is to be used with new fully differential dust-particle detectors [3].

## II. THEORY AND DESIGN

### A. Noise in Charge-Sensitive Amplifiers

The noise of the single input transistor CSA has been extensively studied by literature [7, 8] and will only briefly be reviewed here with application to wideband differential amplifiers. Fig. 2 shows a simplified schematic of a differential CSA, including the dominant noise sources. The detector is represented by the current source $i_{det}$ and the detector capacitances $C_{det}$ and $C_G$. The input referred noise of the amplifier is represented by $V_{n,in}$. The noise of the feedback resistors is represented by $V_{n,Rf}$.

Assuming $R_f \gg \frac{1}{\omega c_f}$ and infinite amplifier gain, the charge transfer function can be calculated inside the observed band to

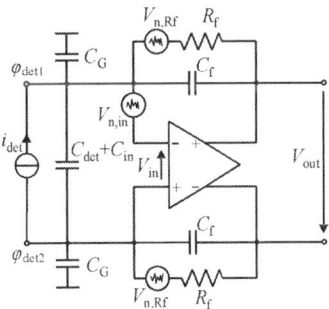

Fig. 2. Schematic of a charge sensitive amplifier and the dominant noise sources. The detector is modeled by the differential detector capacitance $C_{det}$, the detector to ground capacitance $C_G$ and the detector current $i_{det}$.

$$A_Q = \frac{V_{out}}{Q_{in}} = \frac{2}{C_f} \qquad (1)$$

where $Q_{in}$ is the integrated detector current $i_{det}$, $V_{out}$ is the differential output voltage and $C_f$ is the feedback capacitance.

The noise of the CSA is commonly expressed as the equivalent noise charge (ENC), which is defined as the virtual rms charge at the input of the amplifier which would cause the noise that is observed at the output of the CSA. As the filter system is implemented digitally, no pulse shaping is applied to the amplifier and noise calculations are conducted assuming an ideal bandpass from $f_1$=7 Hz to $f_2$=300 kHz. The ENC of the amplifier is derived as:

$$ENC_{rms}^2 = \int_{f1}^{f2} \left( \left( C_{det} + C_{in} + \frac{C_f + C_G}{2} \right)^2 V_{n,in}^2(f) \right. \qquad (2)$$
$$\left. + \frac{1}{2} \left( \frac{4kT}{(2\pi f)^2 R_f} \right) \right) df$$

Where $C_{in}$ is the differential input capacitance of the amplifier without Miller capacitance and $V_{n,in}^2(f)$ is the input referred noise spectral density. If properly designed, the input referred noise of the amplifier is dominated by the input transistors.

To compare this ENC to the value of a classical single input transistor CSA (SI-CSA), a system of two SI-CSAs is assumed, of which one SI-CSA is connected to each electrode. The difference of the output voltages of each amplifier is then used as the output voltage. It can be shown that this results in the same equation for the ENC as (2) with

$$V_{n,in}^2(f) = 2 \, V_{n,in,SI-CSA}^2(f) \qquad (3)$$

Where $V_{n,in,SI-CSA}$ is the input referred noise of the SI-CSA. Therefore both the differential amplifier and the SI-CSA system show the same ENC, if equally sized and biased input transistors are assumed and only input transistor noise is considered. Compared to the SI-CSA system, the differential CSA offers the advantage of suppressing noise from bias current references, which otherwise would have to be attenuated by external capacitors. The differential CSA also shows the ability to realize common mode (CM) suppression in the input stage and thus is less sensitive to CM input charge drifts.

To determine the optimum achievable ENC and input transistor size of the differential CSA for a given process and power consumption, both flicker- and thermal noise have to be considered. The concurrent analytic optimization is not feasible, therefore the following steps focus on the optimization of flicker noise, which dominates at low frequencies. In section III. numerical results of the overall noise optimization are shown.

It is assumed that the input referred noise of a MOSFET is given by:

$$V_G^2(f) = 4kT \frac{2}{3} \frac{1}{g_m} + \frac{K_{fl}}{WLC_{ox}'f} \qquad (4)$$

Assuming that the input transistor noise dominates the overall input referred noise it follows that:

$$V_{n,in}^2(f) = 2 \, V_G^2(f) \qquad (5)$$

By inserting the flicker noise component of (4) in (5) and (5) in (2), the ENC of the differential CSA in dependence of the detector capacitances and the input transistor parameters can be calculated. The minimum of the resulting expression is found at:

$$C_{in,opt,fl} = C_{det} + \frac{1}{2} C_G + \frac{1}{2} C_f \qquad (6)$$

As $C_{in}$ is defined as the differential input capacitance of the amplifier, for optimum flicker noise performance, the input MOSFET gate capacitance should be about as large the sum of twice the differential detector capacitance $C_{det}$ and the detector ground capacitance $C_G$.

### III. AMPLIFIER DESIGN

A crucial step during the design of a CSA, in terms of noise performance, is the selection of the optimum process. This is especially true for low frequency amplifiers like the one presented in this paper. The process has critical influence on the gate-oxide-quality and thus both on the flicker noise and the gate leakage of the input transistor. As classic flicker noise reduction measures like chopping cannot be applied to CSAs, the selection of a low noise process remains the only option to optimize flicker noise besides the dimensioning of the MOSFET.

At the same time the gate leakage has to be as small as possible to reduce the noise due to leakage currents (to be analyzed in Section III.B). To compare different processes, the optimum size input transistor for each process is determined by numeric optimization using the simulated input referred noise, the simulated input capacitance and (2). Four processes with a minimum gate length between 130 nm and 350 nm were evaluated. All provide advanced flicker noise models. The results indicate that the 350 nm process node is best suited for low-frequency CSAs. While the 350 nm PMOS reaches an optimum ENC of 19.4 e⁻ for the given bandwidth in the typical corner (only considering the input transistors) the other tested processes do not reach values below 90 e⁻ for their thin-oxide transistors. This difference is most likely explained by the addition of nitride to the gate oxide after the 350 nm mode, but also by model imprecisions discovered after first measurements. As a byproduct of the process selection the optimum input transistor size is determined. For the detector capacitances $C_{det} = 5$ pF, $C_G = 0$ pF ($C_G$ was included in $C_{det}$) and a feedback capacitance of $C_f = 100$ fF the optimum transistor size and operating point is given in Tab. 1 ($L = 350$ nm).

TABLE I.  INPUT TRANSISTOR PARAMETERS

|         | $W$/mm | $I_D$/mA | $C_{GG}$/pF | ENC/$e^-$ |
|---------|--------|----------|-------------|-----------|
| optimum | 8      | 15       | 10.3        | 19.4      |
| chosen  | 2      | 1.5      | 2.45        | 27.0      |

The resulting optimum gate capacitance of 10.3 pF closely corresponds to the result of (6) and thus confirms the theoretical derivation.

As apparent from fig. 3, the ENC of the input transistor shows only a weak dependence on both the width and the drain current of the transistor. To reduce both the power consumption and the size of the transistor, it is therefore decided to allow for a slightly higher ENC in the final design. In this case the drain current is reduced to 1.5 mA, resulting in an optimum ENC of 27 e⁻.

### A. Core Amplifier

The main design target of the core amplifier is to optimize the noise performance in such a way that the noise optimized input transistor dominates the overall noise power. At the same time a high gain and a low miller effect are desirable to provide a constant closed-loop gain. These goals can be achieved by employing a double cascode design shown in fig. 4. The first cascode transistor (M1) thereby compensates the low intrinsic gain of the minimum gate length (high $g_{DS}$) input transistor. The second cascode transistor (M2) further increases the low frequency gain.

While the cascode transistors do not contribute a significant amount of noise, special care has to be taken during the design of the current sources. As these contain common source transistors, incorrectly designed current sources can easily contribute noise power in the order of the one of the input transistors. In this design the current source noise is reduced by minimizing the effective transconductance of the common source transistors both by resistive degeneration and biasing. The final simulation shows, that approx. 75 % of the amplifier noise power is generated by the input transistors, with the degeneration resistor $R_{\text{deg2}}$ contributing further 12 %. While the differential amplifier ideally suppresses CM noise from bias and supply voltages, detector capacitances with an asymmetry towards ground and transistor mismatch can provide paths for CM noise to enter the differential output signal. Therefore degeneration is also employed inside the current reference, the CMFB circuit and the amplifier tail current source.

### B. Feedback

The feedback network consists both of the feedback capacitor, which is chosen to $C_f = 100$ fF and a feedback resistor $R_f$ to provide the required DC-feedback path. Eq. (2) indicates, that this resistor causes an $\frac{1}{f^2}$ component in the ENC spectral density, which is defined as:

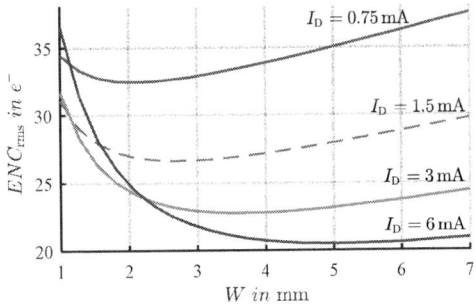

Fig. 3. ENC of the input transistors in dependence of the transistor width and drain current. The transistor length is always $L = 350$ nm. The results indicate that both power consumption and transistor size can be significantly reduced, while allowing only for about 3 dB of additional noise.

$$ENC(f) = \frac{1}{A_Q} \cdot U_{n,out}(f) \qquad (7)$$

This is especially problematic as the presented amplifier is intended to work at very low frequencies. In theory the previously described dust-particle detectors do not contain intrinsic leakage paths due to the capacitor-like design. In practice leakage paths will always exist, but by careful selection of the insulation and PCB materials, the detector can initially be assumed to be leakage-free. Under this assumption the feedback resistor dominates the leakage input noise and can be chosen based on noise considerations alone. Fig. 5 displays the ENC caused by the feedback resistor in dependence of $R_f$. To minimize extra noise caused by the feedback resistor, a value of $R_f = 7.6$ TΩ is targeted, resulting in $ENC_{Rf,rms} = 12.4\ e^-$. This resistance is implemented by a mid-oxide (5 V) PMOS in the subthreshold region, which can be reset by an external reset signal. The bulk is connected to $V_{DD}$. The dimensions are $W = 1\ \mu m$ and $L = 2\ \mu m$.

### IV. MEASUREMENTS

The measurement setup is shown in fig 6. The detector is emulated by a capacitive voltage divider, which presents a 5.4 pF differential capacitance to the input of the amplifier. The output signal of the CSA is high-pass filtered and then amplified using low noise amplifiers. To reduce leakage currents, the PCB and the input capacitors are made of Rogers RO4350B laminate.

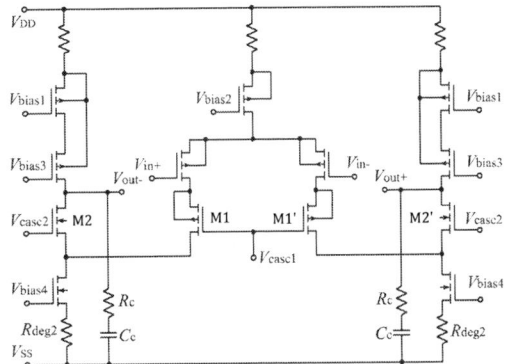

Fig. 4. Schematic of the core amplifier. The amplifier is compensated by a 62 Ohm resistor and a 12 pF capacitance in series towards ground on each output. The CMFB is connected to Vbias2.

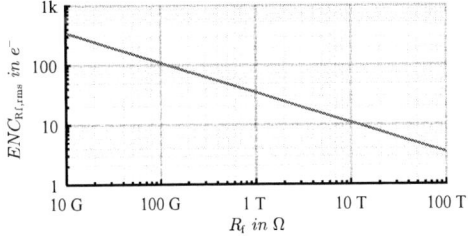

Fig. 5. ENC caused by the feedback resistor in dependence of feedback resistor value. Due to the low frequency operating region of the amplifier, a high feedback resistor is needed to avoid that the feedback resistor dominates the overall noise power.

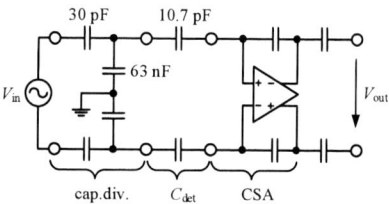

Fig. 6. Simplified measurement setup. The output of the CSA is capacitively coupled to a measurement amplifier.

The noise spectrum is recorded using a real-time oscilloscope and then calculated via discrete Fourier transform. Reset pulses are applied to the feedback MOSFET before each measurement or after about 5 s. The resulting averaged ENC spectral density and the corresponding simulated data is shown in fig. 7. While the typical corner was used during the selection of the input transistor, the measurements show an ENC that is slightly worse (2.2 dB) than the provided worst noise corner model. This difference is most likely explained by unprecise flicker noise models and the extreme W/L-ratio of the input transistor. The integrated ENC in the observed frequency range amounts to 114 electrons. The extra flicker noise prevents precise feedback resistor measurements, but the noise measured at 3 Hz indicates a feedback resistance of at least 2.6 TΩ. The deembedded charge gain of the amplifier shown in fig. 8 corresponds well to the simulation.

## V. Conclusion

The comparison with the state of the art given in tab. 2 shows that the presented amplifier offers the best available noise performance in the low frequency range according to the best knowledge of the authors. Future, improved designs based on worst noise corner optimizations should yield an even better noise performance. A direct comparison with results in other frequency ranges is not possible due to the different requirements in terms of feedback resistance, bandwidth and detector architecture. The most important considerations in terms of output noise of differential CSA have been shown and are implemented in a reference design.

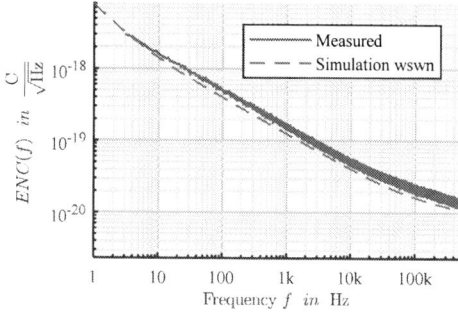

Fig. 7. Measured ENC spectral density vs. simulated data for the worst noise, worst speed (wswn) corner. The overall noise power is dominated by flicker noise.

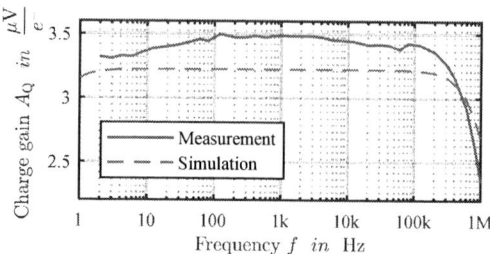

Fig. 8. Charge gain of the CSA. The observed gain is slightly larger, most likely due to process variations of the internal feedback capacitances. The visible noise is caused by the gain deembedding procedure.

## Acknowledgement

The authors would like to thank Mr. Strack and Mr. Srama from IRS at University of Stuttgart for their support and advice.

TABLE II.    Comparison with State of the Art

| Ref. | ENC/e⁻ | Freq. range | $C_{det}$/pF | notes |
|------|--------|-------------|--------------|-------|
| [6] | 95 | 1 kHz … 10 MHz | 5 | JFET |
| [5] | 95 | 10 kHz … 10 MHz | 5 | 180 nm CMOS |
| [4, 9] | 356 | 7 Hz … 10 kHz | 5.4 | comm. CMOS |
| This work | 114 | 7 Hz … 300 kHz | 5.4 | 350 nm CMOS |

## References

[1] Kenneth Hatch, "Aspects of resolution degradation in a nuclear pulse analyzer system," *IEEE Trans. Nucl. Sci.*, vol. 15, no. 1, pp. 303–314, 1968.

[2] S. Auer and F. O. Vonbun, "Highly transparent and rugged sensor for velocity determinations of cosmic dust particles," in *Particle Capture, Recovery and Velocity/Trajectory Measurement Technologies*, 1994.

[3] Y. Li *et al.*, "Instrument concept of a single channel dust trajectory detector," *Advances in Space Research*, vol. 59, no. 6, pp. 1636–1641, 2017.

[4] N. Duncan *et al.*, "The Electrostatic Lunar Dust Analyzer (ELDA) for the detection and trajectory measurement of slow-moving dust particles from the lunar surface," *Planetary and Space Science*, vol. 59, no. 13, pp. 1446–1454, 2011.

[5] R. Srama *et al.*, "A Trajectory Sensor for Sub-micron Sized Dust," *Dust in Planetary Systems*, vol. 643, pp. 213–217, 2007.

[6] S. Auer, "Low-Noise Amplifier for Measuring Dust Charges in the Presence of Plasma and UV Radiation," *Dust in Planetary Systems*, vol. 643, pp. 231–233, 2007.

[7] P. Manfredi, M. Manghisoni, L. Ratti, V. Re, and V. Speziali, "Resolution limits achievable with CMOS front-end in X- and γ-ray analysis with semiconductor detectors," *Nuclear Instruments and Methods in Physics Research Section A: Accelerators, Spectrometers, Detectors and Associated Equipment*, vol. 512, no. 1-2, pp. 167–178, 2003.

[8] P. O'Connor and G. de Geronimo, "Prospects for charge sensitive amplifiers in scaled CMOS," *Nuclear Instruments and Methods in Physics Research Section A: Accelerators, Spectrometers, Detectors and Associated Equipment*, vol. 480, no. 2-3, pp. 713–725, 2002.

[9] J. Xie *et al.*, "Laboratory testing and data analysis of the Electrostatic Lunar Dust Analyzer (ELDA) instrument," *Planetary and Space Science*, vol. 89, pp. 63–70, 2013.

*PRIME 2018, Prague, Czech Republic*

*Session: Analog Circuits I*

# A ZTC-based 0.5 V CMOS Voltage Reference

Y. Wenger and B. Meinerzhagen

Institute for Electron Devices and Circuits, Technical University Braunschweig
Hans-Sommer-Str. 66, 38106 Braunschweig, Germany
Email: y.wenger@tu-bs.de

*Abstract*—This paper presents a voltage reference circuit operational from a 0.5 V supply which is based on the zero-temperature coefficient (ZTC) operating point of a MOS transistor. The ZTC condition is reviewed and it is found that a MOSFET biased below its ZTC point with a PTAT current source can yield a temperature stable output at this low supply voltage. With this idea in mind, a circuit which does not rely on the availability of special devices like Shottky diodes is designed in 130 nm CMOS. Simulations show that this circuit generates an average reference voltage of 318 mV from a 0.5 V supply. The temperature coefficient is 154 ppm/K and the voltage reference has a power supply rejection ratio (PSRR) of 41 dB at DC.

*Index Terms*—Voltage reference, ultra-low voltage, temperature compensation, CMOS, Zero-temperature coefficient.

## I. INTRODUCTION

Driven mainly by the demand for low power digital circuits, CMOS ICs with reduced supply voltage have been studied intensively in recent years. This has created a new challenge for analog designers: Many proven circuit concepts no longer work at supply voltages significantly below 1 V and those which still can be employed show a severely degraded performance. Therefore, new circuit architectures have to be found that can be used in an ultra-low supply voltage environment. One circuit that is affected by this supply voltage reduction is the well-known bandgap reference (BGR) [1]. As this circuit relies on a forward biased PN junction, it is difficult to generate reference voltages below 0.5 V [2].

To reduce the supply voltage of a circuit into the range of 0.5 V and below, different devices or architectures have to be used: One option is to substitute the PN diode by a Shottky diode [3]. These diodes usually have a forward voltage in the range of 0.2 V to 0.6 V, and they can be easily implemented into a standard CMOS process. However, most design-kits do not make Shottky diodes available to the circuit designer, which prevents their use. Another way to make the BGR ultra-low voltage compatible is to use discrete-time techniques like charge pumps [4].

An interesting alternative is to generate the reference voltage using only MOSFETs as active elements. One often used possibility is to rely on the diode-like characteristic of a MOS transistor in weak inversion. A recent example describing a 0.5 V voltage reference based on this idea can be found in [5]. Another option, the one which will be further explored in this work, is to bias a diode-connected MOS transistor near its zero-temperature coefficient (ZTC) operating point. In this operating point at a fixed gate-source voltage depending

on technology and geometry, the transistor's drain current becomes independent of temperature. If biased at a constant current slightly outside the ZTC point, the gate-source voltage will either show CTAT or PTAT behavior. This leaves room for two basic circuit concepts to realize a temperature-independent voltage: The transistor can be biased directly at its ZTC point, as e.g. in [6], or outside this point with a current source that shows the opposite temperature behavior compared to the gate-source voltage. In [7], a Shottky-based PTAT current source is used to bias an nMOS transistor. The design also employs zero-$V_{th}$ transistors.

In this work, we find that the second concept is more suitable for designing an ultra-low voltage reference in CMOS exploiting the ZTC effect without the need for special devices. The rest of the paper is organized as follows: Following this introduction, the ZTC effect will be reviewed and different techniques for lowering the ZTC voltage are discussed. These ideas are then used to design a voltage reference operating at 0.5 V supply voltage. Finally, the circuit performance is discussed based on simulation results before the paper concludes.

## II. REVIEW OF THE ZTC EFFECT

The ZTC effect is usually explained by the mutual compensation of the temperature dependencies of threshold voltage $V_{th}$ and carrier mobility $\mu$ [8]. Measurements show that it can even be found in the most advanced CMOS nodes [9]. Assume that threshold voltage and mobility can be modeled by

$$V_{th}(T) = V_{th}(T_0) + \alpha_{V_{th}}(T - T_0) \qquad (1)$$

$$\mu(T) = \mu(T_0)\left(\frac{T}{T_0}\right)^{\alpha_\mu}, \qquad (2)$$

with a reference temperature $T_0$, usually chosen to be 300 K, and the technology dependent temperature coefficients $\alpha_{V_{th}}$ and $\alpha_\mu$, which commonly lie in the range of $-3$ mV/K to $-0.5$ mV/K and $-2.42$ to $-1.2$, respectively [10], [11].

As derived in [12], these equations together with the square-law characteristic of a strongly-inverted MOSFET in saturation can be used to find a simple model for the ZTC operating point of a diode-connected MOS transistor:

$$V_{GSF} = V_{th}(T_0) - \alpha_{V_{th}}T_0 \qquad (3)$$

$$I_{DF} = \frac{1}{2}\mu(T_0)T_0^2 C_{ox}\frac{W}{L}\alpha_{V_{th}}. \qquad (4)$$

978-1-5386-5388-3/18 $31.00 © 2018 IEEE

*Paper P5*

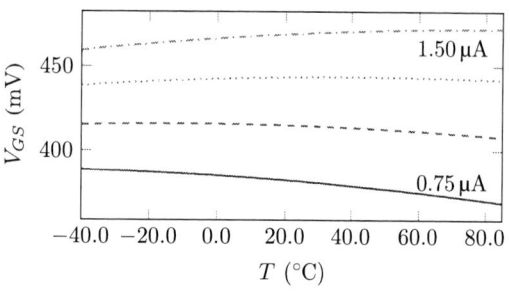
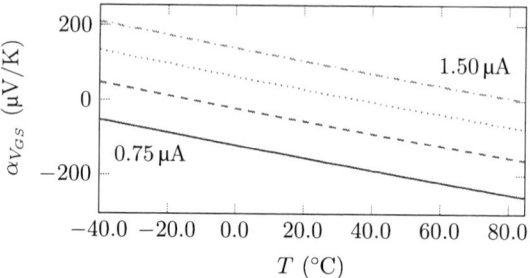

Figure 1: Gate-source voltage and its temperature coefficient of a diode-connected nMOS with $W/L = 1\,\mu m/10\,\mu m$ for different drain currents

Moreover, it can be shown that a transistor biased at a drain current that deviates from the ZTC current $I_{DF}$ by a small difference $\Delta I_D$

$$I_D(T) = I_{DF} + \Delta I_D \qquad (5)$$

has a gate-source voltage with a linear temperature dependence

$$V_{GS}(T, \Delta I_D) = V_{GSF} + \alpha_{V_{GS}}T. \qquad (6)$$

Depending on the sign of $\Delta I_D$, the sign of $\alpha_{V_{GS}}$ can be either positive or negative. In circuit design, $\Delta I_D$ can either be set by changing the current through a MOS diode with a fixed aspect ratio, or more practically by keeping the transistor current $I_D$ constant and varying the aspect ratio and thereby $I_{DF}$.

Fig. 1 depicts the gate-source voltage and its temperature coefficient for a diode-connected nMOS with $W/L = 1\,\mu m/10\,\mu m$ from IHP's (Innovations for High Performance Microelectronics) 130 nm BiCMOS technology. Each of the curves is for a different drain current between $0.75\,\mu A$ and $1.50\,\mu A$ which has been kept constant in the given temperature range. We can see that the behavior deviates visibly from the one predicted by the simple models above. Particularly, the coefficient $\alpha_{V_{GS}}$ is itself temperature-dependent. Nevertheless, the basic conclusions remain the same: There exists an operating point where drain current and gate-source voltage show little temperature dependence (in this case: $I_{DF} \approx 1.25\,\mu A$, $V_{GSF} \approx 440\,mV$) and $\alpha_{V_{GS}}$ is positive for $I_D > I_{DF}$ and negative for $I_D < I_{DF}$.

### III. Low-Voltage Reference Design

The output voltage of the reference circuit should ideally lie at least a saturation voltage $V_{sat}$ away from both supply voltage rails. Independent of technology, the minimum $V_{sat}$ for a transistor in weak or moderate inversion is around five times the thermal voltage (about $125\,mV$ at $300\,K$). For a transistor in strong inversion it is slightly higher [13].

Therefore, the ZTC voltage of the example shown in Fig. 1 is still too high for a supply voltage of $0.5\,V$. If, however, we bias the transistor in his CTAT range below the ZTC point, we come into the voltage range where these low input voltages are possible. To achieve temperature independence, this means biasing the transistor with a PTAT current source.

Another possibility to achieve a lower output voltage can be

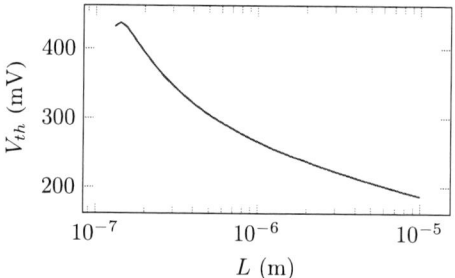

Figure 2: Dependence of threshold voltage on channel length for an nMOS

found from (3). As the ZTC voltage is proportional to the threshold voltage, it is sensible to try to lower $V_{th}$ as much as possible. One way of doing so is to exploit the dependence of threshold voltage on channel length in submicron technologies, resulting from halo implants and charge-sharing effects [11]. As shown in Fig. 2, the nMOS transistors in the 130 nm BiCMOS technology have their $V_{th,max} \approx 440\,mV$ close to minimum channel length, while the threshold voltage of transistors with a channel length of $10\,\mu m$ (the maximum length for which a model is available) is about $0.2\,V$. A similar, yet less pronounced, behavior can be observed for the pMOS transistors. Hence, for the low-voltage design presented here mostly channel lengths around the maximum of $10\,\mu m$ are used.

To reduce the threshold voltages even further, body biasing can be employed. However, one has to be careful that no junctions become forward biased. Especially, as silicon diodes show a strong current increase at higher temperatures. This can result in an upper limit for the range of supply voltages for which the presented design works.

The schematic of the PTAT current source is shown in Fig. 3. The well-known beta multiplier circuit with feedback amplifier [14] is used as it is straightforward to implement and does not pose very stringent requirements on supply voltage. The basic operation is formed by the transistors $M_1$, $M_2$, where $\frac{W_2}{L_2} = K \cdot \frac{W_1}{L_1}$, and resistor $R$, which together set

*PRIME 2018, Prague, Czech Republic*

*Session: Analog Circuits I*

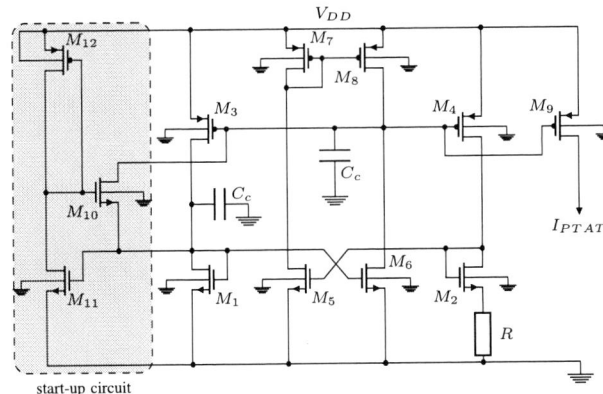

Figure 3: Schematic of the PTAT current source

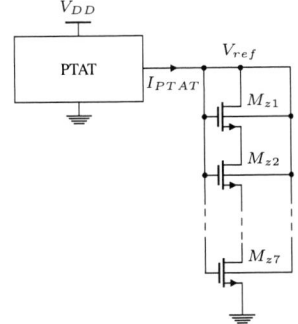

Figure 4: Output voltage generation

the current $I_{PTAT}$. It has been shown [14] that the current's temperature coefficient mainly depends on the temperature behavior of the resistor and the transistor's threshold voltage. If a resistor with a CTAT characteristic is chosen, the output current will be PTAT.

Transistors $M_5$-$M_8$ form a feedback amplifier which together with $M_3$ and $M_4$ forces the voltage over and the current through $M_1$ and $M_2$ to be the same. This feedback loop is stabilized by two capacitors $C_c$.

The obligatory start-up circuit consists of $M_{10}$-$M_{12}$. It prevents a zero current condition by equalizing the gate potentials of $M_{1,2}$ and $M_{3,4}$.

All nMOS except $M_{10}$, which has $\frac{W_{10}}{L_{10}} = \frac{10\,\mu m}{0.13\,\mu m}$, and $M_2$ ($K = 4$) have $W/L = \frac{10\,\mu m}{10\,\mu m}$. The pMOS $M_{12}$ is $0.15\,\mu m$ wide and $10\,\mu m$ long. All other pMOS have $W/L = \frac{5\,\mu m}{2\,\mu m}$. Their bulks are tied to ground reducing the gate-source voltage of $M_7$ and making sure that all transistors in the amplifier are safely in saturation.

The resistor $R = 92\,k\Omega$ is realized via p-doped polysilicon. The result is a PTAT current of $I_{PTAT} \approx 500\,nA$ with a temperature coefficient of $\alpha_I = 3200\,ppm/K$ between $-40\,°C$ and $85\,°C$.

Fig. 4 is the schematic of the final voltage reference circuit. The nMOS diode for output voltage generation is split into seven series-connected transistors $M_{z1}$-$M_{z7}$. In this design, the length of the ZTC transistor has to be significantly larger than its width. Because the maximum length of a single transistor in the target technology is $10\,\mu m$, the shown series connection is used as it behaves similar to a transistor with the added length of all seven MOSFETs. Furthermore, for the same aspect ratio larger widths can be used reducing the sensitivity to process variations. Each transistor has $\frac{W}{L} = \frac{6\,\mu m}{10\,\mu m}$. IHP's $130\,nm$ technology offers the option to use isolated pWells. Although not strictly necessary, it was decided to use these, resulting in a slight reduction of the output voltage by tying the ZTC transistors' bulks to the output voltage. Note that shorting the pWell to ground is not an option as it significantly reduces the PSRR of the voltage reference.

The circuit layout was done in the $130\,nm$ SiGe technology mentioned before. The whole layout occupies a chip area of approximately $170\,\mu m \times 100\,\mu m$.

## IV. SIMULATION RESULTS

The design was verified in post-layout circuit simulations for the temperature range from $-40\,°C$ to $85\,°C$ at a nominal supply voltage of $0.5\,V$.

Fig. 5a shows the variation of the output voltage over temperature at nominal conditions. At $27\,°C$ a voltage of $317.1\,mV$ is generated. Its temperature coefficient is $68.7\,ppm/K$ in the observed temperature range.

The circuit's line regulation can be found from Fig. 5b, which shows the variation of the reference voltage with supply voltage. It is approximately $1.8\,\%/V$ in the range from $0.45\,V$ to $0.7\,V$. Above $V_{DD} \approx 700\,mV$ the temperature stability of the reference starts to deteriorate. As explained before, the reason for this is the forward biasing of the SB and DB diodes within the pMOS transistors in the PTAT current generator in Fig. 3. This gives rise to a significant bulk current at higher temperatures.

The PSRR is shown in Fig. 5c. At DC, it is $42.6\,dB$, and its corner frequency lies around $2.2\,kHz$.

A Monte Carlo simulation with 1000 runs has been performed taking into account process variation as well as mismatch. The results are shown in Fig. 6 and in the rightmost column of Tab. I together with the other performance metrics of the voltage reference. As expected, the average temperature coefficient $\overline{\alpha}_{V_{ref}} = 154\,ppm/K$ is worse compared to the nominal simulation. The average output voltage is approximately $318\,mV$ which is close to the nominal case with $\sigma/\mu$ being $10.7\,\%$. The average PSRR lies above $40\,dB$. It should be noted that the proposed circuit can easily be trimmed by changing the values of $R$ in the PTAT current source (Fig. 3) and shorting the MOSFETs in the stack shown in Fig. 4.

## V. CONCLUSION

In Tab. I the presented voltage reference is compared to other untrimmed ultra-low voltage designs. With respect to the circuits based on Shottky diodes, this work shows very good

Table I: Comparison of CMOS ultra-low voltage references

| Type | [4], exp | [5], sim | [15], exp | [16], exp | [3], exp | [7], sim | This work, sim |
|---|---|---|---|---|---|---|---|
| Technology (nm) | BGR | Subthresh. | Subthresh. | 2T | Shottky | ZTC Shottky | ZTC |
| | 130 | 180 | 180 | 130 | 90 | 130 | 130 |
| Temperature Range (°C) | $0-80$ | $-40-125$ | $-40-130$ | $-20-80$ | $10-100$ | $-55-125$ | $-40-85$ |
| Supply Voltage (V) | $> 0.5$ | $> 0.5$ | $> 0.4$ | $0.5-3.0$ | $0.55-0.65$ | $> 0.45$ | $0.45-0.7$ |
| Reference Voltage (mV) | 256 | 51 | 212 | 177 | 247 | 312 | $\mu = 318, \sigma = 34$ |
| Temperature Coefficient (ppm/K) | 40 | 25.4 | 84.5 | 62 | 270 | 214 | $\mu = 154.1, \sigma = 186.8$ |
| $PSRR_{DC}$ (dB) | 40 | 76 | 40 | 53 | - | 28 | $\mu = 40.9, \sigma = 5.1$ |
| Line Regulation (%/V) | - | 0.065 | 1.2 | 0.033 | 0.96 | 11.4 | 1.8 |
| Power Dissipation (µW) | 0.03 | 0.19 | 0.19 | $2.2 \times 10^{-6}$ | 398 | 5.9 | 1.3 |
| Area (mm$^2$) | 0.026 | - | 0.09 | 0.001 | 0.019 | 0.014 | 0.017 |

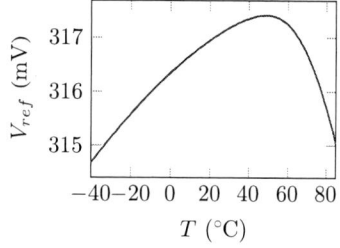

(a) Reference voltage vs. temperature

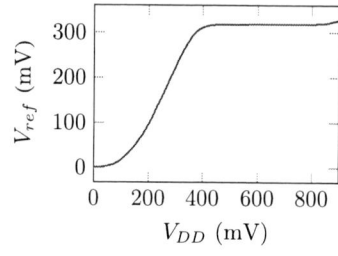

(b) Reference voltage vs. supply voltage

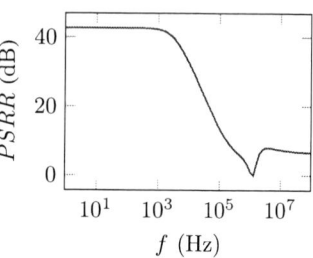

(c) PSRR vs. frequency

Figure 5: Simulation results at nominal conditions

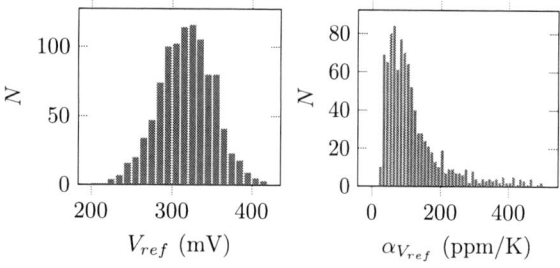

Figure 6: Output voltage and temperature coefficient for 1000 Monte Carlo runs

performance. The recently presented designs from [4], [5], and [15] based on a BGR with charge pumps or subthreshold MOSFETs, however, perform significantly better, especially in terms of power consumption. Unfortunately, no estimation of the area required is given in [5]. The design from [16] needs a process with two types of transistors having distinctly different threshold voltages to achieve a voltage reference with extremely low power consumption.

Nevertheless, the concept presented in this work has the clear potential to cut the power consumption even further. By stacking a higher number of nMOS transistors in Fig. 4, the same output voltage can be achieved with a reduced biasing current. Together with a subthreshold PTAT generator as in [5], a nano-Watt voltage reference becomes possible.

## REFERENCES

[1] K. E. Kuijk, "A precision reference voltage source", *IEEE J. of Solid-State Circuits*, vol. 8, no. 3, pp. 222–226, 1973.

[2] H. Banba, H. Shiga, A. Umezawa, T. Miyaba, T. Tanzawa, S. Atsumi, and K. Sakui, "A CMOS bandgap reference circuit with sub-1-V operation", *IEEE J. of Solid-State Circuits*, vol. 34, no. 5, pp. 670–674, 1999.

[3] P. Kinget, C. Vezyrtzis, E. Chiang, B. Hung, and T. L. Li, "Voltage references for ultra-low supply voltages", in *Custom Integrated Circuits Conference*, 2008.

[4] A. Shrivastava, K. Craig, N. E. Roberts, D. D. Wentzloff, and B. H. Calhoun, "A 32nW bandgap reference operational from 0.5V supply for ultra-low power systems", in *Int. Solid-State Circuits Conference*, 2015.

[5] P. B. Basyurt, E. Bonizzoni, F. Maloberti, and D. Y. Aksin, "A low-power low-noise CMOS voltage reference with improved PSR for wearable sensor systems", in *Int. Symp. on Circuits and Systems*, 2017, pp. 1–4.

[6] J. Jiang, W. Shu, and J. S. Chang, "A 5.6 ppm/°C temperature coefficient, 87-dB PSRR, sub-1-V voltage reference in 65-nm CMOS exploiting the zero-temperature-coefficient point", *IEEE J. of Solid-State Circuits*, vol. 52, no. 3, pp. 623–633, 2017.

[7] D. Cordova, P. Toledo, H. Klimach, E. Fabris, and S. Bampi, "0.5 V supply voltage reference based on the MOSFET ZTC condition", in *Symp. on Integrated Circuits and Systems Design*, 2015.

[8] I. M. Filanovsky and A. Allam, "Mutual compensation of mobility and threshold voltage temperature effects with applications in CMOS circuits", *IEEE Trans. on Circuits and Systems I: Fundamental Theory and Applications*, vol. 48, no. 7, pp. 876–884, 2001.

[9] M. Shin, M. Shi, M. Mouis, A. Cros, E. Josse, G. T. Kim, and G. Ghibaudo, "Low temperature characterization of 14nm FDSOI CMOS devices", in *11th Int. Workshop on Low Temperature Electronics*, 2014, pp. 29–32.

[10] S. M. Sze, *Physics of semiconductor devices*, 2nd ed. New York: Wiley, 1981.

[11] Y. Tsividis and C. McAndrew, *The MOS transistor*, Int. 3rd ed. New York and Oxford: Oxford Univ. Press, 2012.

[12] Y. Wenger and B. Meinerzhagen, "A stable CMOS current reference based on the ZTC operating point", in *Conference on Ph.D. Research in Microelectronics and Electronics*, 2017, pp. 273–276.

[13] P. Kinget, S. Chatterjee, and Y. Tsividis, "Ultra-low voltage analog design techniques for nanoscale CMOS technologies", in *Conference on Electron Devices and Solid-State Circuits*, 2005, pp. 9–14.

[14] R. J. Baker, *CMOS: Circuit design, layout, and simulation*, 3rd ed. Piscataway and Hoboken: Wiley, 2010.

[15] P. B. Basyurt, E. Bonizzoni, D. Y. Aksin, and F. Maloberti, "A 0.4-V supply curvature-corrected reference generator with 84.5-ppm/°C average temperature coefficient within -40 °C to 130 °C", *IEEE Trans. on Circuits and Systems II*, vol. 64, no. 4, pp. 362–366, 2017.

[16] M. Seok, G. Kim, D. Blaauw, and D. Sylvester, "A portable 2-transistor picowatt temperature-compensated voltage reference operating at 0.5 v", *IEEE J. of Solid-State Circuits*, vol. 47, no. 10, pp. 2534–2545, 2012.

PRIME 2018, Prague, Czech Republic

Session: Data Converters

# Generalization of Referenceless Timing Mismatch Calibration Methods for Time-Interleaved ADCs

Arda Uran, Mustafa Kilic and Yusuf Leblebici

Microelectronic Systems Laboratory (LSM)

Swiss Federal Institute of Technology, Lausanne (EPFL) CH-1015 Lausanne, Switzerland

e-mail: arda.uran@epfl.ch

*Abstract*—Calibration of time-interleaved analog-to-digital converters is a problem whose necessity and complexity increase with the number of interleaved channels. In this study, we develop a generic representation of the referenceless timing mismatch calibration scheme for N-channel TI-ADCs. We compare cross-correlation and mean absolute difference based approaches, and investigate the effect of increasing number of channels on the performance. We use both mathematical analyses and simulations to reveal degradation mechanisms, and discuss the extent to which this scheme is applicable.

*Index Terms*—Time-interleaved ADC, calibration, timing mismatch, referenceless, cross-correlation, mean absolute difference.

## I. INTRODUCTION

Time-interleaving is a commonly resorted solution to linearize the power/throughput trade-off curve in high-speed analog-to-digital converter (ADC) design. However, statistical or deterministic intra-die variations cause inter-channel mismatch of critical ADC parameters such as offset, gain, sampling time and input bandwidth. Independent of the quality of a sub-ADC, these variations have detrimental effects on the performance of the overall time-interleaved ADC (TI-ADC), whose mathematical bounds have been investigated extensively [1]–[4]. The focus of this paper is the calibration of inter-channel sampling time mismatch.

Several methods [5], [6] that were proposed to overcome the timing mismatch problem employ an additional channel which samples the input at the same instant as the calibrated channel, providing a reference point for mismatch detection. However, the practical implementation of the additional channel brings additional intractable issues such as unequal crosstalk and power conditions between calibrated stages [7]. On the other hand, the referenceless calibration method [8], relies on the cross-correlation (CC) of the sub-ADC outputs, so it permits independent design of the TI-ADC and its calibration circuit. Nevertheless, the major drawback of this method is misdetection for some certain input signals [9]–[11].

There has been recent effort to mitigate the short-comings and accommodate higher number of channels for the referenceless method. In [10], notch filtering of problematic frequency components is suggested in order to avoid misdetection. In [11], the notches in the frequency response are removed by modifying the sampling sequence of the sub-ADCs, sacrificing output resolution. Furthermore, as cross-correlation is an hardware-expensive operation, there has

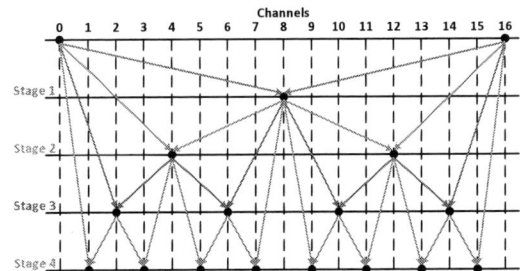

Fig. 1. Calibration stages exemplified on a 16-channel TI-ADC.

been less hardware demanding approaches trading computation accuracy with complexity, such as using mean-absolute-difference (MAD) [9], or logarithmic multipliers [11]. In this paper we elaborate the mathematical basis of the referenceless timing mismatch calibration method when generalized to N-channel TI-ADCs, where N is a power of 2, and compare the performance of the CC-based and MAD-based approaches against the number of channels.

## II. CC-BASED REFERENCELESS DETECTION

The referenceless, cross-correlation based method (originally described for 2-channels in [8]) can be generalized to N-channel TI-ADCs. The output of the $i^{th}$ sub-ADC, $y_i[n]$, sampling an input signal $x(t)$ can be written as

$$y_i[n] = x(n\hat{T}_s + iT_s + \tau_i) + \varepsilon_q \,, \tag{1}$$

where $T_s$ is the TI-ADC sampling period, $\hat{T}_s = NT_s$ is the sub-ADC sampling period, $\tau_i$ is the channel timing error, and $\varepsilon_q$ is the quantization error. For the sake of simplicity in demonstration, we neglect quantization errors for the following analyses.

If we follow the same idea presented in [9] where the number of channels are increased from 2 to 4, $\tau_{N/2}$ can be equalized to $\tau_0$ using $y_0$ and $y_N$, where $y_0$ is accepted as the reference channel output and $y_N$ corresponds to the next output of the reference channel. This will make $y_{N/2}$ a new reference point which can be used to calibrate $y_{N/4}$ and $y_{3N/4}$, and so on. The pseudo-code in Listing 1 summarizes this calibration sequence, and Figure 1 illustrates it on a 16-channel TI-ADC. The inner for loop can be processed in parallel, so calibration can be achieved in $\log_2(N)$ stages.

978-1-5386-5388-3/18 $31.00 © 2018 IEEE

Listing 1. Timing mismatch calibration algorithm for an N-channel TI-ADC

```
for s = 0 : log₂(N)-1
    for p = 0 : 2ˢ-1
        k = N / 2ˢ⁺¹ ;
        ref₀ = pN / 2ˢ ;
        cal = ref₀ + k ;
        ref₁ = cal + k ;
        Calibrate y_cal using y_ref₀ and y_ref₁
```

For any stage $s$, the calibrated channel output $y_{cal}$ and the reference channels $y_{ref_0}$ and $y_{ref_1}$ can be written as

$$y_{ref_0} = x(n\hat{T}_s + \tau_0) \tag{2}$$

$$y_{cal} = x(n\hat{T}_s + kT_s + \tau_1) \tag{3}$$

$$y_{ref_1} = x(n\hat{T}_s + 2kT_s + \tau_0) . \tag{4}$$

The CC of the outputs, $R_{y_{cal}y_{ref_0}}$ and $R_{y_{ref_1}y_{cal}}$, then become the autocorrelation of the input, $R_{xx}$, as

$$R_{y_{cal}y_{ref_0}} = \mathbf{E}[y_{cal} \cdot y_{ref_0}] = R_{xx}(kT_s + \Delta\tau) \tag{5}$$

$$R_{y_{ref_1}y_{cal}} = \mathbf{E}[y_{ref_1} \cdot y_{cal}] = R_{xx}(kT_s - \Delta\tau) , \tag{6}$$

and thus the difference of the CC functions, $D(\Delta\tau)$, becomes

$$\Rightarrow D(\Delta\tau) = R_{xx}(kT_s + \Delta\tau) - R_{xx}(kT_s - \Delta\tau)$$
$$\approx 2\Delta\tau \cdot \left.\frac{dR_{xx}(\tau)}{d\tau}\right|_{\tau=kT_s} . \tag{7}$$

$D(\Delta\tau)$ is proportional to $\Delta\tau = \tau_1 - \tau_0$, thus it can be used as the detection parameter within a calibration algorithm, provided that $dR_{xx}(\tau)/d\tau$ is non-zero at $\tau = kT_s$. This can be investigated by repeating the analysis in [9] for N channels:

$$\left.\frac{dR_{xx}(\tau)}{d\tau}\right|_{\tau=kT_s} = -\int_{-\infty}^{\infty} 2\pi f S_x(f) \sin(2\pi f kT_s) df . \tag{8}$$

The above integral is equal to zero only when the input signal is a sine (or a sum of sines) with frequency integer multiples of $f_s/(2k)$ due to the sifting property of the Dirac function associated with the PSD of sine. As $k_{max} = N/2$, the frequency response of this calibration method has notches at integer multiples of $f_s/N$. This effect is not seen in 2-channel TI-ADCs where $k_{max} = 1$ because the input signal frequency is limited to $f_{in} < f_s/2$ by Nyquist theorem.

Another constraint on the input frequency results from the discrete nature of the autocorrelation estimate. Correct estimation requires more than two distinct samples. If the input is a sine with frequency $f_{in}$ and initial phase $\phi_0$, the sub-ADC outputs are of the form $\sin(2\pi N(f_{in}/f_s)n + \phi_0)$. When $f_{in}$ is an integer multiple of $f_s/(2N)$, the sampled values have the same magnitude. Due to this limitation, notches occur at integer multiples of $f_s/(2N)$, as also stated in [10], [11].

### III. MAD-BASED REFERENCELESS DETECTION

The method presented in [8] can also be implemented without any hardware-expensive multipliers, as demonstrated in [9]. The idea is based on replacing CC in (7) with MAD, which provides a new expression for $D(\Delta\tau)$ as

$$D(\Delta\tau) = \mathbf{E}[|y_{ref_1} - y_{cal}| - |y_{cal} - y_{ref_0}|] . \tag{9}$$

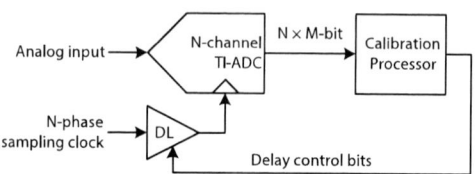

Fig. 2. Conceptual block diagram of a mixed-signal timing mismatch calibration architecture for an M-bit, N-channel TI-ADC.

Here we explore this expression further in order to find a relation between the MAD and CC-based $D(\Delta\tau)$ expressions. The mean square and the mean absolute values of a variable $x$ are related by the Cauchy-Schwarz inequality as

$$\mathbf{E}[|x|] \leq \sqrt{\mathbf{E}[x^2]} . \tag{10}$$

The ratio $\sqrt{\mathbf{E}[x^2]}/\mathbf{E}[|x|]$ is termed as the form factor $k_f$ of $x$, which is a positive constant for a given signal and lower bounded by 1. Substituting $\mathbf{E}[|x|]$ with $\sqrt{\mathbf{E}[x^2]}/k_f$ in (9), an expression for MAD-based $D(\Delta\tau)$ depending on $k_f$ can be derived as

$$D(\Delta\tau) = \mathbf{E}[|y_{ref_1} - y_{cal}| - |y_{cal} - y_{ref_0}|]$$
$$= k_f^{-1}\left(\sqrt{\mathbf{E}[(y_{ref_1} - y_{cal})^2]} - \sqrt{\mathbf{E}[(y_{cal} - y_{ref_0})^2]}\right)$$
$$= \frac{\mathbf{E}[(y_{ref_1} - y_{cal})^2] - \mathbf{E}[(y_{cal} - y_{ref_0})^2]}{k_f\left(\sqrt{\mathbf{E}[(y_{ref_1} - y_{cal})^2]} + \sqrt{\mathbf{E}[(y_{cal} - y_{ref_0})^2]}\right)} . \tag{11}$$

Assuming that the channel output powers are the same and equal to $P_x$, which implies at least wide sense stationarity (WSS) and no offset or gain mismatch, $\mathbf{E}[(y_{ref_1} - y_{cal})^2]$ can be expanded as

$$\mathbf{E}[(y_{ref_1} - y_{cal})^2] = \mathbf{E}[y_{ref_1}^2] + \mathbf{E}[y_{cal}^2] - 2\mathbf{E}[y_{ref_1} \cdot y_{cal}]$$
$$= 2(P_x - R_{xx}(kT_s - \Delta\tau)) . \tag{12}$$

The same expansion of $\mathbf{E}[(y_{cal} - y_{ref_0})^2]$ simplifies the nominator of (11) to $2\mathbf{E}[y_{cal} \cdot y_{ref_0} - y_{ref_1} \cdot y_{cal}]$. Also, if $\Delta\tau$ is small enough compared to $T_s$, the denominator can be approximated as $2k_f\sqrt{2(P_x - R_{xx}(kT_s))}$. Rewriting (11),

$$D(\Delta\tau) = \frac{\mathbf{E}[y_{cal} \cdot y_{ref_0} - y_{ref_1} \cdot y_{cal}]}{k_f\sqrt{2(P_x - R_{xx}(kT_s))}} . \tag{13}$$

This forms the relation between the MAD-based (9) and the CC-based (7) expressions for $D(\Delta\tau)$, based on the aforementioned assumptions. The key observation from (13) is that the relation depends the autocorrelation of the input signal at that calibration stage, $R_{xx}(kT_s)$, for a given form factor, $k_f$.

While the MAD-based method reduces the hardware cost by replacing multiplications in the CC-based method with subtractions, discrete computation of MAD for a sinusoidal input brings additional constraints as discussed in Section IV.

*PRIME 2018, Prague, Czech Republic*

*Session: Data Converters*

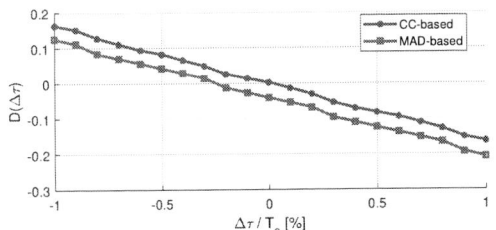

Fig. 3. $D(\Delta\tau)$ versus timing error for CC and MAD based methods (N=16, $N_{samp}$=8, $f_{in}$=0.4$f_s$).

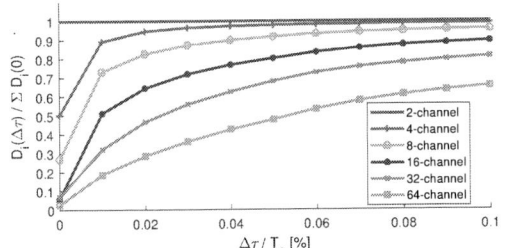

Fig. 4. Dominance of $i^{th}$ channel's $D(\Delta\tau)$ to the sum of magnitudes of $D(0)$ for all channels ($N_{samp}$=8, $f_{in}$=0.4$f_s$).

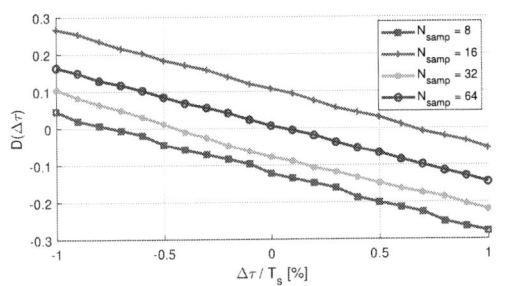

Fig. 5. MAD-based $D(\Delta\tau)$ obtained using various $N_{samp}$ (N=64, $f_{in}$=0.4$f_s$).

## IV. PERFORMANCE EVALUATION

The performance evaluation of both methods are done on a parametrized 10-bit TI-ADC model constructed in Matlab environment, to which sinusoidal inputs are applied. The input frequency ($f_{in}$), the number of channels (N), and the number of samples per channel ($N_{samp}$) for computation are varied in order to explore trade-offs. The two methods discussed in Sections II and III are used for mismatch detection. Discrete delay values are added to the sampling instants in closed-loop to correct channel timing errors, which simulates a delay line (DL) in a mixed-signal calibration scheme as shown in Figure 2. Sampling frequency ($f_s$) is kept the same for all tests, and 0.01% of $T_s$ is chosen as the step size for correction. A simple hill climbing algorithm is preferred over faster gradient descent algorithms such as LMS, in order to avoid an additional block for derivative estimation [11], [12].

### A. Detection performance

Figure 3 shows the monotonic variation of $D(\Delta\tau)$ computed with the CC and MAD-based methods with the skew correction value. Notice that the zero crossing of $D(\Delta\tau)$ does

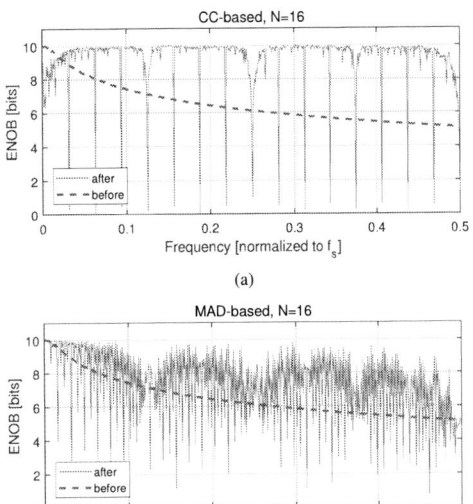

(a)

(b)

Fig. 6. ENOB vs frequency for a) CC-based method b) MAD-based method (N=16, $N_{samp}$=64)

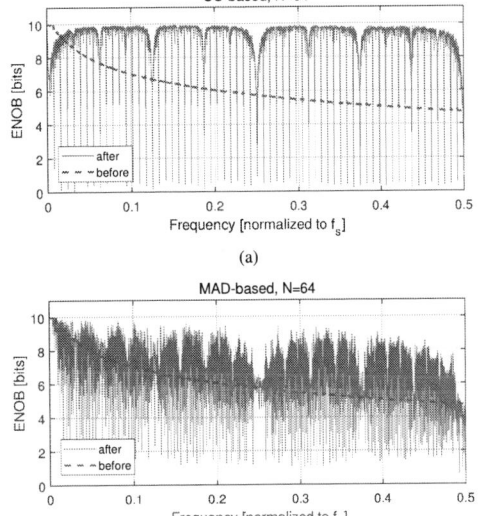

(a)

(b)

Fig. 7. ENOB vs frequency for a) CC-based method b) MAD-based method (N=64, $N_{samp}$=64)

not occur exactly at zero timing error for the MAD-based method. A correction algorithm would try to bring it to zero, so unless $D(\Delta\tau)$ is dominantly higher than the zero mismatch case, misdetection creates additional timing error. This can be better understood from Figure 4, which presents the ratio of $i^{th}$ channel's $D(\Delta\tau)$ to the sum of magnitudes of $D(0)$ for all channels. As the number of channels increase, the detected value becomes less dominant for small timing errors.

978-1-5386-5388-3/18 $31.00 © 2018 IEEE

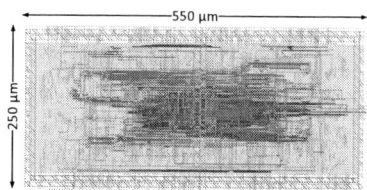

Fig. 8. Layout of the 16-channel, 10-bit, CC-based referenceless timing mismatch calibration circuit synthesized in a 28nm FD-SOI process.

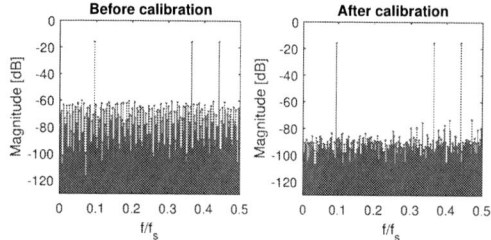

Fig. 9. FFT plots before and after calibration of a 64-channel 10-bit TI-ADC with 16 samples per channel using the CC-based method.

Detection performance of the MAD-based method depends also on the number of distinct samples used for computation. Figure 5 compares $D(\Delta\tau)$ with different number of samples per channel. Increasing the number of samples converges to the ideal response. Nevertheless, this also increases the hardware cost of computation as more number of bits must be used for arithmetic operations and registers.

### B. Calibration performance

The frequency response of the post-calibration effective number of bits (ENOB) for a 10-bit TI-ADC obtained with the CC and MAD-based methods using 64 samples per channel can be seen in Figures 6a and 6b for 16 channels, and in Figures 7a and 7b for 64 channels. Several conclusions can be drawn from these plots. First, the notches at multiples of $f_s/(2N)$ as explained in Section II are clearly visible in CC-based plots. Second, the MAD-based method has additional notches in the frequency response. This is because an integer number of sine periods must be applied for accurate mean estimation and fast Fourier transform (FFT) calculation. For the number of periods which are not relatively prime to the total number of samples, the samples used in the calibration algorithm repeat, reducing the effective $N_{samp}$ and degrading the detection performance as discussed in Section IV-A. Therefore, the best results are obtained when the input and the sampling frequencies follow the coherent sampling relationship [13], which guarantees that each sample is unique. As long as the frequencies outside the notches are selected, ENOB can be recovered to above 9 bits out of 10 bits using both methods.

### C. Hardware implementation

Figure 8 shows the layout of a 16-channel, 10-bit, CC-based referenceless timing mismatch calibration circuit synthesized in a 28nm FD-SOI technology in order to estimate its hardware cost. 8 samples per channel were used for computations. The block occupies 550 μm by 250 μm area and consists of 156k gates. A post-place and route simulation with a multi-tone sinusoidal input with frequency content outside the notches yields the FFT plots seen in Figure 9, proving that multi-tone inputs which are outside the notch frequencies can be calibrated using the CC-based method.

## V. CONCLUSION

We presented a theoretical discussion of the CC-based and the MAD-based referenceless timing mismatch calibration methods for N channels, where N is a power of 2, and evaluated their performance over different parameters using simulations. The notches in the frequency response, which are proportional in number to the number of interleaved channels, are the main drawback of the referenceless scheme. Moreover, the comparison of CC and MAD-based methods show that the input frequency spectrum is more constrained for the latter. Finally, we explored the hardware cost of the CC-based method in a 28nm FD-SOI process.

## REFERENCES

[1] W. C. Black and D. A. Hodges, "Time interleaved converter arrays," *IEEE Journal of Solid-State Circuits*, vol. 15, no. 6, pp. 1022–1029, Dec 1980.

[2] Y. C. Jenq, "Digital spectra of nonuniformly sampled signals: fundamentals and high-speed waveform digitizers," *IEEE Transactions on Instrumentation and Measurement*, vol. 37, no. 2, pp. 245–251, Jun 1988.

[3] N. Kurosawa, H. Kobayashi *et al.*, "Explicit analysis of channel mismatch effects in time-interleaved adc systems," *IEEE Transactions on Circuits and Systems I: Fundamental Theory and Applications*, vol. 48, no. 3, pp. 261–271, Mar 2001.

[4] M. El-Chammas and B. Murmann, *Background Calibration of Time-Interleaved Data Converters*. New York, NY: Springer New York, 2012, pp. 5–30. [Online]. Available: http://dx.doi.org/10.1007/978-1-4614-1511-4_2

[5] M. El-Chammas and B. Murmann, "A 12-GS/s 81-mW 5-bit time-interleaved flash ADC with background timing skew calibration," *IEEE Journal of Solid-State Circuits*, vol. 46, no. 4, pp. 838–847, April 2011.

[6] D. Stepanovic and B. Nikolic, "A 2.8 GS/s 44.6 mW time-interleaved ADC achieving 50.9 dB SNDR and 3 dB effective resolution bandwidth of 1.5 GHz in 65 nm CMOS," *IEEE Journal of Solid-State Circuits*, vol. 48, no. 4, pp. 971–982, April 2013.

[7] A. Buchwald, "A supposedly clever thing I'll never do again," in *2017 IEEE Custom Integrated Circuits Conference (CICC)*, April 2017, pp. 1–8.

[8] B. Razavi, "Problem of timing mismatch in interleaved ADCs," in *Proceedings of the IEEE 2012 Custom Integrated Circuits Conference*, Sept 2012, pp. 1–8.

[9] H. Wei, P. Zhang *et al.*, "An 8 bit 4 GS/s 120 mW CMOS ADC," *IEEE Journal of Solid-State Circuits*, vol. 49, no. 8, pp. 1751–1761, Aug 2014.

[10] Q. Lei, Y. Zheng *et al.*, "A statistic based time skew calibration method for time-interleaved ADCs," in *2014 IEEE International Symposium on Circuits and Systems (ISCAS)*, June 2014, pp. 2373–2376.

[11] A. Salib, B. Cardiff, and M. F. Flanagan, "A low-complexity correlation-based time skew estimation technique for time-interleaved SAR ADCs," in *2017 IEEE International Symposium on Circuits and Systems (IS-CAS)*, May 2017, pp. 1–4.

[12] B. T. Reyes, R. M. Sanchez *et al.*, "Design and experimental evaluation of a time-interleaved ADC calibration algorithm for application in high-speed communication systems," *IEEE Transactions on Circuits and Systems I: Regular Papers*, vol. 64, no. 5, pp. 1019–1030, May 2017.

[13] "IEEE standard for terminology and test methods for analog-to-digital converters," *IEEE Std 1241-2010 (Revision of IEEE Std 1241-2000)*, pp. 1–139, Jan 2011.

*PRIME 2018, Prague, Czech Republic*

*Session: Data Converters*

# A $6.5$-$\mu$W $70$-dB $0.18$-$\mu$m CMOS Potentiostatic Delta-Sigma for Electrochemical Sensors

Joan Aymerich[1], Michele Dei[1], Lluís Terés[1,2] and Francisco Serra-Graells[1,2]

[1]Instituto de Microelectrónica de Barcelona IMB-CNM (CSIC), Spain

[2]Microelectronics and Electronic Systems, Universitat Autònoma de Barcelona, Spain

*Abstract*—**This paper presents the design of a low-power potentiostatic second-order continuous-time (CT) delta-sigma modulator ($\Delta\Sigma$M) for the amperometric read-out and A/D conversion of electrochemical sensors. The proposed architecture reuses the sensor itself as a leaky integrator stage for shaping the quantization noise, resulting in a very compact and energy efficient read-out front end. Low-power CMOS circuits are also presented for the remaining analog blocks of the $\Delta\Sigma$M loop. A design example in $0.18$-$\mu$m CMOS technology is provided with a total area of $0.063$mm$^2$. Post-layout simulations show a dynamic range of 70dB with an overall power consumption of $6.5\mu$W at 1.8-V supply.**

## I. INTRODUCTION

Amperometric electrochemical sensors have undergone a rapid development due to their sensitivity, cost effectiveness, and CMOS compatibility allowing fully integrable and implantable sensing and monitoring systems. They have been extensively used in environmental sensing, health monitoring, and industrial applications. Amperometric sensors generate an output current proportional to the concentration of the measured analyte. The electrochemical cell consists of three electrodes: reference (R), working (W), and counter (C), as shown in Fig. 1(a). The $R_{\mathrm{ct}}$ and $C_{\mathrm{dl}}$ represent the charge-transfer resistance and the double-layer capacitance of the electrode-electrolyte interface [1], respectively. A potentiostat is usually required to measure the sensor output current, and to ensure that no current is flowing through the R terminal and the differential voltage between R and W terminals is kept at a static potential. Recently, new applications demand potentiostats with very low-power consumption while achieving high-accuracy measurement and low-noise performance. Most state-of-the-art potentiostats [2] require multi-OpAmps in addition to the ADC stage for digitization, involving large area and power figures. In [3], a single-bit CT $\Delta\Sigma$M is employed, which reuses the double-layer capacitance of the electrode-electrolyte interface $C_{\mathrm{dl}}$ to realize the noise shaping in order to achieve a compact and energy efficient conversion of the electrochemical signals. $\Delta\Sigma$Ms are particularly well suited to measure slow electrochemical signals because their higher sampling rates result in higher resolution. However, first-order single bit $\Delta\Sigma$Ms present important drawbacks, which have already been addressed in a previous work by these authors [4].

In this paper, an enhanced version of the previous work [4] is presented. The design complexity, area and power

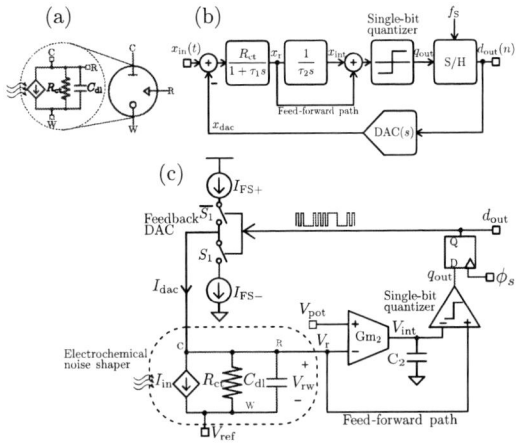

Fig. 1. Electrochemical sensor model (a), signal-processing blocks (b) and electrical circuit (c) of the proposed potentiostatic $\Delta\Sigma$M architecture for the amperometric read-out and A/D conversion.

consumption are improved by removing one of the blocks of the electrochemical $\Delta\Sigma$M potentiostat. Furthermore, a fully circuit design in $0.18$-$\mu$m CMOS technology is also presented, where a rail-to-rail input voltage dynamic comparator for the single-bit quantizer is incorporated.

The paper is organized as follows: Section II presents the potentiostatic amperometric $\Delta\Sigma$M architecture; Section III addresses its design at transistor level; Section IV shows a $0.18$-$\mu$m CMOS design example; the post-layout simulation results are reported in Section V, while conclusions are summarized in Section VI.

## II. ELECTROCHEMICAL $\Delta\Sigma$M ARCHITECTURE

Fig. 1(c) illustrates the architecture of the proposed second-order electrochemical $\Delta\Sigma$M. The input signal estimation ($I_{\mathrm{dac}}$) is subtracted from the chemical input signal ($I_{\mathrm{in}}$). The resulting current error is integrated and converted into voltage ($V_{\mathrm{r}}$) by the sensor impedance itself. A second integrator stage further shapes the quantization error ($V_{\mathrm{r}} - V_{\mathrm{pot}}$) to higher frequencies, as well as it ensures the properly potentiostat operation [4]. The comparator computes the single-bit quantization in $q_{\mathrm{out}}$, while the D-type flip-flop stage implements its sample-and-hold in $d_{\mathrm{out}}$. Finally, this output bit stream is fed back to the current DAC. As a result, $d_{\mathrm{out}}$ is modulated

978-1-5386-5388-3/18 $31.00 © 2018 IEEE

by the chemical input signal. The potentiostatic operation is obtained by the negative feedback of the $\Delta\Sigma M$ loop, which keeps $V_r$ to the desired potentiostat voltage $V_{pot}$, and the high input impedance of the electronic integrator prevents from any current flowing through the reference electrode. Since the working electrode is held to a fixed potential $V_{ref}$, the effective potentiostatic voltage $V_{rw}$ can be programmed by changing the potential $V_{pot}$.

The equivalent signal-processing model of the proposed $\Delta\Sigma M$ is shown in Fig. 1(b), where $\tau_1$ stands for the sensor time constant ($R_{ct}C_{dl}$) and $\tau_2$ is the electronic integrator time constant ($C_2/G_{m_2}$). The feed-forward path introduces a left-half plane (LHP) zero $f_Z = 1/(2\pi\tau_2)$ in the $\Delta\Sigma M$ loop-filter to stabilize its closed operation. In this sense, the loop stability is ensured as long as the condition $f_Z/f_S < 1/(2\pi)$ is satisfied [4].

The summation at the input of the quantizer of Fig. 1(b) increases circuit complexity and power dissipation for feed-forward $\Delta\Sigma Ms$. In the previous work, this adder is done connecting the electrode potential $V_r$ to the other plate of the integrator capacitor $C_2$. However, this approach requires a voltage follower in order to prevent current flowing through the reference electrode, which results in an increase of overall area. The feed-forward (FF) path can be simplified by providing a direct path from $V_r$ to the positive input of the single-bit quantizer and inverting the inputs of the integrator transconductance $G_{m_2}$, such that the differential input of the quantizer is equal to the summation of the integrated ($V_{int}$) and the feed-forward ($V_r$) signals.

## III. Low-Power Circuit Implementation

In order to allow a large flexibility on the selection of $V_{rw} = V_{pot} - V_{ref}$ values for a wide range of sensor applications, the CMOS circuit blocks have been designed to provide wide input and output common-mode swing, as well as low-power consumption.

### A. Gm-C Integrator with Constant Transconductance

Since variations in the electronic integrator time constant $\tau_2$ may compromise the stability of the $\Delta\Sigma$ loop, the integrator transconductance should be designed to be constant over the input common-mode voltage range defined by $V_{pot}$. For this purpose, $G_{m_2}$ can be controlled through the bias of the complementary input stage $M_1 - M_4$ of Fig. 2 operated in weak inversion:

$$G_{m_2} = \frac{I_{biasn} + I_{biasp}}{nU_T} \quad (1)$$

where $I_{biasn,p}$ are the tail currents of the input pairs, $n = (C_{ox} + C_{dl})/C_{ox}$ the slope factor, and $U_T = kT/q$ the thermal voltage. The one-times current mirror, $M_6 - M_7$ together with the current switch $M_5$ [6], maintains the sum of the tail currents constant. Therefore, a decrease of one of the tail currents is compensated by an equal increase of the other tail current. This can be seen as follows. In the input common-mode range close to the positive rail ($V_{dd}$), the current switch $M_5$, is off. Therefore, the bias current flows

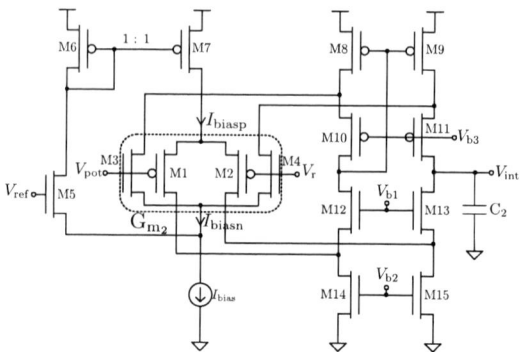

Fig. 2. Gm-C integrator with rail-to-rail input common-mode voltage range to be used for the $G_{m_2}$-$C_2$ stage of Fig. 1(c).

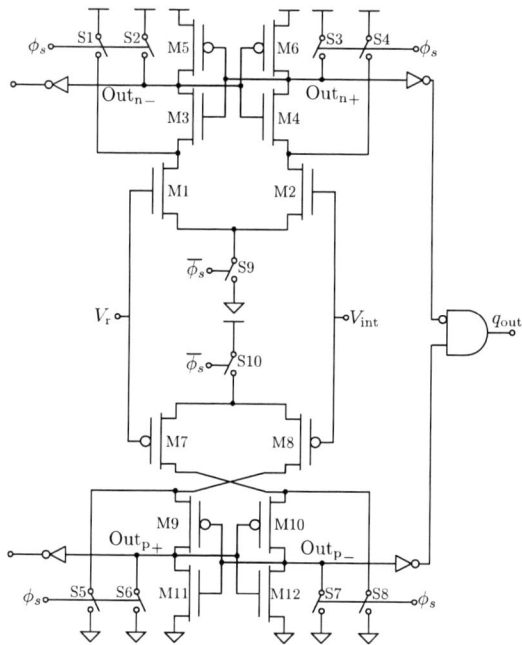

Fig. 3. Low-power rail-to-rail complementary latch comparator for the single-bit quantizer.

entirely into the NMOS pair ($I_{biasn} = I_{bias}$, $I_{biasp} = 0$). As a result, the sum of the tails currents is equal to $I_{bias}$. In the intermediate range, $M_5$ progressively takes the current $I_{bias}$ from the NMOS pair and steers it to the PMOS pair through the one-times current mirror $M_6 - M_7$ ($I_{biasn,p} \approx I_{bias}/2$). The sum of the tail currents is also equal to $I_{bias}$. In the low range of the input common-mode, the bias current flows completely through the PMOS pair ($I_{biasn} = 0$, $I_{biasp} = I_{bias}$). Since the current through NMOS is zero, the sum of both tail currents is still equal to $I_{bias}$. The cascode transistors are designed and biased to ensure a high-gain operation, as well as a wide output swing limited to $2V_{ov} < V_{int} < V_{dd} - 2V_{ov}$, where $V_{ov}$ is the overdrive voltage.

### B. Complementary Latch Comparator

A latch-type dynamic comparator is an ideal option for the implementation of the single-bit quantizer of Fig. 1(c) due to their full output swing, high input impedance and absence of static power consumption. However, traditional latch comparators suffer from limited common-mode input range, especially in those sub-micrometer CMOS technologies where the threshold voltages have not been scaled down at the same ratio as the supply voltages. This issue can be addressed by combining two complementary versions of this circuit with all the NMOS and PMOS transistors swapped, as shown in Fig. 3. During the reset phase ($\phi_s = 1$), switches S2 and S3 pre-charge $\text{Out}_n$ nodes and switches S6 and S7 discharge the $\text{Out}_p$ nodes to $V_{dd}$ and ground, respectively. In the comparison phase ($\phi_s = 0$), the tail switches S9 and S10 are turned on, therefore the input differential voltage ($V_r - V_{int}$) is converted into a differential current and mirrored to the regenerative latch. Positive feedback enables the regeneration of a small differential voltage to a full swing differential voltage. The combinational logic allows to merge both NMOS-PMOS-input comparators; if the PMOS-input is off (input common-mode voltage above $V_{dd} - V_{THp}$), $\text{Out}_p$ nodes remain at the negative rail regardless of $\phi_s$. In this case, the output inverter ensures a logic 1 in one of the AND inputs, and so $q_{out}$ follows the NMOS-input comparison. On the other hand, when NMOS-input is off (input voltage below $V_{THn}$), $\text{Out}_n$ nodes remain charged to the positive supply $V_{dd}$. In that case, two inverters ensure a logic 1, therefore $q_{out}$ follows the PMOS-input comparison. When both comparators are operating simultaneously, $q_{out}$ follows the slowest one.

Concerning the single-bit feedback DAC of Fig. 1(c), it is implemented by a switched-current circuit, with n- and p-type cascode wide-swing current sinks and sources. Finally, the chip bias current is provided by an on-chip all-MOS proportional-to-absolute temperature current reference [5].

## IV. 0.18-$\mu$m CMOS DESIGN EXAMPLE

The proposed potentiostatic $\Delta\Sigma$M has been designed in a 0.18-$\mu$m 1P6M CMOS technology, as illustrated in Fig 4. The overall layout area, excluding the electrochemical sensor and I/O pads, is 0.063mm$^2$.

The $G$m-C integrator occupy most of the area. On the one hand, its input pair transistors are large for matching purposes, since any offset is directly proportional to the potentiostatic error ($V_r - V_{pot}$). On the other hand, a 80-pF triple-MIM capacitor is used for $C_2$ because of the low sampling frequency (kHz-range), which forces a slow integrator time constant $\tau_2$ in order to ensure the loop-stability of the $\Delta\Sigma$M ($\tau_2 \approx 12.5$ms at $f_S = 1$kHz). This also scales down the $G_{m_2}$ transconductance bias current down to a few nA. Within this current range, the input-pair transistors of the complementary input stage $M_1 - M_4$ of Fig. 2 can be easily biased in the weak-inversion region to guarantee the constant $G_{m_2}$ over the rail-to-rail input common-mode voltage, as already explained in Section III. Part of the area is also occupied by the current feedback DAC

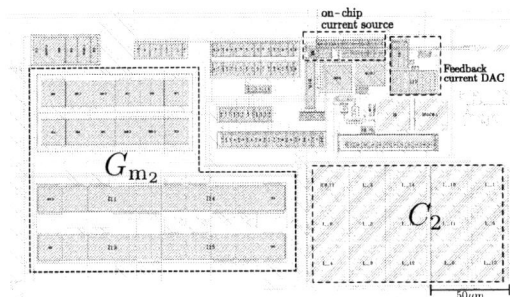

Fig. 4. Layout of the potentiostatic $\Delta\Sigma$M excluding the electrochemical sensor. Overall bounding box is $350\mu$m x $180\mu$m ($0.063$mm$^2$).

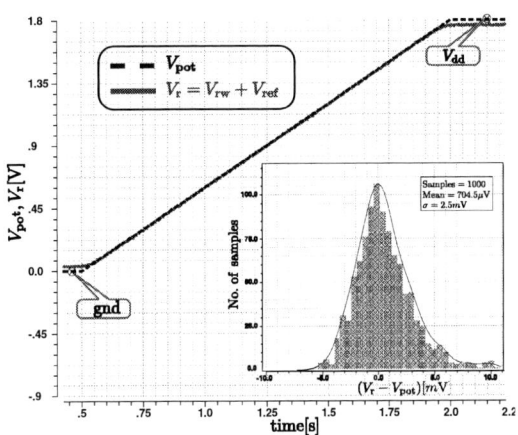

Fig. 5. Simulated potentiostatic performance: $V_r$ voltage swing, sweeping $V_{pot}$ from ground to $V_{dd}$, and Monte Carlo simulation (1000 samples) of the potentiostatic voltage (inset of figure).

due to noise requirements, since the noise coming from this block is not shaped by the $\Delta\Sigma$M loop filter. Indeed, for such low sampling frequency and signal bandwidth the flicker noise can be in practice the main SNDR limiting factor.

The aim of this design is also to provide with a large current full scale in order to cover several electrochemical sensor designs. For that reason, the feedback DAC has a 3-bit digital trimming to extend the current full scale ($I_{FS}$) from 250nA to 3.75$\mu$A. The bias current of the integrator transconductance $G_{m_2}$ can be also trimmed in order to control the loop-filter zero location. This design presents an overall power consumption of 6.5$\mu$W at 1-kHz sampling frequency and a 350nA current DAC full scale.

## V. POST-LAYOUT SIMULATION RESULTS

The post-layout electrical simulation results of the 0.18-$\mu$m CMOS design example of Fig. 4 are presented in Fig. 5 to 7. The proposed design allows nearly rail-to-rail potentiostatic voltage swing, as shown in Fig. 5. The small limitation comes from the output swing of the current feedback DAC or from the output swing of the $G_m$-C, which both require of two over-drive voltages in order to maintain the top and bottom

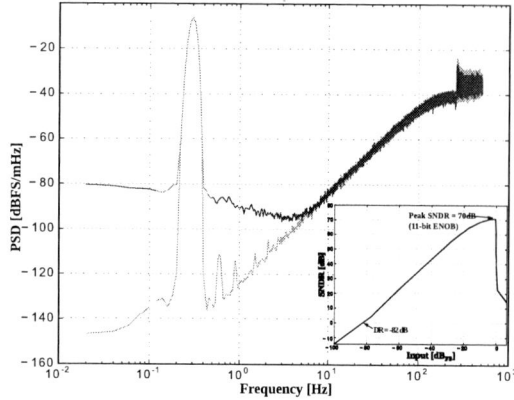

Fig. 6. Simulated $d_{out}$ power spectrum with (a) and without (b) electronic noise at -6dB$_{FS}$, and SNDR vs input signal curve (inset of figure) for $I_{FS} = 1.25\mu A$, $f_s = 1kHz(OSR \approx 500)$, $\tau_1 = 0.15s$ and $\tau_2 = 1.5ms$. Time of observation 50s.

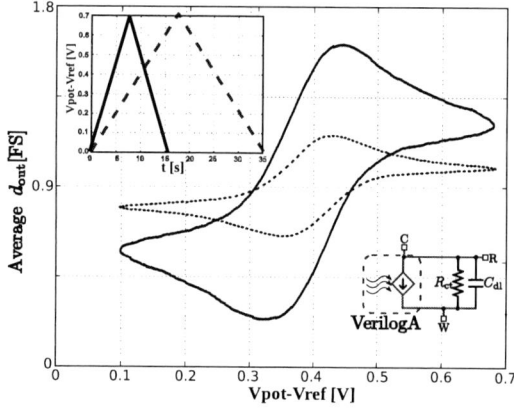

Fig. 7. Simulation of a cyclic voltammetry for a 1mM $K_3[Fe(CN)_6]$ at 0.1M KCL. Scan rate is 40 (solid) and 90 (dotted) mV/s.

cascode transistors in saturation. Since they are biased in weak inversion, the overdrive voltage is limited down to less than $\approx 60$mV at room temperature. The offset of the input pair of $G_{m_2}$, which also represents the variation of the potentiostatic voltage applied to the sensor, has been analyzed using Monte Carlo simulation (inset of Fig. 5) with process and mismatch variations. This result predicts a standard deviation of 2.5mV. Systematic offset is caused mainly because of the $G_{m_2}$ finite open loop-gain.

Fig. 6 compares the output spectrum of the $\Delta\Sigma M$ with and without electronic transient noise for an integrated electrochemical sensor characterized by $R_{ct} = 500k\Omega$ and $C_{dl} = 300nF$ [4]. The flicker noise noticeably increases the noise floor in the low-frequency portion of the spectrum. But for high frequencies approximately above the bandwidth point (BW $\approx 1$Hz for $\tau_1 = 0.15s$) the spectrum is preserved. The SNDR curve (inset of Fig. 6) shows that the electrochemical $\Delta\Sigma M$ achieves a 70-dB peak dynamic range. As already discussed, the main limitation comes from the feedback DAC.

TABLE I
AREA PENALTY AGAINST DYNAMIC RANGE.

| DAC + bias current Area estimation | Total area increased | SNDR | ENOB |
|---|---|---|---|
| $1400\mu m^2$ | Fig. 4 | 70-dB | 11 |
| $5600\mu m^2$ | 7% | 75.5-dB | 12 |
| $22400\mu m^2$ | 33% | 81-dB | 13 |

In this design, the area of the feedback DAC and its current bias (on-chip current reference) is limited to $1400\mu m^2$ due to size requirements, however higher resolution is achievable enlarging the area of these two blocks. Table I shows an estimation of SNDR improvement against area penalty confirmed also by electrical simulation.

In order to validate the potentiostatic function, cyclic voltammetry (CV) simulations were also performed, as shown in Fig. 7. In a CV, a triangular waveform is applied to the reference electrode (inset of figure), while the sensor current is measured simultaneously. To perform these simulations, a VerilogA sensor model was built, which includes two $V_{rw}$-$I_{in}$ DC look-up tables based on two experimental measurements of ferricyanide CVs.

## VI. CONCLUSIONS

This paper has presented the design of a low-power miniaturized potentiostatic and amperometric A/D read-out of electrochemical sensors. The proposed architecture reuses the sensor itself as a leaky integrator stage for shaping the quantization noise to implement a continuous-time mixed electrochemical $\Delta\Sigma M$. The circuit is designed in a 0.18-$\mu$m CMOS technology. The resulting electrochemical $\Delta\Sigma M$ only requires minimalist low-power circuitry, where a rail-to-rail input voltage dynamic comparator architecture for the single-bit quantizer is presented. Post-layout simulations report electrical dynamic range values exceeding 11bit with an overall power consumption of 6.5$\mu$W at 1.8-V supply.

## ACKNOWLEDGMENT

This work has been partially funded by IP4SS (CSIC ref.: 201650E019) and supported by TecnioSpring+ TECSPR16-1-0056.

## REFERENCES

[1] E. T. McAdams, A. Lackermeier, J. A. McLaughlin, D. Macken, and J. Jossinet, "The linear and nonlinear electrical properties of the electrode-electrolyte interface," Biosensors Bioelectron., vol. 10, pp. 67-74, 1995.
[2] S. Martin, F. Gebara, T. Strong, and R. Brown, "A fully differential potentiostat," IEEE Sensors J. vol. 9, no. 2, pp. 135-142, Feb 2009.
[3] S. Sutula, J. Pallarés Cuxart, J. Gonzalo-Ruiz, F. X. Muñoz-Pascual, L. Terés and F. Serra-Graells, "A 25-$\mu$W All-MOS Potentiostatic Delta-Sigma ADC for Smart Electrochemical Sensors," in IEEE Transactions on Circuits and Systems I: Regular Papers, vol. 61, no. 3, pp. 671-679, Mar 2014.
[4] J. Aymerich, M. Dei, L. Terés and F. Serra-Graells, "Design of a Low-Power Potentiostatic Second-Order CT Delta-Sigma ADC for Electrochemical Sensors," 2017 13th Conference on Ph.D. Research in Microelectronics and Electronics (PRIME), pp. 105-108, 2017.
[5] F. Serra-Graells and J. L. Huertas, "Sub-1V CMOS Proportional-to-Absolute-Temperature References," in IEEE Journal of Solid-State Circuits, vol. 38, no. 1, pp. 84-88, Jan 2003.
[6] J. H. Huijsing and D. Linebarger, "Low-Voltage Operational Amplifier with Rail-to-Rail Input and Output Ranges," in IEEE Journal of Solid-State Circuits, vol. SC-20, no. 6, pp. 1144-1150, Dec 1985.

# 11.7b Time-To-Digital Converter with 0.82ps resolution in 130nm CMOS Technology

Rodrigo Granja[1], Mauro Santos[1], Jorge Guilherme[1,3], Nuno Horta[1,2]

[1]Instituto de Telecomunicações, [2]Instituto Superior Técnico Lisboa, Portugal

[3]Instituto Politécnico Tomar, Portugal

r.granja.20@gmail.com; mauro.e.santos@gmail.com; jorge.guilherme@ipt.pt; nuno.horta@lx.it.pt

*Abstract*—**This paper describes a high-resolution 11.7b Time-to-Digital Converter (TDC) designed in a pure digital CMOS 130nm technology. The target architecture comprises a looped delay-line based on an inverter-based pulse-shrinking technique. The proposed technique can achieve a 0.82ps resolution with a dynamic range of 2.918ns, an integral nonlinearity (INL) of -2.4 to 2.11 and a differential nonlinearity (DNL) of -0.91 to 0.87 LSB. In addition, it occupies a low area of 0.148 mm².**

*Keywords— Time-to-Digital Converter, looped delay-line, inverter-based Pulse-Shrinking technique, high resolution and dynamic-range, low area and power consumption, simple and versatile structure.*

## I. INTRODUCTION

The advancements in CMOS technologies are mostly generated by the optimization of digital circuits; therefore, analog approaches are continuously losing their advantage and benefits [1] as technology scaling results in the decrease of the voltage power supply. Hence, voltage-mode circuits, defined by the ratio of the minimum detectable voltage (typically set by the noise floor) and the maximum available voltage, become deeply affected, and scale poorly with those technology advancements. One of the effects of voltage headroom shrinking is a decrease in the Signal-to-Noise Ratio (SNR), meaning that the devices become much more susceptible to noise, leading to an inability to perform proper signal processing [1, 2]. Digital circuits on other hand do not present these disadvantages, having the capability of implementing their functions, in a much smaller area, with lower power consumption, lower susceptibility to noise (resulting from the increase in the operation speed, and even from the type of signal itself), and robustness to process variations. They cannot develop or represent any type of information in the analog voltage domain, but they achieve a very high resolution in the time domain due to the consecutive gate delay reduction, offering exceptional time accuracy surpassing voltage resolutions in analog/mixed-signal approaches [3]. As stated in [1], in a deep-submicron CMOS process, the time-domain resolution of a digital signal edge transition is superior to the voltage resolution of an analog signal.

A time variable possesses a unique duality characteristic, it is an analog variable with continuous amplitude, represented by its pulse duration, and also a digital variable with two distinct logic levels [1]. This time variable can be digitized by a time-to-digital converter (TDC), which is one of the most well-known time-mode circuits. Time-to-Digital converters are a precise stopwatch that converts continuous time domain information into a digital representation. It should be noted that only fully digital TDCs take full advantage of the time domain improvements provided by technology scaling.

With the growing interest in the development of this technology [4, 5], various new architectures emerge increasing the resolutions, dynamic ranges, and lowering the power consumption and conversion times [2, 3, 6].

A looped Pulse-Shrinking TDC is presented in this paper, where instead of using a traditional buffer delay-line, inverter delay-elements were used to perform the time measurement.

This paper is organized as follows: Section II describes the pulse-shrinking technique used to achieve a very precise time resolution, Section III presents the converter architecture, Section IV presents the layout of the converter, the schematic and extracted simulation results for the DNL and INL, and Section V draws the conclusion.

## II. PULSE-SHRINKING TDC

Precise results require advanced electronic techniques leading to extremely high circuit resolutions. Basic TDC topologies have their maximum resolution limited to the gate delay of the technology that is being used, however there are some topologies that enable sub gate delay resolutions.

The operation principle of all delay-line based TDCs is based in the behavior of its basic block, the inverter and its time equations [7, 8]. The parallel scaled delay-line is one of the simplest mechanisms to achieve sub-gate delay resolution, greatly increasing the resolution of the TDC, based on scaled capacitors connected to the output of parallel delay elements, changing its output load capacitance [3]. The working principle behind this technique is described by the following equation,

$$t_{pd} = p + g.h \qquad (1)$$

$$h = \frac{C_{load}}{C_{in}} \qquad (2)$$

where $p$ is the parasitic delay, $g$ the logic effort and $h$ the fan-out of the gate.

Hence, looking at (2), and manipulating the $h$ parameter will lead, theoretically, to any possible propagation delay [2, 9]. The drawbacks that this topology has are related to the huge area consumption, the big offset that exists, the possible parasitic effects that could lead to the change of the modulated capacitors, and mainly the balancing of the start signal net that has to deliver the start signal to all branches at the same time, making the layout construction a very critical process [3].

---

This work was supported in part by the Instituto de Telecomunicações (Research project RAPID UID/EEA/50008/2013, Research project Incentivo/EEI/LA0008/2014) and by the Fundação para a Ciência e Tecnologia (FCT-SFRH/BD/44147/2008).

R. Granja, M. Santos and N. Horta are with Instituto de Telecomunicações and Instituto Superior Técnico, University of Lisbon, Lisboa, Portugal

J. Guilherme is with Instituto de Telecomunicações, Lisboa, Portugal and Escola Superior de Tecnologia de Tomar, Instituto Politécnico de Tomar, Tomar, Portugal.

978-1-5386-5388-3/18 $31.00 © 2018 IEEE

Vernier based TDCs on the other hand, are known for their capability to measure time intervals with a very high resolution, huge versatility and a simple principle of operation. They operate by using two parallel buffer delay-lines, one to receive the start command and the other to receive the stop command [3]. Both lines have the same number of delay elements, but slightly different element sizes. Like the parallel scaled delay-line, the principle behind the Vernier topology is based on the change of the propagation delay of the element gates. However, in the Vernier concept this variation is achieved by changing the transistor sizes of the elements in the two delay-lines, keeping the load capacitance equal in every node [1].

Another solution is the Pulse-Shrinking TDC [9, 10, 11] which consists of a chain of homogenous and inhomogeneous stages. Each homogenous element has similar rising and falling times, and each inhomogeneous stage has slightly different rising and falling times [1]. As depicted in Fig. 1, each of the inhomogeneous stages, $i$, is connected between two homogenous stages, $i$-$1$ and $i$+$1$. This mechanism allows the accurate control of the pulse-shrinking behavior. To quantify the time shrinking made in the pulse, the falling and rising time expressions of the inverters are considered [7, 8]. According to the time equations and applying them to this case, considering that the load capacitance is the input capacitance of the next element, in order to make the pulse-shrinking effect possible when the input pulse travels from the $i$-$1$ to the $i$ gate, it must be assured that $\tau_{PLH,i-1} > \tau_{PHL,i-1}$. Assuming that the NMOS and PMOS transistors have the same threshold voltage $V_T$, the reduction in the width of the pulse is given by,

$$\Delta\tau_{(i-1)\to i} = \tau_{PLH,i-1} - \tau_{PHL,i-1} = \alpha C_i \left( \frac{2}{\beta_{p,i-1}} - \frac{2}{\beta_{n,i-1}} \right) \quad (3)$$

$$\alpha = \frac{2V_T}{(V_{DD} - V_T)^2} + \frac{1}{V_{DD} - V_T} \ln\left( \frac{19V_{DD} - 20V_T}{V_{DD}} \right) \quad (4)$$

where $V_{DD}$ is the supply voltage, $C_i$ the load capacitance and $\beta = \mu C_{ox} \frac{W}{L}$ , where $\mu$ is the MOS charge-carrier effective mobility, $W$ the n-gate width and $L$ is the n-gate length.

Looking at the second stage where the pulse crosses from the element $i$ to the element $i$+$1$ it will have a similar expression,

$$\Delta\tau_{i\to(i+1)} = \tau_{PHL,i} - \tau_{PLH,i} = -\alpha C_{i+1} \left( \frac{2}{\beta_{p,i}} - \frac{2}{\beta_{n,i}} \right) \quad (5)$$

The total pulse-shrinking time is the result of the sum of both factors,

$$T_{LSB} = \Delta\tau = \Delta\tau_{(i-1)\to i} + \Delta\tau_{i\to(i+1)} =$$

$$= \alpha \left[ C_i \left( \frac{2}{\beta_{p,i-1}} - \frac{2}{\beta_{n,i-1}} \right) - C_{i+1} \left( \frac{2}{\beta_{p,i}} - \frac{2}{\beta_{n,i}} \right) \right] \quad (6)$$

So, with this simple structure, high resolutions can be achieved with a basic principle and without having a very complex control logic behind it.

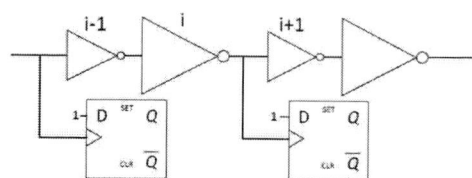

Fig. 1 – Pulse-Shrinking TDC Principle

The area and power consumption are much smaller when compared to the previous architectures while achieving equivalent resolutions. However, the pulse-shrinking architecture suffers from long conversion times and has a limitation on the smallest pulse width that it can measure (typically called an offset), being restricted to pulse widths larger than the propagation delay of the inverter [3].

## III. TDC ARCHITECTURE

The block diagram of the proposed architecture is presented in Fig. 2. The TDC architecture is built around a loop structure with six main blocks:

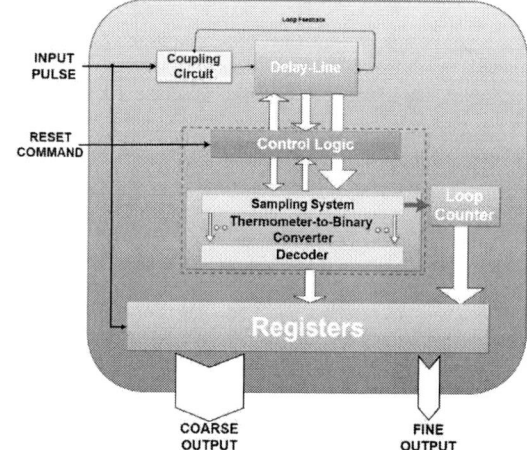

Fig. 2. – Time-to-Digital Converter System Architecture

- *Coupling Circuit* – Responsible for the acquirement of the measurement pulse from an outside source.
- *Delay-Line*– It is the core block, built in a loop structure featuring the pulse-shrinking delay elements.
- *Loop Counter* – Counts the number of times that the pulse looped around the delay-line, delivering the coarse output binary word.
- *Control Logic* – Responsible for the control and command of the system, from the insertion of the pulse into the delay-line, to its reset.
- *Thermometer-to-Binary-Decoder* – Responsible for the acquisition and decoding of the output from the delay-line. It delivers the fine output binary word.
- *Registers* – Responsible for saving the result of the conversion and for delivering it to a posterior circuit. Additionally, this circuit prevents data lost, due to its regenerative topology.

## A. Delay-Line

The system's performance depends mainly on the behavior of this block, so an extra effort was made to achieve the best performance behavior possible. The pulse-shrinking TDC is based on a buffer delay-line [11, 12], and the buffers are built with two inverters in series. The proposed looped delay-line is based on the replacement of those buffer delay elements by inverters, decreasing the area and power consumption, and increasing the maximum resolution. Rewriting (6) in a simpler and more understandable way,

$$\Delta \tau = (t_{PLH,i} - t_{PHL,i}) - (t_{PLH,i+1} - t_{PHL,i+1}) \qquad (7)$$

As referred before, this last equation is the sum of the cumulative effects from both inverters in series and therefore the effect of one buffer. Hence, what is proposed is to divide the equation into two parcels, meaning that every single inverter will shrink the traveling pulse, doubling at least the resolution of the delay-line and decreasing the conversion time taken to obtain the conversion result. The following equations translate the working principle behind this concept,

$$\Delta \tau_i = t_{PLH,i} - t_{PHL,i} \qquad \Delta \tau_{i+1} = t_{PHL,i+1} - t_{PLH,i+1} \qquad (8)$$

Looking at both equations, they look symmetrical. However, for symmetrical rising and falling delays or for asymmetrical but equal inverters, the pulse width will not change; the shrinking will be null. Taking into consideration (8), to obtain a shrink of the $i$ delay elements in Fig. 3, it must be ascertained that $\tau_{PLH,i} < \tau_{PHL,i}$. In the second equation, matching the $i+1$ delay elements is exactly the opposite, to obtain a shrinking element it must be ascertained that $\tau_{PLH,i+1} > \tau_{PHL,i+1}$. In both cases the closer these parameters are the higher the time resolution obtained, however it is always limited by PVT variations. Taking into account (8) and the required relationships between rise and fall times, for the technology used the maximum resolution achievable will be 0.82ps.

After defining the parameters of the time-shrinking elements, it is still necessary to define the length of the delay-line which defines the maximum time measurement that it can perform. Due to the final binary code representation, to simplify the design and maximize the system performance, it is advantageous that the number of delay-elements is a power of two. Looking at the waveforms at the input and output of the inverter, *outi (blue wave)*, *outi+1(red wave)*, depicted in Fig. 4, the propagation-delay of each element, will be 22.8ps. The TDC delay-line was built with 128 elements, leading to a maximum dynamic range of 2.918ns.

## IV. CIRCUIT LAYOUT AND SIMULATION

The complete circuit was designed and simulated in a 130nm UMC process technology in PVT conditions. The layout assembly is critical in a TDC, mainly in its delay-line where the actual measurement occurs. Any asymmetry in the layout can cause a non-linear behavior in the output response. Therefore, it is important to assure that the elements have the same structure, and are equality instantiated to assure that their output nodes have the same parasitic resistances and capacitances. A square structure was used to guarantee that all the elements are placed with the exact same distance from each other, with the exact same connections and metal levels, leading to the exact same parasitics in the component nodes, making it a very homogenous circuit.

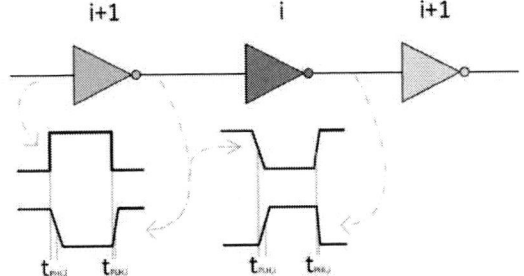

Fig. 3 – Proposed Pulse-Shrinking Delay-Line

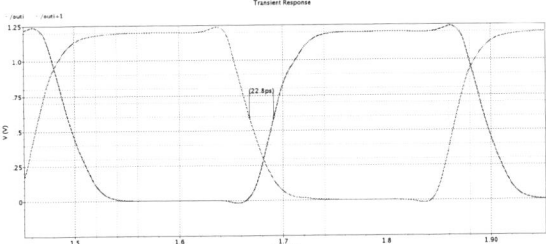

Fig. 4 – Transient Response, Propagation-delay

The proposed layout, depicted in Fig. 6, is divided in six main blocks: the fine registers and fat-tree (yellow box), the one-of-n code (white box), the sampling and control logic (red box), the pulse-shrinking delay-line (green box), loop-counter and coarse registers (orange box) and decoupling capacitors (blue box). To power the system and due to the fact that the circuit has multiple levels, a power grid technique was used, alternating VDD and GND power lines. The total TDC area is 0.148mm$^2$.

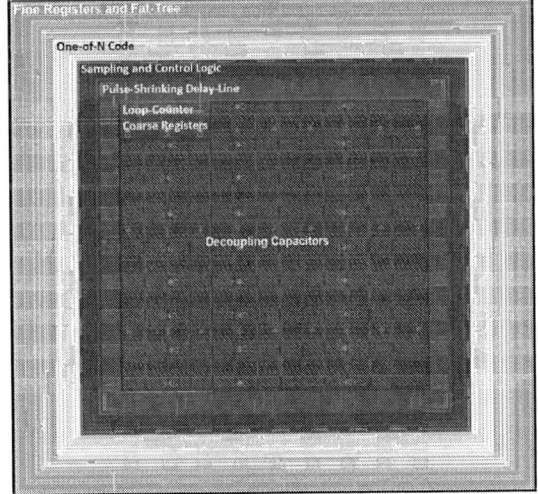

Fig. 6 – TDC Full Layout.

## A. DNL and INL Measurement

The full converter was simulated in PVT corners, the transfer functions for the simulated corners are shown in Fig. 7.

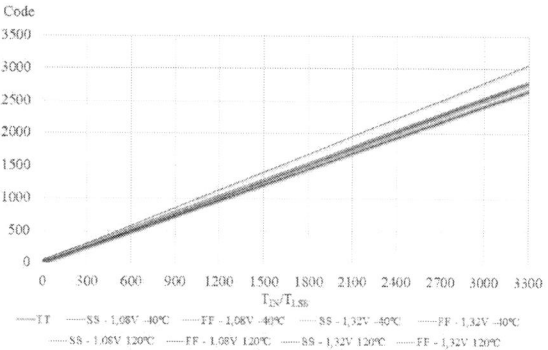

Fig. 7 – TDC Transfer Function

Fig. 8 shows the extracted layout simulation results in typical conditions for the DNL and INL of the full converter. Extracted layout simulation results for corner cases are not shown but show similar performance metrics. The DNL analysis of the extracted circuit shows a variation between -0.91 and 0.87 LSB. The INL analysis shows a variation between -2.4 and 2.11 LSB.

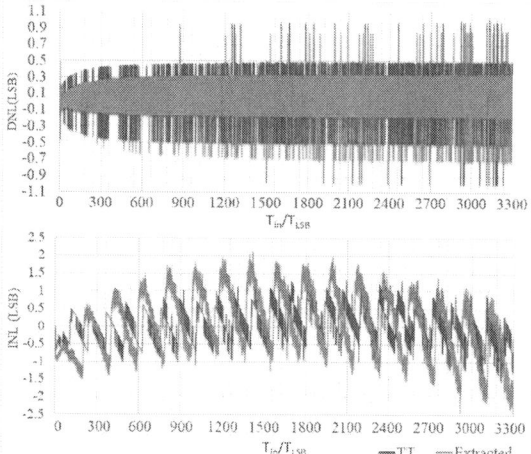

Fig. 8 – DNL and INL of the extracted TDC.

Table I shows a summary of the relevant performance characteristics of the work presented in this paper and other relevant works found in the literature.

TABLE I – PERFORMANCE COMPARISON

| Ref. | [13] | [14] | [15] | [16] | [17] | This work |
|------|------|------|------|------|------|-----------|
| Tech [nm] | 130 | 65 | 90 | 90 | 180 | **130** |
| $T_{LSB}$ [ps] | 6,98 | 7 | 1.25 | - | 1.8 | **0.82** |
| N. Bits | 11 | 7 | 9 | 9 | 9 | **11.7** |
| INL [LSB] | 1.5 | 3.3 | 2 | 1.5 | 8.7 | **2.4** |
| DNL [LSB] | 0.8 | <1 | 0.8 | 1.2 | 1.2 | **0.91** |
| Power [mW] | 0.328 | 1.7 | 3 | 0.014 | 3.4 | **7,5** |
| Area [mm²] | 0.28 | 0.28 | 0.04 | 1 | 0.07 | **0.148** |
| FOM [pJ/conv] | 0.40 | 0.28 | 0.96 | 0.098 | 0.75 | **0.091** |

## V. CONCLUSIONS

An inverter-based pulse-shrinking TDC was presented in this paper. The TDC is based on a simple, direct and modular design allowing the achievement of high-resolutions with a small area (0.148mm$^2$) and power consumption (7.5mW). The 128 pulse-shrinking elements allow the achievement of a resolution of 0.82ps and a dynamic range of 2.918ns. The DNL and INL results show a variation from -0.91 to 0.87 LSB and from -2.4 to 2.11 LSB respectively.

## REFERENCES

[1] F. Yuan, *CMOS Time-Mode Circuits and Systems: Fundamentals and Applications*, 1st ed., Krzysztof Iniewski, Ed. Vancouver, British Columbia, Canada: CRC Press, 2015.

[2] S. Henzler *et al.*, "A Local Passive Time Interpolation Concept for Variation-Tolerant High-Resolution Time-to-Digital Conversion," *IEEE Journal of Solid-State Circuits*, vol. 43, no. 7, pp. 1666-1676, 2008.

[3] S. Henzler, *Time-to-Digital Converters*, 1st ed., Dr. Kiyoo Itoh *et al.*, Eds. Netherlands: Springer Netherlands, 2010.

[4] M. Santos, N. Horta and J. Guilherme, "Logarithmic ad converter with selectable transfer characteristic," in *IEEE Transactions on Circuits and Systems II: Express Briefs*, vol. 63, no.3, pp. 234-238, Mar. 2016.

[5] M Santos, N Horta, J Guilherme, "A survey on nonlinear analog-to-digital converters", Integration, the VLSI Journal 47 (1), 12-22, 2014.

[6] J.D.A. v.d. Broek, "Design and implementation of an Analog-to-Time-to-Digital converter," Master Thesis, Faculty of Electrical Engineering, Mathematics & Computer Science, University of Twente, Box 217, 7500 AE Enschede The Netherlands, Netherlands, 2012.

[7] N. H.E. Weste and K. Eshraghian, *Principles of CMOS VLSI Design : A system Perspective*, 2nd ed., M. C. Varley, Ed. Menlo Park, California, USA: Pearson Education India, 1988.

[8] I. Bombay. (2009, December) National Programme on Technology Enhanced Learning. [Online]. http://nptel.ac.in/courses/117101058/downloads/Lec-16.pdf.

[9] M. S. Kim, "0.18um CMOS Low Power ADPLL with a Novel Local Passive Interpolation Time-to-Digital Converter Based on Tri-State Inverter," Master Thesis, Department of Electrical and Computer Engineering, Northeastern University, Boston, Massachusetts, United States of America, 2012.

[10] P. Chen, S.-I. Liu, and J. Wu, "A CMOS Pulse-Shrinking Delay Element For Time Interval Measurement," *IEEE Transactions on Circuits and Systems II: Analog and Digital Signal Processing*, vol. 47, no. 9, pp. 954-958, 2000.

[11] C. C. Chen, S. H. Lin and C. S. Hwang, "An Area-Efficient CMOS Time-to-Digital Converter Based on a Pulse-Shrinking Scheme," *IEEE Transactions on Circuits and Systems II: Express Briefs*, vol. 61, no. 3, pp. 163-167, 2014.

[12] C. C. Chen, S. H. Lin, and C. S. Hwang, "An Area-Efficient CMOS Time-to-Digital Converter Based on a Pulse-Shrinking Scheme," *IEEE Transactions on Circuits and Systems II: Express Briefs*, vol. 61, no. 3, pp. 163-167, 2014.

[13] Y. K.a.T.W. Kim, "An 11 b 7 ps Resolution Two-Step Time-to-Digital Converter With 3-D Vernier Space," *IEEE Transactions on Circuits and Systems I: Regular Papers*, vol. 61, no. 8, pp. 2326-2336, August 2014.

[14] A. L.a.R.C. L. Vercesi, "Two-Dimensions Vernier Time-to-Digital Converter," *IEEE Journal of Solid-State Circuits*, vol. 45, no. 8, pp. 1504-1512, August 2010.

[15] M. L.a.A.A. Abidi, "A 9 b, 1.25 ps Resolution Coarse–Fine Time-to-Digital Converter in 90 nm CMOS that Amplifies a Time Residue," *IEEE Journal of Solid-State Circuits*, vol. 43, no. 4, pp. 769-777, April 2008.

[16] S. Naraghi, "Time-Based Analog to Digital Converters," PhD Thesis, Department of Electrical Engineering, The University of Michigan, Michigan, United States of America, 2009.

[17] T. Iizuka, T. Koga, T. Nakura, and K. Asada, "A Fine-Resolution Pulse-Shrinking Time-to-Digital Converter with Completion Detection utilizing Built-In Offset Pulse," in *IEEE Asian Solid-State Circuits Conference*, Toyama, Jaoan, 2016, pp. 313-316.

# A High-resolution $\Delta$-Modulator ADC with Oversampling and Noise-shaping for IoT

Ana Correia[1,2], Pedro Barquinha[3], João Marques[2], and João Goes[1]

[1]CTS/UNINOVA, Departamento de Engenharia Electrotécnica (DEE), Faculdade de Ciências e Tecnologia (FCT), Universidade NOVA de Lisboa, 2829-516 Caparica, Portugal
[2]S3 SEMICONDUCTORS, Madan Parque, Rua dos Inventores, 2825-182 Caparica, Portugal
[3]CENIMAT/I3N, Departamento de Ciência dos Materiais (DCM), Faculdade de Ciências e Tecnologia (FCT), Universidade NOVA de Lisboa, and CEMOP/UNINOVA, 2829-516 Caparica, Portugal
E-mail: a.correia@campus.fct.unl.pt

*Abstract*—This paper proposes a novel high-resolution delta-modulator with oversampling and noise-shaping suitable for industrial IoT systems. "Inspired by" a SAR ADC, this architecture uses an integrator in the digital domain, instead of the usual SAR logic. Combining adaptive-step delta-modulation with the adopted low/medium-resolution DAC scheme, an energy efficient A/D converter architecture is proposed, presenting encouraging results in comparison with the reported SAR algorithms. Behavioral simulations demonstrate a dynamic performance of 88 dB of SNDR, corresponding to an ENOB of 14.3 bits, for a 1 MHz input signal bandwidth and using an oversampling-ratio of 64.

*Index Terms*—Industrial IoT systems; High-resolution converters; Delta modulation; Noise-shaping.

## I. Introduction

The 'boom' of the industrial internet-of-things (IoT) has driven a fast re-invention or optimization of conventional industrial systems. Features such as remote access control, effective monitoring, operation of large amount of data, high-level of integration, among others, have strongly contributed for this acceleration, allowing low-cost, high-reliable and multifunctional integrated systems [1].

Analog-to-digital (A/D) converters (ADCs) play a relevant role in industrial IoT systems, spanning from sensor interfaces to radio-frequency (RF) front-ends, where high-resolution and low-power dissipation are generally required. To accomplish high-resolution specifications, the delta-sigma ($\Delta\Sigma$) modulator ($\Delta\Sigma$M) ADC is one of the most popular architectures. However, for a given signal bandwidth (BW), higher effective resolutions are achieved at the expense of larger sampling rates ($F_s$). On the other hand, to fulfill the low-power dissipation requirement, the successive approximation-register (SAR) ADC architecture is widely used, achieving low/moderate resolutions together with a reasonable input signal BWs. Nevertheless, higher effective resolutions are still difficult to achieve relying on a classic SAR ADC, without sacrificing the energy efficiency. Besides the ADC power itself, it will significantly increase the current consumption requirement of

the block driving the ADC. Furthermore, thermal noise will become a relevant issue, compromising the overall dynamic performance (for a fairly good energy-efficiency), and the mismatch of capacitive digital-to-analog converter (DAC) can also contribute for the converter degradation.

In the last years, oversampling and noise-shaping have been introduced in SAR ADCs, taking advantage from the noise-shaping capability typically from $\Delta\Sigma$Ms [2]. Since the effect of mismatch and thermal noise is alleviated through this hybrid scheme, higher resolutions can be accomplished while the low-accuracy circuit blocks and the energy efficiency are kept.

This paper presents a novel delta ($\Delta$) modulator ($\Delta$M) ADC employing oversampling and noise-shaping with high attractiveness for industrial IoT applications. Despite the similarity with existing hybrid SAR ADC architectures, also with oversampling and noise-shaping, the proposed conversion algorithm is rather based on (adaptive) $\Delta$ modulation. Hence, instead of SAR logic, it uses a digital integrator (i.e., an accumulator) that, combined with the adopted DAC structure, can improve significantly the converter's energy-efficiency. In this way, high resolution can be achieved keeping simple the mixed-signal circuitry.

The organization of this paper is as follows. In Section II the architecture is rigorously detailed, showing the block diagram, the associated timing scheme, and providing details of each building block. Additionally, a fair comparison with architectures employing $\Delta$ and $\Delta\Sigma$ modulations is also provided. Section III shows the simulation results using MATLAB®. Finally, the main conclusions are summarized in Section IV.

## II. From SAR ADC to $\Delta$M with Oversampling and Noise-shaping

The proposed $\Delta$M employing oversampling and noise-shaping is schematically presented in Fig. 1. Initially "inspired by" a SAR ADC, and disabling for now, for the sake of simplicity, the noise-shaping section, the proposed architecture uses an integrator $H_1$ in the digital domain (i.e., an accumulator), instead of the typical SAR logic. This modification is crucial since the converter does not act anymore as a SAR but as a $\Delta$M, predicting the input signal ($V_{in}$) based on its change. However, as it happens in SAR ADCs, and in comparison with either single or multi-bit $\Delta\Sigma$Ms, it requires, solely, a

---

This work is funded by FEDER funds through the COMPETE 2020 Programme and National Funds through FCT - Portuguese Foundation for Science and Technology under the project number UID/EEA/00066/2013. The work also received funding from the European Community H2020 program under grant agreement No. 716510 (ERC-2016-STG TREND).

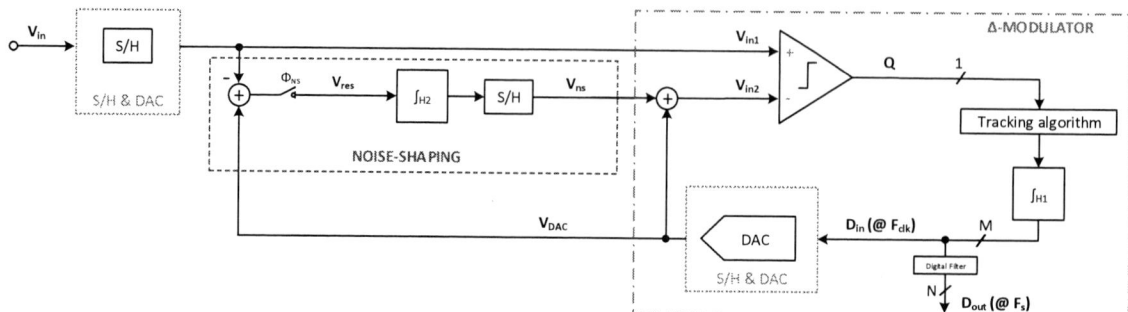

Fig. 1. Diagram of the proposed architecture, a $\Delta$M employing oversampling and noise-shaping.

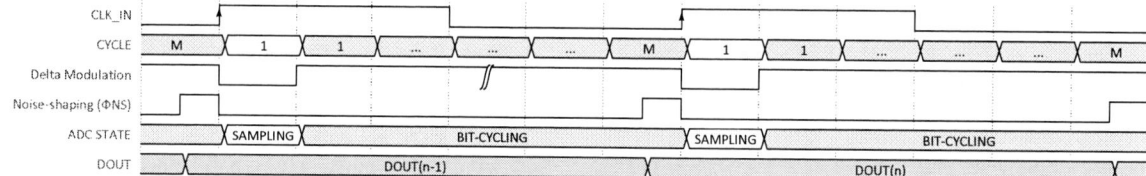

Fig. 2. Timing of the proposed architecture, assuming 1 *clk* cycle for sampling and $M$ for bit-cycling, where $M$ is the DAC resolution.

single comparator and a multi-bit DAC. Notice that although the DAC has higher resolution than in $\Delta\Sigma$Ms, its nominal resolution will be lower than the one of the ADC. In the example, later described, for a 16-bit ADC, a DAC with 10-bit will be used.

The proposed converter also utilizes oversampling and noise-shaping to alleviate the impact of both, thermal and quantization noises. The residual voltage, $V_{res}$, i.e., the difference between the output, $V_{DAC}(n-1)$, and input, $V_{in1}(n-1)$, is integrated and its result, $V_{ns}$, is summed to $V_{DAC}(n-1)$, generating $V_{in2}$. When a new input is sampled, $V_{in1}(n)$, the first comparison is performed using this initial guest, $V_{in2}(n-1)$. During the next cycles of $\Delta$ modulation, $V_{ns}(n-1)$ is always summed to $V_{DAC}(n)$.

The A/D converter's timing is detailed in Fig. 2. As shown, this architecture works using two different frequencies: $F_s$ used for sample-and-hold (S/H) and clock frequency, $F_{clk}$, used for $\Delta$ modulation. Noise-shaping section is activated in the last cycle of $\Delta$ modulation. For now, it is assumed that $F_{clk}$ is $M$ times higher than $F_s$, where $M$ represents the nominal resolution of the DAC.

Fig. 3 presents not only the block diagram of the proposed converter but also the diagrams of usual architectures employing either $\Delta$ or $\Delta\Sigma$ modulations, allowing a fair comparison. In the suggested architecture, and in contrast with the existing ones, sampled $V_{in}$ is an input of comparator and it is compared with the result of $\Delta$ modulation and noise-shaping. Furthermore, while in $\Delta$ and in $\Delta\Delta\Sigma$ structures [3], the integrator in the loop is employed in the analog domain, in this proposal we implement it digitally, contributing for an energy efficient converter. Moreover, and in comparison with the $\Delta\Sigma$M topology, it is worth emphasizing that, in our proposed architecture, the converter's output, $D_{out}$, can be instantly correlated with $V_{in}$.

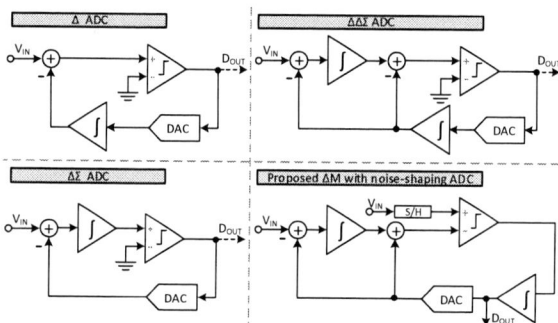

Fig. 3. Block diagram of a $\Delta$M, a $\Delta\Sigma$M, a $\Delta\Delta\Sigma$M [3] and the proposed architecture.

To highlight the advantages and the constraints of the proposed architecture, especially the ones related with energy efficiency and the adopted DAC scheme, detailed information about each building block is provided in the following subsections.

### A. Integrator $H_1$ & Tracking Algorithm

As it is widely known from literature, $\Delta$ modulation is efficient if there is no slope overload and no granular noise, i.e., if the step size is the adequate to follow $V_{in}$. Then, a correct tracking is mandatory to minimize the quantization error and to ensure an instantly correlation between $V_{in}$ and $D_{out}$ [4]. Bearing this in mind, the digital implementation of the integrator $H_1$ will facilitate an adaptive-step approach. In fact, this scheme (shown in Fig. 1 as the "tracking algorithm") adapts the step, $\delta$, during the conversion, minimizing the price of digitalization. The step size is calculated multiplying the comparator's output (+1 or -1), $Q$, by the number of repeated outputs. Otherwise, when a different $Q$ is detected, the step is reset to the minimum (1) defined.

As previously referred, given the domain of this integrator, a simple implementation can be considered, such as:

$$D_{in}(n) = D_{in}(n-1) + \delta Q \qquad (1)$$

### B. Sample-and-Hold & Digital-to-Analog Converter

For practical designing reasons, it is desirable as much as possible a low/medium-resolution DACs implementation. This aspect is a strong concern especially for high-resolution ADCs. However, as in this proposed architecture, noise-shaping reduces the resolution required for DAC, simplifying the design. By this reason, DAC resolutions, $M \leq 12$ bits and, consequently, lower than the nominal resolution of the ADC, (e.g. $N = 16$ bits), are considered in this approach.

Moreover, considering the minimum variation of the integrator $H_1$, there is a strong advantage if the DAC is made using unitary switches and capacitors (rather than binary weighted), with no reset phase. Furthermore, the S/H circuit can also be easily embedded in the DAC [5], minimizing the overall area of the converter, as it is detailed in Fig. 1.

As previously referred, combining the adopted DAC scheme with $\Delta$ modulation, significant energy savings can be achieved. While in the classic SAR algorithm, as it is shown in Table I, even for close successive $V_{in}$'s, the binary search always starts from the beginning (from an explicit DAC reset), different works have been proposed during recent years trying to reduce the energy impact. For instance, the bit-repeating least-significant bit (LSB)-first quantizer (LSBFQ) approach [6], whose an example conversion is detailed in Table II, suggests a method that outperforms the previous algorithm in terms of energy consumption and number of required bit-cycles, for low-activity signals.

In comparison with this last one, the inherent method of the architecture presented in this paper is shown in Table III. Basically, the variation, $\Delta D_{in}$, given by

$$\Delta D_{in} = D_{in}(n) - D_{in}(n-1) \qquad (2)$$

indicates the number of capacitors that will be (un)connected. Then, the switching activity is proportional to the variation obtained from integrator, reducing the energy consumption.

The proposed method can also be straightforwardly extended to an asynchronous version. If a pattern around $\Delta D_{in}$ is detected, such as '+1 -1 +1 -1', the converter can assume that $D_{out}$ was found. Despite the extra logic required, it will allow to save some clock cycles (instead of the usually required $M$ cycles by sample), increasing the conversion speed.

Regarding linearity, since a DAC resolution $M \leq 12$ bits with a unity-weighted topology will be used, data weighted averaging (DWA) can be easily implemented in the design to improve its dynamic linearity. Moreover, behavioral simulations also demonstrate that only the MSB part of the DAC will require DWA (say, the unitary capacitors corresponding to the 5 MSBs).

### C. Integrator $H_2$

The noise-shaping scheme, where the integrator $H_2$ is included, is similar to the one employed in SAR ADC described

TABLE I
EXAMPLE OF SAR ALGORITHM, CONSIDERING A 5-BIT DAC.

| $D_{out}(n\text{-}1) = 15; D_{out}(n) = 18$ | |
|---|---|
| Bit-cycle | D[4:0] |
| Reset phase | 00000 |
| 1 | 10000 |
| 2 | 10000 |
| 3 | 10000 |
| 4 | 10010 |
| 5 | 10010 |

TABLE II
EXAMPLE OF LSBFQ ALGORITHM, CONSIDERING A 5-BIT DAC, WHERE DIR IS THE DIRECTION OF BIT-CYCLING [6].

| $D_{out}(n\text{-}1) = 15; D_{out}(n) = 18$ | | | |
|---|---|---|---|
| Bit-cycle | D[4:0] | DIR | $\Delta$ |
| 1 | 01111 | 0 | |
| 2 | 01111 | 1 | |
| 3 | **10000** | 1 | +1 |
| 4 | 10001 | 1 | +1 |
| 5 | 10011 | 1 | +2 |
| 6 | 10010 | 1 | -1 |

TABLE III
EXAMPLE OF PROPOSED ARCHITECTURE CONVERSION ALGORITHM, CONSIDERING A 5-BIT DAC AND ASSUMING A CORRECT TRACKING.

| $D_{out}(n\text{-}1) = 15; D_{out}(n) = 18$ | | |
|---|---|---|
| Bit-cycle | D[4:0] | $\Delta D_{in}$ |
| 1 | 01111 | |
| 2 | 10000 | +1 |
| 3 | 10010 | +2 |

in [2]. However, given that this integrator will be designed in analog domain, its design should be simple, preferring a quasi-passive structure [7] of an IIR filter. This is possible assuming a function equal to (1), with $\delta = 1$.

### D. Comparator

Given that a single-bit quantizer (a unique comparator) is employed in this architecture, its design is relatively simple. Furthermore, assuming that an $F_{clk}$, around 1.4 GHz may be used (depending on oversampling-ratio (OSR) and BW), a fully-dynamic structure can be used, targeting the highest energy-efficiency in the ADC.

As previously noted, the impact of the input-referred comparator's thermal noise can be alleviated considering the proposed hybrid architecture. Regarding flicker noise, that is especially critical for an input signal BWs below than 10 MHz, a chopping technique can be simply implemented, as proposed by [8], at half of the $F_s$.

### III. RESULTS AND DISCUSSION

Simulations of the described architecture were performed using MATLAB®. The model includes all main static errors, such as, mismatch in capacitors, comparator's offset and thermal noise sources. Regarding flicker noise, it was not introduced in the model since, as stated before, it has been assumed that chopping will be used. Moreover, the dynamic of the only active component, the comparator, has not been modeled since this block can be designed to meet the required comparison-time at the expense of a given power budget.

978-1-5386-5388-3/18 $31.00 © 2018 IEEE

The simulated A/D converter employs a 10-bit DAC, based on unitary capacitors and switches, and with DWA implemented in the 5 MSBs. The integrators' functions, $H_1$ and $H_2$, are described, respectively, by (1) with the detailed adaptive-step ($\delta=1$) and by

$$H_2(z) = \frac{z^{-1}}{1 - z^{-1}} \qquad (3)$$

Additionally, an OSR of 64 and a BW of 1 MHz has been considered, for a 100 kHz input signal frequency (to be able to observe the most important harmonics).

Fig. 4 shows the tracking, implementing the adaptive-step approach. As observed in the insets of Fig. 4, a correct tracking is quickly obtained, decreasing the system's delay, and the output follows instantly the input signal.

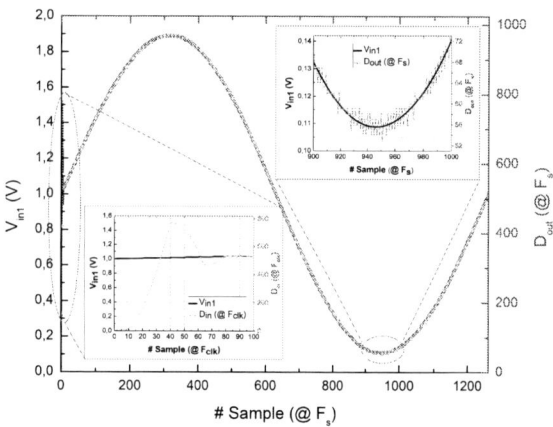

Fig. 4. Comparison between input and output signals, showing the initial tracking and the outputs during conversion, at $F_{clk}$ and $F_s$. The simulated architecture employs a 10-bit DAC.

In Fig. 5, the simulated spectrum of the proposed converter without DWA is presented. With an OSR of 64 and a 10-bit DAC, a peak of 72.8 dB in the signal-to-noise and distortion ratio (SNDR) has been obtained for a 100 kHz input signal and a 1 MHz BW.

Implementing DWA in the 5 MSBs of the DAC, a significant improvement in the overall dynamic performance has been achieved. As shown in Fig. 6, using the same conditions, a peak of 88.1 dB in the SNDR has been obtained, corresponding to an effective resolution (ENOB) of about 14.3 bits. For a 990 kHz input signal, observed in Fig. 7, an SNDR of 87.3 dB was achieved. Since the main static non-idealities were modeled, including noise, it is not expected a degradation higher than 0.5 bits of ENOB when passing to schematic-level.

## IV. Conclusion

A novel high-resolution $\Delta$M with oversampling and noise-shaping has been proposed. Combining, in the feedback-loop, the digital integrator with a DAC structure based on a unity-weighted topology, an energy-efficient ADC with excellent dynamic performance can be designed. Effective resolutions of more than 14 bits can be reached, relying only in a single-comparator and in a 10-bit DAC.

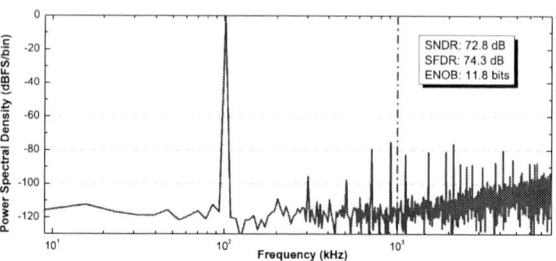

Fig. 5. Simulated spectrum for a 100 kHz input signal and 1 MHz BW. The architecture employs a 10-bit DAC without DWA and an OSR of 64.

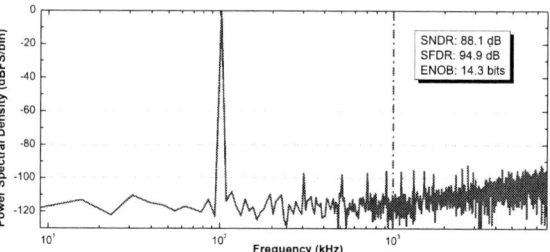

Fig. 6. Simulated spectrum for a 100 kHz input signal and 1 MHz BW. The architecture employs a 10-bit DAC, with DWA in the 5 MSBs, and an OSR of 64.

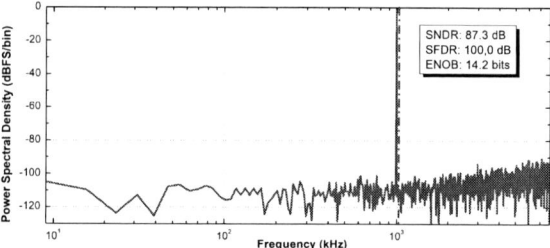

Fig. 7. Simulated spectrum for a 990 kHz input signal and 1 MHz BW. The architecture employs a 10-bit DAC, with DWA in the 5 MSBs, and an OSR of 64.

## References

[1] N. O'Riordan, "Industrial IoT," in *Circuits and Systems for the Internet of Things: CAS4IoT*. IEEE CAS, River Publishers, 2017.

[2] J. Fredenburg and M. Flynn, "A 90MS/s 11MHz bandwidth 62dB SNDR noise-shaping SAR ADC," in *Proc. IEEE Int. Solid State Circuits Conf. (ISSCC)*. IEEE, Feb 2012, pp. 468–470.

[3] J. Ren, S. Sarwana, A. Sahu, A. Talalaevskii, and A. Inamdar, "Low-Pass Delta-Delta-Sigma ADC," *IEEE Trans. Appl. Supercond.*, vol. 25, no. 3, pp. 1–6, Jun 2015.

[4] D. G. Zrilic, *Circuits and systems based on delta modulation : linear, nonlinear, and mixed mode processing*. Springer, 2005.

[5] M. van Elzakker, E. van Tuijl, P. Geraedts, D. Schinkel, E. A. M. Klumperink, and B. Nauta, "A 10-bit Charge-Redistribution ADC Consuming 1.9 $\mu$W at 1 MS/s," *IEEE J. Solid-State Circuits*, vol. 45, no. 5, pp. 1007–1015, May 2010.

[6] A. Waters, J. Leung, and U.-K. Moon, "LSB-first SAR ADC with bit-repeating for reduced energy consumption," in *Proc. IEEE Int. Conf. on Electronics, Circuits and Syst.* IEEE, Dec 2014, pp. 203–206.

[7] J. L. A. de Melo, J. Goes, and N. Paulino, "A 0.7 V 256 $\mu$W $\Delta\Sigma$ modulator with passive RC integrators achieving 76 dB DR in 2 MHz BW," in *Proc. IEEE Symp. VLSI Circuits*, Jun 2015, pp. C290–C291.

[8] P. Harpe, E. Cantatore, and A. van Roermund, "An oversampled 12/14b SAR ADC with noise reduction and linearity enhancements achieving up to 79.1dB SNDR," in *Proc. IEEE Int. Solid State Circuits Conf. (ISSCC)*. IEEE, Feb 2014, pp. 194–195.

*PRIME 2018, Prague, Czech Republic*

*Session: Data Converters*

# A 12.4fJ-FoM 4-Bit Flash ADC Based on the StrongARM Architecture

Abdullah S. Almansouri[1*†], Abdullah Alturki[2†], Hossein Fariborzi[1], Khaled N. Salama[1] and Talal Al-Attar[1]

[1] Computer, Electrical and Mathematical Science & Engineering Division, King Abdullah University of Science and Technology, Thuwal, Saudi Arabia

[2] Electrical and Computer Engineering Division, King Abdulaziz University, Jeddah, Saudi Arabia

* abdullah.almansouri@kaust.edu.sa

† These authors contributed equally to this work

*Abstract*— **This work proposes an efficient 4-bit flash ADC based on the StrongARM comparator architecture. The proposed design eliminates the need for the resistive ladder by systematically modifying the sizing of the input differential pair of each comparator. As a consequence, the area and the power consumed within the ladder is eliminated. Furthermore, a Helpee StrongARM circuit is introduced which enables operation at an input voltage below the threshold voltage of the transistor. An enhanced 1-out-of-15 decoder converts the thermometer code from the StrongARM and the Helpee StrongARM comparators into a 1-out-of-n code. The proposed 4-bit flash ADC architecture, simulated in 90nm standard CMOS technology, consumes 292 µW at 1.6 GHz sampling frequency, has an ENOB of 3.88 and FoM of 12.4 fJ/conv.step.**

*Keywords*— **Flash ADC, Data converter, StrongARM, Helpee StrongARM, Comparator**

## I. INTRODUCTION

The demand for fast and energy efficient data converters is always present, especially with the recent advances in the fields of ultra-low power Integrated Circuits (ICs), Digital Signal Processing (DSP) and Internet-of-Things (IoT). While Flash Analog-to-Digital converter (ADC) in general offers the fastest performance, this architecture suffers from large power- and area- consumption [1]. Figure 1 shows a conventional *n*-bit flash ADC which requires $2^n$ resistors, where *n* represents the number of bits. Furthermore, the current passing through these resistors consumes static power. Typically, the references are generated either by resistor ladder or some types of analog interpolation [2]. However, both are still consuming a considerable amount of area and power. A Threshold Modified Comparator Circuit (TMCC) was proposed to eliminate the need for the resistive/capacitive ladder [3]. However, the TMCC is a single input comparator, which makes it tolerable to noise.

In this work, we propose an energy and area efficient differential-input flash ADC that eliminates the need for the resistive/capacitive ladder and is capable of operating with a rail-to-rail input voltage. The proposed design is based on the enhanced StrongARM comparator introduced in [4] and a modified Helpee StrongARM comparator. The paper is organized as follows: Section II discusses the proposed design, Section III shows the simulation results, and the conclusion is

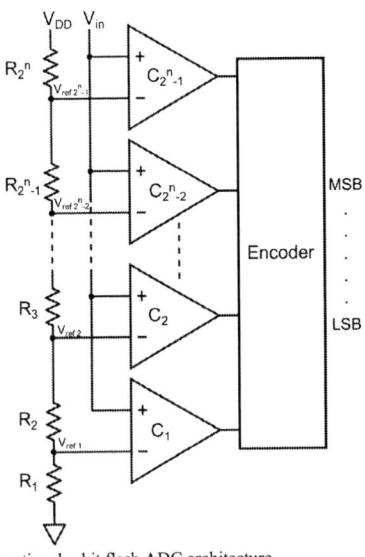

Fig. 1. Conventional *n*-bit flash ADC architecture.

presented in Section IV.

## II. PROPOSED ADC ARCHITECTURE

### A. StrongARM Background

Figure 2 shows the enhanced StrongARM latch comparator. Compared to the conventional design [5] the enhanced StrongARM reduces the clock feedthrough problem, consumes less energy and offers faster performance. Its main working principle is as follows: The gates of the differential pairs $T_5$ and $T_6$ are connected to the input voltages $V_{in}$ and $V_{ref}$, respectively. Once the clock signal goes High, the charging transistors $(CT_{1,2})$ turn OFF allowing the internal nodes (capacitors) to start getting discharged through a tail transistor $(T_7)$. Depending on the values of the Vin and Vref, the internal capacitance (A, A', B and B') discharge at different rates, eventually generating a $V_{DD}$ at one of the nodes (A or A') and a $V_{SS}$ at the other node.

978-1-5386-5388-3/18 $31.00 © 2018 IEEE

*Paper P71*                                                                           *PRIME 2018, Prague, Czech Republic*

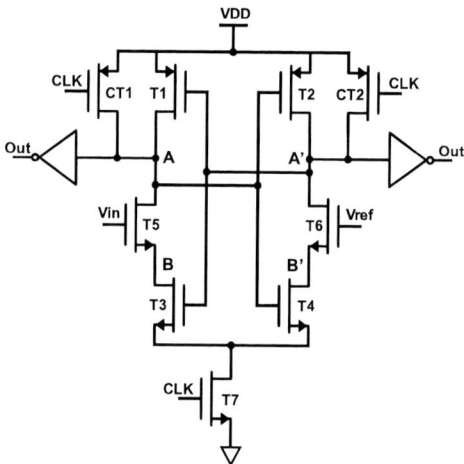

Fig. 2. Enhanced StrongARM circuit.

### B.  Concept

The ultimate need for the resistive/capacitive ladder is to provide a unique reference voltage ($V_{ref}$) for each comparator, as shown in Figure 1. If the StrongARM comparator is used for the flash ADC, different $V_{ref}$ at each comparator results in different current (compared to the other comparators) flowing through $T_6$. Meanwhile, the current flowing through $T_5$ for all the comparators is not influenced, as the current through $T_5$ is mainly controlled by $V_{in}$.

A more efficient approach to achieve similar functionality, while eliminating the need for the resistive/capacitive ladder, is by fixing $V_{ref}$ to a specific value, and modifying the W/L ratio of the differential pair transistors. Since the current is proportional to W/L ($I \propto \frac{W}{L}$), we can achieve different discharging current through $T_6$ by tuning W/L. In other words, even though $V_{ref}$ is equal for all comparators, each comparator represents different code level depending on the W/L ratio of the differential pair. For example, Figure 3 shows the enhanced

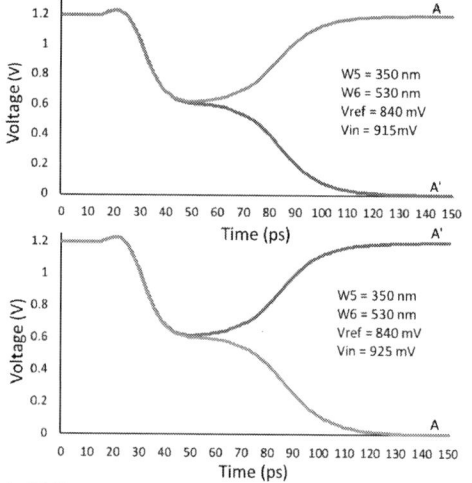

Fig. 3. Shifting the threshold voltage of the StrongARM from $V_{in} = V_{ref}$ to $V_{in} = 920$ mV by changing the sizing of $T_5$

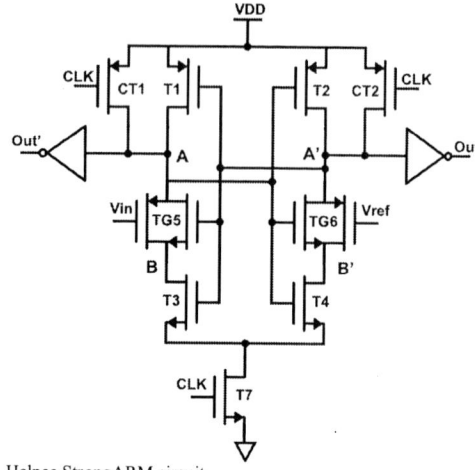

Fig. 4. Helpee StrongARM circuit.

StrongARM comparator with $W_5 = 350$ nm and $W_6 = 530$ nm. The simulation results show that the threshold voltage (i.e., the code level) of the comparator for this configuration is around 920 mV although the reference voltage is fixed at 840 mV. Note that, aside from $W_5$, the sizing for all the other transistors are fixed.

### C.  Full Range ADC

Ideally, the input voltage to the ADC should range from $V_{SS}$ to $V_{DD}$. However, the minimum input voltage for the enhanced StrongARM is determined by the threshold voltage of the input transistors $T_{5,6}$. Therefore, the operation of this architecture is limited to voltages greater than $V_{th}$. This limitation can be solved by utilizing the complementary circuit of the enhanced StrongARM. However, by doing that the complementary circuit requires a different clock (as it operates when the clock signal is Low) and requires additional invertors at the output to obtain a consistent performance with the enhanced StrongARM.

Figure 4 proposes a Helpee StrongARM architecture that covers the lower range of the input voltage ($V_{in} < V_{th}$), operates with the same clock as the enhanced StrongARM and does not require extra invertors at the output. The proposed architecture replaces $T_{5,6}$ with two transmission-gates ($TG_{5,6}$). The gates of the PMOS transistors of the transmission-gates are connected to

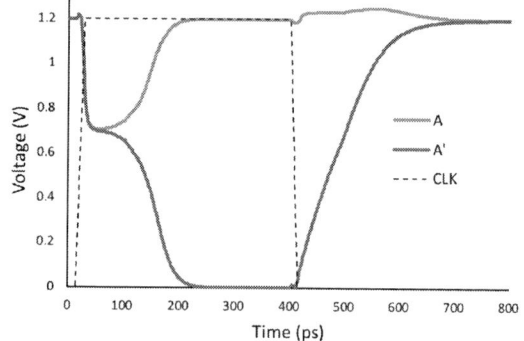

Fig. 5. Operation of the proposed Helpee StrongARM for $V_{in} = 210$ mV and $V_{ref} = 200$ mV.

978-1-5386-5388-3/18 $31.00 © 2018 IEEE              38

the input voltages, while the NMOS transistors of the transmission-gates are cross-coupled to the opposite nodes. Figure 5 shows the simulation results for the Helpee StrongARM when the input voltages $V_{in}$ and $V_{ref}$ are less than the threshold voltage and equals to 210 and 200 mV, respectively. Note that, the polarity of the output in the Helpee StrongARM is opposite compared to the enhanced StrongARM: in the former, $V_{in} > V_{ref}$ results in A = 1 and A' = 0; while in the latter, $V_{in} > V_{ref}$ results in A = 0 and A' = 1.

### D. 1-out-of-n Decoder

The ultimate goal is to use both the StrongARM and the Helpee StrongARM to cover rail-to-rail input range and to output a thermometer code. However, when $V_{in} > V_{DD} - V_{thp}$, the PMOS transistor in the transmission-gate (TG5) remains OFF, resulting in recharging the node B to a different value compared to B'. This systematic difference can cause an incorrect thermometer code for $V_{in} > V_{DD} - V_{thp}$. Nevertheless, this is solved easily by adjusting the next block, namely a thermometer to 1-out-of-15 decoder, to account for this systematic limitation. Figure 6 proposes a 1-out-of-15 decoder for the StrongARM (Figure 6(a)) and the Helpee StrongARM (Figure 5(b)) comparators. These logic blocks compare the output of each comparator ($C_n$) with the next comparator ($C_{n+1}$) and generate 1 at the output only if the $C_n > C_{n+1}$. To eliminate the error of the Helpee StrongARM when $V_{in} > V_{DD} - V_{thp}$, its output is compared with the least StrongARM comparator (in addition to comparing the output to $C_{n+1}$), as shown in Figure 6(b).

## III. RESULTS AND DISCUSSION

Figure 7 shows a schematic of the proposed flash ADC including the 1-out-of-15 decoder. The proposed architecture offset the enhanced StrongARM and the Helpee StrongARM circuits and eliminated the need for the resistive/capacitive ladder. The proposed flash ADC consists of 6 Helpee StrongARM comparators to cover the range where the $V_{in} < V_{thn}$, and 9 StrongARM comparators covering the range of $V_{in} > V_{thn}$. This ratio (i.e., 6:9) of StrongARM and Helpee StrongARM can take a different value. However, since the Helpee StrongARM has a larger internal capacitance due to the extra transmission-gates, reducing the use of the Helpee

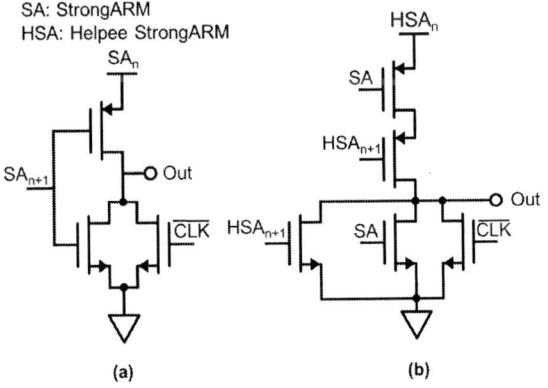

Fig. 6. A 1-out-of-15 decoder for the (a) StrongARM and (b) the Helpee StrongARM comparators.

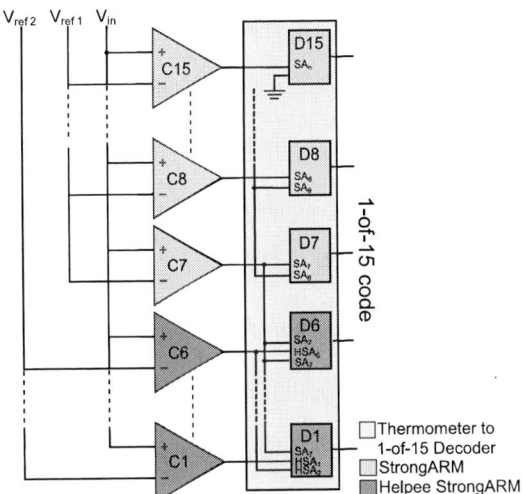

Fig. 7. Proposed flash ADC architecture including the StrongARM, the Helpee StrongARM and the enhanced 1-of-15 decoder.

StrongARM is more efficient. The proposed design requires only two reference voltages $V_{ref1}$ and $V_{ref2}$ for the enhanced StrongARM and the Helpee StrongARM, respectively. The values of $V_{ref1}$ and $V_{ref2}$ are determined to minimize the energy-delay product (EDP) [4]. Nevertheless, it is possible to have only one reference voltage in the ADC (i.e. $V_{ref} = V_{ref1} = V_{ref2} = 0.5 \times (V_{DD} - V_{SS}) + V_{SS}$). By doing so, however, the ADC will operate outside the optimal operating point [4]. In this work, the reference voltages $V_{ref1}$ and $V_{ref2}$ are set to 840 mV and 440 mV, respectively.

Table I shows the threshold voltages of the comparators for the proposed flash ADC and achieved by controlling the sizing of $T_{5,6}$. Aside from $T_{5,6}$, the Length (L) and Width (W) of all the other transistors are fixed for all the comparators. Furthermore,

Table I. The threshold voltages of the comparator for the flash ADC

| Comparator number | Ideal Threshold voltage (mV) | Comparator threshold voltage (mV) |
|---|---|---|
| 1 | 40 | 39 |
| 2 | 120 | 127 |
| 3 | 200 | 208 |
| 4 | 280 | 280 |
| 5 | 360 | 368 |
| 6 | 440 | 441 |
| 7 | 520 | 520 |
| 8 | 600 | 601 |
| 9 | 680 | 687 |
| 10 | 760 | 760 |
| 11 | 840 | 848 |
| 12 | 920 | 917 |
| 13 | 1000 | 1001 |
| 14 | 1080 | 1080 |
| 15 | 1160 | 1162 |

TABLE II.    Performance Comparison

| | [11] FinFET | [1] Time-Based | [10] Interpolation | [9] Digital | [3] TMCC | This Work |
|---|---|---|---|---|---|---|
| Technology (nm) | 16 | 65 | 65 | 90 | 180 | **90** |
| Resolution (Bit) | 4 | 5 | 7 | 7 | 4 | **4** |
| Supply voltage (V) | 0.9 | 1 | 1.2 | 1.2 | 1.8 | **1.2** |
| Input voltage rang (V) | 0.5 | 0.6 | N/A | 0.6 | N/A | **Rail-to-rail** |
| Sampling rate (GS/s) | 4.0625 | 5 | 2 | 1.5 | 0.2 | **1.6** |
| Resistive ladder | Yes | Yes | Yes | No | No | **No** |
| DNL (LSB) / INL (LSB) | 1.444 / 2.364$^2$ | 0.83 / 0.79 | 0.58 / 0.64 | 0.7 / <0.7 | 0.14 / 0.226 | **0.1375 / 0.1125** |
| ENOB | 3.7 | 4.06 | 6.04 | 6.05 | 3.2 | **3.88** |
| Power (mW) | 16.9 | 7.8 | 20.7 | 204 | 0.127 | **0.292** |
| FoM (fJ/Conv.step) | 320 | 94.6 | 157 | 2060 | 69 | **12.4** |

[1] FoM = $Power/(f_s \times 2^{ENOB})$ ; [2] without calibration

the proposed design is capable of handling more than 5% variation in W with minimum change in the performance. Nevertheless, to overcome the mismatch, standard offset cancellation techniques can be used [5]-[8].

The proposed 4-bit flash ADC architecture including the 1-out-of-15 decoder simulated in 90nm standard CMOS technology and using Synopsys simulation tools. It consumes 292 µW, offers an effective number of bits (ENOB) of 3.88 and Figure-of-Merit (FoM) of 12.4 fJ/conv.step when operating at 1.6 GHz sampling frequency. Figure 8 shows the performance of the proposed architecture when $V_{in}$ is varied from 0 to $V_{DD}$. In general, the proposed architecture is capable of operating with a rail-to-rail input voltage while offering an efficient and fast performance.

Table II compares the performance of the proposed design with the state-of-the-art. In general, the proposed design offers the best FoM and consumes relatively low power. [3] has a better power consumption; however, it operates at a much lower sampling frequency.

## IV. CONCLUSION

In this work, we proposed a flash ADC based on the StrongARM architecture. The proposed design systematically varies the sizing of the input differential pair within each comparator to eliminate the need for the resistive/capacitive

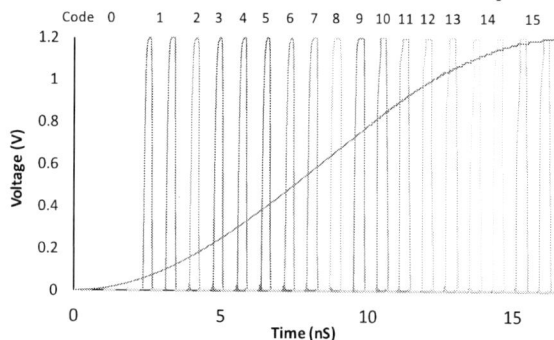

Fig. 8. The performance of the proposed flash ADC when the input voltage is swept from 0 to 1.2 volts (tail-to-rail).

ladder. This work also introduced the Helpee StrongARM that is capable of operating at input voltages less than the threshold voltage. The proposed flash ADC architecture utilizes the StrongARM, the Helpee StrongARM, and an enhanced 1-out-of-15 decoder to improve the performance and enable handling of a rail-to-rail input range.

## REFERENCES

[1] C. H. Chan *et al.*, "A 7.8-mW 5-b 5-GS/s Dual-Edges-Triggered Time-Based Flash ADC," in *IEEE Trans. Circuits Syst. I: Reg. Papers*, vol. 64, no. 8, pp. 1966-1976, Aug. 2017.

[2] S. Weaver, B. Hershberg and U. K. Moon, "Digitally Synthesized Stochastic Flash ADC Using Only Standard Digital Cells," in *IEEE Trans. Circuits Syst. I: Reg. Papers*, vol. 61, no. 1, pp. 84-91, Jan. 2014.

[3] S. Mukherjee, D. Saha, P. Mostafa, S. Chatterjee and C. K. Sarkar, "A 4-bit Asynchronous Binary Search ADC for Low Power, High Speed Applications," *2012 International Symposium on Electronic System Design (ISED)*, Kolkata, 2012, pp. 28-32.

[4] A. Almansouri, A. Alturki, A. Alshehri, T. Al-Attar, H. Fariborzi, "Improved StrongARM latch comparator: Design, analysis and performance evaluation," *2017 13th Conference on Ph.D. Research in Microelectronics and Electronics (PRIME)*, Giardini Naxos, 2017, pp. 89-92.

[5] B. Razavi, "The StrongARM Latch [A Circuit for All Seasons]," in *IEEE Solid-State Circuits Magazine*, vol. 7, no. 2, pp. 12-17, Spring 2015.

[6] G. Van der Plas, S. Decoutere, and S. Donnay, "A 0.16 pJ/conversion-step 2.5 mW 1.25 GS/s 4b ADC in a 90nm digital CMOS process," in *Solid- State Circuits Conference, 2006. ISSCC 2006. Digest of Technical Papers. IEEE International*, 2006, p. 2310.

[7] C. Yanfei *et al.*, "Split capacitor DAC mismatch calibration in successive approximation ADC," *IEICE transactions on electronics*, vol. 93, pp. 295- 302, 2010.

[8] E. Alpman, H. Lakdawala, L. R. Carley, and K. Soumyanath, "A 1.1 v 50mw 2.5 GS/s 7b Time-Interleaved C-2C SAR ADC in 45nm LP digital CMOS," in *Solid-State Circuits Conference-Digest of Technical Papers, 2009. ISSCC 2009. IEEE International*, 2009, pp. 76-77, 77a.

[9] J. Pernillo and M. P. Flynn, "A 1.5-GS/s Flash ADC with 57.7-dB SFDR and 6.4-Bit ENOB in 90 nm Digital CMOS," in *IEEE Trans. Circuits Syst. II: Exp. Briefs*, vol. 58, no. 12, pp. 837-841, Dec. 2011.

[10] J. I. Kim, Dong-Ryeol Oh, Dong-Shin Jo, Ba-Ro-Saim Sung, and Seung-Tak Ryu "A 65 nm CMOS 7b 2 GS/s 20.7 mW Flash ADC With Cascaded Latch Interpolation," in *IEEE J Solid-State Circuits*, vol. 50, no. 10, pp. 2319-2330, Oct. 2015.

[11] L. Wang, M. A. LaCroix and A. C. Carusone, "A 4-GS/s Single Channel Reconfigurable Folding Flash ADC for Wireline Applications in 16-nm FinFET," in *IEEE Trans. Circuits Syst. II: Exp. Briefs*, vol. 64, no. 12, pp. 1367-1371, Dec. 2017.

# A low power Voice Activity Detector for portable applications

1st Gabriele Meoni
*Information Engineering Dep.*
*University of Pisa*
Pisa, Italy
gabriele.meoni@ing.unipi.it

2nd Luca Pilato
*Information Engineering Dep.*
*University of Pisa*
Pisa, Italy
luca.pilato@for.unipi.it

3rd Luca Fanucci
*Information Engineering Dep.*
*University of Pisa*
Pisa, Italy
luca.fanucci@unipi.it

*Abstract*—**Voice Activity Detectors (VADs) are used to enhance performances and to reduce the activation rate of speech recognition and key-word spotting applications. The last aspect is crucial for portable applications because it allows to save energy, increasing battery life. During last decades, VADs have been realized through hardware solutions to increase their speed in processing and to reduce their power consumption. However, the hardware implementation often represents a limit on the choice of the features to use, limiting the performances on recognition. This paper shows a low-power and low-area serial logistic regression classifier which uses the frame-energy, the maximum absolute signal finite difference and the maximum absolute squared signal finite difference over a frame as features. The system has been implemented on IGLOO nano Field Programmable Gate Array (FPGA), leading to power consumption of 0.559 mW and offering acceptable performances for its use as a preprocessor for speech recognition systems or a more sophisticated software VAD.**

*Index Terms*—**Voice Activity Detection, low power, FPGA, logistic regression, machine learning**

## I. INTRODUCTION

Voice Activity Detectors (VADs) are used to detect the presence of human voices in the input audio stream in several applications, like wireless communications, speech recognition and speech enhancement systems.

In particular, in speech recognition / keyword - spotting applications, VADs can increase the performances of recognition by providing information over the statistics of the input data [1]. In addition, since those tasks are computationally intensive and need to run no stop, their implementation on portable systems could be critical for battery life [2]. For that reason, VADs can be used to reduce the activation and the false acceptance rates, leading to a reduction of power consumption. Moreover, for higher processing performances and a reduced power consumption, in last decades VADs have been usually realized by means of hardware implementations [3].

Nevertheless, this fact often represents a limit on the choice of the features which can be used for the VAD probability extraction. In fact, solutions like the one proposed by [4] which uses Linear Prediction Coefficients (LPC) are extremely tricky to be implemented on hardware, requiring complex trade-offs between area occupation / power consumption and the kind of extracted features [3].

For that reason, many solutions rely on features which allow an easier implementation like the zero-crossing extraction [5].

However, zero-crossing is sensible to the microphone offset and its usability depends on the length of the processed frame. The solution presented in [6] uses the Degree of Voice and the QSNR and processes data in parallel, privileging the real time requirements and the VAD recognition performances over the area occupation and power consumption. In this paper we present a serial VAD based on a logistic regression approach for the classification. The design and the used features are oriented to minimize the area occupation and the power consumption, guaranteeing good performances on recognition. In particular, the low-area design allowed the VAD implementation on board Microsemi IGLOO nano Field Programmable Gate Array (FPGA), leading to consumption of fractions of mW. The remainder work is structured as described: in section II, the implementation of the VAD as solution of a logistic regression problem is described. For such aim, the system shall be trained to find the optimum coefficients for the classification; the training method is explained in section III. In section IV, the procedures for the features extraction are explained. In section V, we illustrate the proposed hardware architecture. In section VI the obtained results are presented, and the advantages and disadvantages of the solution are discussed. Finally, in section VII conclusions are shown.

## II. ALGORITHM DESCRIPTION

The input data are sampled at 8 kHz and are quantized by using 16 bits with 2's complement balanced representation in $[2^{15} - 1; -2^{15} + 1]$ range. For the processing, they are divided into non-overlapped frames of 10 ms, everyone containing $N_{frame} = 80$ samples. A vector $x[n]$ of features is extracted for each frame $n$. Such features are used as input of the VAD classifier for the voice activity evaluation.

The VAD system is implemented as a binary classifier which estimates $VAD[n] = P(VAD = 1|x[n]; \theta)$ that is the probability that the voice activity is 1, given the features x[n] with $\theta$ as parameters [7]. Such probability is estimated by

Fig. 1. Division of the input audio frame into non-overlapped frames

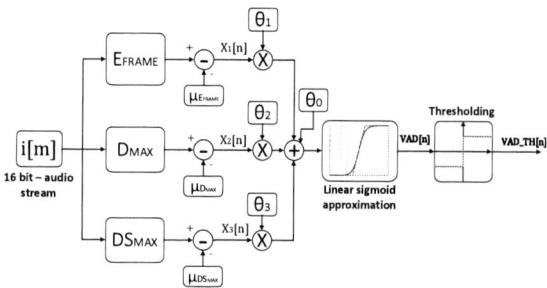

Fig. 2.  Complete VAD algorithm

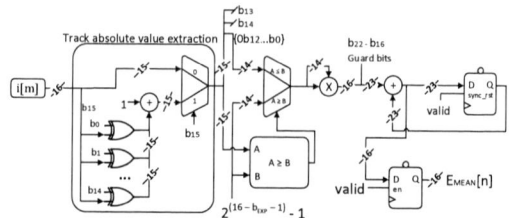

Fig. 3.  Hardware implementation of the frame Energy feature calculator

means of a hypothesis function $h_\theta(x[n]) = h(\theta^T \cdot x[n]) = P(VAD = 1|x[n]; \theta)$.

In particular, in this work we use the sigmoid hypothesis function, described in equation 1:

$$h_\theta(x[n]) = \frac{1}{1 + e^{-\theta^T \cdot x[n]}} \qquad (1)$$

The estimation of the optimum coefficients has been realized through a machine learning approach, as shown in section III. The $n^{th}$ frame features vector $x[n]$ is calculated as described in eq. 2:

$$\begin{cases} x_0[n] = 1, \\ x_1[n] = E_{mean}[n] - \mu_{E_{mean}} \\ x_2[n] = D_{max}[n] - \mu_{D_{max}} \\ x_3[n] = DS_{max}[n] - \mu_{DS_{max}} \end{cases} \qquad (2)$$

where $E_{mean}[n]$ is the energy of the input, $D_{max}[n]$ is the maximum absolute input signal finite difference and $DS_{max}[n]$ is the maximum absolute input squared signal finite difference of the input, every one calculated in the frame $n$. $\mu_{E_{frame}}$, $\mu_{D_{max}}$ and $\mu_{DS_{max}}$ are the sample means of the features calculated on the training dataset. $x_0 = 1$ is used so that the product $\theta_0 \cdot x_0$ generates an additive offset term. The architectures for the calculation of $E_{frame}[n]$, and $D_{max}[n]$, $DS_{max}[n]$ are described in section IV.
The obtained $VAD[n]$ is finally compared with a threshold $thr$ to generate a binary value decision $VAD\_TH[n]$ for each frame. The complete VAD algorithm is shown in figure 1.

### III. Training the system

A training set containing recorded tracks has been created, and each frame has been marked with a *voice* (y = 1) / *no voice* (y = 0) label. Being $m$ the number of frames in the training set, the $(l + 1)$-dimension vector $\theta$ has been estimated by minimizing the cost function $\mathcal{L}(\theta)$ [7]:

$$\mathcal{L}(\theta) = -\frac{1}{m} \sum_{i=1}^{m} [y^i log(h_\theta(x)) + (1 - y^i)log(1 - h_\theta(x))] + \frac{\lambda}{2m} \sum_{j=1}^{l} \theta_j^2 \quad (3)$$

The first term in equation 3 is the negate log-likelihood function; the second one is the regularization term, in which $\lambda$ has been fixed to 0.25. $y^i$ expresses the voice / no voice

label over the $i^{th}$-frame.

The cost function minimization is obtained by the descending gradient algorithm which updates the vector coefficients at every step according to the rule described in equation 4:

$$\theta_{k+1} = \theta_k - \alpha \cdot \nabla \mathcal{L}(\theta_k) \qquad (4)$$

where $\theta_k$ is the weights vector at the step $k$.

The coefficient $\alpha$ has been fixed to 0.01 to guarantee the convergence of the algorithm. Before training the system, the training set features have been preprocessed by subtracting their sample mean. The training procedure produced the following optimum weights:

$$\begin{cases} \theta_0 = -1.072315 \\ \theta_1 = 3.339956 & V^{-2} \\ \theta_2 = 2.247354 & V^{-1} \\ \theta_3 = 3.518937 & V^{-2} \end{cases} \qquad (5)$$

### IV. Features extraction

All the features are represented by using 16 bits, using the 2's complement balanced representation.

#### A. Frame energy

The frame energy of the frame $n$ is calculated as expressed in 6:

$$E_{frame}[n] = \sum_{j=0}^{N_{frame}-1} i_{exp}^2[n \cdot N_{frame} - j] \qquad (6)$$

where $N_{frame} = 80$ is the number of samples in a frame. The term $i_{exp}[m]$ is obtained as described in 7:

$$i_{exp}[m] = \begin{cases} |i[m]|, & \text{if } |i[m]| < 2^{16-b_{exp}-1} - 1 \\ 2^{16-b_{exp}-1} - 1, & \text{otherwise} \end{cases}$$

$$(7)$$

with $b_{exp} = 2$. Such operation makes the input stream $i[m]$ saturate to $2^{16-b_{exp}-1} - 1$, expanding the dynamics for the least significant bits. In that way, it is possible to make a more detailed differentiation among low-power signals with different amplitude for a more effective distinction from the noise. As shown in eq. 6, the use of non-overlapped frames require that the output $E_{frame}[n]$ is valid only every $N_{frame}$ input samples. It makes the implementation of a serial energy

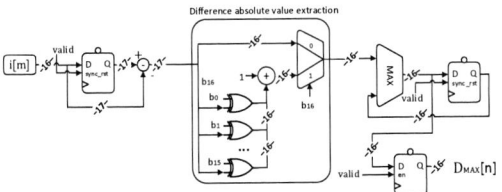

Fig. 4. Hardware implementation of the maximum absolute signal difference feature calculator

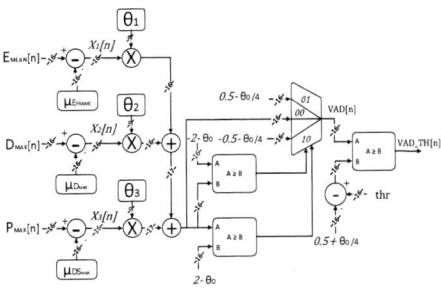

Fig. 5. Implementation of the approximate sigmoid and thresholding system

accumulator possible. The hardware implementation of the frame energy extractor is shown in figure 3.

### B. Maximum absolute signal finite difference over a frame

The maximum absolute signal finite difference on the frame $n$ is calculate as explained in equation 8:

$$\begin{cases} D_{max}[n] = \max_{m \in I} \{|i[m] - i[m-1]|\} \\ I = \{(n-1)\cdot N_{frame} + 1, ..., n\cdot N_{frame}\} \end{cases} \quad (8)$$

The first difference in a frame is calculated by forcing $i[m-1] = 0$. Even for this feature, the output shall be valid once every $N_{frame}$ samples, and a serial implementation has been realized. Every time a new sample is generated as input, the system calculates the $i[m]$ and $i[m-1]$ difference and compares it with the maximum value in the frame. If such difference is higher, it is registered as maximum.

### C. Maximum absolute squared signal finite difference over a frame

The maximum absolute squared signal finite difference over a frame is calculated as described by equation 9:

$$\begin{cases} DS_{max}[n] = \max_{m \in I} \{|i_{exp}^2[m] - i_{exp}^2[m-1]|\} \\ I = \{(n-1)\cdot N_{frame} + 1, ..., n\cdot N_{frame}\} \end{cases} \quad (9)$$

$i_{exp}^2[m]$ is calculated as shown in equation 7 by using $b_{exp} = 1$. The architecture described for the calculation of $D_{max}[n]$, shown in fig. 4, is used to calculate the maximum absolute difference $i_{exp}^2[m] - i_{exp}^2[m-1]$.

### V. Hardware Implementation

The system described in section II has been implemented in HDL. In order to reduce the area and the power consumption, the sigmoid function described in equation 1 has been approximated by using the hypothesis function described in equation 10:

$$\tilde{h}_\theta(x[n]) = \begin{cases} 0, & \text{if } \theta^T\cdot x[n] \leq -2 \\ \frac{\theta^T\cdot x[n]}{4} + \frac{1}{2}, & \text{if } -2 < \theta^T\cdot x[n] < 2 \\ 1, & \text{otherwise} \end{cases} \quad (10)$$

In this way, a look-up table implementation of the sigmoid function can be avoided to exploit a simpler realization which uses one multiplexer and two comparators.

Such approximation, which has been modelled and compared to the original hypothesis function by several simulations, produces negligible errors on the VAD estimation.

The system quantization has been realized by implementing a bit-true simulation. The latter compares the $VAD\_TH[n]$ values produced by the quantized model with the ones generated by the floating point model on a cross-validation set, containing audio tracks. For the simulation, $thr = 0.18$ was used. Such value was fixed by a previous simulation on the cross-validation dataset by using the floating point model. Finally, a 16 bit quantization of $thr$ has been performed for its usage in the quantized model.

The simulation results underlines that the number frames with different $VAD\_TH[n]$ in the floating point and quantized models can be reduced under the 0.1% by exploiting the features extraction architectures shown in section IV, in which 9 bits are used for the coefficients representation.

The synchronization of the systems is realized through a counter which enables the output registers of the features extraction architectures once every $N_{frame}$ input samples.

The architecture of the system including the approximate sigmoid function and the thresholding subsystem is shown in figure 5.

### VI. Results

#### A. Recognition performances results

The training, the bit true simulation and the testing have been carried out by using Matlab®. In order to guarantee good recognition performances with acceptable values of false negatives at the end of a phonation, low values of $thr$ might be used. Such aspect is important if this VAD is used as preprocessor for another VAD or a speech recognition system not to cause a drop in their performances. However, too low values of $thr$ make the system prone to false positive errors in presence of offset or noise. Simulations over the cross-validation set suggests that $thr = 0.18$ offers a good trade off to recognition performance and noise robustness. In order to test the immunity of the system, the approach used in [3] has been used. An amplitude modulated white noise has been added to a input audio track. In addition, an 8-bit offset has been added to test the immunity of the system. Equation 11 shows the the added noise on a frame:

978-1-5386-5388-3/18 $31.00 © 2018 IEEE

TABLE I
FPGA SOURCE OCCUPATION

| Sources | Utilized | Available |
|---|---|---|
| Core Cells | 2845 | 6144 |
| I/O Cells | 52 | 68 |
| Block RAMs | 0 | 8 |

$$n[n] = (1 + \frac{n}{N_{frame}}) \cdot \frac{(2^{15} - 1)}{\sqrt{1000}} \cdot w[n] + 2^7 - 1 \qquad (11)$$

where w[n] is a zero-mean white noise generated by used the Matlab® rand() function.

The obtained $VAD\_TH[n]$ is compared with the one produced by applying the noise-free track to the system. The differences on the $VAD\_TH[n]$ is in the order of 1% which underlines good performances of immunity to noise and offset. Such result is due to the used features (e.g. $D_{max}[n]$ does not change in presence of offset). Nevertheless, the lack of more sophisticated spectral features, like the ones used in [4], makes the system prone to false positives in presence of high impulsive noises. The system behaviour in presence of the $n[n]$ noise can be observed in figure 6 which shows in blue the audio track, in red the $VAD$ probability, in green the threshold $thr$ and in black the binary decision $VAD\_TH$.

### B. FPGA implementation results

The design has been implemented in VHDL onboard a 1.2V - AGLN250V2 IGLOO nano Low Power FPGA in order to minimize the power consumption.
Data on the FPGA source utilization are shown in table I.

Such results underline the low area occupation of the design: no RAM block is required and only around 1380 Logic Elements (LEs) (46% of 3K LEs) are necessary. Finally, for the system power consumption measurement, a post-layout simulation has been performed. The latter is necessary to generate a VCD file containing an estimation of the switching activities of the nodes. Such simulation has been performed by using Questa® advanced simulator. The testbench uses a real audio track as input for the system. The track is sampled at 8kHz, while the system uses a 8MHz clock that is higher than required. In fact, the used serial implementation requires that $T_{CLK} \leq T_{SAMPLE}$ where $T_{SAMPLE} = \frac{1}{8000}$ s to correctly process the input stream.

The power consumption has been extracted by using Smart-Power tool by Libero® SoC. The total power consumption is of 0.559mW.

These results underlines the advantages and disadvantages of the solution: systems like [6] are much faster and process

TABLE II
POWER COMPARISON WITH PREVIUOS WORKS

| | This work | [5] |
|---|---|---|
| Clock frequency [MHz] | 8 | 0.1 |
| FPGA used | AGLN250V2 | Xilinx - Spartan 3 |
| Power supply [V] | 1.2V | 1.8V |
| Power consumption [mW] | 0.559 | 2.10 |

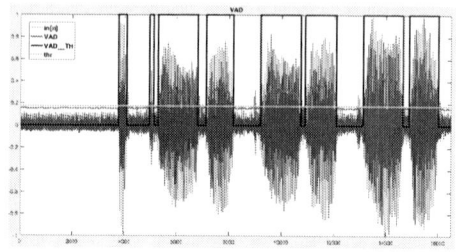

Fig. 6. Output of the VAD on a track with added noise and offset

data in parallel. They can be used when high real time requirements are necessary. However, they are bulky and require FPGA RAMs and DSP blocks, leading to higher power consumption and area occupation. On the contrary, the proposed system is able to produce a VAD response after around $N_{frame} \cdot T_{SAMPLE}$ seconds but offers strong advantages in terms of power consumption and area occupation. For that reason, a power consumption confrontation with such solution is meaningless. On the contrary, a comparison with another serial VAD, described [5], is shown in table II.

### VII. CONCLUSIONS

This work shows a low-power and low area occupation serial Voice Activity Detection system. The serial approach and the chosen features extraction architectures allow to reduce the power consumption to 0.559mW and the source occupation around to 1380 LEs on IGLOO nano FPGAs. The architecture and the obtained performances on recognition make the system suitable to be used as a preprocessor for speech recognition systems or other more sophisticated VAD in portable applications in order to reduce the activation rate, reducing the global power consumption.

### REFERENCES

[1] J. Ramirez, J. M. Górriz, and J. C. Segura, "Voice activity detection. fundamentals and speech recognition system robustness," in *Robust speech recognition and understanding*. InTech, 2007.

[2] Y. Zhang, N. Suda, L. Lai, and V. Chandra, "Hello edge: Keyword spotting on microcontrollers," *arXiv preprint arXiv:1711.07128*, 2017.

[3] M. Oukherfellah and M. Bahoura, "Fpga implementation of voice activity detector for efficient speech enhancement," in *New Circuits and Systems Conference (NEWCAS), 2014 IEEE 12th International*. IEEE, 2014, pp. 301–304.

[4] A. Benyassine, E. Shlomot, and H. S. I.-T. Recommendation, "G729 annex b: A silence compression scheme for use with g729 optimized for v. 70 digital simultaneous voice and data applications," *IEEE Commun. Mag*, vol. 9, pp. 64–73.

[5] H. Noguchi, T. Takagi, M. Yoshimoto, and H. Kawaguchi, "An ultra-low-power vad hardware implementation for intelligent ubiquitous sensor networks," in *Signal Processing Systems, 2009. SiPS 2009. IEEE Workshop on*. IEEE, 2009, pp. 214–219.

[6] J. Jung, S. Jin, D. Kim, H. S. Kim, J. S. Choi, and J. W. Jeon, "A voice activity detection system based on fpga," in *Control Automation and Systems (ICCAS), 2010 International Conference on*. IEEE, 2010, pp. 2304–2308.

[7] J. Bergstra, O. Breuleux, F. Bastien, P. Lamblin, R. Pascanu, G. Desjardins, J. Turian, D. Warde-Farley, and Y. Bengio, "Theano: A cpu and gpu math compiler in python," in *Proc. 9th Python in Science Conf*, 2010, pp. 1–7.

# Fractional-Order Hartley Oscillator

Agamyrat Agambayev[1], Aslihan Kartci[2], Ali H. Hassan[1], Norbert Herencsar[2], Hakan Bagci[1], and Khaled N. Salama[1]

[1]Computer, Electrical and Mathematical Science and Engineering (CEMSE) Division, King Abdullah University of Science and Technology (KAUST), Thuwal 23955, Saudi Arabia

[2]Faculty of Electrical Engineering and Communication, Brno University of Technology, Brno, Czech Republic

agamyrat.agambayev@kaust.edu.sa, kartci@feec.vutbr.cz, ali.h.hassan@ieee.org, herencsn@feec.vutbr.cz, hakan.bagci@kaust.edu.sa, khaled.salama@kaust.edu.sa

*Abstract*—A fractional-order capacitor (FOC) is developed using a Molybdenum disulfide ferroelectric polymer composite. The fabricated FOC exhibits constant phase over five decades between 100 Hz-10 MHz, which is the broadest operating frequency bandwidth reported so far for an FOC. Furthermore, a fractional-order Hartley oscillator is built using this FOC, and provide ten times higher oscillation frequency than the frequency of the conventional Hartley oscillator counterpart.

*Keywords*—*constant phase element, FOC, fractional-order capacitor, Hartley oscillator*

## I. INTRODUCTION

Although, the three basic passive circuit elements, namely, capacitor, resistor, and inductor have been around for three centuries, understanding and design of classical resistor-capacitor, and resistor-inductor, and capacitor-inductor, and resistor-capacitor-inductor circuits have remained fundamentally unaltered, since only integer-order differential equations are used in circuit design [1]. However, great improvements in the study of fractional calculus that deals with fractional-order derivation and integration have been achieved over the last five decades [2]. In the meantime, the fractional-order generalization has been used for circuit/device design and modeling [2] in various field such as biomedicine [3-4], energy-storage and generation [5], agriculture [6], electromagnetics [7], and control systems [8].

Recently, different approaches to realization of fractional-order elements (called constant phase elements in electrochemistry), particularly, fractional-order capacitors (FOCs) [9-13] have been developed. Among these approaches, those make use of composites of different materials have gained traction, since they lead to FOCs that can be easily integrated with microelectronics , and design an FOC with tunable constant phase angle (CPA) [9-10]. However, their narrow band of operating frequency (i.e., maximum two decades) limited their use in building circuits such as filters and oscillators with fractional-order transfer functions. It is also extremely important to develop an FOC with a constant phase zone (CPZ), where FOC exhibits a CPA with only small variation, in the order of ±4° over the full band of operation.

The impedance of an FOC is given by $Z = 1/[(j\omega)^{\alpha} C_{\alpha}]$ where $C_{\alpha}$ is the pseudocapacitance (also known as fractance) and $\alpha$ ($0 < \alpha < 1$) is the fractional order. The phase angle associated with $Z$ is given by $\phi = -\alpha\pi/2$. Table I compares

TABLE I. IMPEDANCE, ORDER, AND PHASE ANGLE OF INDUCTOR, RESISTOR, CAPACITOR, AND FOC.

| Components | Inductor | Resistor | FOC | Capacitor |
|---|---|---|---|---|
| Impedance | $Z = j\omega L$ | $Z = R$ | $Z = 1/[(j\omega)^{\alpha} C_{\alpha}]$ | $1/(j\omega C)$ |
| $\alpha$ | $\alpha = -1$ | $\alpha = 0$ | $0 < \alpha < 1$ | $\alpha = 1$ |
| Phase ($\phi$) | $\phi = 90°$ | $\phi = 0$ | $0 > \phi > -90°$ | $\phi = -90°$ |

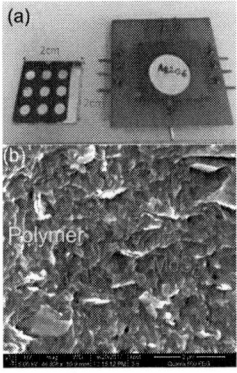

Fig. 1. (a) PCB-compatible FOC fabricated using MoS$_2$-polymer nanocomposite. (b) The cross-sectional SEM image of the nanocomposite.

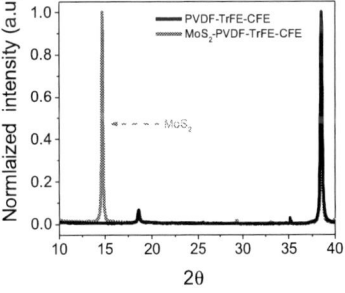

Fig. 2. Normalized XRD patterns of prepared polymer and MoS$_2$-polymer nanocomposite films.

the frequency dependent impedance, order, and as well as phase angle of passive circuit elements, namely, inductor, resistor, capacitor, and ideal FOC.

In this work, we report on an FOC fabricated using a new class of Molybdenum disulfide ($MoS_2$) filled Polyevinelidenefluoride-trifluoroethylene–chlorofluoroethylene (PVDF-TrFE-CFE) composite and exhibits a constant phase over five decades between 100Hz-10MHz. This FOC is then used in building an oscillator, which can, for example, be used for modulating/demodulating frequency between the high-speed radio-frequency (RF) input and the baseband system blocks. The circuit design used here adopts the Hartley oscillator since it has a comparable performance like the other topologies such as CMOS LC and Colpitts oscillators [14-15]. To show the advantages of using an FOC in the design of an oscillator, two Hartley oscillators are built; one with an FOC and other with classical capacitor counterpart, and their oscillation frequencies are compared. Measurement results demonstrate that the output frequency of the fractional-order Hartley oscillator is almost ten times greater than the frequency of its conventional counterpart.

## II. FOC FABRICATION AND CHARACTERIZATION

First, terpolymer namely, PVDF-TrFE-CFE powder is dissolved in a N, N-Dimethylformamide (DMF) solvent, under magnetic stirring for 48 hours to get 0.1 g / ml solution [16]. The $MoS_2$ powder is dispersed in DMF at a concentration of 20 mg/ml and stirred for two hours using a tip-type sonicator. 90 % of the $MoS_2$-DMF solution is collected and DMF is evaporated in a freezer dryer. The resulting $MoS_2$ nanosheets are dispersed again in DMF at the 0.03 g/ml concentration using ultra-sonication bath for two hours and mixed with the PVDF-TrFE-CFE solution in proportion by stirring for 24 hours at room temperature. Finally, the mixture is further stirred for one hour using a tip-type sonicator before drop casting. The composite solutions are casted onto a 200 nm Au covered $SiO_2$ / Si substrate and DMF is evaporated for overnight at 110 °C under vacuum. In order to make the top electrode, Au circular electrode with 1.5 mm radius and 200 nm thickness is deposited onto nanocomposite film using a shadow mask. Nine FOCs are fabricated on a chip with 2 cm × 2 cm area. The chip is bonded using the silver paste on a printed circuit board (PCB) to characterize and use in electrical circuits.

Fig. 1(a) shows the photograph of the nanocomposite film and the PCB-compatible FOC. The cross-section scanning electron microscopy (SEM) image of the nanocomposite film with $MoS_2$ is shown in Fig. 1(b). The composite presents a compact structure where the $MoS_2$ nanosheets are distributed homogeneously. During the mixing process, the polymer chains wrap the $MoS_2$ nanosheets and avoid the aggregation of $MoS_2$.

Furthermore, Fig. 2 compares the X-ray diffraction (XRD) patterns of the terpolymer and nanocomposite films, namely. PVDF-TrFE-CFE and the $MoS_2$:PVDF-TrFE-CFE respectively. It is important to note here that the XRD spectra are normalized with respect to peak at 38° (due to the gold electrode). An intense peak take places at 18.2° XRD spectrum

of the both film which correspond to a (111) plane of PVDF-TrFE-CFE. The XRD spectrum of the nanocomposite shows only one additional peak at 15.1° which belong to the $MoS_2$ nanosheets as expected, and. confirms that extra complicated molecular structures are not formed at the interface of polymer-$MoS_2$ nanosheets.

An Agilent 4994A precision impedance analyzer with the 16048G fixture is used in electrical characterization of fabricated FOCs. The magnitude and phase angle of the FOCs' impedance are measured. Standard calibration tests (open and short circuits) are performed to calibrate the instrument. Note that the thickness of films used for electrical characterizations is 30 µm. A total of 801 measurements are taken in the frequency band between 100 Hz to 10 MHz. Fig. 3(a) plots the measured phase angle versus frequency. The figure clearly shows that the phase angle is nearly constant and stable over a broad frequency range. If the maximum error allowed in phase angle is set to ±4°, the CPA is computed (by averaging) to be −58° between 100 Hz and 10 MHz (five decades). This CPA value corresponds to $\alpha = 0.64$. The phase angle deviation (the difference between the measured phase angle and the CPA) is plotted versus frequency in the inset in Fig. 3(a).

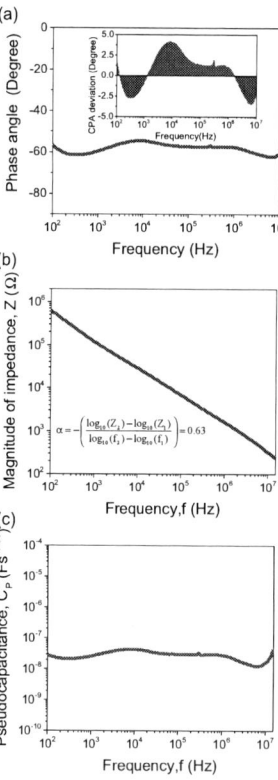

Fig. 3. (a) Measured phase angle versus frequency. (b) Measured magnitude of impedance versus frequency. (c) Pseudocapacitance computed using measured magnitude and phase angle versus frequency.

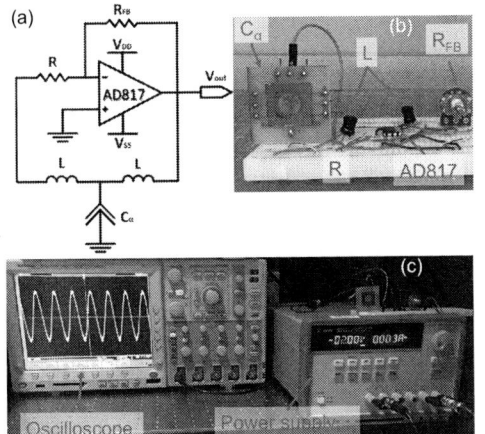

Fig. 4. (a) Hartley oscillator circuit schematic with a fractional-order capacitor (b) Circuit implementation of the Hartley oscillator. (c) Test setup for measuring the output voltage of the oscillator.

Fig. 3(b) plots the measured magnitude of impedance versus frequency. Note that $\log|Z| = -\alpha\log(\omega) - \log(C_\alpha)$ and $\alpha$ (which is the slope in the above equation) is computed by applying (linear) least-squares fit to the samples of $\log|Z|$ in the defined CPZ. This operation yields $\alpha = 0.63$, which confirms the value obtained by averaging the CPA [Fig. 3(a)]. Along $\alpha$, pseudocapacitance $C_\alpha$ also plays key role in designing an FOC. Fig. 3(c) plots $C_\alpha = 1/(\omega^\alpha|Z|)$ versus frequency and shows that $C_\alpha$ is almost stable over the defined CPZ with an average value of $28.3\ \mathrm{nFs}^{-0.36}$.

### III. FRACTIONAL-ORDER HARTLEY OSCILLATOR

This section introduces the fractional-order Hartley oscillator [Fig. 4(a)]. The oscillation frequency of this circuit $f_{osc}$ satisfies the given equation below [17]:

$$(2\pi f_{osc})^{1+\alpha} - \frac{(2r+R)/L}{\tan(\alpha\pi/2)}(2\pi f_{osc})^\alpha - \frac{2/LC_\alpha}{\sin(\alpha\pi/2)} = 0 \quad (1)$$

where $L$ is the inductance, and $r$ is the internal resistance of the inductors, and $R$ is the resistance of the resistor.

Eq. (1) is solved for $f_{osc}$ for values of $\alpha$ changing between 0.1 and 1 while $C_\alpha = 28.3\ \mathrm{nFs}^{-0.36}$, $R = 1\mathrm{k}\Omega$, $L = 1\mathrm{mH}$, and $r = 15\Omega$. Fig. 5 plots $f_{osc}$ versus $\alpha$. The figure shows that $f_{osc}$ increases as $\alpha$ is decreased. As expected, the relation is nonlinear. Especially for small values of $\alpha$, a small change in $\alpha$ results in a large change in $f_{osc}$. Note that setting $\alpha = 1$ in Eq. (1) provides the oscillation frequency of the conventional Hartley oscillator.

The fractional-order Hartley oscillator is built using the fabricated FOC, the resistor with $R = 1\mathrm{k}\Omega$, the inductors with

$L = 1\mathrm{mH}$ and $r = 15\Omega$, and a commercial AD 817 operational amplifier with $\pm\ 2\mathrm{V}$ supplies [Figs. 4(a) and (b)]. In addition, the feedback resistor $R_{FB}$ is equal to 100 k$\Omega$ potentiometer. Fig. 4(c) shows the test set up for Hartley oscillator.

By solving Eq. (1) with these parameters, the oscillation frequency is calculated to be 805.8 kHz for the fractional-order oscillator. For the oscillator with the conventional capacitor ( $\alpha = 1$), the oscillation frequency is found to be 82.8 kHz. These calculated values are in close agreement with the measured output frequencies 793.9 kHz and 81.1 kHz respectively (Fig. 6). The output voltage swing is constant around 2.18 $\mathrm{V}_{pp}$ for both oscillators. Finally, the measured results show that oscillator with the fabricated FOC with $\alpha = 0.64$ exhibits a higher oscillation frequency. It is almost ten times of the frequency of the oscillator with the conventional capacitor ($\alpha = 1$).

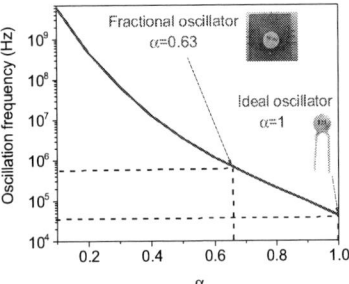

Fig. 5. Calculated oscillation frequency versus $\alpha$. The chip connected to PCB contains nine individual FOCs.

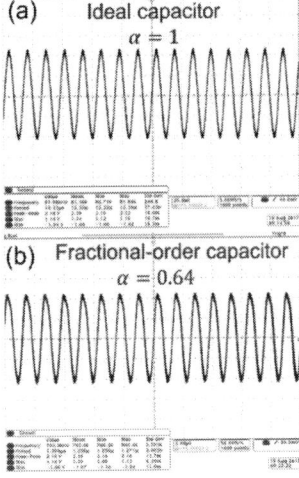

Fig. 6. The measured output voltage of the Hartley oscillator for (a) $\alpha = 1$ (b) $\alpha = 0.64$.

## IV. CONCLUSION

MoS$_2$ reinforced PVDF-TrFE-CFE is used to fabricate an FOC. A straightforward solution-mixing approach to generate the nanocomposites films makes easy the tailoring of the electrical properties of the resulting FOCs. The cross-sectional SEM image of the composite confirms that MoS$_2$ is distributed uniformly inside PVDF-TrFE-CFE without forming an aggregation. Moreover, the structural composition of the PVDF-TrFE-CFE film and the MoS$_2$:PVDF-TrFE-CFE composite are investigated using XRD. Only specific peaks associated with each constituent are observed in XRD spectrum showing that unexpected molecular structure is not formed at the MoS$_2$-polymer interface.

The FOC fabricated using the MoS$_2$:PVDF-TrFE-CFE nanocomposite has a CPA of −58° within the CPZ changing from 100 Hz to 10 MHz (five decades). This is significantly broader than the CPZ of the previously reported nanocomposite FOCs, which changes from 50 kHz to 10 MHz.

A fractional-order Hartley oscillator is designed employing the fabricated FOC. Measurements and analytical predictions show that oscillation frequency of the fractional-order Hartley oscillator is almost ten times than the oscillation frequency of the conventional Hartley oscillator counterpart.

In summary, we believe that FOCs from MoS$_2$-PVDF based polymers nanocomposites paves the way for novel fractional-order electronic systems [18-21].

## REFERENCES

[1] A. Kartci, A. Agambayev, N. Herencsar, and K. N. Salama, "Series-, parallel-, and inter-connection of solid-state arbitrary fractional-order capacitors: theoretical study and experimental verification," IEEE Access, vol. 6, pp. 10933–10943, 2018.

[2] A. S. Elwakil, "Fractional-order circuits and systems: An emerging interdisciplinary research area," IEEE Circuits and Systems Magazine, vol.10, pp.40–50, 2010.

[3] T. C. Doehring, A.H. Freed, E.O. Carew, and I. Vesely, "Fractional-order viscoelasticity of the aortic valve: an alternative to QLV," J. Biomech. Eng., vol. 127, pp. 700–708, 2005.

[4] T. J. Freeborn, "A survey of fractional-order circuit models for biology and biomedicine," IEEE Journal on emerging and selected topics in circuits and systems, vol. 3, pp: 416–424, 2013.

[5] M. E. Fouda, A. S. Elwakil, A. G. Radwan, and A. Allagui, "Power and energy analysis of fractional-order electrical energy storage devices," Energy, vol. 111, pp: 785–792, 2016.

[6] S. Jesus, T. J. A. Machado, and B. J. Cunha, "Fractional electrical impedances in botanical elements," J. Vib. Control 14, pp. 1389–1402, 2008.

[7] F. L. Oustaloup, and B. Mathieu, "Frequency-band complex noninteger differentiator: characterization and synthesis," IEEE Trans. Circuits Syst. I, Fundam. Theory Appl., vol. 47, pp. 25–39, 2000.

[8] G.W. Bohannan, "Analog fractional-order controller in temperature and motor control applications," Journal of Vibration and Control, vol. 4, pp. 1487–1498, 2008.

[9] A. M. Elshurafa, M. N. Almadhoun, K. N. Salama, and H. N. Alshareef, "Microscale electrostatic fractional capacitors using reduced graphene oxide percolated polymer composites," Applied Physics Letters, vol. 102, no. 23, pp. 232901, 2013.

[10] D. A John, S. Banerjee, G. W Bohannan, and K. Biswas, "Solid-state fractional capacitor using MWCNT-epoxy nanocomposite," Applied Physics Letters, vol. 110, no. 16,pp.163504, 2017.

[11] A. Agambayev, K. H. Rajab, A. H. Hassan, M. Farhat, H. Bagci, and K. N. Salama, "Towards fractional-order capacitors with broad tunable constant phase angles: Multi-walled carbon nanotube-polymer composite as a case study," J. of Physics D: App. Physics, vol. 51, pp. 1–6, 2018.

[12] A. K. Gil'mutdinov, P. A. Ushakov, and R. El-Khazali, Fractal Elements and their Applications. Springer, 2017.

[13] A. Agambayev, S. Patole, H. Bagci, and K. N. Salama, "Tunable fractional-order capacitor using layered ferroelectric polymers," AIP Advances, vol. 7, no. 9, pp. 095202, 2017.

[14] I. Chlis, D. Pepe, and D. Zito, "Comparative analyses of phase noise in 28 nm CMOS LC oscillator circuit topologies: Hartley, Colpitts, and common-source cross-coupled differential pair," The Scientific World Journal, vol. 2014, 2014.

[15] A. Kartci, N. Herencsar, J. Katon, L. Brancik, K. Vrba, G. Tsirimokou and C. Psychalinos. "Fractional-order oscillator design using unity-gain voltage buffers and OTAs," IEEE International Midwest Symposium on Circuits and Systems (MWSCAS), Boston, USA, pp. 555–558, 2017.

[16] A. Agambayev, S. Patole, M. Farhat, A. Elwakil, H. Bagci, and K. N. Salama, "Ferroelectric fractional-order capacitors," ChemElectroChem, vol. 4, pp. 2807–2813, 2017.

[17] A. S Elwakil, A. Agambayev, A. Allagui, K. N. Salama, "Experimental demonstration of fractional-order oscillators of orders 2.6 and 2.7," Chaos, Solitons & Fractals, vol. 96, pp. 160–164, 2017.

[18] S. Elmetennani, I. N'Doye, K. N. Salama, and T. M. Laleg- Kirati, "Performance analysis of fractional-order PID controller for a parabolic distributed solar collector," IEEE AFRICON, pp. 440–445, 2017.

[19] I. N'Doye, K. N. Salama, and T. M. Laleg-Kirati "Robust fractional-order proportional-integral observer for synchronization of chaotic fractional-order systems," IEEE/CAA Journal of Automatica Sinica, pp.1–10, 2018.

[20] A. Tepljakov, E. Petlenkov, J. Belikov, and M. Halas, "Design and implementation of fractional-order PID controllers for a fluid tank system," American Control Conference (ACC), pp. 1777-1782, 2013.

[21] A. G. Radwan and K. N Salama, "Fractional-order RC and RL circuits," Circuits, Systems, and Signal Processing, vol. 31, pp. 1901–1915, 2012..

*PRIME 2018, Prague, Czech Republic*　　　　*Session: Modeling, Optimization, and Characterization*

# System Level simulation framework for the ASICs development of a novel particle physics detector

Alessandro Caratelli, Simone Scarfi
*LSM, EPFL, Lausanne, Switzerland*
*CERN, Geneva, Switzerland*

Davide Ceresa, Kostas Kloukinas
*CERN, Geneva, Switzerland*

Yusuf Leblebici
*LSM, EPFL, Lausanne, Switzerland*

*Abstract*—The simulation of the passage of particles through matter using Monte Carlo methods is broadly used in the development of particle detectors for high energy physics experiments. To develop the readout electronics for the Compact Muon Solenoid (CMS) experiment at CERN, and to assist the design of the on-detector ASICs, a simulation framework was build capable to link the physics Monte Carlo simulations platforms with an industry standard EDA simulation tools. This contribution focuses on the implementation of the simulation framework based on the System Verilog language and the Universal Verification Methodology (UVM). The simulation results that guided the development of the ASICs and the choice of the final architecture are presented.

*Index Terms*—Simulation methods and programs, Front-end electronics for particle detectors, ASICs verification, High Energy Physics, UVM

## I. INTRODUCTION

The outer-tracker of the CMS experiment for the High Luminosity Large Hadron Collider (HL-LHC) introduces new challenges for the front-end readout electronics. The increased radiation levels and particle density lead to the requirement of improved radiation tolerance and higher detector granularity. For this reason, the entire tracking system will be replaced with new detectors featuring higher radiation tolerance and ability to handle higher data rates and readout bandwidths together with improved trigger capabilities. The CMS Technical Proposal [1] describes the requirements for the detector upgrade. For the first time data coming from a tracker will be used in the L1 trigger decision of an High Energy Physics experiment. Therefore, the front-end electronics is required to provide primitives of particles with high transverse momentum (called Stubs) which are transmitted for every collision to the Level-1 (L1) trigger system. This new approach involves the development of silicon sensor pixel detectors capable to reject locally signals of low transverse momentum particles ($< 2\,GeV/c$), which are not interesting for the trigger system reconstruction. For this reason the assembly called Pixel-Strip (PS) module [2], combining a pixelated silicon sensor with a silicon-strip layer, requires the development of two different read-out ASICs: the Short-Strip ASIC (SSA) [3] for the readout of the strip sensor and the Macro Pixel ASIC (MPA) [4] for the readout of the pixelated sensor and the particle recognition. Due to the sensor size (5 cm by 10 cm), 16 MPAs and 16 SSAs are needed to readout the entire sensor area. A third ASIC called Concentrator IC (CIC) is required

to aggregate the data and to perform spacial and temporal averaging. Fig. 1 shows a cross-section of the module.

The simulation framework described in this paper provides functional verification and performance evaluation for the design involving several ASICs and communication protocols. It allow to study and compare different system topologies in order to identify the best solution that optimizes the particle recognition efficiency and the data compression algorithm, while minimizing the overall power consumption.

The environment is based on the System Verilog hardware description and verification language and on the Universal Verification Methodology (UVM) [5] from which it inherits the base classes. Implementation details of the simulation framework and examples reported in this paper refer to the specific case of the CMS Outer Tracker ASICs development. Nevertheless, thanks to the layer structure and the modular approach adopted, the tool usability can be extended to different particle pixel sensor readout ASICs.

This paper is organized as follows: in Section II the structure of the simulation environment is presented with regards on the implementation methodology. In Section III the stimuli generation components are described. Section IV focuses on the output monitor components and their integration. Section V shows the modeling of the analog components. Section VI describes the reference model, the scoreboards implementation and the simulation results. Finally, in Section VII the conclusions are drawn.

## II. THE FRAMEWORK IMPLEMENTATION

The main purpose of the tool is to verify and assist the Register Transfer Level (RTL) development work as well as to provide accurate performance evaluation and final design verification. Subsystems and ASICs which compose the module can be integrated in a single simulation environment avoiding the need to develop several test-benches. In particular, a module level simulation provides a single chip verification together with the capability to check the chip-to-chip communication protocols and timing, and the compatibility of several ASICs in an integrated system. Multiple test scenarios allow to configure the environment and address the simulation to specific subsystems while providing capability to evaluate the performance of different algorithms implementations. A block diagram of the simulation framework is shown in Fig. 2.

The environment is composed by 4 main layers:

978-1-5386-5388-3/18 $31.00 © 2018 IEEE

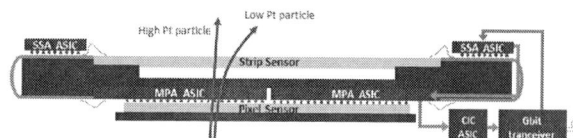

Fig. 1. Block diagram of the CMS Outer-Tracker detector chain

- The verification system that includes the UVM Verification Components (UVCs) related to the stimuli generation, the output monitors, the analysis component and the scoreboards.
- The reference model represents an high level description of the CMS Outer-Tracker readout system functionalities, without taking into account inefficiencies or limitations related to a particular implementation. It receives in input stimuli from the generation UVCs and provides the expected transactions for each component of the module.
- The main test class creates the environment during its build phase (one of the built-in execution phases of the UVM methodology). Several test classes extend the main one, allowing to define multiple scenarios by configuring the UVCs and by constraining the stimuli generation.
- The Design Under Verification (DUV) instantiates the RTL description or the gate-level netlist of the CMS Outer-Tracker ASICs. It defines their interconnectivity and it binds their building blocks to the framework components via interfaces.

The framework components are implemented at the Transaction Level Modeling (TLM) by extending the UVM base classes. At this level of abstraction channels hide the complexity of the protocols by implementing the communication in the form of function calls instead of signals. The communication among verification components, reference model and scoreboards is based on transaction objects. This approach, together with the high level of reconfigurability, allows to achieve a modular implementation easy to extend and reuse for different pixel readout applications, while drastically reducing the simulation time.

## III. STIMULI GENERATION AND CONFIGURATION

Three main types of stimuli can be provided to the DUV:

- For functional verification, constrained random generated stimuli allow to stress the design and reach high coverage values.
- For performance evaluation, physics Monte Carlo simulation events are generated by the CMS tracker physics simulation toolkit and integrated in the System Verilog simulation environment. This allows to perform system studies based on realistic stimuli and to evaluate the system performances in the target application.
- Global clock and control signals are generated for the operation of all ASIC components. This includes the generation of the system clock and reset signals and the

sequences for the high speed control and for the serial configuration ports.

Each stimuli generation UVC is composed by four main components as shown in Fig. 2. The Sequence class, derived from the uvm_sequence, creates the series of transactions at TLM level. The sequencer allows to randomize the sequence items and to transmit the TLM transactions to the driver where are converted into the RTL signals provided in input to the DUV via interfaces. The main test class allows to control the configuration class via the UVM factory mechanism, an object oriented design pattern that provides the ability to configure the verification objects from anywhere else in the code. The configuration class determines the sequence item by constraining the UVC operations and the data randomization.

The stimuli components generate the pixel particle charge deposition matrix and the corresponding strip hits array, emulating the CMS Outer-Tracker double layer silicon sensors. The generated values are transmitted to the reference model in the form of TLM transactions and to the System Verilog models of the ASICs analog Front-Ends in the form of 32-bit signals representing the particle charge deposited on the sensors expressed in $fC$.

In order to reduce the simulation time, the TLM transactions are created only in relation to the generated hit clusters, avoiding to transmit the full matrix for each event. Each transaction carries the information related to the HL-LHC Bunch Crossing reference cycle, the $\phi$ and Z coordinates of the particle clusters, the particle transverse momentum, the charge of the particle and time of arrival.

### A. The generation of expected events

This stimuli generation UVC produces randomized transactions emulating detectors hits due to high transverse momentum particles (High-Pt), namely stubs. This kind of stimuli represents the main primitives expected to be identified by CMS Outer-Tracker. Any missing stub at the output of the DUV may identify either a design error or an inefficiency of the particle recognition algorithm or of the implementation. These exceptions are handled and reported by the framework. The stub generation sequencer allows to randomize the stimuli in terms of cluster centroid position, cluster radius, time of arrival, particle energy and transverse momentum. The density of stubs follows a Poisson distribution configurable per test case. The randomization is constrained by the configuration object and can be controlled at run-time via the UVM factory.

### B. The combinatorial stimuli generation

The combinatorial generation UVC generates randomized stimuli that follow a uniform distribution. It can be activated separately or in addition to the stub generation UVC. In the first case it allows to stress the DUV by generating events that do not represent necessarily the expected ones permitting to reach higher levels of test coverage. In the latter case, when the combinatorial generation UVC is enabled together with the stub generation UVC, it allows to emulate detector hits that do

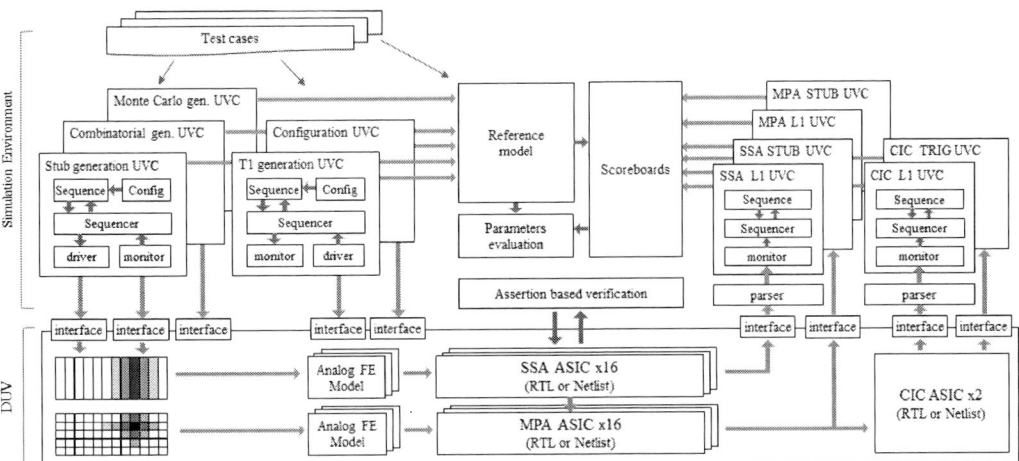

Fig. 2. Simulation framework block diagram of the CMS Outer-Tracker detector chain

not represent valid stubs such as noise, machine background or not interesting particles depositing charge on the sensors.

### C. The Monte-Carlo stimuli generation

In order to evaluate parameters based on the real physics events an additional stimuli generation component, referred in Fig. 2 as Monte Carlo generation UVC, allows to import particle hits from Monte-Carlo (MC) programs for computer simulation of complex interactions in high-energy particle collisions. The data-sets provide event samples for the entire CMS Outer-Tracker [6]. Via the configuration mechanism it is possible to import stimuli related to different module position within the Outer-Tracker geometry allowing to verify the functionalities of the design with several configurations, input activities and module geometries.

### D. The generation of commands

The Outer-Tracker readout ASICs require a fast command input signal, namely T1, in order to control their internal operations and the module functionality (i.e. the complete pixel raw data transmission request, resynchronization request, the shutter control and several others). The environment allows to randomize the commands generation taking into consideration the correlation with the input data, and to configure their average rate from the test cases. This allows to evaluate maximum trigger rate acceptable for a given bandwidth and eventually resize the internal FIFOs.

### E. The configuration UVC

The DUV ASICs implement a large number of control registers to program their functionalities and operating modes. Additional registers for each pixel allow to configure and equalize the analog front-ends, modeled in the environment with a System Verilog description. The configuration UVC allows to verify the control protocol implementation and to evaluate the functionalities of the DUV under different conditions and different operating phases. This component has

the purpose to translate the test class configuration into signals applied to the DUV in the format defined by the communication protocol. The same set of configuration parameters is than transmitted in form of TLM transactions to the stimuli generation components, to the reference model and to the output monitors. The user can define simple test files to setup both the DUV and the simulation framework in a specific operating mode.

## IV. OUTPUT MONITOR COMPONENTS

The simulation framework implements several monitors connected to the output of each of the ASICs and to critical points of the design. The monitors have the purpose to convert the RTL signals into TLM transactions that can be handled by the analysis components as described in Section VI. The CMS Outer-Tracker readout ASICs are required to provide at the same time: stub data (primitives of particles with high transverse momentum which are transmitted for every event) and L1 data (the complete pixel and strip events when requested by Level-1 trigger system). The monitors relative to the stub data path evaluate the signals in output at every clock cycle and decode their information. On the other hand, the monitors relative to the L1 data path implement an event-driven behavior which generates transactions only when triggered by the stimuli generation UVCs or when they detects activity on the monitored signals.

## V. THE MODELING OF ANALOG MODULES

The CMS Outer-Tracker ASICs embed several analog modules that requires to be accurately modeled in the environment (i.e. the analog front-ends necessary to readout the pixel-strip sensors, the biasing circuitry, the DLL for phase alignment and several others). The analog functionalities are modeled in the framework with an accurate System Verilog description in order to drastically reduce the simulation time compared to a mixed signal simulation approach. In the case of the analog front-ends, for example, the model receives in input from the stimuli UVCs a 32-bit signals representing the charge deposition in the sensor expressed in $fC$. It approximates the internal response parameterized with input charge and other factors as peaking time, voltage peak and return to zero time, based on a precise characterization of the analog circuit.

## VI. Reference model and Scoreboards

A TLM reference model implements a high level description of the CMS Outer-Tracker readout functionalities. It receives as input the stimuli from the generation UVCs and generates the expected transactions for each component of the DUV. At this stage no bandwidth limitations are foreseen. Several threads reproduce the different sub-systems functionalities by providing a set of TLM transaction to the scoreboard via analysis exports. In addition, possible sources of inefficiencies are evaluated and the correspondent transactions are flagged for more precise analysis in the scoreboards. The analog functionalities are represented with their simplified behavior.

Several scoreboards allow to perform conformity checks between predicted and actual DUV outputs. The results of the comparison are reported in the simulation log files for further analysis. Every mismatch can either represent an error in the design or a limitation of the ASICs implementation. Discrepancies are categorized as:

- Errors which represent the percentage of losses due to artifacts introduced by the algorithm implementation in the hardware.
- Geometrical losses due to detector construction coming for example from a dead area among ASICs which can alter pixel centroids position value in output of the DUV.
- Bandwidth limitations in the module that can lead to missing stubs at the DUV outputs. High L1 trigger rate compared to the internal FIFO may lead to packet losses in the triggered data acquisition.

Fig. 3 reports the effect of those discrepancy on the system efficiency, by comparing the outputs of the ASICs post place and route netlist with the reference model. Evaluated parameters are reported in the summary file. Moreover, the framework generates several detailed logs. Multiple levels of details can be set in order to monitor stimuli generation, or data outputs at different stages of the system.

Fig. 4 reports the efficiency of the architecture for different values of occupancy and different bandwidths, measured at the output of the ASICs that compose the readout module. The simulation is performed with a constant 3.125 ratio between the number of generated stubs (interesting events to be recognized by the architecture) and combinatorial hits (spurious events representing not interesting particle interactions or background noise). The number of particles interactions and the width of the cluster generated are randomized following a Poissonian distribution while the tracks transverse momentum follows a uniform distribution.

## VII. Conclusions

A simulation framework was build for the CMS Outer Tracker on-detector electronics readout chain, capable to im-

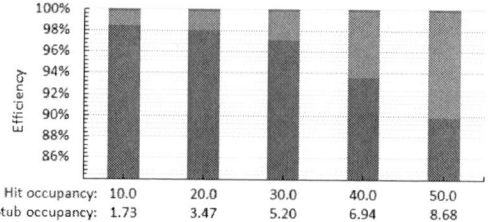

Fig. 3. Efficiency of the particle recognition hardware implementation for different occupancy values. In green the percentage of correctly recognize clusters, in orange the percentage of lost information due to bandwidth limitation, in red the errors due to ASIC implementation inefficiency.

Fig. 4. Efficiency of the implemented system architecture at the output of the CIC ASIC and SSA ASIC for different values of data occupancy and for different transmission bandwidths among the module ASICs.

port test vectors extracted from physics Monte Carlo simulations. The framework was used extensively throughout the entire development phase of the front-end ASICs and assisted to evaluate the overall module performances of the particle recognition. The simulation framework allowed to achieve full ASICs verification and limit data losses to less than 0.002 % of the interesting data on high-pT particles in the operating conditions (200 pile-up).

## References

[1] CMS collaboration *et al.*, "The phase-2 upgrade of the cms tracker," tech. rep., CERN-LHCC-2017-009, CERN, Geneva Switzerland,[CMS-TDR-17-001], 2017.

[2] D. Ceresa, A. Caratelli, J. Kaplon, K. Kloukinas, and S. Scarfi, "Readout architecture for the Pixel-Strip module of the CMS Outer Tracker Phase-2 upgrade," *PoS*, vol. Vertex2016, p. 066, 2017.

[3] A. Caratelli, D. Ceresa, J. Kaplon, K. Kloukinas, Y. Leblebici, J. Murdzek, and S. Scarfi, "Short-Strip ASIC (SSA): A 65nm Silicon-Strip Readout ASIC for the Pixel-Strip (PS) Module of the CMS Outer Tracker Detector Upgrade at HL-LHC," *PoS*, vol. Twepp-17, 2018.

[4] D. Ceresa, J. Kaplon, R. Francisco, A. Caratelli, K. Kloukinas, and A. Marchioro, "A 65 nm pixel readout asic with quick transverse momentum discrimination capabilities for the cms tracker at HL-LHC," *Journal of Instrumentation*, vol. 11, no. 01, p. C01054, 2016.

[5] "IEEE standard for universal verification methodology language reference manual," *IEEE Std 1800.2-2017*, pp. 1–472, May 2017.

[6] S. Viret, "Data transmission efficiency of the phase II tracker front-end system," *CMS Internal Note*, vol. IN-2015, November 2015.

# A Bayesian indicator for Run-to-Run performance assessment in semiconductor manufacturing

1st Taki Eddine KORABI
5th Jacques Pinaton
*STMicroelectronics*
Rousset, France
mohamed-taki-eddine.korabi@lis-lab.fr

2nd Guillaume Graton
*Ecole Centrale Marseille*
LIS lab
Marseille, France
guillaume.graton@centrale-marseille.fr

3rd El Mostafa El Adel
4th Mustapha Ouladsine
*Aix-Marseille Université and LIS lab*
Marseille, France
EL-Mostafa.el-adel@lis-lab.fr
mustapha.ouladsine@lis-lab.fr

*Abstract*—In this paper, Bayesian theory is used to build an indicator for Run-to-Run control. This indicator is used for assessing the performances of the regulation loops in a batch industry. The indicator is using four main inputs which are the output/target error, the dispersion of the output, the out of tolerance rate (oot) and the value of the industrial risk. The efficiency of the proposed Bayesian method has been tested on the deposition area of a semiconductor foundry.

*Index Terms*—Run to Run control, Bayesian detection, performance assessment, industrial risk

## I. INTRODUCTION

The semiconductor industry is a batch industry where we produce slices of semiconductor component called wafers [1]. A wafer goes through a long way of thousands of successive production steps (*i.e.* manufacturing route), before finally becoming a finished product. This route is constituted of a number of steps which are realized in several manufacturing areas such as photo-lithography, etch, chemical mechanical polishing (CMP), chemical vapor deposition (CVD), etc. see for instance [2]–[4].

At each step of the manufacturing route, several inspections and measurement steps have been introduce in order to detect problems early enough to avoid significant loss and damage. As a result, each wafer is regularly and strictly controlled especially when the production step is sensitive or critical [1], [5].

the goal behind the introduction of measurement steps lies in the fact that this significantly reduces the industrial risk. It is true that a measurement step is not an added value, that it is time consuming and very costly. However, this allows early detection of problems which is very important in an industry where the least wafer rework or scrap is extremely expensive and can lead to client dissatisfaction. [6].

Various types of measurement are present in the semiconductor manufacturing field. Among these measures one can find, Physical measurement such as Critical Dimension (CD) and thickness, electrical measurement such as Parametric Test (PT), and defectivity wafer inspections where wafer defaults and contamination are detected. The obtained results after measurement are used to inspect and supervise the production machines and their capability for manufacturing products within specifications.

Statistical Process Control (SPC) is one of the main methods of Advanced Procees Control (APC). It is used to detect abnormal activities in the production machines by detecting product drifts [7]. When analyzing a control chart, if a problem is detected, process control engineers stop the equipment and diagnosis actions are taken in order to detect and fix the origin of the problem. SPC is not an automatic control method in the sense where it does not fix the origin of the problem in an automatic manner. It has only the ability of triggering alarms for diagnosis.

Unlike SPC, Run-to-Run (R2R) control is the only automatic control strategy of APC [8]. A R2R controller is composed of three parts: a mathematical model which needs to be as representative as possible of the real system, a filter or an observer which is generally an EWMA filter and a control law allowing the follow-up of the target. R2R controller uses the obtained metrology results at run $k$ to adjust the input related to run $k+1$ so that its output can meet the target. The success of R2R control lies on its simplicity. This is why it has been successfully deployed in batch industries especially in semiconductor manufactures. R2R permits to improve yield, process capability, and product quality by avoiding process drift in an automatic way.

In order to evaluate the performance of production machines and the presented control techniques above, several capability indicators such as $Cp$, $CpK$, $Cpm$ have been used in the industry [9]. However, in the case of R2R control, we can have unsatisfactory results while the regulator is well optimized, as we can have satisfactory results while the regulator is not optimal [2]. In this case, classical indicators like $CpK$ are no longer efficient.

As a result, other types of indicators such as Control Performance Assessment (CPA) have been used to assess R2R controllers performances [10]. However, to the best of our knowledge, those works have never been deployed in a industrial environment because they make conservative assumptions (see for instance [10] and references therein). This paper proposes a novel R2R controller indicator based on the Bayesian theory [11], [12]. The choice of the Bayes theorem is motivated by the fact that this approach has been widely and successfully used in the industry notably in the semiconductor domain to detect different types of disturbances. The method

978-1-5386-5388-3/18 $31.00 © 2018 IEEE

is also intuitive and easily understandable. Which is very important in an industrial context since it will be used by both engineers and technicians. The inputs of the indicator are, the distance to the target, the dispersion, the out of tolerance rate and the industrial risk value. The indicator is tested on the deposition area of a semiconductor manufacture ( STMicroelectronics foundry in Rousset). In order to demonstrate its efficiency, the indicator is compared with the $CpK$ capability index.

The paper is organized as follows: In section II, an approach description is presented. In order to illustrate the validity of the presented approach, an industrial example is presented in section III. Finally a conclusion is presented in section IV.

## II. Approach description

### A. Basic statement

In this section, an indicator for Run-to-Run control performance assessment is introduced. The indicator is based on the well-known Bayes' theorem [11]:

$$\mathbb{P}(A|B) = \frac{\mathbb{P}(B|A)\mathbb{P}(A)}{\mathbb{P}(B)}, \qquad (1)$$

where $\mathbb{P}(A|B)$ is the conditional probability of $A$ occurring given that $B$ is true. This probability is also called posterior probability, $\mathbb{P}(B|A)$ is the conditional probability of $B$ occurring given that the proposition $A$ is true, $\mathbb{P}(A)$ and $\mathbb{P}(B)$ are the probabilities of observing respectively $A$ and $B$, these probabilities are independent. An alternative form of Bayes' theorem is presented in the form

$$\mathbb{P}(A|B) = \frac{\mathbb{P}(B|A)\mathbb{P}(A)}{\mathbb{P}(B|A)\mathbb{P}(A) + \mathbb{P}(B|\bar{A})\mathbb{P}(\bar{A})}, \qquad (2)$$

equation (2) is obtained from

$$\begin{aligned} \mathbb{P}(B) &= \mathbb{P}(A \cap B) + \mathbb{P}(\bar{A} \cap B) \\ &= \mathbb{P}(B|A)\mathbb{P}(A) + \mathbb{P}(B|\bar{A})\mathbb{P}(\bar{A}), \end{aligned} \qquad (3)$$

in the original equation (1). The second form of the Bayes' theorem (2) is the one which will be used in the rest of the paper. Detection using Bayes theorem has been introduced in [13], the idea was to use the prior probability $\mathbb{P}(A)$ to detect eventual future disturbances on the system, the method was called Bayesian detection and has been able to detect both step and impulse disturbances by calculating the posterior probability $\mathbb{P}(A|B)$.

### B. Problem formulation

In this paper, a similar approach to Bayesian detection is used for Run-to-Run control assessment, the proposed indicator will also use the industrial risk to evaluate the efficiency of the regulation loop. First, we consider that the whole state space $\Phi$ is partitioned in two subspaces $\Phi_N$ and $\Phi_D$, where $\Phi_N$ is the subspace of the normal observations and $\Phi_D$ is the subspace of the abnormal observations. Notice that if the regulator is performing well, abnormal observations are not supposed to be observed, moreover a limitation for normal observations will be presented later in the paper. We introduce

the likelihood function of a single normal observation for a system under Run-to-Run control where Gaussian distribution $\mathcal{N}(0, \sigma^2)$ is assumed

$$\mathbb{P}(y_i|\Phi_N) = \frac{1}{\sqrt{2\pi\sigma^2}} \exp\left[\frac{-y_i^2}{2\sigma^2}\right], \qquad (4)$$

where $\Phi_N$ stands for the normal subspace of the whole state space (i.e, the subspace regrouping all the normal observations), $\sigma$ is the standard deviation for the normal process. By defining the space $Y_k$ regrouping all the observations $y_i$ with $i \in \{1, \ldots, k\}$, $k \in \mathbb{N}$ by

$$Y_k = \{y_1, y_2, \ldots, y_k\}, \qquad (5)$$

and by assuming that all the observations $y_i$, $i \in \{1, \ldots, k\}$ are independent and identically distributed one can write

$$\mathbb{P}(Y_k|\Phi_N) = \prod_{i=1}^{k} \mathbb{P}(y_i|\Phi_N). \qquad (6)$$

Replacing (4) in (6) one obtains

$$\mathbb{P}(Y_k|\Phi_N) = \frac{1}{(\sqrt{2\pi\sigma^2})^k} \Psi(y), \qquad (7)$$

where $\Psi(y) = \exp\left[-\frac{1}{2\sigma^2}\sum_{i=1}^{k} y_i^2\right]$. In parallel to normal observations, Gaussian distribution $\mathcal{N}(d, \sigma^2)$ is also assumed for abnormal observations and the equivalent likelihood function is given by

$$\mathbb{P}(Y_k|\Phi_D) = \frac{1}{(\sqrt{2\pi\sigma^2})^k} \Psi(y - d). \qquad (8)$$

Notice that $d$ stands for the mean of the process when the observations $y_i$ are abnormal,

$$d = \frac{\sum_{i=1}^{k} y_i}{k}, \qquad (9)$$

By replacing the newly introduced notations in the Bayes' theorem (2) one obtain

$$\mathbb{P}(\Phi_D|Y_k) = \frac{\mathbb{P}(\Phi_D)\mathbb{P}(Y_k|\Phi_D)}{\mathbb{P}(\Phi_D)\mathbb{P}(Y_k|\Phi_D) + \mathbb{P}(Y_k|\Phi_N)\mathbb{P}(\Phi_N)}, \qquad (10)$$

where $\mathbb{P}(\Phi_D)$ is the probability of abnormal observations or prior probability, $\mathbb{P}(\Phi_N) = 1 - \mathbb{P}(\Phi_D)$ is the probability of normal observations. By substituting the obtained results (7) and (8) in (10) we obtain

$$\mathbb{P}(\Phi_D|Y_k) = \frac{\mathbb{P}(\Phi_D)}{\mathbb{P}(\Phi_D) + \mathbb{P}(\Phi_N)\exp\left[-\frac{\left(\sum_{i=1}^{k} y_i\right)^2}{2k\sigma^2}\right]} \qquad (11)$$

### C. proof

Replacing $\mathbb{P}(Y_k|\Phi_D)$ and $\mathbb{P}(Y_k|\Phi_N)$ by their expressions (7) and (8) in (10) we obtain

$$\mathbb{P}(\Phi_D|Y_k) = \frac{\mathbb{P}(\Phi_D)\Psi(y - d)}{\mathbb{P}(\Phi_D)\Psi(y - d) + \mathbb{P}(\Phi_N)\Psi(y)} \qquad (12)$$

putting $\mathbb{P}(\Phi_D|Y_k) = \mathbb{P}_l$, $\mathbb{P}(\Phi_D) = \mathbb{P}_s$, and $\mathbb{P}(\Phi_N) = 1 - \mathbb{P}_s$ we obtain

$$\mathbb{P}_l = \frac{\mathbb{P}_s \Psi(y-d)}{\mathbb{P}_s \Psi(y-d) + (1-\mathbb{P}_s)\Psi(y)}, \qquad (13)$$

by setting, $y_i^2 = (y_i - d)^2 - d^2 + 2dy_i$ we obtain

$$\mathbb{P}_l = \frac{\mathbb{P}_s}{\mathbb{P}_s + (1-\mathbb{P}_s)\exp\left[-\frac{1}{2\sigma^2}\sum_{i=1}^{k} 2dy_i - d^2\right]}. \qquad (14)$$

Now, let us remark that,

$$
\begin{aligned}
\sum_{i=1}^{k} 2dy_i - d^2 &= d\left(\sum_{i=1}^{k} 2y_i - \sum_{i=1}^{k} d\right) \\
&= d\left(2\sum_{i=1}^{k} y_i - \sum_{i=1}^{k} d\right) \qquad (15)\\
&= d\left(2\sum_{i=1}^{k} y_i - kd\right),
\end{aligned}
$$

and by replacing $d$ by its equivalent equation in (9), we obtain

$$\sum_{i=1}^{k} 2dy_i - d^2 = \frac{\left(\sum_{i=1}^{k} y_i\right)^2}{k}, \qquad (16)$$

finally, by substituting the obtained result (16) in (14) one obtain

$$\mathbb{P}_l = \frac{\mathbb{P}_s}{\mathbb{P}_s + (1-\mathbb{P}_s)\exp\left[-\frac{1}{2k\sigma^2}\left(\sum_{i=1}^{k} y_i\right)^2\right]}, \qquad (17)$$

which ends the proof.

### D. Discussing the proposed result

The obtained indicator in (11) permits to calculate the probability of abnormal observations under R2R control. In others words, it permits to know if the regulator performs well and if it is optimally tuned. Indeed, obtaining a value relatively close to 0 means that the probability of obtaining undesired values is relatively small (i.e, it will have optimal performances). On the other side, having a value relatively close to 1 means that the probability of having undesired values is relatively high and that the regulator is not optimal. However, one may ask this question: how can we calculate the prior probability $\mathbb{P}_s$? The answer depends on the exigence that we want to impose to the regulator. This one is related to the out of tolerance rate (oot) parameter

$$oot = \frac{\#abnormal\ observations}{\#observations}. \qquad (18)$$

In order to obtain the $oot$, the number of abnormal observations needs to be calculated. Which means that a limit between normal and abnormal observations needs to be fixed. One may use Upper Control Limit (UCL) and Lower Control Limit (LCL) in (19) as a limit between normal and abnormal observations.

$$
\begin{aligned}
UCL &= \mu + 3\sigma \\
LCL &= \mu - 3\sigma,
\end{aligned} \qquad (19)
$$

where $\mu = 0$ stands for the mean of normal observations. However, these limits are not relevant in a case of normal distribution because the number of observations exceeding these limits is too small which means that the related exigence

is small. Thus, smaller limits (tolerance limits) will be used in this paper, these limits are defined as follows

$$
\begin{aligned}
UTL &= \mu + \sigma \\
LTL &= \mu - \sigma,
\end{aligned} \qquad (20)
$$

where UTL is the Upper Tolerance Limit and LTL is the Lower Tolerance Limit. Hence, every single observation is compared to these tolerance limits, and if it exceeds their values, the number of abnormal observations is incremented. The ooc rate is then calculated as shown in (18). The proposed indicator in (11) allows to see if the regulator is optimal or not, but it does not take into account the effect of the industrial risk in this evaluation. In the next section, the industrial risk will be introduced to the indicator and a relationship between the optimal behavior of the indicator and the risk will be proposed.

### E. Industrial risk

The industrial risk is defined by the consequence in term of cost resulting from a certain technical decision. In semiconductor manufacturing field, the industrial risk is defined by the number of wafers scraped after taking a production decision. In the case of our indicator, the industrial risk is related to the number of scrapped wafers if the indicator tells that the Run-to-Run controller is optimal when it is not. In this context, semiconductor industry introduces the so-called Acceptable Risk (AR), which is defined by the maximal "acceptable" number of lost wafers, this number can not be exceeded without any consideration no matter which decisions have been taken. Generally speaking, whatever the result communicated by the indicator, the decision related to this result should not lead to a number of scrapped wafer exceeding the value of the Acceptable Risk. In this paper, the standard industrial value of the industrial risk is fixed to 500 wafers, for confidentiality reasons, the true value used in STMicroelectronics is not disclosed. However, the proposed methodology is not impacted. The goal here is to introduce the industrial risk in the calculation of the performance indicator in (11) in a way that will permit to avoid an exceeding of the AR. The proposition is formulated as follows

$$I = AR.\frac{\mathbb{P}(\Phi_D)}{\mathbb{P}(\Phi_D) + (1-\mathbb{P}(\Phi_D))\exp\left[-\frac{(\sum_{i=1}^{k} y_i)^2}{2k\sigma^2}\right]}, \qquad (21)$$

where, $I \in [0, AR]$. The proposition in (21) can be seen as the number of wafers at risk, these wafers are made under the control of the evaluated regulator. In other words, the indicator evaluates the performances of the regulator in term of the wafers that are in risk regarding the behavior of this regulator. The result of the indicator will be compared with two main values: accident and excursion, the two values are fixed respectively at 200 and 300 wafers at risk.

## III. INDUSTRIAL EXAMPLE

In this section, a test of the proposed indicator is realized on 44 regulators. Generally, a regulator is defined by a combination of different manufacturing contexts such as layer, recipe, tool, product, etc. called threads. In the studied case, the regulator is defined by a combination of two contexts, recipe and tool. In this example we will use the indicator presented in (21) on 44 different threads in order to evaluate their performances and the obtained values are compared with the two fixed values of accident and excursion in order to assess the performances of the regulator. The idea is to highlight all the suboptimal regulators so that necessary actions can be taken for making the regulator more optimal. Note that, the suboptimality of the regulator may be the result of a mismatch between the system and the used model in the R2R control. It also can be the results of big sampling frequencies, or poorly tuned parameter. The impacts of a sub-optimal regulator may disastrous in a semiconductor factory precisely because R2R is implemented in what are rightly called "critical steps" ina manufacturing route. The data have been collected for a period of 3 months in the fab's deposition area. Without loss of information, the set of data has been modified so that they can not be exploited in other applications. Note that in this area, the deposition time is adapted by the regulator in a Run-to-Run manner. If the regulator is sub-optimal, the consequences in term of quality and yield may be significant and this is what motivates the choice of this area. Moreover, many steps like CMP are directly related to the smooth running of the deposition stage. Figure 1 summarizes the obtained values of wafers at risk for this stage using the result in (21).

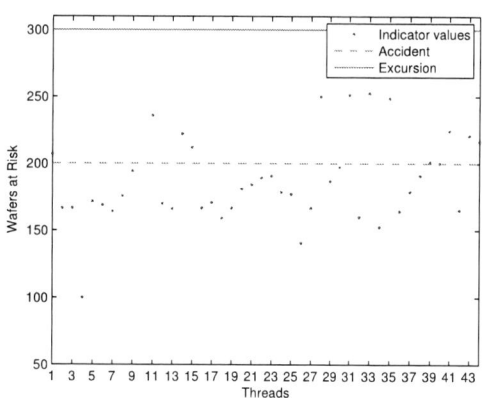

Fig. 1.  Performances evaluation of 50 threads in STMicroelectronics FAB

In Figure 1, the number of threads exceeding the accident limit is 14. Which means that these threads need to be examined and actions have to be taken so that they become more efficient. The First thread has been pointed as suboptimal because the value exceeds the accident limit. For comparison, the capability index $CpK$ has been calculated for Thread 1 and the obtained value is $CpK = 1.73 > 1.67$ which is a very good value of $CpK$ [9]. The main inconvenient of $CpK$ is that it gives a good capability value when the product is under specification, which does not necessarily mean that the regulator has optimal performances. Indeed, specificaiton limits for Thread 1 are relatively large making

the $CpK$ not relevant for assessing the performances of this regulator. Deeper analysis concerning Thread 1 show that the regulator was poorly tuned and that the tuning parameter of the EWMA filter was not adapted to its sampling frequencies. Therefore, a re-tuning of its parameters was necessary to fixed its behaviour. Then, the use of the proposed Bayesian indicator was more efficient in detecting the suboptimality of Thread 1 with comparison to $CpK$.

## IV. CONCLUSION

In this paper, a a performance indicator for Run-to-Run control assessment is proposed. The indicator is based on the Bayesian theory and is using four inputs: error between the target and output, the dispersion of the output, the out of tolerance rate, and the industrial risk. The efficiency of this indicator has been proved using an industrial example with data from the deposition area of STMicroelectronics fab in Rousset. At the end of this paper, the indicator has been compared to the classical capability index, $CpK$ and the results have been discussed. The next step will be the industrialization of the indicator. The method can also be used for adapting the R2R parameters in term of its needs.

## REFERENCES

[1] G. S. May and C. J. Spanos, *Fundamentals of semiconductor manufacturing and process control.* John Wiley & Sons, 2006.

[2] E. Del Castillo and A. M. Hurwitz, "Run-to-run process control: Literature review and extensions," *Journal of Quality Technology*, vol. 29, no. 2, p. 184, 1997.

[3] D. S. Boning, W. P. Moyne, T. H. Smith, J. Moyne, R. Telfeyan, A. Hurwitz, S. Shellman, and J. Tayor, "Run by run control of chemical-mechanical polishing," *IEEE Transactions on Components, Packaging, and Manufacturing Technology: Part C*, vol. 19, no. 4, pp. 307–314, 1996.

[4] A. J. Toprac, D. J. Downey, and S. Gupta, *Run-to-run control process for controlling critical dimensions.* Google Patents, 1999.

[5] J. Moyne, E. Del Castillo, and A. M. Hurwitz, *Run-to-run control in semiconductor manufacturing.* Springer, 2001, vol. 200.

[6] S. J. Qin, G. Cherry, R. Good, J. Wang, and C. A. Harrison, "Semiconductor manufacturing process control and monitoring: A fab-wide framework," *Journal of Process Control*, vol. 16, no. 3, pp. 179–191, 2006.

[7] C. J. Spanos, H.-F. Guo, A. Miller, and J. Levine-Parrill, "Real-time statistical process control using tool data (semiconductor manufacturing)," *IEEE Transactions on Semiconductor Manufacturing*, vol. 5, no. 4, pp. 308–318, 1992.

[8] T. F. Edgar, S. W. Butler, W. J. Campbell, C. Pfeiffer, C. Bode, S. B. Hwang, K. Balakrishnan, and J. Hahn, "Automatic control in microelectronics manufacturing: Practices, challenges, and possibilities," *Automatica*, vol. 36, no. 11, pp. 1567–1603, 2000.

[9] Y. S. Chang, I. S. Choi, and D. S. Bai, "Process capability indices for skewed populations," *Quality and Reliability Engineering International*, vol. 18, no. 5, pp. 383–393, 2002.

[10] C. Bode, B. Ko, and T. Edgar, "Run-to-run control and performance monitoring of overlay in semiconductor manufacturing," *Control engineering practice*, vol. 12, no. 7, pp. 893–900, 2004.

[11] B. P. Carlin and T. A. Louis, "Bayes and empirical bayes methods for data analysis," *Statistics and Computing*, vol. 7, no. 2, pp. 153–154, 1997.

[12] L. Yang and J. Lee, "Bayesian belief network-based approach for diagnostics and prognostics of semiconductor manufacturing systems," *Robotics and Computer-Integrated Manufacturing*, vol. 28, no. 1, pp. 66–74, 2012.

[13] J. Wang and Q. P. He, "A bayesian approach for disturbance detection and classification and its application to state estimation in run-to-run control," *IEEE Transactions on semiconductor manufacturing*, vol. 20, no. 2, pp. 126–136, 2007.

*PRIME 2018, Prague, Czech Republic*                    *Session: Modeling, Optimization, and Characterization*

# UTBB FD-SOI Circuit Design using Multifinger Transistors: A Circuit-Device Interaction Perspective

Arvind Sharma[1], Naushad Alam[2], Anand Bulusu[1]

[1]Indian Institute of Technology Roorkee, Roorkee, India
[2]Aligarh Muslim University, Aligarh, India
[1]E-mail:arvuce22@gmail.com

*Abstract*—This paper examines and models the performance of mlutifinger Ultra-Thin Box and Body Fully-Depleted (UTBB FD-SOI) MOSFETs in the presence of process induced mechanical stress. We model the channel stress and effective drive current ($I_{eff}$) in a multifinger FDSOI MOSFET (Si channel NMOS and SiGe channel PMOS) as a function of number of fingers (NFs). The proposed $I_{eff}$ model predicts the performance (i.e. $I_{eff}$) of Inverter/NAND-2/NAND-3 with a maximum error of 7% compared to Sentaurus TCAD simulations. We show that as the NFs in an NMOSFET increases from 1 to 12, $I_{eff}$ per micrometer width increases by $\approx 36\%$ and subsequently saturates. This is due to the increase in the channel stress with the NFs, which is similar to as observed in the bulk-CMOS technology. However, due to high biaxial stress, a PMOSFET's (SiGe channel) performance changes only marginally with the NFs. This is because the improvement in hole mobility saturates at high biaxial stress values. We also show that the FO4 delay of an inverter reduces by 16.67% when twelve-finger devices are used to design the inverter rather than single finger devices. Finally, using our stress model and $I_{eff}$ for an inverter/NAND/NOR cells, we propose a modified logical effort methodology (LEM) for combinational data-paths which incorporates multifinger effects in FDSOI MOSFETs. A 5-stage data-path designed using our LEM results in 14.3% improvement in total active area and 7% reduction in leakage power without a loss in speed compared to conventional technique.

*Keywords*—*Effective current, Layout dependent effects, Multifinger MOSFETs, Strain Engineering, UTBB FDSOI.*

## I. INTRODUCTION

Ultra-Thin Box and Body Fully-Depleted CMOS (UTBB FD-SOI) is a promising technology over bulk-CMOS below 32 nm technology node due to the absence of random dopant fluctuations and excellent channel control. Because of the planar structure, bulk-CMOS circuit design techniques are also valid for UTBB FDSOI. Therefore, for future technology nodes, circuit design is easy in FDSOI with respect to other technology options [1]. Moreover, ease of body-biasing offers an extra optimization knob that makes FDSOI technology a promising option for low voltage circuit design [1], [2]. Furthermore, process induced stress as performance booster, which is integrated in the process flow of other technologies (e.g. Bulk, FinFET etc), is also applicable to FDSOI [3]–[9]. However, modifications in existing bulk-CMOS circuit design techniques are required to fully exploit the stress induced performance enhancement.

Process induced mechanical stress has been used since its introduction with CMOS 90 nm technology node to enhance a transistor's performance. Thereafter, stress engineering

has become an essential part of CMOS process flow. Stress engineering improves a transistor's performance, but, at the cost of layout dependent performance variations commonly known as layout dependent effects (LDEs) [3], [4], [6]–[8]. LDEs have been extensively studied and modeled for bulk-CMOS and multi-gate devices [10]–[12]. Similar LDEs are also observed for FDSOI devices [3], [4], [6], [7], [9]. Authors in [3] investigated the impact of strain engineering on a transistor's performance with changing length of diffusion, transistor width, and gate pitch. Impact of silicon body thickness and gate length scaling on stress induced performance enhancement is analyzed in [4], [5]. It is concluded that with the silicon body thickness scaling stress induced mobility improvement reduces for electrons but not for holes. The strain induced LDEs in a PMOSFET due to two types of SiGe integration (SiGe-first and SiGe-last) processes are discussed in [6]. Authors in [7], [8] concluded that a SiGe channel PMOSFET's driving capability increases with the reduction in the transistor width, whereas, NMOSFET's performance is not affected. Consequently, narrow-width PMOSFETs are better in driving a load.

The recent strain engineering work is focused towards optimization of stress in UTBB FDSOI devices to ameliorate the performance at device level. Few researchers examined the impact of LDEs on FDSOI circuit level performance [6], [7] and a systematic design methodology in the presence of LDEs is missing. Therefore, an analysis of circuit performance and design techniques in the presence of LDEs for FDSOI technology is essential.

Effective current ($I_{eff}$) is an accurate performance parameter to measure the transient performance of digital circuits [13], [14]. Further, the $I_{eff}$ methodology can be used to design the circuits in the presence of LDEs [11]. In this paper, first, an empirical function is used to model the channel stress in multifinger FDSOI devices. Thereafter, employing the stress model and $I_{eff}$ of an inverter/NAND/NOR gate, a relationship between $I_{eff}$ and NFs in an FDSOI MOSFET is derived. We also discuss the impact of multifinger MOSFETs on the transient performance of Inverter, NAND-2/NOR-2, and NAND/NOR-3 gates. Finally, a modified logical effort methodology (LEM) is proposed for combinational data paths, which also incorporates the effects of strain engineering on multifinfer FDSOI devices. The rest of the paper is organized as follows: Section II presents the details of our simulation setup. In Section III, the channel stress and $I_{eff}$ modeling in multifinger MOSFETs is introduced. Impact of NFs on a transistor and logic cell performance is demonstrated in

978-1-5386-5388-3/18 $31.00 © 2018 IEEE                    57

Section IV. A modified LE methodology is presented in Section V. Section VI concludes the paper.

## II. TCAD SIMULATION SETUP AND CALIBRATION

Synopsys Sentaurus TCAD [15] has been used for this simulation based analysis. First, we generate 2-D UTBB FDSOI NMOSFET/PMOSFET devices using Sentaurus TCAD Process Simulator. The devices are generated using device parameters of STMicroelectronics 14 nm node [8]. Subsequently, we calibrated physics models in TCAD device mixed-mode simulations to match the electrical characteristics with [8] as shown in Fig. 1. For this calibration, we keep the devices structure consistent with the devices discussed in [8]. In the device mixed-mode simulations, appropriate physics models are employed to account for the velocity saturation, high field mobility degradation, generation-recombination, process induced mechanical stress and quantum effects.

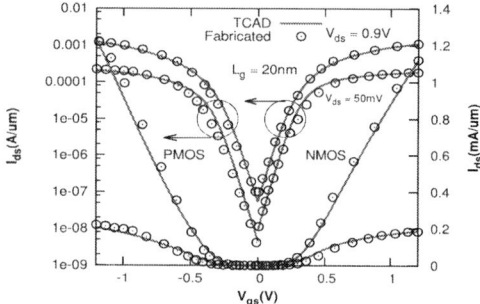

Fig. 1: Calibration of the TCAD models with fabricated data given in [8] for $L_g = 20nm$.

## III. CHANNEL STRESS AND $I_{eff}$ MODELING IN MULTIFINGER FD-SOI MOSFETs

In a circuit design, the drive strength of a transistor can be increased by increasing the width of active region or by using multifinger transistors. Multifinger transistors offer smaller gate resistance and smaller junction capacitance. Therefore, multifinger transistors are widely used in high frequency circuits [16]. However, in stress enabled technologies, channel stress is a function of NFs in a transistor. Consequently, the current does not increases linearly with the NFs [10]–[12]. We plot the normalized average channel stress per finger in a multifinger transistor with the increase in NFs in Fig. 2. Fig. 2 shows that the average stress along the channel direction first increases and subsequently saturates for large number of fingers. This variation can be modeled using following empirical equation, which is similar to that used for bulk-CMOS [10]–[12]:

$$\frac{\sigma(NF)}{\sigma(NF=1)} = P_1 + \frac{P_2}{NF+P} \qquad (1)$$

where, $\sigma(NF)$ represents, $\sigma_{xx}$ (i.e. longitudinal stress) or $\sigma_{zz}$ (i.e. vertical stress) stress component in a multifinger MOSFET and $\sigma(NF=1)$ corresponds to the stress values when single finger transistor is used. $P_1$, $P_2$, and $P$ are model coefficients

extracted for a technology using curve fitting on simulated stress values. The channel stress in a device affects the carrier mobility ($\mu$), threshold voltage ($V_{th}$), and carrier injection velocity ($\nu_{inj}$) in the following fashions [12], [17], [18]:

$$\frac{\triangle\mu}{\mu} = B\left[exp(\frac{\triangle E}{kT}) - 1\right] \qquad (2)$$

where B is a physical constant, k is the Boltzmann constant, and T is the temperature. $\triangle E$ represents the strain-induced change in the conduction band ($\triangle E_C$) or valance band ($\triangle E_V$).

$$q\triangle V_{thp} = m\triangle E_C - (m-1)\triangle E_V$$
$$q\triangle V_{thn} = m\triangle E_V - (m-1)\triangle E_C \qquad (3)$$

where m is the body effect coefficient. Please note that deformation potential model along with our stress model (1) is used to estimate $\triangle E_C$ and $\triangle E_V$ [19].

$$\frac{\triangle\nu_{inj}}{\nu_{inj}} = \alpha * \frac{\triangle\mu}{\mu} \qquad (4)$$

where $\alpha \approx 0.85$ is used in this work.

The vertical stress (i.e. into the wafer direction) is negligible (in the order of 1-50Mpa) compared to the stress along the longitudinal direction (in the order of 1-2 GPa), which can be evidenced from earlier works. Therefore, the vertical stress is not considered [4], [7].

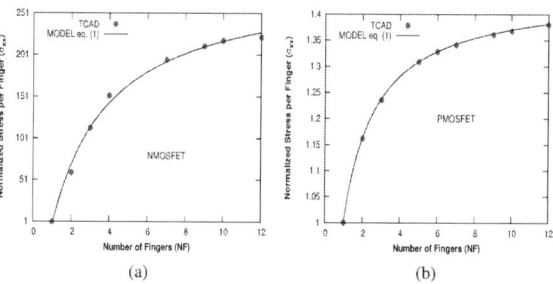

Fig. 2: Longitudinal stress variation with the NFs for a (a) NMOSFET (b) PMOSFET

The change in the effective current of an inverter with the NFs is given as [13]:

$$\frac{I_{eff}(NF)}{I_{eff}(NF=1)} = 1 + \alpha * \frac{\triangle\mu(NF)}{\mu(NF=1)} \qquad (5)$$

Please note, (5) is derived considering that improvement in a transistor current due to stress induced $V_{th}$ decrement is negligible compared to improvement due to $\triangle\mu$. The effective current variation of 2/3-input NAND/NOR gates will be similar to the (5) [14].

## IV. MULTIFINGER FD-SOI MOSFETs BASED LOGIC CELL PERFORMANCE ASSESSMENT

$I_{eff}$ of a logic cell is a measure of its transient performance [13], [14]. Fig. 3-4 shows $I_{eff}$ of an inverter, 2-input, and 3-input NAND gate with the NFs. The impact of stress variation due to the change in NFs is marginal on pull-up network $I_{eff}$ (i.e. PMOSFET performance) as shown in Fig. 3(b). This is due to the less sensitivity of hole mobility on biaxial stress and high stress values (1.8-2.5 GPa) in a PMOSFET at which impact of biaxial stress on hole mobility saturates [7]. However, pull-down network $I_{eff}$ in an inverter/NAND gate (i.e. NMOSFET performance) changes significantly with the NFs as shown in Fig. 3-4. We observe that $I_{eff}$, which is a transient performance evaluation metrics of a logic cell improves significantly with the NFs. From Fig. 3-4, we observe that our model (5) predict the $I_{eff}$ variation with the NFs within 7% of TCAD Sentaurus simulations.

Finally, we analyze the impact of NFs on an inverter cell fan-out-4 (FO4) delay. Fig. 5 shows the FO4 delay of an inverter, when devices with different NFs are used to design the inverter cell. We observe that FO4 delay of an inverter cell designed using two-finger devices is improved by 7%, when compared to an inverter designed with single finger devices (reference inverter). Further, improvement in FO4 delay is increased to 16.67%, when twelve-fingered devices are used to implement an inverter cell compared to the reference inverter, thereafter, saturates. We compare the change in $I_{eff}$ of an inverter (i.e. average of NMOSFET and PMOSFET $I_{eff}$, refer to Fig. 3) with the NFs, It is found that improvement in FO4 delay of an inverter is close to improvement in the $I_{eff}$ of an inverter. Therefore, $I_{eff}$ is an authentic parameter to evaluate a logic cell speed.

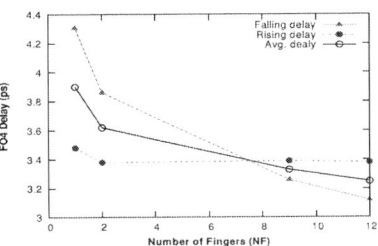

Fig. 5: An inverter FO4 delay variation when devices with different NFs are used.

From above analysis, we conclude that in circuit design, wherever possible, a transistor should be implemented with the maximum NFs (while maintaining total width as a constant). This will improve the circuit performance in two manners: first, as NFs in a transistor increases, stress along the channel direction increases for an N/P-MOSFET, consequently, the transistor current improves. Moreover, if the NFs are increased without changing the total transistor width (i.e. $W = W_F * NF$, is constant), finger width ($W_F$) decreases, consequently, hole mobility increases [7]. This is because, as PMOSFET width reduces biaxial stress converts into uniaxial, which is more beneficial for a PMOSFET. However, NMOSFET performance does not changes significantly with the width [7]. Therefore, the impact of process induced stress will be more when NFs in a transistor are increased without changing its total width.

In this paper, we used our calibrated TCAD Sentaurus simulation setup for all simulations. Although, in our work, we used 2-D devices; this is due to the limitation of TCAD tools to simulate large 3-D devices (i.e. with large NFs). However, the stress modeling along the width direction of a transistor (i.e. transverse stress) will be similar to (1) as evidenced from experimental data [6]. This can be also verified from Fig. 6 in which linear current of a PMOSFET ($V_D = 50mV$, and $|V_G - V_T| = 0.5V$) is shown against transistor width. We observe that linear current of a PMOSFET which is a function of device stress follow the model (5).

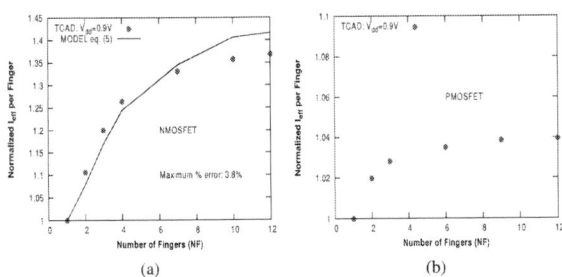

Fig. 3: $I_{eff}$ variation with the NFs for a (a) NMOSFET (b) PMOSFET.

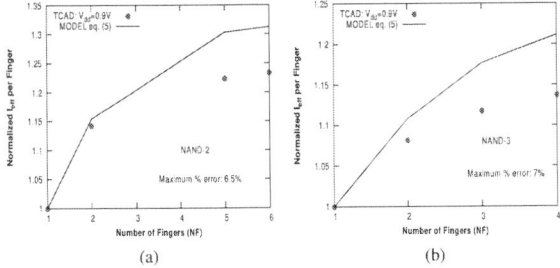

Fig. 4: $I_{eff}$ variation of (a) NAND-2 (b) NAND-3 gate with the NFs.

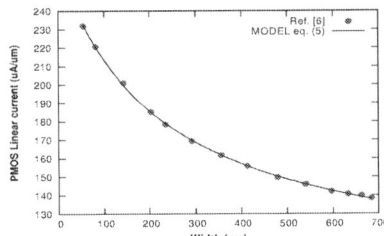

Fig. 6: Linear current of a PMOSFET ($V_D = 50mV$, and $|V_G - V_T| = 0.5V$) with the transistor width [6].

## V. Circuit Design using Multifinger FD-SOI MOSFETs

From last section, we have seen that a transistor performance varies significantly with the NFs, therefore, this has to be considered at an initial design phase to minimize the number of design iterations. Logical effort methodology (LEM) is a simple and accurate first hand method, which is used to optimize a data-path delay [20]. In LEM, logical effort (LE) of a cell is defined as the ration of its $I_{eff}$ to input capacitance ($C_{in}$) compared to a minimum sized inverter called reference inverter for a given technology node [10], [11]. Therefore,

$$LE = \frac{C_{in}}{C_{ref,in}} * \frac{I_{ref,eff}}{I_{in,eff}} \qquad (6)$$

LE of a cell is considered to be independent of transistor sizes and layout of a logic cell. However, as we have seen that $I_{eff}$ of a logic cell is a function of layout parameters and $I_{eff}/C_{in}$ varies with the transistor width. Consequently, LE of a logic cell also varies with the layout and transistor width.

Conventionally, data-paths are designed such that each stage bears equal effort [20]:

$$LE_i h_i = LE_{i+1} h_{i+1} = f \qquad (7)$$

where, $LE_i$ and $h_i$ are logical effort and electrical effort of $i^{th}$ stage, respectively, and 'f' is the stage effort. In our LEM, instead of assuming LE of a logical cell to be a constant, we use (6)-(7) to design a data-path. This will help in minimizing the design iterations to optimize a data-path. To validate our methodology, a 5-stage data-path shown in Fig. 7 is designed using conventional and our methodology. The data-path is designed for a $\frac{C_{out}}{C_{in}} = 500$. We found that our proposed technique results in $7\%$ improvement in leakage power dissipation and 14.3% reduction in total active area compared to conventional methodology [20] without degrading the speed.

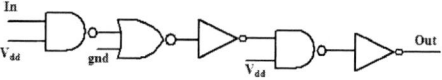

Fig. 7: A Typical circuit to verify the useability of the proposed model.

## VI. Conclusion

In this paper, an effective current ($I_{eff}$) and channel stress model for UTBB FDSOI multifinger MOSFETs is presented. It is found that bulk-CMOS stress modeling is also applicable to UTBB FDSOI technology. The proposed model matches the $I_{eff}$ of an inverter/NAND-2/NAND-3 gate within 7% of TCAD Sentaurus simulations. We have shown that an inverter cell FO4 delay is reduced by 16.67%, when twelve-fingered devices are used instead of single finger devices. We observed that SiGe channel PMOS performance is not sensitive to the NFs; this is due to the hole mobility weak dependence on biaxial stress at higher values. It has been shown that a 2-stack (3-stack) NMOSFETs (i.e. pull-down network in an NAND-2 (NAND-3) gate) $I_{eff}$ increases by $\approx 23\%(13\%)$ for six finger

(four finger) devices compared to single finger devices. Finally, we proposed a modified logical effort methodology (LEM), which also considers the stress induced layout dependent effects. A reduction of 7% and 14.3% in leakage power and active area, respectively, is observed, when a 5-stage data-path designed using proposed LEM compared to conventional logical effort methodology.

## References

[1] D. Jacquet *et al.*, "A 3 GHz dual core processor ARM cortex TM-A9 in 28 nm UTBB FD-SOI CMOS with ultra-wide voltage range and energy efficiency optimization," *IEEE Journal of Solid-State Circuits,* vol. 49, no. 4, pp. 812-826, 2014.

[2] R. Taco, I. Levi, M. Lanuzza, and A. Fish, "Low voltage logic circuits exploiting gate level dynamic body biasing in 28nm UTBB FD-SOI," *Elsevier Solid-State Electronics,* vol. 117, no. 3, pp. 185-192, 2016.

[3] F. Andrieu *et al.*, "Strain and layout management in dual channel (sSOI substrate, SiGe channel) planar FDSOI MOSFETs," in *IEEE European Solid State Device Research Conference (ESSDERC),* 2014, pp. 106-109.

[4] N. Xu *et al.*, "Stress-induced performance enhancement in Si ultra-thin body FD-SOI MOSFETs: Impacts of scaling," in *IEEE Symposium on VLSI Technology (VLSIT),* 2011, pp. 162-163.

[5] M. Cass *et al.*, "Study of piezoresistive properties of advanced CMOS transistors: thin film SOI, SiGe/SOI, unstrained and strained Tri-Gate Nanowires," in *IEEE IEDM,* 2012, pp. 637-640.

[6] R. Berthelon *et al.*, "Performance and layout effects of SiGe channel in 14nm UTBB FDSOI: SiGe-first vs. SiGe-last integration," in *IEEE ESSDERC,* 2016, pp. 127-130.

[7] B. DeSalvo *et al.*, "A mobility enhancement strategy for sub-14nm power-efficient FDSOI technologies," in *IEEE Int. Electron Devices Meeting (IEDM),* 2014, pp. 7.2.1-7.2.4.

[8] Q. Liu *et al.*, "High performance UTBB FDSOI devices featuring 20nm gate length for 14nm node and beyond," in *IEEE Intl. Electron Devices Meeting (IEDM),* 2013, pp. 9.2.1-9.2.2.

[9] R. Berthelon *et al.*, "Impact of the design layout on threshold voltage in SiGe channel UTBB-FDSOI pMOSFET," in *IEEE European Ultimate Integration on Silicon (EUROSOI-ULIS),* 2016, pp. 88-91.

[10] N. Alam, B. Anand, and S. Dasgupta, "Gate-pitch optimization for circuit design using strain-engineered multi-finger gate structures," *IEEE Trans. Electron Devices,* vol. 59, no. 11, pp. 3120-3123, 2012.

[11] A. sharma, N. Alam, S. Dasgupta, and A. Bulusu, "Multifinger MOSFETs' optimization considering stress and INWE in static CMOS circuits," *IEEE Trans. Electron Devices,* vol. 63, no. 6, pp. 2517-2523, 2016.

[12] C. C. Wang, W. Zhao, F. Liu, M. Chen, and Y. Cao, "Modeling of layout-dependent stress effect in CMOS design," in *Proc. IEEE ICCAD,* 2009, pp. 513-520.

[13] M. H. Na , E. J. Nowak , W. Haensch, and J. Cai, "The effective drive current in CMOS inverters," in *Proc. IEEE IEDM,* 2002, pp. 121 -124.

[14] K. V. Arnim *et al.*, "An effective switching current methodology to predict the performance of complex digital circuits," in *Proc. IEEE IEDM,* 2007, pp. 483-486.

[15] *Version J-2014.09,* Synopsys Inc., Mountain View, CA, USA, 2014.

[16] C. C. Ho , C. W. Kuo , Y. J. Chan , W. Y. Lien, and J. C. Guo, "0.13-$\mu m$ RF CMOS and varactors performance optimization by multiple gate layouts," *IEEE Trans. Electron Devices,* vol. 51, no. 12, pp. 2181 -2185, 2004.

[17] A. Khakifirooz and D. A. Antoniadis, "Transistor performance scaling: The role of virtual source velocity and its mobility dependence," in *Proc. IEEE IEDM,* 2006, pp. 667 -670.

[18] S. K. Marella and S. S. Sapatnekar, "The impact of shallow trench isolation effects on circuit performance," *in Proc. IEEE ICCAD,* 2013, pp. 289-294.

[19] G. L. Bir and G. E. Pikus, "Syymmetry and Strain Induced Effects in Semiconductors," New York: John Wiley & Sons, 1974.

[20] I. Sutherland, B. Sproull, and D. Harris, *Logical Effort: Designing Fast CMOS Circuits.* San Mateo, CA, USA: Morgan Kaufmann, 1999.

# Let's make it Noisy: A Simulation Methodology for adding Intrinsic Physical Noise to Cryptographic Designs

Kashif Nawaz, Léopold Van Brandt, François-Xavier Standaert and Denis Flandre
ICTEAM institute, Université catholique de Louvain, Belgium

*Abstract*—Noise in digital circuits for the sake of performance has always been minimized in typical designs. However, for cryptographic applications, increased noise could be beneficial. It can be used effectively to reduce the mathematical SNR (signal-to-noise ratio) further and make it more difficult for the adversary to gather useful information from the side channel leakage data. In this paper, we introduce a methodology to exploit the intrinsic physical noise (i.e. flicker and thermal noise) at the circuit level and use the obtained values in a relevant cryptographic context. Our simulations show that the calculated cryptographic noise values are in close agreement with the noise levels extracted from noisy distributions using transient noise analysis. Consequently, this noise is shown to increase with the number of transistors or the supply voltage.

## I. Introduction

Side channel attacks, such as differential power analysis (DPA) [1], exploit the leakage signal from a cryptographic device to guess the secret keys. Logic styles such as differential dual rail, have been proposed to reduce the signal value, but do not scale well with technology [2] e.g, moving from 65nm to 28nm nodes. Existing countermeasures against side-channel analysis such as shuffling (adding noise in time domain) and masking (or algorithmic noise, adding noise in the amplitude domain) work well only if the SNR has been sufficiently reduced. A possible option to reduce the SNR even further would be to increase the intrinsic physical noise coming from the transistors themselves. In this paper, we explore the design of noisy CMOS implementations, and propose a methodology to exploit the *intrinsic* physical MOSFET noise allowing designers to derive an insight of the impact through gate-level simulations. More specifically, we investigate a methodology to answer the following questions, i.e. can the MOSFET noise be quantified from a cryptographic perspective, how does it scale up with the number of transistors, and how does it behave with low voltage implementations?

This paper is divided into 5 sections. We first review the state-of-the-art existing for this work, then in Section III we discuss our methodology to introduce noise sources, their simulation cost and budget. In Section IV we discuss the results of our methodology. In section V, we present a statistical analysis of our traces and finally wrap up in Section VI with conclusion and perspectives.

## II. State-of-the-art: A Review

A quick state-of-art metric for quantifying the cryptographic leakage from a side channel (a leaky implementation) arises from the classical univariate metric, the Signal-to-Noise ratio. In this paper, we use Mangard's SNR defined in [3] as:

$$\text{SNR} = \frac{\hat{\text{var}}_x(\hat{\text{E}}_i(L_x^i))}{\hat{\text{E}}_x(\hat{\text{var}}_i(L_x^i))}, \quad (1)$$

where $\hat{\text{E}}$ (resp. $\hat{\text{var}}$) denotes the sample mean (resp. variance) operator and $L$ the leakage. In our following simulations, this SNR will be computed for noise-based traces of the current consumption of the digital gates at the supply rail as a function of time, denoted as $I_{DD}(t^*)$ and would include the noise coming from physical *intrinsic* MOSFET noise sources. Using eqn (1), the signal is the "useful" part that is obtained by the adversary as a measure of the information leakage. The lower the signal value, the lower is the "perceived" side-channel leakage. The maximum signal, as a metric to quantify the leakage, in case of noiseless simulations [2], has been used by the authors to show the scaling trends of the signal with respect to technology scaling from 65nm bulk to 28nm FDSOI for standard CMOS and dual rail differential logic styles. As reported, dual rail logic styles lose their advantage with technological scaling and standard CMOS continues to be the design of choice with respect to technological scaling. For CMOS, with lowering of the supply voltage $V_{DD}$, in 28nm, the signal value reduces further compared to 65nm technology, which is highly desirable for the implementation of cryptographic algorithms. These comparisons justify the choice of standard CMOS implementations in scaled down technologies over dual-rail styles and provides the necessary motivation for the design of noisy CMOS implementations.

## III. Transient Noise Analysis: A simulation methodology

### A. Target Designs

Conventional noise analyses in circuit designs mostly use the ac or the harmonic based approaches. However, in digital cryptographic applications, where each point in a transient run is potentially a source of information leakage (from an adversary perspective), it makes it worth analyzing the effect of noise on *each* time sample. From a cryptographic perspective, this is a univariate analysis compared to a bivariate analysis (where multiple time-samples are used). In the scope of this work, we focus on the univariate aspect only. The time sample with the highest value of signal (or SNR) is chosen as the point-of-interest (POI). Using the Transient noise simulations in Eldo software (provided by Mentor Graphics)

[4], we provide a methodology to introduce the *intrinsic* physical noise sources (i.e. noise coming from the transistors themselves and *not* externally) and calculate the resulting "cryptographic"[1] signal and noise (units of $A^2$) (as defined in eqn. 1), then compare them to the values obtained from ac simulations and variance calculations.

Our results are based on transient noise simulations in an Eldo environment using a 28nm FDSOI PDK (process design-kit) provided by an industrial foundry. The sizing of the transistors is kept minimum to maximize the noise produced (flicker noise especially). We use a simple 2-bit XOR, a 4-bit PRESENT Sbox and an 8-bit AES S-box, all custom designed using Cadence Virtuoso software, to show the scaling trends of the signal and noise w.r.t the supply voltage and the number of transistors.

### B. Simulation settings

All the 3 designs are simulated with the Transient Noise analysis built in Eldo (called by the *.noisetran* command) upto 100 transient noise runs. The noise sources correspond to the physical flicker and thermal noises intrinsic to the MOS transistors. They are generated by Eldo in the frequency bandwidth specified by the input parameters of the transient noise analysis. In our simulations, we chose $f_{min} = 1/T$ and $f_{max} = 1/2*dt$, where $dt$ is the minimum time step being used by the simulator or specified by the user. The input data signals to the circuits are a recurring 0 to an arbitrary input for 4 transitions (for the 2-input XOR), 16 and 256 transitions (for the 4-bit PRESENT and 8-bit AES S-boxes repectively) at a clock frequency of 10 MHz. All simulations are done at 298K, $TT$ corner and for a $V_{DD}$ range from 0.5V to 1V.

### C. Simulation cost and Budget

In this section, we investigate the cost of our methodology and quantify the total budget, both in terms of CPU runtime and number of runs required to obtain convergent metrics which estimate the "crypto" signal and noise. The noise transient simulations are indeed well known to be time-and memory-intensive [5]. Basically, our simulation budget can be stated as

$$N_{traces} \cdot \frac{T}{dt}, \qquad (2)$$

where $N_{traces}$ is the number of traces, i.e. noise realizations, $T$ is the simulation duration for one trace, and $dt$ is the time step. These parameters correspond to NBRUN, TSTOP − TSTART and HMAX of Eldo NOISETRAN command [4], respectively. The number of samples for each trace is given by

$$N_s = \frac{T}{dt}. \qquad (3)$$

Since the *.noisetran* analysis specifies a number of input parameters, it is of importance to analyze the effect of each of these on our calculated values of signal and noise.

1  We first analyze the impact of the $f_{max}$ parameter specified in the *.noisetran* analysis on the CPU runtime, as shown in figure 1. Choosing a larger $f_{max}$ also

---

[1]We use the term cryptographic and crypto interchangeably, they both denote the one and same thing

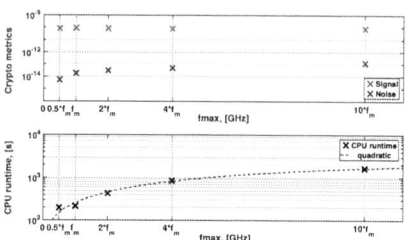

Figure 1: Impact of $f_{max}$ parameter on the "Cryptographic" *Signal* and *Noise* and the CPU runtime for an XOR gate at $V_{DD}$=0.9V and $f_m$ = 12.5 GHz

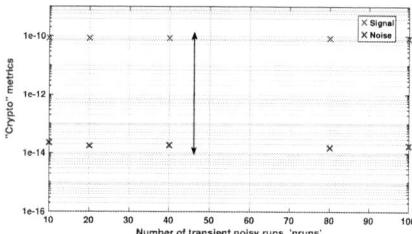

Figure 2: Impact of *nruns* parameter on the "Cryptographic" *Signal* and *Noise* for an XOR gate at $V_{DD}$=0.9V

means increasing very significantly the CPU runtime, as shown in figure 1(bottom). We then propose to choose an optimum value of the $f_{max}$, which trades-off CPU simulation time and convergent signal/noise values. As the fmax parameter is primarily defined by the sampling time, keeping in mind the Nyquist sampling criteria, we see that oversampling may lead to higher obtained noise values, however, at the expense of increased CPU runtime. Moreover, oversampling would simply allow the "cryptographic" adversary to average over a larger number of time samples, thus effectively reducing the noise by averaging and hence, increasing the SNR, which is contrary to our requirements. Hence, an optimum value of fmax has to be chosen which minimizes the simulation time.

2  The simulation time also depends on the number of transient runs, *nruns* as shown by eqn 2. The greater this value, the longer the simulation time; we also define a tolerance value $\eta$ which is defined as

$$\eta = \frac{\sigma}{\mu} \qquad (4)$$

where $\sigma$ is the standard deviation across the observations and $\mu$ is the mean for the observed values. For the purposes of our simulations, we choose an $\eta$ tolerance value of $\lesssim 15\%$. We observe in figure 2 that even with an increasing number of runs, the values of "crypto" signal and noise calculated remain well within our tolerance levels. This justifies the usage of lower number of transient noisy runs to minimize the simulation run-time.

## IV. RESULTS OF THE TRANSIENT NOISE ANALYSIS

In this section, we now discuss the results of our ongoing work. We aim to show our present results for a 2-input XOR, a 4-bit Sbox using the PRESENT cryptographic implementation and an 8-bit AES Sbox implementation. These include a total of 12, 684 and 1884 transistors respectively.

### A. Effect on the Cryptographic Signal and Noise

Thanks to the 40 transient noise runs, we are now able to calculate the maximum signal and the maximum noise, (as per equation 1). Figures 3 and 4 show the scaling of the maximum signal and noise for a range of $V_{DD}$, i.e, from 0.5V to 1V. We can make the following observations from the above 2 plots,

1. By increasing the supply voltage, $V_{DD}$, the value of the maximum signal and noise increase. This can be explained by the fact that as the $V_{DD}$ increases, the power consumption, $P_{dyn}$ increases (hence the increase in $I_{ON}$) which increases the *Signal* value. The increase in the "crypto" *noise* could be explained by the increase in the thermal noise which increases with the increase of the $I_{ON}$ current value. Reciprocally, signal and noise decrease with $V_{DD}$. We can observe the signal decrease is faster than the noise decrease which is of interest for our purpose.

2. For a particular, $V_{DD}$, the signal (and the noise) increase with the increase in design complexity, i.e as the number of transistors increases for the given circuit (e.g. moving from a 2-bit XOR to a 4-bit PRESENT Sbox to an 8-bit AES Sbox)

The increase in the signal can be modeled by the following relation

$$\bar{S}_{V_{DD}}^{circuit} = S_{V_{DD}}^{XOR} N_T^{\beta} \qquad (5)$$

where $\bar{S}_{V_{DD}}^{circuit}$ is the signal for the target cricuit, $S_{V_{DD}}^{XOR}$ is the signal produced by a 2-bit XOR for the same supply voltage, $V_{DD}$, $N_T$ is the ratio of the increase in the number of transistors w.r.t a 2-bit XOR and $\beta$ is a technology factor which varies $0.4 < \beta < 1.5$ for most circuits and depends on the value of the supply voltage, $V_{DD}$.

The increase in the noise can be modeled as

$$\bar{N}_{V_{DD}}^{circuit} = N_{V_{DD}}^{XOR} N_T^{\alpha} \qquad (6)$$

where $\bar{N}_{V_{DD}}^{circuit}$ is the noise for the target circuit, $N_{V_{DD}}^{XOR}$ is the noise calculated for a 2-bit XOR at the same $V_{DD}$, $N_T$ is the ratio of the number of transistors w.r.t a 2-bit XOR and $\alpha$ is a parameter which scales with the supply voltage and is $\lesssim 2$ for most circuits and depends on the value of the supply voltage, $V_{DD}$. Consequently, we observe that noise increases faster with the number of transistors than the signal. This could be related to the fact that the intrinsic MOSFET noise sources are not correlated and hence add on $I_{DD}$, whereas the signal is more proportional to the number of circuit branches connected to $V_{DD}$.

Since the calculated "cryptographic" noise is essentially a mean of the variance across different inputs for *nruns* number of traces, we should be able to relate this noise to the histogram of the measured current. We explore this in the next section.

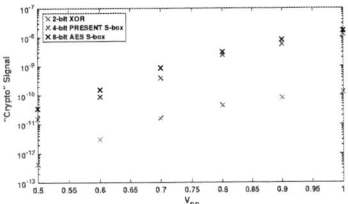

Figure 3: Scaling of "Cryptographic" *Signal* as a function of $V_{DD}$ for different circuits

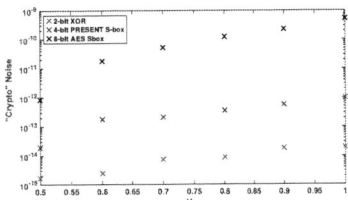

Figure 4: Scaling of "Cryptographic" *Noise* as a function of the number of transistors, $N_T$ for different $V_{DD}$

## V. STATISTICAL PROPERTIES OF THE SUPPLY CURRENT NOISE OF THE XOR GATE

In this section, we study the *first-order statistics* of the noise present in the supply current for one of the above mentioned circuits of interest, i.e. the XOR gate.

Fig. 5 contains noise traces for the set of parameters $N_{traces} = 40$, T = 800$ns$ and $dt = 40ps$ for a total budget of $\sim 8 \times 10^5$ for a supply $V_{DD} = 0.5$V

### A. Statistical characterization of the static region

The static region indicated in Fig. 5 is useful to get insight on the noise behaviour within the circuit and how it affects the supply current. Input voltages of the gate are fixed, and so are all the averages of the branch currents and node voltages within the circuit (since there are noise fluctuations). Hence, the supply current noise is treated as a *wide-sense stationary*

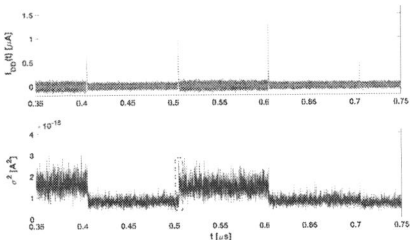

Figure 5: Zoom on the dynamic and static regions (top) and associated variance plot based on all the 40 traces (bottom). Noise in the static region is shown to follow a stationary Gaussian distribution.

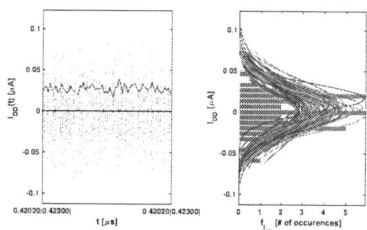

Figure 6: Histogram construction for time t = 0.4202μs - 0.4230μs within the static region in Fig. 5 (left). Extracted Gaussian distribution is also shown (right).

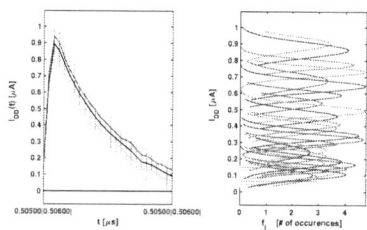

Figure 7: Histogram construction for time t = 0.505μs - 0.506μs within the dynamic region marked red in Fig. 5(below). Extracted Gaussian distribution mixture is also shown(right).

stochastic process, for which a complete definition can be found in [6]. Especially, the probability density function (pdf) does not depend on $t$:

$$f\left(I_{DD}\left(t\right)\right) = f\left(I_{DD}\right), \qquad (7)$$

and the variance is also independent of $t$:

$$\sigma\left(t\right) = \sigma. \qquad (8)$$

As a consequence, every sample $I_{DD}\left(t^{*}\right)$ at some time $t^{*}$ of every trace is understood as a realization of one single random variable $I_{DD}$.

### B. Challenges regarding the dynamic region

In order to accurately capture the dynamic region behaviour, we plot the time-varying histograms of the very *narrow* dynamic region enclosed by a red rectangle in figure 5. The supply current noise now is a *nonstationary* stochastic process. Its distribution is explicitly time-dependent:

$$f\left(I_{DD}\left(t\right)\right) = f\left(I_{DD}, t\right), \qquad (9)$$

as well as the variance $\sigma\left(t\right)$. Since each point in the dynamic region is non-stationary, we observe the *time-varying* histograms for each sample to extract the mean and the variance and show the *envelope* of the $\pm 1\sigma$ over the mean trace as shown in figure 7

Fortunately, our main goal is not the histograms themselves but the extracted variances, for which rough estimations are sufficient. In this case, this is achieved by performing a

nonlinear fitting of the granular histogram. Using the extracted $\sigma$ values from the distributions above and comparing them with the mathematical "cryptographic" noise values calculated by eqn (1), we obtain a good matching between the data, thus validating the fact that the noise present in the mathematical calculations can indeed be traced back to the results of the transient noise analyses including the MOSFET noise sources. This is of importance to further expand, optimize and exploit the methodology of our on-going work.

## VI. CONCLUSION AND OPEN QUESTIONS

In this case study, we have for the first time, to the best of our knowledge, proposed a methodology to analyze and simulate the addition of *intrinsic* MOSFET physical noise on cryptographic implementations using transient noise simulations to compute the first-order univariate security metrics. This can be used to discuss how the physical noise sources from the MOSFETs can be used for effectively deriving the cryptographic SNR value, especially at design stages of a cryptographic implementation. Longer transient runs (higher fmax), which add more noise, however come at the expense of increased simulation run times (high CPU time). The complexity would further increase with the number of gates in case of a full cryptographic implementation such as the AES or the PRESENT block cipher. At this stage our results, while suggesting that the use of intrinsic physical noise in MOSFETs to add more cryptographic noise is an effective method (since such noise sources are predicted to increase significantly with further technology scaling and voltage), its simulation and extension to correlated noise sources especially for larger circuits needs to be optimized and remains an open research question.

**Acknowledgments.** This work has been funded in parts by the ARC Project NANOSEC. François-Xavier Standaert is a research associate of the Belgian Fund for Scientific Research.

## REFERENCES

[1] P. C. Kocher, J. Jaffe, and B. Jun, "Differential power analysis," in *Advances in Cryptology - CRYPTO '99, 19th Annual International Cryptology Conference, Santa Barbara, California, USA, August 15-19, 1999, Proceedings*, pp. 388–397. [Online]. Available: http://dx.doi.org/10.1007/3-540-48405-1_25

[2] K. Nawaz, D. Kamel, F.-X. Standaert, and D. Flandre, "Scaling trends for dual-rail logic styles against side-channel attacks: A case-study," in *Constructive Side-Channel Analysis and Secure Design - 8th International Workshop, COSADE 2017, Paris, France, April 13-14, 2017, Revised Selected Papers*, 2017, pp. 19–33.

[3] S. Mangard, "Hardware countermeasures against DPA ? A statistical analysis of their effectiveness," in *Topics in Cryptology - CT-RSA 2004, The Cryptographers' Track at the RSA Conference 2004, San Francisco, CA, USA, February 23-27, 2004, Proceedings*, 2004, pp. 222–235. [Online]. Available: http://dx.doi.org/10.1007/978-3-540-24660-2_18

[4] Mentor Graphics Corporation, "Eldo User's Manual, Release AMS 2008.2," 2008.

[5] A. Demir and A. Sangiovanni-Vincentelli, "Time-domain non-monte carlo noise simulation," in *Analysis and Simulation of Noise in Nonlinear Electronic Circuits and Systems*. Springer, 1998, pp. 113–161.

[6] A. Papoulis, *Probability, Random Variables, and Stochastic Processes*, ser. McGraw-Hill Series in Electrical Engineering. McGraw-Hill, 1991.

PRIME 2018, Prague, Czech Republic

Session: Circuits for Memories and Security

# Analysis on Sensing Yield of Voltage Latched Sense Amplifier for Low Power DRAM

Suk Min Kim, Byungkyu Song, Tae Woo Oh and Seong-Ook Jung
School of Electrical and Electronic Engineering
Yonsei University
Seoul 03722, Korea
{sukmin_kim, bksong, oto92 and sjung}@yonsei.ac.kr

*Abstract*—Various types of sense amplifiers are widely used in memory products. In this paper, we have studied on the optimization of a voltage latched sense amplifier (VLSA) with 65nm CMOS process for low-power DRAM. In particular, we have classified sensing failure into the offset failure and the latch-delay failure, and have found that the latch-delay failure becomes even worse at low supply voltages below 1.0V. We also found that conventional NMOS-driven sensing operation was no longer effective on VLSA for low supply voltage, and investigated various methods to decrease the latch-delay failure probability.

*Keywords—DRAM, voltage latched sense amplifier, low-power, offset failure, latch-delay failure*

## I. INTRODUCTION

The voltage latched sense amplifier (VLSA) has been used for a long time as the sensing circuit of the data stored in the DRAM cell due to the strong positive feedback and the fast sensing speed [1], [2]. Fig. 1 (a) shows the typical VLSA circuit for commercial DRAM. Two cross-coupled inverters composed of MN1, MN2, MP1 and MP2 amplify the small voltage difference ($\Delta V$) between BLT and BLB according to the cell data polarity (data "0"; D0, data "1"; D1). It is important to note that accurate sensing operation can only occur if $\Delta V$ obtained from charge sharing operation is greater than the offset voltage ($V_{OS}$) of VLSA [2]. If $\Delta V$ is smaller than $V_{OS}$, the VLSA amplifies in the opposite to the stored data polarity, causing a sensing failure. For this reason, many efforts have been made to minimize the influence of $V_{OS}$ resulting from the mismatch of the threshold voltage ($V_{th}$) between the NMOS (NM1, NM2) or PMOS (MP1, MP2) transistors. For example, various studies such as offset cancellation and charge transfer techniques [3], [4] have been reported, but these have not yet been commercialized due to their complicated structure and area penalty.

Rather, steady efforts have been made to optimize the NMOS-driven sensing operation without changing the basic VLSA structure. In previous studies [5]-[7], it was reported that the $V_{th}$ mismatch of the NMOS is a dominant factor on $V_{OS}$ since the NMOS in VLSA inverters is turned on earlier than the PMOS in the initial sensing operation, therefore relatively increasing the NMOS width is advantageous for correct sensing operation by reducing the $V_{th}$ variation and

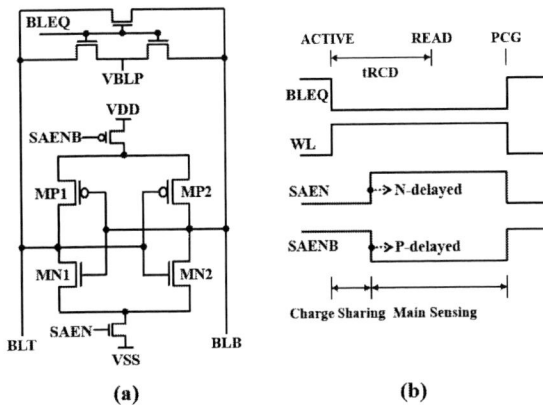

Fig. 1. (a) Schematic and (b) Timing Waveforms of typical Voltage Latched Sense Amplifier for DRAM

lowering the trip point of the inverters. In addition, commercial DRAMs are used to adopt a scheme that separates the sensing start signal into SAEN and SAENB, as shown in Fig. 1 (a), and helps to operate with more reliable NMOS-driven sensing by intentionally activating the SAEN signal first (i.e. P-delayed sensing operation) [8].

In this paper, we find that NMOS-driven (P-delayed sensing) VLSA scheme is no longer useful in low-power DRAMs such as mobile application due to severe latching time degradation, and consequently focus on various methods for resolving sensing failure at low supply voltages (VDD).

## II. SIMULATION

For statistical analysis of VLSA sensing failure, HSPICE Monte Carlo simulations are performed 10,000 times using industry-compatible 65nm model parameters. The random $V_{th}$ variation of each latch transistor affecting $V_{OS}$ is determined by considering the increase in the process variability with technology scaling, and the value of Pelgrom's coefficients ($A_{VT}$) for NMOS and PMOS are used as 3.33 and 3.24 mV·μm, respectively [2].

978-1-5386-5388-3/18 $31.00 © 2018 IEEE

*Paper P89*                                                    *PRIME 2018, Prague, Czech Republic*

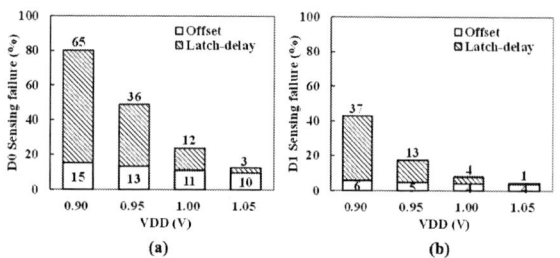

Fig. 2.   Sensing failure rate of (a) D0 and (b) D1 by various VDD

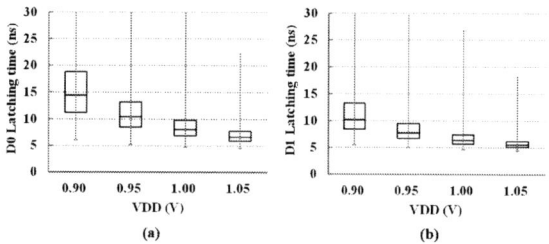

Fig. 3.   Latching time of (a) D0 and (b) D1 by various VDD

In this study, we also implement NMOS-driven or PMOS-driven sensing operation by separating and controlling SAEN and SAENB signals. The active point of SAEN signal is maintained and the SAENB signal is delayed to make the NMOS-driven (PMOS-delayed) sensing operation. Conversely, the SAENB signal is maintained and the SAEN signal is delayed to make the PMOS-driven (NMOS-delayed) sensing operation.

In addition, we classify sensing failure into two types. One is an offset failure when the sensing operation occur opposite to data polarity due to $V_{OS}$ larger than $\Delta V$. The other is a latch-delay failure. Since the read and write operations of DRAM data must be synchronized with the command clock (CLK) of the external system, it is very important to keep a timing specification for each operation. Considering the tRCD (RAS to CAS delay time, RAS; Row Access Strobe and CAS: Column Access Strobe) specification, the latched voltage difference between BLT and BLB at the appointed time after the active command must be sufficiently secured to withstand the data loss of the following read path and to accurately transmit the sensed data. In this paper, the latch-delay failure is defined when BLT~BLB voltage difference at 15ns after WL active is less than 80% of supply voltage even though the latching polarity is correct [9]. To consider worst sensing noise environment, we also applied an isolated data pattern that means different data polarity between adjacent BLs [10].

### III.   SIMULATION RESULTS AND ANALYSIS

First, we investigated the sensing failure characteristics of typical VLSA by various VDD. In this case, considering the NMOS-driven sensing operation of commercial DRAM, the width ratio of NMOS to PMOS of 2:1 and the P-delayed time of 3 ns are applied. As shown in Fig. 2, the increase in the

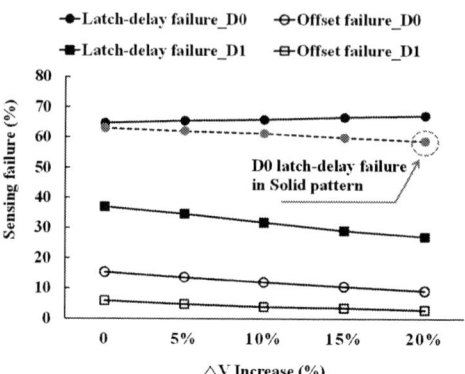

Fig. 4.   Change of sensing failure with $\Delta V$ increase at VDD of 0.9V

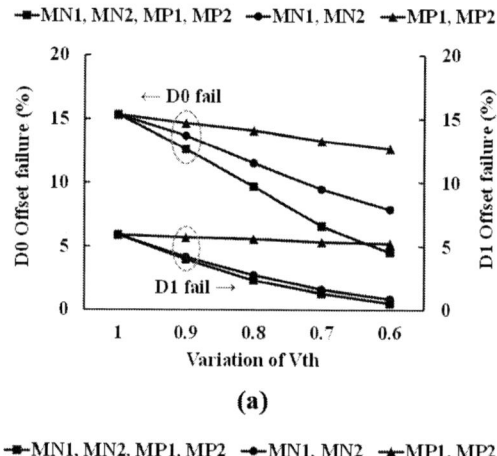

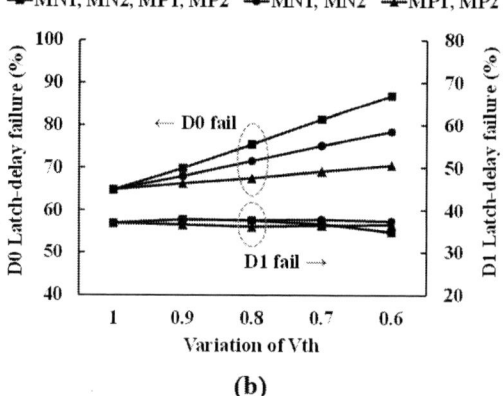

Fig. 5.   Change of (a) offset failure and (b) latch-delay failure with variation of $V_{th}$ decrease at VDD of 0.9V

offset failure is insensitive, while the latch-delay failure increases sharply and becomes a major factor in sensing failure as VDD decreases below1.0V. This means that it is required to

978-1-5386-5388-3/18 $31.00 © 2018 IEEE          66

decrease the offset failure at high supply voltages, but it will be much more important to improve latch-delay characteristic in future low-power VLSA at low VDD. This tendency can be confirmed again in the latching time characteristic, which is the time when the voltage difference between BLT and BLB is developed up to 80% of VDD from when the sensing enable signal is activated. It can be noted that both the dispersion and the median value of latching time significantly increase, as the VDD becomes low in Fig. 3. We investigated whether several methods are effective to improve the latch-delay failure characteristics at low supply voltage as follows.

### A. Charge Shared Voltage between BL and Cell Capacitor

Fig. 4 shows the sensing failure when $\Delta V$ obtained by charge sharing is increased. In order to focus on the effect of the input voltage levels in the static low supply voltage of 0.9V, we controlled $\Delta V$ by changing not the VDD but directly the cell capacitance in this simulation. The offset failure reduces as $\Delta V$ increases due to the increase of sensing margin, while only the latch-delay failure of D0 does not change or rather increases. It is clear that the latch-delay failure should be improved because the sensing speed can be increased by increasing $\Delta V$ as in the case of D1. However, it is assumed that the latch-delay failure of isolated D0 could be worse because adjacent D1 data are sensed more quickly, as shown in Fig. 3 (b), and act as a noise source. Fig. 4 shows that the D0 latch-delay failure in solid pattern (dashed line), which means same data polarity between adjacent BLs, decreases with increasing $\Delta V$ unlike that of isolated pattern, thus supports this assumption of the coupling noise effect.

### B. Random $V_{th}$ variation of VLSA transistors

In order to investigate the effect of $V_{th}$ variation of NMOS and PMOS transistors, we reduced the value of $A_{VT}$ to 90% ~ 60% without changing the length and width. In Fig. 5 (a), it can be confirmed that it is more effective to improve the $V_{th}$ variation of NMOS than that of PMOS for reducing the offset failure, as in the previous study [6]. On the other hand, Fig. 5 (b) shows that the reduction of $V_{th}$ variation does not help the latch-delay failure at all. Conversely, the latch-delay failure of D0 rather increases. This can be analyzed that the latching characteristic of D0 is very bad so that, even if the offset failure is reduced due to the $V_{OS}$ improvement, the latch-delay failure cannot help occurring at the low VDD.

### C. Width ratio of VLSA transistors

As previously mentioned, since the $V_{OS}$ of the NMOS was regarded as a dominant factor of sensing failure in the commercial DRAM so far, it was considerably advantageous to increase the NMOS width relative to the PMOS width. This is because increasing NMOS width can lower the inverter's trip point as well as reduce the $V_{th}$ variation and $V_{OS}$ [5]. However, even though the VDD is scaled down, the previously optimized N:P ratio (e.g., N:P = 2:1) has not been changed much due to the risk of changing the commercially proven VLSA structure.

In this section, we examine whether the effect of larger NMOS width is still valid for improving the latch-delay failure at low supply voltage. Fig. 6 shows simulation results of offset and latch-delay failures according to the N:P width ratio in

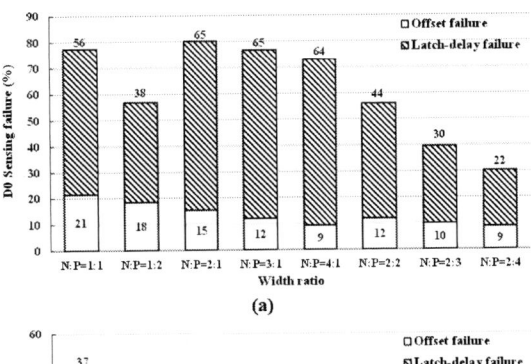

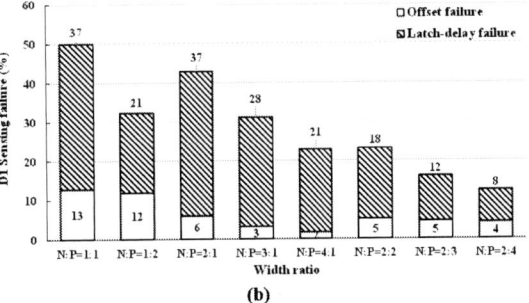

Fig. 6. (a) D0 and (b) D1 sensing failure by various width ratio at VDD of 0.9V

various cases at low VDD of 0.9V. From these results, we can identify several characteristics.

First, increasing the NMOS width is still effective for reducing the offset failure of both D0 and D1, whereas increasing the PMOS width is less effective for the offset failures as mentioned in previous paper [5]. Second, it is observed that the increase of NMOS width also has an effect on the reduction of D1 latch-delay failure, but has no effect on the D0 latch-delay failure. This is because the D1 sensing speed becomes faster due to lower trip point with the increase of NMOS width, whereas, the isolated D0 sensing is affected more by the coupling noise of the adjacent faster D1 sensing like the effect of $\Delta V$ increase mentioned in section A .

Finally, it is more effective to increase the PMOS width for reducing the latch-delay failures of both D0 and D1. In this study, we have examined the offset failure and latch-delay failure separately, and could confirm that the PMOS width of VLSA should not be reduced relatively no longer below supply voltage 1.0V. That is, the upsurge of latch-delay failure at low VDD shown in Fig. 2 is due to the weak strength of PMOS, thus the pull-up strength in latching inverter needs to be improved by increasing PMOS width. For example, based on the sensing failure result of width ratio 2:1, Fig. 6 shows that the width ratio 2:2 is more advantageous than 3:1 with the same area to improve the sensing failure at low VDD.

### D. NMOS-driven or PMOS-driven Sensing Operation

In order to improve the latch-delay failure, PMOS width should be increased. Since it is necessary to consider the VLSA

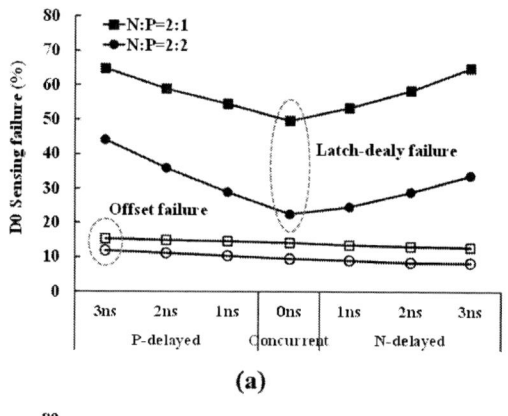

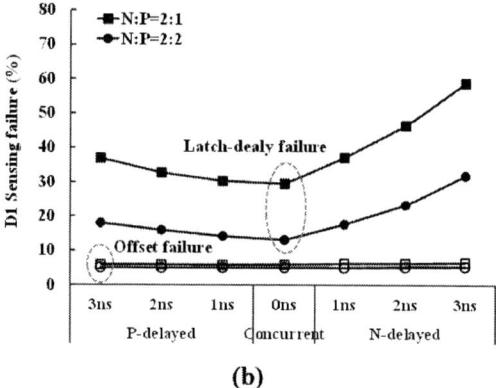

Fig. 7. (a) D0 sensing failure and (b) D1 sensing failure depending on NMOS-driven or PMOS-driven sensing operation at VDD of 0.9V

area, it is inappropriate to drastically increase the PMOS width. In addition, the NMOS-driven VLSA is no longer effective at low VDD. In this section, we investigate whether the latch-delay failure can be further improved by adjusting the timing of SAEN and SAENB signals shown in Fig. 1.

Fig. 7 shows that the probability of the offset failure at VDD of 0.9V is almost constant regardless of NMOS-driven sensing operation or PMOS-driven sensing operation. On the other hand, it is noted that the latch-delay failure varies dramatically depending on the conditions of the sensing start signals. It is considered that latching operation can be most effectively achieved by applying each source power to the VLSA inverters at the same time, rather than shifting the sensing operation in one direction like NMOS-driven type or PMOS-driven type. This tendency does not seem to correlate with the width ratio and data polarity. From Fig. 6 and 7, it can be concluded that it is more effective to additionally adjust the SAEN and SAENB signals in the appropriate width ratio than to increase the only PMOS width further in order to completely improve the latch-delay failure. That is, rather than increasing

the width ratio up to 2:4 in the NMOS-driven sensing mode, increasing the width ratio moderately up to 2:2 and achieving the concurrent sensing mode are advantageous in terms of VLSA area as well as lowering down to a similar sensing failure probability.

## IV. CONCLUSION

In this paper, several methods to improve the latching characteristics of VLSA have been studied based on 65nm CMOS technology for low-power DRAM. Up to now, since the typical VLSA of commercial DRAMs have shown efficient sensing characteristics under NMOS-driven sensing conditions, DRAM's VLSA have focused on improving the offset failure by adopting a relatively large NMOS width and intentional NMOS fast sensing. However, we confirmed that NMOS-driven VLSA is no longer effective in aspects of the latch-delay failure when the supply voltage is below 1.0V. It can be seen that the improvement of the latch-delay failure is affected not only by simply increasing the $\Delta V$ or reducing the $V_{os}$ of the VLSA transistors. It was noted that increasing the PMOS width rather than increasing the area was more effective in improving the latch-delay failure of VLSA at the low supply voltage. In addition, both the offset and latch-delay failures can be improved without excessive increase of PMOS width when the SAEN and SAENB signals are simultaneously activated.

## REFERENCES

[1] B. Wicht, T. Nirschl and D. Schmitt-Landsiedel, "Yield and Speed Optimization of a Latch-Type Voltage Sense Amplifier," IEEE J. Solid-State Circuits, vol. 39, no. 7, pp. 1148-1158, July 2004.

[2] S.-H. Woo, H. Kang, K. Park and S.-O. Jung, "Offset voltage estimation model for latch-type sense amplifiers," IET Circuits Devices Syst., vol. 4, iss. 6, pp. 503-513, July 2010.

[3] Sanghoon Hong, S. Kim, J. Wee and S. Lee, "Low-Voltage DRAM Sensing Scheme With Offset-Cancellation Sense Amplifier," IEEE J. Solid-State Circuits, vol. 37, no. 10, pp. 1356-1360, October 2002.

[4] Choongkeun Lee and H. Yoon, "Highly Robust and Sensitive Charge Transfer Sense Amplifier for Ultra-Low Voltage DRAMs," 5th Asia Symposium on Quality Electronic Design, pp. 227-232, August 2013.

[5] Joyce Yeung and H. Mahmoodi, "Robust Sense Amplifier Design under Random Dopant Fluctuations in Nano-Scale CMOS Technologies," IEEE International SOC Conference, pp. 261-264, September 2006.

[6] Larry Pileggi, G. Keskin, X. Li, K. Mai and J. Proesel, "Mismatch Analysis and Statistical Design at 65nm and Below," IEEE Custom Intergrated Circuits Congerence, pp. 9-12, September 2008

[7] Abhinav V. Deshpande, " Process Variation Induced Mismatch Analysis in Sense Amplifier," Journal of Science, vol. 6, iss. 8, pp. 426-429, 2016

[8] Hee-Bok Kang, et al., "A Sense Amplifier Scheme with Offset Cancellation for Giga-bit DRAM," J. Semiconductor Technology and Science, vol. 7, no. 2, pp. 67-75, June 2007

[9] Tao Zhang, Cong Xu, Yuan Xie and Guangyu Sun, "An Overhead-free Method to Reduce Precharge Overhead for Memory Parallelism Improvement of DRAM System," IEEE 31st International Conference on Computer Design, pp. 138-144, October 2013

[10] Yan Li, H. Schneider, F. Schnabel, R. Thewes and D. Schmitt-Landsiedel, "DRAM Yield Analysis and Optimization by a Statistical Design Approach," IEEE Transactions on Circuits and Systems-□, vol. 58, no. 12, pp. 2906-2918, December 2011

# The key impact of incorporated $Al_2O_3$ barrier layer on W-based ReRAM switching performance

Elmira Shahrabi*, Cecilia Giovinazzo†, Jury Sandrini*, and Yusuf Leblebici*

*Microelectronic Systems Laboratory (LSM), Swiss Federal Institute of Technology (EPFL), Lausanne, Switzerland
Email: elmira.shahrabi@epfl.ch
† Department of applied science and technology (DISAT), Politecnico di Torino, Turin, Italy

*Abstract*—In this article, we inspected the bipolar resistive switching behavior of W-based ReRAMs, using $HfO_2$ as switching layer. We have shown that the switching properties can be significantly enhanced by incorporating an $Al_2O_3$ layer as a barrier layer. It stabilizes the resistance states and lowers the operating current. $Al_2O_3$ acts as an oxygen scavenging blocking layer at W sides, results in the filament path constriction at the $Al_2O_3/HfO_2$ interface. This leads to the more controllable reset operation and consecutively the HRS properties improvement. This allows the $W/Al_2O_3/HfO_2/Pt$ to switch at 10 times lower operating current of 100 $\mu$A and 2 times higher memory window compared to the $W/HfO_2/Pt$ stacks. The LRS conduction of devices with the barrier layer is in perfect agreement with the Poole-Frenkel model.

*Index Terms*—ReRAM, Tungsten, Aluminum oxide, barrier layer, resistive switching

## I. INTRODUCTION

Resistive random access memory (ReRAM) in the past decays has drawn great attention with the prospects of creating a replacement for both embedded storage memory and mass memory applications [1]. Typical ReRAM cells consist of a simple metal-insulator-metal structure, in which the electrical resistance can switch between two resistance states, high resistance state (HRS) and the low resistance state (LRS). Simple structure, great scalability, high device density and CMOS-compatibility are the key features of the ReRAMs for many different applications. For successful embedding of ReRAM in any demands, the main concerns are still to control the device variation and reduce the power consumption [2]. Among different high-k dielectric materials, $HfO_2$ [3], $Al_2O_3$ [4], $TiO_2$ [5] are the most mature candidates for the resistive switching devices in which the resistance levels are controlled by the formation and destruction of the conductive nano filament (CNF) path. The main performance variations are due to the limited control on such nano-scale conductive path. To improve the switching uniformity and device performances, different solutions have been proposed, such as insertion of the oxygen reservoir to facilitate the ion migrations [6], stacking a barrier layer between the oxide layer and electrode to suppress the operating current down to 1 $\mu$A [7], proper selection of the electrodes with respect to the work function and electronegativity to control the interface resistance and the switching performances [8]. Moreover, several other advanced processing techniques like, doping, nano-crystals implanting in the switching layer, operating condition adjustment and

annealing at different atmosphere and temperature were carried out to reduce the switching variations [9] [10]. These massive efforts prove that it is still challenging to realize a simple, time and cost effective way to improve device stability along with the CMOS-compatible stack selection that can be applied directly for the relevant applications.

In this article, we have carefully investigated the role of ultra thin $Al_2O_3$ barrier layer on the switching variability enhancement of the $W/Al_2O_3/HfO_2/Pt$ ReRAM devices. Using W as the electrode is beneficial since it is one the widely accepted material in the microelectronic devices. In our previous study [11], we have applied nano-scale W interconnectors inside the CMOS technology to integrate nano-scale, self-align $WO_x$-based ReRAMs. Due to the CMOS thermal budget, $WO_x$ is not the best candidate to provide stable switching performances with the W electrodes. We have designed W-based ReRAMs using $HfO_2$ as a switching layer and $Al_2O_3$ as a barrier layer. We demonstrated that the insertion of 3 nm $Al_2O_3$ at the $W/HfO_2$ interface can provide enough asymmetry barrier potential to (i) reduce the operating current, (ii) improve the resistance window and (iii) control the variability of operating voltage and HRS. Finally, the switching mechanism for the $W/Al_2O_3/HfO_2/Pt$ is well explained by the Poole-Frenkel mechanism. These studies are highly valuable for the further co-integration of the similar stack on the W via surface of the CMOS periphery.

## II. DEVICE DESCRIPTION

Fig.1 (a) is the schematic representation of the ReRAM cells fabrication. In this process, both bottom electrode (BE) and top electrode (TE) are deposited through a shadow mask membrane. Shadow mask prevents the probable reactive ion etching (RIE) electrode damaging such as ion bombardment, radiation-induces bonding changes and charge buildup. The Si-based shadow masks are fabricated using the conventional photolithography and deep reactive ion etching (DRIE) followed by the grinding to obtain the membrane with the thickness of 300 $\mu$m. For the ReRAM devices, 100 mm $Si/SiO_x$ wafer substrate is used. The BE of W (100 nm) is sputtered at room temperature with the 15 nm of TiN as an adhesion layer. Afterwards, 100 nm of low temperature oxide (LTO) is grown using LPCVD technique at 425 °C to isolate the ReRAM cells and to define the active device area. The LTO passivation layer is patterned via photolithography and BHF wet etching

*Paper P56*                                                                    *PRIME 2018, Prague, Czech Republic*

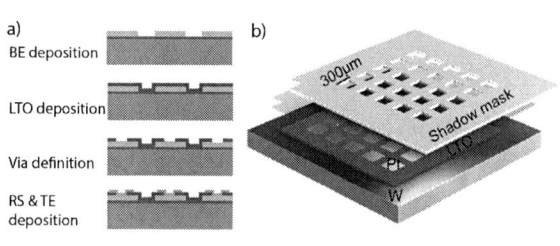

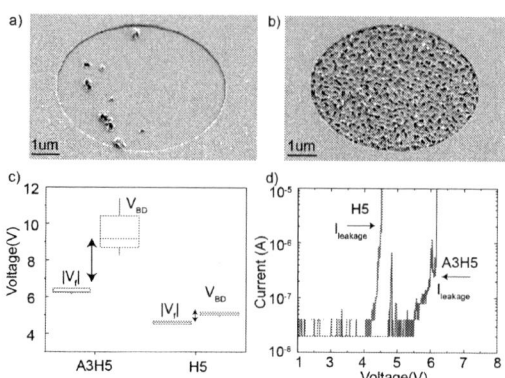

Fig. 1. a) Schematic representation of the stand-alone ReRAM fabrication process flow and b) final device stack.

Fig. 2. SEM micrograph of the same stack after a) soft forming operation with the negative bias and b)breakdown operation with the positive applied voltage. c) $V_{BD}$ and $V_f$ parameters comparison for the H5 and A3H5 cells. d) I-|V| Forming behavior obtained from the H5 and A3H5 cells.

to create the cylinder-shape active area with the diameter of 800 nm, 2 $\mu$m, 3 $\mu$m, 5 $\mu$m and 10 $\mu$m. Then the wafer is diced into the small chips of 1x1 cm$^2$ as shown in Fig.1 (b). Next, the amorphous ALD HfO$_2$ and Al$_2$O$_3$ are deposited as the resistive switching layer (RS) materials. Finally a 100 nm Pt (TE) was sputter-deposited to complete the cell structure. In this paper, we have analyzed two main stacks that are referred as H5 and A3H5 within the text. Table.I summarizes the samples specifications.

TABLE I
SAMPLES SPECIFICATION

| Sample ID | Stack |
|---|---|
| H5 | W 100 nm (BE)/HfO$_2$ 5 nm/ Pt 100 nm (TE) |
| A3H5 | W 100 nm (BE)/Al$_2$O$_3$ 3 nm/ HfO$_2$ 5 nm/ Pt 100 nm (TE) |

### III. FORMING PHENOMENOLOGY

To qualify the impact of stack material engineering on the $V_{\ddot{O}}$ formation energy, we investigated the forming and breakdown operation for the both H5 and A3H5 ReRAM cells by applying the voltage ramp to the TE while the BE is grounded. During the forming operation, the negative voltage sweep from 0 V to -7 V is applied and the current is limited using the Agilent B1500 parameter analyzer to avoid the hard breakdown in the dielectric layers. The memories were able to switch appropriately with the negative forming operations. The breakdown voltage is obtained from the positive voltage sweep from 0 V to 12 V with no current limitation. Fig.2 (a) and (b) show the SEM micrograph of ReRAM cells after employing the forming voltage ($V_f$) and breakdown voltage ($V_{BD}$) respectively. For the $V_{BD}$ with no current elimination, the memory active area goes under massive physical damages due to the Joule heating as explained by Lu et al. [12] while for the regular forming operation with the current compliance, there are only few blow-off zones that caused due to the current overshoots problems with the parameter analyzer limitations. As shown in Fig.2 (c), there is a higher gap between the ($V_{BD}$, |$V_f$|) of the A3H5 samples compare to the H5 samples. A3H5 cells are formed at -6.3 V with the I$_{cc}$ of 100 uA and the forming voltage of -4.5 V with the I$_{cc}$ of 2 mA is achieved for

the H5 cells. Moreover, the forming leakage current (pointed in Fig.2 (d)), for the A3H5 ($3\times10^{-7}$ A) is 10 times less than the one for the H5 ($2.5\times10^{-6}$ A). It is evident that the insertion of the tunnel barrier Al$_2$O$_3$ at the W/HfO$_2$ interfaces immunes the HfO$_2$ from the interactions with the W. Lower density of the vacancy at one interface results in a better conductive path confinement with asymmetrical geometry that can facilitate the conductive filament destruction at lower operating voltage and current. This effect is further discussed in the next sections.

### IV. SWITCHING CHARACTERISTICS

Fig.3 and Fig.4 show the typical DC bipolar resistive switching I-V curve and resistance variations from cycle to cycle of 5 $\mu$m H5 and A3H5 cells at room temperature. The voltage was swept from 0 V→ -1.5 V → 0 V → 2 V→ 0 V on the TE while the BE is kept grounded. During the set operation, the I$_{cc}$ of 2 mA and 100 $\mu$A are applied on the H5 and A3H5 cells respectively. From the DC characteristics of H5, it can be seen that the devices switched to the LRS of 300 $\Omega$ at the negative set voltage of -1 V with 2 mA and in the positive voltage region, the LRS retained up till -2 V and gradually switched back to the HRS of 10 k$\Omega$. It should be noted that the switching performance of the H5 stack for I$_{cc}$ of 100 $\mu$A is strongly attenuated. Therefore, we had to progressively increase the current limit until 2 mA to get more stable and distinguishable resistance levels. For the A3H5 cells, in the negative voltage region, the cell switched to the LRS of 1 k$\Omega$ at the set voltage of -1.1 V and far smaller operating current of 100 $\mu$A and the positive reset associates with several intermediate jumps from LRS of 1 k$\Omega$ to the HRS of 100 k$\Omega$. In the reset part, the first sudden jump occurs at the reset voltage of 0.9 V and another two obvious resistance intermediates states are visible before reaching 2 V.To assess the cycle-to-cycle stability and reproducibility of both H5 and A3H5 devices, the DC endurance test was conducted as demonstrated in Fig.5. The resistance values of both H5 and A3H5 were calculated

978-1-5386-5388-3/18 $31.00 © 2018 IEEE

*PRIME 2018, Prague, Czech Republic*

*Session: Circuits for Memories and Security*

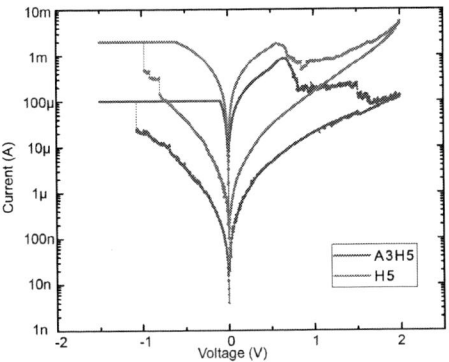

Fig. 3. DC bipolar resistive switching I-V curves of the H5 device (W/ HfO$_2$ (5 nm)/ Pt) and the A3H5 device (W/ Al$_2$O$_3$ (3 nm)/ HfO$_2$ (5 nm)/ Pt) cells with the negative set operation and positive reset operation.

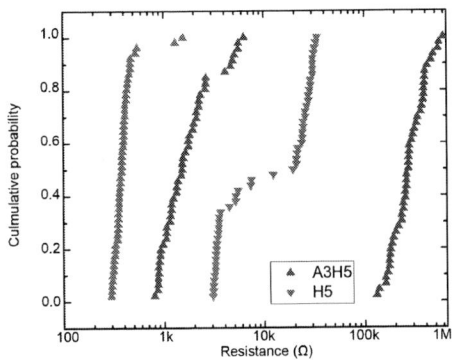

Fig. 4. Cumulative probability distribution of LRS and HRS for both H5 and A3H5 devices measured under DC read voltage of 0.25 V.

at the reading voltage of 0.25 V. For the H5 cells, the memory shows very stable HRS and LRS for 50 cycles, and after that, there is an obvious degradation of HRS from 30 k to 3 k which stabilizes again over contentious cycling. This degradation is expected to be due to the incomplete reset at the W/ HfO$_2$ interface.

During the set operation, the negative bias on the Pt TE pushes the oxygen ions towards the W electrode. W is known as the electrode material with several non-stoichiometric level of oxidation [13]. When W encounters the oxygen ions in the set step, it starts to extract more oxygen ions to grow the thin layer of non-stoichiometric conductive WO$_x$ at the W/HfO$_2$ interface and make the HfO$_2$ more deficient and form the thicker filaments. In order ro reduce the formed WO$_x$ to return the oxygen ions back to the HfO$_2$ layer and break the filaments at the reset part, higher bias voltage (>2 V) is required which makes it unfavorable for the electronic devices. The A3H5 device demonstrates stable switching over 100 cycles by maintaining a resistance window of 10$^2$, which can accomplish the requirement of ReRAM applications. Moreover, both HRS and LRS states are distinctly separated without any explicit

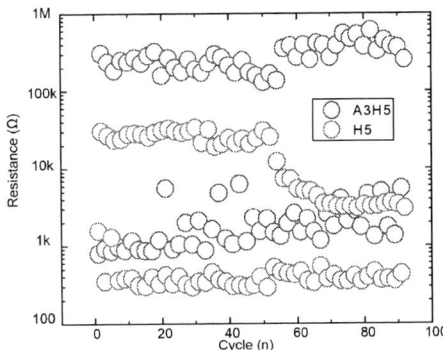

Fig. 5. Endurance characteristics of H5 and A3H5 devices under 100 DC consecutive cycling.

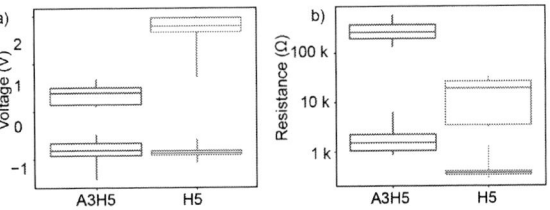

Fig. 6. Voltage (a) and Resistance (b) variation, comparison between H5 and A3H5. The statistical analysis have been carried out on 5 devices for each stack.

degradation. The excellent cycling uniformity and reduced operating current in A3H5 is related to the insertion of Al$_2$O$_3$ at the W interface which has much higher oxygen scavenging immunity compare to the HfO$_2$ [4]. Al$_2$O$_3$ immunity reduces the random formation of the [V$_{\ddot{O}}$] at Al$_2$O$_3$ induced by the W electrode which narrows down the filament path at the Al$_2$O$_3$/HfO$_2$ interface that can be easily ruptured with the lower power consumption. The switching parameters variation of different A3H5 and H5 devices are shown in Fig.6 (a) and (b). It is noticeable that improved switching characteristics are achieved in A3H5 stacks. The ON and OFF resistance window and the resistance stability from device-to-device is much higher compared to the H5 stacks while the operating voltage is decreased.

## V. CONDUCTION MECHANISM

The switching conduction mechanism of the A3H5 stack has been investigated to have a better insight on the device switching properties. Fig.7 (b) exhibits the conductive filament dynamic in the presence of the Al$_2$O$_3$ barrier layer. The different oxygen affinity of the two oxides determines the formation of an asymmetric filament, where the interface Al$_2$O$_3$/HfO$_2$ acts as a constriction zone for conductive filament rupture and formation in reset/set operations. It is assumed that the presence of the Al$_2$O$_3$ increases the barrier potential at the W interface which results in even lower vacancy accumulation at W/Al$_2$O$_3$ compare to the HfO$_2$/Pt interface. The confined

978-1-5386-5388-3/18 $31.00 © 2018 IEEE

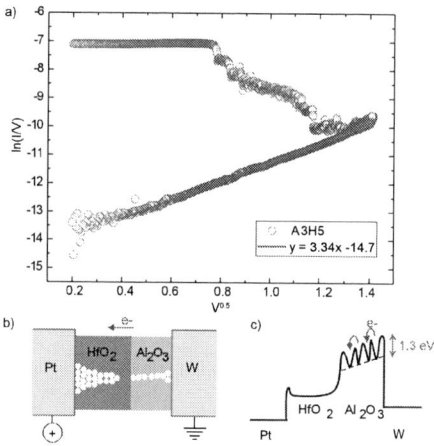

Fig. 7. a) Linear fit on the reset I-V characteristic of A3H5 in $ln(I/V)$-$\sqrt{V}$ scale; b) Scheme of the conductive filament dynamic during reset; c) energetic band diagram for the Poole-Frenkel conductive mechanism.

filament profile improves the stability of the resistance states, particularly the HRS. The narrow filament formation inside the $Al_2O_3$ layer is attributed to the break of Al-O bonds and the creation of $V_{\ddot{O}}$ interstitial defects [4]. Due to the very low self diffusion of $V_{\ddot{O}}$ in amorphous $Al_2O_3$ deposited via ALD, this defects are confined inside the barrier layer, playing a crucial role in the conduction transport mechanism responsible for the reset of A3H5 cells. A trap-led mechanism, as the Poole-Frenkel is proposed. In this model, the current density is defined by:

$$J_{PF} = q\mu N_C E \, exp\left[\frac{-q(\Phi_T - \sqrt{qE \setminus \pi\epsilon})}{kT}\right]$$

where $\mu$ is the mobility of electrons, $N_C$ is the state density in conduction band and $\Phi_T$ is the potential well of the traps. Trapped electrons, thermally excited by the electric field, drift inside the oxide layer by passing from a localized trap state to the adjacent ones (Fig.7 (c)).

In order to verify the accuracy of Poole-Frenkel conduction mechanism, the positive part of IV characteristic is plotted in $ln(I/V)$-$\sqrt{V}$ and a linear fit of data between 0.2 V and 1.5 V is performed (Fig.7 (a)). The calculations have been carried out on 45 consecutive cycles, revealing that the Poole-Frenkel mechanism correctly describes the electron transport of the device. The dimension of the potential well of the traps ($\Phi_T$) has been extracted by the intercept of the linear fit, resulting in 1.29±0.03 eV. The traps in $Al_2O_3$ lower the energy required for the electron motions. This impact is visible in the steep reset current changes for A3H5 devices.

## VI. CONCLUSION

In this study, we have studied the bipolar switching properties of $HfO_2$-based ReRAM using the W as the bottom electrode. We have ascertained the variability improvement

of the resistance switching in the $Al_2O_3/HfO_2$ bilayers with W interface barrier engineering by insertion of an ultra thin 3 nm of $Al_2O_3$. $Al_2O_3$ acts as an oxygen scavenging blocking layer at W sides, results in the filament path constriction at the $Al_2O_3/HfO_2$ interface. This leads to the more controllable reset operation and consecutively the HRS properties improvement. This allows the A3H5 to switch at 10 times lower operating current of 100 $\mu$A and 2 times higher memory window compare to the H5 stacks. The conduction mechanism of the A3H5 device is perfectly explained by the Poole-Frenkel model. This study provides valuable insights in the application of the similar stack for the 1T1R CMOS-ReRAM co integration.

## REFERENCES

[1] H. Akinaga and H. Shima, "Resistive random access memory (reram) based on metal oxides," *Proceedings of the IEEE*, vol. 98, no. 12, pp. 2237–2251, 2010.
[2] W. Banerjee, S. Z. Rahaman, and S. Maikap, "Excellent uniformity and multilevel operation in formation-free low power resistive switching memory using irox/alox/w cross-point," *Japanese Journal of Applied Physics*, vol. 51, no. 4S, p. 04DD10, 2012.
[3] J. H. Yoon, S. J. Song, I.-H. Yoo, J. Y. Seok, K. J. Yoon, D. E. Kwon, T. H. Park, and C. S. Hwang, "Highly uniform, electroforming-free, and self-rectifying resistive memory in the pt/ta2o5/hfo2-x/tin structure," *Advanced Functional Materials*, vol. 24, no. 32, pp. 5086–5095, 2014.
[4] L. Goux, N. Raghavan, A. Fantini, Nigon *et al.*, "On the bipolar resistive-switching characteristics of al2o3-and hfo2-based memory cells operated in the soft-breakdown regime," *Journal of Applied Physics*, vol. 116, no. 13, p. 134502, 2014.
[5] W. Banerjee, X. Xu, H. Lv, Q. Liu, S. Long, and M. Liu, "Variability improvement of tio x/al2o3 bilayer nonvolatile resistive switching devices by interfacial band engineering with an ultrathin al2o3 dielectric material," *ACS Omega*, vol. 2, no. 10, pp. 6888–6895, 2017.
[6] Y. Y. Chen, L. Goux, S. Clima, B. Govoreanu, R. Degraeve, G. S. Kar, A. Fantini, G. Groeseneken, D. J. Wouters, and M. Jurczak, "Endurance/retention trade-off on hfo2 metal cap 1t1r bipolar rram," *IEEE Transactions on electron devices*, vol. 60, no. 3, pp. 1114–1121, 2013.
[7] S.-G. Park, M. K. Yang, H. Ju, D.-J. Seong, J. M. Lee, E. Kim, S. Jung, L. Zhang, Y. C. Shin, I.-G. Baek *et al.*, "A non-linear rram cell with sub-1$\mu$a ultralow operating current for high density vertical resistive memory (vrram)," in *Electron Devices Meeting (IEDM), 2012 IEEE International*. IEEE, 2012, pp. 20–8.
[8] V.-Q. Zhuo, Y. Jiang, M. Li, E. Chua, Z. Zhang, J. Pan, R. Zhao, L. Shi, T. Chong, and J. Robertson, "Band alignment between ta2o5 and metals for resistive random access memory electrodes engineering," *Applied Physics Letters*, vol. 102, no. 6, p. 062106, 2013.
[9] L. Chen, H.-Y. Gou, Q.-Q. Sun, P. Zhou, H.-L. Lu, P.-F. Wang, S.-J. Ding, and D. Zhang, "Enhancement of resistive switching characteristics in al2o3-based rram with embedded ruthenium nanocrystals," *IEEE Electron Device Letters*, vol. 32, no. 6, pp. 794–796, 2011.
[10] S. Park, K. Cho, J. Jung, and S. Kim, "Annealing effect of al2o3 tunnel barriers in hfo2-based reram devices on nonlinear resistive switching characteristics," *Journal of nanoscience and nanotechnology*, vol. 15, no. 10, pp. 7569–7572, 2015.
[11] E. Shahrabi, J. Sandrini, B. Attarimashalkoubeh, T. Demirci, M. Hadad, and Y. Leblebici, "Chip-level cmos co-integration of reram-based non-volatile memories," in *Ph. D. Research in Microelectronics and Electronics (PRIME), 2016 12th Conference on*. Ieee, 2016, pp. 1–4.
[12] Y. M. Lu, M. Noman, W. Chen, P. A. Salvador, J. A. Bain, and M. Skowronski, "Elimination of high transient currents and electrode damage during electroformation of tio2-based resistive switching devices," *Journal of Physics D: Applied Physics*, vol. 45, no. 39, p. 395101, 2012.
[13] J. K. Kim, S. W. Nam, S. I. Cho, M. S. Jhon, K. S. Min, C. K. Kim, H. B. Jung, and G. Y. Yeom, "Study on the oxidation and reduction of tungsten surface for sub-50 nm patterning process," *Journal of Vacuum Science & Technology A: Vacuum, Surfaces, and Films*, vol. 30, no. 6, p. 061305, 2012.

# A Variability-Aware Analysis and Design Guideline for Write and Read Operations in Crosspoint STT-MRAM Arrays

Y. A. Belay, A. Cabrini, G. Torelli

Department of Electrical, Computer and Biomedical Engineering, University of Pavia, Pavia, Italy
Email: yilkalandualem.belay01@universitadipavia.it, alessandro.cabrini@unipv.it, guido.torelli@unipv.it

*Abstract*—Benefiting from emerging resistance-switching memory technologies, crosspoint array has become an attractive array architecture to obtain high storage density. Among the emerging technologies, Spin-Transfer Torque magnetic memory (STT-MRAM) is a potential candidate as storage class memory (SCM) or static/dynamic RAM replacement due to its high write speed, scalability and other interesting characteristics. In this paper, we present a variation-aware comprehensive analysis of the boundary conditions for write and read requirements for the implementation of crosspoint STT-MRAM Arrays. The results of the analysis are very useful as design guide and for choosing a suitable selector device for Crosspoint STT-MRAM arrays.

## I. INTRODUCTION

Emerging resistance-switching memory technologies, such as resistive RAM (RRAM), phase change memory (PCM), and spin-transfer torque magnetic RAM (STT-MRAM) are actively under research and development with the expectation of replacing the mainstream memory technologies that are facing some critical challenges [1], [2]. In these emerging memory technologies, the resistance of the basic storage device is switched between a high resistance (or RESET) state (HRS) and a low resistance (or SET) state (LRS), which are used to store binary data. In particular, STT-MRAM, which features non-volatility, high write speed, area (and) current scalability, and practically infinite endurance, is a potential candidate as storage class memory (SCM) or static/dynamic RAM replacement [1], [3].

Driven by the increasing demand for large data storage capacity and benefiting from the two-terminal structure of the aforementioned memory technologies, crosspoint array has become an attractive array architecture to obtain high storage density and, hence, low cost per bit [4], [5]. In a crosspoint array, the memory cell is formed at the junction of a lower and an upper plane of parallel wires (bitlines, BLs, and wordlines, WLs) running at right angles to each other. If both the width of the wires and the spacing between them are equal to the minimum lithographic feature size, $F$, the memory cell is allocated within the smallest footprint of $4F^2$, thus providing high density. The effective area per memory cell can be further reduced to $4F^2/N$ with $N$-layer 3D stacking [1], [4], [6]. However, crosspoint arrays suffers from sneak current paths and IR drop (along interconnection metal lines). To suppress the sneak current paths, a selector device must be connected in series with each memory element [6], [8], thus giving rise to the 1S1R (one selector, one resistive element) crosspoint array structure as shown in Fig. 1.

In general terms, write and read performance of 1S1R crosspoint arrays depends on the memory element, selector device and the interconnection metal line [9], [10]. We present a variation-aware comprehensive analysis of the boundary conditions for memory write and read requirements that impose constraints on the implementation of crosspoint STT-MRAM Arrays. The analysis is useful both as design guide and as a tool to understand requirements of selector threshold voltage and switching voltage of the STT-MRAM element.

## II. 1S1R CROSSPOINT MEMORY ARRAYS

Fig. 1 shows a circuit schematic of a crosspoint STT-MRAM array with one selected cell located at the lower right corner. The biasing voltages used to operate the memory array are also indicated. The selected WL and BL are connected to write source, $V_W$, and ground, respectively.

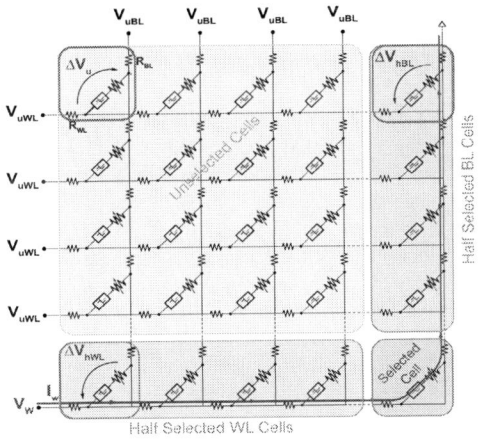

Fig. 1. Circuit schematic of 1S1R crosspoint STT-MRAM array

978-1-5386-5388-3/18 $31.00 © 2018 IEEE

On the other hand, unselected WLs and BLs are connected to intermediate bias voltages $V_{uWL}$ and $V_{uBL}$, respectively. As indicated in Fig. 1, non-addressed cells sharing either the WL or the BL with the selected cell will be referred to as half-selected cells. Whereas, cells connected to unselected WLs and BLs will be referred to unselected cells. The voltage drop across WL and BL half-selected cells and thus, the leakage current through them, is determined by the unselected WL and BL bias voltages. Indeed, the voltage across the cells can be expressed in terms of $V_{uWL}$ and $V_{uBL}$ as follows: $\Delta V_{hWL} = V_W - V_{uBL}$, $\Delta V_{hBL} = V_{uWL}$, and $\Delta V_u = V_{uBL} - V_{uWL}$. Memory write operation requires sufficient current and voltage to be delivered to the selected cell while still avoiding unintended write (or write disturbance) in half-selected and unselected memory cells. In other words, $V_W$ should be large enough to turn-on the selector device and to switch cell while $\Delta V_{hWL}$, $\Delta V_{hBL}$, and $\Delta V_u$ should be less than $V_{th}$ to avoid unintended overwriting of the stored data.

## III. SPIN-TRANSFER TORQUE MAGNETIC RAM (STT-MRAM)

The basic storage element in STT-MRAM is a magnetic tunnel junction (MTJ) device that can be switched using spin-transfer torque effect: depending on the direction of the electric current, the spin torque switches the magnetization of the free layer, which is then oriented either to a parallel or to an anti-parallel direction with respect to the magnetization of the fixed layer [11]. The parallel configuration resulting in LRS, whereas the anti-parallel configuration yields HRS. Spin-transfer-torque (STT)-based switching is intrinsically stochastic. The probability of switching of state is generally a function of the amplitude, and the width the current pulse passing through the MTJ device, $I$ and $t_p$, respectively, and two technology dependent parameters: intrinsic critical current, $I_{c0}$, and thermal stability factor, $\Delta$. It has been demonstrated in the literature that for sufficiently long current pulses ($t_p \geq$ 10ns) with amplitude $I < I_{c0}$, the probability of switching can be expressed as [11], [12]:

$$P_{sw} = 1 - \exp\left\{\frac{-t_p}{\tau_0}\exp\left[-\Delta\left(1 - \frac{I}{I_{c0}}\right)\right]\right\} \quad (1)$$

where $\tau_0 = 1$ns is referred to as thermal attempt time. The thermal stability factor, $\Delta$, is expressed as: $\Delta = \frac{E_{eff}}{K_b T}$. where $E_{eff} = E_0\left(1 - I/I_{c0}\right)$ is the effective energy barrier between the two magnetization states, $k_b$ is Boltzmann constant, and $T$ is absolute temperature.

In crosspoint STT-MRAM arrays, (1) can be used to evaluate read and write disturbance (in half-selected and unselected cells). For example, write/read disturbance rates are indicated in Fig. 2a for combinations of two pulse widths: $t_p = 10$ns and 100 ns and two thermal stability factor values: $\Delta = 45$ and 60. In our analysis, we consider $\Delta = 60$ $t_p = 10$ns, and $I/I_{c0} = 0.4$ (for the read operation), which gives a disturbance probability of $< 10^{-12}\%$. Besides, for the chosen $t_p$ and $\Delta$, it is safe to assume that switching occurs at $I_{c0}$. Accordingly, the switching currents are: $I_{c0} = 50$ $\mu$A and $25\mu$A for the P

to AP and AP to P switching, respectively. The considered switching voltage $V_{sw} = 0.5$ V, as it can be seen from the I-V characterstic of the STT-MRAM device shown in Fig. 2b.

## IV. ANALYSIS OF WRITE AND READ CONSTRAINTS

Each memory cell (or 1S1R) is made up of as a bipolar non-linear selector device serially connected with the STT-MRAM element. The I-V characteristic of some implementations of the two common categories of selector devices are shown in Fig. 3, as reported in the literature [8], [10]. A threshold selector is turned on abruptly at certain threshold voltage, $V_{th}$, and threshold current, $I_{th}$, as shown in Fig. 3. Whereas, in a nonlinear (or exponential) selector the current changes exponentially with the applied voltage. However, threshold selectors have gained more attention recently. Indeed, one study has demonstrated that threshold selectors give a better read performance in crosspoint STT-MRAM arrays [13]. Similarly in this work, we focus on threshold selectors: below $V_{th}$ the device is turned-off while above $V_{th}$ the device is turned-on and provides the high current required for programming the memory element. However, it is worth mentioning that the analysis is equally applicable for exponential selectors with sufficiently high nonlinearity. We also assume that change of state of the memory element occurs at $V_{sw}$. Besides, in the half-selected unselected and cells, since the selector device should be effectively turned-off giving rise to a very high resistance, the total voltage across the 1S1R pair can be approximated by the voltage across selector device neglecting the voltage across the memory element. Accordingly, we assume that the 1S1R device is turned-off when the voltage across it is less than the threshold voltage of selector device. Once the device is turned-on, the total voltage falls partly across the selector device and partly across the memory element.

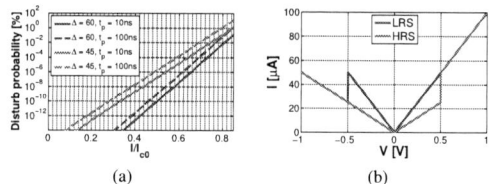

Fig. 2. STT-MRAM: (a) write/read disturbance probability and (b) simplified I-V characteristic

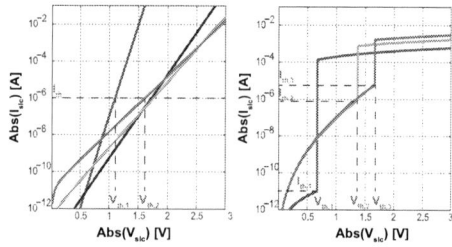

Fig. 3. Selector I-V characteristic: nonlinear selectors(left) and threshold selectors (right)

Based on the aforementioned simplifying assumptions, we can draw some considerations on the boundaries of the biasing voltages as a function of selector device and memory element characteristics. On the one hand, the voltage across selected memory cell must be large enough to turn-on the selector device and switch the memory element. Analytically, this condition can be represented as:

$$V_W \geq V_{th} + V_{sw} \tag{2}$$

where $V_{sw}$ is the voltage required to switch the memory element. On the other hand, the voltages, $\Delta V_{hWL}$, $\Delta V_{hBL}$, and $\Delta V_u$ should be less than $V_{th}$ to avoid unintended over-writing of the stored data. Hence, $V_{uWL}$ must be lower than the selector threshold voltage, $V_{th}$, so as to avoid undesired selection of the half-selected cells along the BL (for example the cell in the upper right corner of Fig. 1). Assuming to operate with only positive voltages, we can write the first boundary condition:

$$V_{uWL} \leq V_{th}. \tag{3}$$

Analogously, in order to prevent undesired selection of half-selected cells along WL (for example the cell at the lower left corner), we must ensure that

$$V_W - V_{uBL} \leq V_{th}. \tag{4}$$

Furthermore, undesired selection of unselected cells (e.g the cell at the upper left corner) is avoided provided that

$$V_{uBL} - V_{uWL} \leq V_{th}. \tag{5}$$

By using (3), (4), and (5) the boundaries of unselected bit-line biasing voltage are calculated as

$$V_W - V_{th} \leq V_{uBL} \leq 2V_{th} \tag{6}$$

which, by combining with (2), gives the design space for write voltage:

$$(V_{sw} + V_{th}) \leq V_W \leq 3V_{th}. \tag{7}$$

However, it is worth mentioning that in practical arrays, the write voltage provided from the driver has to be boosted from $V_W$ to compensate for the IR drop along the selected WL and BL.

Let us further consider cell-to-cell variations on the values of $V_{th}$ and $V_{sw}$. In the worst-case scenario, for the condition in (2) we should take into account the maximum values of $V_{th}$ and $V_{sw}$. Whereas, in (3), (4), and (5) we should consider the minimum $V_{th}$. By denoting the nominal values of $V_{sw}$ and $V_{th}$ as $\overline{V}_{sw}$ and $\overline{V}_{th}$, respectively, we can express the maximum and minimum values in each case. For example, we can express the maximum switching voltage $V_{sw}$ as

$$V_{sw} = (1 + \alpha_{V_{sw}}) \cdot \overline{V}_{sw} \tag{8}$$

and the minimum threshold voltage as

$$V_{th} = (1 - \alpha_{V_{th}}) \cdot \overline{V}_{th}. \tag{9}$$

In the above two equations, $\alpha_{V_{sw}}$ and $\alpha_{V_{th}}$ are equal to $n\sigma_{V_{sw}}/\overline{V}_{sw}$ and $n\sigma_{V_{th}}/\overline{V}_{th}$, respectively, where $\sigma_{V_{th}}$ and $\sigma_{V_{sw}}$ represent the standard deviations of $V_{th}$ and $V_{sw}$, respectively.

Memory read operation requirement also also imposes its own constraints on array implementation. Indeed, by assuming, for example, a current-mode read operation (where the current through the selected cell is compared to a reference current), the cell state can be determined by biasing the selected cell with a read voltage $V_R$ set to

$$V_R \leq V_{th} + V_{safe} \tag{10}$$

where $V_{safe}$ is the voltage across the memory element during reading, which must be sufficiently smaller than $V_{sw}$ in order to prevent undesired switching of the cell state. Without loss of generality it is possible to write

$$V_{safe} = \beta \cdot \overline{V}_{sw} \cdot (1 - \alpha_{V_{sw}}). \tag{11}$$

where $\beta$ is a safety factor less than unity. This means $V_{safe}$ is set considering the minimum switching voltage and some additional margin. Since the selector device and memory element are connected in series, the IR drop across the memory element increases when the IR drop across the selector device decreases. Hence, $V_{safe}$ should be determined by considering the lowest value of selector threshold voltage, i.e., $V_{th} = (1 - \alpha_{V_{th}}) \cdot \overline{V}_{th}$. On the other side, $V_R$ must be set larger than the highest value of the selector threshold voltage, which is equal to $(1 + \alpha_{vth}) \cdot \overline{V}_{th}$. Hence, the design space for the read voltage, $V_R$ can be expressed as:

$$(1 + \alpha_{vth}) \cdot \overline{V}_{th} \leq V_R \leq (1 - \alpha_{V_{th}}) \cdot \overline{V}_{th} + V_{safe} \tag{12}$$

Furthermore, the terms in (7) [along with the aforementioned worst-case scenario considerations for the write operation] and (12) can be rearranged to give a a generic relationship between the acceptable nominal values of the selector threshold voltage and the switching voltage of the memory element; expressed as:

$$\frac{1}{2} \cdot \left( \frac{1 + \alpha_{V_{sw}}}{1 - 2 \cdot \alpha_{V_{th}}} \right) \leq \frac{\overline{V}_{th}}{\overline{V}_{sw}} \leq \frac{1}{2} \cdot \left( \frac{\beta(1 - \alpha_{V_{sw}})}{\alpha_{Vth}} \right) \tag{13}$$

## V. RESULTS AND DISCUSSION

The analytical equations discussed above demonstrate the suitable values of biasing voltages for operating the STT-MRAM crosspoint array during write and read operations by considering the impacts of process spread. The minimum (lower bound) and maximum (upper bound) allowed values for the write voltage, $V_W$, are stated in (7). Fore example the boundaries of $V_W$ are shown in Fig. 4 for a selector device with $V_{th} = 0.5$ V and STT-MRAM device with $V_{sw} = 0.5$ V when each of $\alpha_{V_{sw}}$ and $\alpha_{V_{th}}$ are varied from 0 to 0.4. For a better visibility, contour lines of the difference between the upper bound and the lower bounds (denoted as $\Delta V_W$) is shown on the right side of Fig. 4. The $\Delta V_W = 0$ line marks the intersection of the two surfaces and positive values represent availability of a design space for $V_W$.

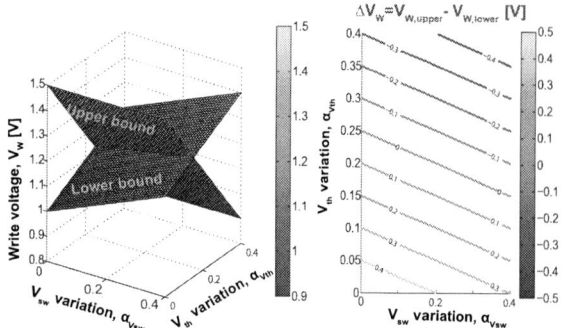

Fig. 4. Upper and lower bounds of write voltage, $V_W$, for $V_{th} = 0.5$ V and $V_{sw} = 0.5$ V.

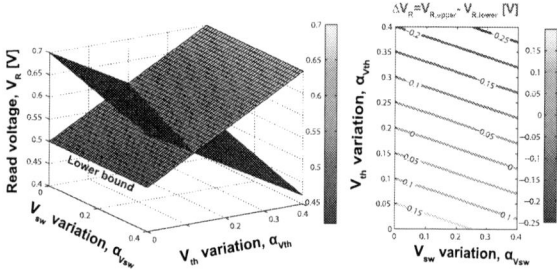

Fig. 5. Upper and lower bounds of read voltage, $V_R$, for $V_{th} = 0.5$ V and $V_{sw} = 0.5$ V and $\beta = 0.4$.

As it is evident from (7), the higher the selector threshold voltage the larger the design space for the write voltage. However, increasing $V_{th}$ is not advantageous for the read operation. As it is evident from (12), assuming similar percentage variation, the higher the $V_{th}$ the narrower the design space for $V_R$. For

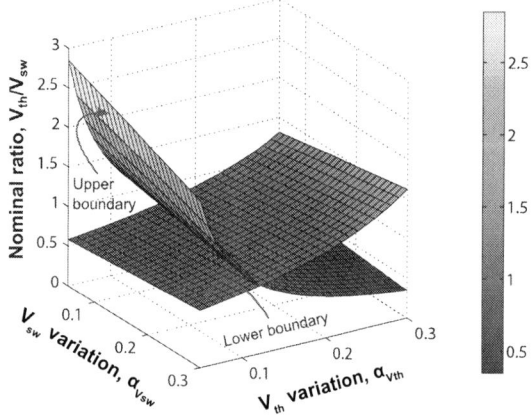

Fig. 6. Lower and upper boundaries of $\overline{V}_{th}/\overline{V}_{sw}$ for different process variations ($\beta = 0.3$).

example, for $V_{th} = 1$ V and $V_{sw} = 0.5$ V , there is no design space for $V_R$ even for small $\alpha_{V_{th}}$ and $\alpha_{V_{sw}}$ values. Whereas, $V_{th} = 0.5$ V gives a sufficient design space for $V_R$ as illustrated in Fig. 5, at the expense of diminishing design space for $V_W$. Hence, the threshold voltage of the selector device should be tuned to a suitable value that compromising write and read performance. In this regard, (13) gives a generic guideline for selecting a suitable selector device for a given STT-MRAM switching voltage, or vice versa. Fig. 6 shows the lower and the upper boundaries of the ratio between $\overline{V}_{th}$ to $\overline{V}_{sw}$, as a function of the amount of variations $\alpha_{V_{sw}}$ and $\alpha_{V_{th}}$.

## VI. CONCLUSION

We have presented a variation-aware comprehensive analysis of the boundary conditions for memory write and read requirements that impose strict constraints on the implementation of crosspoint STT-MRAM Arrays. The results of the analysis could be used as design guide and for choosing a suitable selector device for Crosspoint STT-MRAM arrays.

## REFERENCES

[1] S. Yu and P. Y. Chen, "Emerging Memory Technologies: Recent Trends and Prospects," *IEEE Solid-State Circuits Magazine*, vol. 8, no. 2, pp. 43-56, 2016.

[2] T. Endoh, H. Koike, S. Ikeda, T. Hanyu, and H. Ohno, "An Overview of Nonvolatile Emerging Memories- Spintronics for Working Memories," *IEEE Journal on Emerging and Selected Topics in Circuits and Systems*, vol. 6, no. 2, pp. 109-119, June 2016.

[3] R. Carboni, S. Ambrogio, W. Chen, M. Siddik, J. Harms, A. Lyle, W. Kula, G. Sandhu, D. Ielmini, "Understanding Cycling Endurance in Perpendicular Spin-Transfer Torque (p-STT) Magnetic Memory," *IEEE IEDM*, pp. 21.6.1-21.6.4, Dec., 2016

[4] A. Chen, Z. Krivokapic, and M. R. Lin, "A comprehensive Model for Crossbar Memory Arrays," *70th Annual Device Research Conference*, pp. 219-220, June 2012.

[5] A. Chen, "Analysis of Partial Bias Schemes for the Writing of Crossbar Memory Arrays," *IEEE Transactions on Electron Devices*, vol. 62, no. 9, pp. 2845-2849, Sept. 2015.

[6] R. Aluguri and T. Y. Tseng, "Overview of Selector Devices for 3-D Stackable Cross Point RRAM Arrays," *IEEE Journal of the Electron Devices Society*, vol. 4, no. 5, pp. 294-306, Sept. 2016.

[7] B. Govoreanu, L. Zhang, and M. Jurczak, "Selectors for High Density Crosspoint Memory Arrays: Design Considerations, Device Implementations and Some Challenges Ahead," *International Conference on IC Design and Technology (ICICDT)*, pp. 1-4, Nov. 2015.

[8] L. Zhang, "Study of the Selector Element for Resistive Memory," *Ph.D. Dissertation*, Dept. Elec. Eng., Univ. of KU Leuven, Leuven, Belgium, 2015.

[9] L. Zhang, S. Cosemans, D. J. Wouters, G. Groeseneken, M. Jurczak, and B. Govoreanu, "Selector Design Considerations and Requirements for 1S1R RRAM Crossbar Array," *IEEE 6th International Memory Workshop*, pp. 1-4, 2014.

[10] P. Narayanan, G. W. Burr, K. Virwani, and B. Kurdi, "Circuit-Level Benchmarking of Access Devices for Resistive Nonvolatile Memory Arrays," *IEEE Journal on Emerging and Selected Topics in Circuits and Systems*, vol. 6, no. 3, pp. 330-338, Sept. 2016

[11] Z. Diao, Z. Li, S. Wang, Y. Ding, A. Panchula, E. Chen, L. Wang, Y. Huai, "Spin-Transfer Torque Switching in Magnetic Tunnel Junctions and Spin-Transfer Torque Random Access Memory," *Journal of Physics: Condensed Matter*, vol. 19, no.16, p. 165209(13pp), Apr. 2007.

[12] J. Harms and F. Ebrahimi, "SPICE macromodel of spin-torque transfer-operated magnetic tunnel junction ," *IEEE Trans. on Electron Devices*, vol. 57, no. 6, pp. 1425-1430, Jun. 2010.

[13] J. Woo and S. Yu, "Comparative Study of Cross-Point MRAM Array With Exponential and Threshold Selectors for Read Operation," *IEEE Electron Device Letters*, vol. 39, no. 5, pp. 680-683, May 2018.

# A Simulated Approach to Evaluate Side Channel Attack Countermeasures for the Advanced Encryption Standard

Luca Sarti, Luca Baldanzi, Luca Crocetti, Berardino Carnevale, Luca Fanucci
Department of Information Engineering
University of Pisa, Pisa, Italy

*Abstract*—**Modern networks have critical security needs and a suitable level of protection and performance is usually achieved with the use of dedicated hardware cryptographic cores. Although the Advanced Encryption Standard (AES) is considered the best approach when symmetric cryptography is required, one of its main weaknesses lies in its measurable power consumption. Side Channel Attacks (SCAs) use this emitted power to analyze and revert the mathematical steps and extract the encryption key. In this work we propose a simulated methodology based on Correlation and Differential Power Analysis. Our solution extracts the simulated power from a gate-level implementation of the AES core and elaborates it using mathematical-statistical procedures. An SCA countermeasure can then be evaluated without the need for any physical circuit. Each solution can be benchmarked during an early step of the design thereby shortening the evaluation phase and helping designers to find the best solution during a preliminary phase. The cost of our approach is lower compared to any kind of analysis that requires the silicon chip to evaluate SCA protection.**

## I. INTRODUCTION

The amount of data flowing over communication networks is continuously increasing and any disclosure or modification of such private data can have severe consequences. Furthermore, in the Internet of Things (IoT) world every object will interact with each other, providing malicious attackers a wide surface with multiple entry points. Security countermeasures aim to ensure that the data is hidden to unwanted parties and unauthorized modifications of the content can be immediately identified. A typical security solution for communications is symmetric cryptography [1]. This is based on the encryption of the data bit-stream performed on the transmission side and the decryption on the reception side. The encryption/decryption functions receive as input the data and a specific parameter called the key. In symmetric cryptography the key is the same for encryption and decryption. A malicious entity monitoring the channel would notice only a meaningless flow of information. The most common symmetric encryption algorithm is the Advanced Encryption Standard, also known as Rijndael [2], released by the National Institute of Standards (NIST) in 2001 [3]. The AES can be implemented both in software and hardware depending on the security needs and the performance required. Hardware implementations are often considered to be more efficient, reaching higher throughputs and lower latencies. The main drawback of hardware approaches is that they are prone to power emission Side Channel Attacks (SCA)

[4]. SCAs are based on the idea that a chip implementing the AES emits power that is statistically related to the key used in the encryption process. This means that by repeatedly observing the power traces, the attacker could extract the key and understand the data flow. Extracted power traces are elaborated using different mathematical techniques like Differential Power Analysis (DPA) [5] and Correlation Power Analysis (CPA) [6].

SCA countermeasures aim to increase the time required for an attack to be successful, making it infeasible. However, every countermeasure must be physically implemented in order to be evaluated. Evaluation of countermeasures is required to benchmark the solutions and choose the one that is more appropriate for the final application and its required context. Implementing the required hardware can often be extremely expensive and requires time to produce a suitable circuit prototype [7]. A simulated approach can thus represent a good workaround to evaluate a countermeasure before its implementation and limit the effort in terms of time and cost.

To solve the power emission vulnerability, various SCA countermeasures have been proposed based on modifying the AES circuit in order to change the power emission [8]. Additionally, several authors have proposed creating equivalent circuits for AES implementation to extract simulated power traces [9]. However, very few have investigated the usage of gate-level circuits for AES models in order to guarantee a more accurate VLSI implementation model [10].

We present an approach to characterize an SCA countermeasure using simulated data alone. The methodology simulates all the steps that a potential attacker performs, including power trace extraction and statistical elaboration. A similar work was presented in [11], in which the authors showed a comparison of the robustness against SCA of the same implementation of the AES realized using two different technologies, CMOS and MCML. The authors used a transistor level simulation methodology using SPICE software while managing the coordinated simulation of sub-portions of the circuit, to reduce the high simulation time. In our work we present instead a methodology to compare and benchmark different countermeasures realized with the same technology. The simulation environment provides a quick and cheap solution that can be used to make estimations without the need for any hardware prototype.

The remainder of this paper is organized as follows: Section

2 describes SCAs, Section 3 explains material and methods of our approach, Section 4 analyses and discusses the results, and finally Section 5 gives the conclusions.

## II. SIDE CHANNEL ATTACKS

The AES algorithm is based on the iteration of some mathematical steps called *rounds*. The number of rounds is 10, 12 and 14 for 128-, 192- and 256-bit master keys, where longer keys guarantee more secure results. Each round uses as input the output of the previous round and for each round a *round-key* derived from the master one is used. Figure 1 depicts the AES algorithm, highlighting the difference between the rounds. As shown, there is a preliminary key addition step and the last round is less complex than the others, being the MixColumn stage missing. Hardware solutions can implement

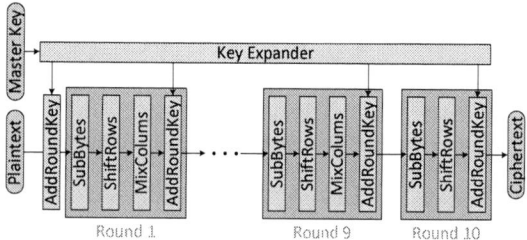

Fig. 1. The AES algorithm in the case of 128-bit keys. For 192- and 256-bit keys the number of rounds increases to 12 and 14, respectively.

one or more rounds in a clock cycle. When a single round per clock cycle is executed, the implementation is called single-round pipelined, otherwise multi-round pipelined. Furthermore, the pipeline stage can be placed between each round (inter-round pipelined solution) or inside the rounds (intra-round pipelined solution) [12]. We focus on single-inter-round pipelined implementations, but the approach can be extended to the other pipelined approaches with minor mathematical modifications.

A SCA is based on the concept that the power emitted into the external environment by an integrated circuit performing AES operations is related to the processed data. The same approach can be applied to the circuit implementing the AES round where the input data is the output of the previous round and the round key is derived from the encrypting/decrypting master key. Therefore, an entity monitoring the output and the emitted power of the AES core for a sufficient number of encryptions can disclose the key. An attacker who wants to carry out an SCA needs to collect several power traces relative to different plaintext encryptions. In particular, in a single-intra-round pipelined AES core, the rounds can be easily identified on the output because each of them corresponds to a peak of the emitted power [13]. Furthermore, the easiest round on which the attacker can focus is the last one, due to its reduced mathematical complexity. Indeed, the last AES round skips the MixColumns stage, and the other stages are byte-oriented: this allows an attacker to focus individually on each

byte of the key, rather than the whole 128-bit key, therefore reducing the computational cost of the attack.

We define the vector of power traces as $T_{extracted} = [t_0, t_1, \cdots, t_{N-1}]^T$, where every $t_X$ element is the mean power consumed during the clock cycle in which the last AES round is performed while encrypting one of the N 128-bit blocks. This set of collected data then needs to be processed with specific statistical approaches to extract the bytes of the key. We focus on the CPA and the DPA approaches, which are both based on finding a statistical relation between the $T_{extracted}$ vector and a set of expected vectors each of which is related to a key guess. The vectors with the best relation represent the *guessed key* that matches the last round key if the attack has been successful.

## III. MATERIAL AND METHODS

The proposed methodology has two main steps: power extraction and statistical analysis. The following sub-sections explain the methodology by analysing in detail the power extraction and the two alternative ways of performing the statistical analysis.

### A. Power extraction

The first step in our simulated SCA is based on the extraction of the power emitted by the circuit. We implemented an AES core in Verilog and then synthesized the circuit on a 65 nm standard-cell CMOS technology. The gate-level netlist obtained as a result of the synthesis represents an approximation of a real physical implementation of the same core. The switching activity of the implementation can be stored in a Value Change Dump (VCD) file during the simulation of the gate-level implementation. Finally, the VCD file together with the standard-cell library leads to the power consumed for each clock cycle (i.e. each AES round). This allowed us to simulate a set of encryptions of N plaintexts and obtain a simulated $T_{extracted}$ vector which we call $T_{simulated}$. Figure 2 shows the flow of the implemented power extraction in which synthesis, simulations and power extractions were performed using Synopsys Design Compiler [14], VCS [15], and PrimeTime [16], respectively. As shown in Figure 2, the extracted power can be elaborated using the DPA or the CPA analysis. The following two subsections highlight the main differences between the two techniques.

### B. CPA statistical analysis

The CPA is based on finding the statistical correlation between the $T_{simulated}$ vector and a set of predicted data for each byte of the key. Therefore, let us define the predicted data as a set of 16 matrices $M_0, ..., M_{15}$ where each matrix is relative to one byte of the key. We write the matrices in the following form:

$$M_B = \begin{bmatrix} m_{B,0,0} & \cdots & m_{B,0,255} \\ \vdots & \ddots & \vdots \\ m_{B,N-1,0} & \cdots & m_{B,N-1,255} \end{bmatrix} = \begin{bmatrix} M_{B,0} \cdots M_{B,255} \end{bmatrix}$$

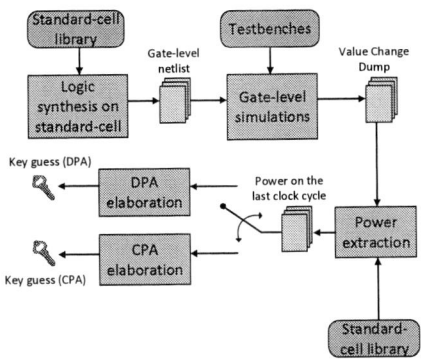

Fig. 2. The complete flow of the proposed approach.

where $B = 0, \cdots, 15$. Each column of the matrix relates to a byte guess starting from $\{00000000\}$ to $\{11111111\}$. On the other hand, each row relates to a plaintext from the 0-th to the (N-1)-th.

The problem now is how to extract the $m_{x,y,z}$ element of the matrix, i.e. the predicted power trace for a specific key guess, plaintext and byte of the key. To estimate this value we use another hypothesis:

**Hypothesis 1.** Each round of a single-inter-round pipelined AES dissipates a power proportional to the Hamming Distance (HD) between the value of the 128-bit string before and after the round has been executed

The HD between two strings $a$ and $b$ is the number of bit differences between them and can be written as follows:

$$HD_{a,b} = HW(a \; XOR \; b)$$

where the weight function $HW()$ returns the number of 1's in the string. Therefore, in our methodology the elements of the matrix can be computed by simply reverting the round of interest, which in our case is the last one. We can write the following equation:

$$m_{x,y,z} = HD_{y,z}(C_x, I_x)$$

where $C_x$ and $I_x$ are the values of the $x$-th byte of the 128-bit block before and after the last AES round for the $z$-th key guess and for the $y$-th plaintext. Indeed, during the last round of the AES each byte of the 128-bit block is processed only by one byte of the key, and there is a unique correspondence between them. Therefore, considering that the $m_{x,y,z}$ values of each matrix relates to a single byte of the key, we have that $0 \leq m_{x,y,z} \leq 8$, as the HD between two bytes can never be larger than 8. Once the set of predicted data has been computed, the methodology simply entails finding the column in each matrix that has the highest correlation factor with the

$T_{simulated}$. The correlation factor used in our approach is the Pearson coefficient defined as follows:

$$p = \frac{\sum\limits_{i=0}^{n-1} (x_i - \bar{x})(y_i - \bar{y})}{\sqrt{\sum\limits_{i=0}^{n-1} (x_i - \bar{x})^2} \sqrt{\sum\limits_{i=0}^{n-1} (y_i - \bar{y})^2}}$$

where $\bar{x} = \frac{1}{n} \sum\limits_{i=0}^{n-1} x_i$ and $\bar{y} = \frac{1}{n} \sum\limits_{i=0}^{n-1} y_i$. In our case $x$ is the $T_{simulated}$ vector and $y$ is each column of the selected $M$ matrix. We call Best Guess (BG), the 16 bytes that relate to the 16 columns (one for each matrix) with the highest Pearson coefficient:

$$BG = BG_0 \mid BG_1 \mid \ldots \mid BG_{15} \mid$$

$$BG_B = t_x \in \left[ M_{B,0} \cdots M_{B,255} \right] \qquad B = 0 \cdots 15$$

The BG represents the result of our methodology and therefore the result of our simulated attack.

*C. DPA statistical analysis*

The proposed DPA methodology differs from the typical DPA in terms of the number of samples required. The classical approach performs a bit-by-bit attack that needs a power sampling frequency higher than the clock frequency. The sampling of the power emission requires a smaller resolution and no more than one power sample per clock cycle. This makes the technique suitable for simulated power based on switching activities and standard-cell libraries in which the power resolution is the same as the clock frequency.

The proposed DPA methodology works similarly to the CPA. The $T_{simulated}$ and the set of 16 $M$ matrices are computed exactly in the same way, but the BG is estimated differently. This procedure takes the $T_{simulated}$ vector and, for each column of the first matrix, divides the $T_{simulated}$ values in two groups: $T_l$ and $T_s$. The selection factor to divide the elements between the $T_l$ and $T_s$ is the following:

$$\begin{cases} T_{li} = t_x : m_{1,x,i} > 4 \\ T_{si} = t_x : m_{1,x,i} < 4 \end{cases} \qquad \text{where } i = 0, \cdots 255$$

This means that the selection function for $T_{simulated}$ is whether or not the corresponding value in the $M_k$ is greater or smaller than 4. The sub-division is performed for each column of the matrix (i.e. for each guess on the byte of the key).

Finally, the differences between the averages of TL-TS couples is computed for each byte guess:

$$D_i = |\overline{TL}_i - \overline{TS}_i| \qquad \text{with } i = 0, \cdots, 255$$

and the best one is selected as the "highest difference of averages". The procedure is then repeated for the remaining $M$ matrices, thus giving the BG for the DPA approach.

## IV. RESULTS AND DISCUSSION

The proposed solution was characterized to quantify the number of AES executions such that BG is equal to the encryption key. Figure 3 shows an execution of the technique using CPA analysis with a fixed key and a set of 8 thousand plaintexts. In the particular case showed, 15 bytes of the BG correctly correspond to the encryption key bytes. The average number of samples to extract the complete key is around 15 thousand for DPA and 10 thousand for CPA: as expected from literature, CPA requires less samples than the DPA. The number of samples required to extract the key is not

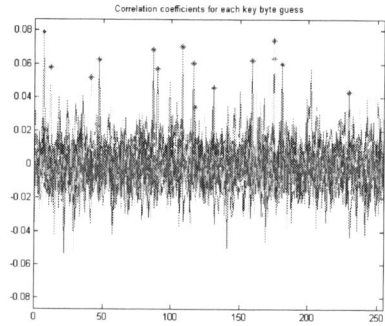

Fig. 3. Example of CPA results for 8000 samples in which each colour represents a different byte.

a meaningful value per se, without knowing the relationship between real and simulated samples. Let's call $N_s$ and $N_r$ the number of samples required to extract the key from the original AES circuit in a simulated SCA and in a physical SCA, respectively.

We can hypothesize the following simplified relationship:

$$N_s = \beta N_r$$

This relationship has to be specifically characterized in the real cases, as parasitic capacitance and routing resistance heavily affect power emission. The extraction and characterization of the $\beta$ will be addressed in our future works.

Assuming the above expression, let's suppose that we have a set of $M$ SCA countermeasures that increase the complexity of an attack. We expect that the number of power traces that the attacker would thus have to extract in order to carry out an SCA is increased. Therefore, for each countermeasure we have $N_{s_i} = \beta N_{r_i}$. The ratio between the number of samples required to perform a real physical SCA with and without the countermeasure is the same as the ratio between the samples required to perform a simulated attack. It is thus possible to define a benchmark factor $E$ that measures the effectiveness of each solution:

$$E_i = \frac{N_{r_i}}{N_r} = \frac{\beta N_{r_i}}{\beta N_r} = \frac{N_{s_i}}{N_s}$$

The higher the $E$ value for a specific countermeasure, the higher the level of security it guarantees against SCAs. The

designer can choose the best solution to be integrated into the system by comparing the $E$ parameter of different approaches and evaluating the specific trade-off of the host system.

## V. CONCLUSIONS AND FUTURE WORKS

We have described a simulated SCA that does not require any hardware prototype. The SCA results would obviously be different in a real-world scenario, but our techniques could be easily used to characterize countermeasures before prototyping. The results are therefore valid as a benchmark for comparing SCA countermeasures and measuring their effectiveness. Our approach gives a quick result thus reducing the time required to choose the right solution for a given context and shortening the characterization phase. Future works will include a characterization of the relationship between simulated and physical results and the benchmarking of several countermeasures implementations using this methodology.

## REFERENCES

[1] S. Chandra, S. Bhattacharyya, S. Paira, and S. S. Alam, "A study and analysis on symmetric cryptography," in *Science Engineering and Management Research (ICSEMR), 2014 International Conference on*, Nov. 2014, pp. 1–8.

[2] J. Daemen and V. Rijmen, *The design of Rijndael: AES-the advanced encryption standard.* Springer Science & Business Media, 2013.

[3] N. F. Pub, "197: Advanced encryption standard (aes)," *Federal Information Processing Standards Publication*, vol. 197, pp. 441–0311, 2001.

[4] M. Z. Rahaman and M. A. Hossain, "Side channel attack prevention for AES smart card," in *Computer and Information Technology, 2008. ICCIT 2008. 11th International Conference on*, Dec. 2008, pp. 376–380.

[5] Z. Martinasek, V. Clupek, and T. Krisztina, "General scheme of differential power analysis," in *Telecommunications and Signal Processing (TSP), 2013 36th International Conference on*, Jul. 2013, pp. 358–362.

[6] H. Pahlevanzadeh, J. Dofe, and Q. Yu, "Assessing cpa resistance of AES with different fault tolerance mechanisms," in *2016 21st Asia and South Pacific Design Automation Conference (ASP-DAC)*, Jan. 2016, pp. 661–666.

[7] W. F. Lee, "Asic design flow," *Verilog Coding for Logic Synthesis*.

[8] A. G. Bayrak, F. Regazzoni, D. Novo, P. Brisk, F. X. Standaert, and P. Ienne, "Automatic application of power analysis countermeasures," *IEEE Transactions on Computers*, vol. 64, no. 2, pp. 329–341, Feb. 2015.

[9] K. Iokibe, K. Maeshima, T. Watanabe, and Y. Toyota, "Security simulation against side-channel attacks on advanced encryption standard circuits based on equivalent circuit model," in *2015 IEEE International Symposium on Electromagnetic Compatibility (EMC)*, Aug. 2015, pp. 224–229.

[10] M. Nagata, D. Fujimoto, and D. Tanaka, "Power current modeling of cryptographic VLSI circuits for analysis of side channel attacks," in *Electromagnetic Compatibility (APEMC), 2013 Asia-Pacific Symposium on*, May 2013, pp. 1–4.

[11] F. Regazzoni, T. Eisenbarth, A. Poschmann, J. Großschädl, F. Gurkaynak, M. Macchetti, Z. Toprak, L. Pozzi, C. Paar, Y. Leblebici *et al.*, "Evaluating resistance of mcml technology to power analysis attacks using a simulation-based methodology," in *Transactions on Computational Science IV*. Springer, 2009, pp. 230–243.

[12] D. Kotturi, S.-M. Yoo, and J. Blizzard, "Aes crypto chip utilizing high-speed parallel pipelined architecture," in *IEEE International Symposium on Circuits and Systems 2005*. IEEE, 2005, pp. 4653–4656.

[13] E. Oswald, S. Mangard, N. Pramstaller, and V. Rijmen, "A side-channel analysis resistant description of the aes s-box," in *International Workshop on Fast Software Encryption*. Springer, 2005, pp. 413–423.

[14] P. Kurup and T. Abbasi, *Logic synthesis using Synopsys®*. Springer Science & Business Media, 2012.

[15] V. Synopsys, "Verilog simulator," *Avaliable HTTP: http://www. synopsys. com/products/simulation/simulation. html*, 2004.

[16] G. Yip, "Expanding the synopsys primetime solution with power analysis," *Synopsys, Inc., http://www. synopsys. com*, 2006.

# Enabling Secure Boot Functionality by Using Physical Unclonable Functions

Kai-Uwe Müller*, Robin Ulrich*, Alexander Stanitzki*, and Rainer Kokozinski*[†]
*Fraunhofer IMS, 47057 Duisburg, Germany
[†]Universität Duisburg-Essen, Germany
Email: kai-uwe.mueller@ims.fraunhofer.de

*Abstract*—A firmware encryption for embedded devices can prevent the firmware from being read out to clone the device to a counterfeited one or to steal the intellectual property of the software developer. Also the integrity is ensured to hinder an attacker from manipulating the firmware to a malicious one. In this work, a cryptographic concept to implement a Secure Boot functionality using the intrinsic properties of a specific hardware device is shown. After describing the Physical Unclonable Function and the cipher used for the implementation, the key generation algorithm is explained. Further, the function of the crypto-module inside the system architecture and the secure boot sequence are described.

*Index Terms*—Physical Unclonable Functions, PUF, Secure Boot, Firmware Encryption

## I. INTRODUCTION

The development of firmware is a major matter of expense in the design of embedded systems. Also a manipulation of firmware data may cause severe damage to industrial production lines or other decentralized systems. To protect the firmware from being copied and used on another hardware device or being manipulated, two general approaches can be chosen. The first is to make the firmware unreadable for an attacker, which could be achieved by encrypting the firmware with a cryptographic cipher using a hard-coded key. This would make it impossible for an attacker to read out the firmware and steal the intellectual property from the device owner if the device is powered off and the secret key is not accessible. A second approach is to associate the firmware to a specific hardware device and make it non-executable on any other one.

In this work, the two approaches are combined by using a Physical Unclonable Function (PUF) to generate a hardware-specific identifier which functions as a key for a lightweight hardware-implemented block cipher to decrypt and encrypt the firmware for providing a Secure Boot functionality for an embedded system.

## II. RELATED WORK

A common way of implementing hardware-based security in PCs and embedded systems is the trusted platform module (TPM) [1]. The standard (ISO/IEC 11889) describes a crypto-processor whose main purpose is to ensure the integrity of a platform (i.e. an embedded system). The security is based on RSA, SHA or AES using secret keys, which are saved onto the device by the manufacturer (Endorsement Key, EK) or

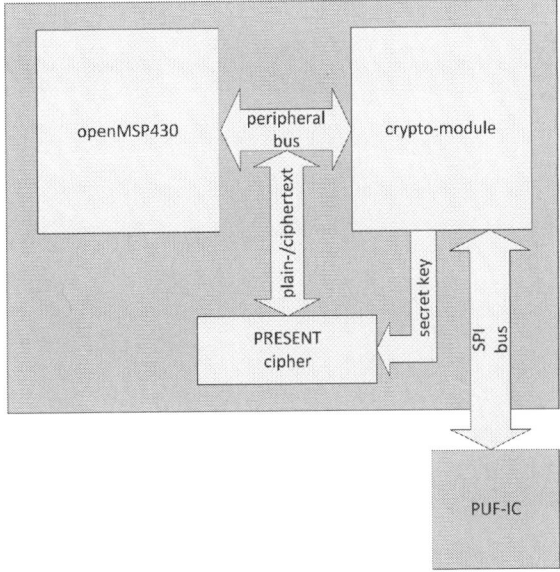

Fig. 1. System architecture

generated internally on the device (Storage Root Key, SRK). The secret keys are not externally accessible and never have to leave the device. Unfortunately, the keys may be accessible via hardware attacks, as shown in [2] for Flash EEPROM memories. Additionally, the Endorsement Key may be saved by the manufacturer or given to federal organizations.

## III. EMBEDDED SYSTEM ARCHITECTURE

The example embedded system consists of a microcontroller device which could be connected to several sensor devices in an industrial environment. An open-source derivative of the Texas Instruments MSP430 family was used as a 16bit microcontroller, namely the openMSP430 [3]. Advantages of this design are the compatibility to the standard MSP430 toolchain and its low area usage (only 8,000 gates are needed for the implementation). It is also well suited for embedded systems because of its ultralow-power architecture to maximize the operating hours of battery powered devices. The microcontroller and the peripheral parts for key generation and

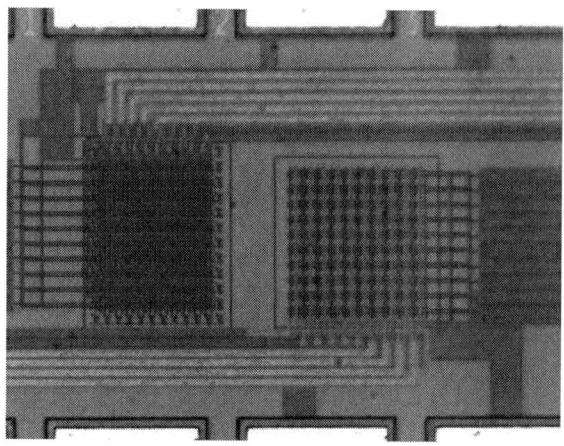

Fig. 2. Chip photograph of the PUF array

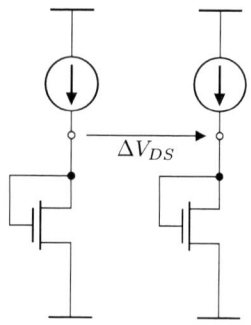

Fig. 3. PUF-pair

encryption, which are described later, are implemented on a Virtex 5 field programmable gate array (FPGA). An external IC providing the PUF functionality is connected over an SPI bus (see Fig. 1).

## IV. Physical Unclonable Function

A hardware-specific identifier is created by using a Physical Unclonable Function (PUF). A PUF describes a physical one-way function that extracts intrinsic properties of an integrated circuit to output a specific sequence of bits which should be random and unpredictable, but at the same time reproducible for the exact same hardware device. For PUF, these properties can be divided into the metrics Randomness, Uniqueness, Bias and Stability [4]. For this specific implementation, the integrated differential readout-circuit for an array-based analog PUF which has been described in [5] is used.

The PUF primitives are build as diode-connected nMOS transistors with a gate-area clearly under the specified minimum dimensions of the used 350 nm CMOS process, namely with a size of W=320 nm and L=200 nm, which leads to an increased mismatch of the devices. The primitives are

arranged as a 16 by 16 array resulting in 256 primitives in total. A chip photograph of different array versions is shown in Fig. 2. The transistors are driven by a constant current source as shown in Fig. 3. The current-voltage characteristic of these devices varies randomly due to variations of several process parameters, such as the threshold voltage $V_{th}$ or the transconductance, and matching errors between the individual transistors. These variations effect the drain-source voltage of the transistor. A pair building is used to form a voltage $V_{DS}$ to only extract the matching errors instead of measuring also the overall process or environmental variations.

## V. Used Cipher and Key Generation Algorithm

A lightweight cipher is needed to realize the decryption of the firmware on the embedded system. The PRESENT cipher [6] was chosen since it is a standardized lightweight block cipher in the ISO/IEC 29192-2 standard and has shown to be well implementable in hardware. The cipher needs a 128 bit key which can be exactly provided by building 128 element pairs out of the 256 elements which comprise the array of the integrated PUF circuit. The pair building is based on an approach called Sequential Pairing Algorithm (SPA), which is described briefly for the use in a Ring-Oscillator PUF implementation in [7]. In this case, the algorithm (see Algorithm 1) works as follows: at first the output voltages $V_{DS}$ of all $n$ elements in the array are measured. Then the elements are sorted in an descending order of the output voltages and indexed from 1 to $n$. From this sorted list, each element from $i = 1$ to $i = n/2$ is combined with element $j = i + n/2$ to build $n/2$ pairs if every voltage difference is greater than a specified threshold $V_{DS_{th}}$. The outputs of the devices in a pair combination now have a sufficient distance to generate a reliable bit out of the decision which one has the higher output value, i.e. whether the generated voltage $\Delta V_{DS}$ is positive or negative.

---

**Algorithm 1** Sequential Pairing

---
1: **procedure SortItems( )**
2:     Sort voltages in descending order:
3:     $V_{DS_1} > V_{DS_2} > ... > V_{DS_n}$
4: **procedure BuildPairs**
5:     $i \leftarrow 1$
6:     $j \leftarrow \frac{n}{2} + 1$
7:     **while** $i < \frac{n}{2} + 1$ **do**
8:         **if** $V_{DS_i} - V_{DS_j} > V_{DS_{th}}$ **then**
9:             Build Pair $\{Element_i, Element_j\}$
10:         $i \leftarrow i + 1$
11:         $j \leftarrow j + 1$

---

The pair combinations can be saved to the internal memory without leaking any information about the key or the general intrinsic properties. As an additional safety feature, the pairing list is protected by a Message Authentication Code (MAC) which also uses the generated key as a secret to hinder an adversary from manipulating the code in order to extract the secret key.

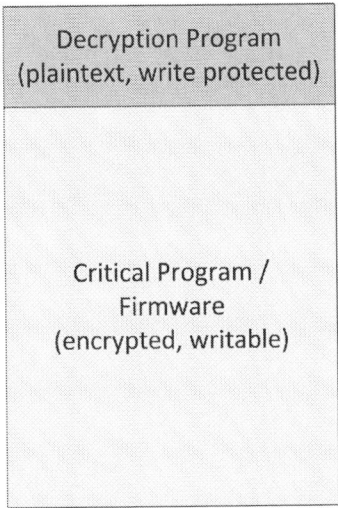

Fig. 4. Program memory organization

## VI. Implementation of a Crypto-Module

The key generation algorithm is implemented in a separate module which also comprises an SPI Master Interface to communicate with the integrated PUF circuit. The PRESENT cipher described in section V is also implemented as a hardware module, which gets the secret key directly from the crypto-module and the encrypted data over the peripheral bus. By using a separate module to perform the key generation and decryption, the secret key is never exposed to the microcontroller. The PRESENT cipher can be seen as a black box on the peripheral bus of the microcontroller which has the ciphertext as input and outputs the plaintext.

## VII. Secure Boot Sequence

The program memory is divided into two parts as shown in Fig. 4. The upper part is write-protected and contains an executable program which can perform the decryption sequence. The other part is not write-protected and contains the encrypted program and the above mentioned MAC-protected transistor pairing list. The block random-access memory (BRAM) of the FPGA is initialized with the configuration data on every startup, including the microcontroller design with an initialized program memory.

The Secure Boot Sequence is shown in Fig. 5. After powering up the system, the Decryption Program is being executed. The first step is to read the $MAC_{enc}$, which was generated in the encryption phase to protect the pairing list from manipulation, and the pairing list out of the header of the Critical Program. Using the information of the pairing list, the Secret Key is generated by checking if the output voltage $\Delta V_{DS}$ (see Fig. 3) for every pair $P_i$ from $i = 1$ to $i = 128$ is positive ($bit_i = 1$) or negative ($bit_i = 0$) and concatenating the outputs to a 128bit key. The check can be

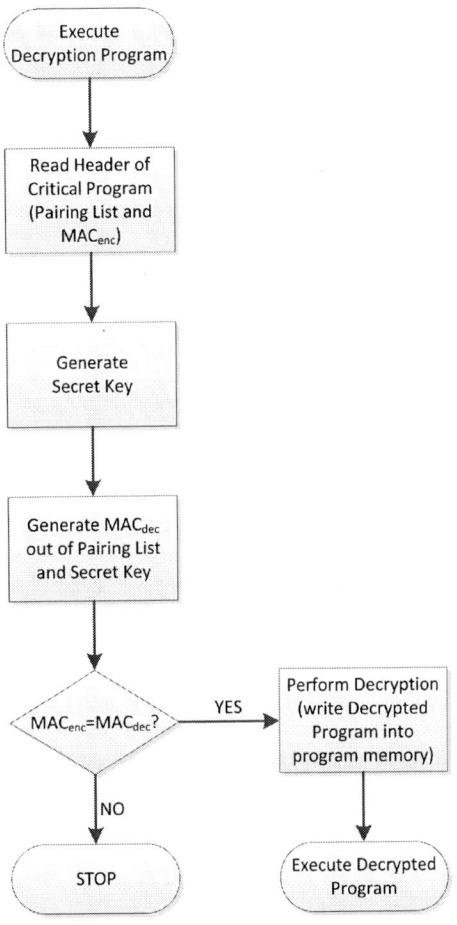

Fig. 5. Secure Boot Sequence

done by using a simple comparator block or in this case by using the most significant bit of the analog-to-digital converter which is integrated on the PUF-IC and accessed over the SPI interface. The key is used to perform the generation of $MAC_{dec}$ over the pairing list. $MAC_{enc}$ and $MAC_{dec}$ are compared to detect if the pairing list has been changed by an attacker. After the verification of the pairing list, the Critical Program is decrypted using the hardware implemented block cipher and can be executed.

This sequence serves two major functions. The first is to hinder the system from executing malicious program code which changes the functionality of the device. If an attacker wants to implement his own executable program into the writable memory, it has to be encrypted with a key corresponding to a specific pairing list. This combination of key and pairing list can only be generated if the attacker knows the intrinsic properties of the device, namely the PUF responses. A possible attack to get knowledge of these properties would be to change

978-1-5386-5388-3/18 $31.00 © 2018 IEEE

the pairing list to a version which implies the actual secret key. An example would be to use the same transistor pair for every key bit. This attack is prevented by adding a validation sequence which checks if every transistor pair is only used once in the key building sequence.

The second major functionality is to prevent the software from being executed on another hardware device since the secret key derived from the PUF is device-specific. This serves as an anti-counterfeiting mechanism.

## VIII. CONCLUSION AND FURTHER WORK

A cryptographic concept for using Physical Unclonable Functions as a hardware security anchor for a Secure Boot functionality has been developed and implemented into an FPGA. It was shown that the concept is suitable to ensure the integrity of the system and to bind a given software to a certain hardware device. A key generation algorithm to ensure a stable PUF-based key without the need of additional helper data or error correction codes was presented.

As further work the design will be implemented on an ASIC together with the PUF readout-circuit to have an embedded system in which it is impossible to get information about the PUF responses, which in this first implementation may be accessible on the SPI bus.

## REFERENCES

[1] TrustedComputingGroup, "Trusted platform module (tpm) summary," April 2008. [Online]. Available: https://trustedcomputinggroup.org/trusted-platform-module-tpm-summary/

[2] F. Courbon, S. Skorobogatov, and C. Woods, "Reverse engineering flash eeprom memories using scanning electron microscopy," in *CARDIS*, 2016.

[3] O. Girard, "openmsp430," June 2009. [Online]. Available: https://opencores.org/project,openmsp430

[4] M. D. Yu, R. Sowell, A. Singh, D. M'Raïhi, and S. Devadas, "Performance metrics and empirical results of a puf cryptographic key generation asic," in *2012 IEEE International Symposium on Hardware-Oriented Security and Trust*, June 2012, pp. 108–115.

[5] B. Willsch, K. U. Müller, Q. Zhang, J. Hauser, S. Dreiner, A. Stanitzki, H. Kappert, R. Kokozinski, and H. Vogt, "Implementation of an integrated differential readout circuit for transistor-based physically unclonable functions," in *2017 Austrochip Workshop on Microelectronics (Austrochip)*, Oct 2017, pp. 58–63.

[6] A. Bogdanov, L. R. Knudsen, G. Leander, C. Paar, A. Poschmann, M. J. B. Robshaw, Y. Seurin, and C. Vikkelsoe, "Present: An ultra-lightweight block cipher," in *Cryptographic Hardware and Embedded Systems - CHES 2007*, P. Paillier and I. Verbauwhede, Eds. Berlin, Heidelberg: Springer Berlin Heidelberg, 2007, pp. 450–466.

[7] C. E. D. Yin and G. Qu, "Lisa: Maximizing ro puf's secret extraction," in *2010 IEEE International Symposium on Hardware-Oriented Security and Trust (HOST)*, June 2010, pp. 100–105.

# Encryption of test data: which cipher is better?

Mathieu Da Silva, Emanuele Valea, Marie-Lise Flottes, Sophie Dupuis, Giorgio Di Natale, Bruno Rouzeyre

LIRMM (Université Montpellier - CNRS), Montpellier, France

{mathieu.da-silva,emanuele.valea,flottes,dupuis,dinatale,rouzeyre}@lirmm.fr

*Abstract*—Testing is a mandatory step in the Integrated Circuit (IC) production because it ensures the required quality of the devices. The most common solution for easing IC testing is the scan chain insertion. This way, a tester can control and observe the internal states of the circuit through dedicated pins. However, a malicious user can exploit this infrastructure in order to extract secret information stored inside the chip. This is the case for cryptographic circuits where partially encrypted results can be observed by shifting out the scan chain content and exploited to retrieve secret keys. Existing countermeasures consist in encrypting the scan content, ensuring the confidentiality of the exchanged messages between the circuit and the tester. The encryption techniques that have been proposed so far rely on the use of two different ciphers: stream ciphers and block ciphers. In this paper, we present pros and cons of both solutions in terms of security and performance. The purpose is to provide an overview of the state-of-the-art in test data encryption and to give elements of comparison between the two ciphers.

*Keywords—Test and Security; Test data encryption; Block cipher; Stream cipher*

## I. Introduction

Steady advances in the semiconductor technology have resulted in devices with hundreds of millions of transistors. The consequence is an increasing probability of physical defects in manufactured Integrated Circuits (ICs), each possibly leading to the failure of the system. Typical defects are shorts or opens involuntarily created during IC manufacturing. All ICs are thus tested after production in order to sort out faulty devices. The circuits that pass the manufacturing test are then packaged. A second test is performed on the packaged devices to eliminate those that may have been damaged during the packaging process or assembled into faulty packages. Finally, other tests are performed after the assembly of ICs on boards. They are used to ensure the final quality of the IC before going to market. Testing is therefore an important step along the IC production, representing half of the cost of the final product. Test costs include development costs (test sequence computation), implementation costs (design practices for high testability) and application costs (time needed to test every single IC), the latter representing a recurrent cost since every single IC must be tested before shipping.

Design-for-Testability (DfT) is a domain of paramount importance. Its goal is to maximize the capability in detecting faults at test time, possibly to perform diagnosis, while minimizing the test time and the required number of additional pins. The most popular DfT technique for dealing with sequential circuits is the scan chain insertion. It consists in replacing the Flip-Flops (FFs) of the IC by Scan Flip-Flops (SFFs). These SFFs are serially connected to form one or several shift-registers, the so-called scan chain(s). In mission mode, the SFFs behave as regular FFs while in test mode they can be serially written or read through scan-in and scan-out pins. Doing so, a tester can control the internal states of the IC by shifting-in test vectors into the scan chain(s), and it can observe internal states stored into the scan chain(s) by shifting-out test responses. An Automatic Test Pattern Generator (ATPG) is used to produce test vectors depending on the target fault models, expected fault-free responses are computed as well. The test procedure consists in serially shifting the test vectors inside the scan chain, and collecting the corresponding responses. The tester compares then the actual test responses with the expected ones in order to identify the presence of faults within the circuit.

Note that the scan-based test procedure introduces numerous cycles for each test data since test patterns and test responses must be serially propagated in the scan chain(s). Fortunately, while a test response is shifted out for observation, a new test vector can be concurrently scanned-in the same scan chain. Test time is thus financially affordable thanks to the simultaneous scan-in and scan out operations.

Unfortunately, scan chains jeopardize the security of the data processed by the IC. The observability feature provided by the test infrastructures can indeed be a source of information leakage, useful to retrieve secret keys of devices implementing cryptographic primitives, such as the Advanced Encryption Standard (AES) [1]. The target of the attack in that case is the AES round register that store partial encrypted results. The scan attack consists in scanning out the content of this round register after the execution of the first AES round. It has been shown in [2] that the scan attack to retrieve the whole 128-bit secret key of an AES crypto-processor can be completed applying on average 512 plaintexts.

Attacks performed on the scan chains are called "scan-based attacks" [2]–[5]. These hardware attacks do not require any invasive handling nor sophisticated equipment. On the other hand, a countermeasure consisting in disconnecting test IOs after manufacturing is not entirely satisfactory as it restrains debug and diagnostic during the IC life cycle.

Beside the security threat involving the scan chains, standard test interfaces can also be maliciously exploited. The IEEE 1149.1 [6] standard, named JTAG, was originally used for testing printed circuit boards. With the increasing number of cores implemented inside System-on-Chips (SoCs), the IEEE 1500 [7] standard was proposed to facilitate their testing. Nowadays, complex ICs integrate a great variety of instruments to ease test and diagnosis. Interfacing this large number of embedded instruments with the user is a challenge that has been addressed by introducing Reconfigurable Scan

978-1-5386-5388-3/18 $31.00 © 2018 IEEE

Networks (RSNs). The RSNs have been standardized in the IEEE Std. 1687 [8], named also IJTAG. They provide a flexible and scalable access to the instruments. These test infrastructures usually allow the access to the scan chains after IC packaging, since the dedicated pads on the die are only accessible during manufacturing test. Moreover, standardized infrastructures are not strictly limited to test purposes, they also allow the access to the circuit for debugging. An attacker can use these structures to steal the contents of on-chip memories or to modify the firmware. The test infrastructure usually requires also the connection of the devices to a network, organized in a daisy chain structure. This represents another threat inside the chip. Indeed, if a malicious device is connected to the test daisy chain, this device can read and/or modify the test data shifted through it, in order to steal confidential data, or to force the device into an illegal behavior. In order to prevent misuse of these test infrastructures, several countermeasures have been proposed in the literature, protecting the access to the scan chains as well as the debugging features.

One of these countermeasures consists in the encryption of test data shifted to and from the test interface. An authorized user encrypts the test vectors off-chip using the established secret key. The encrypted test vectors are shifted in the circuit through the test interface. On-chip decryption is performed before the test vectors are applied to the circuit. The resulting test responses are then encrypted before scanning them out of the circuit. The authorized tester collects the encrypted test responses, and decrypts these data in order to compare them with the expected ones. This solution has the advantage to preserve testing and debugging facilities, while preventing malicious users from accessing the test infrastructure. Since the test communication is encrypted, a user with no knowledge of the secret key is not able to set the circuit in an undesired state, nor to read its internal states. Another advantage of the test data encryption is to not affect the fault coverage achieved with classical scan design, since the same test data are applied and collected once the encryption/decryption is processed.

The test communication encryption is performed by a symmetric cipher. A cipher transforms a plain message into a ciphered version using a secret key. In the same manner, the inverse transformation is performed in order to retrieve the plaintext from the ciphertext, by using the same secret key.

Two ciphers can be used in the test infrastructures, the stream cipher and the block cipher. The stream cipher performs a bit-to-bit encryption of a serial bitstream, while the block cipher encrypts an $n$-bit block of plaintext into a ciphertext block of $n$ bits.

In this paper, we give an overview of the existing countermeasures based on test data encryption. We also compare the solutions based on stream ciphers with the ones based on block ciphers.

The remainder of this paper is organized as follows. In Section II we provide a background on block and stream ciphers, as well as the state-of-the-art on the encryption for securing the test infrastructures. In Section III we compare the test data encryption based on stream cipher with the one based on block cipher. Section IV finally draws some conclusions.

## II. BACKGROUND

We give in this Section a brief overview on both stream and block ciphers. We introduce then the existing countermeasures based on these ciphers, in order to compare them in the next Section.

### A. Ciphers

A cipher ensures the confidentiality of a communication, executing an encryption function $E$ on a message $m$, to produce a ciphertext $c$ using a secret key $k$, such that $E(m, k) = c$. Only a receiver knowing the key $k$ can properly apply the inverse function $D$ to retrieve the original message, i.e. $D(c, k) = m$.

#### 1) Stream ciphers

The stream cipher performs a bitwise XOR operation between the plaintext and a pseudo-random bitstream, called keystream, generated from a seed. In some stream ciphers (e.g. TRIVIUM [10] cipher), the seed is composed of the key $k$ and the initial value $IV$. The key needs to be secret, while the $IV$ can be public, but it is supposed to be different for each encryption session. The generated keystream is denoted as $S(k, IV)$. The encryption and decryption functions $(E, D)$ of the stream cipher are thus respectively defined as $E(m, k) = m \oplus S(k, IV) = c$, and $D(c, k) = c \oplus S(k, IV) = m$.

The first requirement that must be fulfilled in order to consider a stream cipher secure is to produce an unpredictable keystream. This way, it is impossible to retrieve the plaintext from the ciphertext without knowing the keystream. The second requirement is to never use the same keystream more than once. In the case where two different plaintexts $m_1$ and $m_2$ are encrypted with the same keystream $S(k, IV)$, an attacker can exploit the XOR of the two respective ciphered messages $c_1$ and $c_2$. Indeed, this operation leads to remove the encryption: $c_1 \oplus c_2 = (S(k, IV) \oplus m_1) \oplus (S(k, IV) \oplus m_2) = m_1 \oplus m_2$. The XOR between two messages can then be exploited in a differential attack, such as it is the case for scan attacks [1-4]. That's why it is important to use a different seed, i.e. a different $IV$ and/or secret key, to initialize the stream cipher between different encryption sessions.

#### 2) Block ciphers

The block cipher executes iterative transformations based on substitutions and permutations on fixed-length groups of bits, called blocks. The transformation function depends on a secret key. The block encryption results in the diffusion and confusion of the plaintext on the ciphertext at each iteration of the execution. The iterations performed by the block cipher are called rounds.

The most used block cipher is defined by a standard, named AES [1]. Nevertheless, AES may induce a large area overhead to the device under test (see Tab. 1). For this reason, lightweight block ciphers implying a lower area cost have been studied, such as PRESENT [9]. This block cipher guarantees a lower security level than AES, but cryptoanalysis studies show that it is enough for most applications. The encryption is performed on block size of 64 bits in 31 rounds with two possible lengths for the secret key, 80 bits or 128 bits.

### B. Test data encryption

Several countermeasures to the scan attacks have been reported in the literature [10]–[15] to ensure the confidentiality of the exchanged test data between the tester and the circuit, while preserving the use of the test interface for authorized users. These solutions are applied to the existing test infrastructure by inserting two ciphers in the circuit. One is placed at the serial input of the test interface in order to decrypt the encrypted test data sent by the user. The other is placed at the serial output in order to encrypt the test responses before being shifted out of the circuit.

The decryption performed at the scan-in of the test interface takes the controllability of the circuit away from an unauthorized user, who is unable to apply chosen data. The encryption performed at the scan-out of the test interface prevents him/her from observing the plain scan content. An attacker is thus not able to perform scan attacks, nor illegally debugging the circuit. The confidentiality established between the protected circuit and the tester ensures also a protection from threats placed inside the chip, such as malicious devices connected to the test daisy chain. The encryption prevents these devices from making sense of the sniffed encrypted data, or from modifying them in a controlled way.

The encryption of test data proposed in the literature is based on stream ciphers as well as on block ciphers. The stream cipher used to encrypt the test communication is the TRIVIUM [10], because of its low silicon footprint. The pseudo-random sequence, used as keystream, is generated with a Non-Linear Feedback Shift Register (NLFSR) from an 80-bit secret key and an 80-bit IV. The TRIVIUM stream cipher encrypts the test interfaces of JTAG in [10], IEEE 1500 in [12] and IJTAG in [13], while the PRESENT block cipher is used to encrypt the scan chain in [14][15]. We will see the pros and cons of each encryption method in the next section.

### III. COMPARISON: STREAM VS BLOCK ENCRYPTION

We evaluate the stream-based countermeasures and the block-based ones according to several cost functions: the area and power overheads, the impact on the testing cost and the provided security level.

### A. Area and power consumption overhead

Tab. 1 shows the area and power consumption for the AES [1], the PRESENT [9] block cipher with 128-bit secret key, and for the TRIVIUM [10] stream cipher.

Stream ciphering is the technique that has been preferred so far in the literature [10]–[13]. The choice of the stream cipher is motivated by the lower impact on area and power costs. Block ciphers imply the usage of a larger area footprint than stream ciphers, as is the case for the AES block cipher. However, some modified versions of the AES have been

| Ciphers | Area (Gate Equivalent) | Power consumption @ 10 MHz (µW) |
|---|---|---|
| *Block ciphers* | | |
| AES-128 | 22 535 | 134.2 |
| PRESENT-128 | 2 139 | 26.26 |
| *Stream cipher* | | |
| TRIVIUM | 2 016 | 36.35 |

Tab. 1 Area and power consumption for block and stream ciphers

designed to be lightweight, such as PRESENT. PRESENT block cipher and TRIVIUM stream cipher have similar costs in terms of area and power consumption, as shown in Tab. 1.

A more realistic estimation of area and power costs takes into account the number of ciphers that have to be implemented. In fact, depending on whether the encryption is performed with block or stream ciphers, a different number of ciphers must be placed inside the circuit. The block-based solution [14][15] requires two ciphers, one dedicated to the decryption of the test patterns, the other dedicated to the encryption of the test responses (both test vector in and test response out are concurrent operations).

Conversely, the stream-based solutions [10]–[13] can use only one stream cipher to generate both the decryption and encryption keystream. Therefore, if a lightweight block cipher is used, which has a cost comparable to the stream cipher, the block-based solutions implies twice more area and power overhead, due to the duplication.

### B. Testing cost

The impact on test coverage is also important to evaluate the countermeasures applied to the test infrastructure. The set of faults that are detected by the test sequences, originally generated by the ATPG, must not decrease because of the insertion of the security countermeasure. The encryption of test data assures this condition. The content of the applied test vectors and the produced responses is not disrupted by the additional encryption/decryption steps. Concerning the test of the ciphers themselves, authors in [14][15] showed that the extra logic introduced for encryption is tested in the same time as the test data are processed. Therefore, there is no impact on the original test coverage even with the implementation of the ciphers.

Nevertheless, the decryption/encryption of the test data shifted through the test interface adds a cost in terms of test time. It is important to increase as less as possible the test time, since this overhead has an impact on the cost of each sample of the circuit. If this is not taken into account, the time to test an entire product chain can increase significantly.

Concerning the stream cipher, an additional initialization time is required. This overhead is 1152 clock cycles for the TRIVIUM, representing a marginal cost compared to the millions of clock cycles needed to test an entire SoC. Moreover, since both the testing interface and the stream cipher have a serial access, no additional timing overhead is required.

Contrariwise, the parallel interface of the block cipher requires padding the test data acquired serially into a multiple of the block size. The padding of test data results in additional clock cycles to complete the shifting operations, implying a test time overhead on each pattern. This results in higher overhead than the stream-based solutions. However, an optimization is proposed in [14][15], based on an alternative DfT approach that makes the scan chain length multiple of the block size. Block ciphers have thus to be adapted to cope with the serial interface of the testing infrastructures.

### C. Security

As shown in Section II.B, the state-of-the-art countermeasures based on the test communication encryption protect against the aforementioned threats. However, the stream-based solutions [10]–[13] show a vulnerability due to the mismanagement of the seed generating the keystream. In this case, the attacker has the possibility to provoke the generation of the same keystream to encrypt different test data. The requirement on the use of the stream cipher, stated in Section II.A.1), is therefore not respected, circumventing the encryption in the case of differential scan attacks [1-4].

This security flaw is not present on the block-based countermeasures, representing therefore a more secure encryption solution than the stream-based ones. Tab. 2 resumes the pros and cons of both solutions.

| Test data encryption | Stream cipher | Block cipher |
|---|---|---|
| Security | - | + |
| Area | + | - |
| Power | + | - |
| Test time | + | - |

Tab. 2 Comparison overview

The security of the stream-based countermeasures can be improved, making sure that the stream cipher does not generate the same keystream for multiple encryptions. To produce a different keystream, the stream cipher has to change its seed between each cipher initialization, i.e. the secret key and/or the initial value $IV$. A possible solution is to generate a random $IV$ at each circuit reset. As a result, the stream cipher is initialized with a different seed between each encryption. The differential scan attacks [2]–[5] are thus no longer feasible. However, the issue with this solution is to share the random $IV$ to the authorized users in order to perform the encryption/decryption of the test data.

## IV. CONCLUSION

Granting access to the internal states of the ICs is fundamental for testing during production, as well as for debugging and diagnosis in the field. Test infrastructures composed of scan networks meet these needs, but compromise the security at the same time. To prevent attacks that exploit the scan side-channel, several countermeasures exist. Some of them are based on the encryption of the test data. Two types of ciphers can be used to encrypt the test communication: the stream cipher and the block cipher. From our study, it comes out that stream ciphers can be preferred due to their smaller overhead and their easy adaptation to the serial test interface. Nevertheless, as implemented in [10]–[13], the stream-based solutions present a vulnerability, due to a misuse of the stream cipher, while the block-based solutions prove to be secure in all cases.

### ACKNOWLEDGMENT

This project has been funded by the French Government (BPI-OSEO) under grant FUI#20 TEEVA (Trusted Execution EVAluation).

### REFERENCES

[1] J. Daemen and V. Rijmen. "The Design of Rijndael". Springer-Verlag New York, Inc., Secaucus, NJ, USA, 2002.

[2] B. Yang, K. Wu and R. Karri. "Secure scan: a design-for-test architecture for crypto chips". In Design Automation Conference (DAC), pp. 135-140, 2005.

[3] J. DaRolt, G. Di Natale, M.-L. Flottes and B. Rouzeyre. "Scan Attacks and Countermeasures in Presence of Scan Response Compactors". In European Test Symposium (ETS), pp. 19-24, 2011.

[4] J. Da Rolt, G. Di Natale, M.-L. Flottes and B. Rouzeyre. "Are advanced DfT structures sufficient for preventing scan-attacks?". In VLSI Test Symposium (VTS), pp. 246-251, 2012.

[5] Sk Subidh Ali, Ozgur Sinanoglu, Samah Mohamed Saeed, and Ramesh Karri. "New scan-based attack using only the test mode". In International Conference on Very Large Scale Integration (VLSI-SoC), pp. 234-239, 2013.

[6] Committee, I. S. (1990). IEEE Standard Test Access Port and Boundary-Scan Architecture. IEEE Std (Vol. 2001).

[7] IEEE Standard Testability Method for Embedded Core-based Integrated Circuits. (2012). IEEE Std 1500-2005.

[8] The IEEE Standards Association. (2014). IEEE Standard for Access and Control of Instrumentation Embedded within a Semiconductor Device.

[9] A. Bogdanov, L.R. Knudsen, G. Leander, C. Paar, A. Poschmann, M.J.B. Robshaw, Y. Seurin, and C. Vikkelsoe, P. Paillier and I. Verbauwhede. PRESENT: An Ultra-Lightweight Block Cipher. CHES 2007, LNCS 4727, pp. 450–466, Springer-Verlag Berlin Heidelberg 2007

[10] C. De Canniere and B. Preneel. "TRIVIUM Specifications". ECRYPT Stream Cipher Project, Report, 30, 2005.

[11] K. Rosenfeld and R. Karri. . "Attacks and defenses for JTAG". In IEEE Design and Test of Computers, 27(1), 36–47, 2010.

[12] K. Rosenfeld and R. Karri "Security-aware SoC test access mechanisms". In IEEE VLSI Test Symposium (VTS), pp. 100–104, 2011.

[13] S. Kan, J. Dworak and J. G. Dunham. "Echeloned IJTAG data protection". In IEEE Asian Hardware Oriented Security and Trust Symposium (AsianHOST), 2016.

[14] M. Da Silva, M.-L. Flottes, G. Di Natale, B. Rouzeyre, P. Prinetto and M. Restifo. "Scan chain encryption for the test, diagnosis and debug of secure circuits". In IEEE European Test Symposium (ETS) pp. 1–6, 2017.

[15] M. Da Silva, M.-L. Flottes, G. Di Natale and B. Rouzeyre. "Experimentations on scan chain encryption with PRESENT". In International Verification and Security Workshop (IVSW), pp. 45–50, 2017

# Increasing EM Robustness of Placement and Routing Solutions based on Layout-Driven Discretization

Steve Bigalke and Jens Lienig
Institute of Electromechanical and Electronic Design
Dresden University of Technology
✉ steve.bigalke@outlook.com, jens@ieee.org

Thorben Casper and Sebastian Schöps
Institut für Theorie Elektromagnetischer Felder
Technische Universität Darmstadt
✉ casper@gsc.tu-darmstadt.de, schoeps@gsc.tu-darmstadt.de

*Abstract*—Nowadays, electromigration (EM) is mainly addressed in the verification step. This is no longer possible due to the ever increasing number of EM failures in the future. An EM-aware physical synthesis could reduce the number of critical locations but the layout complexities prevent this from already being used. To solve this problem, we propose a novel method to discretize placement and routing solutions to enable a fast EM analysis. In addition, we suggest adjustments in the placement and routing step to enhance the EM robustness based on early analysis results. In contrast to the standard approach of running a numerical simulation outside the physical design step and after the synthesis, we perform most of the analysis steps within our placement and routing tools to consider the results; thus enabling early and specialized EM-robust solutions. Particularly, our methodology exploits layout structures to enable an efficient discretization inside the geometrical representations of synthesis tools. We demonstrate how to reduce the discretization effort significantly while achieving sufficient accuracy to improve EM robustness.

*Index Terms*—Reliability, Electromigration, Placement, Routing

## I. INTRODUCTION

The semiconductor industry continuously upgrades the performance of very large scale integration (VLSI) circuits with every new technology node. This leads to an ongoing downscaling of the transistor dimensions and a subsequent shrinking of interconnects.

Electromigration (EM) is a process of material migration caused by the momentum exchange between moving electrons and lattice atoms. The main driving force behind EM is current density, but it also depends on temperature, among other things. Both factors are increasing over time, escalating the EM impact. The International Technology Roadmap for Semiconductors (ITRS) predicts EM damage to worsen significantly in the future, while at the same time claiming that there are no known solutions [1].

To assure the EM robustness of a layout, circuit designers use rule-based verification tools or spatial discretization methods, e.g., the finite element method (FEM). The former are based on simplified models and the latter are typically performed on only parts of the layout due to its complexity. The usual goal of the FEM is to develop a simplified model, which can then be incorporated in a verification tool. Examples of this methodology are presented in [2], [3] and [4].

In this paper, we present a method to close the gap between the aforementioned simplified EM analysis and a complex FEM simulation. We suggest performing most of the analysis steps within existing electronic design automation (EDA) tools and to incorporate the EM results in order to synthesize EM robust solutions (compare Fig. 1). This is becoming increasingly important because of the increasing number of EM problems in the verification step. In the future, fast analyses integrated within the synthesis tools are needed to estimate the layout robustness as early as possible to take appropriate countermeasures.

As illustrated in Fig. 1, we integrate basic analysis steps such as the discretization and the boundary condition (BC) assignment step into the placement and routing tools based on their layout representations. While these steps are carried out using an in-house tool, we exploit an external tool only for solving the already discretized EM equations.

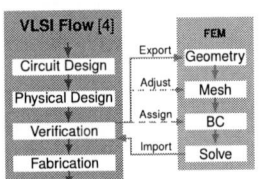 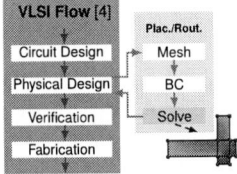

| (a) Standard method | (b) Proposed method |

Fig. 1. Illustrating the difference between the (a) standard method and (b) the proposed method. Our suggestion is performing the discretization and boundary condition assignments (analysis steps) within the placement and routing tools and during synthesis. This generates fast EM results and enables synthesis algorithms to improve their results.

To integrate the analysis steps and to reduce the discretization time (which can take up to 80% of the overall analysis time [6]), we also develop algorithms to discretize placement and routing structures. These algorithms are incorporated in the placement and routing step, respectively.

Summarized, our main contributions are

- a new method for an EM analysis based on algorithms to discretize placement and routing solutions within EDA tools achieving a comparable result quality to an FEM simulation while significantly reducing the problem complexity, and
- procedures on how to consider these EM results within the placement and routing algorithms in order to synthesize EM-robust layouts.

978-1-5386-5388-3/18 $31.00 © 2018 IEEE

## II. ELECTROMIGRATION

### A. Physical Model

The physical EM model contains the electrical, mechanical and thermal domains as depicted in Fig. 2.

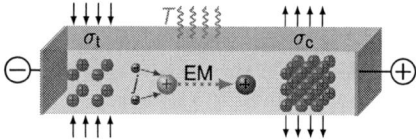

Fig. 2. Because of the collision of the electrons (blue) with the atoms (red), atoms migrate in the electron flow direction. This depletes atoms at the cathode and accumulates them at the anode. The change of concentration introduces tensile ($\sigma_t > 0$) and compressive stresses ($\sigma_c < 0$), respectively. At the same time, the current flow heats the interconnect due to Joule heating.

To describe the effect of EM, the governing equations are Fick's first and second law. In the presence of an external force, such as a current density or a stress gradient, Fick's first law (describing the atomic flux $\vec{J}$ of chemical diffusion (CD)) is extended by the atomic flux of EM and stressmigration (SM) given by

$$\vec{J} = \vec{J}_{CD} + \vec{J}_{EM} + \vec{J}_{SM} = -D\nabla C + \frac{CD}{kT}eZ^*\rho\vec{j} + \frac{CD}{kT}\Omega\nabla\sigma, \quad (1)$$

where the concentration gradient $\nabla C$, the stress gradient $\nabla\sigma$ and the current density $\vec{j}$ are the main driving forces of the atomic flux [7]. It also depends on the diffusion coefficient $D$, Boltzmann's constant $k$, the temperature $T$, the charge of an electron $e$, the effective charge number $Z^*$, the resistivity $\rho$ and the atomic volume $\Omega$. Fick's second law (the mass balance equation) given by

$$\frac{\partial C}{\partial t} = -\nabla \cdot \vec{J} + G, \quad (2)$$

expresses the change of the atom concentration over time depending on the divergence of the atomic flux and a generation/annihilation rate $G$. Korhonen et al. [8] presented a one dimensional mechanical stress equation derived from Eq. (2) by neglecting the chemical diffusion and the generation/annihilation term and using Hooke's law ($\partial\sigma = -B\partial C/C$) with the bulk module $B$ to express the stress caused by EM as

$$\frac{\partial\sigma}{\partial t} = \frac{\partial}{\partial x}\left[\frac{DB}{kT}\left(eZ^*\rho j + \Omega\frac{\partial\sigma}{\partial x}\right)\right]. \quad (3)$$

To calculate the EM stress in Eq. (3), one must also solve the electrokinetic and heat equation for static problems to calculate the current density and temperature, respectively.

### B. Numerical Techniques

To obtain a solution for the problems presented in Sec. II-A, analytical means are not suitable due to the complexity of the physical equations and the geometries that are investigated. Therefore, numerical techniques such as FEM, finite difference (FD) or finite volume (FV) schemes are required. Since we separated the discretization from the solving step, we can utilize any solver. We calculate the heat map by an in-house tool using the finite integration technique (FIT) [9], which is similar to FD and FV schemes. The commercial FEM solver from *ANSYS* solves Eq. (3).

## III. DISCRETIZATION

This section presents the algorithms to discretize placement and routing solutions to model cell placements and interconnect structures. To estimate EM effects, the placement discretization enables a fast temperature simulation based on the placement. The interconnect discretization enables a fast EM estimation on routing results due to a small number of elements.

### A. Discretization of Placement Structures

We developed the algorithm to enable a fast discretization of the current placements in order to calculate the heat distribution within a chip. We suggest using the regular placement grid as the discretization template with optional subdivision. Therefore, we discretize only once before the first placement solution is calculated and then, we reuse this discretization after a new placement has been calculated. Furthermore, we use an *rtree* from the *boost geometry* library to map our elements to the newly calculated cell positions. This means that we can iterate through our elements and query the *rtree* to determine simulation parameters like power dissipation or material properties at the corresponding locations. Fig. 3a-c show this procedure. By adding the third dimension to Fig. 3a, one can model the entire chip as shown in Fig. 3d to consider Joule heating within interconnects if a routing solution is provided. Since the placement discretization might be too coarse for interconnects, one has to scale the current densities within the elements covering a part of an interconnect to keep the power generation constant.

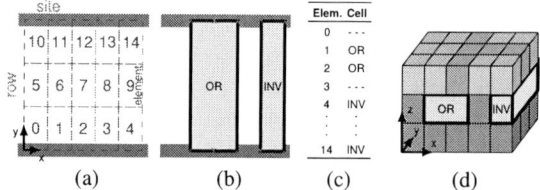

Fig. 3. (a) Generated elements (blue) align with the placement sites (red). (b) Geometric placement representation for fast querying. (c) Mapping of element properties to cell locations. (d) Procedure can be repeated for all layers.

### B. Discretization of Routing Structures

To perform a fast EM analysis on interconnects while routing, we propose an algorithm to discretize routing structures, which aligns elements according to the geometry of the interconnects. Fig. 4 shows an exemplary interconnect structure to demonstrate the algorithm.

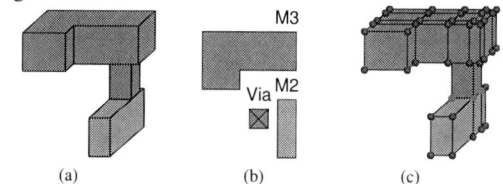

Fig. 4. (a) Exemplary interconnect. (b) Decomposition of (a) into layers M2, Via and M3. (c) Generated elements on each layer. The colors represent continuous material and the red circles electrically coupled nodes.

The algorithm to discretize interconnects operates in the following three steps (Fig. 5): (1) Remove vias from layer, (2) generate maximum rectangles and (3) recursively decompose the maximum rectangles into non-overlapping rectangles. In the first step, the vias are subtracted from the connecting routing layers and stored as one element. The second step creates maximum rectangles within the resulting structure on each layer. A maximum rectangle is a rectangle which extends in the $x$ and $y$ directions until it touches the structure borders at both sides. The third step decomposes recursively the maximum rectangles into overlapping and non-overlapping rectangles. The former are gathered for a re-run and the latter become elements.

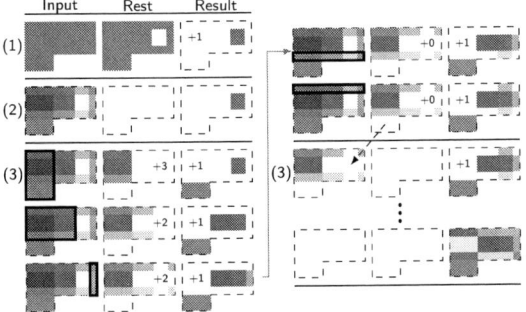

Fig. 5. Discretization of the routing structures. (1) Subtraction of vias from routing layer. (2) Construction of maximum rectangles. (3) Decomposition of maximum rectangles into overlapping and non-overlapping rectangle (become elements). The numbers +1, +2 and +3 represent the number of added rectangles per column and step.

To address the state of the art *dual-damascene-techniques (DDT)*, elements represent continuous materials within each copper layer (brown and orange color in Fig. 4c). Between the different copper layers, we couple the electrical domains of coincident nodes (red nodes in Fig. 4c). This blocks the migration between the layers as given by the DDT.

## IV. SIMULATION RESULTS

### A. Case Study

In this section, we assume that we have identified one net connecting three cells and driving a relatively high current compared to the other nets in the *multiplier plus adder* design from [10]. Therefore, this net is potentially EM critical. We demonstrate how the placement and routing step can reduce the EM risk of this net by running and considering a thermal analysis in the placement and a stress analysis in the routing step. As mentioned in Sec. I, we use our in-house placement and routing tools to perform the discretization, to assign the BCs and to call the external solver (Fig. 1).

### B. Placement with Temperature Consideration

Since most of the placement algorithms iteratively calculate cell locations, we suggest performing temperature simulations in between these iterations and consider high temperature locations as inconvenient locations for EM critical cells. Our proposed strategy is to perform the discretization step before

the first placement iteration to create elements aligned to the placement sites as shown in Fig. 3. In between the different iterations, we pause the computation of the cell locations to run the thermal analysis. We suggest considering the current placement solution by performing geometry queries with all element locations to determine the properties such as material or power generation (given by the corresponding data sheets of the physical design kit). These queries are typically very fast. At the end, we transfer the element property information together with the appropriate boundary conditions to a solver to calculate the temperature map. Here, we use an in-house FIT tool solving the heat equation for static problems. In the next placement iteration, we consider the heat results (Fig. 6) and move critical cells out of hot areas by adding pseudo nets and pins, which drag these cells to less critical areas.

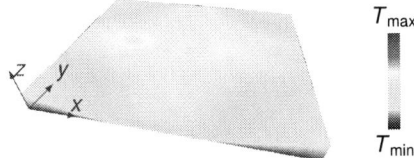

Fig. 6. Three-dimensional heat map of the case study layout. Our algorithm discretizes the placement solution and we solve the static heat equation with appropriate BCs by an in-house FIT solver.

The temperature map in Fig. 6 shows that sites close to the corners of the chip are favorable for EM critical cells (Fig. 7) because of the lower temperatures compared to the center.

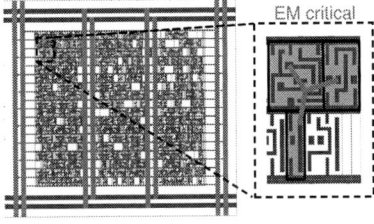

Fig. 7. Case study layout with zoom in to the EM critical cells. Locations close to the corners of the layout are favorable for EM critical cells.

### C. Routing with Stress Consideration

To estimate the EM effects while routing, we run current density and stress analyses. Since EM moves atoms and changes atom concentrations, hydrostatic stress builds up within the interconnect, which we refer to as stress. Our analysis includes the effect of SM, which counteracts EM to reduce the stress gradient.

In this work, the interconnect dimensions are taken from the process design kit of [10]. The current values are given in look up tables (depending on the parasitics) in [11], which are assigned as boundary conditions to the end of each interconnect.

Our idea of an EM-aware routing is the following: First, we calculate a routing solution. Second, we use our algorithm to discretize the interconnects within the router. Third, we transfer the elements and boundary conditions to an external solver that calculates stress results by solving Eq. (3)

in all three dimensions. In the last step, we control the commercial router from *Cadence* through its interface function `setAttribute` with the argument `{bottom,top}_preferred_routing_layer` to avoid the critical routing layer with the maximum stress.

To investigate the result quality of our approach, we simulate the interconnect structure of the critical net from Fig. 7 with the *ANSYS* FEM tool which we refer to as the standard method. The top row of Fig. 8 shows the discretization and stress values of the standard and our method. Both stress results of the initial routing solution identify the maximum stress on metal 2 and therefore, we instruct the router to prefer metal 3. As a result, we generated an improved solution (bottom row of Fig. 8), which experiences around 25% less tensile stress than the standard solution and thus, significantly prolongs the lifetime of the interconnect. Although we choose a much coarser discretization, our method still leads to the avoidance of the same metal layer without time-consuming discretization.

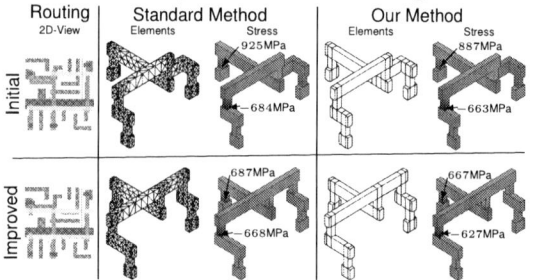

Fig. 8. Comparison between the initial and improved routing solution with stress results calculated by the standard and our method.

Tab. I shows that the principle stress results are qualitatively in good agreement between the standard and our method. However, our approach uses fewer elements on straight lines. Therefore, we trade result quality for a smaller problem size since the current density (main driving force of EM) is almost constant within a straight line. This is a reasonable inaccuracy to enable fast EM estimations on different routing solutions.

TABLE I

COMPLEXITY AND RESULT COMPARISON OF THE IMPROVED ROUTING SOLUTION BETWEEN THE STANDARD AND OUR METHOD.

| Characteristic | Standard Method | Our Method | Relative Difference |
|---|---|---|---|
| Elements | 1596 | 36 | 98% |
| Runtime | 69s | 11s | 84% |
| Max. Stress | 687 MPa | 667 MPa | 3% |
| Min. Stress | -668 MPa | -627 MPa | 6% |

### D. Discretization Performance

To show the performance of our discretization algorithms, we test it on four benchmarks with different problem sizes. They are synthesized with the technology and cell library from [10] and [11], respectively. The problem sizes range from a small analog design to a large digital design. Our benchmark suite contains a *clock divider, counter, multiplier plus adder* from [10] and the *OpenRISC 1200* core from [12]. Tab. II demonstrates that our algorithms are capable to discretize large designs in a reasonable amount of time on a single core of an Intel Xeon E5-2620 at 2.40 GHz.

TABLE II
ELEMENT NUMBER AND DISCRETIZATION TIME

| Design | Nets | Placement-driven | | Interconnect-driven | |
|---|---|---|---|---|---|
| | | Elements | Time | Elements | Time |
| Divider | 5 | 660 | <1s | 44 | <1s |
| Counter | 15 | 137K | <1s | 13K | 1s |
| Mult/Add | 344 | 467K | 3s | 34K | 7s |
| Or1200 | 15K | 30M | 123s | 13M | 1h57m |

## V. SUMMARY AND CONCLUSION

We presented a novel method to improve the EM robustness of layouts. With our approach, temperature and stress results are effectively calculated and widely considered within the placement and routing steps. To achieve this, we proposed algorithms to discretize placement and interconnect solutions, which can be used within a placement or routing tool.

Our approach is the first to show how to perform most of the analysis steps within placement and routing tools. In addition, we describe how EM results can be effectively achieved and considered in these tools. Our methodology has the advantage of being fast, accurate and flexible due to the separation of discretization and solving strategies. Fast and accurate, because our developed discretization algorithms are specialized to layout structures; thus generating relatively few elements. Flexible, because the discretization step is independent of the solving process; thus enabling the use of any solver in the background. The simulation results show that we are able to find less EM critical cell locations in the placement step and robust interconnect structures in the routing step.

We believe that our EM-aware synthesis will become highly important in the near future due to increasing reliability concerns. Our approach is the starting point for future work on EM-aware synthesis to increase EM robustness of a layout.

## REFERENCES

[1] International Technology Roadmap for Semiconductors 2.0 (ITRS 2.0), "More moore," http://www.itrs2.net/itrs-reports.html, 2015 edn (2016).

[2] P. Gibson, M. Hogan, and V. Sukharev, "Electromigration analysis of full-chip integrated circuits with hydrostatic stress," in *2014 IEEE Int. Reliability Physics Symposium*, June 2014, pp. IT.2.1–IT.2.7.

[3] S. Bigalke and J. Lienig, "Load-aware redundant via insertion for electromigration avoidance," in *Proceedings of the 2016 on International Symposium on Physical Design (ISPD '16)*, 2016, pp. 99–106.

[4] V. Sukharev, A. Kteyan, J. H. Choy, S. Chatterjee, and F. N. Najm, "Theoretical predictions of EM-induced degradation in test-structures and on-chip power grids with analytical and numerical analysis," in *2017 IEEE Int. Reliability Physics Symposium*, April 2017, pp. 6B/5.1–5.10.

[5] A. B. Kahng, J. Lienig, I. L. Markov, and J. Hu, *VLSI Physical Design: From Graph Partitioning to Timing*. Springer, Dordrecht, 2011.

[6] P. T. Boggs, A. Althsuler, A. R. Larzelere, E. J. Walsh, R. L. Clay, and M. F. Hardwick, "Dart system analysis," Sandia National Laboratories, Tech. Rep., 2005.

[7] R. Kirchheim, "Stress and electromigration in Al-lines of integrated circuits," *Acta Metall. et Mater.*, vol. 40, no. 2, pp. 309–323, 1992.

[8] M. A. Korhonen, P. Børgesen, K. N. Tu, and C.-Y. Li, "Stress evolution due to electromigration in confined metal lines," *Journal of Applied Physics*, vol. 73, no. 8, pp. 3790–3799, Apr. 1993.

[9] M. Clemens, E. Gjonaj, P. Pinder, and T. Weiland, "Self-consistent simulations of transient heating effects in electrical devices using the Finite Integration Technique," *IEEE Trans. Magn.*, vol. 37, no. 5, pp. 3375–3379, Sep. 2001.

[10] J. E. Stine, I. Castellanos, M. Wood, J. Henson, F. Love, W. R. Davis, P. D. Franzon, M. Bucher, S. Basavarajaiah, J. Oh, and R. Jenkal, "FreePDK: An open-source variation-aware design kit," in *2007 IEEE Int. Conf. on Microelectronic Systems Education (MSE'07)*, June 2007, pp. 173–174.

[11] NanGate, *45nm Open Cell Library*, http://www.nangate.com, 2011.

[12] OpenCores, http://www.opencores.org, 2017.

# Characterising Soft-Failures in Component-Level ESD Testing

Patrick Schrey

Graz University of Technology, Institute of Electronics
Inffeldgasse 12/I, 8010 Graz, Austria
e-mail: patrick.schrey@tugraz.at

*Abstract*—The gap between component-level and system-level Electrostatic Discharge (ESD) tests is an ongoing problem for microelectronics. Without powering the Device Under Test (DUT) during component-level tests, there is no way to identify soft-failures of a component during operation of the final system, before the first prototype is available. In this work, detection of the component-level soft-failures by monitoring the response of the DUT while stressing with Transmission Line Pulse (TLP) is discussed on the example of an LM741C operational amplifier. Parameters used for soft-failure detection are included in the quasi-static Current over Voltage Characteristic. Additionally, a Wunsch-Bell-like characteristic for TLP induced soft-failure is presented. With these new characteristics, Integrated Circuit (IC) designers can provide system designers with valuable information on DUT soft-failure susceptibility and its response to TLP, without revealing any sensitive technology information.

## I. INTRODUCTION

Electrostatic Discharge (ESD) is one of the major failure mechanisms for an Integrated Circuit (IC). ESD can physically damage an IC, as shown in Figure 1. But also electronic systems struggle to fulfil their ESD requirements. The common misconception in the field of ESD that more robust ICs yield a more robust system was proven wrong by case studies [1]. The ESD-requirements for components, like ICs, and systems differ completely and can seldom be compared to each other. While components are expected to make it through packaging and assembly without suffering permanent damage by ESD, systems are required to operate properly even if an ESD occurs. Components are therefore tested to withstand permanent damage or so-called hard-failures and are un-powered during these component-level ESD tests, e.g. IEC 60749-26:2013. Systems, on the other hand, are tested for any disturbance in operation, e.g. a display blackout, or so-called soft-failures, and are powered and un-powered during these system-level ESD tests, e.g. IEC 61000-4-2:2008. To summarize, standardised component-level tests are incapable of detecting soft-failures, which are of major interest for system-level tests.

To find a link between component- and system-level ESD tests, component-level tests must be able to detect soft-failures. This requires the component to operate during ESD tests. Recent investigations deal with soft-failures induced by ESD into tablet computers [2], micro-controller boards [3] and D flip-flops [4], [5], all of which are digital components, where soft-failures can be identified easily.

Analogue components do not show an easily identifiable soft-failure like a flipping bit in digital components.

This paper presents how an analogue IC such as an operational amplifier can be characterised by Transmission Line Pulse (TLP) while being powered. A new characterisation for analogue components is proposed, featuring different levels of soft-failure criteria. This allows system designers to estimate the impact of a component's ESD susceptibility on the overall system. The change in output voltage of an operational amplifier serves as an indicator for soft-failure and seconds the leakage current introduced by Barth *et al.* [6]. Characterising ICs this way allows IC designers to provide crucial information on the DUTs ESD response to system designers without revealing sensitive technology data.

In Section II, the powering of components and coupling of the stress pulses are detailed. Section III presents how soft-failures of components can be categorized. Measurement results are discussed in Section IV.

## II. POWERING COMPONENTS DURING TLP TESTS

TLP is a powerful tool for ESD characterisation and was introduced by Maloney and Khurana [7]. TLP generates square current pulses with tens of amperes, nano- to micro-seconds pulse width, and short rise- and fall-times by discharging a transmission line.

As suggested in [4], the coupling network of the Direct Power Injection (DPI) measurement technique defined in IEC 62132-4:2006 can be used to power the DUT during TLP testing. A schematic coupling network is given in Figure 2. A DC block capacitor is used to isolate the DC potential from the TLP generator. According to IEC 62132-4:2006, the impedance $Z_L$ of the inductor has to be greater than $400\,\Omega$ for the used frequency range. Since TLP generates a square current pulse, the frequency range is theoretically infinite.

A typical TLP does not generate a perfect square pulse, but has a certain rise- and fall-time. These can either be limited by the measurement setup or introduced on purpose to investigate the impact of $du/dt$-triggering [8], [9].

The spectral density of a trapezoidal pulse with an amplitude $\hat{\imath}$, a full-width half-maximum pulse width $T_P$, and equal rise- and fall-time $t_r = t_f$ as a function of frequency $f$ can be calculated by

$$I(f) = 2 \cdot \hat{\imath} \cdot T_P \cdot \frac{\sin\left(\pi \cdot f \cdot T_P\right)}{\pi \cdot f \cdot T_P} \cdot \frac{\sin\left(\pi \cdot f \cdot t_r\right)}{\pi \cdot f \cdot t_r} \quad (1)$$

Fig. 1. Power bus of an IC damaged by ESD. Visible are the individual layers down to the tungsten plugs at the silicon surface.

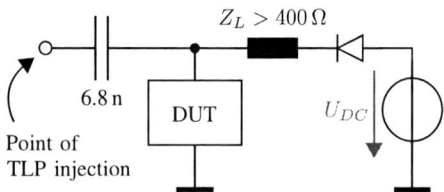

Fig. 2. Schematic for pulse injection while powering the DUT. The circuit is an adaptation of IEC 62132-4:2006 and [4].

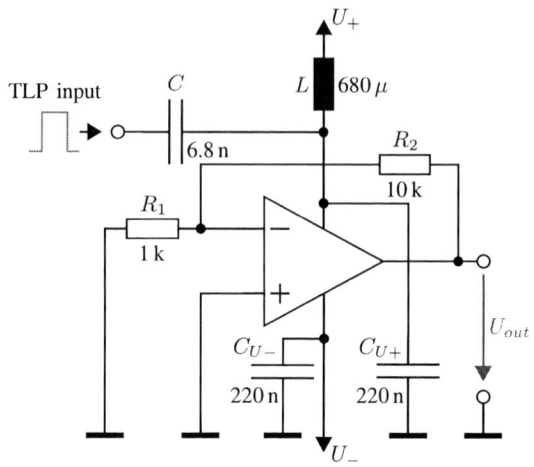

Fig. 3. Test circuits for the LM741C operational amplifier. The three inductors are represented by the single $680\,\mu$H one. The inductors are only required if the pin must be connected to an external voltage source, like the VDD pin.

The envelope of the spectral density has two corner frequencies $f_1$ and $f_2$ if rise- and fall-time are the same $t_r = t_f$. For a $T_P = 1000$ ns, $t_r = 1$ ns TLP, the two corner frequencies are

$$f_1 = \frac{1}{\pi \cdot T_P} = \frac{1}{\pi \cdot 1000\,\text{ns}} \approx 318\,\text{kHz} \qquad (2)$$

$$f_2 = \frac{1}{\pi \cdot t_r} = \frac{1}{\pi \cdot 1\,\text{ns}} \approx 318\,\text{MHz} \qquad (3)$$

For designing the inductor in the coupling network, the longest available pulse width $T_P$ has to be used. The in-house TLP system can generate stress with a pulse width of 1000 ns. Selecting a minimum frequency $f_{min} = 100$ kHz which is below $f_1 = 318$ kHz for a $T_P = 1000$ ns pulse, yields an inductor of $L \approx 640\,\mu$H. The next higher E12-series preferred value for inductors is $680\,\mu$H. Due to parasitic capacitances, the inductor will be split into three individual components; a $680\,\mu$H inductor, a ferrite bead with three wire turns, and a ferrite bead with two wire turns. This provides a total impedance of $Z_L > 400\,\Omega$ over the frequency range.
The DC block capacitor $C = 6.8$ nF is selected according to the standard IEC 62132-4:2006.
The DUT for this work is the operational amplifier LM741C. For the test circuit, the LM741C is operated as an inverting amplifier with a gain $G = -10$. The test circuit is shown in Figure 3 for a pin requiring an external voltage source.

## III. CRITERIA FOR SOFT FAILURES

Defining a soft-failure criterion is virtually impossible for an IC designer, as the soft-failure criterion depends on the system in which the IC is going to be built into. A general categorisation of failure criteria is given in IEC 61000-6-2:2005. The standard gives three criteria for a DUT, i.e. a system, while being tested.

**Criterion A** DUT undisturbed during and after the test.
**Criterion B** DUT undisturbed after the test.
　　　　　　　During the test, disturbance is allowed.
**Criterion C** Temporary disturbance is allowed,
　　　　　　　provided the function can be restored.

These system-level criteria serve as guidelines to define soft-failure criteria for component-level TLP testing while the DUT operates. To generate a quasi-static Current over Voltage Characteristic ($I - U$ characteristic), an averaging window is selected [10]. Similar to generating the $I - U$ characteristic, the above listed criteria are used to set a second averaging window for the output voltage. For criterion A, the monitored parameters must not change more than a specified limit, e.g. $\pm 10\,\%$, for the entire recorded time. This means, the averaging window for soft-failure detection starts at the beginning of the TLP stress and reaches to the end of the measurement record. For criterion B, the monitored parameters must not change more than a specified limit after the TLP stress. This means, the averaging window starts at the end of the TLP stress and reaches to the end of the measurement record. For criterion C, the monitored parameters are allowed to exceed the defined limits during the measurement. For example, the TLP stress can cause the circuit to oscillate with an amplitude above the defined limits. In a worst-case, the circuit has to be reseted manually. As long as the circuit operates normally after a reset, criterion C is fulfilled. If circuit operation cannot be restored, the DUT suffered a hard-failure and has to be replaced.
The actual levels which are considered as a failure have to be defined by either customer, specification, regulations, or similar sources. For example, the specification for the output voltage of an operational amplifier could define $\pm 5$ V for criterion A, $\pm 2$ V for criterion B.

## IV. TLP Test of Powered Components

Depending on the DUT and application, a number of quantities can be used as soft-failure indicators. For this work, a soft-failure is detected by monitoring the maximum deviation of the operational amplifier's output voltage from its equilibrium potential caused by TLP stress. During TLP measurements, the change of output voltage $U_{out}$ during and after the stress is monitored. This allows to plot a soft-failure $I - U$ characteristic, similar to the conventional one with the leakage current presented in [6]. The new soft-failure $I - U$ characteristic is printed in Figure 4 for soft-failure criteria A and B. As the DUT passed criterion C, the corresponding plots are omitted.

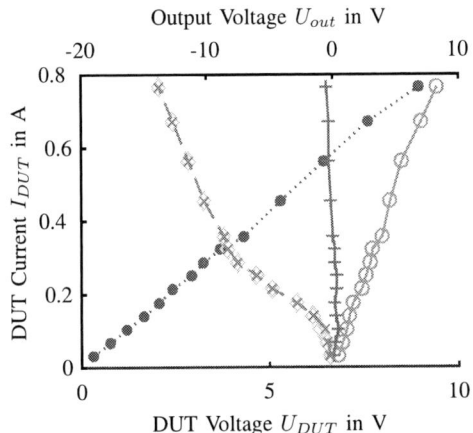

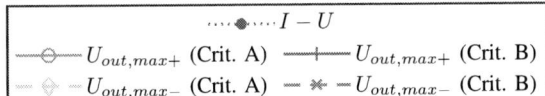

Fig. 4. Soft-failure $I - U$ characteristic of an LM741C operational amplifier, VDD stressed towards GND. Test performed with a constant-current 500 Ω TLP system with a pulse width $T_P = 100$ ns, rise-time $t_r = 1$ ns, and charge voltages ranging from 40-800 V. Averaging window for the $I - U$ characteristic ranges between 50-90% of the pulse width. Averaging window for criterion A is 900 ns wide and ranges between the beginning of the stress pulse and the end of measurement record. Averaging window for criterion B is 800 ns wide and ranges between the end of the stress pulse and the end of measurement record. The DUT fully met criterion C and its plots are omitted.

The soft-failure $I - U$ characteristic contains a lot of information on the DUT response to TLP stress for the selected soft-failure criteria. Comparing the individual soft-failure curves reveals two things: First, criteria A and B show the same maximum negative output voltage depicted by the dashed lines in Figure 4, i.e. the negative output voltage occurs after the TLP stress. Second, the maximum positive output voltage given by the solid lines in Figure 4 is close to 0 V for criterion B, while the maximum positive output for criterion A increases. This means, the output voltage has a positive pulse, which occurs while the TLP stress is applied to the DUT.

The test is repeated for a pulse width of 50 ns, 200 ns, 500 ns, and 1000 ns. The soft-failure $I - U$ characteristic for 1000 ns is given in Figure 5.

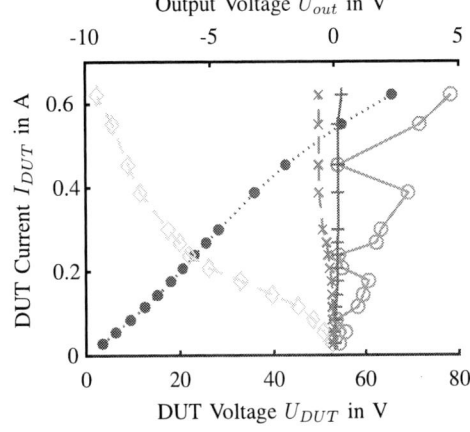

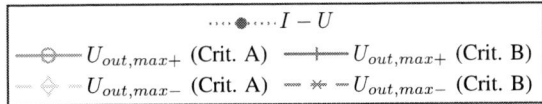

Fig. 5. Soft-failure $I - U$ characteristic of an LM741C operational amplifier, VDD stressed towards GND. Test performed with a constant-current 500 Ω TLP system with a pulse width $T_P = 1000$ ns, rise-time $t_r = 1$ ns, and charge voltages ranging from 40-800 V. Averaging window for the $I - U$ characteristic ranges between 50-90% of the pulse width. Averaging window for criterion A is 9 μs wide and ranges between the beginning of the stress pulse and the end of measurement record. Averaging window for criterion B is 8 μs wide and ranges between the end of the stress pulse and the end of measurement record. The DUT fully met criterion C and its plots are omitted.

The DUT responds differently compared to the previous 100 ns pulses. First, for 1000 ns the maximum negative output voltage is no longer the same for criteria A and B as it was for 100 ns. The underlying output voltage curves reveal that the DUT tries to compensate the TLP stress. For 100 ns pulses, the DUT responds too slowly and the averaging window for criterion B starts before the additional output voltage caused by the TLP stress vanished. For 1000 ns, the DUT manages to compensate the TLP stress before the averaging window for criterion B starts.

The zigzagging line for positive maximum output voltages, denoted by the circle markers in Figure 5, is not due to DUT response. The corresponding pulse is a narrow voltage spike, which is not sufficiently sampled by the oscilloscope. The actual maximum voltage will be higher than the detected one, and the data cannot be used for further analysis.

A system designer can now define soft-failure criteria for the output voltage according to system specifications. The soft-failure $I - U$ characteristics are used to determine the power-to-failure $P_F$ for each pulse width $T_P$ and plot a Wunsch-Bell-like characteristic [11]. This would allow a system designer to compare different components, select the most suitable component to meet the ESD goals, and to estimate the system's susceptibility towards an arbitrary transient pulse [12], [13].

For instance, if the output of the operational amplifier is connected to a Schmitt trigger, the output voltage must not exceed the switching thresholds or absolute maximum ratings of the Schmitt trigger. Selecting, for example, $\pm 5\,\mathrm{V}$ for criterion A and $\pm 2\,\mathrm{V}$ for criterion B results in the Wunsch-Bell-like pulse susceptibility characteristic for soft-failures shown in Figure 6.

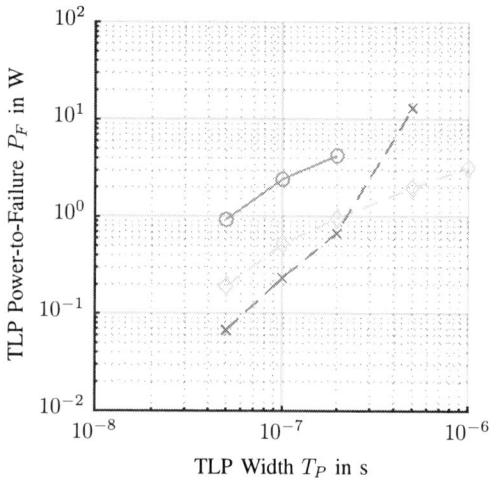

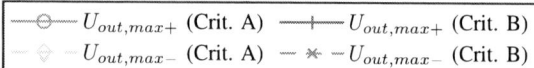

Fig. 6. Pulse susceptibility characteristic for soft-failures of an LM741C operational amplifier, VDD stressed towards GND. Test performed with a constant-current $500\,\Omega$ TLP, rise-time $t_r = 1\,\mathrm{ns}$, and charge voltages ranging from 40-800 V. Averaging window for the $I - U$ characteristic ranges between 50-90% of the pulse width. The output voltage must not change more than $\pm 5\,\mathrm{V}$ for criterion A and $\pm 2\,\mathrm{V}$ for criterion B. Averaging window for criterion A ranges between the beginning of the stress pulse and end of measurement record. Averaging window for criterion B ranges between the end of the stress pulse and end of measurement record. The DUT fully met criterion C.

Contrary to the original Wunsch-Bell characteristic, the power-to-failure $P_F$ increases for longer pulse width $T_P$. This is due to the capacitors $C$ and $C_{U+}$, which become charged by the applied TLP stress. Thus, the sensed TLP voltage increases while the TLP stress is applied. The longer the applied pulse, the higher the voltage and power become.

The pulse susceptibility characteristic in Figure 6 indicates that the DUT fails the maximum negative output voltage criteria before failing the positive ones. For $U_{out,max+}$ criterion A, the data at 500 ns and 1000 ns has been removed due to insufficient sampling, as discussed above. If the DUT met the requirements, the point in the characteristic is missing. Most notably, the line for $U_{out,max+}$ criterion B is missing. This means the DUT passed this criterion for all applied pulses.

Depending on the system, in which the operational amplifier is intended to be used, system designers can select soft-failure indicators to be monitored, set soft-failure criteria, and plot the $I - U$ characteristics and pulse susceptibility characteristic. With these, the system designer can identify problems related to susceptibility against arbitrary transient pulses.

## V. CONCLUSION

ESD induced soft-failures of an operational amplifier are described by $I - U$ characteristics and a Wunsch-Bell-like pulse susceptibility characteristic. For this work, a soft-failure is detected by monitoring the maximum deviation of output voltage from its equilibrium potential due to TLP stress. Which parameters are monitored, depends on the system's requirements. The individual $I - U$ characteristics provide information on the DUT response to TLP stress. Defining soft-failure criteria allows to plot the pulse susceptibility characteristic, and to estimate the impact of arbitrary pulses. The pulse susceptibility characteristic provides information on the power of TLP stress at a certain pulse width that caused a violation of the set soft-failure criteria.

Both types of plots enable more detailed ESD modelling of components with respect to soft-failures and a deeper understanding of the impact of a single components ESD response on the surrounding system. IC manufacturers can provide such plots without revealing sensitive information.

## REFERENCES

[1] Industry Council on ESD Target Levels, *White Paper 1: A Case for Lowering Component Level HBM/MM ESD Specifications and Requirements*, Sep. 2011.

[2] A. Hosseinbeig, O. H. Izadi, S. Solanki, T. D. Lingayat, B. P. Subramanya, A. K. Vaidhyanathan, J. Zhou, and D. Pommerenke, "Methodology for Analyzing ESD-Induced Soft Failure Using Full-Wave Simulation and Measurement," *IEEE Trans. Electromagn. Compat.*, vol. PP, no. 99, pp. 1–9, 2018.

[3] S. C. Yener, S. Frei, and S. Scheier, "Behavioural Model Based Simulation of the ESD-Soft-Failure-Robustness of Microcontroller Inputs," in *Proc. Int. Symp. Electromagn. Compat. - EMC EUROPE*, Sep. 2016, pp. 541–545.

[4] N. Monnereau, F. Caignet, N. Nolhier, M. Bafleur, and D. Tremouilles, "Investigation of Modeling System ESD Failure and Probability Using IBIS ESD Models," *IEEE Trans. Dev. Mat. Rel.*, vol. 12, no. 4, pp. 599–606, Dec. 2012.

[5] G. Shen, S. Yang, V. V. Khilkevich, D. J. Pommerenke, H. L. Aichele, D. R. Eichel, and C. Keller, "ESD Immunity Prediction of D Flip-Flop in the ISO 10605 Standard Using a Behavioral Modeling Methodology," *IEEE Trans. Electromagn. Compat.*, vol. 57, no. 4, pp. 651–659, 2015.

[6] J. E. Barth, K. Verhaege, L. G. Henry, and J. Richner, "TLP Calibration, Correlation, Standards, and New Techniques," *IEEE Trans. Electron. Packag. Manuf.*, vol. 24, no. 2, pp. 99–108, Apr. 2001.

[7] T. J. Maloney and N. Khurana, "Transmission Line Pulsing Techniques for Circuit Modeling of ESD Phenomena," in *Proc. of the EOS/ESD Symp.*, vol. 7, 1985, pp. 49–54.

[8] Y. Cao, W. Simburger, and D. Johnsson, "Rise-Time Filter Design for Transmission-Line Pulse Measurement Systems," in *2009 German Microwave Conference*, Mar. 2009, pp. 1–5.

[9] C. Musshoff, H. Wolf, H. Gieser, P. Egger, and X. Guggenmos, "Rise-time Effects of HBM and Square Pulses on the Failure Thresholds of GGNMOS Transistors," in *Proc. ESREF Symp.*, 1996, pp. 1743–1746.

[10] W. Simbürger, D. Johnsson, and M. Stecher, "High Current TLP Characterisation: An Effective Tool for the Development of Semiconductor Devices and ESD Protection Solutions," *ARMMS RF & Microwave Society*, 2012.

[11] D. C. Wunsch and R. R. Bell, "Determination of Threshold Failure Levels of Semiconductor Diodes and Transistors Due to Pulse Voltages," *IEEE Trans. Nucl. Sci.*, vol. 15, no. 6, pp. 244–259, Dec. 1968.

[12] F. Lafon, F. de Daran, and S. Lecointre, "ESD analysis methodology, From IC behavior to PCB prediction," in *6th Int. Workshop Electromagn. Compat. Integr. Circuits - EMC COMPO*, vol. 7, 2007.

[13] B. Deutschmann, F. Magrini, and F. Klotz, "Internal IC Protection Structures in Relation to New Automotive Transient Requirements," in *8th Int. Workshop Electromagn. Compat. Integr. Circuits - EMC COMPO*, Nov 2011, pp. 47–52.

# Torus Topology based Fault-Tolerant Network-on-Chip Design with Flexible Spare Core Placement

P. Veda Bhanu*, Pranav Kulkarni*, Soumya J*, Linga Reddy Cenkarmaddi†, and Henning Idsøe†

*Department of EEE, Birla Institue of Technology and Science-Pilani, Hyderabad, Telangana, India - 500078

*{vedabhanuiit2010, kulkarni.pranav2, soumyatkgp}@gmail.com

†Department of Information and Communication Technology, University of Agder, Norway

†{linga.cenkeramaddi, henning.idsoe}@uia.no

*Abstract*—The increase in the density of the IP cores being fabricated on a chip poses on-chip communication challenges and heat dissipation. To overcome these issues, Network-on-Chip (NoC) based communication architecture is introduced. In the nanoscale era NoCs are prone to faults which results in performance degradation and un-reliability. Hence efficient fault-tolerant methods are required to make the system reliable in contrast to diverse component failures. This paper presents a flexible spare core placement in torus topology based fault-tolerant NoC design. The communications related to the failed core is taken care by selecting the best position for a spare core in the torus network. By considering this we propose a meta-heuristic based Particle Swarm Optimization (PSO) technique to find suitable position for the spare core that minimizes the communication cost. We have experimented with several application benchmarks reported in the literature by varying the network size and by varying the fault-percentage in the network. The results show significant reduction in terms of communication cost compared to other approaches.

*Keywords*—Fault-tolerance, Network-on-Chip, Particle Swarm Optimization, Torus topology, Communication cost, Spare core.

## I. INTRODUCTION

In the nano-scale era more number of IP cores being fabricated onto a chip has been increased rapidly. This has led to an increase in the communication complexity on a chip. To achieve high communication efficiency and to overcome communication complexity, authors have proposed a packet-based switching technique namely Network-on-Chip (NoC) [1]. NoCs consist of three components namely Network Interface (NI), Routers or switches and Interconnection links which facilitate communication between different processing elements. The cores are connected to the routers and the information is exchanged in terms of packets known as flits. NoC has to be designed by taking into consideration of many aspects such as mapping of the cores, routing algorithms, power consumption, reliability and so on. Many performance parameters like latency and throughput in turn depend on these various factors.

The NoCs are scalable unlike the traditional buses and are designed to handle the high inter-core bandwidth requirements. The high bandwidth requirements translates to high switching activity leading to higher heat dissipation. The non-uniform heat dissipation might lead to failure of the components. It is therefore essential to introduce fault tolerant techniques to build reliable and robust NoC systems subject to minimizing the communication costs. The design of an NoC system for satisfying incoming applications with high inter-core bandwidths requirements necessitates proper application mapping strategies to develop reliable systems. To cater the core failures, this paper mainly focuses on flexible placement of a spare core in the given network. The paper is structured as follows. The literature survey has been outlined in Section II. Section III is devoted to detailed explanation of PSO. The placement of spare core using DPSO technique has been detailed in Section IV. The experimental results are presented in Section V followed by the conclusion in Section VI.

## II. RELATED WORK

Quite recently, considerable attention has been paid to fault-tolerant methodologies in NoC design. In the literature for the fault-tolerant NoC design, most of the publications reported have considered redundancy in router and link. To make the system reliable, most of the works have been concentrated on different routing techniques proposed for fault-tolerant NoC. However, to the best of our knowledge, very few publications can be found in the literature that address the issue of fault-tolerant application mapping in NoC. As reported by [2], energy and reliable aware application mapping approach proposes a cost model by combining energy and reliability. In this work it was observed that branch-and bound-based mapping technique for regular topologies has been used. However, they have assumed fault-free cores and link faults in this work. FoToNoC, folded torus like NoC based multi-core SoC in the dark silicon era has been presented in [3], they have considered hierarchical mapping which includes strategies of cluster management and mapping of the tasks. The study of optimizing application performance on a 5-D torus network via the topology-aware task mapping technique has been presented in [4]. Their work focuses on placement of processes on computing nodes. However approaches in [3] and [4] does not provide a solution in the event of core faults. In [5], task-remapping for fault-tolerant NoC has been performed by

978-1-5386-5388-3/18 $31.00 © 2018 IEEE

using an Integer Linear Programming (ILP) based technique. For the problem of online task remapping the same group of authors in [5] have developed heuristics. In [6], a fault-tolerant multi-application mapping has been presented. It works in two stages, initially an application is mapped onto non-faulty processing cores and in the next stage the spare core is placed onto the fault free cores. The authors in [6] has developed heuristic for mapping phase and design space exploration algorithm has been proposed for spare core placement. In [7] it was shown that spare core has been placed at the center of the mesh network using a mapping algorithm. Authors in [8] have considered the faults in links and developed a design using spare link. The faults in routers have been considered and developed a design using spare routers [9]. To tolerate router faults the authors in [9] have proposed a fault-tolerant NoC design in [10] using spare links and double network interface routers. As per the user requirements the authors in [11], have proposed a cost function by using application mapping method. Although several studies have indicated that faults in cores, links and routers can be addressed using different techniques proposed in the literature, little attention has been paid to faults in application cores that has to be mapped onto NoC.

### III. DISCRETE PARTICLE SWARM OPTIMIZATION

Particle Swarm Optimization (PSO) [12] is a population based stochastic technique designed and developed by Eberhart and Kennedy in 1995. The following optimization problem requires a discrete variant of PSO - DPSO [13]. In a DPSO system, multiple candidate solutions coexist and collaborate simultaneously. Instead of using genetic operators, each individual particle (solution) adjusts its flying in problem space according to its own flying experience as well as its companions flying experience. In the search space of DPSO, each individual particle has a fitness value, which determines its quality. Each particle is treated as a point in n-dimensional space. In every run, the fitness function is evaluated by taking the position of the particle in solution space. The position of $i^{th}$ particle at $k^{th}$ iteration is denoted as $p_k^i$. Each particle keeps track of its best value obtained so far. This is called as personal best or local best of $i^{th}$ particle ($pbest^i$). Similarly, the best fitness value across the whole swarm is called global best ($gbest_k$) for the generation $k$. The new position of the $i^{th}$ particle can be calculated by

$$p_{k+1}^i = (x_1 * I \oplus x_2 * (p_k \to pbest^i) \oplus x_3 * (p_k \to gbest_k))p_k^i \quad (1)$$

In the equation (1), $y \to z$ denotes the minimal length of swap sequence to be applied on the components of $y$ to transform it to $z$. For example, if $y = <6, 7, 8, 9>$ and $z = <9, 6, 7, 8>$, $y \to z = <$ swap (1, 4), swap (2, 4), swap (3, 4)$>$. The fusion operator ($\oplus$) is enforced on two swap sequences. The sequence of swaps in $y$ is followed by the sequence of swaps in $z$ and it is given by $y \oplus z$. The constants $x_1, x_2, x_3$ are the inertia, self-confidence and swarm confidence values, respectively. Swap sequences such as $<$swap (1, 1),

swap (2, 2), swap (3, 3) ... swap (n, n)$>$ are known as identity swaps which is denoted by $I$. It corresponds to the inertia of the particle to maintain its current configuration. To generate a new particle $p_{k+1}^i$, the final swap sequence is applied on particle $p_k^i$ mentioned in above equation (1). From [14], PSO convergence condition is given by,

$$(1 - \sqrt{x_1})^2 \leq x_2 + x_3 \leq (1 + \sqrt{x_1})^2 \quad (2)$$

The experiment has been carried out for different values of $x_1$, $x_2$ and $x_3$. Out of them based on the following values of $x_1$= 0.8, $x_2$=0.4 and $x_3$= 0.6, the results are reported in this paper. This completes an overview of PSO.

### IV. SPARE CORE PLACEMENT VIA DPSO

We present our formulation for mapping the incoming core graph onto the torus architecture along with the spare core. The input torus floorplan gives the number of available position onto which the cores can be mapped to. The input torus floorplan should have enough number of available positions to accommodate all the cores in the core graph along with the spare core. In our formulation, the spare core has been numbered the last. However, any entry in the particle can be taken.

#### A. Particle formulation and Fitness Function

Each particle is a specific arrangement of cores which represents their association with routers. The quality or fitness of a particle depends on the association of the cores to the routers. The fitness of a particle is the communication cost calculated by the following formula:

$$Communication\ cost = \sum_{\forall Edges} (Number\ of\ Hops * Bandwidth) \quad (3)$$

For the problem of spare core placement, each particle is modeled as single dimensional array wherein the index of the array determines the router number and the value at the index determines the core number. This scheme conveys the association of the core with the router. The formulation takes the failed core as user input and in the event of a core failure, the communications associated with the failed core are taken care by the spare core.

#### B. Local and Global bests

During the course of evolution, the particles undergo modification which changes the association of the cores with the routers. This leads to a change in the quality/fitness of the particles. Each particle keeps track of the best set of core positions that it has encountered resulting in minimum fitness known as local best (lbest). Across the generations, the swarm keeps track of the set of core positions which it has encountered resulting in minimum fitness known global best (gbest). Both these values help in guiding the particles towards minimum fitness during the process of evolution. These are updated if newer values are lesser than the previously stored values during evolution.

## C. Evolution of generation

Over the generations the particles are evolved and transformed into new particles having fitness closer to the best value. Initially, the particles are generated randomly and quality of each particle is evaluated. In the first generation, local best *(lbest)* of each particle is initialized as the particle itself and the global best is evaluated. The remaining generations are evaluated according to the eqn. 1. The evolution centers around the particles being swapped with a probability towards *lbest* and *gbest* set of core positions. The termination condition is reached when the number of generations reach a preset number of generations or if there is no improvement in the solution for a certain number of generations.

## V. EXPERIMENTAL RESULTS

We present the results obtained by simulating our technique on a PC embedded with Intel Xeon processor operating at 3.5 GHz with 32 GB internal memory. The DPSO was coded in a high level language *C++* and simulated for several benchmark applications reported in the literature. Communication cost is calculated by equation (3). For fair comparison we have extended the work followed in approach [7] to torus topology. We compare the results generated by DPSO with the approach followed by [7]. A sample application mapping with spare core placement using approach [7] and proposed approach has been shown in Fig. 1.

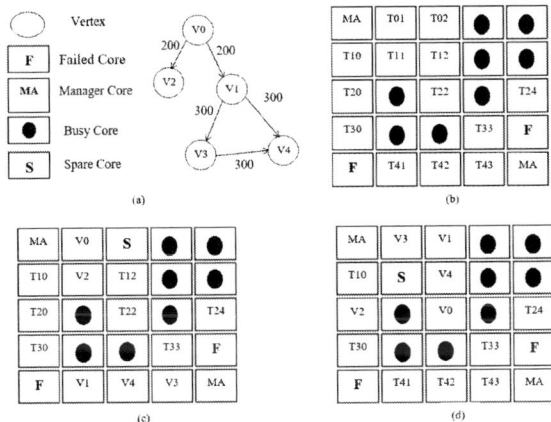

Fig. 1: Placement of a spare core in 5x5 torus network (a) Sample application core graph, (b) Available positions in the 5x5 torus network, (c) Spare core placement using proposed approach, (d) Spare core placement using approach [7].

Fig. 1(a) shows an example core graph, let us assume core V1 as failed core input given by user, hence the spare core has to be placed by taking V1 into consideration. It can be observed from Fig. 1(a), that the communication bandwidth between core V1 and V0, V3, V4 are 200, 300 and 300 Mbps respectively. The spare core should be placed such that in the consequence of application core failure (V1), the spare core (represented by S in Fig. 1(c) and 1(d)) should

TABLE I: Calculation of the communication cost for Fig. 1

| Source Core | Destination Core | Communication Bandwidth | Approach [7] | | Proposed approach | |
|---|---|---|---|---|---|---|
| | | | Number of hops (N) | Communication cost (N x BW) | Number of hops (N) | Communication cost (N x BW) |
| V0 | V2 | 200 | 2 | 400 | 1 | 200 |
| V0 | V1 | 200 | 2 | 400 | 1 | 200 |
| V1 | V3 | 300 | 1 | 300 | 2 | 600 |
| V1 | V4 | 300 | 1 | 300 | 1 | 300 |
| V3 | V4 | 300 | 2 | 600 | 1 | 300 |
| Communication Cost | | | | 2000 | | 1600 |

take over the communication associated to V1 with minimum communication cost. In the Table I, proposed approach could give a better communication cost of 1600 against 2000 given by the approach [7].

To test for the scalability and efficiency of proposed approach compared to approach [7], the size of the torus network was increased (5x5, 10x10 and 15x15). The simulation results showing the percentage improvement in communication cost in proposed approach compared to approach [7] are tabulated in Table II.

TABLE II: Comparison of communication cost by varying the torus network size to 5x5, 10x10 and 15x15 with 30% faults in the network.

| Application | 5x5 | | | 10x10 | | | 15x15 | | |
|---|---|---|---|---|---|---|---|---|---|
| | I | II | III | I | II | III | I | II | III |
| MPEG | 5751 | 3850 | 27.47% | 6431 | 3946 | 27.47% | 7880 | 5213 | 33.84% |
| MWD | 1568 | 1184 | 12.24% | 1824 | 1664 | 22.80% | 2464 | 1792 | 27.27% |
| 263Encoder | 30.85 | 23.04 | 17.34% | 36.48 | 29.28 | 28.00% | 43.58 | 28.01 | 35.72% |
| MP3Encoder | 24.21 | 17.05 | 26.76% | 39.58 | 17.97 | 44.11% | 24.24 | 19.99 | 17.52% |
| 263Decoder | 29.29 | 19.94 | 30.62% | 30.88 | 24.53 | 32.40% | 65.49 | 24.01 | 63.33% |
| VOPD | 6374 | 4105 | 29.55% | 5819 | 4708 | 7.92% | 11790 | 7962 | 32.47% |

I = Approach in [7]; II = Proposed approach; III = % Improvement by proposed approach;

A major drawback of approach [7] is that spare core is likely to be placed at the center of the torus network while ignoring the communication cost of the failed core. On the other hand, in the proposed approach, flexibility is given for placement of the spare core among the available positions in the input torus floorplan. The flexibility helps in reducing the communication cost. From Table II, it can be observed that as the size of the torus network is increasing there is an improvement in the communication cost by 23.99%, 26.32% and 35.02% for 5x5, 10x10, and 15x15 respectively. Such improvement in the communication cost is due to the search space for the placement of spare core increases which in turn finds the optimum position for the spare core. This shows the scalability of the proposed approach.

Fault percentage is termed as number of non-available positions to total number of available positions in the torus network. The communication cost comparison between proposed approach and approach followed in [7] has been shown in Table III, by varying the fault percentage (0%, 15%, 30% and 50%) for different application benchmarks in the 10x10 torus network. A maximum of 45.14% improvement in terms of communication cost has been achieved when compared to the approach followed in [7] . From the Table III, it can be observed that the average percentage improvement (20.72, 21.15, 24.24 and 30.77) increases as the fault percentage (0%, 15%, 30% and 50%) increases, which shows the applicability of proposed approach in serious fault environments. Fig.

TABLE III: Comparison of communication cost by varying the faults from 0% to 15%, 30%, 50% in 10x10 torus network.

| Application | 0% | | | 15% | | | 30% | | | 50% | | |
|---|---|---|---|---|---|---|---|---|---|---|---|---|
| | I | II | III | I | II | III | I | II | III | I | II | III |
| MPEG | 4314 | 3570 | 17.24% | 5823 | 3780 | 35.08% | 6431 | 3946 | 41.38% | 6773 | 4487 | 33.75% |
| MWD | 1440 | 1408 | 2.22% | 1856 | 1536 | 17.24% | 1824 | 1664 | 8.77% | 2464 | 1664 | 32.46% |
| 263Encoder | 29.28 | 25.52 | 12.84% | 33.17 | 25.52 | 23.06% | 36.48 | 29.28 | 19.73% | 43.17 | 25.11 | 41.89% |
| MP3Encoder | 22.73 | 17.07 | 24.90% | 21.58 | 17.74 | 17.79% | 39.58 | 17.97 | 31.98% | 31.23 | 17.13 | 45.14% |
| 263Decoder | 35.88 | 24.39 | 32.03% | 31.47 | 25.43 | 19.19% | 30.88 | 24.53 | 20.56% | 45.15 | 38.10 | 15.69% |
| VOPD | 7160 | 4647 | 35.09% | 5508 | 4708 | 35.09% | 5819 | 4708 | 23.05% | 7724 | 6514 | 15.66% |

I = Approach in [7]   II = Proposed approach   III = % Improvement by proposed approach

2 represents the communication cost calculated for MPEG application by taking failed core as an input from the user in proposed approach and approach followed in [7]. On X-axis failed core has been taken and on Y-axis the communication cost is considered. We can observe significant improvement in communication cost for each core failure in the MPEG application by proposed approach. However, experiments have been carried out with all other benchmarks reported in Table I, but only MPEG application has been reported by considering 30% faults in 5x5 torus network. This shows that proposed approach can be efficiently applicable to any core failures given as an input by the user in different benchmark applications.

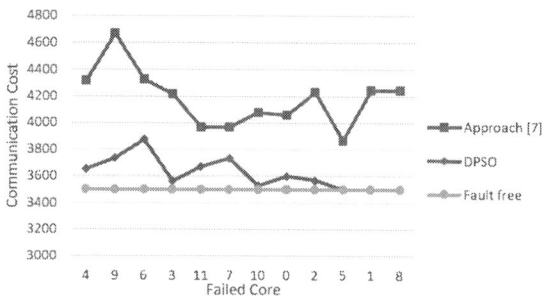

Fig. 2: Communication cost with varying core faults

## VI. Conclusion

In summary, a DPSO based meta-heuristic technique has been proposed to select the optimal position for the spare core to be placed in the torus network which reduces communication cost. Based on the results, it can be concluded that proposed approach has achieved minimal communication cost by flexible spare core placement in the torus topology. Future work involves flexible placement of multiple spare cores from the available positions in the torus based floorplan and proposing exact methods like Integer Linear Programming.

## Acknowledgements

This work is partially supported by the research project No. ECR/2016/001389 Dt. 06/03/2017, sponsored by the SERB, Govt. of India

## References

[1] W. J. Dally and B. Towles, "Route packets, not wires: on-chip interconnection networks," in *Proceedings of the 38th Design Automation Conference (IEEE Cat. No.01CH37232)*, 2001, pp. 684–689.

[2] L. Liu, C. Wu, C. Deng, S. Yin, Q. Wu, J. Han, and S. Wei, "A flexible energy- and reliability-aware application mapping for noc-based reconfigurable architectures," *IEEE Transactions on Very Large Scale Integration (VLSI) Systems*, vol. 23, no. 11, pp. 2566–2580, Nov 2015.

[3] L. Yang, W. Liu, W. Jiang, M. Li, P. Chen, and E. H. M. Sha, "Fotonoc: A folded torus-like network-on-chip based many-core systems-on-chip in the dark silicon era," *IEEE Transactions on Parallel and Distributed Systems*, vol. 28, no. 7, pp. 1905–1918, July 2017.

[4] A. Bhatele, N. Jain, K. E. Isaacs, R. Buch, T. Gamblin, S. H. Langer, and L. V. Kale, "Optimizing the performance of parallel applications on a 5d torus via task mapping," in *2014 21st International Conference on High Performance Computing (HiPC)*, Dec 2014, pp. 1–10.

[5] O. Derin, D. Kabakci, and L. Fiorin, "Online task remapping strategies for fault-tolerant network-on-chip multiprocessors," in *Proceedings of the Fifth ACM/IEEE International Symposium*, May 2011, pp. 129–136.

[6] F. Khalili and H. R. Zarandi, "A fault-tolerant low-energy multi-application mapping onto noc-based multiprocessors," in *2012 IEEE 15th International Conference on Computational Science and Engineering*, Dec 2012, pp. 421–428.

[7] B. N. K. Reddy, M. H. Vasantha, and Y. B. N. Kumar, "A gracefully degrading and energy-efficient fault tolerant noc using spare core," in *2016 IEEE Computer Society Annual Symposium on VLSI (ISVLSI)*, July 2016, pp. 146–151.

[8] N. Chatterjee, N. Prasad, and S. Chattapadhya, "A spare link based reliable network-on-chip design," in *18th International Symposium on VLSI Design and Test*, July 2014, pp. 1–6.

[9] N. Chatterjee, S. Chattopadhyay, and K. Manna, "A spare router based reliable network-on-chip design," in *2014 IEEE International Symposium on Circuits and Systems (ISCAS)*, June 2014, pp. 1957–1960.

[10] N. Chatterjee and S. Chattopadhyay, "Fault tolerant mesh based network-on-chip architecture," in *2015 IEEE International Symposium on Circuits and Systems (ISCAS)*, May 2015, pp. 417–420.

[11] N. Chatterjee, S. Reddy, S. Reddy, and S. Chattopadhyay, "A reliability aware application mapping onto mesh based network-on-chip," in *2016 3rd International Conference on Recent Advances in Information Technology (RAIT)*, March 2016, pp. 537–542.

[12] J. Kennedy and R. Eberhart, "Particle swarm optimization," in *Neural Networks, 1995. Proceedings., IEEE International Conference on*, vol. 4, Nov 1995, pp. 1942–1948 vol.4.

[13] K.-P. Wang, L. Huang, C.-G. Zhou, and W. Pang, "Particle swarm optimization for traveling salesman problem," vol. 3, 2003, pp. 1583–1585, cited By 218. [Online]. Available: https://www.scopus.com/inward/record.uri?eid=2-s2.0-1542316003partnerID=40md5=f0d49a515af704754d12beb9bbba63ca

[14] G. Luo, H. Zhao, and C. Song, "Convergence analysis of a dynamic discrete pso algorithm," in *2008 First International Conference on Intelligent Networks and Intelligent Systems*, Nov 2008, pp. 89–92.

# Design and Analysis of Energy Efficient Self Correcting Latches considering Metastability

Chaudhry Indra Kumar, *Student Member, IEEE*, and Anand Bulusu, *Member, IEEE*

*Abstract*— In modern CMOS technologies, meta-stability is becoming an important issue for designing sequential systems, especially in the near/sub-threshold regime. This is because, with a reduction in supply voltage, mean time between failures (MTBF) increases exponentially. This paper presents a detailed analysis of the meta-stability in design of near/sub-threshold resilient flip-flops. We show that a proper transistor sizing, a selection of optimum number of fingers in the main path inverter of the slave stage, and an insertion of a small sized inverter in the output stage of the resilient flip flops can result a significant reduction in the resolution constant ($\tau$). The post-layout simulation results show that the meta-stability-power-delay-product (MPDP) improves using the modification as compared to an equivalent conventional design. This improves the resilient flip-flop's resolution constant while incurring a design trade-off between power dissipation and performance.

*Index Terms*— Energy efficient latches. metastability, mean time between failures, near threshold circuits, resilient design.

## I. INTRODUCTION

Modern digital CMOS ICs require ultra-low power and energy efficient circuit operation [1]. To design modern energy efficient CMOS circuits, one of the important criteria is to operate them in the near/sub-threshold region [2], [3]. However, the impact of variations and meta-stability in the near/sub-threshold regime is more pronounced compared to the super-threshold regime [2]. This is due to the exponential dependence of the MOSFET current on the power supply, threshold voltage, and temperature.

Sequential elements such as latches and flip-flops are an essential part of CMOS digital circuits, since; they synchronize the flow of an input data with respect to the clock signal. Meta-stability arises during such synchronization operations [4]-[6]. Meta-stability can be defined as a phenomenon in which a bi-stable element (latch/flip-flop) enters an unpredictable third state. In a meta-stable state, output remains at an intermediate voltage level between "0" and "1" logic voltage levels. Consequently, the output may become unstable and take indefinite amount of time to settle to a stable state. The meta-stability occurs due to the change in input data after the setup time or before the hold time of a latch/ flip-flop [7].

As CMOS technology is scaling down, process, voltage, and temperature (PVT) variations on a chip increased. Further, the increased PVT variations along with the high clock frequencies enhance the possibility of meta-stability failures in pipelined systems. Figure 1(a) shows the block diagram of a CMOS static D latch. When clock signal (CLK) is high, the output (Q) takes the value of the input data signal (D), while when CLK is low the data is retained at the output node by back-to-back inverters. In this case, the latch operates as a memory cell, statically storing the previous data. Figure 1(b) shows a meta-stability window ($t_{window}$), which is defined by the setup and hold time around the rising/falling edge of the clock signal. If the input data (D) changes outside of the $t_{window}$, the output (Q) becomes equal to D. If the input data transitions happen in the $t_{window}$, then the output enters in a meta-stable region resulting in a long time for Q to settle to a stable state.

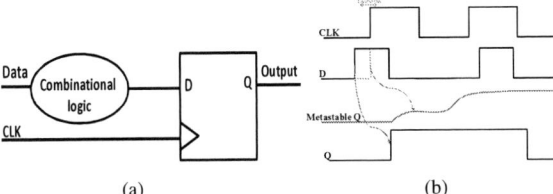

(a)  (b)

Fig. 1. (a) Block diagram of CMOS static D-Flip Flop  (b) Flip-flop waveforms in the case of stable and metastable outputs

In this paper, existing self correcting flip flops meta-stability is analyzed from transistor sizing, circuit and layout architecture perspective. We introduced a designing approach for improving the meta-stability in the near/ sub-threshold region. From new design approach, we compare the meta-stability of various self correcting flip flops power and performance with conventional design approach.

This paper is organized as follows: background information related to meta-stability is presented in Section II. Design consideration for meta-stable hardened resilient latches/filp-flops is discussed in Section III. Section IV illustrates the post-layout simulations with or without modifications on resilient flip-flops and the conclusions with observations drawn are discussed in Section V.

## II. A REVIEW OF META-STABILITY ISSUES

Past research shows that the latch/ flip-flop delay is increases exponentially with respect to a time offset from a metastability event [8]. The behavior of a latch delay in the meta-stable region is modeled by two parameters ($\tau$ and $T_0$) [7]. These two parameters can be extracted from Hspice simulation. To analyze the meta-stability, meta-stability window $\delta$ is used:

$$\delta = T_0 \, e^{\frac{-t_{st}}{\tau}} \qquad (1)$$

Meta-stability window, can be defined such that if the input data switches within $\delta$ then output will not be resolved within a given settling time, $t_{st}$. Where $\tau$ is the resolution time constant and $T_0$ is the asymptotic width of the window with no settling time. The resolution time constant is the inverse of the gain-bandwidth product of an inverter pair in a latch (slave stage) and defines the latch's ability to settle the transitional voltage level of the output. Equation (1) shows that the impact of $\tau$ is more compared to $T_0$ [7]. This is because $\delta$ is exponentially dependent on $\tau$. Therefore, a metastable hardened latch requires a reduction of the value of $\tau$. A small value of $\tau$, results in fast settling time in the meta-stable region and thus reduces the meta-stability window $\delta$.

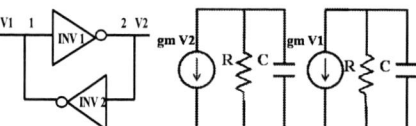

Fig. 2. Small signal models for cross coupled inverter

If an input data switches randomly with respect to the clock signal, the probability that the input data will switch during the meta-stability window $\delta$ is simply $F_D*\delta$ ($F_D$ is the input data switching frequency). The mean time between failures (MTBF), is depicted by [8],

$$ \text{MTBF} = \frac{1}{F_D F_{CLK} T_0 e^{-t_{st}/\tau}} \qquad (2) $$

where, $F_{CLK}$ is the clock signal frequency. From (2) we see that MTBF is exponentially dependent on $\tau$. A meta-stable hardened latch/flip flop would require a lower value of $\tau$ [9]. A lower value of $\tau$ signifies a fast settling time, and consequently improves MTBF. The resolution time constant, $\tau$ is defined as the time required for a metastable state to settle into a stable state in a latch/flip-flop. Figure 2 shows the small-signal model of a cross coupled inverter pair with positive feedback in a latch. In first-order model of MOS transistor, the NMOS transistor is modelled as a current source ($g_m$) and the PMOS transistor is modelled as a load resistance $R$. Using small-signal analysis (Fig. 2), $\tau$ of an inverter pair can be modelled by equation (3), based on the capacitances and loop trans-conductance, $\tau$ is:

$$ \tau = \frac{C_{total}}{g_m} \qquad (3) $$

where, $C_{total} = C_{parasitic} + C_{miller}$. From (3), we observe that a meta-stable hardened latch should have maximum trans-conductance while minimum capacitance around the critical node in a latch to improve the settling time.

## III. DESIGN CONSIDERATION FOR META-STABLE HARDENED RESILIENT LATCH/FLIP FLOP

In this section, we discuss design modifications required to improve the meta-stability in a conventional resilient latch/flip-flop operating in the near/ sub-threshold regime.
1) Ratio of inverter ($k = W_p/W_n$): From (3), we observe that resolution time $\tau$, is the ratio of load capacitance ($C_L$) and trans-conductance ($G_m$) of an inverter in the closed loop system. Therefore, a minimum value of $C_L/G_m$ ratio is required to minimize the meta-stability. We find the trans-conductance of an inverter from its ac small signal analysis for several values of inverter $W_p/W_n$ ratios at different power supplies. Figure 3 (a) and (b) depict the input low-to-high (lh) and high-to-low (hl) transition $C_L/G_m$ ratio versus inverter size

ratio ($W_p/W_n$), respectively. As shown in Fig. 3 that minimum $C_L/G_m$ ratio occurs for inverter ratio, k=1 in 65nm technology at 0.35V, 0.4V and 0.45V power supply. Physically, it can be interpreted that higher values of k results in larger $C_L/G_m$ ratio due to the larger junction capacitance, while lower values of k lowers the current driving capability of PMOS transistor, which means too small a trans-conductance ($G_m$).

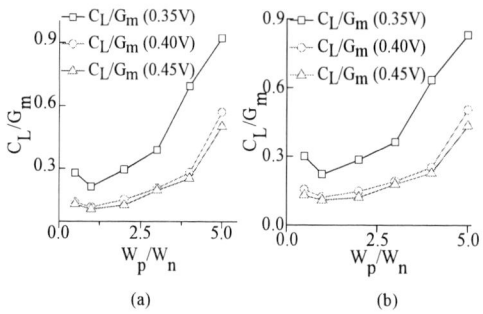

(a)        (b)

Fig. 3. $C_L/G_m$ v/s inverter P/N ratio of CMOS D-latch for (a) low to high input transition and (b) high to low input transition.

2) Optimum number of fingers: In circuit design, number of fingers (NF) is used to increase the drive strength of a logic gate. However, due to the inverse narrow width effect (INWE) and process induced stress, the transistor current does not increase linearly with the NFs [10]-[12]. Therefore, an optimum NF in a transistor can improves a circuit performance. INWE occurs due to the shallow Trench Isolation (STI) in deep sub-micron technologies [12]. INWE reduces the trans-conductance of the cross coupled inverter, consequently, improves the meta-stability issues in a latch as shown in Fig.4.

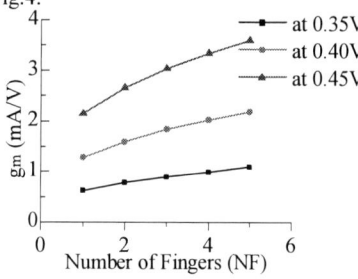

Fig.4. Trans-conductance of an inverter vs Number of fingers (NF) for a fixed size of transistor

3) Using small sized inverter: Meta-stability of a resilient latch/ flip flop can be improved by adding a minimum seized inverter at the output of the forward path of a latch (salve stage) as shown in Fig. 5. Insertion of an inverter improves the settling time of the output state to getting a stable state from an unstable state. Although, the addition of an extra inverter increases the delay in the data-path, however, because of a typical large load capacitance using an extra inverter stage reduces the D-Q delay in the data-path. Therefore, addition of an inverter at the output of a latch improves the resolution time with the nearly same D-Q delay as well as CLK-Q delay. A small sized inverter is used because it minimizes the

parasitic load capacitance of an inverter at the output node and provides a higher resolution capability at the output of a latch.

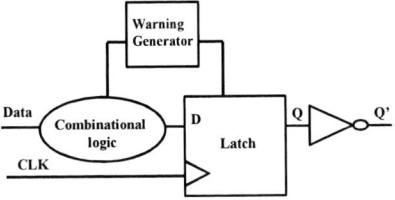

Fig. 5 Self-correcting D-latch (slave stage) with an extra added inverter

## IV. EXPERIMENTAL RESULTS

### A. EXPERIMENTAL SETUP

In our simulation setup, each self-correcting latch receives input data from the standard data-path, which is adopted from [3]. Further, we extract timing parameters, meta-stability parameter, and energy for each resilient latch topology. All the simulations are done in an industrial 65nm CMOS technology PDK at 25°C. For near threshold region analysis three different power supply voltages are used: 0.35V, 0.4V, and 0.45V. The clock frequency is 2MHz used in the simulations. This frequency is used because output has enough time to getting a stable value when it is in meta-stable state.

### B. CONSIDERED SELF-CORRECTING FLIP FLOP TOPOLOGIES

In this work, three self correcting latches/FF are considered for analysis. The analysis is limited only to these three latches because these are widely used latches in near threshold region. Also these latches are based on the selection of sequential block through multiplexing when timing violations occurs. Error Resilient Flip Flop (ERFF) [13], Soft Edge Flip Flop (SEF) [14], Variation Aware Flip Flop (VAFF) [15], and Conventional D Latch chosen for analysis of meta-stability in this paper. In order to make a fair comparison, the self correcting latches are designed in a same manner.

### C. META-STABILITY ANALYSIS

For meta-stability analysis we use resolution time period as a performance parameter. Resolution time period is defined as the time interval when a latch enters into a meta-stable state to reaching a stable state i. e. either logic "0" or logic "1". In this paper, we compare our design approach for designing the self correcting latches to the conventional design approach. In our designing approach we use all design modification that we discussed in Section III The results of the HSPICE simulations at 0.35V, 0.40V and 0.45V power supply are shown in Table I. Simulation results show that the resilient latch/FF designed using our methodology results in a significant improvement in the meta-stability parameter ($\tau$) as compared to conventionally designed latch/FF. This is because $\tau$ is inversely proportional to $g_m$ and our design techniques improve the value of $g_m$ as compared to the conventional designing. Meta-stable hardened latch/FF requires minimum $\tau$. From the simulations, resolution time constant ($\tau$) is extracted from the plot of C-Q delay v/s arrival time. Slope of this curve gives the value of $\tau$ [8].

TABLE I
Meta-stability analysis of considered resilient Flip Flops

| Techniques | $V_{dd}$ (V) | Design Method | Resolution constant (nsec) | Percentage Improvement |
|---|---|---|---|---|
| ERFF | 0.35 | Conventional | 32.89 | 9.09 % |
| | | Modified | 29.90 | |
| | 0.40 | Conventional | 20.56 | 12.88 % |
| | | Modified | 17.91 | |
| | 0.45 | Conventional | 17.60 | 13.58% |
| | | Modified | 15.21 | |
| VAFF | 0.35 | Conventional | 38.60 | 5.69 % |
| | | Modified | 36.40 | |
| | 0.40 | Conventional | 36.94 | 10.20 % |
| | | Modified | 33.17 | |
| | 0.45 | Conventional | 30.04 | 11.55 % |
| | | Modified | 26.57 | |
| SEF | 0.35 | Conventional | 37.70 | 8.19 % |
| | | Modified | 34.61 | |
| | 0.40 | Conventional | 35.30 | 13.17 % |
| | | Modified | 30.65 | |
| | 0.45 | Conventional | 19.60 | 13.47 % |
| | | Modified | 16.96 | |
| Static D-Latch | 0.35 | Conventional | 36.01 | 7.77% |
| | | Modified | 33.21 | |
| | 0.40 | Conventional | 30.20 | 11.35% |
| | | Modified | 26.77 | |
| | 0.45 | Conventional | 20.53 | 12.42% |
| | | Modified | 17.98 | |

### D. PERFORMANCE ANALYSIS

Post layout simulations shows that the resilient latches designed with modifications have nearly equal performance in terms of power and delay when compared to the conventional design approach. This is because INWE increase the charging/ discharging current of the forward path inverter, hence, improves the performance of a latch. In general the power and delay of the modified flip-flops (described in section III) is within 1% of the delay and power of the conventional FF. Post layout results in Table II show that the PDP of ERFF technique with design modification is 0.44% more as compared to conventional designed ERFF technique at 0.4V power supply. While in VAFF and SEF techniques with modifications, PDP is 0.42% and 1.17% more as compared to conventional approach respectively at 0.4V supply voltage.

The presence of extra inverter at the output stage of resilient latch/FF improves the latch/FF meta-stability performance by minimizing the value of $\tau$. Improvement in meta-stability, comes at the cost of additional clock-output delay and increase in power dissipation. In [7] meta-stability of latch/FF is analyzed by the meta-stability-power-delay product (MPDP), which is given in equation (4). For a good trade-off between meta-stability, power and performance MPDP is a useful merit:

$$MPDP = \tau \times Power \times Delay \qquad (4)$$

Therefore, in this paper we use MPDP as the overall Figure-of-merit (FOM). , for comparing $\tau$, power and delay of the various resilient latch/FF designed with conventional and proposed design modifications. Figure 6 shows the meta-stability Power Delay Product (please note that maximum CLK-Q delay is taken for these MPDP calculations) of the ERFF, VAFF, and SEF technique at 0.35V, 0.4V, and 0.45V power supply. In Fig. 6 percentage data represents the improvement of the modified design over conventional design approach.

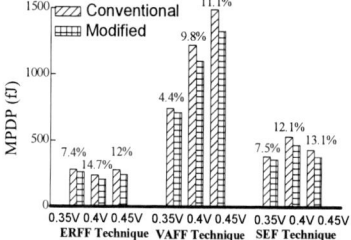

Fig. 6. Metastability Power Delay Product of ERFF, VAFF, and SEF technique at 0.35V, 0.4V, and 0.45V power supply.

## V. CONCLUSION

In this paper, we have presented a methodology for designing high-performance, meta-stable-hardened, and low-power resilient latch/ flip-flops in the sub/near-threshold region. Metastability in resilient latches is analyzed from transistor sizing, increasing the number of fingers in the main path inverter, and adding a small size inverter at the output stage of a latch. We have shown that our methodology improves the resolution time constant of the resilient latches up to 8%, 12%, and 13% at 0.35V, 0.4V, and 0.45V power supply, respectively. We also observed at 0.4V supply voltage that compared to conventional design techniques our design approach improves MPDP for ERFF, VAFF, and SEF technique by 14.7%, 9.8%, and 12.1%, respectively. Similar results are also found at 0.35V and 0.45V power supply. The post-layout results show that our methodology for designing meta-stable hardened resilient latch/flip flops reduces the meta-stability parameter $\tau$ without degrading the clock-output delay.

TABLE II
Performance analysis of resilient Flip Flops designed employing our methodology

| $V_{dd}(V)$ | Parameter | Resilient Technique | | | | | |
|---|---|---|---|---|---|---|---|
| | | ERFF | | VAFF | | SEF | |
| | | Conventional | Modified | Conventional | Modified | Conventional | Modified |
| 0.35 | Power Dissipation (nW) | 26.57 | 26.82 | 46.02 | 46.23 | 24.54 | 24.81 |
| | CLK-Q delay (nS) | 321.00 | 324.00 | 419.00 | 423.00 | 416.00 | 414.00 |
| | PDP (fJ) | 8.52 | 8.68 | 19.28 | 19.55 | 10.20 | 10.27 |
| 0.40 | Power Dissipation(nW) | 38.79 | 38.94 | 81.14 | 81.29 | 37.37 | 37.64 |
| | CLK-Q delay(nS) | 290.00 | 290.11 | 407.00 | 408.00 | 405.00 | 406.90 |
| | PDP (fJ) | 11.24 | 11.29 | 33.02 | 33.16 | 15.13 | 15.31 |
| 0.45 | Power Dissipation (nW) | 56.47 | 56.68 | 123.50 | 123.81 | 55.38 | 55.59 |
| | CLK-Q delay (nS) | 281.06 | 282.10 | 401.00 | 402.00 | 400.70 | 401.00 |
| | PDP (fJ) | 15.87 | 15.98 | 49.52 | 49.77 | 22.19 | 22.29 |

## REFERENCES

[1] D. Markovic, C. C. Wang, L. P. Alarcon, T. Liu and J. M. Rabaey, "Ultralow-Power Design in Near-Threshold Region", in *Proc. IEEE*, Feb. 2010, pp. 237-252.

[2] H. Kaul, M. Andres, S. Hsu, A. Agarwal, R. Krishnamurthy, and S. Borkar, "Near-Threshold Voltage (NTV) design opportunities and challenges" in *IEEE DAC*, June 2012, pp. 1149 – 1154.

[3] Chaudhry Indra Kumar, Arvind Sharma, Sandeep Miryala and Anand Bulusu, "A Novel Energy-Efficient Self-Correcting Methodology Employing INWE," *IEEE SMACD*, June 2016.

[4] H. J. M. Veendrick, "The behavior of flip-flops used as synchronizers and prediction of their failure rate," IEEE J. Solid-state Circuits, vol. SC-15, no. 2, pp. 169-176, Apr. 1980.

[5] D. Li, D. Rennie, P. Chuang, D. Nairn, and M. Sachdev, "Design and analysis of metastable-hardened and soft-error tolerant high-performance, low-power flip-flops," in Proc. 12th Int. Symp. Quality Electron. Design, Mar. 2011, pp. 583–590.

[6] L. Kim and R. Dutton, "Metastability of CMOS latch/flip-flop," J. Solid-State Circuits, vol. 25, pp. 942–951, Aug. 1990.

[7] D. Rennie, D. Li, M. Sachdev, B. L. Bhuva, S. Jagannathan, W. ShiJie, R. Wong, "Performance metastability and soft-error robustness trade-offs for flip-flops in 40 nm CMOS", *IEEE Trans. Circuits Syst. I Reg. Papers*, vol. 59, no. 8, pp. 1626-1634, 2012.

[8] C. Portmann and T. Meng, "Metastability in CMOS Library Elements in Reduced Supply and Technology Scaled Applications," *JSSC*, vol. 30, pp. 39–46, January 1995.

[9] T. Sakurai, "Optimization of CMOS arbiter and synchronizer circuits with submicrometer MOSFET's," *IEEE J. Solid-State Circuits*, vol. 23, no. 4, pp. 901-906, Aug. 1988.

[10] C. Pacha, B. Martin, K. Arnim, R. Brederlow, D. Landsiedel, P. Seegebrecht , J. Berthold, and R. Thewes, "Impact of STI-induced stress, inverse narrow width effect, and statistical $V_{th}$ variations on leakage currents in 120 nm CMOS," in *IEEE ESSDERC*,2004, pp. 397-400.

[11] L.Xinfu, L. Kheeyong, W. Zhihua, X. Zhibin, D. Yongping, N. Hao, W. Yanping, S. Yanping, T. Bin, L. Louis, C. Sally, Y. Xing, H. Feng, and S. Yang, "A study of inverse narrow width effect of 65nm low power CMOS technology," in *IEEE ICSICT*, Oct. 2008, pp.1138-1141.

[12] Hyeokjae Lee, et al., An Anomalous Device Degradation of SOI Narrow Width Devices Caused by STI Edge Influence, IEEE Transactions on Electron devices. Vol.49, April 2002, pp605-611.

[13] C. M. Huang, T. Liu, and T. Chiueh, "An energy-efficient resilient flip-flop circuit with built-in timing-error detection and correction", in *IEEE VLSI-DAT*, Apr. 2015, pp. 1-4.

[14] S. Dillen, D. Priore, A. Horiuchi, and S. Naffziger, "Design and implementation of soft-edge flip-flops for x86-64 AMD microprocessor modules," *Proc. of IEEE Custom Integrated Circuits Conference*, Sept. 2012.

[15] Youngkyu Jang, Yoon Changnoh, Kim Jinsang, and Won-Kyung Cho, "Low-power variation-aware flip flop." in *IEEE ISCAS*, May 2012, pp. 488-491.

# A CMOS gate driver with ultra-fast dV/dt embedded control dedicated to optimum EMI and turn-on losses management for GaN power transistors

Plinio Bau[1,2], Marc Cousineau[2,] Bernardo Cougo[1], Frédéric Richardeau[2], Davy Colin, Nicolas Rouger[2]

[1] IRT Saint Exupéry, 3, rue Tarfaya - CS 34436, 31405 Toulouse cedex 4, France
Email : plinio.bau@irt-saintexupery.com - nicolas.rouger@laplace.univ-tlse.fr
[2] LAPLACE, Université de Toulouse, CNRS, Toulouse, France

*Abstract* — **In this paper, a CMOS gate driver in 180nm technology is presented. The gate driver implements an integrated and independent ultra-fast dV/dt control circuit dedicated to manage switch-on transients for GaN HEMT technology. In order to mitigate a detrimental effect in EMI spectrum for wide bandgap transistors, a novel method to reduce dV/dt without increasing so much switching losses is proposed. A comprehensive benchmark with the classical method is also presented, where the gate driver resistance is typically adjusted. Simulations are conducted to show the feasibility of the proposed method and the amount of switching energy that can be saved. Time responses of a feedback loop lower than 200ps are expected. The preliminary characterization of the integrated CMOS circuit is shown.**

*Keywords—Active Gate Driver, dV/dt, EMI, Electromagnetic Interference, HEMT Transistors, Wide Bandgap Transistors, GaN.*

## I. INTRODUCTION

To improve the performance of power converters, a more recent technology of wide bandgap devices can be used. Furthermore, power transistors based on GaN or SiC technologies are expected to represent a growing market in the next few years [1]. These components require less gate charge to switch on or off, and have overall better figures of merit than classical silicon based power devices. However, since these power transistors turn on faster, a larger amount of electromagnetic emission (EMI) is generated and can interfere with the systems close by. Therefore, several techniques to reduce dV/dt (and so undesired EMI effects) are being investigated [2-4]. In [2], a method using discrete transistors was proposed to actively drive silicon IGBT devices. In [3] a similar discrete approach was adapted for HEMT GaN transistors. This paper proposes a novel method to control turn-on switching speed based on CMOS integration offering very large bandwidth analog circuits. An improved solution, where all required devices are integrated into one ASIC is presented and discussed. Table I compares different methods of dV/dt reduction, $\Delta t_{MIN}$ is the time to switch all the $V_{HVdc}$ voltage.

TABLE I.    COMPARISON OF DIFFERENT METHODS OF DV/DT REDUCTION

|  | **Published by [2]** | **Published by [3]** | **This work** |
|---|---|---|---|
| Reduction in \| dV/dt \| | 1.8 to 0.5 V/ns | 27 to 8V/ns | 48 to 6V/ns |
| $\Delta t_{MIN}$ | 333 ns | 11 ns | < 1 ns |
| Embedded in an ASIC | No | No | Yes |

This work is a partnership between the research lab Laplace, and the research institute IRT Saint Exupéry. The authors also would like to thank to S. Vinnac and D. Flumian for their contribution to this work.

First, the well-known trade-off between the switching losses during the power transistor's turn-on ($E_{ON}$) and its related dV/dt (slew rate) value is briefly reminded. A method to improve this trade-off is then presented. Analytical study of the proposed CMOS circuits is also explained. In order to perform the slew rate measurement under high voltage condition, a double pulse test-bench is provided using a power switching-cell with two GaN HEMT transistors, implemented in a power PCB with the CMOS gate driver. Simulations are conducted and preliminary characterization of the integrated CMOS circuits is described.

## II. CLASSICAL APPROACH TO ADJUST THE LOSSES VS. SLEW RATE TRADE-OFF

The classical trade-off between the output voltage switching speed of a power device (dV/dt) and switching losses ($E_{ON}$) is presented in Fig. 1. Classically, the gate resistor $R_G$ is adjusted to obtain a good balance between EMI performance (related to switching speed) and switching losses. A low $R_G$ value reduces switching losses but increases the dV/dt value (point A. in Fig. 1 compared to point B). High dV/dt of several hundreds of V/ns have been already achieved [5]. As a consequence, the Common Mode Transient Immunity (CMTI) of signal and power isolators are pushed further, and EMI issues can dramatically degrade the operation of power converters. Higher $R_G$ values are then used to reduce the dV/dt, leading consequently to higher losses (point B in Fig. 1). Thus, innovative techniques are required to reduce at the same time the switching losses and the dV/dt. Using the proposed approach, presented hereinafter, point C of Fig. 1 can be reached, where dV/dt is reduced with a limited increased of $E_{ON}$.

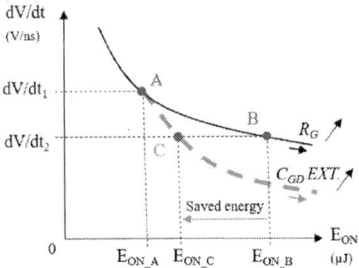

Fig. 1.   Trade-off $E_{ON}$ vs dV/dt function of $R_g$ (solid line) or $C_{GD}$ (dashed line) values for constant $V_{HVdc}$, $I_L$ and temperature. The proposed method aims to reach the line dV/dt₂ without increasing the external gate resistor $R_G$.

When a power transistor is turned ON (or OFF), some amount of charges has to flow in (or out of) the gate of the component. During the classic turn-on of a power MOSFET transistor, the derivative of the output voltage (dV/dt) depends on the gate current $I_{GM}$ and the gate to drain capacitance $C_{GD}$, as shown in (1):

$$\frac{dV_{DS}}{dt} = -\frac{I_{GM}}{C_{GD}} = -\frac{V_{DRV}-V_{GM}}{R_G C_{GD}} \qquad (1)$$

where $V_{DRV}$ is the power supply voltage of the driver and $V_{GM}$ is the Miller Plateau voltage.

As presented earlier, one can increase the gate resistor $R_G$ value to reduce the turn-on dV/dt. However, the switching losses will be increased, as shown in Fig. 2. This reduction in dV/dt corresponds to the transition from point A to B in Fig. 1, where the absolute value of dV/dt has been reduced from 15V/ns to 6.3V/ns, and the switching energy increased from 1.46µJ to 2.34µJ.

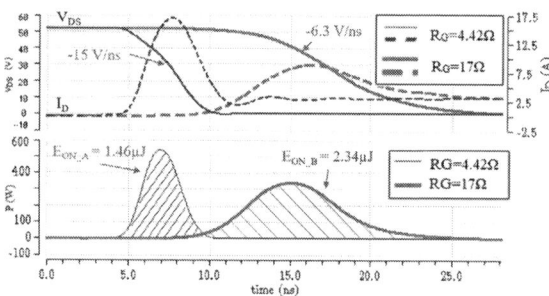

Fig. 2. Two classic simulated turn-on waveforms $V_{DS}$ and $I_D$ for different dV/dt with different $R_G$ values. Component GaN EPC 2001C switching $I_L$=3A, $V_{HVdc}$=50V at $V_{DRV}$=4V.

### III. ACTIVE GATE DRIVER ARCHITECTURE FOR DV/DT CONTROL

The principal limit of the previous techniques to control the slew rate is that both dV/dt and dI/dt phases are affected by a change in the gate resistor value. As a consequence, a new technique is required to control independently the dI/dt sequence and the dV/dt during the switching transitions. The main idea of our technique is to control separately the dI/dt by a fixed low $R_G$ value and the dV/dt by a local ultra-fast active feedback circuit.

A system-level diagram is shown in Fig. 3: the derivative of the voltage $V_{DS}$ (in the range of 50V up to 1200V) is sensed and converted into a small current $i_{dVdt}$, then amplified by a gain stage. This feedback current $I_{FB}$ sinks a part of the driver current $I_{RG}$ during the Miller Plateau, leading to a reduced $I_{GD}$. As a consequence, the dV/dt is reduced and controlled by this feedback loop. The reduction of dV/dt is proportional to the subtracted current from the gate terminal. The following expression is obtained:

$$\frac{dV_{DS}}{dt} = -\frac{V_{DRV}-V_{GM}}{R_G C_{GD}} + \frac{I_{FB}}{C_{GD}} = -\frac{V_{DRV}-V_{GM}}{R_G(C_{GD}+G C_S)} \qquad (2)$$

where G is the gain of the selected current mirror explained after.

The maximum current $I_{FB}$ that can be subtracted is the same as the maximum gate current obtained during the Miller Plateau

$I_{GM}$, defined in (1). We notice that if $I_{FB}$ is higher than $I_{RG}$ the output voltage $V_{DS}$ will oscillate.

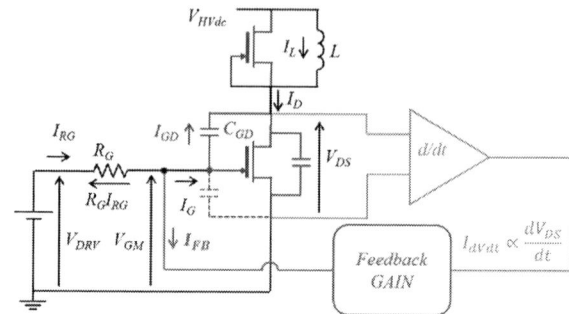

Fig. 3. System-level diagram of the proposed method in a switching arm setup for double pulse test-bench (operating during Miller Plateau).

In order to sense the derivative of the voltage $V_{DS}$, a small capacitor $C_S$ is used as a sensor. Note that this sense capacitor $C_S$ has to sustain the high voltage of the DC bus voltage. Therefore, a high voltage capacitor is needed. To amplify the current generated into the sense capacitor $C_S$ by the dV/dt, a large-bandwidth gain current mirror is proposed. However, the main challenge of this approach is to propose a circuit able to provide at the same time a high feedback current with a high bandwidth. This feedback current $I_{FB}$ can be as higher than 1A, and the turn-on time $T_{ON}$ during the dV/dt phase can be as low as 1ns.

Details of the feedback loop implementation is shown in Fig. 4. The feedback loop is composed by the four low voltage transistors M1 to M4, and the external high voltage sense capacitor $C_S$. The current buffer (or also called main stage driver) is made by a PMOS M5 and a NMOS M6 to source and sink high gate currents. This buffer M5-M6 has a split output architecture, making it possible to adjust separately the turn-on and turn-off switching losses vs. EMI trade-off. Capacitors $C_{GD}$, $C_{GS}$ and $C_{DS}$ are intrinsic to the GaN power transistor. Diode $D_P$ is used only as a protection at the current mirror input and is reverse biased during normal operation.

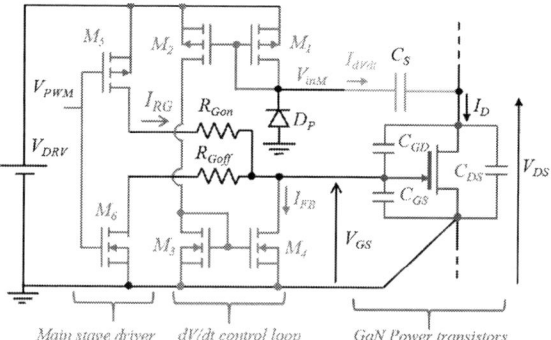

Fig. 4. Transistor-level schematic of the CMOS gate driver with a dV/dt feedback control loop for switch-on control.

## A. Transistor sizing and CMOS design

The power device chosen is the GaN EPC2001C. The maximum voltage $V_{DS}$ for this component is 100V. To ensure a safety margin of overvoltage signals during turn-off, the value of 50V is chosen for the $V_{HVdc}$ bus voltage of the test-bench.

A test-chip integration approach has been adopted, where several current mirror gains have been integrated on the same chip with the output buffer. The current mirror gain values are 5, 10, 20 and 50. Each gain is provided by the two cascaded current mirrors M1 to M4. For the circuit with gain 20 for example, the length of M2 ($L_2$) is 3 times $L_1$ and the length of M4 is 6.6 times larger than M3 providing a total gain of 20.

To calculate the gate width of M1, the current mirrors are designed to make the input voltage sweep from $V_{DRV}$ to 0.5V (safety margin towards 0V) when the input current goes from 0A to the maximum value obtained at maximum dV/dt. Note that a constant 1-10µA biasing current source can be added to pre-charge M1 and make it ready to respond faster. This current source must be placed in parallel with the diode $D_P$ (not shown in Fig. 4).

Another dimensioning parameter is the ratio between M2 and M3. When dV/dt is maximum, $I_{dVdt}$ is also maximum. In that case the width of M2 and M3 are defined so that a voltage equal to $V_{DRV}/2$ is obtained on their drains.

The propagation delay from the input to the output of the current mirror is simulated. A value of less than 200ps is obtained in the worst case using a gain of 50, where the transistors are wider, with larger intrinsic capacitances.

Notice that the size of the transistor M4 is about the same size of M5 and M6 in the main stage driver. This is because the order of magnitude of the currents $I_G$ and $I_{FB}$ are similar. Note that the transistors used to control M5 and M6 are not shown in Fig. 4 for the sake of simplicity.

## B. Simulations Considering PCB Parasitic Components

Fig. 5 shows simulation results of the dV/dt control circuit. The dV/dt is reduced by the proposed active gate driver (case C of Fig. 1), compared with a fixed small $R_G$ value (case A). The $I_{GD}$ and $V_{GS}$ waveforms clearly show the turn-on sequence: the Miller plateau's duration is extended with our technique, without modifying the dI/dt sequence. Circuits are simulated with Cadence Spectre simulator using the design-kit libraries of the AMSHI18A6 technology. One can notice the fast transient response of the control loop where $V_{DS}$ voltage is switched in few nanoseconds.

Table II shows the simulated values related to the points presented in Fig. 1. Reduction in dV/dt is achieved using the proposed method with less switching energy $E_{ON}$.

TABLE II.    SIMULATING DV/DT AND EON IN OPEN AND CLOSED LOOP

| Point* | $R_G$ (Ω) | $E_{ON}$ (µJ) | \|dV/dt\| (V/ns) |
|--------|-----------|---------------|------------------|
| A | 4.42 | 1.46 | 15 |
| B | 17 | 2.34 (+60%) | 6.3 (-58%) |
| C | 4.42 | 1.93 (+32%) | 6.3 (-58%) |

*Same points as defined in Fig. 1

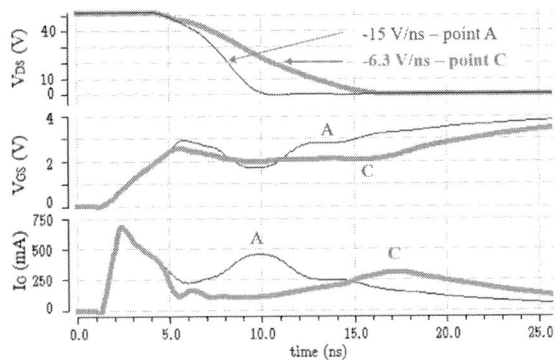

Fig. 5.  Cadence simulation waveforms of $V_{DS}$, $V_{GS}$ and $I_G$, with (Case C) and without (Case A) the proposed method, including PCB parasitic elements.

Fig. 6 shows the three currents involved at the gate node of the power device. The main idea of the method can be seen in this graph: the dV/dt generates a current $I_{dVdt}$ in the high voltage capacitor $C_S$, which is further amplified to $I_{FB}$ by the current mirror and then subtracted from $I_{RG}$. The resulting gate current $I_G$ during the Miller plateau is then reduced only during the dV/dt part of commutation. The second part of Fig. 6 shows the input voltage of the current mirror $V_{inM}$. This waveform starts from 3.6V (and not $V_{DRV}$=4V) because of the 2µA biasing current. A dV/dt signal makes a current flow through $C_S$ and the left terminal of this capacitor is dropping to a value towards zero when the current mirror is active.

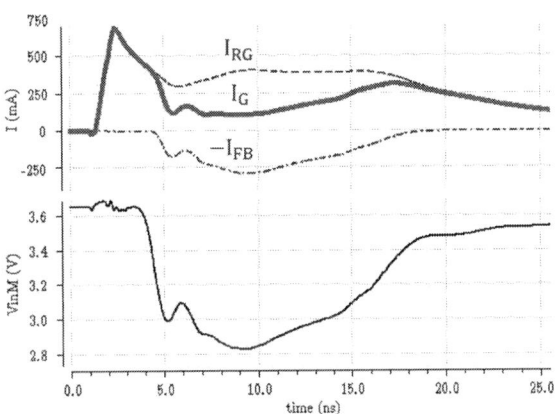

Fig. 6.  Current waveforms of the driver $I_{RG}$, the feedback loop $I_{FB}$, the transistor gate $I_G$ and the waveform of the input voltage of the current mirroir (case C of Fig.1)

## IV. PRELIMINARY CHARACTERIZATION

The circuit is fabricated in the 180nm AMS bulk CMOS technology and a microscope picture is shown in Fig. 7. Fig. 8 shows the characterization of the current mirrors with a wafer prober test Cascade MPS150 and a Keithley 2612B source meter unit. Linear pulsed-mode current sweeps have been used to prevent self-heating during the measurements. Tests are conducted with a biasing current of 2µA and $V_{DRV}$=4V. One can note that the output currents are in the same order of magnitude than $I_G$ achieved during Miller plateau.

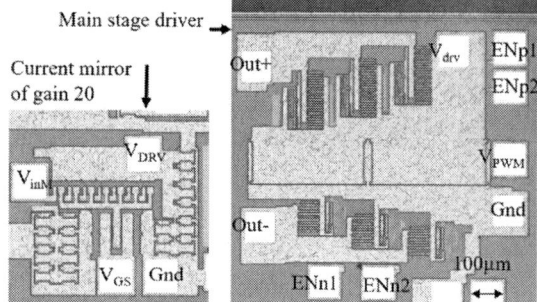

Fig. 7. Microscope picture of parts of the CMOS gate driver.

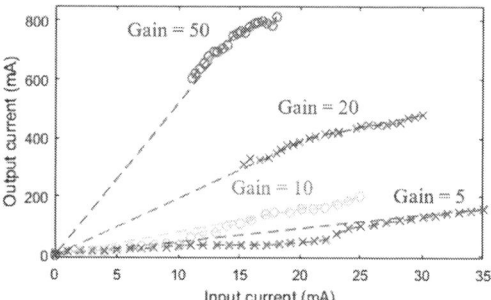

Fig. 8. Current mirrors characterization using pulsed DC method with a probe station.

The output buffer and the current mirrors have been packaged in QFN24 6mmx6mm. Decoupling capacitors of 2.2µF are placed close to the power supply input pins of the chip (both for $V_{DRV}$ and current mirrors). Decoupling capacitors for the $V_{HVdc}$ bus voltage are 6 capacitors of 1µF (in parallel to reduce parasitic inductance) and one of 10µF. Protection buffers for $V_{PWM}$ signal and enable bits of the main stage driver (ENn1, ENn2, ENp1 and ENp2, see Fig. 7) are added as showed in Fig. 9. A 4-layer board is used to reduce the parasitic inductance of the power loop. Fig. 10 shows the waveforms of the measured and simulated $V_{DS}$ signal during turn-on for case A. Those two waveforms are confronted to extract PCB parasitic components, and are shown in this same plot.

## V. CONCLUSION

A gate driver with a built-in ultra-fast active dV/dt control loop has been designed and implemented in a CMOS technology. Thanks to the proposed method, negative EMI effects of the fast transients occurring during the turn-on of new wide-bandgap transistors can be drastically attenuated. The active gate driver is described. Simulations and preliminary characterizations are shown. Experimental measurements are ongoing in order to demonstrate the performances of the ultra-fast close-loop active gate driver.

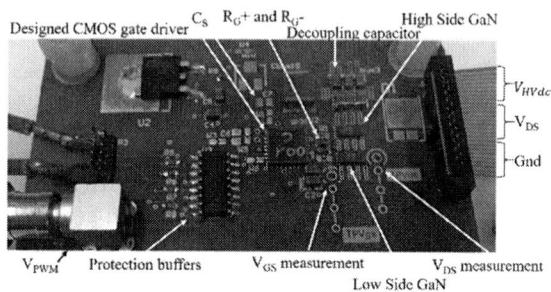

Fig. 9. Hardware for double pulse test-bench with the proposed dV/dt control circuit.

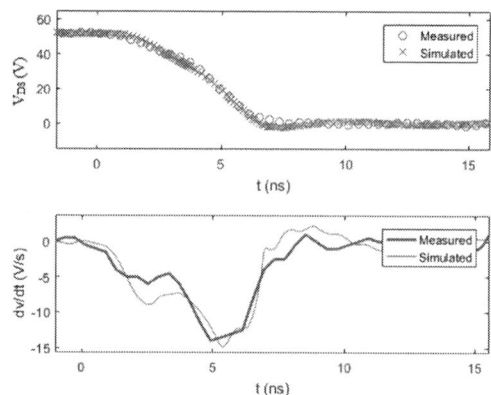

Fig. 10. Measured and simulated dV/dt during turn-on transients for $R_G$=4.45Ω (case A of Fig. 1).

## REFERENCES

[1] A. Bindra, "Wide-Bandgap Power Devices: Adoption Gathers Momentum," in *IEEE Power Electronics Magazine*, vol. 5, no. 1, pp. 22-27, March 2018.

[2] Shihong Park and T. M. Jahns, "Flexible dV/dt and di/dt control method for insulated gate power switches," in IEEE Transactions on Industry Applications, vol. 39, no. 3, pp. 657-664, May-June 2003.

[3] B. Sun, R. Burgos, X. Zhang and D. Boroyevich, "Active dV/dt control of 600V GaN transistors," 2016 IEEE Energy Conversion Congress and Exposition (ECCE), Milwaukee, WI, 2016, pp. 1-8.

[4] Pierre Lefranc, Dominique Bergogne. "State of the art of dv/dt and di/dt control of insulated gate power switches". Proceedings of the Conference Captech IAP1, Power Supply and Energy Management for Defence Applications, Jun 2007, Bruxelles, Belgium. pp.1-8, 2007.

[5] S. Moench, P. Hillenbrand, P. Hengel and I. Kallfass, "Pulsed measurement of sub-nanosecond 1000 V/ns switching 600 V GaN HEMTs using 1.5 GHz low-impedance voltage probe and 50 Ohm scope," *2017 IEEE 5th Workshop on Wide Bandgap Power Devices and Applications (WiPDA)*, Albuquerque, NM, 2017, pp. 132-137.

[6] T. Nieminen, O. Viitala, M. Voutilainen and J. Ryynaenen, "A Scalable Low-Voltage Signaling (SLVS) Driver for a Low-Power MIPI M-PHY Serial Link in 40 nm CMOS," *PRIME 2012; 8th Conference on Ph.D. Research in Microelectronics & Electronics*, Aachen, Germany, 2012, pp. 1-4.

[7] D. Colin and N. Rouger, "High speed optical gate driver for wide band gap power transistors," *2016 IEEE Energy Conversion Congress and Exposition (ECCE)*, Milwaukee, WI, 2016, pp. 1-6.

PRIME 2018, Prague, Czech Republic

Session: Power Circuits and Harvesting

# Design of a SIBO DC-DC Converter for AMOLED Display Driving

Filippo Boera, Arunkumar Salimath, Edoardo Bonizzoni, Franco Maloberti

Dept. of Electrical, Computer and Biomedical Engineering

University of Pavia

Pavia, Italy

E-mail: filippo.boera01@universitadipavia.it

*Abstract*—This paper describes a Single Inductor Bipolar Outputs (SIBO) DC-DC converter for Active Matrix Organic Light Emitting Diode (AMOLED) displays. The circuit is able to generate with a single inductor both the positive and the negative voltage necessary to turn on the AMOLED pixels and does not require the addition of post-regulation techniques. The circuit works with an input battery voltage ranging from 2.4 V to 4.9 V, compliant with Li-Ion batteries, and is able to generate +5 V and −6 V. The maximum output current delivered to the load is 0.6 A. The switching frequency is 2 MHz and the control loop has been implemented in Voltage Control Mode (VCM). The effectiveness of the single inductor based scheme has been verified at the behavioral level in Matlab-Simulink.

## I. INTRODUCTION

During the last decade, AMOLED displays are increasingly replacing the traditional Liquid Crystal Displays (LCDs). While at first only in portable devices, AMOLED displays can now be found also in larger size devices, like televisions. The advantages of AMOLED displays over LCDs are many: higher image quality (in terms of contrast ratio, refresh rate, and viewing angles), thinner displays, and lower power consumption, [1]. Flexible displays are realizable using the AMOLED technology, [2], which led to the production of curved displays in recent smartphones, already available on the market, and foldable displays are expected in the near future. The core advantage of AMOLED displays is the usage of thin layers of an organic (i.e., carbon-based) material that have the property of producing luminescence if stimulated by an electrical current. Since the luminescence comes from the organic layer itself, there is no need for a backlight panel, while LCDs use a LED backlight panel to provide light to the pixels.

In an OLED pixel, the organic layer is placed between two electrodes that work as anode and cathode, providing the current required to turn on the pixel when the voltage is applied. The single pixel is modeled as a diode and, in order to build a full screen, all the pixels are arranged in a matrix. A distinction has to be specified between Passive Matrix OLED (PMOLED) and AMOLED. Fig. 1(a) shows a PMOLED panel: multiple electrodes are placed orthogonally on the top and bottom of the organic layer, defining the rows and columns of the matrix. Each interception of the matrix is a pixel. Fig. 1(b) shows the simplified schematic diagram of a single pixel of a PMOLED display. As shown in the figure,

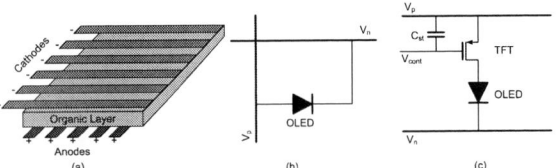

Fig. 1. Simplified schematic diagrams of a) a PMOLED panel, b) a PMOLED pixel, and c) an AMOLED pixel.

the pixel consists of a diode and two supply lines: one positive ($V_p$) and one negative ($V_n$). In order to draw a full image, the pixels of the matrix need to be turned on sequentially, [3]. The main problem in the PMOLED technology is that, when line $N$ is turned on, line $(N − 1)$ is turned off, and the pixels on that line do not have a way to store their charge, so they will be turned off.

An active matrix addressing scheme is used in AMOLED displays: the electrodes supply all the pixels at the same time, and the on/off state of each pixel is controlled by a matrix of thin-field transistors (TFTs). The schematic diagram of a single AMOLED pixel is shown in Fig. 1(c). A storage capacitor $C_{st}$ is added, so that, when the next line is activated, the pixels can maintain their charge, [4]. The active matrix addressing scheme ensures a consistent improvement in the performance of the display. Since PMOLEDs need more brightness and higher voltage than AMOLED, the lifetime of a PMOLED screen is reduced. AMOLED can also provide a higher refresh rate. These are the main reasons why AMOLED is the only dominant OLED technology in the smartphone and tablet market, while the use of PMOLED is limited to smaller, low-resolution displays, [5].

Each pixel of Fig. 1(c) needs three different voltages to be turned on: +5 V for the positive supply, −6 V for the negative supply, and a proper gate voltage at the TFT to turn on a particular pixel. The typical Li-Ion battery of a smartphone provides a single voltage ranging from 2.4 to 4.9 V. In order to generate the two supply voltages in AMOLED displays, two separate DC-DC converters are typically used: a boost converter for the positive voltage, and an inverting buck-boost for the negative one.

In this paper, a DC-DC converter that can provide both

978-1-5386-5388-3/18 $31.00 © 2018 IEEE 109

+5 V and −6 V using only one inductor is presented. The use of a single inductor instead of two leads to an obvious and significant reduction in the bill of material and of the overall system area, making the proposed scheme particularly attractive.

## II. Review of the State of the Art

The typical approach of using two DC-DC converters to generate $V_p$ and $V_n$ in Fig. 1 is well consolidated, easy to design and suitable for time-to-market. Its obvious limitation is the need of two inductors: the bulkiness and the integration difficulty of inductors are well known issues in integrated circuits, especially when dealing with portable devices, where the area consumption is a critical parameter. Another problem is the need for a Low-Dropout Regulator (LDO) for the positive output, which is power consuming and, hence, degrades the overall system efficiency, [6]. Different attempts have been proposed in order to spare one inductor in the generation of the two required voltages. In [7] an interesting way of generating both outputs with only one inductor is presented: the positive output is obtained with a boost converter, while the negative one is generated by means of a reverse charge pump. However, the complexity of the proposed loop control techniques mandates the use of a large active area, and the ripple at the positive output would still require some post-regulation techniques. A less area-consuming solution for generating the two desired voltages is the use of the Single Inductor Bipolar Outputs (SIBO) configuration: the two output nodes are connected to the complementary terminals of the inductor. This will result in outputs having opposite polarity, [8]. Some examples of SIBO DC-DC converters for OLED displays can be found, [9], [10], but again the stringent requirements on the positive output ripple for portable devices still require the addition of post-regulation techniques like LDOs [9]. The next Section presents a non-conventional design of a SIBO DC-DC converter for AMOLED displays that overcomes the above mentioned drawbacks.

## III. Proposed SIBO DC-DC Converter

### A. Power Stage

Fig. 2 shows the schematic diagram of the power stage of the proposed SIBO DC-DC converter. $V_p$ indicates the positive output voltage, $V_n$ the negative output voltage, $V_{bat}$ the input battery voltage. Two equal filter capacitors, $C_1$ and $C_2$, are connected to the two terminals of the shared output load, $R$.

Within each switching period, $T$, three phases of operation are needed: an inductor charge phase with duty cycle $D_1$ and two discharge phases with duty cycles $D_2$ and $D_3$, respectively. Fig. 3 shows how the circuit is reconfigured during each phase. In the charge phase of Fig. 3(a), $SW_1$ and $SW_2$ are closed to charge the inductor, $L$. The current, $I_L$, flows from $V_{bat}$ to ground, energising the inductor. In this phase, $C_1$ and $C_2$ supply the load current. In the first discharge phase, Fig. 3(b), $SW_1$ and $SW_2$ are opened, while $SW_3$ and $SW_4$ are closed. The inductor discharges to transfer charge into $C_1$. The voltage at the terminals of the inductor is $−V_p$, while the inductor current

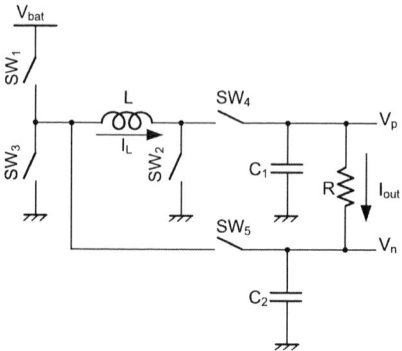

Fig. 2. Schematic diagram of the proposed SIBO converter.

$I_L$ flows into $C_1$. In the second discharge phase, Fig. 3(c), $SW_3$ and $SW_4$ are opened, while $SW_5$ and $SW_2$ are closed. The voltage across the inductor is $V_n$. Since $V_n$ is connected to the complementary terminal of the inductor, and the inductor current can not change its direction, during this phase, the inductor current discharges $C_2$. Fig. 4 shows the inductor current in Continuous Conduction Mode (CCM) together with the three duty cycles identifying the three phases of operation.

### B. Control Strategy

The Voltage Mode Control (VCM) technique is chosen to regulate the SIBO DC-DC converter. The block diagram of the control loop is shown in Fig. 5. To regulate both outputs, two different error signals must be calculated. The two error signals, $V_{err1}$ and $V_{err2}$, are obtained by suitably scaling both the positive output $V_p$, and the difference of the outputs $V_p − V_n$. The scaled-down voltages are subtracted from a reference voltage $V_{ref}$. The two branches of the control loop process the sum ($V_{err1} + V_{err2}$) and the difference ($V_{err1} − V_{err2}$) of the two error signals, respectively. The linear combinations of the two error signals are, after compensation, compared with a sawtooth signal, $V_{saw}$. The two Pulse-Width Modulated (PWM) signals at the output of the comparators are used to generate the required duty cycles $D_1$, $D_2$, and $D_3$.

### C. Loop Analysis and Compensation

VCM regulation requires error compensation to ensure the stability of the loop. An equivalent small signal model of the SIBO converter is obtained using the perturbation and linearisation approach, [11]. The control-to-output transfer function, $G_{vd}(s)$, is calculated considering the following parasitic effects: the equivalent series resistances (ESR) of the capacitors $R_{C1} = R_{C2} = R_C$, the inductor series resistance, $R_L$, and the on-resistance of the switches, $R_{on}$, kept equal for all the switches. It results:

$$G_{vd}(s) = \frac{D_2 V_{bat} - I_L R_{loss}}{D_2(D_2^2 R + R_{loss})} \frac{(1 - \frac{sLI_L}{D_2 V_{bat} - I_L R_{loss}})(1 + sCR_C)}{1 + s\frac{L + C[R_{loss}(R+R_C)+D_2^2 RR_C]}{D_2^2 R + R_{loss}} + s^2 \frac{LC(R+R_C)}{D_2^2 R + R_{loss}}}$$
(1)

*PRIME 2018, Prague, Czech Republic*  *Session: Power Circuits and Harvesting*

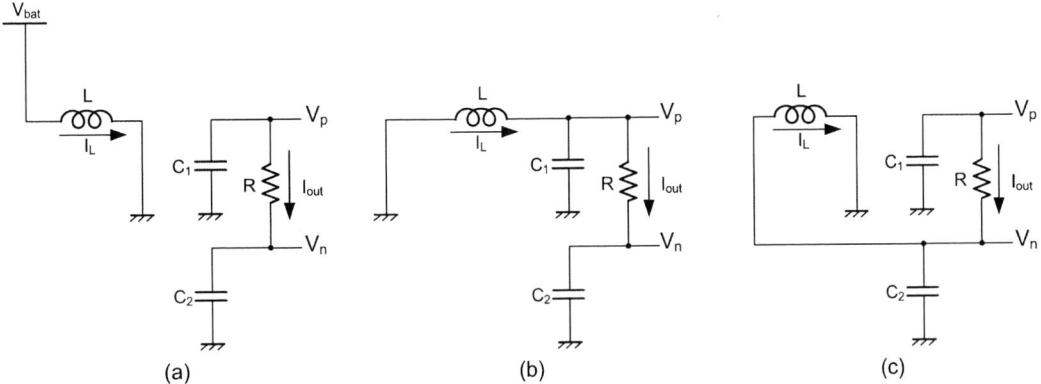

Fig. 3. Circuit configuration for a) inductor charghe phase $D_1$, b) dicharge into the positive output phase $D_2$, c) discharge into the second output phase $D_3$.

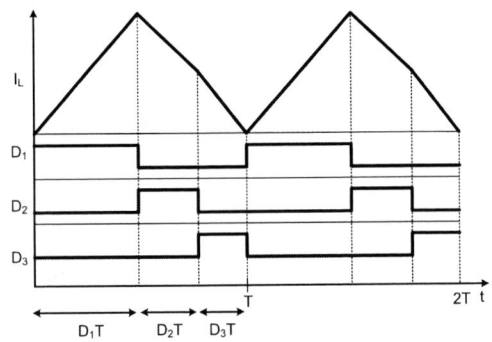

Fig. 4. Steady-state switching cycle of the power stage in CCM.

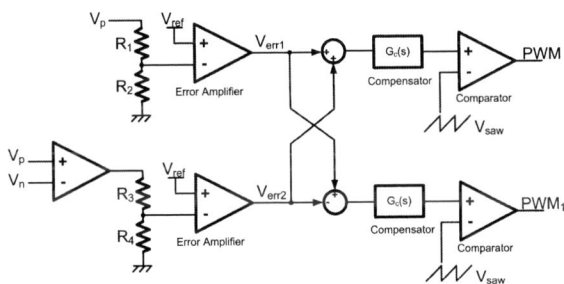

Fig. 5. Bock diagram of the control loop.

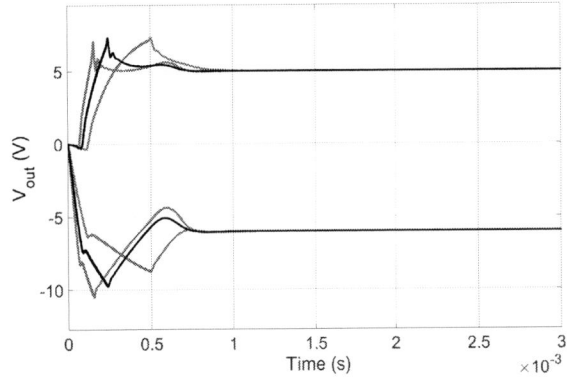

Fig. 6. Simulated positive and negative output voltages for $V_{bat}$ = 2.4 V (green), 3.6 V (black), 4.9 V (red).

50 kHz in the maximum battery voltage condition. Stability at other line and load conditions is verified by simulations.

## IV. SIMULATION RESULTS

The proposed SIBO DC-DC converter has been modeled and simulated at the behavioural level in Matlab-Simulink. The considered values of the inductor, $L$, and of the two filter capacitors, $C_1$ and $C_2$, are 470 nH and 25 µF, respectively. The above values are adequate for modern smartphone applications. The chosen switching frequency is 2 MHz. The load resistance, $R$, is 36 Ω in order for the output current to be set at around 0.3 A. The switches are chosen to be all equal, with the lowest on-resistance that can be achieved with a reasonable area with modern CMOS technologies (50 mΩ). The ESR of the capacitors is 5 mΩ, and the inductor series resistance is equal to 20 mΩ. The saturating current of the inductor is considered to be equal to 6 A.

The SIBO DC-DC converter steady-state behaviour has been verified in the three conditions of maximum, minimum, and typical battery voltage value (4.9 V, 2.4 V, and 3.6 V,

where $R_{loss}$ is the total parasitic resistance seen by the inductor, equal to $R_L + 2R_{on}$ in each of the three phases. The transfer function has two poles and two zeroes. These are compensated by the compensator $G_c(s)$ (see Fig. 5), at the lowest frequency locations of the poles and zeroes. The lowest frequency of poles and zeroes in (1) is for the minimum battery voltage value and the maximum considered load current. A PID scheme is used to compensate for the poles and zeroes in $G_{vd}(s)$. The resulting crossover frequency is 10 kHz, reaching

978-1-5386-5388-3/18 $31.00 © 2018 IEEE

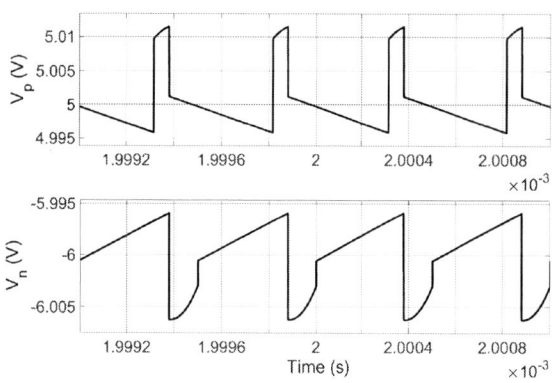

Fig. 7. Detail of the output ripples.

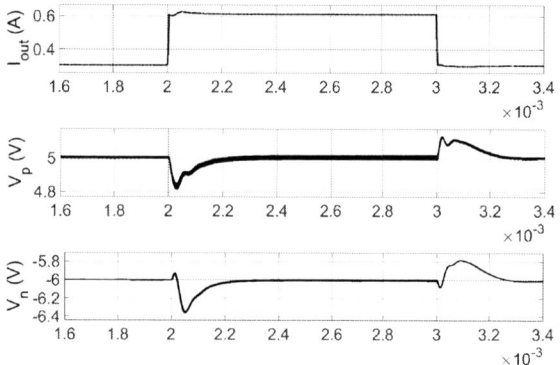

Fig. 8. Simulated load transient response.

respectively). Fig. 6 shows the achieved results. The circuit is correctly regulating the desired +5 V and −6 V in all the range of $V_{bat}$. The initial overshoot is due o a forced startup 90% duty cycle square wave. Fig. 7 shows the detail of the ripples on the positive and negative output. The ripple on $V_p$ is 15 mV while the ripple on $V_n$ is 5 mV. Both the achieved values are well below what required (generally, 20 mV) for display driving of modern smartphones. An estimation of the efficiency based on conduction and switching losses leads to an efficiency above 80% in all the battery conditions.

The load transient response was simulated by changing the load current with a slope of 60 mA/μs from 300 mA 600 mA, which is the maximum value allowed for the output current. The shape of the load current, as well as the load transient response of $V_p$ and $V_n$, is reported in Fig. 8. The circuit exhibits a very good load transient response: 0.1 mV/mA for the positive output and 0.18 mV/mA for the negative output. Table I summarises the achieved performance.

## V. Conclusion

In this paper, the design of a SIBO DC-DC converter for AMOLED display driving has been discussed. The circuit

generates the two output voltages (+5 V and −6 V) required to turn on an AMOLED pixel using only one inductor and without requiring post-regulation techniques. The input battery voltage ranges from 2.4 V to 4.9 V, the typical values of a Li-Ion battery for portable devices. The switching frequency of the DC-DC converter is 2 MHz. The steady-state behaviour has been verified: the two output voltages are correctly regulated and their ripple is below the maximum allowed by the design specifications of modern smartphones displays. The simulated load transient response is also adequate for the target application.

### TABLE I
PERFORMANCE SUMMARY

| Battery voltage range | from 2.4 V to 4.9 V | |
|---|---|---|
| Inductor | 470 nH with ESR= 20 mΩ | |
| Filter capacitors | 25 μF with ESR= 5 mΩ | |
| Switching frequency | 2 MHz | |
| Inductor current ripple | 2.2 A | |
| Estimated efficiency | >80% | |
| Output voltages | +5 V | −6 V |
| Output ripples | 15 mV | 5 mV |
| Load regulation | 0.1 mV/mA | 0.18 mV/mA |
| Line regulation | 66 mV/V | 72 mV/V |

### REFERENCES

[1] 4D Systems, "Introduction to OLED Displays Design Guide for Active Matrix OLED (AMOLED) Displays", www.4dsystems.com.au.

[2] Y. Nakaima, T. Takei, G. Motomura, T. Tsuzuki, H. Fukagawa, M. Nakata, H. Tsuji, T. Shimzu, K. Morii, M. Hasegawa, Y. Fuisaki, T. Kurita, and T. Yamamoto, "Flexible AMOLED Display Using an Oxide-TFT Backplane and Inverted OLEDs ", *IEEE Photonics Conference*, October 2014.

[3] S. Forest, P. Burrows, M. Thompson , "The dawn of organic electronics", *IEEE Spectrum*, vol. 37, no. 8, pp. 29–34, August 2000.

[4] I. Pappas, S. Siskos, and C. A. Dimitriadis, "Active-Matrix Liquid Crystal Displays - Operation, Electronics and Analog Circuits Design", *New Developments in Liquid Crystals*, Georgiy V Tkachenko (Ed.), InTech, DOI: 10.5772/9686.

[5] OLED-info, "PMOLED vs AMOLED–what's the difference?", July 2013, www.oled-info.com.

[6] STOD13AS: 250 mA Dual DC-DC Converter For Powering AMOLED Displays. ST Microelectronics, 2012, www.st.com.

[7] C. S. Chae, H. P. Le, K. C. Lee, G. H. Cho and G. H. Cho, "A Single-Inductor Step-Up DC-DC Switching Converter With Bipolar Outputs for Active Matrix OLED Mobile Display Panels,"*IEEE Journal of Solid-State Circuits*, vol. 44, no. 2, pp. 509-524, Feb. 2009.

[8] D. Kwon and G. A. Rincon-Mora, "Single-InductorMultiple-Output Switching DCDC Converters", *IEEE Transactions on Circuits and Systems II: Express Briefs*, vol. 56, no. 8, pp. 614-618, Aug. 2009.

[9] TPS65120-4: Single-Inductor Quadruple-Output TFT LCD Power Supply. Texas Instruments, Dallas, TX, 2004. www.ti.com.

[10] B. C. Kwak, S. K. Hong and O. K. Kwon, "A Highly Power-Efficient Single-Inductor Bipolar-Output DC-DC Converter Using Hysteretic Skipping Control for OLED-on-Silicon Microdisplays," in IEEE Transactions on Circuits and Systems II: Express Briefs.

[11] R. W. Erickson, D. Maximovic "Fundamentals of Power Electronics", Springer, 2001.

*PRIME 2018, Prague, Czech Republic*  ·  *Session: Power Circuits and Harvesting*

# A human body powered sensory glove system based on multisource energy harvester

Alfiero Leoni, Vincenzo Stornelli, Giuseppe Ferri
*Dept.of Industrial and Information Engineering and Economics*
P.le Pontieri 1, 67100, Monteluco di Roio
L'Aquila (Italy)
email: vincenzo.stornelli@univaq.it

Vito Errico, Mariachiara Ricci, Antonio Pallotti, Giovanni Saggio
*Dept.of Electronic Engineering*
via del Politecnico 1, 00133, Roma (Italy)
Email: saggio@uniroma2.it

*Abstract*— **In this work we present and evaluate a multi-source power management system, based on human body energy harvesting, to extend the battery lasting of an electronic sensory glove, used to measure flexion/extension, abduction/adduction movements of fingers of the hand. The system exploits heat of the human forearm and pressure impressed by the foot heel during walking, so to gather additional energy. The aim is to allow hours of energy-autonomy for the user working with the sensory glove. Such a glove is equipped with a number of flex sensors which furnish data from finger movements, acquired and pre-processed by a microcontroller, and wireless sent to a Personal Computer for analysis, visualization and storage purposes. The multi-source harvester is based on vibrational and thermic sources. Prototype discrete element boards were designed and tested for the microelectronics integration. Measurement results demonstrate how the overall system extends the battery lasting time up to 20%.**

*Keywords—Energy harvesting, power management, Sensory glove*

## I.  INTRODUCTION

Human hands are the most important way of interaction with the world around us, so much so that many other parts of the body are used very often for the sole purpose of placing our hands in a specific point of the space. Therefore, the possibility to track and measure the human hand motions can be beneficial for many applications, like sign language recognition [1,2], post-trauma therapy [3,4] or Virtual Reality (VR) [5] for serious-game purposes [6]. In this scenario, a sensory glove can be advantageously adopted, since it can sense all the fingers and wrist movements [6,7]. Typically, the hand sensory system is composed by a couple of light, elastic gloves equipped with flex sensors able to provide electric signals related to specific movements of hand fingers. These gloves overcome limitations of traditional optical measurements which force the user to stand under the camera-lights. Generally speaking, the hand movements are represented by 27 degrees of freedom (DOFs), according to a commonly adopted kinematic hand model [8], including flexion and extension of the fingers joints and

rotation/bending of the wrist. Because of the complexity, usually only a subgroup of DOFs is considered, leading to lack of accuracy. Recently, a combination of flex sensors and inertial measurement unit (IMU) sensors for a complete 27 DOFs hand tracking glove system has been designed and evaluated by some of the authors (Figure 1).

Performance and capabilities of this system have been largely analyzed and discussed elsewhere [9-12], with evidence of good values in repeatability and reproducibility of the measurements. Nevertheless, one of the main issues regards the portability and power-supply autonomy of the whole system, including the electronic circuits, which need a local, adequate energy source to guarantee a long-life wireless functionality, as well as lightness and low encumbrance. Traditional supplying systems, including large size embedded batteries or wired systems, are not appropriate, in terms of human comfort. For this reason, here we considered the development of an unwired, self-powered, size- and weight-reduced system which, based on energy harvesting techniques, can allow gathering "environmental" energy.

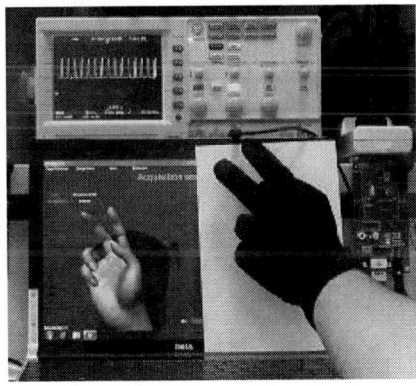

*Figure 1:  27 DOFs sensory glove*

Despite remarkable efforts dedicated to developing power harvesting techniques to scavenge power either from the environment or from the human body, these techniques are not yet considered reliable and feasible, but research is steadily progressing [13]. Currently, Energy Harvesting (EH) [14-18] and its applications for smart glove and generally array of sensor systems or wireless sensors networks are more and more gaining in interest [19-21]. The source-power required by the sensory glove can be harvested by means of different energy sources, depending on the application and the environment, such as electromagnetic, solar, thermal, acoustic, vibrational, and so ahead. Generally, it is not a trivial matter to integrate different harvesters within the same system, especially considering that they have to be simultaneously available and furnish an approximatively equal energy level. In this perspective, here we propose the way to extend the battery life-time of the sensory glove by gathering energy directly from the human heat and vibrations. The here proposed multi-harvester system integrates Thermo-Electric Generators (TEGs) - that can be applied to the forearm - and piezoelectric disks, placed in the heel of a shoe, so to scavenge energy from the pressure generated by feet during walking or a running session. In order to demonstrate the feasibility of the overall system, we have realized and tested a prototype board.

## II. GENERAL SYSTEM OVERVIEW

Nowadays wireless sensors connectivity is desirable in many applications, such as in medical health care, smart home, gaming, remotely controlled devices, and so on. Wireless connectivity adds freedom in person life who is wearing it (freedom of moving, freedom to work from anywhere within the possible range, etc.) and reduces the constraint in natural environment. Our overall system (Figure 2) is made by four different main blocks: 1) the glove, 2) the multi-harvester block, 3) the transmitting block, 4) the receiving block.

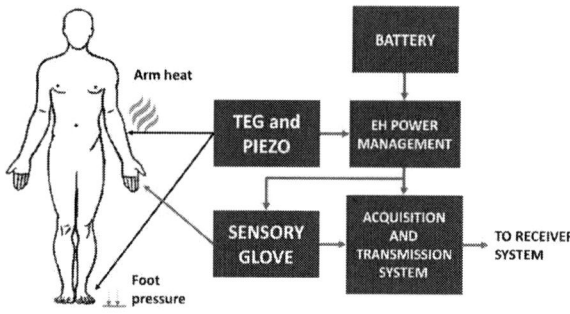

*Figure 2: The complete system block scheme*

Figure 3 shows a typical application scenario. The specific working principle and electronic interface of the sensory glove is discussed elsewhere [6] and, for the sake of shortness, not replayed here. Differently, this work is mainly focused on the multisource harvester block, here as a discrete prototype (while in the future it can become an integrated solution reducing occupation area). The transmitting and receiving blocks are made by commercial discrete components too.

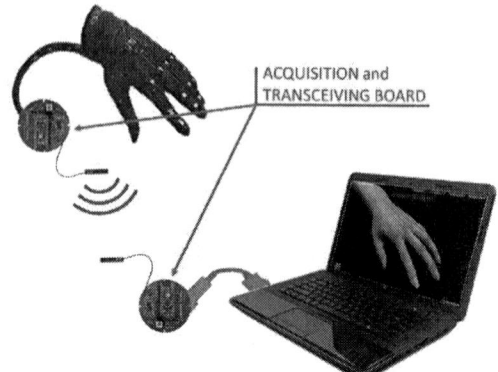

*Figure 3: Data glove acquisition and visualization process*

## III. THE MULTISOURCE ENERGY HARVESTING SENSORY GLOVE AND THE TRANSMITTING ARCHITECTURE

Figure 4 shows our double source energy harvesting system, with piezoelectric and TEG generators. The harvester outputs contribute in parallel to directly power the system circuit. Each harvester sub-system is terminated with an off-chip dual Schottky diode to avoid reverse current flow. Of course, this solution affects the overall harvester efficiency dissipating energy over the diode (even if this phenomenon is mitigated by a low reverse voltage drop of the Schottky barrier) but, usefully, the total amount of collected energy is greater than power coming from a unique energy source.

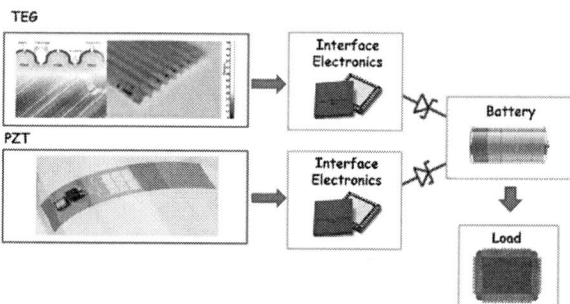

*Figure 4: Multi-harvester block scheme*

As shown in figure 5, the EH system is conceived to provide the required current directly from the gathered energy, which is converted in a regulated voltage output. When the energy source is absent or not sufficient to supply the glove, an extra current is provided by the battery. In this discrete prototype, the commercial component LTC3107 from Linear Technology is used to gather energy from a set of TEGs, and to arbitrate the power path with the battery, while the LTC3588 is used to convert the energy obtained by the human walk. In detail, the energy coming from the human heat is collected by means of a set of standard 3cm x 3cm Peltier cells (see Figure 6a for the prototype harvester implementation).

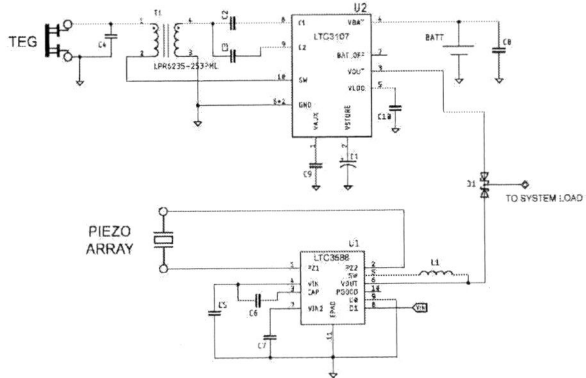

Figure 5: Harvesting system electrical scheme

This is because of the distribution of the heat along the human arm, which is not uniform, thus each cell cannot provide exactly the same output voltage and a parallel connection could cause a dispersion current between cells, reducing the overall efficiency. Since the TEG cells, in this configuration, can provide an output voltage ranging from 20mV (static arm, rest condition) to 180 mV (movement), a 1:100 transformer has been employed for the fly-back converter, integrated in the LTC3107, in order to allow the DC-DC converter so guaranteeing its functionality with input voltages as low as 20 mV. Concerning the piezo harvesting branch (Figure 6b shows the prototype harvester implementation with the connection of multiple stacked disk in parallel), preliminary simulations proved how the amplitude of the voltage signal, obtained with a pressure of 75 Kg over a single piezo disk surface, can reach up to 25 Vpp, while the frequency ranges from about 1 Hz (for a normal walk) to 2.5 Hz (for a running session). The first input stage of the LTC3588 chip is composed by a full wave rectifier bridge, followed by a 20V clamping Zener diode.

## IV. MEASUREMENT RESULTS

A first discrete prototype board (Figure 6c), with discrete components mounted both on the top and bottom layer, was designed and fabricated, in order to demonstrate the feasibility of the proposed system. Figure 7 shows the TEG and PZT harvester physical implementation. For the TEG harvester, Table I shows measurement results about conversion efficiency, for a 3.3V regulated output voltage and different loads, without any battery. Results show a relatively high conversion efficiency at lower loads, while the minimum start-up voltage increases as the current requirements rises as well. This problem is mitigated with the introduction of the battery, thus enabling the power path management, which provides as much current as the TEGs can provide by varying the equivalent load seen from the TEG harvester. Also, experimental results proved that best performance can be achieved by connecting all the cells in series.

Table I: TEG System Measurement results

| Load [kΩ] | Minimum start-up Voltage [mV] | Conversion efficiency [%] |
|---|---|---|
| 100 | 67 | 38 |
| 82 | 69 | 25 |
| 68 | 70 | 22 |
| 47 | 76 | 37 |
| 39 | 79 | 30 |
| 33 | 80 | 37 |
| 27 | 84 | 43 |
| 10 | 120 | 56 |
| 4.7 | 177 | 43 |

Concerning the piezo harvesting channel, Figure 8 shows a typical waveform of a walk, captured with a digital oscilloscope at 20V/div without any load applied to the system. The peak-to-peak voltage value reaches 75V, however the amplitude drops down to about 18V – for the best case – when a load current is applied, thus avoiding a voltage limitation due to the internal zener diode of the LTC3588 chip. In Figure 9 the overall harvested power is illustrated, considering an equivalent resistive load. Results show a pieak of maximum power transfer around 1 kΩ of equivalent resistive load. Another minor peak is located around 82 kΩ of equivalent load. In order to test the capabilities of the whole harvesting system, a low-power acquisition and transmission board was implemented (Figure 10). The system includes an ARM Cortex M0 and a Si4463 transceiver set up to short 5 dBm transmission range. The idle current of the system is about 4 mA, while is as-low-as 10 mA during transmission. According to [9], the acquisition and transmitting rate can go down to 15 ms and 5 ms, respectively, for a mean current consumption of 5.5 mA. In this perspective, results demonstrate that the harvester can extend the battery life-time (ion lithium battery has been used) up to 20% of the normal duration.

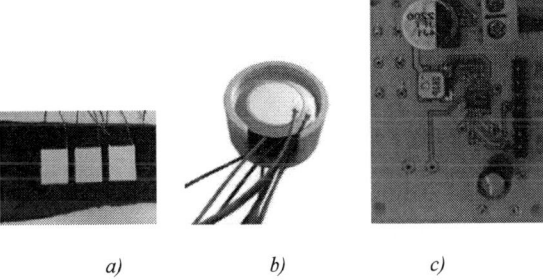

| a) | b) | c) |
|---|---|---|

Figure 6: a) TEG Harvester; b) PZT Harvester; c) Electronic board

## V. CONCLUSION

We have presented and discussed a human body powered sensory glove system, based on a multi-source power harvester architecture to extend the source battery life time. The prototype boards, that integrates circuitry to scavenge energy form Thermo-Electric Generators (TEGs) and piezoelectric disks, demonstrated the system feasibility towards IC applications.

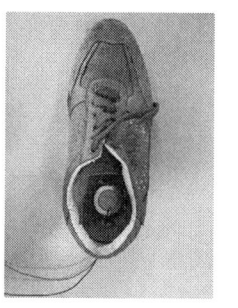 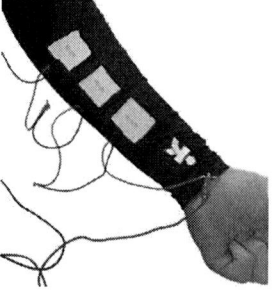

*Figure 7: PZT (left) and TEG (right) harvester physical implementation*

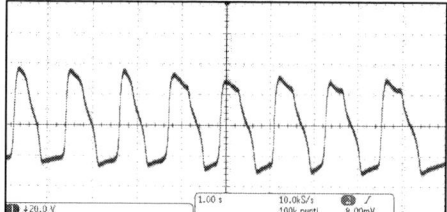

*Figure 8: Piezo harvester output with no load*

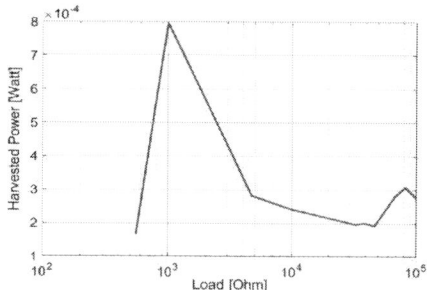

*Figure 9: Overall piezo harvested power*

*Figure 10: Acquisition and transmission board*

## REFERENCES

[1] G. Saggio, F. Cavrini, C.A. Pinto, "Recognition of arm-and-hand visual signals by means of SVM to increase aircraft security", Studies in Computational Intelligence, 669, pp. 444-461, 2017.

[2] G. Saggio, G. Latessa, F.D. Santis, et al., "Virtual reality implementation as a useful software tool for e-health applications",

IEEE International Symposium on a World of Wireless, Mobile and Multimedia Networks and Workshops, 2009.

[3] R. Gentner, J. Classen, "Development and evaluation of a low-cost sensor glove for assessment of human finger movements in neurophysiological settings", J. Neurosci. Methods, 178, pp. 138-147, 2009.

[4] G. Saggio, L. Sbernini, A. De Leo, M. Awaid, N. Di Lorenzo, A. L. Gaspari, "Assessment of hand rehabilitation after hand surgery by means of a sensory glove", 9th International Conference on Biomedical Electronics and Devices (BIODEVICES), pp. 187-194, 2016.

[5] Lin J., Wu Y. and Huang T. S., "Modeling the constraints of human hand motion", IEEE Workshop on Human Motion, pp. 121-126, 2000.

[6] R. Mugavero, G. Saggio, V. Sabato, M. Bizzarri, "The multisensory integrated modules for training", International Carnahan Conference on Security Technology, 2014.

[7] L. Dipietro, A. Sabatini, P. Dario, "Evaluation of an instrumented glove for hand-movement acquisition", Journal of Rehabilitation Research and Development, N. 2, pp. 181-191, 2012.

[8] G. Saggio, F. Riillo, L. Sbernini, L.R. Quitadamo, "Resistive flex sensors: A survey", Smart Materials and Structures, 25(1), 2015.

[9] G. Saggio, M. De. Sanctis, E Giannini, "Long Term Measurement of Human Joint Movements for Health Care and Rehabilitation Purposes" Wireless VITAE, pp. 674-678, 2009.

[10] G. Saggio, S. Bocchetti, C.A. Pinto, G. Orengo, F. Giannini, "A novel application method for wearable bend sensors", 2nd International Symposium on Applied Sciences in Biomedical and Communication Technologies, ISABEL 2009.

[11] G. Saggio, M. Bizzarri,"Feasibility of teleoperations with multi-fingered robotic hand for safe extravehicular manipulations", Aerospace Scienceand Technology, vol. 39, pp. 666–674, 2014.

[12] G. Saggio, S. Bocchetti, G. Orengo, C. Pinto, "Electronic interface and signal conditioning circuitry for data glove systems useful as 3D HMI tools for disabled persons", 4th International Conference on Health Informatics (HEALTHINF, part of BIOSTEC), Rome (Italy), 2011.

[13] V. C. Gungor and G. P. Hancke, "Industrial Wireless Sensor Networks: Challenges, Design Principles, and Technical Approaches", IEEE Transactions on Industrial Electronics, vol.56, no.10, pp. 4258–4265, 2009.

[14] C. Lu, C. Ying Tsui, W. Hung Ki, "Vibration energy scavenging system with maximum power tracking for micropower applications," IEEE Transactions on Very Large Scale Integration Systems, vol. 19, no. 11, pp. 2109-2119, 2011.

[15] C. Park, P. Chou, "Ambimax: Autonomous energy harvesting platform for multi-supply wireless sensor nodes," IEEE Communications Society on Sensor and Ad Hoc Communications and Networks, pp. 168-177, 2006.

[16] C. Lu, S. Phill Park, V. Raghunathan, K. Roy, "Analysis and design of ultra low power thermoelectric energy harvesting systems," IEEE International Symposium on Low Power Electronics and Design, pp. 183-188, 2010.

[17] Q. Huang, C. Lu, M. Shaurette, R. Cox, "Environmental thermal energy scanning powered wireless sensor network for building monitoring," International Symposium on Automation and Robotics in Construction, pp. 1376-1380, 2011.

[18] H. Xiao, H. Shao, K. Yang, F. Yang, W. Wang, "Multiple Timescale Energy Scheduling for Wireless Communication with Energy Harvesting Devices", Radioengineering Journal, vol. 21, 2012.

[19] M. Piñuela, P. D. Mitcheson and S. Lucyszyn, "Ambient RF Energy Harvesting in Urban and Semi-Urban Environments," in IEEE Transactions on Microwave Theory and Techniques, vol. 61, no. 7, pp. 2715-2726, 2013.

[20] P. Di Marco, V. Stornelli, G. Ferri, L. Pantoli, A. Leoni, "Dual band harvester architecture for autonomous remote sensors", Sensors and Actuators A: Physical, Volume 247, pp. 598-603, 15 August 2016. ISSN 0924-4247.

[21] L. Pantoli, A. Leoni, V. Stornelli, G. Ferri, "An IC architecture for RF energy harvesting systems". Journal of Communication Software and Systems, vol. 13, pp. 96

# HW platform for BMS algorithm validation

Luca Buccolini, Federico Garbuglia, Matteo Unterhorst, Massimo Conti
Department of Information Engineering
Università Politecnica delle Marche, Ancona, Italy

*Abstract*—Lithium batteries are more and more used for energy storage. They need to be controlled by some battery management system (BMS) to maintain the batteries working in a safety range, estimate the state of charge, the state of health and to maximize the energy stored. Software simulation allows to improve the BMS algorithms but they need to be validated on real BMS and cells. This paper presents an HW platform composed by a custom open-source BMS with a standard Arduino Uno compatible pinout, a battery charger, six lithium cells, and the software that manages the test. The platform is validated by using a passive cells balancing algorithm and a charging algorithm that interacts between them. The test shows as during the passive balancing the algorithm must take into account also the temperature of the board.

*Keywords*—*BMS; battery management; lithium batteries; test platform, cell balancing; BMS validation; Arduino Uno shield;*

## I. INTRODUCTION

Nowadays lithium batteries are the most used energy storage systems either for electrical vehicles and stationary storage solution. The researchers focus their works to improve the lithium batteries performances and solve the main drawback of this technology. The battery pack is made up of several basic electrochemical units, i.e. the cells that could be connected in series and or in parallel to obtain certain voltages capacity or powers. The most evident advantages of the lithium batteries, if compared with other technologies, are the higher energy and power density, higher charge/discharge efficiency and longer lifetime. In addition, there are several types of lithium-based chemistries available therefore the right technology can be used for the particular application needed. On the other hand, some disadvantages exist that need to be balanced against the benefits. Because lithium is more reactive, special safety precautions are needed to prevent physical or electrical abuse and to maintain the cell within its design operating limits in terms of cells voltage, temperature and current [1]. To this purpose, the batteries must embed an electronic device, named battery management system (BMS) with the aim to measure the cell voltages, the cell temperatures and actuate the safety precautions, such as the stop of the charge, discharge or the opening of the main contactor. A battery is made up of several cells that can have different total capacities, internal resistances, self-discharge rates and age. Thus a battery stack is limited in performance by the lowest capacity cell in the stack; once the weakest cell is depleted, the entire stack is effectively depleted. The BMS overcomes for this cell-to-cell mismatch by acting cell balancing. The balancing technique is passive, when the charge is drained from the cells with the highest State of Charge (SOC) and dissipating excess charge in some bleed resistors, and active when the charge is transferred among cells by using more complex methods [2]. Different hardware and algorithms are under investigations by researchers [3]. Another important feature of the BMS in which the researchers are focused too is the estimation of the state of charge (SOC) and the state of health

(

in the battery and for the battery's capability to deliver the desired power. Since the time needed to test the BMS's algorithms is high, many researchers use computer simulations to reduce the algorithm development time [6],[7]. Nevertheless, these algorithms have to be validated on a real system because some side effects such as the BMS board temperature, could be difficult to be modelled. To the knowledge of the authors, in fact, the issues inherent to the high temperatures of the BMSs have yet not been addressed by the researchers. In this paper we propose a platform composed by a custom open-source BMS, a battery charger, real lithium-cells and a PC that holds the software, written in MATLAB, that manages the test and visualize and log the data in real-time. The BMS, being an Arduino compatible shield, has the possibility to be connected with a huge variety of microcontroller or FPGA boards giving the possibility to exploit different solutions. Furthermore, different spare analog and digital inputs/outputs pins lead to freely expand the board and permit to connect several BMS in series to build a high voltage battery. Few platforms were proposed, but whilst in [8] the platform is not expandable and it is not possible to connect several BMS in series, respect to [9] this work focuses on the passive balancing hardware and highlight the ability to control the whole platform (BMS and battery charger) by adding interaction between the charging algorithm and the balancing algorithm. The presented platform allows to develop and test the whole battery model by using a PC simulation software and to use the same algorithm implemented in the model to control the real hardware thus validating the model.

## II. THE ARCHITECTURE OF THE PLATFORM

This section describes the design and the development of a controlled HW/SW platform composed by a six-cell battery management system named BMSino, a controlled battery charger and a host PC. A MATLAB® architecture was developed to control both the hardware instruments and to display the cells voltages, cells temperatures, the balancing status, the battery current and the BMS board temperature at run-time allowing to develop battery management algorithms with high level programming techniques. The proposed hardware[1] and software[2] architecture is completely open-source and could be controlled by an external host-PC, via serial ports, for development purposes, or it could work also as standalone system for on-field tests by logging all the information on SD-cards through its expansion connectivity. The platform is expandable, for example, by adding an active load controllable by the same host PC, thus developing also management techniques during discharging phases. Furthermore, BMSino allows to control different relays, cooling fans, etc. by using spare digital inputs and outputs or to measure different physical quantities through its analog inputs. These features allow the maximum flexibility. Fig. 1 illustrates a basic block diagram of the HW platform proposed.

tps://circuitmaker.com/Projects/Details/Erwins-Fluffy/BMSino
tps://github.com/Lucast85/BMSino-Matlab-test

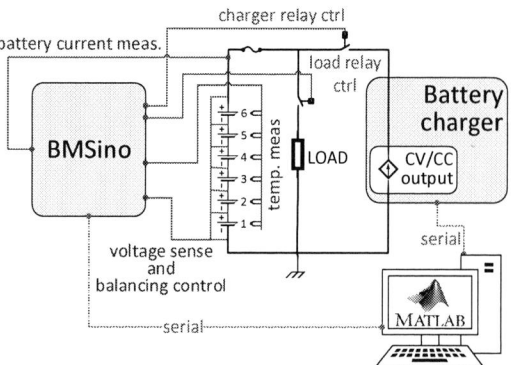

Fig. 1. The block diagram of the open-source HW platform based on BMSino

## A. BMS design

The battery management system for lithium batteries must ensure that the battery works under safety conditions. To this aim the BMS monitors the voltage and the temperature of all the cells and the current of the whole battery [1]. In addition, to improve the exploitable capacity, a balancing method is required [3]. We have implement passive balancing circuitry because this is the most used method being the cheapest and the safer among the methods. The circuit allows to study on the improvements of the performances of passive balancing algorithms, by reducing the energy losses or speed up the charging phase. Different communication buses allow to communicate with external devices such as the battery charger and they're also required when it is necessary to plot the real time data or to save the data on external SD-CARD. The BMS is designed as an expansion board (shield) for the single board microcontrollers compatible with the standard Arduino Uno pin headers. This increases the flexibility of the platform and allows to choose different microcontrollers-based or FPGA-based boards compatible with the Arduino pinout. We have tested the BMS controlling it through an Arduino Uno board, thus in Fig. 1 the block named BMSino is composed by the BMS and Arduino Uno boards. The following paragraphs describe the block schemes of BMSino which is shown in Fig. 2. Fig. 3 shows the complete prototype.

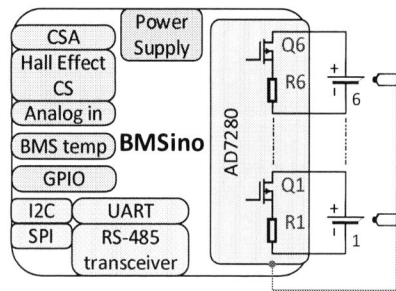

Fig. 2. The block diagram of BMSino: a BMS shield based on AD7280 ic.

### 1) BMS integrated circuit

An AD7280 ic by Analog Devices was used to measure up to 6 cells temperatures, through NTCs, and 6 cells voltages. The resolutions are 976 uV in the range 1-5 V for cells voltages and 1.22 mV in range 0-5 V for temperature measurement. The maximum sample rate is 11.6 kHz. The same ic controls the gate of the balancing mosfets that acts as switch and connect the high voltage cell during the charging cycle to a 10 ohm, 2 W resistor, that waste the cell energy in heat.

### 2) Current sensing capabilities

The BMS has 3 different methods to measure the battery current. All these methods are based on the ADC inputs of the microcontroller board. In the case the board is Arduino, the ADC input are in the range 0-5 V, with a resolution of 10 bits. For high currents, in the order of tenths or hundreds of amperes, the shield can fit an open loop hall effect sensor of the HAIS family by LEM. The HAIS 50-P has a range of ±150A that translate into a resolution of 0.39 A. When using low currents, an INA170 current sense amplifier (CSA) by Texas Instruments could be used to sense for the voltage drop across a shunt resistor of 0.22 ohm. The current sense range is ±11.3 A and a resolution of 22.2 mA. The third option for current sense is to use an external sensor connected to a dedicated input of the BMS board. The resolution and the accuracy depends on the external sensor chosen.

### 3) BMS temperature sensor

During the balancing time, the temperature of the balancing resistors increases quickly, thus a 10 kohm NTC is placed in the middle of the resistors group to monitor the temperature of hottest area on the board. The charge and balance have to be stopped when the temperature is too high.

### 4) Communication

Different communication buses are available on the board. An SPI bus and a chip select pin permit to control external SPI devices such as external SD CARDs, that memorize all the logged information of the battery. A UART bus permits to share the information with a PC using the standard serial communication bus available on the Arduino boards. The Arduino programming USB cable was used to control the balancing mosfets of BMSino and to read back all the cells voltages, temperatures and the battery current. The same UART bus could be used to communicate with the RS-485 differential bus by mounting a dedicated transceiver. Finally, an optionally isolated I2C bus is available in order to communicate with other BMSino connected in series thus giving the possibility to build and experiment on high voltage batteries.

### 5) General purpose I/O

Many general purpose pins are available from BMSino itself. There are two analog inputs with RC anti-aliasing filters, two digital inputs and two power low side outputs capable to control inductive loads (i.e. security contactors) with a power supply up to 50 V and 2.5 A.

*6) Power supply logic*

The shield could be powered either: by the microcontroller shield during programming, debug or test mode or by the cell stack itself to make it work as a standalone battery without the control of an external PC. If a pin is shorted with the battery voltage, BMSino, through its DC/DC converter, feeds the microcontroller board.

*7) Firmware*

As BMSino can be fitted in every Arduino compatible microcontroller or FPGA boards, the FW can be customized by the user. We have designed a FW Arduino-compliant, whose purpose is to define a simple communication interface between a device with serial port and BMSino. The main purpose of the FW developed is to provide a SW layer to connect the BMS to a host PC which is running the SW with the algorithms being validated. The firmware was developed in C++ , defining a class object for almost every block shown in Fig. 2. Thus, to define every object physically connected to the board such as a specified NTC or a specified current sensor, it is sufficient to define an instance of the relative class that holds the characteristics of the object (look up tables for NTCs or proportional factors for current sensors, etc.). The firmware developed for BMSino provides essential security features. In the main loop the firmware is continuously monitoring the cells voltages, the cells temperatures, the battery current and the PCB temperature with a period of 14 ms. If any of these parameters goes out from the security range, the BMSino stop the charge and opens all the balancing mosfets leaving the cells disconnected from the load. While monitoring the cells status, BMSino could be queried by simple serial commands received from the UART port. Commands are available to read the cells voltages, cells and PCB temperatures, battery current and the status of the balancing mosfets. With another serial command it is possible to control the single balancing mosfet.

*B. Battery charger design*

The battery charger is a DC/DC converter based on the LM2596 step-down regulator from Texas Instruments. The acceptable input voltage is up to 40 V, whilst the output voltage is in the range [0.01 – 35] V with a max output current of 3 A. It is controllable from the serial port, setting it in constant current mode with a resolution of 1 mA or in constant voltage mode with a resolution of 10 mV. Also, a command to enable or disable the output is available, as well as precise auto-calibration feature.

*C. Software design*

The SW is built in MATLAB by using the OOP paradigm. The test, as Fig. 4 depicts, is controlled by a timer that execute four different tasks: "start task", which initialize the timer and instantiate the charger and the battery objects; "tick task", which contains the main code to control the battery and display the data in real time; "error task" and "end task", that contain respectively the code that handles the errors and the code that stops the test execution and save the data. The flow chart is represented in Fig. 5. After an initial phase, where all the measurements were performed, a security check is done to control that all the parameters are within the safety range. Later, the balancing is computed and applied to each cell with the following criteria: if the cell voltage are above the "*start of balancing voltage*", indicated in TABLE I. ,then if that cell voltage is close for less than "*End of charge delta voltage*" to the highest cell voltage then that cell is balanced else it is not balanced. The current setpoint estimation algorithm follow a CC-CV logic and has the following criteria: if the highest cell voltage is below the "*Reduction of charge set-point*" then the current is fixed to "*Max charging current*" else, the current setpoint linearly depends from the delta voltage between the highest cell voltage and the "*End of charging voltage*".

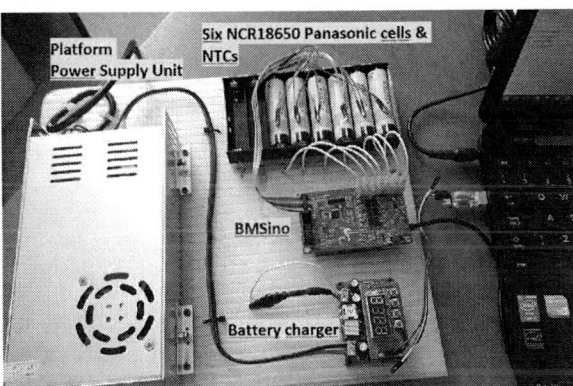

Fig. 3. The test platform composed by the custom BMS, the battery charger, the cells and a power supply unit.

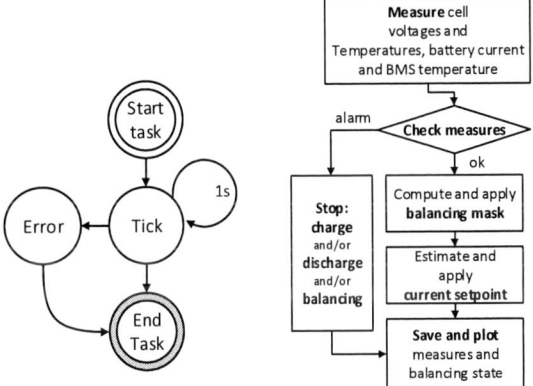

Fig. 4. Flowchart of the test alghoritm developed in MATLAB

Fig. 5. Flowchart of the Tick task

## III. ARCHITECTURE VALIDATION AND EXPERIMENTAL RESULTS

The platform was validated using the test setup represented in Fig. 1 and Fig. 3.during the charge. We used six Panasonic NCR18650 cells; their main characteristics and the main parameters used to charge and balance the battery are summarized in TABLE I.

TABLE I.        THE SETPOINT USED DURING THE CHARGE TEST

| Nominal voltage | 3.6 V |
|---|---|
| Discharging end voltage / End of charging volt. | 2.5 V / 4.2V |
| Rated capacity | 3200 mAh |
| Max charging current | 1.5 A |
| Start of balancing voltage | 4.0 V |
| End of charge delta voltage | 0.03 V |
| Reduction of charge set-point | 4.0 V |
| Max. PCB temperature | +75 °C |

As an example of the test developed, Fig. 6 shows, on the left side, the values of the voltages of the six cells and, on the right site, the current feeding the battery pack, during charge. The HW validation test starts with the cell unbalanced. The lower cell voltage is 3.51 V and the cell with the highest voltage has an open circuit voltage of 3.74 V. According to the algorithm show in Fig. 5, the battery charger is set to the maximum charging current. The charge setpoint remains constant upon the cell with the highest voltage reach 4 V, then the balancing mosfets starts to switch on as shown in Fig. 8 consequently the BMS reaches very quickly the max. PCB temperature (Fig. 7). In meantime, the current setpoint starts to decrease linearly with the highest cell voltage until the end of charge. At the end of charge, the current drops due to the temperature safety protection and the balancing mosfets are activated less frequently.

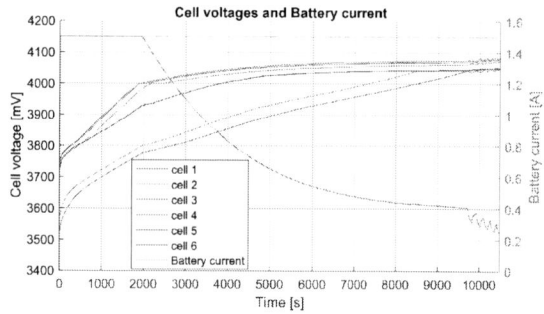

Fig. 6.    The cell voltages during the charging and balancing validation test.

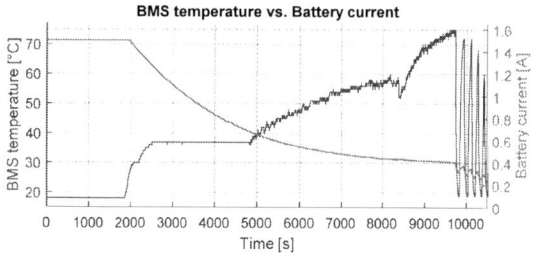

Fig. 7.    The BMS temperature during the experiment.

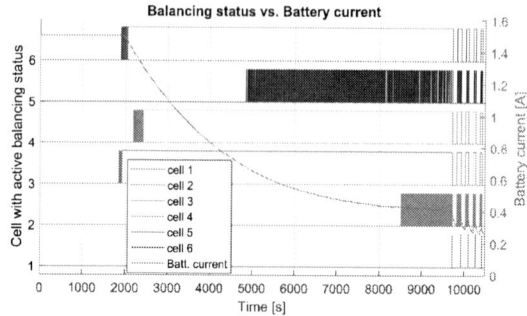

Fig. 8.    The status of the balancing mosfets during the validation test.

## IV. CONCLUSIONS

A test platform helpful for the development of BMS algorithms was presented. The platform is composed by a custom open-source BMS, a battery charger and a software that control the test and embeds models and algorithms. A battery charging algorithm interacts with a passive balancing algorithm and tests were conducted to validate the hardware/software platform pointing out the potentiality of the system and highlighting the difficult thermal management of the board that is a topic not yet addressed by other researchers. The platform allows to research in BMS charge and in passive balancing algorithms with the goal to reduce the charging time, reduce the energy loss or to estimate SOC and SOH by managing also the increasing of BMS and cells temperature. The BMS algorithms could be tested with the help of the simulation SW in the PC and then validated using the HW platform developed.

## REFERENCES

[1]    J. Nguyen and C. Taylor, "Safety performance for phosphate based large format lithium-ion battery," 2004 10th Int. Work. Comput. Electron. (IEEE Cat. No.04EX915), pp. 146–148, 2004.

[2]    J. Qi and D. Dah-Chuan Lu, "Review of battery cell balancing techniques," 2014 Australas. Univ. Power Eng. Conf. AUPEC 2014 - Proc., no. October, pp. 1–6, 2014.

[3]    M. Daowd, N. Omar, P. Van Den Bossche, and J. Van Mierlo, "Passive and active battery balancing comp. based on MATLAB sim.," Veh. Power Propuls. Conf., 2011.

[4]    R. Xiong, J. Cao, Q. Yu, H. He, and F. Sun, "Critical Review on the Battery State of Charge Estimation Methods for Electric Vehicles," IEEE Access, vol. 6, pp. 1832–1843, 2018.

[5]    C. Lin, A. Tang, and W. Wang, "A Review of SOH Estimation Methods in Lithium-ion Batteries for Electric Vehicle Applications," Energy Procedia, vol. 75, pp. 1920–1925, 2015.

[6]    S. Orcioni, L. Buccolini, A. Ricci, M. Conti, "Lithium-ion Battery Electrothermal Model, Parameter Estimation, and Simulation Environment," Energies 2017, Vol. 10, p. 375, vol. 10, n. 3, 2017.

[7]    C. Scavongelli, et al, "Battery Management System Simulation using SystemC", Int. Workshop on Intelligent Solutions in Embedded Systems WISES2015, pp.151-156, Ancona, Italy, October 29-30, 2015.

[8]    N. Lotfi, P. Fajri, S. Novosad, J. Savage, et al., "Development of an Experimental Testbed for Research in Lithium-Ion Battery Management Systems," Energies, vol. 6, no. 10, pp. 5231–5258, Oct. 2013.

[9]    B.Alvarez, S.Garcia, and C.Ramis, "Developing an active balancing model and its Battery Management System platform for lithium ion batteries," IEEE Int. Symp. on Industrial Electronics, 2013, pp. 1–5.

[10]  J. Cao and A. Emadi, "Batteries Need Electronics," IEEE Ind. Electron. Mag., vol. 5, no. 1, pp. 27–35, 20

*PRIME 2018, Prague, Czech Republic*                                    *Session: Analog Circuits II*

# Ultra Low Frequency Low Power CMOS Oscillators for MPPT and Switch Mode Power Supplies

Francarl Galea[1], Owen Casha, Ivan Grech, Edward Gatt and Joseph Micallef

Department of Microelectronics and Nanoelectronics
University of Malta
[1]Email: francarl.galea@um.edu.mt

*Abstract*—This paper presents the design of two low power consumption analog oscillators implemented in a $0.35\,\mu m$ CMOS technology. These oscillators were designed for a power conditioning circuit with an analogue Perturbation and Observation (P&O) Maximum Power Point Tracker (MPPT) to maximize the scavenged power generated by energy harvesting devices. The nominal frequency of the two oscillators is $15\,Hz$ and $200\,kHz$, respectively. The $15\,Hz$ oscillator is used to clock the MPPT, whereas the second oscillator generates a sawtooth wave required for the Pulse Width Modulation (PWM) of the switch mode converter. Both oscillators work with a supply voltage of $1\,V$ and use a reference current generated by a self-biasing zero temperature coefficient circuit. All the circuitry was designed to operate in the sub-threshold region in order to keep its power consumption to a minimum. The frequency of the $15\,Hz$ oscillator varies by $7.1\,\%$ over a temperature variation from $-40\,^{\circ}C$ to $125\,^{\circ}C$. The total power consumption including the current reference circuit is $30\,nW$ at $27\,^{\circ}C$ and reaches a maximum of $90\,nW$ at $80\,^{\circ}C$. The frequency of the $200\,kHz$ oscillator varies by $33\,\%$ over a temperature variation from $-40\,^{\circ}C$ to $125\,^{\circ}C$. The sawtooth generator, together with the current reference circuit, consume $63\,nW$ across this temperature range.

## I. INTRODUCTION

Various applications require low frequency oscillators as part of their circuitry. Cardiac pacemakers, navy and submarine radio communication, MPPT circuits, sensor monitoring circuits, and low power switch mode converters all require a low frequency clock signal which has to be generated by ultra low frequency oscillators [1], [2].

The conventional CMOS ring oscillator generates a frequency in the order of MHz [3], [4]. To produce a lower frequency signal of the order of hundreds of Hz, frequency dividers have to be employed. Although this method is capable to accurately generate its designated frequency, its main drawback is the high static power consumption. Alternative circuit topologies, such as relaxation oscillators, exist but they are not capable of generating low frequency signals unless very large off-chip passive components are used [5].

An ultra low frequency ring oscillator using the concept of CMOS thyristors, was designed in [6]. The drawbacks of this oscillator are that it requires a minimum supply voltage of 2.5 V, occupies a considerable area since it requires three capacitors and dissipates a static power of $5.7\,\mu W$.

This paper presents the design of two low power consumption analog oscillators to be employed in a charge conditioning MPPT circuit for energy harvesting applications [2]. Both

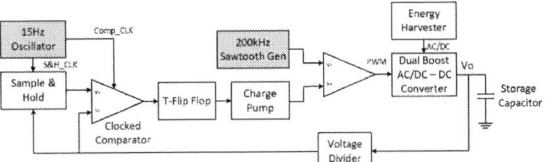

Fig. 1: Schematic diagram of the MPPT Charge Controller [2].

oscillators require only one capacitor and work with a supply voltage of 1 V. Such a low supply voltage is beneficial so that the MPPT can operate once the voltage generated by the energy harvester exceeds 1 V. The schematic diagram of the charge controller is shown in Fig. 1. The 15 Hz oscillator provides a clock signal to both a sample and hold (S&H) circuit and to a clocked comparator, which are essential to implement the P&O MPPT algorithm. The 200 kHz sawtooth generator is required to generate the PWM of an improved version dual boost converter, which is able to directly convert an AC input voltage to a DC output voltage without full wave rectification [7]. The power consumption of the sawtooth generator is very low and may also be used to generate PWM signals for other types of switch mode converters and power supplies. The frequency and duty cycle of the designed oscillators is controlled by means of a reference current, generated by a sub-threshold zero temperature coefficient self biasing circuit.

## II. DESIGN OF THE 15 Hz OSCILLATOR

For the application described above [2], a low precision low power consumption oscillator is required. This oscillator is implemented by looping two Schmitt inverters to a current starved inverter, as shown in Fig. 2. It generates a clock signal of 15 Hz. The current starved inverter charges and discharges the 1 pF timing capacitor $C1$ and uses $1149\,\mu m^2$ of area. The current which charges and discharges capacitor $C1$ is set via two current mirrors.

The frequency and duty cycle of the generated clock pulses are determined by the current of the inverter, the capacitance of $C1$ and the hysteresis values of the Schmitt inverter. The charging time of $C1$ determines the clock pulse width, whereas the discharging time of $C1$ determines the periodic time of the clock. In order for the clock frequency not to drift with temperature and voltage variations, a self biasing circuit generating a reference current was employed.

978-1-5386-5388-3/18 $31.00 © 2018 IEEE          121

*Paper P20*

*PRIME 2018, Prague, Czech Republic*

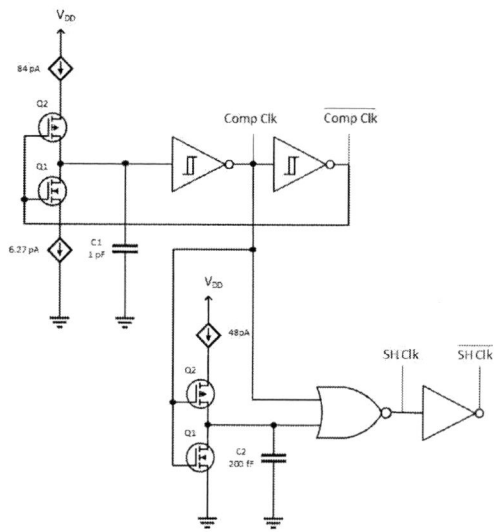

Fig. 2: 15 Hz MPPT Oscillator.

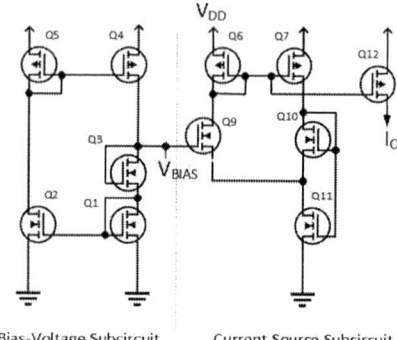

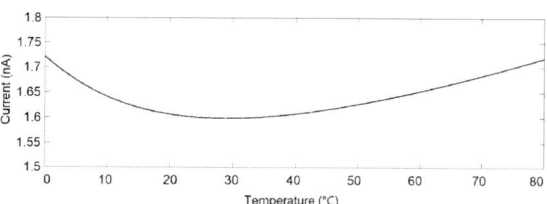

Fig. 3: Self-biasing temperature independent current reference circuit.

Fig. 4: Variation of the current reference output with temperature.

A zero temperature coefficient CMOS current reference circuit consisting of subthreshold transistors was designed [8]. This circuit consists of a self biased voltage reference subcircuit which generates the voltage $V_{BIAS}$ to bias the current-source subcircuit. The former exhibits a positive temperature coefficient, whereas the current source subcircuit exhibits a negative temperature coefficient, resulting in a net zero temperature coefficient output current. Practical results show that this circuit consumes a power of $1\,\mu W$ [8]. The MPPT circuit without the reference current consumes $250\,nW$ [2] and therefore the reference current circuit proposed in [8] consumes too much power for this application. In order to consume less power, the circuit was modified as shown in Fig. 3. The bias-voltage subcircuit was redesigned to require only a self-biased Wilson current mirror [9]. To reach the required bias voltage for Q9 with smaller currents, the diode connected transistor Q3 was added. Analysis of this circuit yields:

$$I_O = I_{D_0} K_9 \left( \frac{K_1 K_{11}}{2 K_2 K_{10}} \right)^2 e^{\frac{2 V_{GS3} + \delta V_{TH}}{n V_t}} \qquad (1)$$

where $\delta V_{TH} = V_{T_2} + V_{T_{11}} - V_{T_3} - V_{T_9} - V_{T_{10}}$ and K is the transistor aspect ratio.

From Eq. 1, it can be deduced that the thermal coefficient of the output current can be minimized through appropriate adjustment of the threshold voltages of Q2, Q3, Q9, Q10 and Q11. This can be achieved by appropriate sizing. The reference current circuit designed for the oscillator consumes a power of $7.6\,nW$ at a temperature of $27\,°C$. The current output variation with temperature is presented in Fig. 4.

The clock pulses supplied to the clocked comparator should be of at least $2\,ms$ for the comparator and the TFF to carry out their operation. The comparator is designed with low bias currents for minimal power dissipation resulting in a low slew

rate. In order to give allowance due to temperature and process variations, the oscillator was designed to have a pulse width of $5\,ms$. The hysteresis limits of the designed Schmitt inverter are of $180\,mV$ and $600\,mV$. Hence using $I = C \frac{\delta V}{\delta t}$ the charging current and discharging current are calculated to be $84\,pA$ and $6.27\,pA$, respectively.

This $15\,Hz$ oscillator also generates the clock pulses to the S&H circuit as shown in Fig. 1. This clock pulse is applied exactly once the comparator's clock pulse is over and should be of at least $500\,\mu s$ wide for the S&H's capacitor to charge or discharge to the input voltage.

This pulse is generated by the lower part of the circuit in Fig. 2. When Comp_Clk is at $1\,V$, capacitor C2 discharges through transistor Q1 and so the output of the NOR Gate which drives the S&H goes low. As soon as Comp_Clk is low, the NOR Gate's output goes high initiating the S&H clock pulse. Consequently, capacitor C2 starts charging with a current of $48\,pA$ via another current mirror from the same reference current circuit. Once it charges to $340\,mV$, which is considered a high signal by the NOR gate, the NOR Gate's output goes back to low.

The Schmitt inverter, shown in Fig. 5, uses four stacked transistors to invert the input signal and two additional transistors, Q5 and Q6, to feedback the output voltage [10] [11] to alter the threshold voltage of the inverter depending on the current state of the inverter. The aspect ratio of transistors Q2 and Q6 where adjusted to obtain the upper limit voltage ($V_{IH}$) required and the aspect ratio of transistors Q5 and Q4 was adjusted to achieve the lower limit voltage ($V_{IL}$) required.

The oscillator was simulated at various temperature con-

978-1-5386-5388-3/18 $31.00 © 2018 IEEE

*PRIME 2018, Prague, Czech Republic*     *Session: Analog Circuits II*

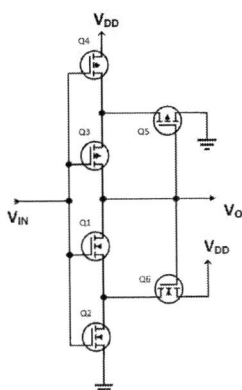

Fig. 5: The Schmitt Inverter.

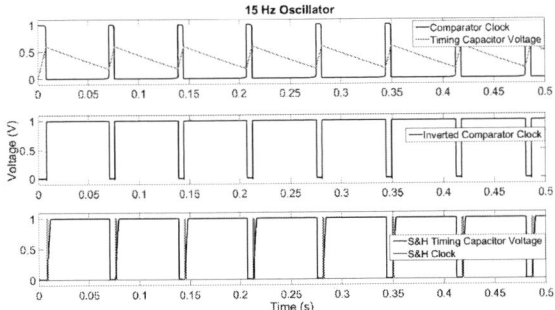

Fig. 6: 15 Hz Oscillator Timing Capacitors Voltages and Comparator and Sample & Hold Clock Pulses Plots

ditions. The results of the optimized design and its power dissipation (including reference current circuit) are listed in Table I. Corner analysis and Monte Carlo simulations were carried out to obtain a robust design. The frequency may drift from 14 Hz to 20 Hz when simulating with both temperature and process variations. Monte Carlo simulations resulted in a standard deviation of 6 Hz. The offset of the oscillator output frequency from the nominal 15 Hz are not expected to affect the operation of the MPPT circuit. Table II compares the performance of the proposed oscillator with the state of the art (SOA). The temperature coefficient and voltage coefficient of

TABLE I
PERFORMANCE OF THE 15 HZ OSCILLATOR ACROSS DIFFERENT TEMPERATURES.

| Temp. | Frequency | Comp. Pulse | S&H Pulse | Cons. |
|---|---|---|---|---|
| 0 °C | 16 Hz | 6.3 ms | 1.5 ms | 15 nW |
| 27 °C | 15 Hz | 5.5 ms | 1.2 ms | 30 nW |
| 80 °C | 16 Hz | 3.8 ms | 0.8 ms | 90 nW |

TABLE II
COMPARISON OF THE 15 HZ OSCILLATOR WITH THE SOA.

| Reference | Technology | Min. Supply Voltage | Frequency | Power |
|---|---|---|---|---|
| [1] | 180 nm CMOS | 2.5 V | 0.303 Hz | 6.6 nW |
| [6] | 250 nm CMOS | 2.5 V | 8.94 Hz | 5.7 µW |
| [12] | 2 µm CMOS | 2 V | 100 Hz | 0.3 µW |
| This Work | 0.35 µm CMOS | 1 V | 15 Hz | 30 nW |

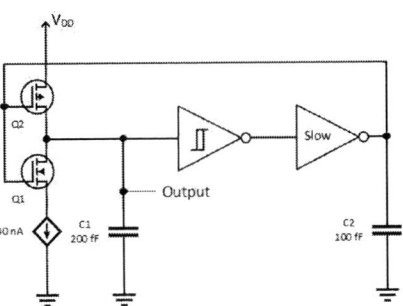

Fig. 7: 200 kHz Sawtooth Generator.

this oscillator are $0.27\,\%/°C$ and $-0.069\,\%/°C$ respectively. The area usage of this oscillator including the 200 fF capacitor required to generate the S&H clock pulse is of $1645\,\mu m^2$.

### III. DESIGN OF THE 200 kHz SAWTOOTH GENERATOR

The sawtooth generator, shown in Fig. 7, is required to generate a PWM signal to control the switch mode boost converter. It generates a sawtooth waveform with a signal swing which goes from 0 V to 1 V at a frequency of 200 kHz. The circuit consists of a current starved inverter followed by a Schmitt inverter and a NOT gate designed to have a low output slew rate. The current starved inverter uses a current source in order to precisely control the rate at which the 200 fF timing capacitor $C1$ is discharged.

The sawtooth output voltage wave is generated across the timing capacitor $C1$. The implemented Schmitt inverter circuit is shown in Fig. 5. It is designed to have a $V_{IL}$ as close to 0 V as possible and, since the NOT gate has a low slew rate and a capacitive load of 100 fF, the current starved inverter would continue discharging capacitor $C1$ for 50 ns after the Schmitt inverter changes its state. The timing capacitor is charged in 20 ns which is less than 1 % of the full cycle. Hence, the frequency of the oscillator is mainly determined by the discharging current, the timing capacitor and the hysteresis values of the Schmitt inverter, which are 0 V and 1 V.

A discharging current of 40 nA is required to achieve an oscillation frequency of 200 kHz. This current is set by mirroring a current from a self biased temperature independent reference current circuit, similar to the one shown in Fig. 3 and consumes 2.5 nW. It was ensured that the sawtooth waveform reaches the maximum value of the charge pump output voltage (1 V) at all operating temperatures and corner conditions, in order to prevent the output duty cycle from reaching 100%. Additionally, it was ensured that, across various process and temperature variations, the sawtooth generator does not have a dead band at 0 V, as this limits the lowest PWM duty cycle which can be obtained, thus limiting the operation of the boost converters at specific input and output voltages which require a low duty cycle. It is preferable that the sawtooth wave does not reach 0 V rather than having a dead-band at 0 V. The Schmitt inverter, similar to the one shown in Fig. 5, was designed to have $V_{IL} = 90\,mV$ in typical conditions, by adjusting the aspect ratios of transistors $Q4$ and $Q5$. $V_{IH}$ was set to

978-1-5386-5388-3/18 $31.00 © 2018 IEEE     123

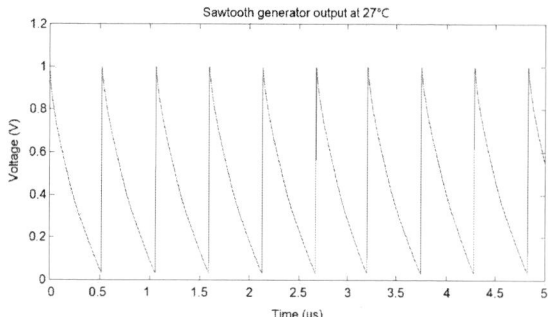

Fig. 8: Transient response of the Sawtooth Generator.

TABLE III
PERFORMANCE OF THE 200 kHz SAWTOOTH OSCILLATOR ACROSS
DIFFERENT TEMPERATURES.

| Temp. | Frequency | Dead-band @ 0 V | Minimum Voltage | Maximum Voltage |
|---|---|---|---|---|
| 0 °C | 180 kHz | 0 μs | 10.7 mV | 998 mV |
| 27 °C | 190 kHz | 0 μs | 37.5 mV | 998 mV |
| 80 °C | 200 kHz | 0 μs | 80 mV | 999 mV |

800 mV in typical conditions. This was achieved by adjusting the aspect ratios of transistors $Q2$ and $Q6$. The output of the sawtooth generator simulated at 27 °C with typical conditions is shown in Fig. 8.

The design of the oscillator was simulated and tested across various temperature conditions. The simulated results are presented in Table III, showing that no dead-band occurs neither at 1 V nor at 0 V. Corner analysis and Monte Carlo simulations were also carried out. Corner analysis with temperature variations resulted in the oscillator frequency to fluctuate from 140 kHz to 220 kHz. Monte Carlo simulations calculated a standard deviation of 34 kHz. Although the 200 kHz oscillator frequency is the ideal switching frequency for the boost converter in the MPPT circuit, variations in this frequency would only affect the switching losses and the ripple currents of the boost converter. The temperature coefficient and voltage coefficient of the proposed oscillator are 0.17 %/°C and 0.029 %/°C respectively and the circuit consumes an area of 436 μm². Table IV compares the specifications of the proposed oscillator with that of existing designs.

## IV. CONCLUSION

Simulation results have shown that the frequency of the 15 Hz oscillator varies by 7.1 % over a temperature variation from −40 °C to 125 °C. The total power consumption, including that of the self-bias zero temperature coefficient current reference circuit, is 30 nW at 27 °C and reaches a

TABLE IV
COMPARISON OF THE 200 kHz OSCILLATOR WITH THE SOA.

| Reference | Technology | Min. Supply Voltage | Frequency | Power |
|---|---|---|---|---|
| [5] | 65 nm CMOS | 1.5 V | 18.5 kHz | 120 nW |
| [13] | 0.35 μm CMOS | 1 V | 80 kHz | 1.14 μW |
| [14] | 65 nm CMOS | 1.05 V | 100 kHz | 41 μW |
| [15] | 0.35 μm CMOS | 1.2 V | 200 kHz | 84 μW |
| This Work | 0.35 μm CMOS | 1 V | 200 kHz | 63 nW |

maximum of 90 nW at 80 °C. The frequency of the 200 kHz sawtooth oscillator varies by 33 % over a temperature variation from −40 °C to 125 °C. The sawtooth generator together with the current reference circuit consume 63 nW, which remains constant throughout all temperature range. The variations due to temperature, in the sawtooth wave particularly the minimum voltage, as listed in Table III were found to be negligible and do not affect the operation of the switch mode converter and its controller.

## ACKNOWLEDGEMENTS

The research work disclosed in this publication is funded by the ENDEAV-OUR Scholarship Scheme (Malta). The scholarship may be part-financed by the European Union European Social Fund (ESF) under Operational Program II Cohesion Policy 2014-2020, Investing in human capital to create more opportunities and promote the wellbeing of society.

## REFERENCES

[1] V. R. Kanth and K. N. Kumar, "Low power and low frequency cmos ring oscillator design," *European Journal of Advances in Engineering and Technology*, vol. 2, no. 7, pp. 82–87, 2015.

[2] F. Galea, E. Gatt, O. Casha, I. Grech, and J. Micallef, "A CMOS MPPT Power Conditioning Circuit for Energy Harvesters," *ICECS 2017*, 2017.

[3] J. Eusébio, L. B. Oliveira, L. M. Pires, and J. P. Oliveira, "A 0.5 V Ultra-low Power Quadrature Ring Oscillator," in *Doctoral Conference on Computing, Electrical and Industrial Systems*. Springer, 2014, pp. 575–581.

[4] W. S. T. Yan and H. C. Luong, "A 900-MHz CMOS low-phase-noise voltage-controlled ring oscillator," *IEEE Transactions on circuits and systems II: analog and digital signal processing*, vol. 48, no. 2, pp. 216–221, 2001.

[5] A. Paidimarri, D. Griffith, A. Wang, A. P. Chandrakasan, and G. Burra, "A 120nW 18.5 kHz RC oscillator with comparator offset cancellation for±0.25% temperature stability," in *Solid-State Circuits Conference Digest of Technical Papers (ISSCC), 2013 IEEE International*. IEEE, 2013, pp. 184–185.

[6] A. K. Mahato, "Ultra low frequency CMOS ring oscillator design," in *Engineering and Computational Sciences (RAECS), 2014 Recent Advances in*. IEEE, 2014, pp. 1–5.

[7] G. D. Szarka, S. G. Burrow, and B. H. Stark, "Ultralow power, fully autonomous boost rectifier for electromagnetic energy harvesters," *IEEE Transactions on Power Electronics*, vol. 28, no. 7, pp. 3353–3362, 2013.

[8] K. Ueno, "CMOS Voltage and Current Reference Circuits consisting of Subthreshold MOSFETs–Micropower Circuit Components for Power-Aware LSI Applications–," in *Solid state circuits technologies*. InTech, 2010.

[9] G. Palumbo, "Design of the Wilson and improved Wilson MOS current mirrors reach the best settling time," in *Circuits and Systems, 1994. ISCAS'94., 1994 IEEE International Symposium on*, vol. 5. IEEE, 1994, pp. 413–416.

[10] N. Lotze and Y. Manoli, "A 62 mV 0.13 μm CMOS Standard-Cell-Based Design Technique Using Schmitt-Trigger Logic," *IEEE journal of solid-state circuits*, vol. 47, no. 1, pp. 47–60, 2012.

[11] C. Zhang, A. Srivastava, and P. Ajmera, "Low voltage CMOS Schmitt trigger circuits," *Electronics Letters*, vol. 39, no. 24, pp. 1696–1698, 2003.

[12] C. Hwang, S. Bibyk, M. Ismail, and B. Lohiser, "A very low frequency, micropower, low voltage CMOS oscillator for noncardiac pacemakers," *IEEE Transactions on Circuits and Systems I: Fundamental Theory and Applications*, vol. 42, no. 11, pp. 962–966, 1995.

[13] G. De Vita, F. Marraccini, and G. Iannaccone, "Low-voltage low-power CMOS oscillator with low temperature and process sensitivity," in *Circuits and Systems, 2007. ISCAS 2007. IEEE International Symposium on*. IEEE, 2007, pp. 2152–2155.

[14] F. Sebastiano, L. J. Breems, K. A. Makinwa, S. Drago, D. M. Leenaerts, and B. Nauta, "A low-voltage mobility-based frequency reference for crystal-less ULP radios," *IEEE journal of solid-state circuits*, vol. 44, no. 7, pp. 2002–2009, 2009.

[15] K. Lasanen and J. Kostamovaara, "A 1.2-V CMOS *RC* oscillator for capacitive and resistive sensor applications," *IEEE transactions on instrumentation and measurement*, vol. 57, no. 12, pp. 2792–2800, 2008.

PRIME 2018, Prague, Czech Republic

Session: Analog Circuits II

# Analysis of Gain and Bandwidth Limitations of Operational Amplifiers in Sigma-Delta Modulators

Tobias Saalfeld, Alexander Meyer, Eva Schulte Bocholt, Ralf Wunderlich, Stefan Heinen

Integrated Analog Circuits and RF Systems Laboratory, RWTH Aachen University
Kopernikusstrasse 16, D-52074 Aachen, Germany
Email: mailbox@ias.rwth-aachen.de

*Abstract*—In Sigma-Delta modulator system design the loop filter coefficients are calculated based on ideal circuit behavior. Advancing to an actual circuit implementation based on operational amplifiers, which show finite gain bandwidth products, a deviation of these coefficients is observed. These influences which were not considered during initial coefficient development can lead to performance degradation or even instability. This paper presents an analysis of the influence of the 3 dB bandwidth and the DC gain of an opamp to the signal and noise transfer function of a third order Sigma-Delta modulator. Based on a mathematical analysis of the transfer functions the opamp's influences on a complex integrator and a Tow-Thomas biquad are discussed in order to give a target specification for a circuit implementation.

## I. INTRODUCTION

Short range wireless systems with low power consumption are increasingly used for communication of IoT devices and sensor networks. The communication of these networks usually takes place on three bands, which are also used for a various range of common standards. WiFi, Bluetooth or Bluetooth LE are located in the 2.4 GHz band whereas IEEE 802.15.4 specifies multiple physical layers not only for 2.4 GHz, but also for 868 MHz and even for the 433 MHz ISM band. In the last decade the trend is going towards multi-mode, multi-standard transceivers which offer a wide range of functionality. On the receiver side the Low-IF architecture with Sigma-Delta modulators ($\Sigma\Delta$M) for analog to digital conversion are a well-fitted solution due to their high reconfigurability.

The system design of a filter for a $\Sigma\Delta$M is an iterative process. After a decision on the filter architecture has been made, coefficients for the filter have to be developed and verified depending on the system specifications. With advancing to the actual circuit implementation, multiple non-idealities of the selected circuit components influence the coefficients. Accordingly, the coefficients have to be tuned and verified again in order to guarantee system functionality and performance.

The development of filter coefficients has been extensively investigated and new methodologies advance further like the simulation based approach presented by Wagner et al. [1]. Besides, established design practices like the MATLAB toolbox for discrete-time noise transfer function (NTF) design by Schreier et al. [2] with a transformation to continuous time designs exist. Likewise a broad range of circuit equivalents for different filter topologies have been proposed.

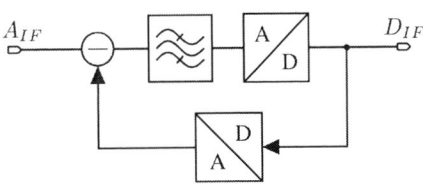

Fig. 1. Simplified block diagram of a sigma delta modulator.

Advancing towards a circuit implementation the non-ideal component behavior reveals a not negligible influence on the filter and system transfer function (TF). Several studies towards the gain band width product ($GBW$) limitations of operational amplifiers (opamp) towards stability and effective resolution of $\Sigma\Delta$Ms exist. Nevertheless, the proposed $GBW$ limits for stability vary from $GBW \geq f_s$ [3] to $GBW \geq 5f_s$ [4] or even $GBW \geq 2\pi f_s$ [5]. Within a system using a sample frequency of $f_s = 96\,\text{MHz}$ this leads to a wide $GBW$ range from 96 MHz to 603 MHz. Furthermore, these $GBW$ limits are mostly derived for a specific filter architecture and can only be used as a coarse estimation for arbitrary architectures. Consequently, an investigation of the $GBW$ influences on the selected circuit implementation has to be carried out.

The following sections present an analysis of the filter's TF with respect to the 3 dB bandwidth, DC gain, and $GBW$ of a single-pole opamp model. Consequently, a brief overview of the target system together with an introduction to the used filter architecture is given. The mentioned influences of non-ideal behavior is investigated and discussed. Finally, this paper concludes with a summary of the assumptions which can be drawn from this investigation regarding the opamp design specifications.

## II. SYSTEM OVERVIEW

The investigated filter has been developed for use in a $\Sigma\Delta$M of a multi-standard, multi-band transceiver like the one presented by Scholl et al. [6]. The $\Sigma\Delta$M, which is shown in Fig. 1, is used in a Low-IF receiver with a reconfigurable RF front-end to enable multiple intermediate frequencies (IF) and bandwidth (BW) settings. It consists of the loop filter, which is analyzed in the upcoming sections, a reconfigurable multi-bit analog-to-digital converter and a corresponding digital-to-analog converter in the feedback path. The $\Sigma\Delta$M directly

978-1-5386-5388-3/18 $31.00 © 2018 IEEE

*Paper P69*                                          *PRIME 2018, Prague, Czech Republic*

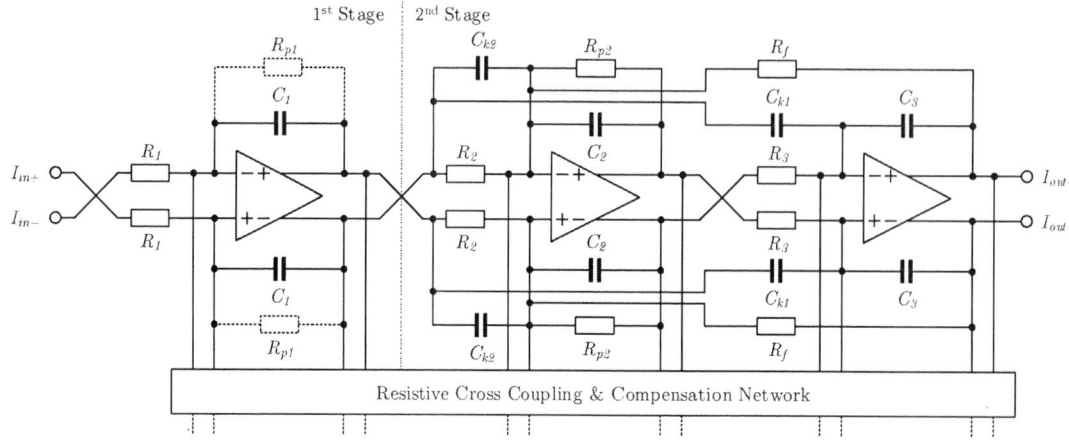

Fig. 2. Schematic of the *in-phase* path of the CICFF quadrature bandpass filter architecture as presented by Kim et al. [5].

samples analog baseband signal $A_{IF}$ at $\omega_{IF}$ and provides the digital signal representation $D_{IF}$ for further digital down conversion and demodulation. The required $\Sigma\Delta$M's dynamic range (DR) of 60 dB is achieved with a 3rd order loop filter and a sampling rate $f_s$ of up to 96 MHz. The following sections focus on a loop filter configuration with a 1 MHz BW and a center frequency of $\omega_c = \omega_{IF} = 1$ MHz.

A poly-phase quadrature bandpass filter architecture has been chosen as the loop filter. It consists of an in-phase ($I$) and quadrature ($Q$) path. Fig. 2 shows the schematic of the $I$ path. The chain of integrators with capacitive feed-forward summation (CICFF) architecture is based on two stages. The first stage, a complex valued integrator with input resistors $R_1$ and integrating capacitor $C_1$, is responsible for a pole at $\omega_1 = \frac{1}{R_1 C_1}$. The shift to a center frequency $w_c \neq 0$ is performed with a resistive cross coupling network omitted on the bottom of Fig. 2. A Tow-Thomas architecture has been chosen for the second stage. Resistors $R_{p2}$ are used to compensate the non ideal opamps and the influences of the resistive cross coupling. Based on the desired $\omega_c$ the cross coupling resistors are chosen equal for the Tow-Thomas biquad and the first stage's integrator.

Each integrator is based on a fully differential opamp resulting in a total amount of six opamps for a third order bandpass filter. Therefore, the power dissipation of the filter is dominated by the consumption of the opamps. Hence, a lower power consumption of the opamps will result in a higher efficiency of the overall system. However, the high $GBW$ and low power consumption are, to some extend, contrary to each other in terms of circuit implementation. Furthermore, stable $\Sigma\Delta$M operation is achieved with less effort by using a high $GBW$ opamp. The $\Sigma\Delta$M performance like Signal-to-Noise Ratio (SNR) or DR of the ADC is also dependent on the $GBW$. In conclusion a balance between $GBW$ and power consumption has to be found based on performance and stability concerns.

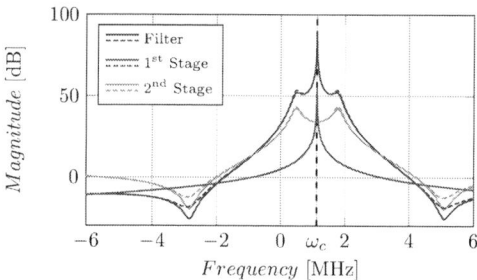

Fig. 3. Comparison of mathematical derived transfer function without $R_{p1}$ compensation (solid) and simulation results (dashed).

The effects of finite $GBW$ on the input signal of the modulator is minimal since $\omega_{IF} \pm \frac{BW}{2} << GBW$ for $GBW$ specifications which achieve loop stability. Consequently, the effects on the $\Sigma\Delta$M's signal transfer function (STF) are negligible. In order to investigate the influences on the NTF a mathematical description of the filter's TF has been derived. Equation 1 is the TF derived from the schematic of the first stage's complex integrator. The circuit topology of the integrator is similar to the one presented by Kim et al. [5].

$$H_1(s) = \frac{\omega_1}{\frac{1}{A_D(s)}(s + \omega_1 + \omega_c) + (s - j \cdot \omega_c)} \quad (1)$$

with the coefficient $\omega_c = \frac{1}{R_C C_1}$ where $R_C$ represents the cross coupling resistor. In order to consider an opamp with finite $GBW$, $A_D(s)$ is given for a single pole opamp by equation 2.

$$A_D(s) = \frac{A_{DC}}{\frac{s}{w_{3dB}} + 1} = \frac{GBW}{s + \omega_{3dB}} \quad . \quad (2)$$

The opamps $A_{DC}$ and $\omega_{3\,dB}$ can be set individually to observe the influences on the system's TF.

978-1-5386-5388-3/18 $31.00 © 2018 IEEE            126

*PRIME 2018, Prague, Czech Republic*                                                    *Session: Analog Circuits II*

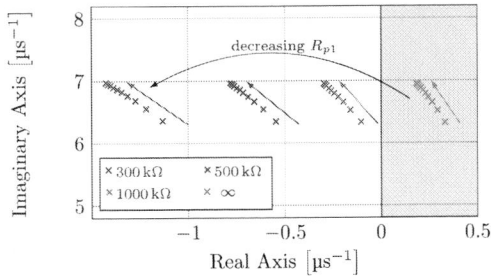

Fig. 4. Movement of critical pole $p_1$ by adding $R_{p1}$ to the integrator stage. The arrows indicate increasing $GBW$ of the used opamp.

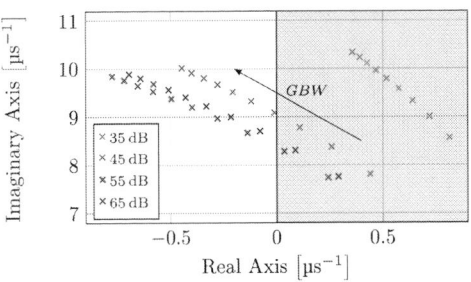

Fig. 5. Movement of the dominant second stage pole $p_2$ for different DC gain specifications. The arrow denotes the direction of increasing $GBW$.

For an ideal opamp $A_D \to \infty$ applies which reduces equation 1 to the ideal complex integrator's TF equivalent. Regarding the second stage, the Tow-Thomas biquad reduces to equation 3 for this case.

$$H_2(s) = \frac{c_0 + c_1(s - j\omega_c) + c_2(s - j\omega_c)^2}{b_0 + b_1(s - j\omega_c) + b_2(s - j\omega_c)^2} \quad (3)$$

$$c_0 = \frac{1}{R_2} \qquad c_1 = C_2 \frac{\omega_{p2}}{\omega_{k1}} \qquad c_2 = \frac{R_2}{\omega_2 \omega_{k1}} \qquad \omega_2 = \frac{1}{R_2 C_2}$$

$$b_0 = \frac{1}{R_f} \qquad b_1 = C_2 \frac{\omega_{p2}}{\omega_3} \qquad b_2 = \frac{R_2}{\omega_2 \omega_3} \qquad \omega_3 = \frac{1}{R_3 C_3}$$

$$\omega_{k1} = \frac{1}{R_3 C_{k1}} \qquad \omega_{p2} = \frac{1}{R_{p2} C_2}$$

Consequently, the overall TF can be calculated using the TF of equation 1 and 3, resulting in:

$$H_{CICFF}(s) = H_1(s) \cdot H_2(s) \quad (4)$$

Figure 3 shows the magnitude plot in solid of the derived description of the first and second stage's TF as well as the combined CICFF filter. Furthermore, the transfer characteristic from a simulated schematic with ideal parts is shown as a dashed curve. Hence, the derived TF actually depicts the implemented schematic.

### III. INFLUENCES OF OPAMP NON-IDEALITIES

*A. Stability and Pole Placement*

The first stage's integrator introduces the pole $p_1$ to the system which is placed in the right half plane of the pole-zero plot and shown in equation 5.

$$p_1 = -\frac{a_0}{2} + \sqrt{\left(\frac{a_0}{2}\right)^2 + i \cdot a_1 - a_2} \quad (5)$$

$$a_0 = \frac{k_1}{R_1 C_1} + w_{3db} + GBW \qquad a_1 = w_c \cdot GBW$$

$$a_2 = \frac{w_{3db} \cdot k1}{R_1 C_1} + \frac{GBW}{R_{p1} \cdot C_1} \qquad k_1 = 1 + \frac{R_1}{R_{C1}} + \frac{R_1}{R_{p1}}$$

Fig. 4 shows this critical pole $p_1$ examined across a $GBW$ range from $150\,\text{MHz}$ to $600\,\text{MHz}$ with a step size of $50\,\text{MHz}$. In order to increase the stability of the system this pole has to be investigated. Since an increase of $GBW$ of the used opamps does not significantly move this pole towards the left half plane, other compensation techniques have to be taken into account. Equation 5 shows that, $p_1$ can be moved towards

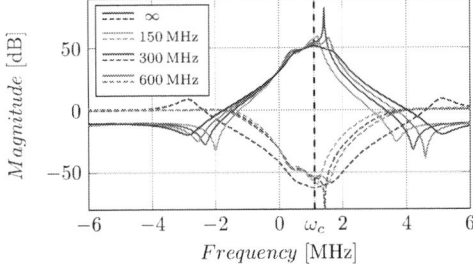

Fig. 6. Filter transfer characteristic (solid) and NTF (dashed) for different opamp $GBW$ specification with constant $A_{DC} = 45\,\text{dB}$.

the left half plane by adding the resistor $R_{p1}$ as suggested by Kim et. al. [5]. This does affect the overall system performance in regards of SNR and, thus, the actual value of $R_{p1}$ has to be carefully chosen to trade between area consumption and SNR.

The dominant pole $p_2$ of the second stage is shown in Fig. 5. It is examined across the same $GBW$ range as $p_1$. This pole is influenced by the opamp's $GBW$ and it can be seen that at least a DC gain of $A_{DC} = 45\,\text{dB}$ is needed to achieve pole positions in the left half plane.

All other poles and zeros of the filter are located far in the left half plane and can be considered non critical for the stability of the system. Hence, the system's stability is ensured by a broad range of bandwidth and gain specifications for the opamp while only a lower boundary for $GBW$ and $A_{DC}$ is given by the pole placement.

*B. Effects of Limited 3 dB Bandwidth*

Based on the $GBW$ limitations due to stability concerns the actual margins for a specification for a transistor based opamp implementation have to be defined. One possible approach is to observe the DR of the whole system which requires numerous simulations and needs a decent configuration to start with. Additionally the influence on the $\Sigma\Delta$M's noise shaping character and, therefore, the NTF can be used in order to set a specific target $GBW$ margin. A common observation is that higher $GBW$ increases stability and the SNR or DR of the $\Sigma\Delta$M system. Accordingly, the actual influence on the noise shaping character of the filter will be investigated in this

978-1-5386-5388-3/18 $31.00 © 2018 IEEE                                      127

*Paper P69*

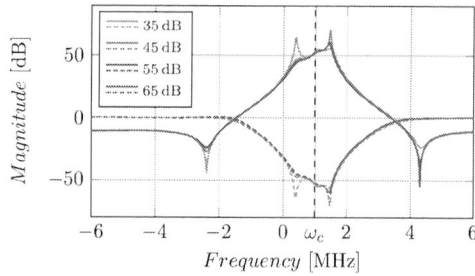

Fig. 7. Filter transfer characteristic (solid) and NTF (dashed) for $A_{DC}$ sweep with constant $GBW = 350\,\text{MHz}$.

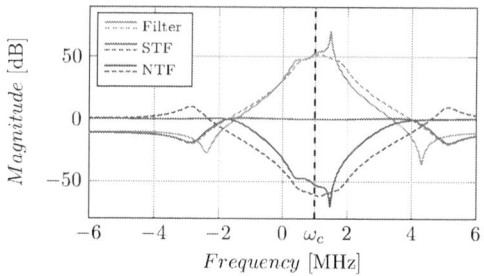

Fig. 8. Comparison of the finite $GBW$ (solid) and ideal (dashed) filter.

section. Furthermore, it is necessary to observe the influence of the opamps $\omega_{3dB}$ and $A_{DC}$ separately.

Fig. 6 shows the TF of the filter in solid lines and the corresponding NTF in dashed lines. The plot shows these characteristics for different $GBW$ setups. During these sweeps the DC gain $A_{DC}$ of the opamp has been kept constant at 45 dB. Hence, the resulting set of curves show the influence of an increasing $\omega_{3dB}$ of the opamps. Apparently, the filter's TF improves to fit the ideal curve plotted in red with higher opamp bandwidths.

Furthermore, it is observable that with decreasing $GBW$ the actual center frequency $\omega_c$ of the filter shifts to lower frequencies. Fig. 6 implies that only the bandwidth of the filter decreases. However, a shift of $\omega_c$ towards DC is also influenced by $GBW$ which can be observed by the real-valued fraction of $\frac{1}{A_D(s)}$ in equation 1 as shown by [5].

### C. Contribution of DC Gain

The set of curves shown in Fig. 7 depicts the influence of the DC gain on the TF of the filter. Accordingly, a sweep over $A_{DC}$ with constant $GBW = 350\,\text{MHz}$ is shown. It can be observed that for higher opamp gain the peaks in the TF are reduced. This behavior corresponds to the poles moving further away from the right half plane in Fig. 5. Hence, the results for $A_{DC} = 35\,\text{dB}$ show a tendency towards instability which has already been shown in section III-A. Additionally, the influence of the DC gain alone on the actual NTF of the system is negligible.

### D. Implications for Opamp Design

The $GBW$ influence on the TF of the investigated filter is mainly based on $\omega_{3\,dB}$ of the implemented opamp and less dependent on the $A_{DC}$. Additionally, a finite $GBW$ is responsible for decreased BW and a shift of $\omega_c$ towards DC. Hence, these factors have to be taken into account during filter design and, therefore, a center frequency of $\omega_c = 1.14\,\text{MHz}$ as well as a BW of 1.5 MHz can be seen with the ideal opamp model. In this regard Fig. 8 shows the CICFF TF, STF and NTF for the elaborated $GBW = 350\,\text{MHz}$ and $A_{DC} = 45\,\text{dB}$ target specification. As already expected the STF is marginally influenced by the non-idealities of the opamps while the NTF differs in terms of BW and $\omega_c$ from the ideal characteristic.

## IV. CONCLUSION

It has been shown that $GBW$ specifications within the limits set by various publications lead to stable operation but are dependent on the selected coefficients and compensation techniques. Furthermore, the actual filter bandwidth is dependent on the opamp's $\omega_{3\,dB}$, whereas a higher $A_{DC}$ is responsible for a pole shift towards stable operation. The investigation on actual pole placements during filter design has shown that improvements of SNR can be achieved at the cost of power or stability margins, assuming that a lower noise floor of the noise transfer function directly translates to a higher SNR. Additionally, instability can be avoided by using higher $GBW$ as well as tuning of filter coefficients like the feedback resistor $R_{p1}$. Depending on the selected circuit implementation a margin for the $GBW$ specifications of the opamps can be drawn to decrease power consumption. Finally, different $GBW$ setups for each stage are a possible solution to further decrease power consumption during circuit design phase.

## ACKNOWLEDGMENT

The authors acknowledge the financial support by the German Federal Ministry of Education and Research (FKZ 16 ESEO154) and the Electronic Components and Systems for European Leadership Joint Undertaking under grant agreement No. 737434.

## REFERENCES

[1] J. Wagner and M. Ortmanns, "Designing CT $\Sigma\Delta$ modulators with www.sigma-delta.de," in *Proceedings of the IEEE International Conference on Electronics, Circuits and Systems*, Dec. 2016, pp. 442–442.

[2] R. Schreier and G. C. Temes, *Understanding delta-sigma data converters*. New York, NY: Wiley, 2005.

[3] M. Ortmanns, F. Gerfers, and Y. Manoli, "Influence of finite integrator gain bandwidth on continuous-time sigma delta modulators," in *Proceedings of the 2003 International Symposium on Circuits and Systems*, vol. 1, May 2003, pp. I–925–I–928.

[4] A. Atac, R. Wunderlich, and S. Heinen, "A variable bandwidth and IF, continuous time $\Delta\Sigma$ modulator for low power Low-IF receivers," in *Proceedings of the IEEE International New Circuits and Systems Conference*, Jun. 2011, pp. 362–365.

[5] S. Kim, S. Joeres, R. Wunderlich, and S. Heinen, "A 2.7 mW, 90.3 dB DR continuous-time quadrature bandpass sigma-delta modulator for GSM/EDGE Low-IF receiver in 0.25 $\mu$m CMOS," *IEEE Journal of Solid-State Circuits*, vol. 44, no. 3, pp. 891–900, Mar. 2009.

[6] M. Scholl *et al.*, "A multistandard, triple band wireless transceiver in a 130 nm CMOS technology with integrated PAs for IoT applications," in *Proceedings of the IEEE Radio and Wireless Symposium*, Jan. 2018, pp. 88–90.

*PRIME 2018, Prague, Czech Republic*                                    *Session: Analog Circuits II*

# Design of a Low-power Ultrasound Transceiver for Underwater Sensor Networks

Gönenç Berkol, Peter G. M. Baltus, Pieter J. A. Harpe, and Eugenio Cantatore

Department of Electrical Engineering, Mixed-Signal Microelectronics Group,
Eindhoven University of Technology, Eindhoven, The Netherlands
g.berkol@tue.nl

*Abstract*—This paper presents an ultrasound (US) transceiver including a transmitter and a receiver for underwater wireless sensor nodes, where low-power operation is desired to extend the life-time of the network. A system-level analysis of the underwater communication has been performed by taking into account the underwater propagation and the medium characteristics to show their impact on the overall performance. In addition, a low-noise amplifier using an inverter-based topology has been introduced to ensure power efficiency of the receiver, where a bulk-feedback method is proposed to stabilize the output bias point of the inverter. Simulation results show that the proposed transceiver has a scalable power consumption from 1.95μW to 10.4μW while achieving 100μV to 20μV sensitivity at a $10^{-3}$ BER level.

## I. INTRODUCTION

Underwater wireless sensor networks (UWSNs) are useful for various emerging applications such as environmental monitoring, early detection of disasters, and localization in pipes [1]. A low-power transceiver is required in these applications to achieve sufficiently long life-time of the network since the sensor nodes are required to operate continuously. US signals are commonly preferred over RF signals for the data exchange among sensor nodes because of their relatively low attenuation in underwater conditions [2]. On the other hand, US data-rate is limited due to the low speed of US signals, and the communication bandwidth is bounded by that of the transducer, which is much lower compared to terrestrial RF applications [3]. The limited bandwidth of the transducer limits the choice of the communication schemes to low-complexity amplitude or frequency modulation methods, which can be sensitive to e.g. multi-path propagation and interference in the communication among different sensor nodes. In this work, a design procedure for a low-power US Transmitter and US Receiver for UWSNs will be described. In order to determine the power trade-off between transmitting and receiving the US data, a system level analysis will be performed, where a top-down approach is used to capture the effects of the transducer and the underwater medium as well as the circuit parameters. The proposed US transceiver is shown in Fig.1. An on-off keying (OOK) scheme has been chosen for the communication due to its simplicity and low power operation [4]. The receiver comprises a low-noise amplifier (LNA), a variable gain amplifier (VGA), an envelope-detector (ED), a low pass filter (LPF), and a comparator to reconstruct the incoming data. To minimize power consumption of the amplifiers, an inverter-based topology has been used, where a bulk feedback method is

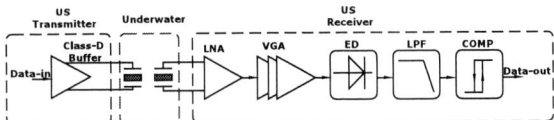

Figure 1. The proposed architecture of the US transceiver. The transducer has a band-pass transfer centered around its resonance frequency.

proposed to stabilize the output bias point of the inverter. On the transmitter side, a power-efficient two-stage class-D buffer has been used thanks to relaxed linearity constraints of the OOK scheme. The paper is organized as follows: In Section II, the system-level analysis of the US transceiver for underwater communication is performed. The details of the circuit design are presented in Section III. Post-layout simulation results are given in Section IV. The performance of the US transceiver with respect to system-level analysis is discussed in Section V. Finally, Section VI concludes this study.

## II. SYSTEM LEVEL ANALYSIS OF THE UNDERWATER COMMUNICATION

Piezoelectric transducers are considered in this work as they do not require external bias [3]. These transducers can be simply modelled by their parallel capacitance, $C_p$, in transmitting mode [5]. As a result, the dominant power is spent by the transmitter circuit to charge and discharge $C_p$. Furthermore, since an OOK based scheme is chosen, linearity is not a relevant factor. Thus, it is possible to use a class-D output buffer [3]. As a result, the power consumption of the transmitter, $P_{Tx}$, can be calculated as

$$P_{Tx} = \frac{C_p \cdot V_{drive}^2 \cdot f}{\alpha}, \quad [W] \quad (1)$$

where $V_{drive}$ is the amplitude of the driving signal, $f$ is the data rate, $\alpha$ is the efficiency of the buffer. Assuming a single omni-directional transducer working in its linear regime, the applied voltage, $V_{drive}$, will generate a source level pressure, $SL$, which can be found as

$$SL = S_v + 20\log(V_{drive}), \quad [dB\ re\ \mu Pa] \quad (2)$$

where $S_v$ is the transmit sensitivity of the transducer. The passive sonar equations [2], [6] can be used to estimate the pressure level at the receiver. In this case, the received

978-1-5386-5388-3/18 $31.00 © 2018 IEEE            129

sound pressure level, $SPL$, can be calculated as

$$SPL = SL - TL, \qquad [\text{dB } re \, \mu\text{Pa}] \qquad (3)$$

where $TL$ is the transmission loss of the environment. The corresponding received voltage level, $V_{rec}$, is given by

$$SPL = |S_r| + 20\log(V_{rec}), \qquad [\text{dB } re \, \mu\text{Pa}] \qquad (4)$$

where $|S_r|$ is the receive sensitivity of the transducer. As a result, the relationship between $V_{drive}$ and $V_{rec}$ can be obtained by substituting equations (3) and (4) into (2) to write

$$
\begin{aligned}
V_{drive} &= V_{rec} \cdot 10^{\frac{|S_r| - S_v + TL}{20}} \\
V_{drive} &= V_{rec} \cdot K, \qquad [\text{V}]
\end{aligned}
\qquad (5)
$$

where the constant $K$ captures the combined effects of the transducer, the communication frequency and distance, and the medium of interest. The level of $V_{rec}$ is important for the estimation of the power consumption of the receiver, since it determines the signal-to-noise ratio (SNR). Moreover, assuming that the VGA provides sufficiently large gain to suppress the noise and non-linearity of the ED, the noise factor of the LNA basically determines the input noise level of the front-end, which makes its power consumption dominant in the receiver chain. To estimate the power consumption of the LNA, the noise-efficiency factor (NEF) of an amplifier can be used. It is defined as

$$NEF = V_{in,noise}\sqrt{\frac{2I_{amp}}{\pi \cdot U_t \cdot 4kT \cdot BW}}, \qquad (6)$$

where $V_{in,noise}$ is the total input-referred rms noise voltage, $U_t$ is the thermal voltage, $k$ is the Boltzmann constant, $T$ the temperature, $I_{amp}$ and $BW$ are the current consumption and the bandwidth of the LNA. The ratio between $V_{rec}$ and $V_{in,noise}$ needs to be larger than 12dB to have $10^{-3}$ bit error rate in OOK [4], and BW can be assumed to be equal to the transducer bandwidth. Therefore, given that NEF of 2 is achievable with inverter-based LNA topologies [7], $I_{amp}$ and thus the power consumption of the LNA can be calculated. Fig.2 shows the estimated amount of power consumption in the transmitter and the LNA with respect to the amplitude of the received voltage, assuming a commercial transducer [8] is used in a shallow water[1]. Further assumptions are: 80% efficiency of the transmitter, $C_p$ of 1nF, the communication distance is $10m$, the bandwidth is $50kHz$, the data rate is $250bps$, and the supply voltage of the LNA is $0.6V$. As a result, the power of the LNA dominates when a higher sensitivity must be achieved, whereas the transmitter power dominates when sensitivity is relaxed. It should be noted that in this discussion no duty-cycling is considered. The methodology described in this section indicates that there is an optimum power consumption for the underwater US transceiver when the performance of the receiver and the transmitter are considered together, rather than focusing on the performance of the individual blocks.

---

[1]For a communication distance of $10m$ and a frequency of $200kHz$, the value of $TL$ in shallow water is calculated to be 30dB at $27\,^{\circ}C$ temperature [2], and around $20dB$ margin is added to $TL$ to account for the additional losses and the variation of the underwater medium [6].

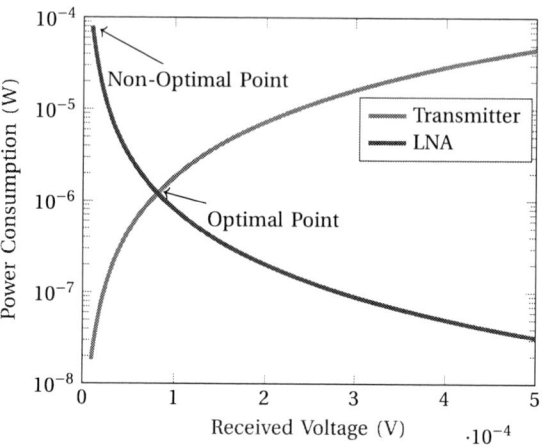

Figure 2.    Power consumption behaviour of Transmitter and LNA

## III. Circuit Design

### A. Low Noise Amplifier and Variable Gain Amplifier

A voltage amplifier with high input resistance is desired for the piezo transducers due to their relatively low impedance at the resonance frequency [5]. A closed-loop capacitor feedback architecture as shown in Fig.3 is selected for the LNA. An inverter-based topology is chosen for the core amplifier to increase the transconductance and improve the NEF. No current sources in series with the

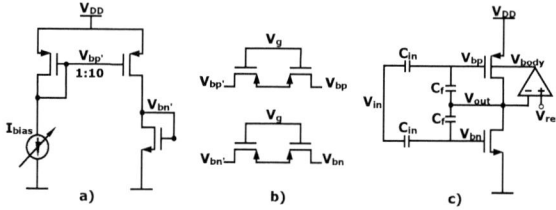

Figure 3.    Circuit implementation of the LNA a) Biasing current mirror b) Pseudo-Resistors c) Capacitive-feedback amplifier with bulk-feedback.

inverter are used to bias it. This makes possible to lower aggressively the supply voltage, $V_{DD}$. The inverter transistors are biased in weak inversion and use relatively large aspect ratio to maximize the current efficiency. To reduce the effects of parasitic capacitors on the in-band gain, the capacitor splitting method [9] is applied and the transistors are biased separately. The bias current, $I_{bias}$, is multiplied 10 times by the the current mirror. The gate voltages for the input transistors ($V_{bp}$ and $V_{bn}$) are provided through pseudo-resistors ($PRs$) having a large resistance (100M$\Omega$ to 10G$\Omega$) to set the high-pass corner below 1kHz. The gate voltage ($V_g$) of the PRs is provided separately to control their resistance against process variations. An important point in the inverter-based topology is to stabilize the output bias point against process variations and mismatches in the biasing networks. This can be achieved by applying a negative feedback to control the pull-up current provided by a transistor in series with the inverter [7], or to control the gate bias of one of the transistors in the inverter [9]. The former option limits the minimum $V_{DD}$ level, whereas the latter results in extra parasitic capacitance to the gate of one of the input transistors, which degrades the gain and the noise performance. Although this problem can be solved via a dynamic bias loop [9], this technique is not applicable to continuous operations since it requires a pre-determined on and off time of the LNA. To solve this problem, a bulk-feedback method is proposed, where the negative feedback is applied to the bulk terminal of the p-type transistor to control its threshold voltage via the body-effect. The output DC point is compared to a mid-rail reference voltage ($V_{ref}$) and the error signal is fed back to the bulk terminal by an error amplifier (EA), thereby avoiding parasitic loading on the gate of the input transistors. On the other hand, this method requires higher EA gain compared to the method in [9] due to reduced bulk transconductance, and the EA output voltage should be around $V_{DD}$. As a result, a supply voltage ($V_{DDH}$) higher than $V_{DD}$ is used for the EA, which is built as a conventional folded-cascode amplifier with diode-connected load [10]. The power overhead is negligible since the bias current of the EA is kept to $10nA$ to slow down its operation. A similar topology is used for the amplifiers of the VGA; here five inverter-based stages like the one depicted in Fig.3 are cascaded to provide high gain, as shown in Fig.4. A reduced closed-loop gain with respect

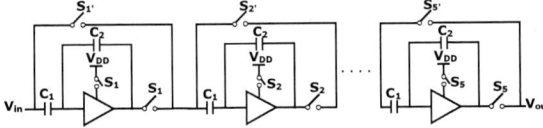

Figure 4. Schematic diagram of the VGA.

to the LNA as well as reduced bias current is used for the VGA stages since these amplifiers are not noise-limited. By appropriately selecting the switches ($S_1$ to $S_5$), the number of cascaded stages, and thus the total gain can be changed.

*B. Envelope Detector, Low Pass Filter, and Comparator*

The amplitude of the modulated signal can be extracted by using the ED, where a common drain topology [11] has been chosen as shown in Fig.5a. The transistor is biased in weak-inversion to provide the non-linearity. A $5pF$ capacitance $C_L$ is used as hold capacitor. The ED is followed by the LPF given in Fig.5b, which is implemented as a $Gm-C$ filter for low-power operation. A simple inverter is used as a transconductance block. After the signal is squared and filtered, a hysteresis comparator as shown in Fig.5c has been used to compare the amplitude with an external reference ($V_{ref2}$) and obtain a digital output signal.

*C. Transmitter*

As discussed in section II, a class-D buffer as shown in Fig.6 is used as a transmitter. It is formed by a chain of inverters sized to drive large capacitance load of 1nF. In this circuit a level shifter and inverters using 3.3.V transistors

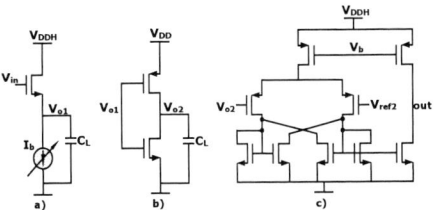

Figure 5. The schematic of the a) ED b) LPF c) Comparator (The biasing network of the circuits are not shown for simplicity)

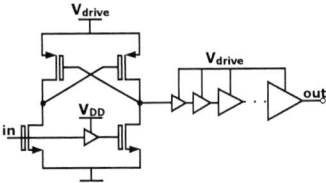

Figure 6. The schematic of the Transmitter

have been used, to enable to use a separate supply level, $V_{drive}$, higher than the one used in the receiver.

## IV. SIMULATION RESULTS

A 65nm commercial CMOS technology is used for the implementation. The $I_{bias}$ of the LNA is provided from outside of the chip, and can be tuned to achieve various power and noise performances, as summarized in Table I. $V_{DD}$ can be lowered down to 0.4V for low-power operation. The current drawn from the $V_{DD}$ (including LNA bias and core amplifier), $I_{total}$, varies from $0.33\mu A$ to $7.1\mu A$ and corresponds to a total power consumption of $0.2\mu W$ to $4.26\mu W$. The functionality of the bulk-feedback method is shown by the output DC voltage, $V_{dc,out}$, which is set to half of the $V_{DD}$ ($V_{ref}$) with a maximum error of $4mV$. The capacitance ratio, $\frac{C_{in}}{C_f}$, is set to 10. The simulated closed-loop gain ($A_v$) is 19dB in all cases. Since large aspect ratios are utilized in the inverter, the 1dB discrepancy in the closed-loop gain is due to the parasitic capacitors at the inverter input. A 0.6V supply is used for the VGA, where the capacitance ratio is set to achieve $12dB$ closed loop gain of each stage. The maximum power consumption of the VGA is $1.1\mu W$ when the five stages are all activated, and it can be reduced down to $0.45\mu W$, depending on the input signal level. A 1.2V supply voltage is used for the EA, ED, and the comparator where the power consumption of these circuits at their typical biasing conditions is $0.012\mu W$, $0.036\mu W$, and $0.038\mu W$, respectively. Total power consumption of the US receiver is scalable from $1\mu W$ to $7\mu W$, while achieving an input-referred noise density ($IRND$) from $28.4nV/\sqrt{Hz}$ to $8.5nV/\sqrt{Hz}$ at $200kHz$ and a sensitivity from $100\mu V$ to $20\mu V$. On the transmitter side, the maximum available voltage in the technology used is 3.3V, which results in a power consumption of $3.4\mu W$ for the 1nF transducer considered in this work. This can be lowered down to

| Table I. | PERFORMANCE OF THE LNA |

| $V_{DD}$ (V) | $I_{bias}$ (nA) | $I_{total}$ (μA) | $V_{dc,out}$ (mV) | $A_v$ (dB) | IRND@200kHz ($nV/\sqrt{\text{Hz}}$) |
|---|---|---|---|---|---|
| | 10 | 0.33 | 299 | 19 | 26.2 |
| 0.6 | 100 | 3.69 | 299 | 19 | 10.2 |
| | 200 | 7.1 | 299 | 19 | 8.5 |
| 0.4 | 10 | 0.41 | 198 | 19 | 28.4 |
| | 100 | 2.68 | 196 | 19 | 11.4 |

| Table II. | PERFORMANCE COMPARISON OF THE PROPOSED US TRANSCEIVER |

| Specs | [4] | [12] | This Work* | |
|---|---|---|---|---|
| Frequency (kHz) | 40.6 | 41 | 200 | |
| CMOS process (nm) | 65 | 250 | 65 | |
| Modulation | OOK | OOK | OOK | |
| Data Rate (bps) | 250 | 250 | 250 | |
| Supply Voltage (V) | 0.6 | 0.3 | 1.2 - 0.6 - 0.4 | |
| Off-chip L&C | Yes | Yes | No | |
| On-CHIP Rx&Tx | No | No | Yes | |
| Communication Distance (m) | 8.6 (air) | 6.3 (air) | 10 (shallow underwater) | |
| | | | Range | Optimal |
| Tx Power Consumption (μW) | 16000 | 1000 | 0.45 - 3.4 | 1 |
| Rx Power Consumption (μW) | 4.4 | 1 | 0.95 - 7 | 1.2 |
| Sensitivity (μV) | 20 | 20 | 100 - 20 | 70 |

*Based on post-layout simulations

$0.45\mu W$ and $1\mu W$ when $V_{drive}$ is set to 1.2V and 1.8V, respectively.

## V. DISCUSSION

The power consumption for the total US transceiver, including transmitter and receiver, with respect to the received voltage level, is shown in Fig.2. This trend shows an optimum where the power used in the transmitter and receiver are similar. When a more challenging scenario is to desired, e.g. when a longer communication distance is needed in harsh environmental conditions, the optimal point moves to higher power levels. This problem is addressed in this work, and a flexible transceiver is proposed to cover a range of performance, as summarized in the bottom of Table II. The flexibility is achieved by varying $V_{drive}$ at the transmitter and $I_{bias}$ at the receiver. As a result, considering the 10m underwater communication scenario described in Sec.II, it can be shown that a voltage sensitivity around $70\mu V$ leads to an optimum power consumption of $2.20\mu W$, which is achievable with the proposed transceiver when $V_{drive} = 1.8V$, and $I_{bias} = 10nA$.

The performance of the proposed US transceiver and its comparison with relevant integrated US transceivers from recent literature is given in Table II. Each work uses the same modulation scheme and data-rate, but the prior art focuses on transmission in air, while we exploit US for underwater communication, where the US signal attenuation is much smaller. A super-regenerative receiver is proposed in [12], where a supply voltage of 0.3V is used and a power consumption of $1\mu W$ on the receiver side is achieved for $20\mu V$ sensitivity. The work in [4] achieves a $4.4\mu W$ receiver power consumption for $20\mu V$ sensitivity. Our work relaxes the receiver sensitivity requirements thanks to the underwater environment. To reach 10m communication distance, only $70\mu V$ receiver sensitivity is needed, which brings the power consumption in our receiver down to $1.2\mu W$. Besides, exploiting an holistic optimization of both transmitter and receiver power for the low-loss underwater environment, we achieve a total power consumption for the 10m communication link of only $2.2\mu W$.

## VI. CONCLUSION

The underwater medium offers various emerging applications demanding low-power circuit design to enable long-term continuous operations of the sensor nodes. This work presents a system-level design approach of a US transceiver, where it is shown that the minimum power consumption can be achieved when transmitter and receiver performance

are optimized together. The circuit implementation of the overall transceiver is described, and post-layout simulations show the suitability of the proposed circuits for use in underwater wireless sensor networks achieving extremely competitive, $\mu W$ power budgets for communication in the 10m distance range.

## ACKNOWLEDGEMENT

This work has been funded by the European Union's Horizon 2020 research and innovation programme under grant agreement No 665347.

## REFERENCES

[1] (2018) Phoenix project. [Online]. Available: https://www.phoenix-project.eu/

[2] R. J. Urick, *Principles of underwater sound for engineers.* Tata McGraw-Hill Education, 1967.

[3] H.-Y. Tang, D. Seo, U. Singhal, X. Li, M. M. Maharbiz, E. Alon, and B. E. Boser, "Miniaturizing ultrasonic system for portable health care and fitness," *IEEE transactions on biomedical circuits and systems*, vol. 9, no. 6, pp. 767–776, 2015.

[4] K. Yadav, I. Kymissis, and P. R. Kinget, "A 4.4 $\mu W$ wake-up receiver using ultrasound data communications," in *VLSI Circuits (VLSIC), 2011 Symposium on.* IEEE, 2011, pp. 212–213.

[5] R. J. Przybyla, H.-Y. Tang, A. Guedes, S. E. Shelton, D. A. Horsley, and B. E. Boser, "3D ultrasonic rangefinder on a chip," *IEEE Journal of Solid-State Circuits*, vol. 50, no. 1, pp. 320–334, 2015.

[6] M. C. Domingo, "Overview of channel models for underwater wireless communication networks," *Physical Communication*, vol. 1, no. 3, pp. 163–182, 2008.

[7] P. Harpe, H. Gao, R. van Dommele, E. Cantatore, and A. van Roermund, "A 3nW signal-acquisition IC integrating an amplifier with 2.1 NEF and a 1.5 fJ/conv-step ADC," in *Solid-State Circuits Conference-(ISSCC), 2015 IEEE International.* IEEE, 2015, pp. 1–3.

[8] BII7519FB. (2018) Benthowave instrument inc. communication transducer. [Online]. Available: http://www.benthowave.com/products/BII-7510communicationtransducer.html

[9] C. Chen, Z. Chen, Z.-y. Chang, and M. A. Pertijs, "A compact 0.135-mW/channel LNA array for piezoelectric ultrasound transducers," in *European Solid-State Circuits Conference (ESSCIRC), ESSCIRC 2015-41st.* IEEE, 2015, pp. 404–407.

[10] W. M. Sansen, *Analog design essentials.* Springer Science & Business Media, 2007, vol. 859.

[11] X. Huang, G. Dolmans, H. de Groot, and J. R. Long, "Noise and sensitivity in rf envelope detection receivers," *IEEE Transactions on Circuits and Systems II: Express Briefs*, vol. 60, no. 10, pp. 637–641, 2013.

[12] H. Fuketa, S. O'uchi, and T. Matsukawa, "A 0.3-V $1\mu W$ super-regenerative ultrasound wake-up receiver with power scalability," *IEEE Transactions on Circuits and Systems II: Express Briefs*, vol. 64, no. 9, pp. 1027–1031, 2017.

# On the Design of a Linear Delay Element for the Triggering Module at CERN LHC

Jordan Lee Gauci*, Edward Gatt*, Owen Casha*, Giacinto De Cataldo†, Ivan Grech* and Joseph Micallef*

*Department of Microelectronics and Nanoelectronics, University of Malta
†Istituto Nazionale di Fisica Nucleare, Sezione di Bari, Italy
*E-mail: jordan-lee.gauci.10@um.edu.mt

*Abstract*—This paper presents an analytical model of a linear delay element circuit to be employed in the triggering module for the High Momentum Particle Identification Detector (HMPID) at the CERN Large Hadron Collider (LHC). The aim of the analytical model is to facilitate the design of the linear delay element circuit, while maximizing its linearity and delay range. The analytical model avoids the need of time consuming parametric sweeps on the aspect ratios of the various transistors of the delay element in order to optimize it. In addition, the analytical model can be used to predict the variation of the delay with the input tuning voltage. The proposed analytical model is verified via the simulation of the delay element circuit using the $0.18\,\mu m$ X-FAB technology.

*Index Terms*—Delay Lines, Delay Range, Linearity, Jitter, Modelling

## I. INTRODUCTION

Precise delay generation is an active research area due to the employment of delay generators in high-energy physics, time-of-flight experiments, time-to-digital converters and time interval measurement circuits. The core element of the delay generator is the delay line, which is a device capable of delaying the input signal by a predefined value. This delay can either be fixed or variable. In the case of variable delay lines, the delay can be tuned by both an analog or a digital mechanism. Delay lines have four important characteristics which determine their performance [1]–[3]:

- **Delay Step**: The finest incremental time delay step that can be produced by the delay line.
- **Delay Range**: The maximum time by which a signal can be delayed.
- **Linearity**: The ability to achieve equal and uniform delay steps.
- **Jitter**: Variation in the delay of the output signal due to noise, which has a direct effect on the smallest delay step that may be generated.

Some analog delay lines have fixed delays, such as those employing transmission gate delay elements and inverter based delay elements. In these cases, the generated delay mainly depends on the dimensions of such devices. However, the delay may be tuned either by modulating the power supply voltage or through the use of additional circuitry such as a current-starved inverter, where by tuning the quiescent current, the delay is increased or decreased accordingly. Although this type of delay element has a non-linear transfer characteristic, techniques were proposed to linearize it [4]. The work

in [5], proposes to add diode connected transistors to the inverter in order to generate very linear delays. Thyristor-based delay elements can also be used, where long delay ranges can be achieved at the expense of area and higher power consumption [6]. This paper presents an analytical model of the delay element proposed in [7]. The aim of this model is to facilitate the design of the delay element circuit, while maximizing its linearity and delay range. Compared to the design methodology employed in [7], this analytical model avoids the need of time consuming parametric sweeps on the aspect ratios of the various transistors of the delay element in order to optimize it. The analytical model can also be used to predict the variation of the delay with the input tuning voltage.

## II. BACKGROUND

This work will aid the design of the trigerring module (better known as the Fan-In/Fan-Out module) to be used by the High Momentum Particle Identification Detector (HMPID) at the CERN Large Hadron Collider (LHC). HMPID is a triggered detector, where the signals on the detector pads are read on receiving a trigger signal from the Central Trigger Processor [3]. Currently, during Run1 and Run2, the HMPID receives the Level 0 (L0) trigger signal, after approximately $1.2\,\mu s$, which corresponds exactly to the peaking time of the signal on the pads. A digital delay generator is used to coarsely adjust the delay on the trigger signal in steps of $25\,ns$. Because of the upgrade that will occur during the second long shutdown (2019-2021), the HMPID will use the Level Minus (LM) trigger instead of L0. This signal arrives at approximately $700\,ns$ after a collision. This means that a new delay generator is required to fine-tune the delay on the trigger. In [3], an improved digital delay generator was proposed as a preliminary solution. However, results have shown the issue of the creation of a random offset in the generated delay profile which is not deterministic [3]. An analogue delay generator is therefore being considered to overcome this issue, and an analytical model to facilitate the design of the chosen delay element is being proposed.

## III. DELAY ELEMENT CIRCUIT

The current-starved inverter architecture suffers from a non-linear relationship between the delay and the tuning voltage while having an input range limited to the from $V_t$ to $V_{DD}$, where $V_t$ is the threshold voltage and $V_{DD}$ is the supply

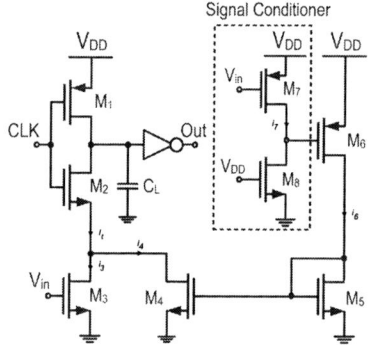

Fig. 1. Linear Delay Element Circuit [7].

voltage. The delay time ($T_d$) between the input and the output signal of a current-starved inverter can be modelled by [7, 8]:

$$T_d \propto \frac{C_L}{I_{cp}} V_{DD} \qquad (1)$$

where $C_L$ represents the capacitive load of the inverter and $I_{cp}$ is the charging/discharging current through the capacitive load. Eq. 1 shows that the delay may either be varied through the control of $C_L$, $I_{cp}$, or $V_{DD}$. Eq. 1 further illustrates how the delay is non-linear, as it is inversely proportional to the current. Furthermore, the constant of proportionality is equal to $\ln(2)$ [8].

To overcome the linearity problem, the work in [7] presents a low power linear delay element circuit, shown in Fig. 1. The circuit is based on a current-starved inverter architecture and can obtain both a linear delay and rail-to-rail operation. This is achieved through the addition of transistors M4-M8. If the tuning voltage, $V_{in}$, is applied directly through transistor $M_6$, the delay response of the circuit would be highly non-linear and non-monotonic. For this reason, an inverting common-source amplifier is added consisting of $M_7$ and $M_8$ and this helps in achieving a monotonic and quasi-linear relationship in the delay response of the circuit.

The current, $i_2$, through transistor $M_2$ determines the delay time of the circuit, and it consists of the currents $i_3$ and $i_4$. When $V_{in} < V_{tn}$, $M_3$ works in the sub-threshold region, while $M_6$ works in saturation. The delay in this case is therefore primarily dependent on the current through $M_6$. For values of $V_{in}$ in the region $V_{tn} < V_{in} < V_{DD} - V_{tp}$, both M3 and M6 are on and a linear delay-voltage characteristic may be achieved. When $V_{in} > V_{DD} - V_{tp}$, $M_7$ is switched off and thus the current through $M_6$ would be saturated. In this case, the delay would depend on the $M_3$. Transistor sizing optimization was performed by means of a parameteric sweep, in order to obtain the most linear voltage-delay characteristic, particularly when both $M_3$ and $M_6$ are turned on [7].

In particular, the sizing of $M_3$ and $M_4$, were chosen to be as small as possible such that the maximum voltage-to-delay gain is achieved. The size of $M_5$ and $M_6$ was

minimized to reduce the current. The sizing of the transistors of the inverting common-source amplifier, $M_7$ and $M_8$, was chosen specifically to maximize the gain without affecting the linearity. As such, $M_7$ is chosen as large and $M_8$ as small. However, if $M_8$ is chosen too small, there would be a large voltage swing on the gate of $M_6$ and this will have a negative effect on linearity [7].

## IV. ANALYTICAL MODEL

An analytical model for the current of the delay element shown in Fig. 1 is presented in this section. This model is valid for the range $V_{tn} \leq V_{in} \leq V_{dd} - V_{tp}$. Two versions of the model are presented in this section: a first-order model and a second-order model. In addition, an approximation method is used in order to linearize the delay transfer characteristic of the circuit. Let $i_3$ be the current through M3, $i_4$ be the current through M4, and so on. The total current, $i_t$, is equal to:

$$i_t = i_3 + i_4 \qquad (2)$$

$$= \frac{K_n'}{2} \frac{W_3}{L_3}(V_{in} - V_{tn})^2 + \frac{\frac{W_4}{L_4}}{\frac{W_5}{L_5}} i_5 \qquad (3)$$

The current $i_6 = i_5$ is equal to

$$i_6 = \frac{K_p'}{2} \frac{W_6}{L_6}(V_{DD} - V_x - V_{tp})^2 \qquad (4)$$

where

$$V_x = i_7 r_{ds8} = \frac{K_p'}{2} \frac{W_7}{L_7}(V_{DD} - V_{in} - V_{tp})^2 r_{ds8} \qquad (5)$$

and $r_{ds8}$ is the drain to source resistance for $M_8$. To simplify some terms,

$$\mu_1 = \frac{K_n'}{2} \frac{W_3}{L_3} \qquad (6)$$

$$\mu_2 = \frac{\frac{W_4}{L_4}}{\frac{W_5}{L_5}} \frac{K_p'}{2} \frac{W_6}{L_6} \qquad (7)$$

This means that $i_t$ is equivalent to:

$$i_t = \mu_1(V_{in} - V_{tn})^2 + \mu_2(V_{DD} - V_x - V_{tp})^2 \qquad (8)$$

Eq. 8 may be expanded to a fourth order polynomial and therefore the current equation may be expressed as:

$$i_t = \alpha_4 V_{in}^4 + \alpha_3 V_{in}^3 + \alpha_2 V_{in}^2 + \alpha_1 V_{in} + \alpha_0 \qquad (9)$$

where $\alpha_4, \alpha_3, \alpha_2, \alpha_1, \alpha_0$ are functions of the transistor aspect ratios $\frac{W}{L}$, the process parameters $K_p'$ and $K_n'$, the supply voltage and the threshold voltages of the NMOS and PMOS transistors, $V_{tn}$ and $V_{tp}$, respectively. This implies that the delay would be equal to

$$T_d \propto \frac{1}{i_t} = \frac{1}{\alpha_4 V_{in}^4 + \alpha_3 V_{in}^3 + \alpha_2 V_{in}^2 + \alpha_1 V_{in} + \alpha_0} \qquad (10)$$

The above equation assumes that $r_{ds8}$ has a constant value. In reality, this is not true since it depends on the current $i_8$ and the drain-source voltage $V_{ds8} = V_x$. Thus a better model of $V_x$ should be taken. To achieve this, we know that the current through $M_7$ is equal to the current through $M_8$. Thus,

$$i_7 = \frac{K_p'}{2} \frac{W_7}{L_7} (V_{dd} - V_{in} - V_{tp})^2 \qquad (11)$$

$$= K_n' \frac{W_8}{L_8} ((V_{dd} - V_{tn})V_x - \frac{V_x^2}{2}) \qquad (12)$$

From Eq. 11 and Eq. 12, $V_x$, can be found (Eq. 13).

$$V_x = V_N - \sqrt{(V_N)^2 - 2\frac{K_p}{K_n}(V_{dd} - V_{in} - V_{tp})^2} \qquad (13)$$

where $V_N = V_{dd} - V_{tn}$, $K_p = K_p' \frac{W_7}{L_7}$ and $K_n = K_n' \frac{W_8}{L_8}$. Eq. 10 shows that the time delay that can be achieved is a highly non-linear function. However linearity may be increased through the choice of values of $\alpha_4, \alpha_3, \alpha_2, \alpha_1,$ and $\alpha_0$. The Lagrange Polynomial approximation was used to transform Eq. 10 into a second order polynomial equation. The second order equation is a good approximation to the equation as it will provide two degrees of freedom, in terms of the coefficients of $V_{in}$, and $V_{in}^2$. The former would enable to control the range, while the latter would permit to control the linearity. Thus, Eq. 10 may be rewritten in the form of:

$$T_d \approx AV_{in}^2 + BV_{in} + C \qquad (14)$$

where the coefficients $A, B,$ and $C$ can be directly related to the parameters of the transistors. A closer examination of the above coefficients shows that the coefficient $A$, can be minimized through an increase in the dimensions of transistor $M_5$, a decrease in the sizing of $M_3$, and an increase in $r_{ds8}$ (decrease in $\frac{W_8}{L_8}$). This however will have an effect on the second coefficient, $B$. When the aforementioned parameters are changed, the gradient (and therefore the range) changes significantly. This is due to the fact that the denominator of the three coefficients is the same, and it is a function of the transistor ratios $M_3 - M_8$. This therefore implies that in this architecture there is a trade-off between delay range, resolution, and linearity.

## V. MODEL VERIFICATION

The circuit illustrated in Fig. 1 was implemented and simulated in Cadence using the $0.18$ $\mu$m X-FAB technology, with the transistor aspect ratios, as implemented in [7] (refer to Table I). A scaling factor was added to the dimensions such that any channel-length modulation effects are minimized. In addition, each individual transistor was characterized to find the process parameters $K_n'$ and $K_p'$ together with the transistors' respective threshold voltages.

An input pulse was applied to the $CLK$ terminal while varying the tuning voltage, $V_{in}$. The delay was calculated by measuring the time difference between the input and output waveforms, when the voltage reaches 50% of the final value. A MATLAB script was written in order to test the analytical model. The script contains the equations for the currents in each branch of the circuit. Through a DC analysis, with an input voltage of 1 V, it was found that there is a good relationship between the values, thus showing that the

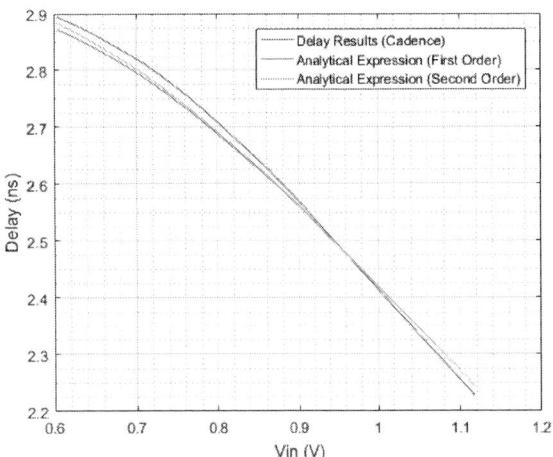

Fig. 2. Comparison of the simulated results with those obtained from the first order and second order analytical models.

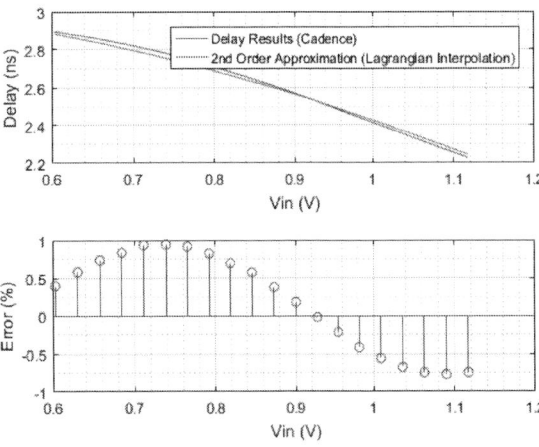

Fig. 3. Comparison of the simulated delay variation with that obtained from a second order Lagrangian polynomial.

TABLE I
TRANSISTOR ASPECT RATIOS.

| Transistor Name | Width ($\mu$m) | Length ($\mu$m) |
|---|---|---|
| M1 | 6.9 | 0.18 |
| M2 | 3.5 | 0.18 |
| M3 | 0.225 | 0.25 |
| M4 | 0.69 | 0.18 |
| M5 | 2.08 | 0.18 |
| M6 | 2.21 | 0.18 |
| M7 | 3.18 | 0.18 |
| M8 | 1.94 | 0.18 |

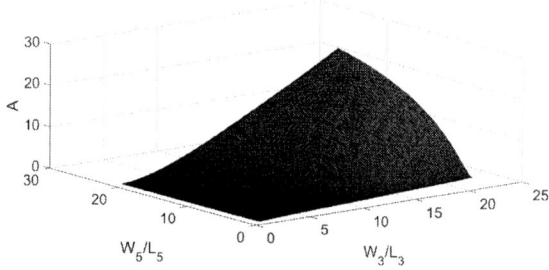

Fig. 4. Variation of the quadratic coefficient $A$ with the aspect ratio of transistors $M_3$ and $M_5$.

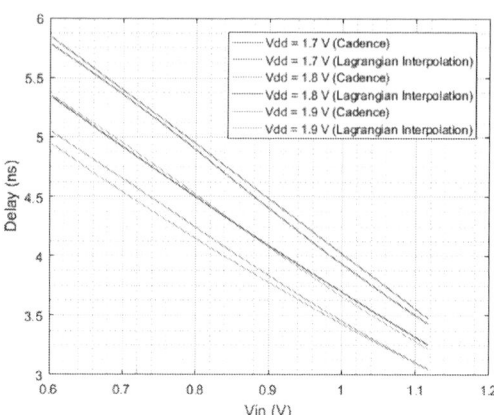

Fig. 5. Plot of the delay versus the input tuning voltage for the optimised linear delay element circuit. The plot presents both the simulation results and the second order approximation results obtained by means of the analytical model, for different values of $V_{dd}$.

equations describing the current are correct. A voltage sweep on the tuning voltage, in the range $V_{tn}$ to $V_{DD} - V_{tp}$ was then performed and the delay results are illustrated in Fig. 2, where the delay is shown in blue. The first order analysis is illustrated in orange, and the improved analysis is illustrated in yellow. It can be seen that there is a good relationship between the analytical results and the simulated delay. There is, however, a discrepancy between the analytical results and the obtained result. This can be attributed to parasitic effects.

The Lagrangian interpolation method was used to approximate the delay curve by a second-order polynomial. The results are shown in Fig. 3, where the simulated delay results from Cadence are plotted in blue, and the approximation is plotted in orange. The error is also plotted in the subplot, where the error illustrated reaches a maximum of 0.94%. This therefore shows that the proposed model can be used to effectively approximate the delay transfer charactertistic with a second-order polynomial. To this end, different values for the transistors aspect ratios can be modified to improve linearity and range. As stated in Section IV, the coefficient of the term in $V_{in}^2$, can be minimized through an increase in $r_{ds8}$, an increase in the sizing of $M3$, or a decrease in the sizing of $M_5$. The aspect ratio of $M_8$ was decreased to $700\,\text{nm}/180\,\text{nm}$, which yields an $r_{ds}$ value of around $1.7\,\text{k}\Omega$. A sweep on the dimensions of $M_3$ and $M_5$ was then performed with the coefficient in $V_{in}^2$ worked out on each iteration. The results are presented in Figure 4. From this plot, the values of the aspect ratios of $M_3$ and $M_5$, that yields the lowest value of the coefficient of $V_{in}^2$ can be selected. These aspect ratios were calculated to be 0.9 and 20.8, respectively.

Transient simulations were performed in order to obtain the variation of the delay with the input tuning voltage for different supply voltages. The results are reported in Fig. 5 and are compared to those estimated using the Lagrangian Interpolation analytical model. There is good agreement between the plots, which shows that the Lagrangian Interpolation for Eq. 14 can be successfully used to set the transistor aspect ratios in order to obtain the most linear response.

## VI. Conclusion

This paper presented an analytical model for the current-starved inverter proposed in [7]. A model was derived from first principles, where the delay was expressed in terms of the current and the input tuning voltage. This resulted in an inverse polynomial equation. The equation was approximated using the Lagrangian interpolation method and by means of this approximation, the transistor aspect ratios were found in order to achieve a linear delay transfer characteristic. The design was validated via simulation using the X-FAB $0.18\,\mu\text{m}$ technology.

## Acknowledgements

The research work disclosed in this publication is funded by the ENDEAVOUR Scholarship Scheme (Malta). The scholarship may be part-financed by the European Union - European Social Fund (ESF) under Operational Programme II - Cohesion Policy 2014-2020, *"Investing in human capital to create more opportunities and promote the well being of society."*

## References

[1] B. Abdulrazzaq, I. A. Halin, S. Kawahito, R. M. Sidek, S. Shafie, and N. A. Yunus, "A Review on High-Resolution CMOS Delay Lines: Towards sub-picosecond Jitter Performance," *SpringerPlus*, vol. 5, no. 1, pp. 1–32, 2016.

[2] M. Maymandi-Nejad and M. Sachdev, "A Monotonic Digitally Controlled Delay Element," *IEEE Journal of Solid-State Circuits*, vol. 40, no. 11, pp. 2212–2219, 2005.

[3] J. L. Gauci, E. Gatt, G. De Cataldo, O. Casha, and I. Grech, "An analytical model of the delay generator for the triggering of particle detectors at cern lhc," in *CAS (NGCAS), 2017 New Generation of.* IEEE, 2017, pp. 69–72.

[4] G. Jovanović and M. Stojčev, "Current starved delay element with symmetric load," *International journal of electronics*, vol. 93, no. 03, pp. 167–175, 2006.

[5] B. Markovic, S. Tisa, F. A. Villa, A. Tosi, and F. Zappa, "A high-linearity, 17 ps precision time-to-digital converter based on a single-stage vernier delay loop fine interpolation," *IEEE Transactions on Circuits and Systems I: Regular Papers*, vol. 60, no. 3, pp. 557–569, 2013.

[6] N. R. Mahapatra, S. V. Garimella, and A. Tareen, "An empirical and analytical comparison of delay elements and a new delay element design," in *VLSI, 2000. Proceedings. IEEE Computer Society Workshop on.* IEEE, 2000, pp. 81–86.

[7] H. Rivandi, S. Ebrahimi, and M. Saberi, "A low-power rail-to-rail input-range linear delay element circuit," *AEU-International Journal of Electronics and Communications*, vol. 79, pp. 26–32, 2017.

[8] E. Zafarkhah, M. Maymandi-Nejad, and M. Zare, "Improved accuracy equation for propagation delay of a cmos inverter in a single ended ring oscillator," *AEU-International Journal of Electronics and Communications*, vol. 71, pp. 110–117, 2017.

# A 4W 37.5-42.5 GHz Power Amplifier MMIC in GaN on Si Technology

Ferdinando Costanzo[1], Rocco Giofrè[2], Alessandro Salvucci, Giorgio Polli, Ernesto Limiti
Electronic Engineering Department, University of Roma Tor Vergata
Via del Politecnico 1, 00133, Roma, Italy
[1]costanzo@ing.uniroma2.it, [2]giofr@ing.uniroma2.it

*Abstract* — the design of a Q-band high power amplifier (HPA) in Microwave Monolithic Integrated Circuit (MMIC) technology is presented. The HPA is fabricated in a 100nm gate length Gallium Nitride on Silicon (GaN-Si) technology. The HPA, based on a four-stage architecture, was designed accounting for the de-rating rules foreseen for spatial use and to work in continuous wave (CW) conditions. Nevertheless, the realized HPA can provide a saturated output power larger than 36.5dBm with a gain and a power added efficiency higher than 22dB and 30%, respectively, in the operative band from 37.5GHz to 42.5GHz. The chip area is 3.54 x 3.5 mm². Such results are in line with others state-of-art HPAs realized in more expensive GaN processes based on Silicon Carbide, thus demonstrating that high resistivity Silicon substrate can be efficiently adopted also in such a peculiar application.

*Keywords* — *Power amplifier, Gallium nitride, Q-band, MMIC, high efficiency, silicon.*

## I. INTRODUCTION

Modern satellite systems are demanding for higher data throughput and greater network capacity [1]. In order to meet these goals, researchers are looking for new coding techniques and more complex modulation schemes to be applied on wider bandwidth. On the other hand, due to the low frequency spectrum crisis, there is a common trend to move towards higher frequency. Ka-band is extensively used for data downlink purpose [2], but many studies have already started to evaluate the potentialities of using new set of frequencies such as Q/V-band [3-4]. This scenario exasperates the design of high power amplifiers (HPAs), whose performances, especially in terms of efficiency, is tied to the used technology. Gallium Arsenide (GaAs) and Gallium Nitride (GaN) are the most popular technologies usually adopted for space systems. However, comparing them in terms of power capability and ruggedness, the GaN technology is nowadays the solution of choice to implement solid state power amplifiers (SSPAs) for space applications. Indeed, GaN devices can offer higher power density, a higher impedance and can operate at higher voltages as compared with GaAs HEMT [4]. Some studies have already proven that GaN SSPAs can be efficiently adopted to replace TWTA in several applications [5].

Nowadays, GaN technology is available in two flavors: on Silicon Carbide (SiC) and on pure Silicon (Si) substrates.

Gallium Nitride on Silicon Carbide (GaN-SiC) technology has proven to be the best semiconductor platform to implement high power, efficient HPAs at microwave frequencies, i.e., up to Ku-band. On the other hand, the emerging Gallium Nitride on Silicon (GaN-Si) processes can represent, in some cases, an interesting alternative. Indeed, GaN-Si allows to exploit the key features of most common GaN-SiC technologies, while assuring some benefits especially in terms of production costs, thanks to larger wafers.

After a discussion about pros and cons of GaN-Si and GaN-SiC technologies, this paper presents the design of a four stage Q-band high power amplifier on 100nm gate length GaN-Si, for space applications. The HPA achieves an output power higher than 4W, with a power added efficiency (PAE) and a gain around 30% and 25 dB, respectively, in 37.5-42.5 GHz band.

## II. GAN TECHNOLOGY: SI VS. SIC

The selected process is the D01GH available at OMMIC foundry. Such GaN process is available on both Si and SiC substrates. Consequently, this unique feature allows carrying out a fair and significant comparison between the two substrates.

The fields of application are microwave and millimeter-wave circuits as power amplifier, low noise amplifier and other general functions. The process uses AlN/GaN/AlGaN active layer for Double Heterostructure Field Effect Transistors (DHFET) with gate length of 100nm, leading to a cut-off frequency higher than 100GHz. The suggested drain bias voltage for power transistor is $V_{DD}$=12V. Two types of Metal-Insulator-Metal (MIM) capacitors are available: one to implement high capacitance values (400 pF/mm²) and the other for small values (49pF/mm²). Resistors with high sheet resistance values are obtained with a GaN layer, while smaller values (40 Ω/□) are realized through nickel-chrome (Ni-Cr) layer.

For both GaN-Si and GaN-SiC options, load-source pull simulations were carried out on different gate peripheries with the aim to estimate the maximum power density, efficiency and gain, while assuring a device junction temperature ($T_j$) lower than 160°C, i.e., the maximum allowable value for space application.

978-1-5386-5388-3/18 $31.00 © 2018 IEEE

As expected, the thermal resistance ($R_{th}$) of GaN-Si devices results to be more than double as compared to GaN-SiC counterparts, due to the poor thermal conductivity of the Si. This clearly implies a different limit in the maximum dissipated power and, consequently, in the achievable RF power density ($P_D$). As an example, Fig. 1 shows the non-linear simulated performance of a single 4x75µm FET in both GaN-SiC and GaN-Si processes at 40GHz and assuming a backside temperature for the device of $T_{BP}$=80°C. The former was biased at $V_{DD}$=12V and exhibits a maximum output power of 1.4W, thus resulting in a power density of $P_D$=4.2W/mm, with an associated PAE and gain around 45% and 7dB, respectively. Moreover, such performances were achieved guaranteeing a maximum junction temperature of the device always lower than 160°C. In the GaN-Si case, to keep $T_j$ under this limit, it was necessary to de-rate the drain bias voltage up to $V_{DD}$=8V, which corresponds a maximum output power of 0.7W only, thus a $P_D$=2.1W/mm, even if PAE and gain values are almost the same as in the GaN-SiC case.

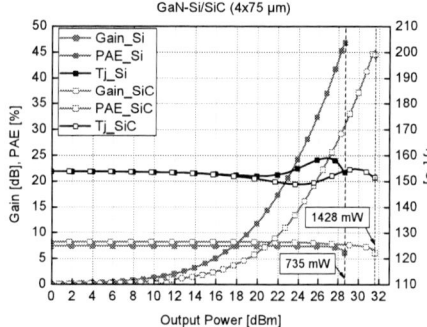

Fig. 1: Gain, PAE and Junction temperature of a 4x75µm device in both GaN-Si (filled symbols) and GaN-SiC (empty symbols) processes as functions of output power at f=40 GHz. Bias point is: $V_{DD}$= 8V for GaN-Si, $V_{DD}$= 12V for GaN-SiC and $V_{GG}$=-1.1 V for both.

A second key difference between SiC and Si substrates is related with the losses of passive structures, especially transmission lines. Fig. 2 shows the simulated insertion loss of a transmission line with characteristic impedance of $Z_0$=50Ω and 1mm length in both processes. At 40GHz, the GaN-Si line shows an insertion loss that is 0.13dB higher than the one in GaN-SiC.

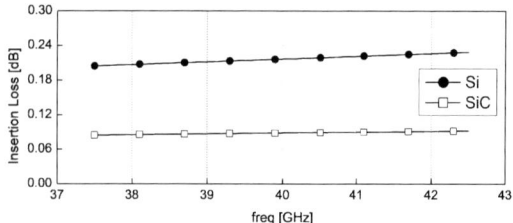

Fig. 2: Simulated insertion loss of a transmission line with characteristic impedance of $Z_0$=50Ω and 1mm length when realized on GaN-Si (filled symbols) and GaN-SiC (empty symbols) substrates.

## III. HPA DESIGN

The MMIC requirements are listed in Table 1. Accounting for the above-mentioned constraints of the GaN-Si technology, as well as, the specified Psat and the unavoidable losses in the output combiner, eight 4x75µm devices resulted to be required in the final stage of the HPA. The driver stage uses the same devices in a 1:2 ratio, whereas the second and first stages are composed by two 4x50um and two 2x50um FETs, respectively. Exploiting the symmetry, Fig. 3 shows both the MMIC schematic and related topology. The selected bias point is $V_{DD}$=8V and $V_{GG}$=-1.1V for all the stages. Moreover, the MMIC was designed to directly compensate the parasitic introduced by the input and the output bond wires used to interconnect it with the rest of the system.

Table 1: HPA requirements.

| Feature | Symbol | Unit | Value |
|---|---|---|---|
| Bandwidth | BW | GHz | 37.5-42.5 |
| Gain | G | dB | >19.5±0.5 |
| Sat. Output power | Psat | dBm | >36 |
| Return Loss | RL | dB | >15 |
| Power Added Efficiency | PAE | % | >20 |
| Junction temperature | Tj | °C | <160 |
| Backside temperature | $T_{BP}$ | °C | 80 |

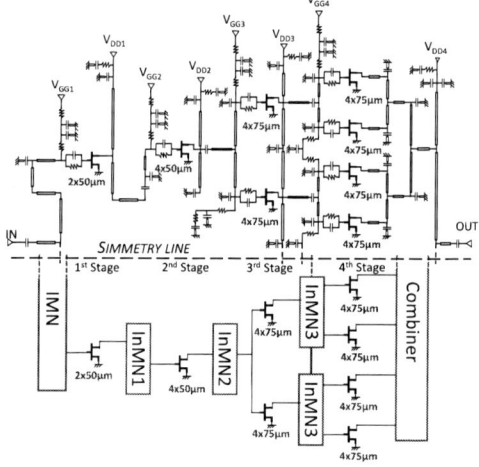

Fig. 3: MMIC schematic and related topology.

Each active device was stabilized in band through an R-C parallel network connected in series to gate. Instead, the addition of a resistor in the gate bias line was necessary to reach out-of-band unconditionally stability from DC to 150GHz.

The output combiner was designed with the aim to synthesize the optimum load required across each device while reducing as much as possible insertion loss and sizes.

Then, the inter-stage matching network (InMN3) between the last and the third stage was implemented. This network must transform the input loads of two devices in the last stage in the optimum load requested by a single 4x75μm of the third stage, minimizing the insertion loss and allowing the biasing of the devices. Following the same approach, the other inter-stage matching networks (i.e. InMN2 and InMN1) were derived. Final step was the design of the input matching network (IMN). This network transforms the input impedance of the two-stabilized device in the first stage into the conjugate impedance seen at the input of the series composed by bond wires and 50Ω standard termination. Thus, as for the output combiner, it provides on board compensation of the bond wires used to interconnect the MMIC with the rest of the system.

The simulated behaviors of the junction temperature of devices are shown in Fig. 4 as a function of the input power and for all the frequencies in the band. It is worth noting that in all the stages of the HPA, the devices are working safe with a Tj lower than 160°C.

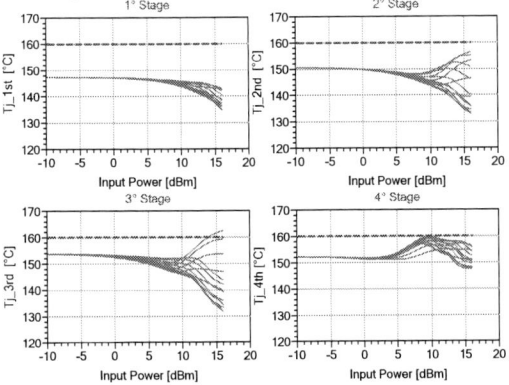

Fig. 4: Junction temperature of each device (one for stage) as a function of the input power and for all the frequencies in the band (from 37.5GHz to 42.5GHz in steps of 0.5GHz).

## IV. PERFORMANCE

Fig. 5 reports a picture of the realized MMIC. Sizes are 3.54 x 3.5 mm². The HPA was tested under small signal excitation at the nominal bias condition i.e., $V_{DD}$=8V and $V_{GG}$=-1.1V.

The comparison between measured and simulated small signal performances is reported in Fig. 6, showing a good agreement. In particular, measured S21 is higher than 23.4dB in the overall band from 37.5GHz to 42.5GHz. In the same frequency range, measured input and output return losses are better than 10 dB and 5dB, respectively.

Simulated large-signal performances as function of input power and for the lower, middle and upper frequency in the operative band (i.e., 37.5 GHz, 40 GHz and 42.5 GHz) are shown in Fig. 7. Saturated output power is about 36dBm, whereas the PAE and gain are in the range of 30% and 22dB, respectively.

The same features are shown in Fig. 8 as functions of frequency and for a constant input power of 14dBm. In the overall bandwidth the output power ripple is almost negligible being lower than 0.2dB. The PAE is close to 30%.

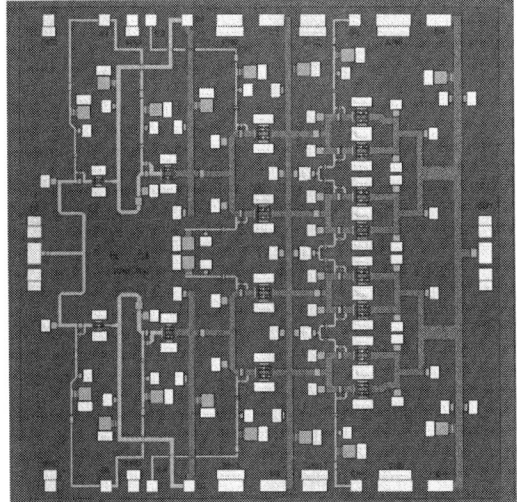

Fig. 5: Picture of the HPA MMIC. Sizes are 3.54 x 3.5 mm².

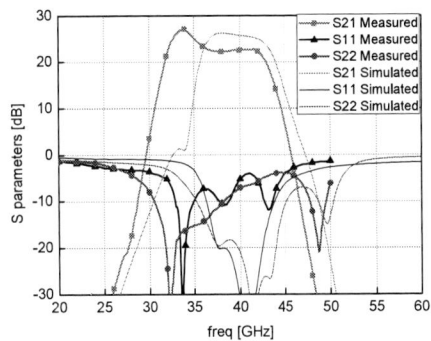

Fig. 6: Measured and simulated HPA scattering parameters.

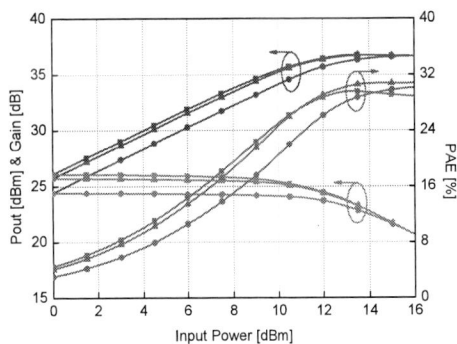

Fig. 7: Output power, PAE, and gain behaviors as functions of input power at 37.5GHz (square symbol), 40GHz (triangle symbol) and 42.5GHz (circle symbol).

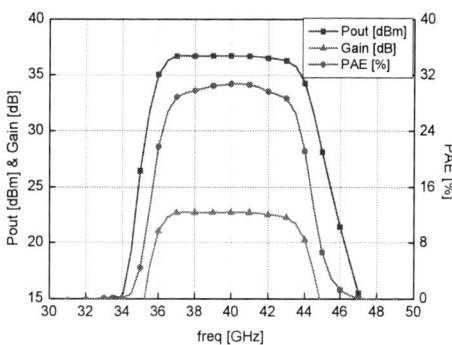

Fig. 8: Output power, PAE, and gain behaviours as functions of the frequency for a constant input power of 14dBm.

Finally, Table 2 shows a comparison between this MMIC and the actual state-of-the-art in Ka/Q band. Even if all the other HPAs are realized in more performing GaN-SiC technologies, this MMIC compares pretty well with them, proving that GaN-Si can be a feasible alternative with respect to GaN-SiC, despite the higher losses and thermal resistance.

## V. CONCLUSION

This paper reported the design and related performances of a Q-band MMIC HPA in 100nm GaN on Si technology. To the best of the Authors' knowledge, this MMIC is the first HPA designed on a GaN-Si able to provide 4W output power while respecting the space de-ratings. Moreover, the HPA shows a PAE and associated gain higher than 30% and 22dB, respectively, from 37.5GHz to 42.5GHz. The comparison between measured and simulated Scattering parameters have shown a good agreement.

Table 2: Comparison of Q-band MMICs PAs performance.

| Gate length-Process Technology | Year paper | Operative condition | Number of stages | Freq. band (GHz) | Small Gain (dB) | Sat. Power (dBm) | Avg. PAE (%) | Die Area (mm²) | Reference |
|---|---|---|---|---|---|---|---|---|---|
| 0.15µm-GaN-SiC | 2005 | CW | 2 | 27.5-34.5 | 15 | 36 | 23.8 | - | [6] |
| 0.15µm-GaN-SiC | 2006 | CW | 2 | 26-36 | 12.5 | 36 | 23 | - | [7] |
| 0.15µm-GaN-SiC | 2016 | pulsed | 3 | 34-36 | 26 | 41.9 | 30 | 9.5 | [8] |
| 0.20µm-GaN-SiC | 2016 | pulsed | 3 | 30-39 | 25 | 37.7 | 38 | 4.2 | [9] |
| 0.10µm-GaN-SiC | 2016 | CW | 1 | 35-40 | 9 | 24 | 18 | 2.3 | [10] |
| 0.15µm-GaN-SiC | 2017 | CW | 2 | 32-38 | 17.5 | 37.1 | 29 | 3.6 | [11] |
| 0.15µm-GaN-SiC | 2017 | CW | 3 | 32-38 | 27 | 40.5 | 33 | 9.9 | [11] |
| **0.10µm-GaN-Si** | **2018** | **CW** | **4** | **37.5-42.5** | **22** | **36.5** | **30** | **12.4** | **This work** |

## REFERENCES

[1] H. Fenech, Amos, Tomatis and Soumpholphkakdy, "KA-SAT and future HTS systems," *2013 IEEE 14th International Vacuum Electronics Conference (IVEC)*, Paris, 2013, pp. 1-2.

[2] P. Blount, S. Huettner and B. Cannon, "A High Efficiency, Ka-Band Pulsed Gallium Nitride Power Amplifier for Radar Applications," *2016 IEEE Compound Semiconductor Integrated Circuit Symposium (CSICS)*, Austin, TX, 2016, pp. 1-4.

[3] A. Tyagi, C. Choudhary, N. Upadhyay, "Future of V Band in Satellite Communication," *International Journal of Science, Engineering and Technology*, vol.1, issue 1, Nov. 2013.

[4] R. Emrick, P. Cruz, N. B. Carvalho, S. Gao, R. Quay and P. Waltereit, "The Sky's the Limit: Key Technology and Market Trends in Satellite Communications," in IEEE Microwave Magazine, vol. 15, no. 2, pp. 65-78, March-April 2014.

[5] R. Giofrè, P. Colantonio, L. Gonzalez, F. De Arriba, L. Cabria, "A 300W Complete GaN Solid State Power Amplifier for Positioning System Satellite Payloads" Proceedings of int. microwave symposium, mtt-s, San Francisco, California, May 22-27, 2016

[6] M. Micovic *et al.*, "GaN MMIC technology for microwave and millimeter-wave applications," *IEEE Compound Semiconductor Integrated Circuit Symposium, 2005. CSIC '05.*, 2005, pp. 3 pp.-.

[7] A. M. Darwish, K. Boutros, B. Luo, B. D. Huebschman, E. Viveiros and H. A. Hung, "AlGaN/GaN Ka-Band 5-W MMIC Amplifier," in *IEEE Transactions on Microwave Theory and Techniques*, vol. 54, no. 12, pp. 4456-4463, Dec. 2006.

[8] X. Yu, H. Tao and W. Hong, "A Ka band 15W power amplifier MMIC based on GaN HEMT technology," *2016 IEEE International Workshop on Electromagnetics: Applications and Student Innovation Competition (iWEM)*, Nanjing, 2016, pp. 1-3.

[9] P. Blount, S. Huettner and B. Cannon, "A High Efficiency, Ka-Band Pulsed Gallium Nitride Power Amplifier for Radar Applications," *2016 IEEE Compound Semiconductor Integrated Circuit Symposium (CSICS)*, Austin, TX, 2016, pp. 1-4.

[10] P. Feuerschütz, C. Friesicke, R. Quay and A. F. Jacob, "A Q-band power amplifier MMIC using 100 nm AlGaN/GaN HEMT," *2016 11th European Microwave Integrated Circuits Conference (EuMIC)*, London, 2016, pp. 305-308.

[11] S. Chen, S. Nayak, C. Campbell and E. Reese, "High Efficiency 5W/10W 32 - 38GHz Power Amplifier MMICs Utilizing Advanced 0.15µm GaN HEMT Technology," *2016 IEEE Compound Semiconductor Integrated Circuit Symposium (CSICS)*, Austin, TX, 2016, pp. 1-4.

# An Integrated Power Detector for a 5GHz RF PA

Valdrin Qunaj
ESAT-MICAS
KU LEUVEN
Leuven, Belgium
Email: valdrin.qunaj@esat.kuleuven.be

Umut Celik
ESAT-MICAS
KU LEUVEN
Leuven, Belgium
Email: umut.celik@esat.kuleuven.be

Patrick Reynaert
ESAT-MICAS
KU LEUVEN
Leuven, Belgium
Email: patrick.reynaert@esat.kuleuven.be

*Abstract*—**A fully integrated power detector is presented that detects both instantaneous RF voltage and current. Both signals are multiplied using an on-chip low-noise, high linearity mixer to measure the real power delivered to the load of an integrated PA operating at 5GHz in predictive 45nm CMOS technology. The dynamic range of the detector is 23.72dB with an accuracy of $< \pm 0.5$dB. Even under antenna load mismatch the detector is able measure the real power delivered to the load up to a voltage standing wave ration of 2.7:1 with an error $< \pm 0.6$dB. The power detector has no effect on the PA performance and can be integrated under the large output matching transformer, resulting in a cost efficient design.**

## I. INTRODUCTION

In the past 20 years the telecommunications market has grown tremendously, this mainly due to the success of mobile phones and rise of the Internet. The total number of connected devices will grow to 27 billion by 2021. This exponential increase in devices will boost the data usage, by 2021 there will be a 12 fold increase in data traffic compared to 2015 [1]. For manufacturers this means that the main focus is low cost and low power consumption. This implies that integration is becoming more important and hence CMOS, due to its low cost and integration with digital circuits, is the preferred technology of choice enabling true SoC integration. The whole transceiver, including PA, is integrated into one single CMOS chip [2].

The PA has the highest power consumption out of all building blocks in a transceiver [3], PA efficiency is thus of great importance. The usage of higher order modulation requires the PA to be highly linear. In order to reduce the EVM, the PA can be operated in back-off where it behaves more linearly. However, in this region the efficiency and output power is lower.

PA's are typically designed for a fixed antenna load of $50\Omega$, the efficiency and linearity changes when the load deviates from the nominal $50\Omega$. This paper will evaluate an architecture that can measure the real power that is delivered to the load and hence detect these changes under antenna load mismatch. Previous publications have introduced power detectors that only detect voltage or current [4] [5], to determine the RF power, however these detector types only give the true RF power for a fixed load. By measuring both the RF current and RF voltage, the RF power can be detected even under antenna load mismatch.

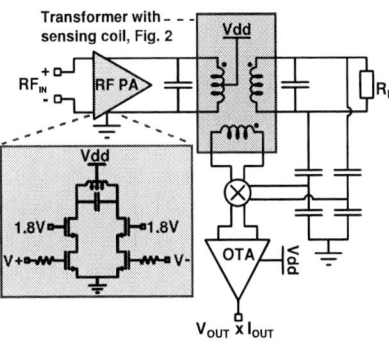

Fig. 1: Schematic representation of the power detector architecture

## II. POWER DETECTOR

From electrical network theory, the AC power is defined as

$$S = \frac{V \cdot I^*}{2} = \frac{V_{out} I_{out}}{2} e^{\phi_v - \phi_i} \tag{1}$$

$$= \frac{V_{out} I_{out}}{2} [\cos(\phi) + j \sin(\phi)] \tag{2}$$

$$= P + jQ \tag{3}$$

where $\phi = \phi_v - \phi_i$ is the phase difference between voltage and current, $P$ is the real power and $Q$ is the reactive power.

The architecture in Fig. 1 detects the instantaneous RF output voltage $v_{out}(t)$ and current $i_{out}(t)$ to calculate the power delivered to load by taking the product of these two signals. The instantaneous RF power $P_{out}(t)$ is defined as

$$P_{out}(t) = v_{out}(t) \cdot i_{out}(t) \tag{4}$$

$$= V_{out} cos(\omega t + \phi_v) \cdot I_{out} cos(\omega t + \phi_i) \tag{5}$$

$$= V_{out} I_{out} \left( \frac{cos(\phi)}{2} + \frac{cos(2\omega t + \phi)}{2} \right) \tag{6}$$

From (6) it is clear that $P_{out}(t)$ has a DC term and a harmonic component at two times the operating frequency. And indeed, it can be concluded that the real power delivered to the load equals the DC component in (6)

$$\Re(S) = P = \frac{V_{out} I_{out}}{2} cos(\phi) \tag{7}$$

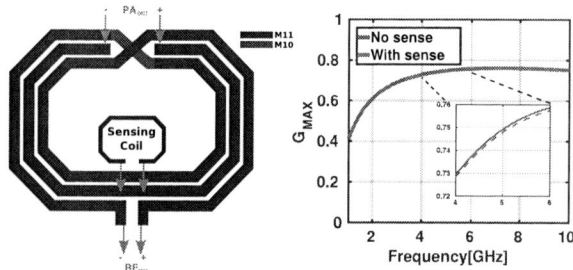

Fig. 2: Output transformer of the PA with sensing coil and it's effect on efficiency

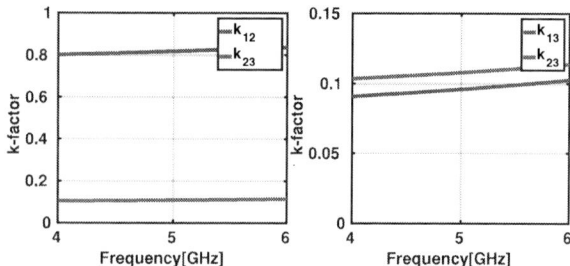

Fig. 3: The coupling factors of the output matching transformer from 4GHz to 6GHz

From equation (7) it is clear that under antenna load mismatch, meaning that there is a phase difference between the voltage and current signal($\phi \neq 0$), the proposed architecture is able to measure the real power delivered to the load $R_L$.

### A. Sensing Coil

The instantaneous RF current $i_{out}(t)$ is detected using a sensing coil in the center of the output matching transformer, as given in Fig. 2. The efficiency of the transformer is simulated with and without sensing coil. The sensing coil has minimal effect on the efficiency of the output matching transformer, as seen in Fig. 2.

The coupling factor of a transformer, between primary winding $L_1$ and secondary winding $L_2$ is defined as

$$k = \frac{M}{\sqrt{L_1 L_2}} \qquad (8)$$

where $M$ is the mutual inductance. For an ideal transformer, the coupling factor $k$ is equal to one. The sensing coil $L_3$ and the secondary winding $L_2$ form a transformer, hence a coupling factor $k_{23}$ between the two can be defined. The sensing coil however is not only coupled to the secondary winding, but also the primary. A second coupling factor $k_{13}$, between the primary winding $L_1$ and the sensing coil $L_3$, can thus be defined. Fig. 3 shows the coupling factors of the output transformer.

Since the sensing coil is coupled to both $L_1$ and $L_2$, the transformer with sensing coil can be represented as a three port

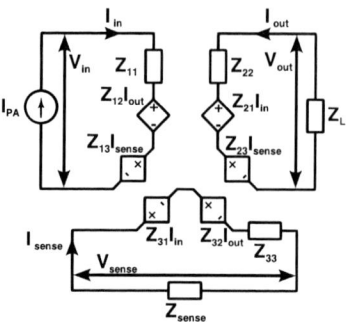

Fig. 4: A three port Z-parameter representation of the output transformer with sensing coil

network, as given in Fig. 4. From the Z-port representation the sensing current $I_{sense}$ can hence be described as

$$I_{sense} = \frac{Z_{32}}{Z_{33} + Z_{sense}} I_{out} - \frac{Z_{31}}{Z_{33} + Z_{sense}} I_{in} \qquad (9)$$

The turn ratio between the primary and secondary winding is defined as [6]

$$n = \frac{V_{out}}{V_{in}} = \frac{I_{in}}{I_{out}} = \sqrt{\frac{L_{out}}{L_{in}}} \qquad (10)$$

Using (9) and (10), the instantaneous RF power can be described as

$$P_{out}(t) = v_{out}(t) i_{out}(t) \cdot A \qquad (11)$$

$$A = n \cdot \left( \frac{-Z_{32}}{Z_{33} + Z_{sense}} + \frac{-Z_{31}}{Z_{33} + Z_{sense}} \right) \qquad (12)$$

where A is a constant characterized by the designed output transformer, this factor A can be calibrated out of the detected output power by doing a simple calibration after the fabrication of the chip. Using (6) which was derived previously, and (11)

$$P_{out} = A \cdot \frac{V_{out} I_{out}}{2} \cos(\phi) \qquad (13)$$

The OTA used after the mixer is designed to have high gain and a very limited bandwidth, this will in turn, filter out the harmonic at twice the operating frequency. Clearly the detected power from (13) is equal to (7) up to the constant A. Indeed the proposed architecture detects the real power delivered to the load.

### B. Mixer

*1) Conversion gain and linearity:* A double balanced Gilbert cell mixer is used to multiply the detected RF voltage $v_{out}$ and RF current $i_{out}$. In order to improve the linearity and conversion gain(CG) of the Gilbert cell mixer, current bleeding and source degeneration is used. Because the RF input transistors are biased to operate in the saturation region, the CG and IIP3 can be defined as

$$CG = \frac{2}{\pi} R_L \sqrt{K_n I_{DS}}, \quad IIP3 = 4\sqrt{\frac{I_{DS}}{3 K_n}} \qquad (14)$$

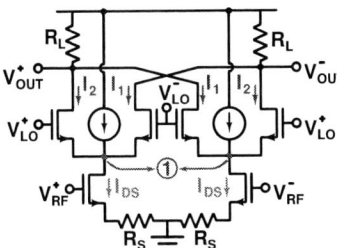

Fig. 5: Schematic of the Gilbert cell with current bleeding and source degeneration

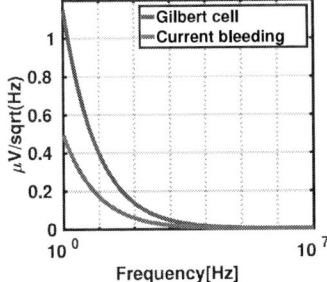

Fig. 6: Output noise of mixer without & with current bleeding

were $I_{DS}$ the current through the RF transistors, $R_L$ the load resistance of the mixer and $K_n = 2\mu C_{ox}W/L$. From these two equations it is clear that both IIP3 and CG are proportional with $\sqrt{I_{DS}}$. Consequently increasing the bias current will improve both conversion gain and the linearity. However, by increasing $I_{DS}$, the voltage drop across $R_L$ will increase as well, which will in turn affect the operation of the switching transistors. Thus in order to improve both CG and IIP3, $I_{DS}$ should be increased without increasing the current through the switching transistors. This can be done by adding a current source that injects a current $I_{CH}$ in node one, indicated in Fig. 5. With current bleeding, $I_{DS}$ is equal to $I_1 + I_2 + I_{CH}$, thus the current is higher, meaning that both CG and IIP3 are higher while at the same time the current through the load remains low, keeping the voltage drop across it small.

*2) Flicker noise:* Active mixers suffer substantially from flicker noise at their output, especially when the output frequency approaches zero. Because the down conversion mixer is multiplying two signals operating at the same frequency, it will produce a DC signal. Using current bleeding, the noise of the mixer will be reduced substantially [7]. This can be clearly seen on Fig. 6.

## III. SIMULATIONS RESULTS

### A. Antenna Load Mismatch

In the antenna load mismatch simulations, the load impedance of the PA is swept across the Smith chart. It was

established that the detected voltage $V_{det}$ coming from the OTA is proportional to the power at the load

$$P_{load} = A \cdot V_{det} \cdot \epsilon \qquad (15)$$

$$10\log_{10}(P_{load}) = 20\log_{10}(A) + 20\log_{10}(V_{det}) + 20\log_{10}(\epsilon) \qquad (16)$$

where $\epsilon$ is the error. By calibrating once with a load impedance of $50\Omega$ the constant $20\log_{10}(A)$ is found, this constant is then simply added to the detected voltage $V_{det}(dBV)$. By then taking the difference between the detected voltage $V_{det}(dBV)$ and $P_{load}(dBm)$ we can hence compute the error $\epsilon(dB)$ when sweeping the load impedance.

The CG of a mixer was introduced. When the relation of CG with respect to the error is plotted, it is clear that they are related to each other. Fig. 7 reveals that due to strong nonlinearity of the CG in low load impedance regions, the error increases tremendously. Ideally if the CG is kept constant, the error of the power detector will be reduced. Using current bleeding, the CG remains constant if load impedance of the PA change and hence the error is reduced.

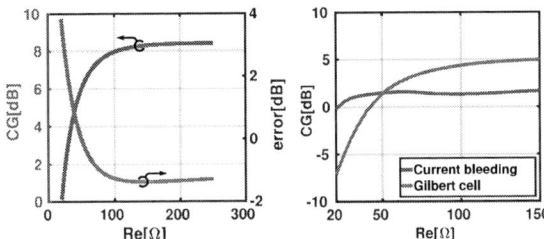

Fig. 7: The conversion gain and the absolute error of the power detector when the real part of the load is swept

Fig. 8 shows the absolute error of the power detector under antenna load mismatch. The absolute error is very large in the area where the real impedance of the load is small. The maximum error for the Gilbert cell mixer is $\pm 7dB$. The absolute error of the power detector using current bleeding and source degeneration on the Gilbert cell mixer is plotted. Across the simulated area of the Smith chart, up to a VSWR of 2.7:1, the maximum error is $< \pm 0.66dB$, as seen on the contour plot in Fig. 9. Current bleeding and source degeneration clearly reduces the error of the power detector.

### B. Dynamic Range Simulations

In order to simulate the dynamic range(DR) of the power detector, the load of the PA is kept constant at $50\Omega$. The input power of the PA is swept from -30dBm to +15dBm, afterwards the output power delivered to the load $P_{load}$ and the output voltage of the power detector $V_{det}$ is evaluated. The DR of the detector is 23.72dB for a linearity error $< \pm 0.5dB$, as can be seen in Fig. 10. In the operating bandwidth of the PA, from 4.5GHz to 5.4GHz, the performance of the power detector is maintained.

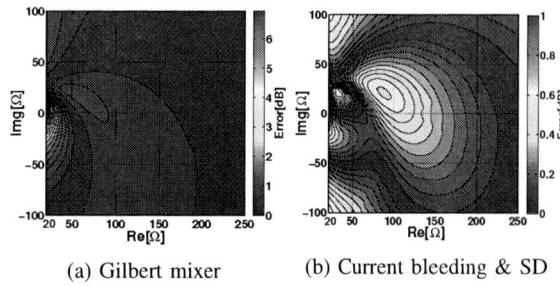

(a) Gilbert mixer   (b) Current bleeding & SD

Fig. 8: The absolute error of the power detector under antenna load mismatch

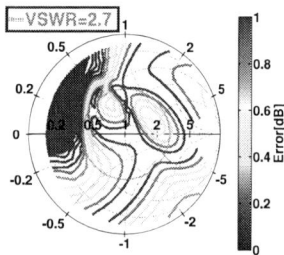

Fig. 9: The absolute error of the power detector under antenna load mismatch using current bleeding and source degeneration

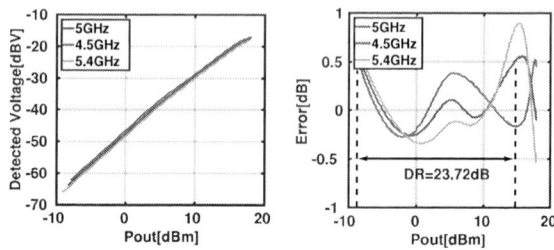

Fig. 10: The detected output voltage $V_{det}(dBV)$ and linearity error for different operating frequencies

TABLE I: Final simulations results of the power detector in comparison with other publications

| | This work | [8] | [4] | [5] |
|---|---|---|---|---|
| Technology | 45nm | 40nm | 90nm | 65nm |
| Integration of PA | Yes | Yes | Yes | No |
| Coupling principle | Capacitive & Inductive | Capacitive & Inductive | Capacitive | Capacitive |
| Detection principle | Voltage & Current | Current & Voltage | Voltage | Voltage |
| Dynamic range(dB) | 23.72 | 32.5 | 27 | 25 |
| Linearity error(dB) | ±0.5 | ±0.5 | ±0.5 | - |
| Maximum error under antenna load mismatch(dB) | ±0.6 | ±1 | - | - |
| VSWR | 2.7:1 | 2.5:1 | - | - |
| Area overhead | No | No | Yes | Yes |
| Power consumption(mW) | 0.118 | 0.31 | 0.3-0.63 | 0.06 |

## IV. CONCLUSION

This paper explored the design and simulations of a fully integrated transformer based power detector with a PA operating at 5GHz in 45nm CMOS predictive technology. By detecting both the instantaneous RF current and RF voltage, the real power delivered to the load is calculated through a simple on-chip mixer. Using current bleeding, the linearity and noise of the mixer was substantially improved. The detector is integrated under the large output matching transformer of the PA, thus there is no area overhead in this architecture. Under antenna load mismatch, the detector is able to measure the real power delivered to the load accurately.

The detector accurately detects the real power delivered to the load with an accuracy of $< \pm 0.5 dB$ over a dynamic range of 23.72dB. The power consumption of the detector is 0.0118mW. Even under antenna load mismatch, with a VSWR of 2.7:1, the error is $< \pm 0.6 dB$. The final results of the power detector are given on Table I.

## REFERENCES

[1] Ericsson, "Ericsson mobility report 2016," URL: https://www.ericsson.com/res/docs/2016/ericsson-mobility-report-2016.pdf, last checked on 2017-04-24, 2016.

[2] D. Su, M. Zargari, P. Yue, S. Rabii, D. Weber, B. Kaczynski, S. Mehta, K. Singh, S. Mendis, and B. Wooley, "A 5 GHz CMOS transceiver for IEEE 802.11a wireless LAN," in 2002 IEEE International Solid-State Circuits Conference. Digest of Technical Papers (Cat. No.02CH37315), vol. 1, Feb 2002, pp. 92–449 vol.1.

[3] Y. Li, B. Bakkaloglu, and C. Chakrabarti, "A comprehensive energy model and energy-quality evaluation of wireless transceiver front-ends," in IEEE Workshop on Signal Processing Systems Design and Implementation, 2005., Nov 2005, pp. 262–267.

[4] H. Nakamoto, M. Kudo, K. Niratsuka, T. Mori, and S. Yamaura, "A real-time temperature-compensated cmos rf on-chip power detector with high linearity for wireless applications," in 2012 Proceedings of the ESSCIRC (ESSCIRC), Sept 2012, pp. 349–352.

[5] J. Gorisse, A. Cathelin, A. Kaiser, and E. Kerherve, "A 60ghz 65nm cmos rms power detector for antenna impedance mismatch detection," in 2009 Proceedings of ESSCIRC, Sept 2009, pp. 172–175.

[6] I. Aoki, S. D. Kee, D. B. Rutledge, and A. Hajimiri, "Distributed active transformer-a new power-combining and impedance-transformation technique," IEEE Transactions on Microwave Theory and Techniques, vol. 50, no. 1, pp. 316–331, Jan 2002.

[7] J. Park, C. H. Lee, B. S. Kim, and J. Laskar, "Design and analysis of low flicker-noise cmos mixers for direct-conversion receivers," IEEE Transactions on Microwave Theory and Techniques, vol. 54, no. 12, pp. 4372–4380, Dec 2006.

[8] B. Francois and P. Reynaert, "3.3 a transformer-coupled true-rms power detector in 40nm cmos," in 2014 IEEE International Solid-State Circuits Conference Digest of Technical Papers (ISSCC), Feb 2014, pp. 62–63.

*PRIME 2018, Prague, Czech Republic*  　　　　　　　　*Session: Radio Frequency Circuits and Systems I*

# Design of an E-band Doherty Power amplifier

Md Najmussadat[1], Raju Ahamed[1], Dristy Parveg[2], Mikko Varonen[2], Kari A. I. Halonen[1]

[1]Department of Electronics and Nanoengineering, Aalto University, Espoo, Finland

[2]VTT Technical Research Centre of Finland Ltd, Espoo, Finland

Email: md.najmussadat@aalto.fi

*Abstract*—This paper demonstrates the design of an E-band Doherty power amplifier (PA) based in an 130nm SiGe BiCMOS process. This design includes main and auxiliary amplifiers, lange coupler and a pre-amplifier. The designed power amplifier exhibits a saturated output power of 14.4 dBm and output referred P1dB of 11.7 dBm. The peak power added efficiency (PAE) of this amplifier is 19.2%. This PA shows PAE of 17% at P1dB and 11.6% at 6-dB output power back off. The peak power gain of this Doherty PA is 23 dB at 75 GHz with a 3-dB bandwidth from 60 to 80 GHz. The designed Doherty PA consumes DC power of 52 mW with a chip area of 900 $\mu$m x 800 $\mu$m without RF pads.

*Index Terms*—BiCMOS, Doherty Power Amplifier, Microstrip, Power Backoff, SiGe.

## I. INTRODUCTION

The need for high data speed communication systems is growing very fast. The millimeter -wave spectrum, with a wide frequency range available for high data rate communication, is drawing increasing attention. High frequency bands are used for satellite communication, automotive radar, point-to-point communication as well as imaging and radar applications [1]. However, advanced communication systems use different modulation schemes that acheives high peak-to-average power ratios (PAPRs) to obtain high data rates and high spectral efficiency [2]. The PA linearity requirement for these modulation schemes is also high to avoid distortion of the signal. It has been seen that in the most of the high-performance mm-wave PA design, much importance was given for high peak PAE. However, these mm-wave PAs show that the efficiency drop considerably when backed off from output saturation power (Psat) [7]. So, for enhancement of the back-off efficiency, different methods such as Doherty, outphasing and envelope tracking can be used. Among them Doherty is the mostly used for its ease of implementation [8].

Traditionally, mm-wave PAs were mainly fabricated using III-V semiconductors such as InP, GaN, GaAs, SiGe, because these processes offer both high output power and high efficiency [3]. Although mm-wave circuits can be designed in modern Si-based technologies due to higher cutoff frequencies, it has never been a favorable choice for PA design because of lower drain-gate and drain-source breakdown voltages and lower quality factor of passive element. However, recent studies indicate that Si is becoming a realistic competitor in mm-wave PA designs [4]- [6]. In this work, a Doherty PA is presented in a 130nm SiGe BiCMOS process.

## II. MM-WAVE DOHERTY PA DESIGN

The basic Doherty PA is shown in a block diagram presented in Fig.1. The Doherty PA is composed of an input power divider, two power amplifiers in parallel namely main PA and auxiliary PA. The main PA is designed to work in all levels of input power and the auxiliary PA works only in higher levels of input power. This is done by operating the main PA in either Class- A or AB and the auxiliary PA in Class-C [2]. The $\lambda/4$ transmission line which is connected after the main PA has characteristic impedance two times higher than load impedance and it acts as an imepedance inverter [7]. The quarterwave transmission line that is connected in between the output of power divider and the input of the auxiliary PA, used as a phase controller so that both the signals from main PA and auxiliary PA combine with same phase at the output. At low input power, the main PA is in on-state and the auxiliary PA is in off-state. In the main PA the output current increases and the impedance at the output of the main PA is four times the load impedance. At higher input power, the main PA saturates with its maximum efficiency and auxiliary PA turns on. Now auxiliary PA controls the impedance at the output of the main PA according to the active load-pull principle [8] and the impedance seen by main PA will start to decrease. When the auxiliary PA is in on-state, the main PA will remain at its peak value for the dynamic load modulation as described in [8]. Therefore, the efficiency of the main PA remains at its peak and overall PA efficiency stays almost unchanged when auxiliary amplifier reaches its peak output power.

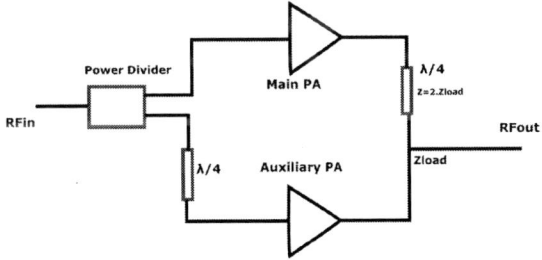

Fig. 1: Basic Doherty amplifier.

Transistor modeling needed careful attention in mm-wave PA design as it has a significant effect on device characteristics [10]. In this PA design, it was observed that the connections from the transistor through via's have a negative effect on the

978-1-5386-5388-3/18 $31.00 © 2018 IEEE　　　　145

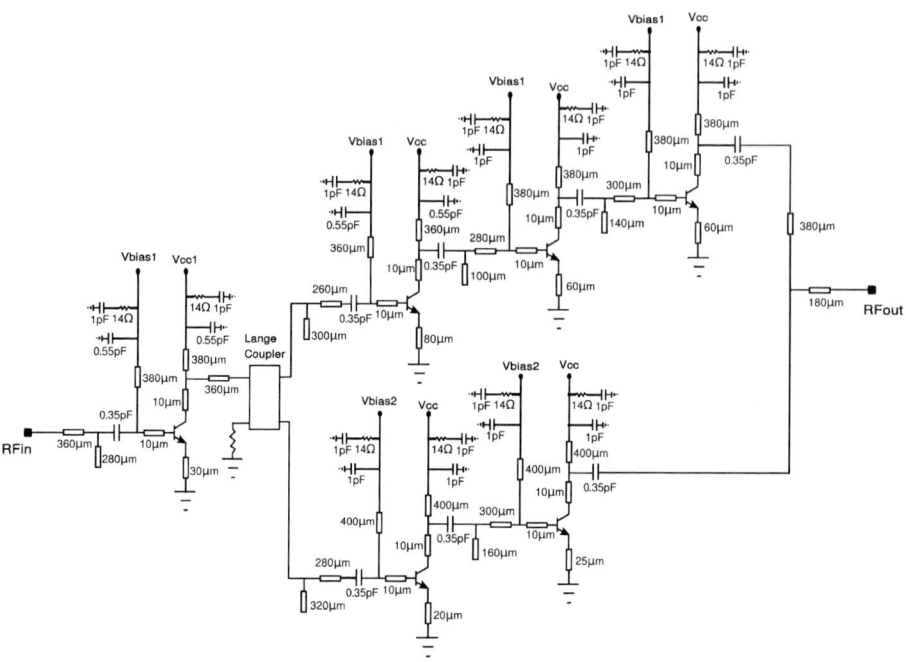

Fig. 2: Complete schematic of the Doherty power amplifier.

maximum gain of the transistors. So these connections were modified for optimizing the maximum gain of the transistors. The designed Doherty amplifier has three stages in the main PA and two stages in the auxiliary PA. For the power requirements, the number of parallel emitters are increased in different stages. To determine the optimal output power, the load pull simulation was done for each stage by varying the emitter's number. A total of twenty emitters (two transistors each having ten emitters in parallel) are used to achieve desired output power in the last stage of the main PA. In the first stage and second stage of the main PA, four and ten emitters are used respectively. From the load pull simulation the best load for each transistor is determined. The load point for the output stage is chosen for the maximum efficiency and for other stages this is selected for the maximum gain. In auxiliary PA, two stages are used with the first stage consisting of transistor with four emitters and the second stage consists of transistor with ten emitters in parallel.

Fig.2 shows the complete schematic of the designed Doherty PA. For the design simplicity, each stages of the main PA, the auxiliary PA and the pre- amplifier uses common emitter topology. Fig.3 shows a individual stage of an amplifier which is used in this design. Each stage of the main PA is operating in class-AB and the auxiliary PA in class-C. Both class-AB and class-C were driven by 2.0 V supply. A pre amplifier is used to increase the overall gain of the PA before the Lange coupler. To ensure for unconditional stability of the amplifier, emitter degeneration is used in each stage trading gain for

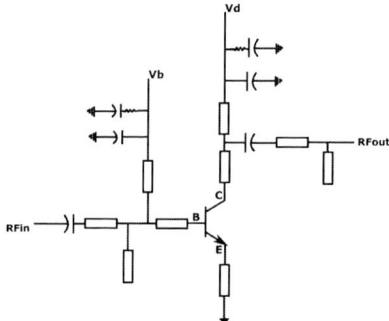

Fig. 3: Single stage amplifier.

stability. The input signal was fed through the pre-amplifier and the output of the pre amplifier is connected to a Lange coupler. The phase shift needed in the auxiliary amplifier is achieved here by the use of Lange coupler as it provides 90° phase shift between coupled port and direct port [5].

In this design series transmission lines and parallel stubs and MIM (metal-insulator-metal) capacitors are used for designing the matching networks. Microstrip transmission lines are also used in the biasing network of the transistors. Microstrip lines in this design are realized with topmost metal as signal lines and metal3 as ground lines. The RF Pad capacitance was taken into account as a part of the matching network. In each DC biasing line of the transistor, a series RC network is placed in parallel with MIM capacitor for better stability at lower frequencies.

### III. SIMULATION RESULTS

The power amplifier in this work was designed in IHP's 130nm SiGe BiCMOS technology with fT of 300 GHz and $f_{max}$ of 450 GHz and seven metal layers. Transmission lines are realized with topmost metal as signal layer and metal3 as ground layer. All the transmission lines in matching networks and biasing networks, MIM capacitors, PAD capacitance were simulated in Momentum EM simulator of Advanced Design System (ADS). Ground-Signal-Ground(GSG) configuration with pitch of 150 $\mu$m and pad diameter of 80 $\mu$m was selected as RF input and output pads.

#### A. Small signal performances

The small signal performances are shown in Fig.4. This PA exhibits peak small signal gain ($S_{21}$) of 23.6 dB at 75 GHz with a 3-dB bandwidth of 20 GHz from 60 GHz to 80 GHz.The PA achieves $S_{11}$ of -9.8 dB and $S_{22}$ of -17 dB at 75 GHz. The stability factor (K) is always greater than 1 from 1 GHz to 200 GHz which is shown in Fig.5.

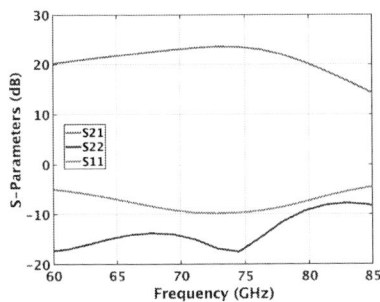

Fig. 4: Small signal performance of the Doherty PA.

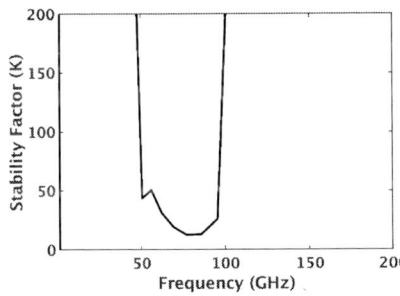

Fig. 5: Stability factor of the designed Doherty PA

#### B. Large signal performances

The PAE, output power and power gain at 75 GHz are shown in Fig 6. The simulated performance of the PA shows

the output saturated power ($P_{sat}$) of 14.4 dB and the output 1-dB compression point ($P_{1dB}$) of 11.7 dBm. The PA exhibits the paek PAE of 19.2% and PAE at $P_{1dB}$ of 17%. Fig.7 shows the PAE against output power. The 6-dB back off PAE is 11.6%. Fig.8 and Fig.9 shows the results of Pout and PAE against frequency respectively. From these figures, it can be seen that P1dB is always greater than 10 dBm from 60-80 GHz and PAE at P1dB is always greater than 15% from 70-80 GHz. The 6-dB backoff PAE is also greater than 10% from 70-80 GHz range.

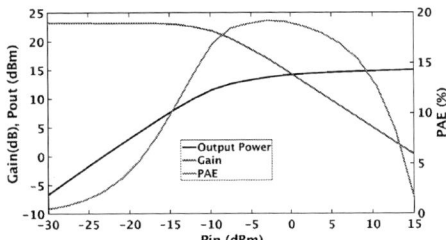

Fig. 6: PAE, Pout and Gain vs Input power.

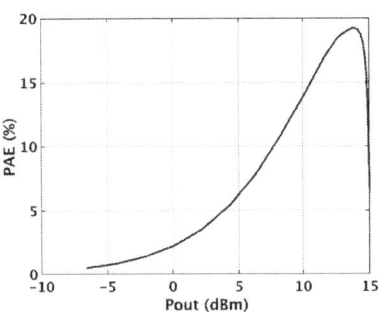

Fig. 7: PAE vs Pout of the Doherty amplifier.

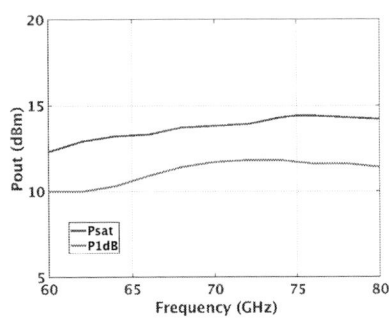

Fig. 8: Pout vs frequency

Simulated performance of the designed PA is compared with the other published designs in SiGe BiCMOS and CMOS

978-1-5386-5388-3/18 $31.00 © 2018 IEEE

TABLE I: Performance summary and comparison

| References | Technology | Type | Frequency (GHz) | Gain (dB) | Bandwidth (GHz) | Psat (dBm) | OP1dB (dBm) | PAE,max (%) | PAE@ OP1dB (%) | PAE_6dB backoff (%) | Area (mm2) |
|---|---|---|---|---|---|---|---|---|---|---|---|
| [2] | 40-nm CMOS | Doherty | 80 | 18.1 | 15.2 | 20.9 | 17.8 | 22.3 | 12 | 3.3 | 0.19 |
| [2] | 40-nm CMOS | Doherty | 73 | 25.3 | 7.6 | 22.6 | 18.9 | 19.3 | 10.7 | 3.3 | 0.19 |
| [9] | 90 nm SiGe | 8-way power combined | 76 | 21.2 | 14 | 24 | 21.3 | 11.6 | 8.2* | 2.5* | 3.52 |
| [10] | 180nm SiGe | 2-stage cascode | 83 | 25 | 9.6 | 14.7 | 12.5 | 8.1 | 5.7* | 1.5* | 0.34 |
| [11] | 130nm SiGe | Transformer coupled | 84 | 27 | 8.6 | 18 | 16 | 9 | 7.5* | 2.5* | 0.68 |
| [12] | 65-nm CMOS | Doherty | 60 | 18 | 5.2 | 14.9 | 13* | 16.8 | 15* | 8.7 | 0.195 |
| This work** | 130 nm SiGe | Doherty | 75 | 23 | 20 | 14.4 | 11.7 | 19.2 | 17 | 11.6 | 0.72 |

\* estimated from measurement plots ** post layout simulated result

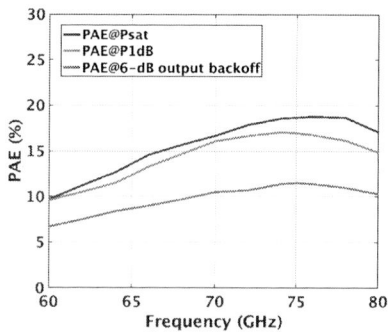

Fig. 9: PAE vs frequency

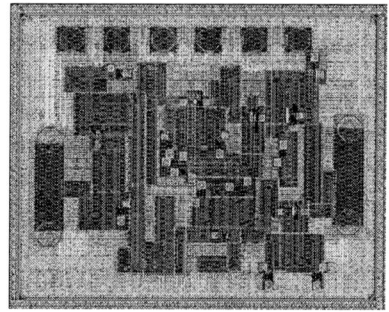

Fig. 10: Layout of the designed Doherty PA

processes in Table I. The designed PA shows 19.2% peak PAE and 23 dB power gain which are considerably high in the E-band. The PA consumes 52 mW of DC power. The designed PA achieves the highest 6-dB back-off PAE among all the other published works in 70-80 GHz. The layout of the designed Doherty PA is shown in Fig.10. The realised PA occupy an area of 900 $\mu$m x 800 $\mu$m excluding RF pads.

IV. CONCLUSION

A Doherty power amplifier in 130nm SiGe BiCMOS technology is presented in this paper. Lange coupler is used for input power dividing and to remove the phase shifting $\lambda/4$ transmission line at the begining of the auxiliary PA. The simulated results of the PA shows state-of-the-art results

providing peak PAE of 19.2%, power gain of 23 dB and highest 6-dB backoff PAE of 11.6% among the other PA at 75 GHz.

ACKNOWLEDGMENT

The authors would like to thank the Finnish Funding Agency for Innovation (Business Finland) for supporting this work through 5WAVE project.

REFERENCES

[1] A. Komijani , A. Hajimiri,A Wideband 77-GHz, 17.5-dBm Fully Integrated Power Amplifier in Silicon,IEEE J. Solid-State Circuits, vol. 41, Issue:8, pp. 1749-1756, Aug. 2006.
[2] E. Kaymaksut, D. Zhao and P. Reynaert, "Transformer-Based Doherty Power Amplifiers for mm-Wave Applications in 40-nm CMOS," in IEEE Transactions on Microwave Theory and Techniques, vol. 63, no. 4, pp. 1186-1192, April 2015.
[3] M. Micovic, A. Kurdoghlian, A. Margomenos, D. F. Brown, K. Shinohara, S. Burnham, I. Milosavljevic, R. Bowen, A.J. Williams, P. Hashimoto, R. Grabar, C.Butler, A. Schmitz, P. J. Willadsen, D. H. Chow, 92 -96 GHz GaN Power Amplifiers, 2012 IMS Digest, June 2012.
[4] ETSI, Uwe Rddenklau, "mmWave Semiconductor Industry Technologies:Status and Evolution", white paper No. 15, July 2016.
[5] D. Parveg et al., "CMOS I/Q Subharmonic Mixer for Millimeter-Wave Atmospheric Remote Sensing," in IEEE Microwave and Wireless Components Letters, vol. 26, no. 4, pp. 285-287, April 2016.
[6] A. Vahdati, M. Varonen, D. Parveg, D. Karaca and K. A. I. Halonen, "Design of an 85-95-GHz differential amplifier in 28-nm CMOS FD-SOI," 2016 Global Symposium on Millimeter Waves (GSMM)& ESA Workshop on Millimetre-Wave Technology and Applications, Espoo, 2016, pp. 1-4.
[7] A. Agah, H. T. Dabag, B. Hanafi, P. M. Asbeck, J. F. Buckwalter and L. E. Larson, "Active Millimeter-Wave Phase-Shift Doherty Power Amplifier in 45-nm SOI CMOS," in IEEE Journal of Solid-State Circuits, vol. 48, no. 10, pp. 2338-2350, Oct. 2013.
[8] S. C. Cripps, RF Power Amplifiers for Wireless Communications, 2nd ed. Norwood, MA, USA: Artech House, 2006.
[9] H. C. Lin and G. M. Rebeiz, "A 70-80 GHz SiGe Amplifier With Peak Output Power of 27.3 dBm," in IEEE Transactions on Microwave Theory and Techniques, vol. 64, no. 7, pp. 2039-2049, July 2016.
[10] A.Y.K. Chen et al., "An 83-GHz High-Gain SiGe BiCMOS Power Amplifier Using Transmission-Line Current-Combining Technique ," in IEEE Transactions on Microwave Theory and Techniques, vol. 61, no. 4, pp. 1557-1569, April 2013.
[11] Yi Zhao, John R. Long, "A Wideband, Dual-Path, Millimeter-Wave Power Amplifier With 20 dBm Output Power and PAE Above 15% in 130 nm SiGe-BiCMOS," in IEEE Journal of Solid-State Circuits, vol. 47, no. 9, pp. 1981-1997, Aug. 2013.
[12] D. Chen, C. Zhao, Z. Jiang, K. M. Shum, Q. Xue and K. Kang, "A V-Band Doherty Power Amplifier Based on Voltage Combination and Balance Compensation Marchand Balun," in IEEE Access, vol. 6, pp. 10131-10138, 2018.

# A Novel Multi-level CMOS Switching Mode Amplifier for Mobile Communication Signals

Robert Bieg, Martin Schmidt, Markus Grözing, Manfred Berroth
Institute of Electrical and Optical Communications Engineering
University of Stuttgart, 70569 Stuttgart, Germany

*Abstract*—**This paper presents simulation results of a CMOS switching mode power amplifier (SA) in a 65 nm technology with adjustable output voltage swing. The output stage is built in a stacked design to prevent dielectric breakdown of the transistors. Inverters at the top and bottom of the stack provide the supply voltage for the stack. The configuration offers a variable output voltage swing between one, two or three times the nominal transistor supply voltage. This paper demonstrates the advantages over a power amplifier with fixed output levels for signals with high peak to average output power ratio (PAPR).**

## I. INTRODUCTION

The power consumption of RF transmitters is one of the cost drivers for operators of mobile communications. Switching mode power amplifiers (SA) offer very good power efficiencies and thus are one of the promising alternatives to linear power amplifiers (PA). Today's most common used architecture is the linear Doherty PA, which reaches good power added efficiencies of 45 % [1]. The drawback of the Doherty PA is the needed matching network at the output, which limits its bandwidth. SA are advantageous considering bandwidth because their high bandwidth is limited mainly by the bandpass filter at the output, which can be set according to the application. Regarding efficiencies digital amplifiers reach competitive results [2].

A disadvantage of a common SA is, that it switches between the positive and negative supply voltage. Quantization of the input signal into two output levels leads to quantization noise. In [3] the resulting signal-to-noise ratio ($SNR$) after quantization assuming an equal distribution of noise and a sinusoidal input waveform is calculated by:

$$SNR = n \cdot 6.02 \, \text{dB} + 1.76 \, \text{dB}, \qquad (1)$$

with $n$ being the number of bits. For the common used SA this results in an $SNR$ of 7.78 dB without oversampling or delta sigma modulation.

The design of the amplifier proposed in this paper is able to output four different voltage levels. This corresponds to $n = 2$ increasing the $SNR$ by about 6 dB.

This paper is organized as follows: in section II the proposed SA topology with its decoder unit is presented. The simulation results of the SA will be discussed in section III. First, the SA's performance with excitation with a rectangular square wave with 50 % duty cycle will be shown. Then, the SA will be driven with a QAM-16 modulated pseudorandom binary sequence (PRBS) and results will be discussed. Section IV closes this paper with a conclusion.

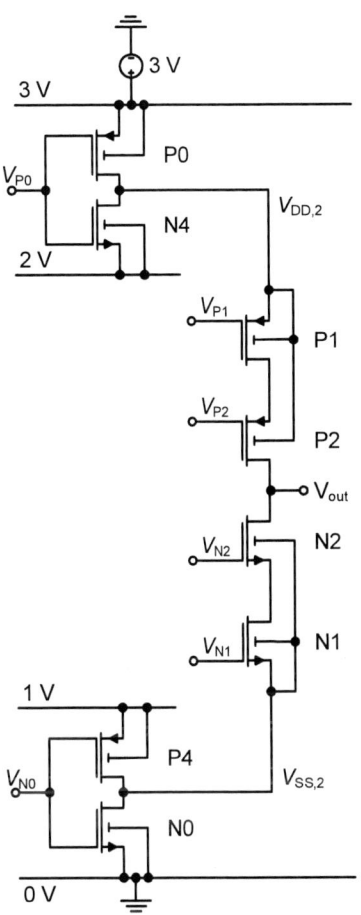

Fig. 1. Output stage for adjustable output level.

## II. ADJUSTABLE 3 V SWITCHING MODE AMPLIFIER

### A. Stacked MOSFET Output Stage Topology

The simulated design uses the 65 nm CMOS technology from ST Microelectronics. Since the breakdown voltage of the transistors is only 1 V the stacked topology from [4] is modified to prevent breakdown of the transistors. In [4] three PMOS-transistors are stacked over three NMOS-transistors. In order to render the additional output levels possible, the PMOS-transistor connected to VDD and the NMOS-transistor

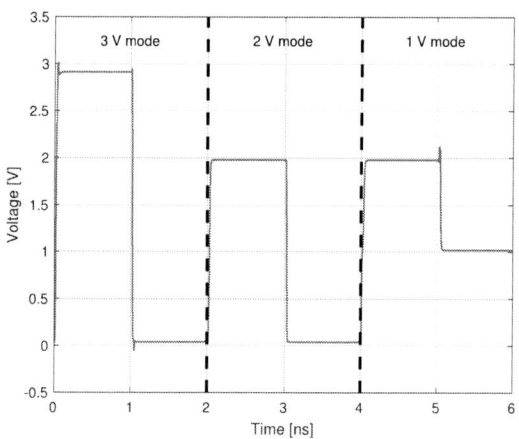

Fig. 2. Output waveform for all three possible modes.

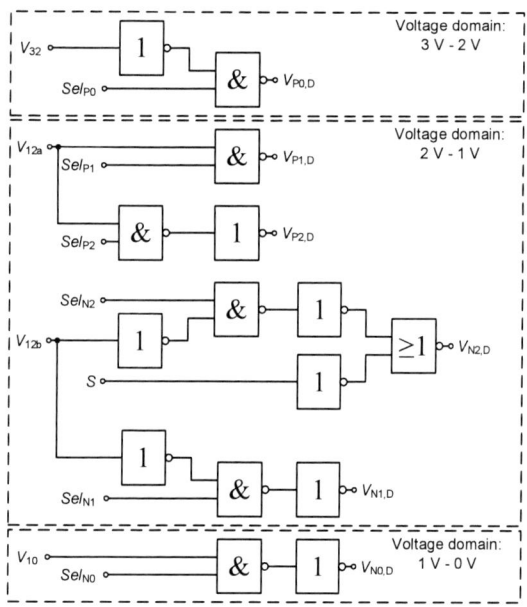

Fig. 3. Proposed decoder to generate driving signals for output stage.

connected to VSS are replaced by CMOS-inverters to set the voltages of the nodes $V_{DD,2}$ and $V_{SS,2}$. The width of the inverter-transistors are dimensioned equal to the width of the remaining stacked transistors to reduce static power loss in the stacked transistors for all output levels. Fig. 1 depicts the resulting output stage. This output stage supports three different modes. Fig. 2 shows the waveform of the output voltage while the SA traverses the three modes: 3 V operation, 2 V operation and 1 V operation. In this simulation the SA drives a 50 Ω load into a 1.5 V supply voltage.

### B. Decoder Circuit

The proposed SA is built of two inverters and four remaining transistors in the stack, which all have their own

TABLE I.    SELECT SIGNALS FOR THE THREE MODES.

| Mode | Select signals | | | | | | |
|------|------|------|------|------|------|------|------|
| | $Sel_{P0}$ | $Sel_{P1}$ | $Sel_{P2}$ | $Sel_{N2}$ | $Sel_{N1}$ | $Sel_{N0}$ | $S$ |
| 3 V | 1 | 0 | 1 | 1 | 0 | 1 | 1 |
| 2 V | 0 | 1 | 0 | 0 | 0 | 1 | 0 |
| 1 V | 0 | 1 | 0 | 0 | 1 | 0 | 1 |
| 3 V | 3 V | 1 V | 2 V | 2 V | 1 V | 1 V | 2 V |
| 2 V | 2 V | 2 V | 1 V | 1 V | 1 V | 1 V | 1 V |
| 1 V | 2 V | 2 V | 1 V | 1 V | 2 V | 0 V | 2 V |

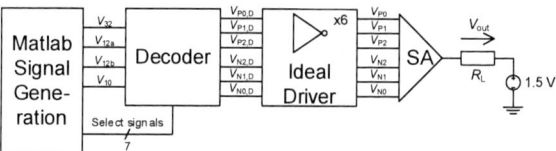

Fig. 4. Block diagram of the used testbench for simulation.

driving signals depending on the desired output level. In order to reduce circuit complexity, a decoder is developed which generates the correct driving signals for the output stage depending on the mode. The driving signals are based on the signals for the SA in [4]. The decoder is depicted in Fig. 3. The input signals of the decoder are the input signals $V_{10}$, $V_{12a,b}$ and $V_{32}$. The indices of the voltages indicate the voltage domain of the signals i.e. $V_{10}$ is between 0 V and 1 V. The input signals $V_{12a,b}$ are inverted compared to the outer input signals. The select signals $Sel_{N0}$ to $Sel_{P0}$ and $S$ are used to set the mode of the output stage.
$V_{P0,D}$ to $V_{N0,D}$ are the resulting signals for the output stage. A following driver is necessary to drive the large output stage. Table I summarizes the settings for the three modes and the select signals with resulting voltages depending on the voltage domain.

### III.    SIMULATION RESULTS

The used testbench is shown in Fig. 4. The select signals and RF-pulses are generated with Matlab-code. The gates of the decoder are built of CMOS-transistors according to Fig. 3. The resulting signals $V_{P0,D}$ to $V_{N0,D}$ go into an ideal driver to keep the signal slopes steep despite the large transistors of the SA. $R_L$ is dimensioned to 50 Ω in all simulations. In order to keep simulation time low, a DC-block capacitor is realized by connecting the resistive load $R_L$ to a DC-voltage source with 1.5 V. All simulations are performed with typical corners using the schematics of the shown circuits.

### A. Excitation with Rectangular Square Wave

The SA is driven by a rectangular square wave with 50 % duty cycle. The performance of the amplifier for the different modes is shown depending on the signal frequency. As the SA generates a lot of power in the harmonics, the pulse efficiency and overall signal power are considered. According to [4] the pulse efficiency is defined by:

$$\eta_p = \frac{P_{out}}{P_{DC}} = \frac{\frac{V_{out,RMS}^2}{R_L}}{P_{DC}}. \tag{2}$$

$P_{DC}$ is the sum of the output power and the static and dynamic

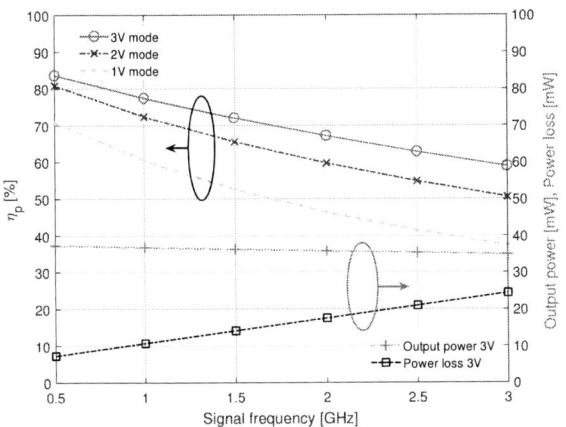

Fig. 5. Pulse efficiency for all SA modes and pulse output power with pulse power loss for the 3 V mode versus signal frequency.

TABLE II.    IDEAL AND SIMULATED VOLTAGE LEVELS AT SA OUTPUT

| | Voltage [V] | | | |
|---|---|---|---|---|
| Ideal levels | 0 | 1 | 2 | 3 |
| Simulated levels | 0.0760 | 1.0345 | 1.9548 | 2.8260 |

power losses. Fig. 5 depicts the pulse efficiencies for the three modes. The pulse efficiency for the 3 V mode is higher than for the lower output voltages because there, the output power is much higher while many loss sources are the same for all modes. The SA achieves higher pulse efficiency with higher output power being generated.

Fig. 5 shows the output power and power loss for the 3 V-mode. The power loss rises with increasing signal frequency due to the increasing dynamic power loss. For higher frequency the SA has to switch the output node and recharge the parasitic capacitances more frequently. As the output power stays constant over the signal frequency, the pulse efficiencies decrease.

### B. Excitation with Delta-Sigma Modulated QAM-16 Data

For simplicity reasons a Delta-Sigma-Modulator (DSM) of second order, which shapes the noise at the carrier frequency is used. The input signal of the DSM is normalized to 3 V peak-peak voltage in all simulations. The modulator quantizes the transmit signal depending on the operating mode. Two simulations are performed to analyze the adjustable SA: in the first simulation the SA is kept in the 3 V mode exclusively. The second simulation uses all available modes of the SA and the DSM chooses between four voltage levels. Due to voltage drop over the conducting transistors in the stack, the output levels of the SA are slightly different from the ideal ones. To improve the $SNR$ and error vector magnitude ($EVM_{\mathrm{RMS}}$) the simulated output levels are used instead of the ideal ones for DSM according to Table II. For the outer levels the voltage drop is higher because of the higher current that is driven by the SA.

The input signal is a PRBS $2^8 - 1$ sequence, which is mapped into QAM-16 symbols. The symbol rate is 100 MBaud, so the resulting bit rate is 400 MBit/s. As pulse shaping filter a raised-cosine (RC) filter is used. The carrier frequency is chosen to

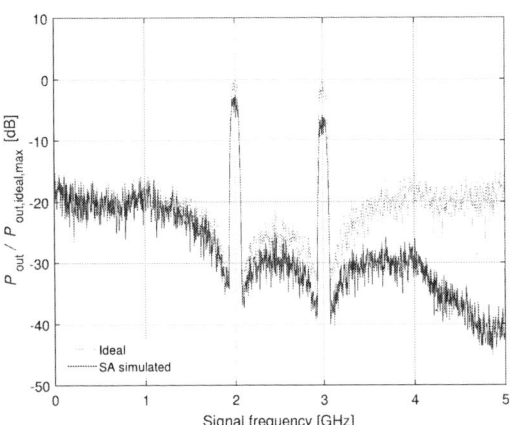

Fig. 6. Calculated FFT from the ideal Matlab signal and the resulting voltage output waveform for simulation with multi-level.

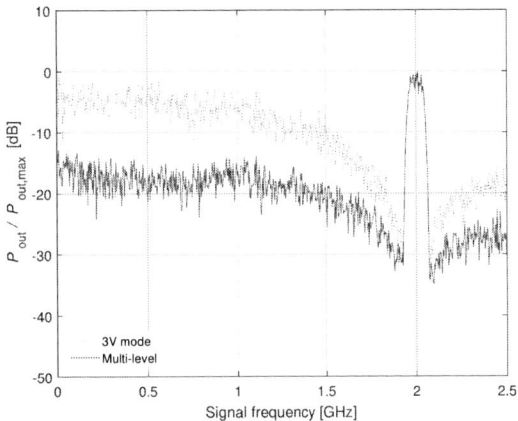

Fig. 7. Calculated FFT from the resulting waveforms for simulations with the 3 V operation mode and multi-level.

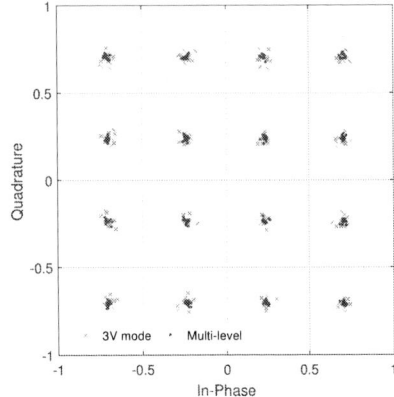

Fig. 8. IQ-diagram for both simulation cases 3 V only and multi-level.

TABLE III.    SUMMARIZATION OF THE SIMULATION RESULTS

| | QAM16 PAPR(QAM16) = 7.9 dB | | |
|---|---|---|---|
| | $\eta_C$ [%] | $EVM_{RMS}$ [%] | $SNR$ [dB] |
| 3 V only | 11.24 | 4.71 | 26.55 |
| Multi-level | 45.52 | 2.12 | 33.46 |
| | QPSK PAPR(QPSK) = 6.8 dB | | |
| | $\eta_C$ [%] | $EVM_{RMS}$ [%] | $SNR$ [dB] |
| 3 V only | 14.55 | 4.76 | 26.46 |
| Multi-level | 47.57 | 1.94 | 34.26 |
| | QPSK$_{RC,off}$ PAPR(QPSK$_{RC,off}$) = 2.9 dB | | |
| | $\eta_C$ [%] | $EVM_{RMS}$ [%] | $SNR$ [dB] |
| 3 V only | 33.65 | 8.18 | 21.74 |
| Multi-level | 49.93 | 7.06 | 23.02 |

$f_C = 2\,\text{GHz}$ and the DSM works with a sampling rate of $f_S = 5\,\text{GS/s}$. The PAPR of the input signal without DSM is 7.9 dB. A reconstruction filter at the output is not used and the power of all harmonics is fed into the $50\,\Omega$ load resistor.

First the spectrum of the ideal Matlab signal is compared with the spectrum of the simulated output waveform of the SA. The results are depicted in Fig. 6. Compared to the ideal waveform the power of the signal is decreased by about 3 dB due to losses in the switching stage. In the second Nyquist band from 2.5 GHz to 5 GHz the limited bandwidth of the SA is seen by the increasing difference between the ideal and simulated waveform.

The spectrum at the load resistor is depicted in Fig. 7 for 3 V only / multi-level simulations. As can be seen, the noise in the exclusive 3 V mode is higher than with the use of all modes. This results in a higher code efficiency $\eta_C$ of 45.53 % for multi-level compared to 11.24 % for 3 V only.

Fig. 8 shows the resulting IQ-diagram after demodulation. The variance of the exclusive 3 V mode is higher leading to a higher $EVM_{rms}$ of 4.71 % compared to 2.12 % for multi-level. According to [5] the $SNR$ can be calculated by:

$$SNR \approx 10 \cdot \log_{10}\left(\frac{1}{EVM_{RMS}^2}\right). \tag{3}$$

This results in an $SNR$ of 26.55 dB for the exclusive 3 V operation mode and 33.46 dB for multi-level.

The simulation results are summarized in Table III. Additionally the simulation results for a QPSK-modulated input signal are shown. The PAPR of the QPSK-signal is 1.1 dB lower resulting in an improved $\eta_C$ for the exclusive 3 V mode. To stress the advantage of the proposed topology, a signal with a low PAPR is generated for comparison. Therefore the same simulations are performed with a theoretical QPSK-signal without RC-filter and a PAPR of 2.9 dB. In this case the results of the 3 V only and multi-level operation modes of the SA are similar, since the 1 V and 2 V voltage levels are barely used. Due to the high PAPR of mobile communication signals, the multi-level SA is advantageous to the 2-level SA.

## IV.    CONCLUSION

This article presents a SA with a multi-level output voltage swing. The maximum output swing is 3 V. The proposed design is interesting for a digital transmitter and a transmit signals with a high PAPR. The output voltage swing of the amplifier can be set according to the required power of the transmit signal. Compared to a SA with a fixed output voltage swing of 3 V the code efficiency as well as the $EVM_{RMS}$ and the $SNR$ can be improved by the proposed design. When using a signal with high PAPR and all available voltage levels, $\eta_C$ increases by 34.28 %, $EVM_{RMS}$ decreases by 2.59 % and $SNR$ increases by 6.91 dB.

## REFERENCES

[1] S. Maroldt and M. Ercoli, "3.5-GHz ultra-compact GaN class-E integrated Doherty MMIC PA for 5G massive-MIMO base station applications," in *2017 12th European Microwave Integrated Circuits Conference (EuMIC)*. IEEE, 2017, pp. 196–199.

[2] T. Hoffmann, A. Wentzel, F. Huhn, and W. Heinrich, "Novel digital microwave PA with more than 40% PAE over 10 Db power back-off range," in *2017 IEEE MTT-S International Microwave Symposium (IMS)*. IEEE, 2017, pp. 2037–2040.

[3] P. E. Allen and D. R. Holberg, *CMOS analog circuit design*, 2nd ed., ser. The Oxford series in electrical and computer engineering. New York: Oxford Univ. Press, 2002. [Online]. Available: http://www.loc.gov/catdir/enhancements/fy0612/2002020034-d.html

[4] R. Bieg, M. Schmidt, and M. Berroth, "A CMOS switching mode amplifier with 3 V output swing for continuous-wave frequencies up to 4 GHz," in *2015 Asia-Pacific Microwave Conference (APMC)*. IEEE, 2015, pp. 1–3.

[5] R. A. Shafik, M. S. Rahman, and A. R. Islam, "On the Extended Relationships Among EVM, BER and SNR as Performance Metrics," in *International Conference on Electrical and Computer Engineering, 2006*. Dhaka, Bangladesh: IEEE, 2006, pp. 408–411.

*PRIME 2018, Prague, Czech Republic*     *Session: Radio Frequency Circuits and Systems I*

# Down-converter solutions for 77-GHz automotive radar sensors in 28-nm FD-SOI CMOS technology

Claudio Nocera, Andrea Cavarra,
Egidio Ragonese, Giuseppe Palmisano
DIEEI, University of Catania
Catania, Italy
claudio.nocera@unict.it
andrea.cavarra@studium.unict.it
egidio.ragonese@dieei.unict.it
giuseppe.palmisano@dieei.unict.it

Giuseppe Papotto
STMicroelectronics
Catania, Italy
giuseppe.papotto@st.com

*Abstract—* **This paper presents a review of 77-GHz down-converter solutions for automotive radar sensors in 28-nm FD-SOI CMOS technology. A comparison of two different topologies based on common source (CS) and common gate (CG) stages is reported. The comparison is carried out at a power supply as low as 1-V and at 15-mA current consumption. CS-based and CG-based down-converters achieve a conversion gain of 27.5 and 21.3 dB over a −3-dB bandwidth of 16 GHz and 22 GHz, respectively, while exhibiting a noise figure of 7.8 dB and 9.1 dB.**

*Keywords—Down-converters, common gate stage, common source stage, integrated transformers, mixers, CMOS technology, EM simulations, W-band.*

## I. INTRODUCTION

The 77-GHz automotive radar sensor in CMOS technology is a very challenging application for integrated circuit designers. Indeed, this application needs a very scaled technology with low power supply to achieve a System-on-Chip (SoC) implementation, capable of including a mm-wave front-end, a wide-band analog base-band, and a digital processing circuit. To satisfy the performance of the radio front-end such as gain, noise, and linearity, high bias currents have to be adopted, which along with the low power supply make very difficult fulfilling the application requirements. Actually, most design problems at mm-wave frequencies come from voltage headroom limitations due to both high currents and low voltages. To this purpose, extensive use of inductive components and folded circuit topologies are usually employed for mm-wave circuits [1], [2], which however increase power consumption and silicon area. On the other hand, both consumption and area have to be contained since the implementation of a single sensor node needs multiple receivers and transmitters in the same die that has to include a digital signal circuitry besides the radio front-end. The most critical circuit in the receiver section is the down-converter that in a large extent has to guarantee several requirements of the overall receiver. Various solutions for the down-converter were published in literature in the last years [3]-[5]. The use of RF amplification before frequency down conversion would have the advantage of greatly reducing the noise requirements of the mixer. However, linearity requirements make very difficult increasing gain at the mm-wave side. Therefore, most promising approaches are mixer-first receiver architectures, which allow avoiding RF amplification, thus guarantying high linearity

[6]-[8]. In this paper, two 77-GHz down-converters with high-gain voltage-to-current (VI) converters followed by intermediate frequency (IF) trans-impedance amplifiers are presented. Common source (CS) and common gate (CG) based topologies are compared through extensive simulations, which accurately account for process worst cases and layout parasitic effects. The comparison aims at identifying the best approach to fulfill the very challenging requirements of the mm-wave receiver of a long-range automotive radar sensor. The comparison is carried out by using a high-performance 28-nm fully depleted (FD) silicon-on-insulator (SOI) CMOS technology at a power supply as low as 1 V.

## II. CIRCUIT DESCRIPTION

Traditional down-converters that use resistances as passive IF load are not suitable for mm-wave applications in nano-meter CMOS technologies due to both high bias currents and low power supply, which drastically reduce output voltage swing and gain. In such cases, an active intermediate frequency (IF) load based on a trans-impedance amplifier can profitably be used to overcome voltage swing and gain limitations at a cost of higher complexity and slightly increase in noise. A simplified schematic of the proposed architecture is shown in Fig. 1. The two down-converters share almost all the components with the exception of the VI converter that is based on the CS or the CG stage. The VI converter uses transformers at the input and output terminals. The input transformer, $T_1$, is connected to the antenna and guarantees both single-differential signal conversion and electrostatic discharge protection. The secondary winding of $T_1$ is instead exploited for input matching. Of course, the noise figure ($NF$) is slightly increased due to the transformer insertion loss ($IL$). Transformer $T_2$ is used to connect the VI converter to the power supply with the primary coil and to the Gilbert quad with the secondary coil. In both CS and CG topologies the drain-source bias voltage is set to the power supply, thus maximizing the operative transition frequency. In this way, the frequency behavior of the input pair is optimized, which means higher gain and lower noise factor. The IF output current of the Gilbert quad feeds the trans-impedance amplifier, which performs the current-to-voltage (IV) conversion through the feedback resistances, $R_F$. Since the two down-converters share the Gilbert quad and trans-impedance amplifier, the attention in this comparison will be focused only on the VI converter, which

978-1-5386-5388-3/18 $31.00 © 2018 IEEE     153

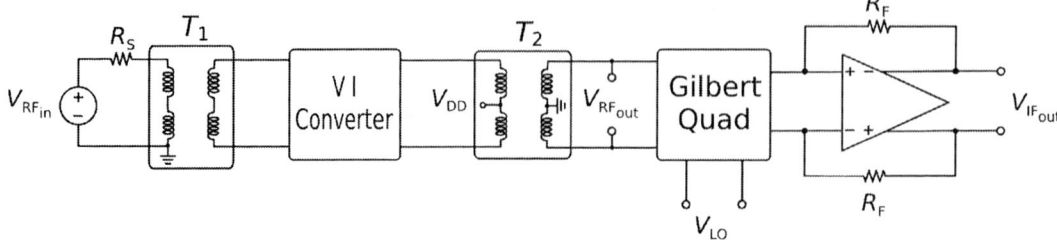

Fig. 1. Simplified schematic of the down-converter architecture based on the trans-impedance amplifier.

mainly determines the performance of gain, bandwidth, linearity, power consumption, and greatly contributes to the down-converter $NF$. The comparison will be carried out by optimizing the VI converter at the same supply current (15 mA).

### A. CS down-converter

The VI converter for the CS down-converter is shown in Fig. 2. Similarly to a conventional CS low-noise amplifier, the real-part input matching is achieved with the degeneration inductances, $L_S$, at the source terminals, whereas the secondary coil inductance of the input transformer is used for the imaginary-part input matching, as mentioned before. Minimum $NF$ requires optimum gate-source voltage ($V_{gsopt}$), which is achieved with the optimum current density (i.e., the drain current divided by the aspect ratio). The gate area and hence the final aspect ratio of the transistors is instead set to make the optimum source resistance for the noise factor equal to the resistance at the secondary coil of $T_1$, which is the real input-pair source resistance (i.e., near to 50 Ohm). A gate-source capacitance, $C_{gs}$, is included [9], which avoids that the optimum source resistance leads to a very high aspect ratio and hence to high current consumption. The frequency response of the down-converter has a band-pass shape whose bandwidth is mainly determined by the resonance of transformer $T_1$ and $T_2$. The bandwidth is large enough to avoid signal attenuation in the radar operating frequency range (i.e., 76 GHz-81 GHz). The down-converter voltage gain is given by:

$$A_{vCS} = \frac{2}{\pi} \frac{1}{R_S} \frac{\omega_T}{\omega_0} \sqrt{\frac{L_1}{L_2}} R_F$$

where $L_1$, $L_2$ are the primary and secondary inductance of $T_2$, respectively, and a unity transformer ratio was set for $T_1$. Moreover, negligible losses are assumed for $T_1$ and $T_2$.

### B. CG down-converter

The VI converter for the CG down-converter is shown in Fig. 3. The real-part input matching is here achieved by properly setting the trans-conductance of transistors $M_1$ and $M_2$. However, this usually leads to very high bias current for the input pair and hence excessive power consumption. To overcome this drawback, feedback capacitors, $C_F$, are included, which determine an increase in the transistor pair input

conductance by a factor 2. Indeed, thanks to capacitor $C_F$, a signal at the gate of $M_1$ ($M_2$) is produced that is equal but with opposite sign to the one at the source of $M_1$ ($M_2$), as can be seen by inspecting Fig. 3. Therefore, the variation of the gate-source voltage of $M_1$ ($M_2$) is doubled with the same input current, i.e. the resistance at the source terminal of $M_1$ is $1/2g_{m1,2}$ instead of $1/g_{m1,2}$ and hence the differential input resistance is $1/g_{m1,2}$ instead of $2/g_{m1,2}$. This means that the value of $g_{m1,2}$ useful for the input matching is reduced by a factor 2, thus saving power consumption. Capacitors $C_F$ also allow a slight reduction of the

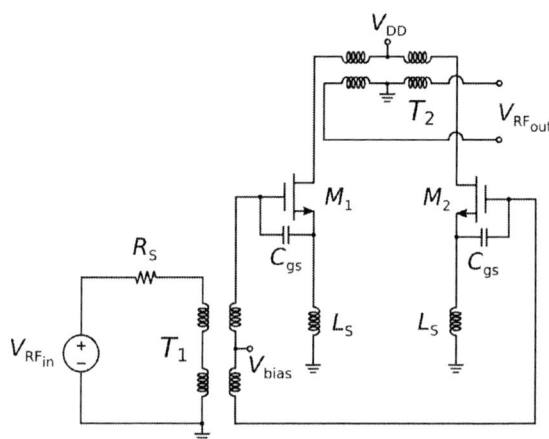

Fig. 2. VI converter for the CS down-converter.

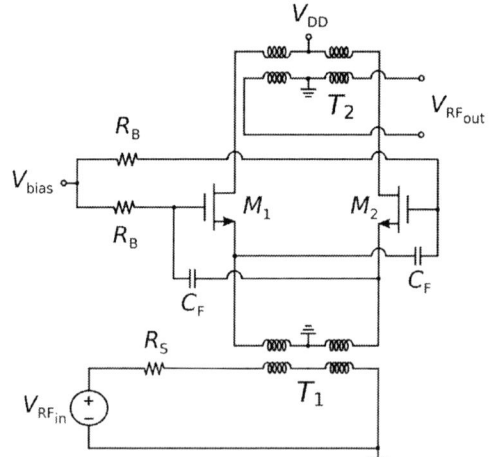

Fig. 3. VI converter for the CG down-converter.

input pair noise. Indeed, the equivalent noise voltage, $V_{n,s1}$ ($V_{n,s2}$), at the source of $M_1$ ($M_2$) transfers to the gate of $M_2$ ($M_1$) thanks to $C_F$ and then to the source of $M_2$ ($M_1$) attenuated by 2 (see Fig. 3) due to matching conditions. Therefore, the noise power spectral density of $V_{n,s1}$ ($V_{n,s2}$) reflects on the source of $M_2$ ($M_1$) multiplied by a factor 1/4. This means that a common-mode noise component of 1/4 arises thanks to $C_F$ giving a noise reduction by a factor 3/4 (1.2 dB). Differently from the CS VI converter, the trans-conductance of the CG converter is constrained by the real-part matching condition and hence it cannot be used for the optimization of NF that becomes almost independent from the trans-conductance. The down-converter voltage gain is given by:

$$A_{vCG} = \frac{2}{\pi}\sqrt{\frac{2g_{m1,2}}{R_S}}\sqrt{\frac{L_1}{L_2}}\,R_F$$

The design parameters adopted for CG and CS down-converters are reported in Table I.

### III. SIMULATION RESULTS

Both down-converters were designed in a 28-nm FD-SOI CMOS technology by STMicroelectronics [10], which provides eight copper metals in addition to a top aluminum layer. As shown in the Fig. 1, they adopt the same bias current for the VI converter, as well as the same Gilbert quad and trans-resistance amplifier. The latter exploits a 50-dB operational amplifier and was designed for a 20-MHz closed-loop bandwidth.

The layout of the VI converter is crucial for the overall down-converter performance. Therefore, a meaningful comparison between down-converter topologies requires an accurate layout design to minimize the parasitic effects. To this aim, extensive electromagnetic (EM) simulations in ADS Momentum were used. In the adopted architecture of Fig. 1, both frequency response and losses are highly dependent from the integrated transformers, $T_1$ and $T_2$. In particular, the most critical component is $T_1$, since its IL directly contributes to the overall receiver NF. A low IL is generally related to high $Q$-factors as well as high coupling factor, $k$, between windings. Typically, stacked transformers are preferred to interleaved ones to implement an input balun thanks to higher $k$. However, they suffer from a considerable $k$ degradation at very low inductance values [11]. Therefore, in this work an advanced transformer structure, namely interstacked [12], [13], was evaluated and compared to the stacked one. Fig. 4 depicts a 3 D view of an interstacked transformer. Each transformer coil is built by shunting two spirals of top metal layers (i.e., metal 7 and metal 8), using complementary structures for primary and secondary windings. Therefore, it takes advantage of both interleaved and stacked magnetic couplings. Indeed, the outer (inner) spiral of primary winding is stacked to the outer (inner) spiral of the secondary winding and interleaved with the inner (outer) spiral of the secondary winding at the same time. Table II compares the geometrical and electrical parameters of

TABLE I. DESIGN PARAMETERS

|  | $W_{1,2}$ | $L_{1,2}$ | $V_{gsopt}$ | $C_{gs}$ | $C_F$ |
|---|---|---|---|---|---|
| CG | 36 μm | 28 nm | 500 mV | - | 60 fF |
| CS | 36 μm | 28 nm | 500 mV | 15 fF | - |

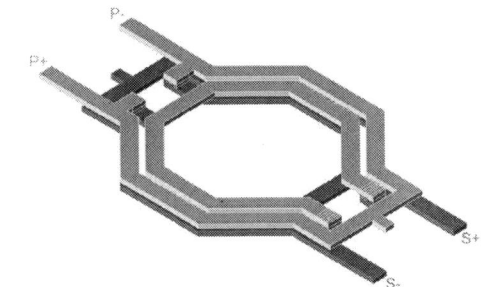

Fig. 4. 3D-view of an interstacked transformer.

TABLE II INPUT TRANSFORMER PARAMETER COMPARISON

| Parameters | Stacked | Interstacked | Units |
|---|---|---|---|
| Metal width | 5.5 | 3.3 | [μm] |
| Inner diameters P/S | 44/44 | 41/37 | [μm] |
| Primary coil inductance @ 77 GHz | 90 | 72 | [pH] |
| Secondary coil inductance @ 77 GHz | 96 | 72 | [pH] |
| Primary coil $Q$-factor @ 77 GHz | 25.6 | 18 | - |
| Secondary coil $Q$-factor @ 77 GHz | 17.3 | 18 | - |
| Self-resonance frequency | 170 | 174 | [GHz] |
| $k$ @ 77 GHz | 0.62 | 0.68 | - |
| IL @ 77 GHz | 5 | 4.8 | [dB] |

stacked and an interstacked transformers designed to implement the input transformer, $T_1$. The interstacked configuration exhibits the lowest IL thanks to a higher magnetic coupling, as well as high enough quality factors. Therefore, it was preferred for the input of the VI converter, as $T_1$ transformer. On the other hand, transformer $T_2$ was implemented by means of a traditional interleaved structure to maximize $Q$-factors, while allowing down-converter gain to be increased, thanks to an inherently lower than 1 ratio between primary and secondary coils. The performance comparison between CG and CS down-converters is shown in Fig. 5. Specifically, Fig. 5(a) reports the conversion gain of both solutions as a function of the intermediate frequency, IF. As apparent, the CS down-converter achieves better conversion gain performance than the CG solution (i.e., 27.5 dB against 21.3 dB). Moreover, as displayed in Fig. 5(b), the CS solution at 1-MHz of IF frequency exhibits a NF as low as 7.8 dB, which is 1.3 dB lower than the NF of the CG topology. Finally, Fig. 5(c) reports the conversion gain at 1-MHz of IF with respect to the input RF frequency. The down-converters exhibit a band-pass frequency response mainly set by transformers $T_1$ and $T_2$. Note that the CG solution guarantees a −3-dB bandwidth of 22 GHz which is 6 GHz wider

TABLE III. DOWN-CONVERTER PERFORMANCE COMPARISON

|    | $V_{DD}$ | $I_{DD}$ | Conversion gain @ 77 GHz | −3 dB bandwidth | NF @ 1 MHz |
|----|----------|----------|--------------------------|-----------------|------------|
| CS | 1 V      | 15 mA    | 27.5 dB                  | 16 GHz          | 7.8 dB     |
| CG | 1 V      | 15 mA    | 21.3 dB                  | 22 GHz          | 9.1 dB     |

than that one of the CS solution.

Table III summarizes the performance of the designed down-converters. According to the achieved results, the CS-based down converter is the best solution for low-noise and high-gain requirements, whereas the CG-based down converter is more suitable for wide-band requirements.

## IV. CONCLUSION

Down-converter solutions in 28-nm FD-SOI CMOS technology using two different input stages have been accurately analyzed and compared in terms of conversion gain and noise figure performance. The simulated results show that the CS down-converter achieves better performance in terms of NF and conversion gain with values of 7.8 dB and 27.5 dB, respectively. On the other hand, the CG down-converter exhibits a −3-dB bandwidth of 22 GHz, which is 6 GHz wider than that one of the CS solution. Therefore, the CS-based down-converter is the best solution for low-noise and high-gain applications, whereas the CG-based down-converter is more suitable for wide-band applications.

## REFERENCES

[1] A. Medra, *et al.*, "An 80 GHz low-noise amplifier resilient to the TX spillover in phase-modulated continuous-wave radars," *IEEE J. Solid-State Circuits*, vol. 51, pp. 2299-2311, May 2016.

[2] M. Vigilante, and P. Reynaert, "On the design of wideband transformer-based forth order matching networks for E-Band receivers in 28-nm CMOS," *IEEE J. Solid-State Circuits*, vol. 52, pp. 2071-2082, Aug. 2017.

[3] H. Jia *et al.*, "A 77 GHz frequency doubling two-path phased-array FMCW transceiver for automotive radar," *IEEE J. Solid-State Circuits*, vol. 51, pp. 2299-2311, Oct. 2016.

[4] J. Lee, Y. A. Li, M. H. Hung and S. J. Huang, "A fully-integrated 77-GHz FMCW radar transceiver in 65-nm CMOS technology," *IEEE J. Solid-State Circuits*, vol. 45, pp. 2746-2756, Dec. 2010.

[5] T. Mitomo, *et al.*, "A 77 GHz 90 nm CMOS transceiver for FMCW radar applications," *IEEE J. Solid-State Circuits*, vol. 45, pp. 928-937, April 2010.

[6] T. Fujibayashi *et al.*, "A 76- to 81-GHz multi-channel radar transceiver," *IEEE J. Solid-State Circuits*, vol. 52, pp. 2226-2241, Sept. 2017.

[7] S. Trotta *et al.*, "An RCP packaged transceiver chipset for automotive LRR and SRR systems in SiGe BiCMOS technology," *IEEE Trans. Microwave Theory and Tech.*, vol. 60, pp. 778-794, March 2012.

[8] D. Kissinger, B. Sewiolo, H. P. Forstner, L. Maurer and R. Weigel, "A fully differential low-power high-linearity 77-GHz SiGe receiver frontend for automotive radar systems," in *Proc. IEEE 10th Annual Wireless and Microwave Technology Conference*, April 2009, pp. 1-4.

[9] G. Girlando and G. Palmisano, "Noise figure and impedance matching in RF cascode amplifiers," *IEEE Transactions on Circuits and Systems II: Analog and Digital Signal Processing*, vol. 46, pp. 1388-1396, Nov. 1999.

[10] A. Cathelin, "Fully depleted silicon on insulator devices CMOS: The 28-nm node is the perfect technology for analog, RF, mmW, and mixed-signal system-on-chip integration," *IEEE Solid-State Circuits Magazine*, vol. 9, pp. 18-26, Fall 2017.

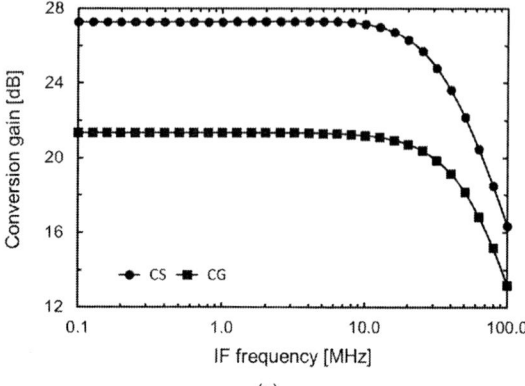

(a)

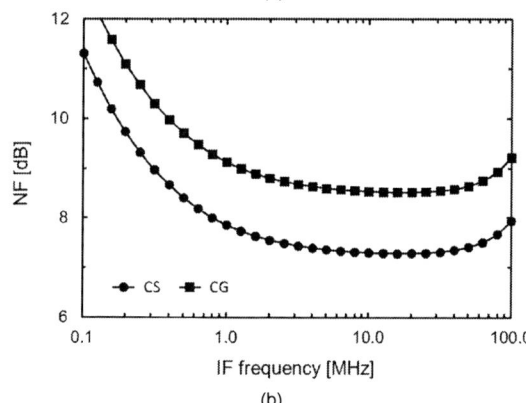

(b)

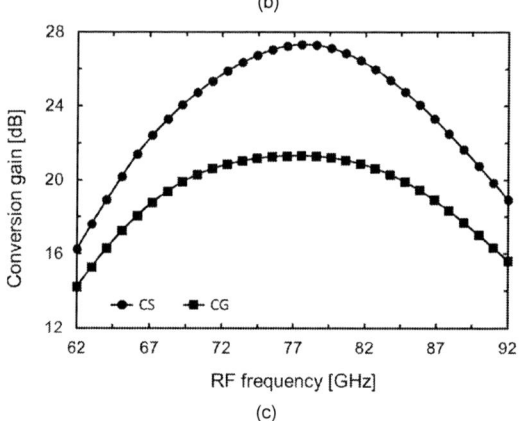

(c)

Fig. 5. Performance comparison of CG and CS down-converters.

[11] T. Biondi, A. Scuderi, E. Ragonese, and G. Palmisano, "Analysis and modeling of layout scaling in silicon integarted stacked transformers," *IEEE Trans. Microwave Theory and Tech.*, vol. 54, pp. 2203-2210, May 2006.

[12] E. Ragonese, G. Sapone, V. Giammello, and G. Palmisano, "Analysis and modeling of interstacked transformers for mm-wave applications," *Springer Analog Integrated Circuits and Signal Processing*, vol. 72, pp. 121-128, June 2012.

[13] V. Giamello, E. Ragonese, and G. Palmisano, "A transformer-coupling current-reuse SiGe HBT power amplifier for 77-GHz automotive radar," *IEEE Trans. Microwave Theory and Tech.*, vol. 60, pp. 1676-1683. June 2012.

PRIME 2018, Prague, Czech Republic

Session: Digital Circuits and Sub-Systems

# A Novel Very Low Voltage Topology to implement MCML XOR Gates

Davide Bellizia[1], Gaetano Palumbo[2], Giuseppe Scotti[1], Alessandro Trifiletti[1]

1- DIET Department, University of Rome "La Sapienza" Rome, Italy.

2- DIEEI Department. University of Catania, Catania Italy.

**Abstract—A new very low-voltage topology to implement MOS current mode logic (MCML) XOR gates is proposed in this paper. Instead of stacking several level of transistors to implement a two inputs XOR gate, a p-type differential pair is used to steer the current in n-type differential pairs through current mirrors. The proposed topology allows to reduce the minimum supply voltage of MCML XOR gates while guaranteeing a fully current mode behavior as in the conventional XOR gate. The proposed topology has been compared against the conventional and triple tail MCML XOR gates. Simulation results referring to a 40nm CMOS technology for $V_{DD}$=1V confirm that the XOR gate presented in this work exhibits a lower propagation delay than the previously published low voltage MCML XOR gate. Furthermore both theoretical analysis and simulation results in a 40nm process show that the proposed topology is able to work with a $V_{DD}$ as low as 0.65V whereas state of the art topologies are not usable below 0.8V.**

*Keywords—MCML; low voltage; nanometer CMOS; current mode; XOR Gate.*

## I. INTRODUCTION

High-speed digital communications and chip-to-chip interconnections are becoming more and more popular in the nanometer CMOS era. These applications require high performance logics suitable to be implemented with CMOS processes and able to guarantee high operating frequencies and low power consumption. Optical communications as well as, very high sampling frequency ADCs demand the use of logic styles able to operate at tens of Gb/s: these speeds cannot currently be achieved using the conventional CMOS logic style [1-3].

Mixed-signal integrated circuits have nowadays become pervasive due to the continuously increasing demand of systems on chip integrating digital and analog functions with more and more stringent specification requirements. The conventional CMOS logic style exhibits a large switching noise that lowers the performance of the analog building blocks and therefore it cannot be used on the same substrate with very high-resolution analog circuits [4-5]. MOS current mode logic (MCML) in contrast to the standard CMOS logic provides higher speed, better power efficiency at high frequencies, lower switching noise and lower sensitivity to process variations [6].

The exclusive-OR (XOR) gate is a key component to build high speed digital circuits such as comparators, adders, multipliers, test pattern generators etc. [7-10].

The most challenging drawback in the use of nanometer CMOS processes is the very low supply voltage available,

(which is usually less than 1V). This constraint on the supply voltage does not allow stacking several levels of transistors in a MCML gate, so that the traditional MCML XOR gate is not usable in practice anymore as will be pointed out in the next sections of this paper.

Low voltage CML topologies based on the triple tail cell concept have been introduced in [8] and analyzed in [9] referring to bipolar technologies. Recently, a new low-voltage MCML XOR gate topology has been proposed in [10]. It exploits the triple-tail cell concept, provides differential signaling and does not need any additional circuitry when compared against previous low voltage topologies. A MCML triple tail XOR gate topology exploiting multi-threshold CMOS processes has been introduced in [11]. Exploiting transistors with different threshold voltages, level shifters are not required, and supply voltage can be reduced [11].

In this work we present a novel low voltage MCML XOR gate topology in which the high performance of p-type MOS transistors in nanometer technologies is exploited to develop very low voltage logic gates by using a "folded" approach instead of the conventional "stacking" approach. A similar approach has been adopted in [12] for the implementation of low voltage D-Latches. In *Section II* we review the conventional MCML XOR gate and the triple tail XOR gate and compute the minimum supply voltage of the two circuits from a theoretical perspective. In Section III we introduce the proposed MCML XOR gate topology showing its advantages in terms of minimum supply voltage. Simulation results and comparisons are shown in Section IV. Conclusions are finally reported in section V.

## II. Review of MCML XOR Gate Topologies

### A. Conventional MCML XOR Gate

The circuit schematic of the conventional MCML XOR gate is depicted in Fig. 1. The transistor $M_1$ implements the bias current source of $I_{SS}$ value. MOS devices $M_2$ to $M_7$ implement the conditional pulldown network, whereas the active load is implemented through p-type transistors $M_8$ and $M_9$. The upper level and lower level differential couples are driven by differential inputs A and B respectively. When input B is low, $M_2$ is off, the tail current $I_{SS}$ flows through $M_3$ and is steered either to $M_6$ or $M_7$ according to the differential input A. The tail current $I_{SS}$ is steered to the load transistors $M_8$ and $M_9$ to provide the differential output voltage [6].

978-1-5386-5388-3/18 $31.00 © 2018 IEEE          157

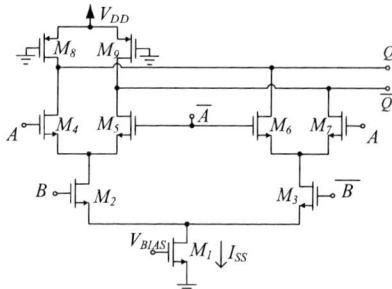

Fig. 1    Conventional MCML XOR Gate

The output differential voltage swing $V_{SWING}$ of the conventional MCML XOR gate can be written as:

$$V_{SWING} = 2 \cdot I_{SS} \cdot R_P \tag{1}$$

where $R_P$ is the equivalent linear resistance of the p-type devices $M_8$ and $M_9$ computed as in [5], [10] and [13].

In order for this circuit to properly operate, all the NMOS transistors have to work in saturation and this condition set a limitation on the minimum supply voltage $V_{DD}$.

Denoting as $V_{CMA}$ and $V_{CMB}$ the common mode voltage of differential input A and B respectively, the following conditions have to be fulfilled in order to guarantee saturation operation for all NMOS devices in the XOR gate circuit in Fig. 1:

$$V_{DS1} = V_{CMB} - V_{GS2,3} > V_{DS1,sat} \tag{2}$$

$$V_{DS2,3} = V_{CMA} - V_{GS4-7} - (V_{CMB} - V_{GS2,3}) > V_{DS2,3,sat} \tag{3}$$

$$V_{DS4-7} = V_{DD} - \frac{V_{SWING}}{2} - (V_{CMA} - V_{GS4-7}) > V_{DS4-7,sat} \tag{4}$$

where $V_{DS}$, $V_{GS}$, and $V_{DS,sat}$ are the drain source, gate source and saturation drain source voltage of the different transistors in Fig. 1 respectively. From equations (2) and (3) it is evident that, setting $V_{CMA} = V_{CMB}$ results in $V_{DS2,3} = 0$, if $M_{2,3}$ and $M_{4-7}$ are equally sized. In the past, MCML designers used source followers to lower the common mode voltage level $V_{CMB}$ with respect to $V_{CMA}$ thus allowing $M_2$ and $M_3$ to work in saturation [5], [6], [8], [11]. However, source follower stages are not power efficient and consume large silicon area footprint. Furthermore by using a source follower (implemented trough a common drain transistor $M_{10}$) the implemented voltage shift is equal to $V_{GS10}$.

By using these assumptions the minimum supply voltage can be computed from (1) as follows:

$$V_{DS1} = V_{CMB} - V_{GS2,3} = V_{CMA} - V_{GS10} - V_{GS2,3} \tag{5}$$

Assuming that the differential signal A is the output of a conventional MCML inverter (differential pair) we can write:

$$V_{CMA} = V_{DD} - \frac{V_{SWING}}{4} \tag{6}$$

Finally, substituting (6) in (5) and using (2), we can express the minimum supply voltage for the conventional MCML XOR gate as follows:

$$V_{DD,min,conv} = \frac{V_{SWING}}{4} + V_{GS10} + V_{GS2,3} + V_{DS1,sat} \tag{7}$$

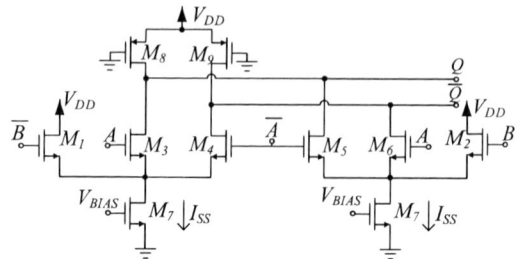

Fig. 2    Triple Tail MCML XOR Gate

From (7), it is evident that the conventional MCML XOR gate is not suited to be used for $V_{DD}$ in the range of 1V assuming typical device parameters for nanometer CMOS technologies.

B. Triple Tail MCML XOR gate

The low-voltage triple tail XOR gate introduced in [10] is reported in Fig. 2. It is made up of two triple-tail cells ($M_3$, $M_4$, $M_1$) and ($M_5$, $M_6$, $M_2$) with two tail current sources of $I_{SS}$ value. The MOS devices $M_1$ and $M_2$ are connected between the supply node and the common source node of differential pairs $M_3$–$M_4$ and $M_5$–$M_6$ respectively. When the differential input B is high, $M_2$ is on and the differential pair $M_5$–$M_6$ is off. In the same condition, $M_1$ is off and the differential pair $M_3$–$M_4$ is on: the output voltage is generated according to the differential input A.

The minimum supply voltage, $V_{DD,min,TT}$ for the triple tail XOR gate is computed by following the same approach followed for the conventional XOR gate as:

$$V_{DD,min,TT} = \frac{V_{SWING}}{4} + V_{GS3,4} + V_{DS7,sat} \tag{8}$$

where $V_{DS}$, $V_{GS}$, and $V_{DS,sat}$ are the drain source, gate source and saturation drain source voltage of the different transistors in Fig. 2 respectively.

It has to be pointed out that, the performances of the triple tail XOR gate strongly rely on the sizing of MOS devices $M_1$ and $M_2$ with respect to the other devices. Furthermore, the common-mode voltage and the voltage swing of inputs A and B have to be carefully chosen to provide proper operation. In particular, MOS devices $M_1$ and $M_2$ have to be sized much larger than the other transistors in order to keep the output swing high and guarantee a good noise margin [10].

According to [10] the output differential voltage swing for the triple tail XOR gate can be expressed as follows:

$$V_{SWING,TT} = 2R_P I_{SS} \frac{M}{1+M}. \tag{9}$$

where $M$ is the ratio between the aspect ratios of $M_{1,2}$ and $M_{3-6}$.

In order to keep the output swing high and guarantee a good noise margin, values of $M$ in the range of 8-10 have to be chosen when adopting short channel MOS technologies. When $M$ is set to be greater than 8 the input capacitance of $M_1$ and $M_2$ is very large and the gates (or buffers) driving $M_1$ and $M_2$ are heavily loaded: this results in a strong limitation of the speed performance of the triple tail XOR gate topology.

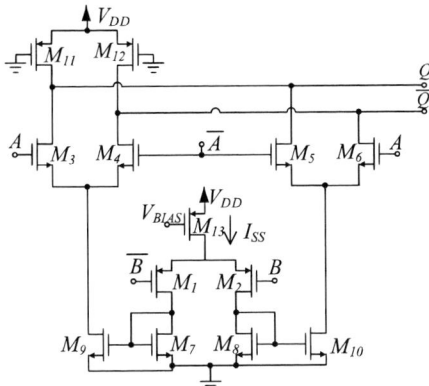

Fig. 3    Proposed Folded MCML XOR Gate

### III.    Proposed MCML XOR gate topology

In this paragraph we introduce a novel, low voltage, MCML XOR gate. The proposed XOR gate exploits the idea of folding one level of the conventional XOR gate. The proposed topology is shown in Figure 3. When input B is low the P-channel differential couple $M_1$-$M_2$ steers the current $I_{SS}$ to the current mirror $M_8$-$M_{10}$, whereas, when input B is high $I_{SS}$ is steered towards $M_7$-$M_9$. The differential pair made up of devices $M_3$-$M_4$ is active when B is high, and the XOR gate generates the output according to the differential input A. When B is low only the differential couple $M_5$-$M_6$ is active and the XOR gate generates the output according to not A.

#### A.  Minimum supply voltage

Denoting as $V_{CMA}$ and $V_{CMB}$ the common mode voltage of differential input A and B respectively, the following conditions have to be fulfilled in order to guarantee proper operation of the proposed XOR gate (i. e. all transistors from $M_1$ to $M_{10}$ working in saturation):

$$V_{DS9,10} = V_{CMA} - V_{GS3-6} > V_{DS9,10,sat} \qquad (10)$$

$$V_{DS3,4} = V_{DD} - \frac{V_{SWING}}{2} - (V_{CMA} - V_{GS3-6}) > V_{DS3,4,sat} \quad (11)$$

$$|V_{DS13}| = V_{DD} - (V_{CMB} - |V_{GS1,2}|) > |V_{DS13,sat}| \qquad (12)$$

Assuming that the differential signal A is the output of a conventional MCML inverter we can write $V_{CMA} = V_{DD} - \frac{V_{SWING}}{4}$ as in (6). The differential input B in Figure 3 can be generated from a p-type MCML inverter (made up of a PMOS differential pair with NMOS load). In this cases $V_{CMB}$ can be easily set to the optimum value which allows to minimize the supply voltage $V_{DD}$:

$$V_{CMB} = \frac{V_{SWING}}{4} \qquad (13)$$

By using the expressions in (6) and (13) for $V_{CMA}$ and $V_{CMB}$ respectively in (10) and (12) the minimum supply voltage of the proposed folded XOR gate can be written as:

$$V_{DD,min,folded} = \frac{V_{SWING}}{4} + V_{GS3-6} + V_{DS9,10,sat} \qquad (14).$$

#### B.  Threshold lowering for very low voltage XOR gates

Equation (14) can be rewritten in terms of the threshold voltages $V_{TH}$ and overdrive voltages $V_{OV} = V_{GS} - V_{TH}$ as follows:

$$V_{DD,min,folded} = \frac{V_{SWING}}{4} + V_{TH} + 2V_{OV} \qquad (15)$$

Referring to a 40nm CMOS technology $V_{TH}$ is typically in the range of 0.4V, whereas $V_{OV}$ and $\frac{V_{SWING}}{4}$ can be both set in the range of 0.1V, resulting in about 0.7V of minimum $V_{DD}$.

To further reduce the minimum supply voltage a threshold lowering technique based on moderate forward body bias can be exploited by using the following body connections for the circuit in Fig. 3: Body of $M_1$ and $M_2$ connected to ground; Body of $M_7$, $M_8$, $M_9$ and $M_{10}$ connected to the common source of $M_1$ and $M_2$; Body of $M_3$, $M_4$, $M_5$ and $M_6$ connected to $VDD$.

With this arrangement the $V_{TH}$ can be reduced to about 0.3V and a minimum supply voltage in the range of 0.6V becomes feasible. It has to be noted that in this configuration all transistors are forward body biased by the same bulk source voltage $V_{SB} = \frac{V_{SWING}}{4} + V_{TH} + V_{OV}$ which is about equal to 0.5V in a 40nm CMOS technology.

### IV.    DESIGN AND COMPARISONS

The triple tail XOR gate and the proposed XOR gate have been compared by means of transistor level simulations referring to a commercial 40nm CMOS technology within Cadence Virtuoso environment.

A supply voltage $V_{DD}$ = 1V, an output swing $V_{SWING}$ = 800mV and a reference bias current $I_{SS}$ = 25μA have been assumed for the comparisons. All the comparisons have been carried out referring to the same power consumption for both circuits.

Bias settings and devices sizing for the reference tail current $I_{SS}$ are shown in Table I. Values in Table I have then been scaled with the current according to [6] for $I_{SS}$ ranging from 25 to 100μA.

It has to be noted that, for values of $V_{SWING}$ lower than 800mV the triple tail XOR gate is not able to properly operate ($M_1$ and $M_2$ in Fig. 2 are not able to properly steer $I_{SS}$ even for $M$=8), whereas the proposed folded XOR gate is able to work with $V_{SWING}$ set to 400mV. The possibility to operate at lower $V_{SWING}$ allows the proposed XOR gate to operate at a lower minimum supply voltage than the triple tail XOR gate.

The two XOR gate topologies have been simulated referring to two testbenches (TB1 and TB2) and two fan-out settings (FO1 and FO4). MCML inverter/buffer cells have been used as driving and load cells in all the simulations. In TB1 ideal voltage sources feed the differential A and B inputs of the gate under test, whereas in TB2 an MCML inverter/buffer cell, biased and sized at $I_{SS}$, drives each one of the A and B inputs.

Simulation results for $V_{DD}$=1V and for the reference $I_{SS}$ are summarized in Table I, and show that the proposed topology outperforms the triple tail one under testbench 2 conditions.

The propagation delay of the two compared XOR gates is reported in Figures 4 and 5.

978-1-5386-5388-3/18 $31.00 © 2018 IEEE

Table I: Bias settings and devices sizing for the compared XOR gate.

|  | Triple Tail XOR Gate | Proposed Folded XOR Gate |
|---|---|---|
| $V_{DD}$ | 1V | 1V |
| L | 40nm | 40nm |
| Reference $I_{SS}$ | 25 $\mu A$ | 25 $\mu A$ |
| $R_P$ | 16 K$\Omega$ | 16 K$\Omega$ |
| $W_{1,2}$ | 5.76 $\mu m$ | 0.72 $\mu m$ |
| $W_{3,4,5,6}$ | 0.72 $\mu m$ | 0.36 $\mu m$ |
| $W_{7,8,9,10}$ | - | 0.36 $\mu m$ |

Table II: comparisons in terms of propagation delay at $I_{SS}=25\mu A$

|  | Triple Tail | | | Folded | | |
|---|---|---|---|---|---|---|
|  | CML Buf (ps) | XOR (ps) | TOT (ps) | CML Buf (ps) | XOR (ps) | TOT (ps) |
| TB1-FO1 | - | 27.2 | **27.2** | - | 32.7 | **32.7** |
| TB1-FO4 | - | 49.2 | **49.2** | - | 39.9 | **39.9** |
| TB2-FO1 | 58.8 | 27.2 | **86.0** | 14.1 | 32.7 | **46.8** |
| TB2-FO4 | 61.3 | 49.2 | **110.5** | 14.1 | 39.9 | **54.0** |

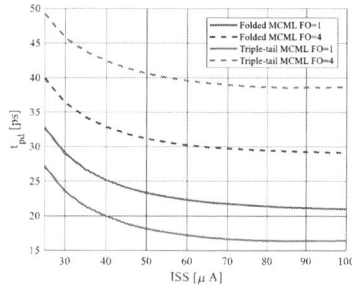

Fig. 4 Propagation delay comparisons for testbench1, in FO1 and FO4 conditions and $V_{DD}$=1V.

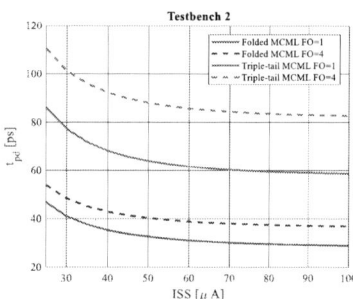

Fig. 5 Propagation delay comparisons for testbench2, in FO1 and FO4 conditions and $V_{DD}$=1V.

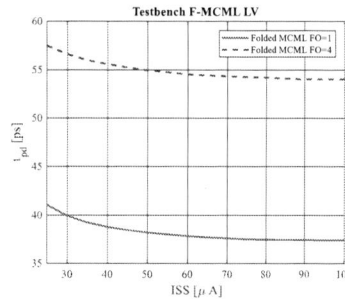

Fig. 6 Propagation delay as a function of $I_{SS}$ for the proposed Folded XOR gate for $V_{DD}$=0.65V, testbench1, FO1 and FO4.

Finally, the proposed topology has been tested with $V_{DD}$ set at its minimum value of 0.65V and the propagation delay as a function of $I_{SS}$ is shown in Figure 6.

V. CONCLUSION

In this paper we have introduced a novel low voltage topology to implement MCML XOR gates. The proposed XOR gate and the low voltage triple tail XOR gate have been compared referring to a 40nm CMOS process. Simulation results for $V_{DD}$ = 1V have demonstrated that the proposed topology allows a speed improvement of about 50% with respect to the triple tail one in the most common testbench conditions. Due to the capability to work with lower $V_{SWING}$, and exploiting a threshold lowering technique, the proposed circuit has been able to work at the lowest minimum supply voltage. Simulation results have shown the capability to operate at $V_{DD}$=0.65V whereas the triple tail XOR gate is not able to properly work for $V_{DD}$ lower than about 0.8V.

REFERENCES

[1] Guanghua Shu; Woo-Seok Choi; Saurabh Saxena; Mrunmay Talegaonkar; Tejasvi Anand; Ahmed Elkholy; Amr Elshazly; Pavan Kumar Hanumolu, "A 4-to-10.5 Gb/s Continuous-Rate Digital Clock and Data Recovery With Automatic Frequency Acquisition," IEEE Journal of Solid-State Circuits, vol. 51, no. 2, 2016, pp. 428-439.

[2] Sangwoo Han; Taegyu Kim; Jintae Kim; Jongsun Kim, "A 10 Gbps SerDes for wireless chip-to-chip communication" International SoC Design Conference (ISOCC), 2015 pp. 17 – 18.

[3] Siliang Hua; Qi Wang; Hao Yan; Donghui Wang; Chaohuan Hou, "A high speed low power interface for inter-die communication," International Conference on Solid-State and Integrated Circuit Technology (ICSICT), 2010 pp. 1916-1918.

[4] M. Anis, M. Allam, and M. Elmasry, "Impact of technology scaling on CMOS logic styles," IEEE Transactions on Circuits and Systems II, vol. 49, no. 8, pp. 577–589, 2002.

[5] M. Alioto and G. Palumbo, "Design strategies for source coupled logic," IEEE Transactions on Circuits and Systems I, vol.50, no. 5, pp. 640–654, 2003.

[6] M. Alioto, G. Palumbo, "Power Aware Design Techniques for Nanometer MOS Current Model Logic Gates: a Design Framework, IEEE Circuits and Systems Magazine, 2006, pp. 41-59.

[7] M. Alioto, R. Mita, G. Palumbo, "Design of High-Speed Power-Efficient MOS Current-Mode Logic Frequency Dividers," IEEE Transactions on Circuits and Systems II: Express Briefs, vol. 53, no.11, 2006, pp.1165-1169.

[8] B. Razavi, Y. Ota, R. Swartz, Design techniques for low-voltage high speed digital bipolar circuits, IEEE Journal of Solid-State Circuits, vol. 29, no. 3, 1994, pp. 332–339.

[9] M. Alioto , R. Mita, G. Palumbo, "Performance Evaluation of the Low-Voltage CML D-Latch Topology", Integration – The VLSI Journal, Vol. 36, No. 4, , 2003, pp. 191-209.

[10] K. Gupta, N. Pandey, M. Gupta, "Analysis and design of MOS current mode logic exclusive-OR gate using triple-tail cells," Microelectronics Journal vol. 44, no. 6, 2013, pp. 561–567.

[11] N. Pandey, K. Gupta, G. Bhatia, B. Choudhary, "MOS Current Mode Logic Exclusive-OR Gate using Multi-Threshold Triple-Tail Cells", Microelectronics Journal. vol. 57, no. 11, 2016, pp. 13–20.

[12] G. Scotti, D. Bellizia, A. Trifiletti and G. Palumbo, "Design of Low Voltage High Speed CML D-Latches in Nanometer CMOS Technologies," ," IEEE Transactions on VLSI, vol. PP, no. 99, pp. 1–12, 2017.

[13] M. Alioto, G. Palumbo, and S. Pennisi, "Modeling of source coupled logic gates," Int. J. Circuit Theory Applicat., vol. 30, no. 4, pp. 459–477, 2002

*PRIME 2018, Prague, Czech Republic*                                    *Session: Digital Circuits and Sub-Systems*

# VLSI Design of Frequent Items Counting Using Binary Decoders Applied to 8-bit per Item Case-study

Katsumi Inoue*, Trong-Thuc Hoang[†], Xuan-Thuan Nguyen[†], Hong-Thu Nguyen[†], and Cong-Kha Pham[‡]

*Advanced Original Technologies Co., Ltd (AOT), Tokyo, Japan

[†‡]The University of Electro-Communications (UEC), 1-5-1 Chofugaoka, Chofu-shi, Tokyo 182-8585, Japan

Email: *ugg44151@nifty.com; [†]{thuc, xuanthuan, hongthu}@vlsilab.ee.uec.ac.jp; [‡]phamck@uec.ac.jp

*Abstract*—In this paper, the Very-Large-Scale Integration design of Frequent Items Counting (FIC) is proposed. The fundamental idea is to use binary decoders to generate a matrix of binary values of all input items, with each column represents for one items binary value. Then, the sums are executed on the rows of the matrix to retrieve the input items counting results. The proposed design is applied to the case-study of 8-bit/item. That means 256 different types of items in total. For storing the counting results, various options of count-register are also presented. The proposed architecture is implemented with seven option of count-register from 8-bit counter to 32-bit counters, with the incremental of 4-bit at a time. The design was implemented on the Altera Arria V SoC Development Kit. After successful built and verified on Field Programmable Gate Array (FPGA), the design was synthesized using Synopsys tools with the process of SOTB (Silicon on Thin Buried-oxide) 65nm. The FPGA results achieved the average speed of 3,883.1 and 4,638.62 million item-counting per second for the 32-bit and 8-bit count-register options, respectively. Compared to our previous work and the software-based application, the achieved speed results are more than three times and more than 150 times faster, respectively. The SOTB-65nm builds achieved the theory speed about 75% of the average practical results of FPGA implementations.

## I. INTRODUCTION

Nowadays, the active development of hardware architecture is opening up new opportunities for implementing basic but essential data mining tasks such as Frequent Items Counting (FIC). The FIC problem is the counting of the frequently appeared item in the itemset. Such a task is the most foundation function of data mining because it is required in almost every data mining algorithms such as the queries of frequent elements [1], [2], iceberg queries [3], [4], Top-k queries [5], quantile queries [6], and iceberg data cubes [7]. The FIC performances in those applications require not only fast respond but sometimes be restricted to one data pass as in data streaming application.

Although FIC plays a vital role in the data mining research area, it is hard to design a parallel FIC in software-based applications [2] due to limitations of CPU-based systems. As a result, there were many database systems with software approach spend most of their runtime to execute only the FIC task. Optimizations at the algorithmic or software level will become irrelevant when the limitations of CPU-based architecture are reached. As a result, an efficient hardware implementation had become necessary in the era of big data. Moreover, Very-Large-Scale Integration (VLSI) designs

have the substantial advantages of low power, highly parallel processing, high throughput, and can be free from significant memory buffers in the designs, unlike CPU-based approaches.

In this paper, a VLSI architecture of FIC is proposed using binary decoders. The key design of the proposed architecture is the use of binary decoders to generate columns of input items' binary values. Then, summing functions are done on the rows of the binary values' matrix. The 8-bit/item case-study was taken for the implementation. That means input items can be categorized into 256 different types. For storing the counting results, seven options of count-register were also given in the paper. The count-register can vary from 8-bit counter to 32-bit counters, with the incremental of 4-bit at a time. A full FPGA system was built to verify the functionality of the design, and then the FIC core module was further synthesized in Application-Specific Integrated Circuit (ASIC) with the process of SOTB (Silicon On Thin Buried-oxide) 65nm. The chosen FPGA board is the Altera Arria V SoC Development Kit with the 5ASTFD5K3F40I3 chip series. In FPGA system, the Direct Memory Access (DMA) was used to exploit the substantial advantages of highly parallel design fully. The DMA's data-width was chosen based on the maximum capacity of the bus transferring that the development kit can support, which is 256-bit. Therefore, with 256-bit input and 8-bit/item, 32 items can be read by the DMA at each operating clock. Hence, the system can categorize and count 32 input items at a time.

The experimental results showed the best and the worst timing results of the FPGA implementations are 8-bit and 32-bit count-register versions, respectively. For specific, 8-bit and 32-bit count-register FPGA implementations had 149.44 MHz and 125.10 MHz maximum operating frequencies, and the average Millions Item-counting Per Second (MIPS) of 4,638.62 and 3,883.1, respectively. In comparison with our previous work [8] of 1,280 MIPS, the FPGA speed results were about three to three and a half times faster. Moreover, they were also more than 150 times faster than the 25 MIPS performance of the software-based approach with the similar setting. For ASIC performances, 8-bit and 32-bit count-register circuits were also held the best and the worst timing implementations with 4,416 and 3,104 MIPS, respectively. Therefore, the SOTB-65nm synthesis had the speed results are about 75% compared to those of the corresponding FPGA implementations.

978-1-5386-5388-3/18 $31.00 © 2018 IEEE                161

The remainder of this paper is organized as follow. Section II review of the recent works. Section III proposes the architecture. Section IV gives the experimental results on FPGA and SOTB-65nm. And finally, Section V concludes the work.

## II. RELATED WORKS

For software-like approaches, the space-saving method was the primary method used to solve the FIC problem, and it was well studied in the literature [2], [5], [9], [10]. First introduced by A. Metwally *et al.* [5] in 2006, the space-saving method was proven to be the best software-based algorithm. The space-saving algorithm is an approximation calculation with the main idea is not to monitor all the item types but only a few best-selected candidates. The selection strategy and the maximum number of candidates are different depending on various aspects of the specific application. There are several worth-mention space-saving-based FIC implementations on CPU-like systems such as the works of G. S. Manku and R. Motwani [1] in 2002 and S. Das *et al.* [10] in 2008. In 2012, the state-of-the-art multi-core CPU-based FIC system was presented by P. Roy *et al.* [9]. The efficient design in [9] exploited the skew of input data by combining a proposed pre-filtering stage with the conventional space-saving method. As a result, the best throughput rate can reach up to 250 MIPS if there are highly skew input data. For mid-range and low-range data skew, the performances were about 100 MIPS and less than 10 MIPS, respectively.

For hardware implementations, the space-saving algorithm was also used such as the Xilinx Virtex-5 FPGA-based FIC systems proposed by Jens Teubner *et al.* [11] in 2011. The two proposed architectures in [11] were Parallel-Lookups and Pipelining models which utilized parallel item search technique and array algorithm, respectively. Although the two models were proven to be independent with input data skew and gave well experimental results, they cannot produce the full FIC table or extract the exact counting results. The reason is that the space-saving method which they were built upon is an approximation algorithm. Therefore, in 2017, our previous work [8] was proposed to overcome such an issue. In the previous work, an FPGA-based FIC implementation was proposed based on the idea of using Equality Comparators (EC) matrix. The EC matrix compares all of the input items with themselves to produce the counting results instantly within a single operating clock. The implementation can count 32 8-bit items at each clock and achieved the counting speed of 1,280 MIPS, which is about 50 times faster than that of the software-based application with the same approach.

## III. PROPOSED IMPLEMENTATIONS

The goal of the proposed system is to produce the completed FIC table. Hence, the implementations did not deploy an approximation method such as the popular space-saving algorithm like the majority of software-based approaches. Our completed proposed systems were built and tested on FPGA development kit. After the successful function verification, the FIC core modules were further synthesized on ASIC level with the process of SOTB-65nm. The proposed implementations were described in the case-study of 8-bit/item, i.e., 256 possible different item categories in total, with various bit-width of count-register. There were seven different options of count-register including 8-bit, 12-bit, 16-bit, 20-bit, 24-bit, 28-bit, and 32-bit counters.

### A. System Overview

The overview of the proposed FPGA system is given by Fig. 1. In the system, depends on application requirement, memories resources can be on-chip memories only or include a high-speed off-chip RAM. Also, memories can store the completed itemset or work as buffers and store a part of itemset at a given time. Due to the 256-bit data-width limitation of the Avalon bus in the Arria V SoC Development Kit, the data-width of the FIC IP and its DMA are also set to 256-bit. Thus, with 8-bit/item, the FIC IP can process 32 items by each clock.

A typical operation of the system begins with the Database Processor (DBP) when the FIC IP is informed by the DBP about the itemset in the memories. Then, the IP controller in the FIC IP gives the information to the DMA and starts the data transferring. The itemset, or a part of the itemset, is read by the DMA and transferred to the FIC core module. The FIC core module then generates the counting results of the input items. The counting process is done when the data transferring through the DMA is completed. Finally, the completed count table, which is the final answer to the FIC task, is given back to the DBP and the operation is finished.

### B. FIC Core Module

Fig. 2 describes the architecture of the FIC core module. The architecture is based on the tuple-scan approach which maintains an array of count-registers to keep the record of the counting results. Tuple-scan approach allows the completed FIC table can be generated by a single pass of the itemset. As seen in Fig. 2, with 8-bit/item, there are 256 count-registers needed to be deployed to produce 256 counting results from *CountResult0* to *CountResult255*. The architecture in Fig. 2 is implemented with seven options of count-register bit-width such as 8-bit, 12-bit, 16-bit, 20-bit, 24-bit, 28-bit, and 32-bit per counter.

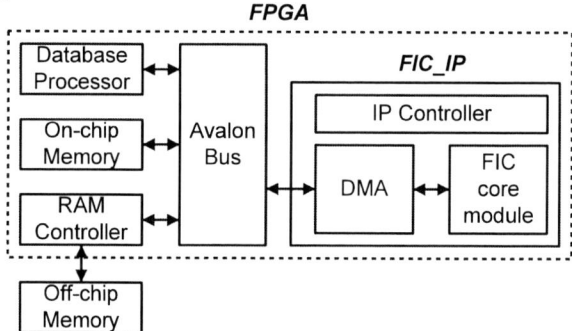

Fig. 1: The system overview of the proposed FIC implementation on FPGA.

*PRIME 2018, Prague, Czech Republic*                    *Session: Digital Circuits and Sub-Systems*

As shown in Fig. 2, 32 8-bit input items are transferred through the 256-bit Avalon bus at a time by the DMA. Then, corresponding to the input items, 32 8-to-256 binary decoders from *Decoder0* to *Decoder31* are deployed to generate the matrix of binary values $I_{x,y}$, where $x$ is from 0 to 31, and $y$ is from 0 to 255. One binary decoder creates one column of the $I$ matrix. When the matrix is formed, the population-count $\sum$ modules are used to sum a row of the matrix into one 6-bit counting result. The counting results after $\sum$ modules are 6-bit wide because the maximum number they can hold is 32. Finally, the current 256 counting results are added up with the previous counting results from the array of count-registers as described in Fig. 2. The array of count-registers will keep track of the total counting numbers untill the end of the operation. They are reset to the initial values of zero at the beginning of the operation or by the DBP command.

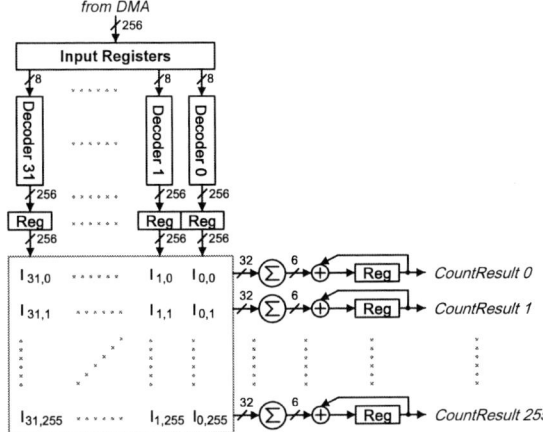

Fig. 2: FIC core module architecture.

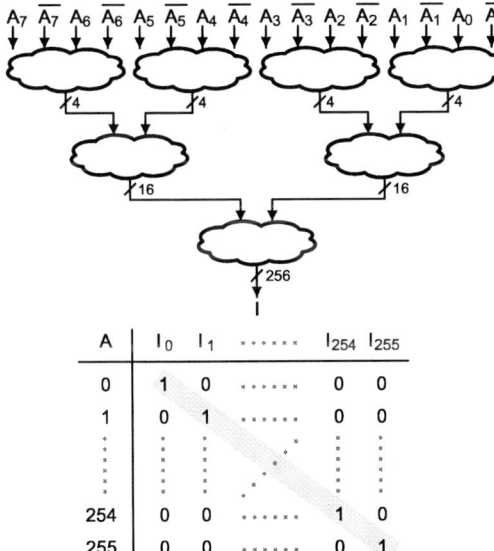

Fig. 3: 8-to-256 binary decoder design.

The design of an 8-to-256 binary decoder is given in Fig. 3. The binary decoder works in the same way of addresses decoder in a conventional RAM design. With 8-bit input data varies from 0 to 255, the 256-bit output result is generated with only one bit is asserted and 255 bits are de-asserted. The asserted bit corresponds to the value of the 8-bit input data as seen in the truth table in Fig. 3. With 8-bit input data $A$, the binary decoder implementation is divided into three stages of the combination as shown in the figure. The first stage is the combination of every two input bits with their inverted values. For example, the $(A_0, \bar{A}_0, A_1, \bar{A}_1)$ makes the combination values of $(A_0\ AND\ A_1)$, $(A_0\ AND\ \bar{A}_1)$, $(\bar{A}_0\ AND\ A_1)$, and $(\bar{A}_0\ AND\ \bar{A}_1)$. Similarly, in the second stage of combination, four distinct values are combined with other four distinct values and make the total of 16 output values. In the final stage, the 256-bit binary result $I$ is produced by the combination of 16-bit and 16-bit in the previous stage.

## IV. EXPERIMENTAL RESULTS

The proposed FIC architecture was implemented in the case of 8-bit/item. That means the total categories of input items are 256. There were seven different versions of the implementations corresponding to the bit-width of count-register as given in the Table I. The proposed designs were built and verified on the Altera Arria V SoC Development Kit with the FPGA chip series of 5ASTFD5K3F40I3. The FIC core modules were further synthesized at ASIC level with the SOTB-65nm process. The experimental results were reported in Table I along with the results of our previous work in 2017 [8] and an equivalent software-based application.

The software was utilized the well-known data analytic tool R-Studio using the function count() in the plyr library. There was no other optimization applied in the program, and the program run on one core of CPU. The reason for such a software development is because other software implementations in other literature [9]–[11] cannot make a fair comparison with the proposed FIC architecture in this paper. That is because their systems used an approximation approach such as the space-saving method which cannot produce a completed FIC counting table, unlike our approach.

To verify the functionality and measure the throughput rate of the proposed FIC systems, the same test data had been built and applied to both FPGA systems and the R-Studio program. The test itemset was generated randomly with various sizes from 1MB to 1GB. After the successful function verification, the FIC core modules were synthesized by the Synopsys tools with the SOTB-65nm process library. The Design Compiler tool was used to synthesize the core, and then the IC Compiler tool made the physical layout of the core. After the IC Compiler step, the ASIC results such as standard cells, area, estimation power, and the number of CMOS were reported. Then, the RC information was extracted from the layouts by the StarRC tool. The maximum operating frequencies of those layouts were retrieved by using the PrimeTime tool to test the layouts with their RC information. Because the FIC core modules were only synthesized to retrieved the ASIC results,

978-1-5386-5388-3/18 $31.00 © 2018 IEEE          163

TABLE I: The experimental results of the proposed FIC architecture with FPGA implementations and ASIC synthesis.

| FPGA implementation (Altera Arria V SoC: 5ASTFD5K3F40I3) | | | | | | | | |
|---|---|---|---|---|---|---|---|---|
| | Proposed architecture (Binary Decoders) | | | | | | | Previous work [8] (Equality Comparators) |
| Count-Register | 8-bit | 12-bit | 16-bit | 20-bit | 24-bit | 28-bit | 32-bit | 32-bit |
| ALUTs | 27,290 | 29,164 | 30,095 | 31,625 | 32,225 | 33,155 | 36,692 | 51,094 |
| Registers | 10,458 | 11,478 | 12,498 | 13,518 | 14,538 | 15,558 | 16,578 | 8,417 |
| $F_{Max}$ (MHz) | 149.44 | 137.75 | 137.02 | 134.47 | 130.60 | 125.30 | 125.10 | 40.85 |
| Average MIPS | 4,638.62 | 4,275.76 | 4,253.1 | 4,173.95 | 4,053.82 | 3,889.31 | 3,883.1 | 1,280 |
| ASIC synthesis (SOTB-65nm) | | | | | | | | Software-based |
| Count-Register | 8-bit | 12-bit | 16-bit | 20-bit | 24-bit | 28-bit | 32-bit | (R-Studio) |
| Standard cells | 70,349 | 75,290 | 82,178 | 91,046 | 98,649 | 116,919 | 125,279 | Windows 10 PC, |
| Area ($\mu m \times \mu m$) | 569×551 | 595×580 | 622×605 | 648×630 | 674×652 | 709×691 | 732×713 | 4GHz Core-i7 |
| Power (mW) | 7.3178 | 8.1773 | 9.1761 | 9.9226 | 11.0924 | 11.6577 | 12.5734 | Intel CPU, |
| Number of CMOS | 813,595 | 889,099 | 975,160 | 1,050,372 | 1,134,045 | 1,232,400 | 1,348,297 | 32GB RAM |
| $F_{Max}$ (MHz) | 105 | 97 | 97 | 96 | 95 | 94 | 94 | Average MIPS |
| Theoretically MIPS | 3,360 | 3,104 | 3,104 | 3,072 | 3,040 | 3,008 | 3,008 | 25 |

they did not have the practical MIPS performances as FPGA implementations. Therefore, their MIPS results are calculated based on their $F_{max}$.

For the FPGA implementations, as can be seen in Table I, it is clear that with the smaller count-register bit-width option, the fewer resources costs and the better timing performances. For specific, when changing from 8-bit counters option to 32-bit counters option, the costs of ALUTs and registers were increased about 35% and 60%, respectively; while the $F_{max}$ and average MIPS results were decreased about 16%. In comparison with our previous work [8] at the same count-register setting of 32-bit, the ALUTs requirement was decreased about 30% (36,692 compared to 51,094), the registers number was increased about twice (16,578 compared to 8,417), and the average MIPS was about three times better (3,883.1 compared to 1,280). In comparison with the R-Studio program, the 8-bit and 32-bit counters versions gain the better MIPS performances about 185 times (4,638.62 compared to 25) and 155 times (3,883.1 compared to 25), respectively. For ASIC performances, the results were similar with the FPGA implementations, the 8-bit and 32-bit counters versions were also held the best and the worst timing responds. The 8-bit and 32-bit layouts achieved 105 MHz and 94 MHz maximum operating clocks, and 3,360 and 3,008 MIPS performances, respectively. The MIPS results of the SOTB-65nm's layouts were about 75% in comparison with those of the FPGA systems at the same option of count-register.

## V. CONCLUSION

A high-speed VLSI architecture of the FIC was proposed in this paper. The proposed design was implemented with the case-study of 8-bit/item and various options of count-register bit-width. The full systems were implemented on the Altera Arria V SoC Development Kit, and the FIC core modules were further synthesized at ASIC level with the process library of SOTB-65nm. The fundamental idea of the proposed architecture is the using of the binary decoders to generate input items' binary values in columns, then applies the sum functions on the rows to compute the counting results. The

implementations can produce a full FIC table at the high-speed of 4,638.62 and 3,883.1 MIPS for the 8-bit and 32-bit count-registers versions of FPGA systems, respectively. The ASIC layouts also achieve the high-speed results of 3,360 and 3,008 MIPS for the 8-bit and 32-bit counters builds, respectively. To conclude, the achieved throughput rates in this paper were far more than those of the best-published results in both terms of software and hardware approaches.

## REFERENCES

[1] G. S. Manku and R. Motwani, "Approximate Frequency Counts Over Data Streams," in *Int. Conf. on Very Large Data Bases Endowment (PVLDB)*, 2002, pp. 346–357.

[2] S. Das, S. Antony, D. Agrawal, and A. El Abbadi, "Thread Cooperation in Multicore Architectures for Frequency Counting Over Multiple Data Streams," in *Int. Conf. on Very Large Data Bases Endowment (PVLDB)*, vol. 2, no. 1, Aug. 2009, pp. 217–228.

[3] M. Fang, N. Shivakumar, H. Gracia-Molina, R. Motwani, and J. D. Ullman, "Computing Iceberg Queries Efficiently," in *Int. Conf. on Very Large Data Bases Endowment (PVLDB)*, 1998, pp. 299–310.

[4] K. AlSabti, "Efficient Computing of Iceberg Queries Using Quantiling," *Journal of King Saud University - Computer and Information Sciences*, vol. 18, pp. 53–75, 2006.

[5] A. Metwally, D. Agrawal, and A. E. Abbadi, "An Integrated Efficient Solution for Computing Frequent and Top-k Elements in Data Streams," *ACM Trans. on Database Systems*, vol. 31, no. 3, pp. 1095–1133, Sep. 2006.

[6] M. Greenwald and S. Khanna, "Space-Efficient Online Computation of Quantile Summaries," in *ACM SIGMOD Int. Conf. on Management of Data*, vol. 30, no. 2, 2001, pp. 58–66.

[7] J. Han, J. Pei, G. Dong, and K. Wang, "Efficient Computation of Iceberg Cubes with Complex Measures," in *ACM SIGMOD Int. Conf. on Management of Data*, vol. 30, no. 2, 2001, pp. 1–12.

[8] Trong-Thuc Hoang, Xuan-Thuan Nguyen, Hong-Thu Nguyen, Nhu-Quynh Truong, Duc-Hung Le, Katsumi Inoue, and Cong-Kha Pham, "FPGA-based Frequent Items Counting Using Matrix of Equality Comparators," in *IEEE Int. Midwest Symp. on Circuits and Systems (MWCAS)*, Boston, USA, Aug. 2017, pp. 285–288.

[9] P. Roy, J. Teubner, and G. Alonso, "Efficient Frequent Item Counting in Multi-core Hardware," in *18th ACM SIGKDD Int. Conf. on Knowledge Discovery and Data Mining*, 2012, pp. 1451–1459.

[10] S. Das, D. Agrawal, and A. E. Abbadi, "CAM Conscious Integrated Answering of Frequent Elements and Top-k Queries Over Data Streams," in *4th Int. Workshop on Data Management on New Hardware*, 2008, pp. 1–10.

[11] J. Teubner, R. Muller, and G. Alonso, "Frequent Item Computation on a Chip," *IEEE Trans. on Knowledge and Data Engineering*, vol. 23, no. 8, pp. 1169–1181, Aug. 2011.

*PRIME 2018, Prague, Czech Republic*　　　　　　　　　*Session: Digital Circuits and Sub-Systems*

# Multi-Stage Complex Notch Filtering for Interference Detection and Mitigation to Improve the Acquisition Performance of GPS

Syed Waqas Arif, Adem Coskun and Izzet Kale

Applied DSP and VLSI Research Group
Department of Engineering
University of Westminster
waqas.arif@my.westminster.ac.uk, a.coskun@westminster.ac.uk, kalei@westminster.ac.uk

*Abstract*— **Continuous Wave Interferences (CWIs) can degrade the accuracy of a Global Positioning System (GPS) receiver and moreover it can completely deteriorate receiver's normal operation. In this paper a low-cost anti-jamming system design is presented for the mitigation and detection of CWIs for GPS receivers. The anti-jamming system comprises of parameterizable Complex Adaptive Notch Filter (CANF) module which is able to detect and excise single or multiple CWIs. The CANF module is composed of a first, second and third order infinite-impulse response filter with an Auto-Regressive Moving Averager structure. The proposed CANF detects the existence of the CWI and estimates JNR level of incoming signal by using the statistical value of the adaptive parameter $b_0$. The impact of the CANF module on the acquisition is analyzed. Moreover, a simple and innovative system level model is proposed which can utilize each CANF efficiently with threshold setting of JNR estimation within the adaptation block. Threshold setting parameters provide trade-off between effective excision of CWI, order of the filter and power consumption. This results in a parameterizable CANF module and provide effective solution for the mitigation of interferences with a high-power profile for GPS based applications.**

*Keywords—Anti-Jamming, Complex Notch Filter, GPS receivers, complex baseband processing*

## I. INTRODUCTION

Jamming and Anti-Jamming of the Global Navigation Satellite System (GNSS) have become a hot research topic over years due to the fast evolving GNSS technology and its rapid growth for the consumer based applications. Due to extremely low power of the GNSS signal received on the surface of the earth GNSS signal is prone to slightest of the RF interference directed to any GNSS receiver. Its effect on the received GNSS signal is unpredictable and severely degrade the quality of the signal making it impossible for the receiver to acquire and track it. It is essential to differentiate between types of interference either as wide or narrowband interference. An interference can be wide band for the civilian C/A code, but if same interference is to be compared with long encrypt P(Y) code will be referred as a narrowband interference. Being wide or narrowband merely depend on the bandwidth of the GNSS signal. Both C/A and P(Y) have respective signal bandwidth of 2.046 MHz and 20.46 MHz [1]. Intentional jamming is categorized as spoofing and jamming. Spoofing is transmission of a fake version of GNSS signal with more strength in order to deceive the receiver and hence

GNSS receiver will report incorrect position and timing data. Vast analysis of jammer signals have shown that any jamming signal can be modelled by linear frequency modulation which means jamming signal instantaneous frequency sweeps a range of frequencies in very small duration usually microseconds targeting the entire GNSS band to jam signal. The ideal GPS L1 signal at receiver front-end can be expressed as

$$s(t) = \sqrt{2P_i}\, D_s(t - \tau_0) C(t - \tau_0) \cos(2\pi f_{L1+D} t + \theta) \quad (1)$$

where $P_i$ is the signal power, $D_s$ is the navigation data bit with chip rate of 50 Hz, and $C$ is coarse C/A code running at chip rate of 1.023MHz and $\tau_0$ is the code phase delay. Frequency parameter $f_{L1+D}$ represents the L1 carrier frequency with a Doppler shift of $f_D$ (where the sum of these two frequency components results in $f_{L1+D} = 1575.42MHz \pm f_D$)[2]. It is assumed that sufficient bits are available from the ADC which provide enough dynamic range to accommodate interference signal without ADC saturation.

## II. JAMMING SIGNAL

### A. Modelling Jamming Signal

Jammers can radiate a variety of interferences signal. Main focus of this research is on Continuous Wave Interference (CWI) signals in the form of Complex Sinusoidal Wave Interference (CSWI) and Complex Chirp-Type Wave Interference (Single or Multi Saw-tooth function). These interferences are the most common type of interferences emitted by the commercially available jammers in the market.

### B. Jamming to Noise Ratio (JNR)

In order to evaluate the performance of any anti-jamming algorithm, it is important to determine the Jamming to Noise ratio (JNR) and it is defined as

$$JNR = 10\log_{10}\left(\frac{P_i}{P_n}\right) \quad (2)$$

$P_i$ And $P_n$ are the respective powers of jamming signal and the noise [3]. Interference's power with respect to noise variance is defined by the JNR which can mathematically be represented by

$$\frac{P_i}{P_n} = \frac{\frac{1}{2}A_i^2}{\sigma_n^2} = \frac{A_i^2}{F_s N_0} = \frac{A_i^2}{2B_{IF}N_0} \quad (3)$$

978-1-5386-5388-3/18 $31.00 © 2018 IEEE　　　　　165

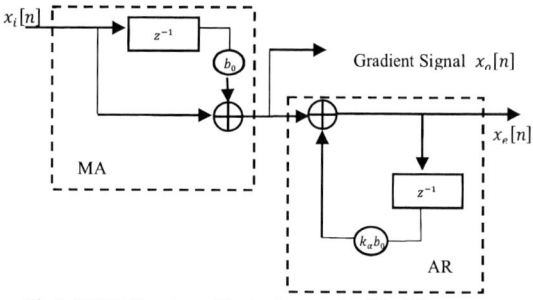

Fig 1. ARMA Structure of first order complex filter [4].

where $A_i$ is amplitude of narrowband CWI, $N_0$ is the noise per unit bandwidth, $F_s$ is the sampling frequency and $B_{IF}$ is the bandwidth of the front-end filter. The variance of the noise $\sigma_n^2$ is equal to $B_{IF}N_0$. This section presents modelling of the narrowband CWI through acquisition chain and its implications on the Cross Ambiguity Function (CAF) of the acquisition scheme. The fixed frequency complex sinusoidal CWI is present as an interference, it can be expressed by

$$J_{int}(t) = \sqrt{P_i}\exp\left(j2\pi f_i T_s + \theta_i\right) \tag{4}$$

where $P_i$ is the power of the narrowband CWI ,$f_i$ is the frequency of the interferer and $\theta_i$ is the initial random phase of the interference. Hence GPS L1 signal at the RF front-end of the receiver is formulated by, $x_i(t) = s(t) + n(t) + J_{int}(t)$.

## III. HIGH ORDER COMPLEX NOTCH FILTER

The structure of first order complex adaptive notch filter is shown in Fig.1, it comprise of two sections Moving Averaging (MA) and Autoregressive (AR). The z-domain transfer function of whole filter can be represented by (5) while K = 1. Our observation on the excision of the CWI and complex sinusoidal signal using first order CNF are;

- The first order CNF filter under performs if the value of JNR is above 25 dB hence GPS L1 signal is unable to be acquired.
- The mitigation of high power CWI and complex sinusoidal signal requires notch filtering with deeper attenuation and tighter bandwidth.

These observations became motivation to undertake further analysis on higher order CNFs and incorporate them together such that, depending on the JNR power level system can decide which CNF order should be used for efficient excision of CWI interference without harming or distorting the actual signal

$$H_{CNF}(z) = \prod_{K=1}^{K} \frac{1-b_0 z^{-1}}{1-k_\alpha b_0 z^{-1}} \tag{5}$$

where the parameter $b_0$, is the complex coefficient of CNF and is also the zero location at the interference frequency and variable $k_\alpha$ determines the width of the notch. The Normalized Least Mean Square (NLMS) is used in this paper as it is less complex and suitable for real-time system for the practical

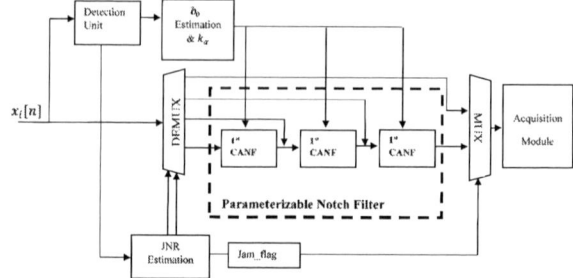

Fig 2. Proposed system level model.

integration into the GPS receivers. NLMS algorithm is used to minimize the cost function (6) and parameter $b_0$ is adapted by minimizing output of moving average part $x_0[n]$ , given by the following equation [5].

$$grad(C[n]) = \nabla_{zo}\{|x_0[n]|^2\} \tag{6}$$

The $b_0$ coefficient of Complex Adaptive Notch Filter (CANF) can be updated using (7) where μ is the normalized step-size the value of which is very critical in determining the rate that the algorithm converge to a solution. A normalized step-size is used to avoid any misadjustment due to signal power level.

$$b_0[n] = b_0[n-1] - \mu \times grad(C[n]) \tag{7}$$

As given in [6], differentiation of cost function with respect to complex parameter $b_0$ in (6) leads to following equation:

$$\nabla_{b_0}\{|x_0[n]|^2\} = -4x_0[n](x_e^*[n-1]) \tag{8}$$

Substituting (8) into (7) gives following adaptation equation for the coefficient $b_0$ of the CANF filter.

$$b_0[n] = b_0[n-1] + \mu \times 4x_0[n](x_e^*[n-1]) \tag{9}$$

## IV. PROPOSED SYSTEM LEVEL MODEL

Proposed system level model is illustrated in Fig.2. Here the goal is to develop a paramterizable multi-stage notch filter that can switch from a lower to a higher order notch filter depending on the input JNR level. As the useful signal with interference enters the system the presence of the interference must be detected. Once interference is detected, system simultaneously estimate the zero location $b_0$ and JNR level. Depending on the level of estimated JNR, the signal $x_i[n]$ will be passed on to 1st, 2nd or 3rd order CANF via a DEMUltipleXer (DEMUX) which is controlled by JNR estimation block in Fig.2. Jam_flag in Fig.2 determines whether incoming signal has to be filtered through CANF or pass directly to acquisition block when no interference is present. There is no point of using higher notch filter for low level of JNR which will consume more power. A simple and less power hungry first order notch filter can be used for mitigating interference with low JNR level. The working thresholds for 1st, 2nd and 3rd order is estimated to be 26dB, 46dB and 51dB respectively as marked with circles in Fig 3. The GPS L1 signal is known to be acquirable if output SNR at acquisition is 10dB or above [7], as represented by the yellow dashed line in Fig.3.

978-1-5386-5388-3/18 $31.00 © 2018 IEEE

*PRIME 2018, Prague, Czech Republic*  *Session: Digital Circuits and Sub-Systems*

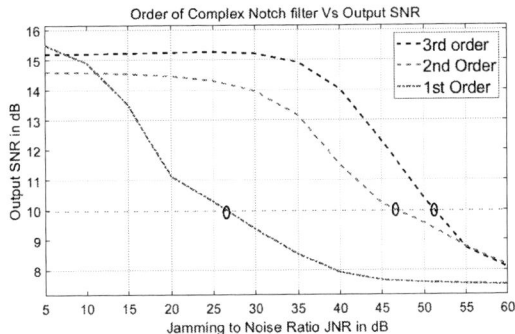

Fig 3. Output SNR vs. Order of Notch filter with same $b_0$ value for all the notch filter.

## V. PROPOSED JNR LEVEL ESTIMATION

---

**ALGORITHM**: *JNR Level Estimation* $(N_p, |b_0|, \beta_1, \beta_2$ and $\beta_3)$

---

Collect $N_p$ number of samples of adaptive parameter $|b_0|$

Step 1: Compute mean of $|b_0|$ over $N_p$ number of samples

Step 2: Determine $\sigma^2_{|b_0|}$ the variance of $|b_0|$ using (10)

$$\sigma^2_{|b_0|} = \frac{1}{N_p} \sum_{n=0}^{N_P-1} abs(b_0[n])^2 - mean(|b_0|) \qquad (10)$$

Now

   **If** $\beta_0 < \sigma^2_{|b_0|} < \beta_1$
        **then** DEMUX$_{ctrl}$  00 ( No CANF filtering)
   **elseif** $\beta_1 < \sigma^2_{|b_0|} < \beta_2$
        **then** DEMUX$_{ctrl}$  01 ( 1$^{st}$ Order)
   **elseif** $\beta_2 < \sigma^2_{|b_0|} < \beta_3$
        **then** DEMUX$_{ctrl}$  10( 2$^{st}$ Order)
   **elseif** $\sigma^2_{|b_0|} > \beta_3$
        **then** DEMUX$_{ctrl}$  11( 3$^{st}$ Order)

---

The $\beta_1, \beta_2$ and $\beta_3$ represent the threshold levels for the DEMUX to pass on the signal either to 1$^{st}$, 2$^{nd}$ or 3$^{rd}$ CANF. $\beta_0$ is the detection threshold for incoming interference. If mean of $|b_0|$ approximate to unity, it indicates the presence of CWI in the useful signal. The absolute value of $b_0$, i.e.$|b_0|$, fluctuates about its mean value whenever CWI is present in the useful signal and these fluctuation differ for different level JNR and hence the variance of $|b_0|$ can be used to estimate the power level of JNR. The range of fluctuation about unity strongly depend on the level of JNR. Fig.4illustrate the convergence curves of $|b_0|$ for different values of JNR. Different level of fluctuation of the magnitude of $|b_0|$ are more prominent in Fig.4 (a) which shows how $|b_0|$ eventually converges to unity for different levels of JNR but have different variance about unity. Another observation from Fig.4 (a) is, when JNR level is 0 dB (the pink curve), the modulus of parameter $b_0$ does try to converge to unity but the level of fluctuation is higher than any of the curve in Fig.4 (a). As the level of JNR increases the respective fluctuation for each successive JNR level damp down and decreases in amplitude. The simulation results in Fig.4 (b) gives an indication that rate of convergence of modulus of $b_0$ is different for different levels of JNR when rest of the parameters in simulation are kept unchanged.

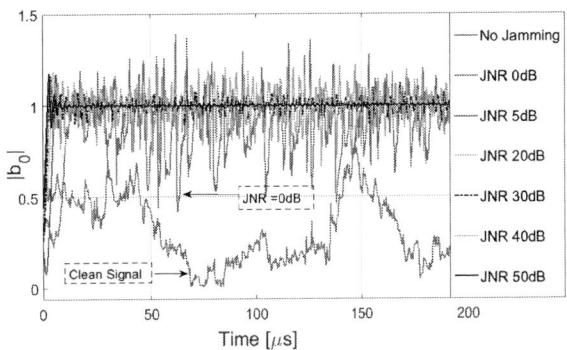

Fig 4. Convergence of magnitude of $|b_0|$ for different level of JNR. (a) shows behavior of $|b_0|$ from zero to 160 µs. (b) The zoomed in version of fig 4(a).

When interfering power is more, less number of iterations taken by NLMS algorithm to lock on the target frequency and the modulus of $b_0$ converges faster. Keeping the rest of the parameters the same or constant, it can be understood from the simulation results in this section that , the convergence of modulus of $b_0$ primarily depend on two factors, the power of narrowband CWI interference and the pole contraction parameter $k_\alpha$. As $|b_0|$ fluctuate about is mean value whenever CWI interference is present in the signal and these fluctuation differ for different level JNR, the variance of $|b_0|$ can be used to estimate the power level of JNR. Proposed interference power estimation is based on the variance of the modulus of $b_0$ about unity. Fig 5(b) shows how variance of modulus of parameter $b_0$ can be used as an estimate of power of jamming signal and JNR. The labels $\beta_1, \beta_2$ and $\beta_3$ in Fig. 5(b) represent three different threshold levels for 1$^{st}$, 2$^{nd}$ and 3$^{rd}$ CANF to be used in proposed system level model.

$$\tilde{J}NR_{Level} = f\left(\sigma^2_{|b_0|}\right) \qquad (11)$$

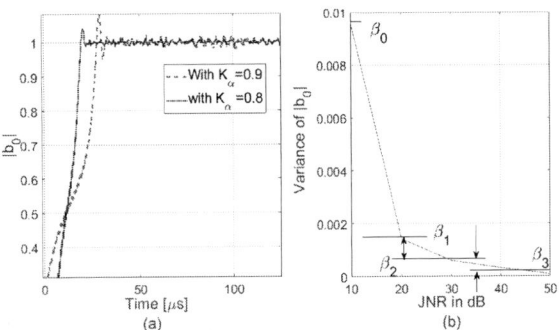

Fig. 5. (a) Convergence of the $|b_0|$ of 1$^{st}$ order notch filter for value of $k_\alpha$ 0.8 and 0.9. (b) Setting up threshold to activate required CANF filter depending on variance of the magnitude of $b_0$

## VI. SIMULATION RESULTS

CANF is employed to remove and excise the CWI and complex chirp-type interference from GPS L1 signals. This section presents the simulation results for CWI in the form of time-frequency representation of signal before and after

978-1-5386-5388-3/18 $31.00 © 2018 IEEE

mitigation, 3D plot of the Cross Ambiguity Function (CAF) plot [8] at the output of the acquisition module after excision

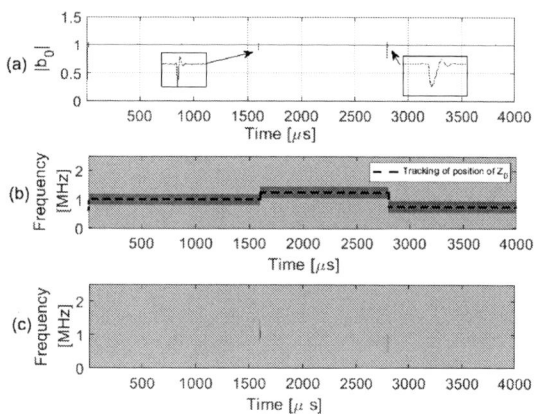

Fig 6. (a) Shows the convergence of $|b_0|$ over 4ms length of data with CSWI at 0.75MHz, 1MHZ and 1.25MHz with $k_a = 0.85$ and JNR level of 20dB. (b) Shows the time-frequency representation of GPS L1 signal with interference as indicated by thickline, and black dotted line represents the tracking position of $b_0$. (c) Shows the cleaned signal after mitigation of interference.

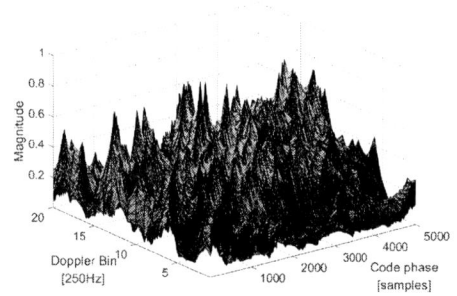

Fig 7. Evaluation of CAF at output of acquisition with interferences.

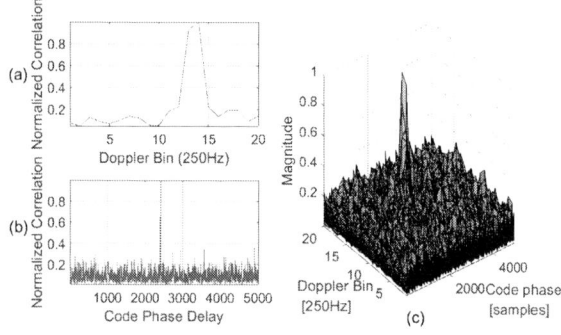

Fig 8. Output of acquisition block after mitigating CWI. (a) Doppler frequency estimation. (b) C/A code delay phase estimation. (c) CAF evaluation after mitigation of CWI at 0.75MHz, 1MHZ and 1.25MHz.

of the interference. Fig. 6(a) shows the convergence of adaptive parameter for first 4000μs and the glitches in Fig. 6(a) shows the adaptation of $b_0$ to new interference frequency. Fig.6 (b) shows the time-frequency representation of CSWIs at 0.75MHz, 1MHZ and 1.25MHz and dotted brown line represents the CANF notch's center frequency following and adapting the interference precisely. The plot in Fig. 6(c) represent the cleaned signal after filtered through CANF. Fig.8 and Fig.8 (c) shows evaluation of CAF before and after the mitigation of CSWIs at 0.75MHz, 1MHZ and 1.25MHz. The correlation peak is in Fig.8 (c) prominent and visible as compare to multiple peaks in Fig.7, hence blindfold the receiver operations. Fig 8 (a) and (b) shows the estimated values of Doppler Shift and code phase delay respectively which are further use to determine the location.

## VII. CONCLUSION

A simple and innovative system level model is proposed utilizing multi-stage CANF efficiently with threshold setting of JNR estimation. Threshold setting parameters provide trade-off between effective excision of CWI and order of the filter. Hence power consumption and results in a parameterizable CANF which provide an effective solution for interference mitigation for GNSS based applications. Different aspects and properties of the parameter $b_0$ are exploited which could be beneficial for the estimation of JNR levels. Variance of the magnitude of parameter $b_0$ is proposed to set as threshold setting variable for estimation of level of JNR. It can be employed to toggle between different orders of the filter depending on level of JNR.

## REFERENCES

[1] A. Ucar, E. Cetin, and I. Kale, "A low complexity DSP driven analog impairment mitigation scheme for Low-IF GNSS receivers," in *2008 IEEE/ION Position, Location and Navigation Symposium*, 2008, pp. 865–870.

[2] E. D. Kaplan and C. Hegarty, *Understanding GPS Principles and Applications.* Boston: Boston : Artech House, 2006.

[3] M. Abdizadeh, J. T. Curran, and G. Lachapelle, 'New decision variables for GNSS acquisition in the presence of CW interference', *IEEE Transactions on Aerospace and Electronic Systems*, vol. 50, no. 4, pp. 2794–2806, Oct. 2014.

[4] H. Xiong, W. Zhang, Z. Du, B. He, and D. Yuan, 'Front-End Narrowband Interference Mitigation for DS-UWB Receiver', *IEEE Transactions on Wireless Communications*, vol. 12, no. 9, pp. 4328–4337, Sep. 2013.

[5] D. Borio, "Loop analysis of adaptive notch filters," *IET Signal Processing*, vol. 10, no. 6, pp. 659–669, 2016.

[6] S. S. Haykin, *Adaptive Filter Theory.* (3rd ed ed.) Englewood Cliffs, NJ ; London: Englewood Cliffs, NJ ; London : Prentice Hall, 1996.

[7] W. L. Mao, A. B. Chen, Y. F. Tseng, F. R. Chang, H. W. Tsao, and W. S. Huang, 'Design of Peak-finding Algorithm on Acquisition of Weak GPS Signals', in *2006 IEEE International Conference on Systems, Man and Cybernetics*, 2006, vol. 3, pp. 1820–1825.

[8] D. Akopian, 'Fast FFT based GPS satellite acquisition methods', *Sonar and Navigation IEE Proceedings - Radar*, vol. 152, no. 4, pp. 277–286, Aug. 2005.

*PRIME 2018, Prague, Czech Republic*    *Session: Digital Circuits and Sub-Systems*

# Stall-Aware Fixed-Point Implementation
# of LMS Filters

Darjn Esposito, Gennaro Di Meo, Davide De Caro, Ettore Napoli, Nicola Petra, Antonio G. M. Strollo

*Dept. of Electrical Engineering and Information Technology, University of Napoli "Federico II", Italy
darjn.esposito@unina.it

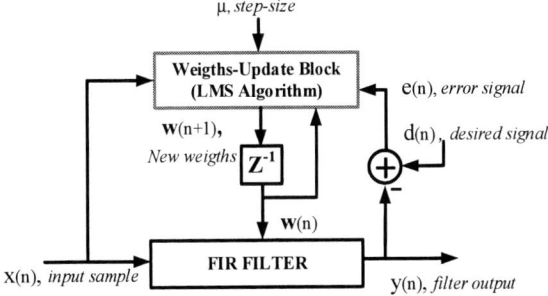

Fig. 1 Adaptive LMS filter.

*Abstract*—**Least-Mean-Square (LMS) is the most popular adaptive filtering algorithm, due to its numerical stability, satisfactory steady-state error and relatively low computational complexity. LMS algorithm is commonly employed in a multitude of tasks such as channel equalization, adaptive noise cancellation, system identification and frequency tracking.**

**Owing to its computational complexity, LMS algorithm is commonly implemented in hardware, in which, due to speed and energy constraints, a fixed-point arithmetic is usually adopted. As shown in the literature, the effects due to fixed-point arithmetic alter the behavior of the algorithm, causing a significant increase in the steady-state error, due to the so-called stall phenomenon.**

**In this paper we propose a novel fixed-point scheme which significantly limits the stalling. This gives increased flexibility to the designer in the choice of the step-size, that controls the tradeoff between convergence rate and steady-state-error. Implementation results in a 40nm CMOS technology are presented in the paper, to assess the impact of the proposed scheme in terms of area, speed and power.**

*Keywords—LMS filters; fixed-point circuits; DSP; stalling.*

## I. INTRODUCTION

Adaptive Least-Mean-Square (LMS) algorithm is employed in numerous applications such as channel equalization, adaptive noise cancellation, system identification and frequency tracking [1].

LMS algorithm owes its popularity to good numerical stability, satisfactory steady-state error and relatively simple computational complexity. This last feature allows the LMS algorithm to be implemented in hardware. In such context, due to speed and energy constraints, fixed-point arithmetic is preferred to floating-point one.

In virtue of its popularity, the effects due to fixed-point arithmetic in LMS algorithm have been extensively studied in [2]-[5], while hardware implementation of adaptive LMS filters has been proposed in [6]-[10]. An adaptive LMS filter is usually composed by a FIR filter, whose weights are updated, in a recursive manner, using the LMS algorithm, as shown in Fig. 1.

The LMS algorithm is an approximation of the Wiener filter (that is not implementable in hardware), which provides the optimum value of the coefficients, that minimize the mean-squared-error (MSE) between the output $y(n)$ of the FIR filter and the desired signal $d(n)$ (see Fig. 1). In the LMS algorithm, this minimization happens in an approximate way and, as a result, the steady-state MSE deviates from the optimum one given by the Wiener filter. The performance of the adaptive LMS filter is quantified by the so-called misadjustment parameter, that indicates the percentage deviation of the LMS steady-state error with respect to the optimal one [1]. The convergence rate of the LMS algorithm is controlled by the step-size parameter $\mu$ (Fig. 1): higher $\mu$ values allow the algorithm to converge faster, at the price of increased misadjustment [1]. When the LMS is implemented in fixed-point arithmetic, the tradeoff between convergence rate and steady-state MSE is altered [1].

In [2] Caraiscos et Liu investigated the round-off effects in fixed-point LMS algorithm, showing that the quantization noise increases with smaller $\mu$ values, limiting the range of achievable misadjustement. Moreover, they showed that, due to finite-precision effects, the weight updating is stopped when $\mu$ is chosen too small: this phenomenon is dubbed as "stalling" [1]. In [2] an expression of the steady-state error in presence of quantization effects is proposed. The model of [2] is extended in [3], introducing an analytical expression for the transient error. In [4] a novel model is proposed, accounting for different rounding types. Ghanassi et al. [5] define an intermediate working region, in which the stalling effect reduces the accuracy of the algorithm while still allowing the weights to update. An expression for the steady-state MSE is provided for this novel region.

With reference to hardware implementation, in [6] automatic generation of IPs for LMS are discussed with the aim of optimizing the fixed-point word-length while meeting the desired accuracy constraint. The usage of distributed arithmetic is considered in [7]-[8] to reduce hardware complexity. In [9] critical path analysis of LMS algorithm is

978-1-5386-5388-3/18 $31.00 © 2018 IEEE    169

discussed, and a low-complexity, pipelined, LMS implementation is proposed. The usage of approximate multipliers in the FIR part of adaptive LMS filters is investigated in [10], to reduce the power dissipation.

In this paper we propose a novel fixed-point scheme, that reduces the effect of fixed-point arithmetic on the algorithm behavior, by drastically limiting the stalling phenomenon. This gives increased flexibility to the designer in the choice of the step-size, that controls the tradeoff between convergence rate and steady-state-error. Implementation results in a 40nm CMOS technology are presented in the paper, to assess the impact of the proposed scheme in terms of area, speed and power.

## II. LMS Algorithm

In this section the LMS algorithm is briefly reviewed and the fixed-point implementation adopted in the literature is discussed.

### A. Floating-point algorithm

Let us assume that the FIR filter of Fig. 1 is composed by $T$ taps. Its input-output relation is given by

$$y(n) = \sum_{k=0}^{T-1} w_k(n) \cdot x(n-k) \tag{1}$$

where $w_k(n)$ with $k \in \{0, 1, ..., T-1\}$ is the *k-th* tap weight and $x(n)$ is the input. In (1) the weights are function of the time instant $n$, allowing the FIR filter impulse response to change over the time in order to minimize the MSE between $y(n)$ and the desired output $d(n)$ (Fig. 1). To compute the weights, the LMS algorithm senses the error $e(n)$

$$e(n) = d(n) - y(n) \tag{2}$$

and uses a stochastic gradient descent algorithm. The weight $w_k(n)$ is updated using the following equation

$$w_k(n+1) = w_k(n) + \mu \cdot e(n) \cdot x(n-k) \tag{3}$$

in which the term $\mu \cdot e(n) \cdot x(n-k)$ is the approximate *k-th* gradient component of the MSE cost-function [1]. Note that the effect of the step-size $\mu$ is to modify the magnitude of the gradient components, affecting the convergence rate and the steady-state error of the algorithm. Due to the approximation in the gradient computation, the steady-state MSE of the LMS algorithm is larger compared to one provided by the Wiener filter. By indicating as $J_{WH}$ the MSE of the Wiener filter, the steady-state MSE of the LMS algorithm can be written as:

$$J_{LMS} = J_{WH}(1+M) \tag{4}$$

where the misadjustment $M$ is introduced.
The misadjustment is linearly dependent on $\mu$ and $T$ [1]:

$$M \propto \mu \cdot T \tag{5}$$

The settling time $t_s$ of the algorithm exhibits, instead, an inverse proportionality with the step size [1]:

$$t_s \propto \mu^{-1} \tag{6}$$

The equations (5) and (6) highlight conflicting requirements in obtaining low misadjustment (that requires

small $\mu$ values) and fast convergence (calling for high $\mu$ values).

### B. Fixed-point algorithm

The fixed-point implementation of the LMS algorithm requires the quantization of the inputs $x(n)$ and $d(n)$ and of the internal operations (Fig. 2 (a)). In the following we will assume that the step-size $\mu$ is a power of two, as common done in order to reduce hardware complexity [9]. In fixed-point arithmetic the equation (1) becomes:

$$y_q(n) = Q\left[\sum_{k=0}^{T-1} w_{k_q}(n) \cdot x_q(n-k)\right] \tag{7}$$

where the subscript $q$ indicates the quantized variables, while $Q[\cdot]$ is the quantization operation. Since the equation (2) is the subtraction of two quantized variables ($y_q(n)$ and $d_q(n)$), the error $e(n)$ is automatically quantized.

The equation (3) requires the computation of the gradient component $\mu \cdot e(n) \cdot x(n-k)$. Note that the product $\mu \cdot e(n)$ is used to update the $w_k(n)$ weights, therefore it is convenient to evaluate this product beforehand. Since $\mu$ is a power of two, this requires a simple hardwired shift; we indicate the result of such operation as: $\mu e_q(n)$.

The standard fixed-point implementation of equation (3) is as follows [1]-[6] (see Fig. 2 (b)):

$$w_{k_q}(n+1) = w_{k_q}(n) + Q\left[\mu e_q(n) \cdot x_q(n-k)\right] \tag{8}$$

where the *k-th* gradient component is quantized to the same bit-width of $w_{k_q}(n)$ (note that, to prevent overflow, some guard bits can be added to $w_{k_q}(n)$, this is equivalent to scale down the desired signal).

Due to the quantization operation on the stochastic gradient component (8), the quantized gradient component can be zero if $\mu$ is too small. This happens when

$$\left|\mu e_q(n) \cdot x_q(n-k)\right| < LSB_w \tag{9}$$

being $LSB_w$ the LSB of the quantized weights. When condition (9) occurs the weight adaptation is stopped (*stalling phenomenon* [1], [2], [5]). It is worth noting that (even when (9) is not verified) the smaller is the step-size the lower will be the precision of the weight update computation (8), since more LSBs of the gradient component are discarded by the quantization operation. Because of this effect, the total output MSE of the fixed-point implementation (in addition to the LMS floating point contribution (4)) presents an error contribution that is inversely proportional to the step-size parameter [1]-[2]. This results, as opposed to (5), in increased misadjustment with smaller step-size $\mu$, limiting the steady-state performance achievable in fixed-point.

## III. Proposed Fixed-Point Implementation

We propose a novel scheme, to overcome the limitations due to the standard fixed-point implementation of weight update. In our scheme, the equation (7) is unchanged, while the precision of the weight update computation is improved by

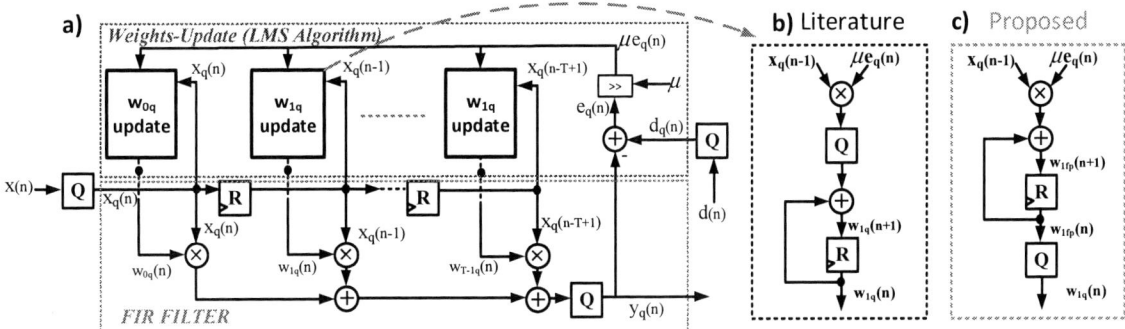

Fig. 2 Adaptive LMS filter. (a) Fixed-point schematic; (b) Standard weight update quantization; (c) Proposed weight-update quantization.

avoiding the quantization of the gradient. It follows that equation (8) becomes:

$$w_{k_{fp}}(n+1) = w_{k_{fp}}(n) + \mu e_q(n) \cdot x_q(n-k) \quad (10)$$

where the subscript *fp* indicates that the weights are considered in full-precision (i.e. no LSBs are discarded). The quantization (needed to decrease the hardware complexity of the FIR filter) is performed after computing the next weights in full-precision using (10), as follows:

$$w_{k_q}(n) = Q\left[ w_{k_{fp}}(n) \right] \quad (11)$$

Performing weight update in full-precision requires an overhead in terms of (i) increased size of the adder in (10) (ii) an additional carry-propagate adder to perform quantization with rounding (in the standard approach of Fig. 2 (b) the adder for the rounding can be fused with the multiplier computing the gradient component) (iii) increased number of flip-flops to store the full-precision coefficients of (10).

On the other hand, we have advantages related to (i) higher precision of the algorithm (ii) drastic limitation of the stalling. In our implementation, stalling occurs when:

$$\left| \mu e_q(n) \cdot x_q(n-k) \right| < LSB_{w_{fp}} \quad (12)$$

since $LSB_{w_{fp}} < LSB_w$, as a result the equation (12) is more hardly verified compared to (9).

## IV. ALGORITHM BEHAVIOR

The effect of the proposed fixed-point scheme on the algorithm behavior has been assessed in a system identification application, see Fig. 3. In this scenario, the LMS updates the weights of the FIR filter, to mimic the impulse response of the system to identify.

The input $x(n)$ is a Gaussian distributed signal with zero mean and 0.3 standard deviation, with values in the range $[-1.500, 1.447]$. The system to identify is an IIR band-stop Butterworth filter, with normalized -3 dB bandwidth equal to $0.2\pi$ rad/sample and stopband attenuation of 50 dB. In our simulations we observed that a $T$=20 taps FIR filter is sufficient to mimic, with excellent fidelity, the harmonic response of the system to identify.

We developed Matlab scripts describing (i) the optimum Wiener filter (ii) the floating-point LMS algorithm (described in the Section II A) (iii) the standard fixed-point LMS algorithm of Section II B (iv) the proposed fixed-point implementation discussed in Section III. The quantization operations have been implemented as rounding.

The standard and the proposed fixed-point implementations have been quantized with 14-bits ($x_q(n) = d_q(n) = w_{k_q}(n)$ =14-bits).

We performed Monte Carlo simulations in Matlab environment, with 99% confidence level and 5% relative error, with step-size parameter $\mu$ assuming values in the set $\left\{ 2^{-8}, 2^{-7}, 2^{-6}, 2^{-5}, 2^{-4}, 2^{-3}, 2^{-2}, 2^{-1} \right\}$. The corresponding results are reported in Fig. 4, where the steady-state MSE in logarithmic scale vs the step-size parameter is reported. Note that the MSE of the Wiener filter is independent on the step-size parameter (i.e. constant line).

The Fig. 4 shows that the floating-point implementation reduces the misadjustement with smaller step-size values, showing performance comparable to the optimum Wiener solution. On the other hand, the standard fixed-point LMS implementation, based on (8), exhibits a dramatic increase of the steady-state MSE when the step-size is reduced, due to the onset of stalling. When the step-size increases above $\mu=2^{-2}$, the MSE increases again according to (4)-(5). This highlights

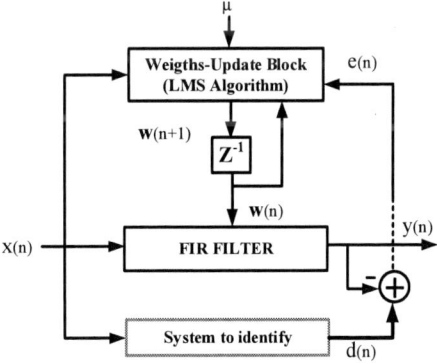

Fig. 3 System identification.

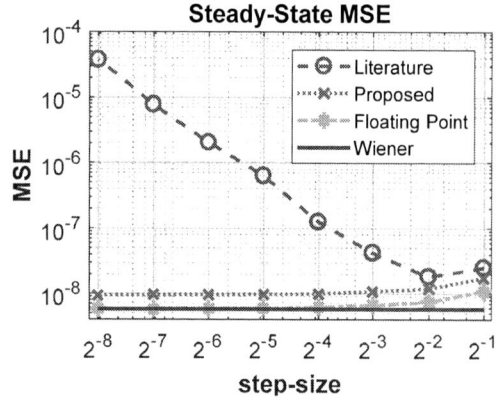

Fig. 4 Steady-state MSE vs step-size parameter. The proposed solution follows the behavior of the floating-point implementation, showing reduced MSE with smaller step-size.

that in the standard fixed-point LMS implementation the step-size $\mu$ is a critical parameter to be defined.

The proposed fixed-point scheme, as expected, is more precise than the standard approach and shows a behavior analogous to the floating-point implementation, with a steady-state MSE that reduces with smaller $\mu$ values (owing to the drastic limitation of the stalling). As a result, the choice of the step-size is much less critical and the design space is enlarged, allowing to trade the settling time (6) for an increased precision, choosing smaller step-size values.

## V. VLSI IMPLEMENTATION

We have implemented the LMS filter described in the previous section using both the proposed fixed-point scheme and the standard implementation. The circuits have been described in Verilog HDL and synthesized in a Global Foundry 40 nm technology, using Cadence RTL Compiler. The power dissipation has been evaluated from VCD and SDF-based post-synthesis simulations. Carry-save optimizations have been performed as much as possible with the help of the synthesis tool. For the standard implementation, involving truncation between the multiplier and the accumulator (see Fig. 2 (b)), a carry-save over

truncation transformation has been designed, as shown in [11].

The Table I and Table II report the obtained results. The increase in area ranges from 4% to 8%, while the increase in power (both dynamic and static) is between 8% and 14%. The minimum clock period of the proposed architecture is only slightly larger than the literature solution. These performance worsening are more than compensated by a marked improvement in the steady-state MSE. Compared to the literature solution, the proposed approach yields a reduction of MSE of 1.5X for $\mu=2^{-2}$ and 12.6X for $\mu=2^{-4}$.

## VI. CONCLUSIONS

A novel fixed-point scheme for adaptive LMS filters has been proposed. Compared to the standard approach, the proposed solution drastically reduces the stalling phenomenon, substantially improving the steady-state mean-squared-error, with a limited hardware overhead.

## REFERENCES

[1]   S. Haykin, *Adaptive Filter Theory,* Prentice-Hall, 2002.

[2]   C. Caraiscos, B. Liu, "A Roundoff Error Analysis of the LMS Adaptive Algorithm", *IEEE Transactions Acoustic, Speech, Signal Processing,* vol ASSP-32, no.1, february 1984.

[3]   S. Alexander, "Transient weight misadjustment properties for the finite precision LMS algorithm," in *IEEE Transactions on Acoustics, Speech, and Signal Processing,* vol. 35, no. 9, pp. 1250-1258, Sep 1987.

[4]   R. Rocher, D. Menard, O. Sentieys and P. Scalart, "Accuracy evaluation of fixed-point LMS algorithm," *2004 IEEE International Conference on Acoustics, Speech, and Signal Processing,* 2004, pp. V-237-40 vol.5.

[5]   M. Ghanassi, B. Champagne, P. Kabal, "On the steady-state mean squared error of the fixed-point LMS algorithm", *Signal Processing,* 87(12), pp.3226-3233, 2007.

[6]   R. Rocher, N. Herve, D. Menard and O. Sentieys, "Fixed-point configurable hardware components for adaptive filters," *2006 IEEE International Symposium on Circuits and Systems,* Island of Kos, 2006, pp. 4 pp.-3304.

[7]   D. J. Allred, Heejong Yoo, V. Krishnan, W. Huang and D. V. Anderson, "LMS adaptive filters using distributed arithmetic for high throughput," in *IEEE Transactions on Circuits and Systems I: Regular Papers,* vol. 52, no. 7, pp. 1327-1337, July 2005.

[8]   M. T. Khan, S. R. Ahamed and F. Brewer, "Low Complexity and Critical Path Based VLSI Architecture for LMS Adaptive Filter Using Distributed Arithmetic," *2017 30th International Conference on VLSI Design and 2017 16th International Conference on Embedded Systems (VLSID),* Hyderabad, 2017, pp. 127-132.

[9]   P. K. Meher and S. Y. Park, "Critical-Path Analysis and Low-Complexity Implementation of the LMS Adaptive Algorithm," in *IEEE Transactions on Circuits and Systems I: Regular Papers,* vol. 61, no. 3, pp. 778-788, March 2014.

[10]  D. Esposito, G. Di Meo, D. De Caro, N. Petra, E. Napoli and A. G. M. Strollo, "On the Use of Approximate Multipliers in LMS Adaptive Filters," *2018 IEEE International Symposium on Circuits and Systems (ISCAS),* Florence, Italy, 2018, pp. 1-5.

[11]  Taewhan Kim, W. Jao and S. Tjiang, "Arithmetic optimization using carry-save-adders," *Proceedings 1998 Design and Automation Conference. 35th DAC. (Cat. No.98CH36175),* San Francisco, CA, USA, 1998, pp. 433-438.

TABLE I

ELECTRICAL PERFORMANCE AND STEADY-STATE MSE –STEP-SIZE $=2^{-2}$

| Circuit | Area [μm²] | P_LEAKAGE [μW] | P_DYNAMIC [mW/MHz] | Min. T_clk [ns] | MSE |
|---|---|---|---|---|---|
| Literature | 95418 | 478 | 0.552 | 1.928 | $1.81\times10^{-8}$ |
| Proposed | 99256 (+4%) | 516 (+8%) | 0.594 (+8%) | 1.944 (+0.8%) | $1.18\times10^{-8}$ |

TABLE II

ELECTRICAL PERFORMANCE AND STEADY-STATE MSE – STEP-SIZE $=2^{-4}$

| Circuit | Area [μm²] | P_LEAKAGE [μW] | P_DYNAMIC [mW/MHz] | Min. T_clk [ns] | MSE |
|---|---|---|---|---|---|
| Literature | 92379 | 449 | 0.536 | 1.930 | $1.25\times10^{-7}$ |
| Proposed | 100108 (+8%) | 513 (+14%) | 0.611 (+14%) | 1.974 (+2%) | $9.89\times10^{-9}$ |

*PRIME 2018, Prague, Czech Republic*

*Session: Digital Circuits and Sub-Systems*

# A SpaceFibre multi lane codec System on a Chip: enabling technology for low cost satellite EGSE

Pietro Nannipieri, Gianmarco Dinelli
Information Engineering Department
University of Pisa
Pisa, Italy

Daniele Davalle
IngeniArs S.r.l.
Pisa, Italy

Luca Fanucci
Information Engineering Department
University of Pisa
Pisa, Italy,
IngeniArs S.r.l.
Pisa, Italy

Email: pietro.nannipieri@ing.unipi.it
gianmarco.dinelli@ing.unipi.it

Email: daniele.davalle@ingeniars.com

Email: luca.fanucci@unipi.it

*Abstract*—In the last few years, data rate requirement on on-board satellite communication systems significantly grown. The need of high speed networks led to the birth of the SpaceFibre protocol, which is able to run at several Gigabit per second and operates over both optical fibre and copper cables. A key feature of SpaceFibre is the possibility to have multi lane link, which increases the overall achievable data rate and link reliability. The growing complexity of satellite payload communication systems requires the definition an accurate monitoring and testing system. In this paper a multi lane SpaceFibre interface integrated in a System on a Chip is presented as enabling technology for an electrical ground segment equipment.

*Index Terms*—SpaceFibre, EGSE, Multi-lane, SoC, Satellite, Communication

## I. INTRODUCTION

The future space missions will require very high speed communication network to fulfil the growing requirement in term of data rate. In fact, the newest spacecraft payloads, such as SAR (Synthetic Aperture Radar) and multi-spectral imaging systems, require data rate in the order or Gbps (Gigabit per second).Communication protocols currently used by the European Space Agency (ESA) missions have lower performances. In example, SpaceWire has a maximum data rate of 200 Mbps and is no longer able to fulfil future mission data rate requirement. For this reason, ESA started to develop a new protocol named SpaceFibre. This standard [1], which is currently under public ECSS review, is a communication protocol specifically developed for spacecraft on-board communication systems. SpaceFibre supports data rate up to 5 Gbps, which can be incremented over 20 Gbps with multi laning. It is able to operate over both copper and optical fibre cables, and it is backward compatible with SpaceWire at packet level. A brief description of the standard togheter with some single lane SpaceFibre implementations, can be found at [2].

One of the SpaceFibre key features is the possibility to establish multi lane communication. There are other space qualified protocols which include multi lane feature, such as RapidIO [3], [4]. The possibility of introducing a multi lane link may be interesting for space applications not only

because it increases the maximum achievable data rate but also because it empowers system robustness and reliability introducing redundancy.

The described ongoing technical evolution of satellite payload communication needs to be preceded by the development of appropriate control and testing equipment. In Figure 1 a schematic description of an EGSE (Electrical Ground Segment Equipment) is presented. The aim of an EGSE is to intensively perform a series of tests by injecting a large number of stimuli to the DUT (Device Under Test), which in our case is the satellite payload. Thus, a generic EGSE will be composed by a monitoring and controlling block, responsible for creating both the appropriate control signals and the input stimuli for the DUT. The output of the DUT shall be analysed and may be stored on-board and used by the control section. This is a very general block scheme, which may vary depending on the application needs. Anyway, a key point of all EGSEs are the interfaces: it is fundamental for these systems to be able to communicate

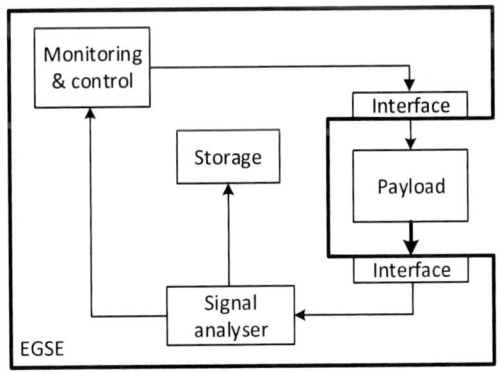

Fig. 1. Generic Electrical Ground Segment Equipment block scheme

978-1-5386-5388-3/18 $31.00 © 2018 IEEE

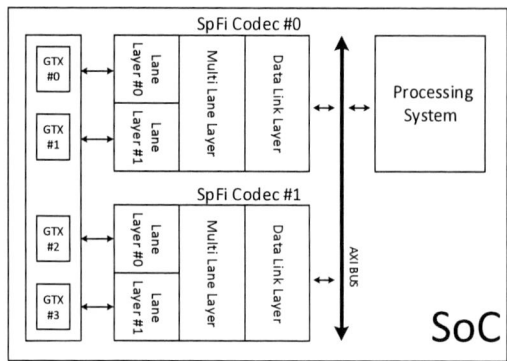

Fig. 2.  Architecture of the system on the Zinq FPGA

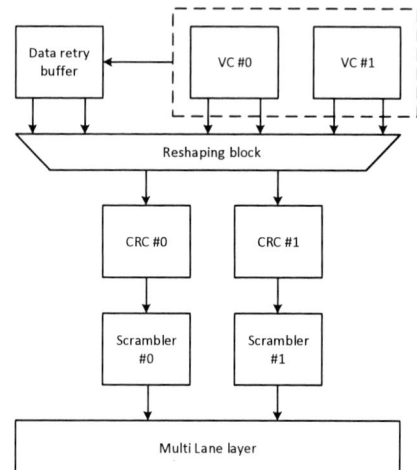

Fig. 3.  Data link layer transmitting side with 2 VC and 2 lanes

with the DUT with its communication protocols. For detailed information about ground segments, please refer to the appropriate standard [5].

On the market there are different EGSE solutions, both standard such as the PXI family from National Instrument and custom, able to operate with SpaceFibre [6]. The problem of such solutions is primarily the cost: it may be required from the user a lower cost and more flexible EGSE. Moreover, the available systems do not implement multi lane communication. The aim of this paper is to present a a multi lane SpaceFibre codec integrated on a SoC, which enables the development of a flexible and customizable EGSE.

In section II the architecture of the proposed system is shown, and in section III the hardware set-up used to test and demonstrate the system functionality is illustrated. Finally, an overview of the complexity of the presented solution is shown in section IV.

## II. SYSTEM ARCHITECTURE

The architecture of the SoC is shown in Figure 2. It includes a general purpose processing unit (PU) plus two hardware SpaceFibre codecs for the high speed serial interfaces, connected to the PU trough a AXI bus. This is a generic application that may be customized by the user depending on the final application. It can be implemented on various hardware platform which embeds both programmable logic, high speed serialiser/deserialiser and a processing system. The codec, completely realised in hardware with VHDL description, contains the data link, the multi lane and the lane layers of the SpaceFibre protocol. For a detailed protocol description please refer to the standard [1]. The number of communication lanes may be selected by the user between 1 and 2. The implementation of a codec with larger number of lanes (up to 16, to be fully compliant with the standard) is currently undergoing. In order to prove the functionality of the single codec, as an EGSE interface enabling technology, a prototype

has been designed. A brief overview on the codec architecture follows.

The codec is an implementation of the SpaceFibre protocol on a Xilinx FPGA. The lanes are interfaced with the PHY trough the Xilinx GTX transceiver, as will be better explained in section III. The data link is interfaced, by means of a configurable number of VCs (Virtual channels), with an AXI bus, which is then accessed also by the processing system, which embeds an hard core processor, responsible for the communication to and from a generic host (i.e via Ethernet or PCI-Express).

In the implemented system, represented schematically in Figure 2, two codecs, each one with two lanes, have been fitted into the FPGA. The idea is to connect externally the two codecs and check the received data stream to see if it matches. Moreover, both the SpaceFibre codec have a configuration and a status interface, to check the status of the codec in each layer and to configure some operating parameters in real time (i.e. turn on the link). In the following paragraph, the most important features, necessary to realize a multi lane SpaceFibre codec, are described.

### A. Multi lane architecture highlights

A schematic architecture of the data link layer transmission side is shown in Figure 3. The VCs carry independent flow of information and are scheduled with different priority. The data link layer requires to read N words per clock cycle, where N is equal to the number of lanes. Thus, each VC shall have N buffers that operates in parallel. The data retry buffer is responsible for the retransmission of corrupted data frames. The *Reshaping block* is responsible for positioning words on the active lanes if one or more lanes fail. In fact, a lane may become inactive in case of failure, becoming no longer able to transmit and receive data. Two parallel *CRC* (Cyclic

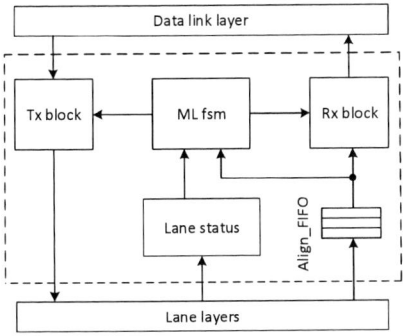

Fig. 4. Multi Lane layer architecture

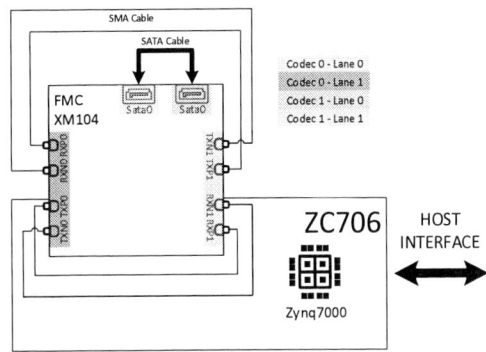

Fig. 5. Hardware set-up and external connections

Redundancy Check) and *Scrambler* blocks are instantiated, one per each lane, as SpaceFibre standard requires. Obviously, the receiving side performs similar operations in reverse order, thus its description is omitted.

A schematic representation of the multi lane layer is shown in Figure 4. A summarized list of the blocks requirements follows. The *Tx_block* shall transmit data from the data link layer to the lane layers. The *Rx_block* shall receive data from the lane layers and send them to the data link layer. The *Lane Status* block shall inform in real-time the *ML fsm* (Multi Lane finite state machine) about the status of each lane. The *align_FIFO* shall buffer 3 or 4 words to ensure receiving row alignment. More details about lanes alignment will follow.

The *ML fsm* shall:

- Allow the *Tx block* to receive words from the data link layer and transmit them to the lane layers (one word per lane).
- Allow the *Rx block* to receive words from the lane layers and transmit them to the data link layer.
- Inform the data link about the status of each lane: in case of one or more lane failures, data stream shall be redirected on the remaining active lanes.
- Check the alignment status of the link. It may happen that separate lanes may have up to a few clock cycle delay respectively, due to different transmission medium: this may lead to data row misalignment. The *ML fsm*, thanks to a proper synchronisation protocol, compensates different delays reading out the *align_FIFO* words belonging to the same row.

The lane layer maintains the same architecture of the single lane codec.

## III. HARDWARE SET-UP

In this section the hardware set-up adopted to implement the system will be described. The *Xilinx ZC706* evaluation board [7] has been chosen for its flexibility and great number of resources. It mounts the *Zynq-7000 XC7Z045-2FFG900C*, a SoC that embeds both a processing system and programmable

logic on the same die. The FPGA can host up to 16 high speed GTX transceivers. An expansion board, the *Xilinx FMC XM104* [8], is used together with the *ZC706* to increase the the the number of high speed ports. The processing system, consisting in a *Cortex ARM A9* hard core, is responsible for the host interface, which may be customized by the user (i.e. via Ethernet or PCI-Express). In Figure 5 the system is described at a block diagram level, with particular attention to serial interfaces: both a couple of SATA ports and a couple of differential SMA ports have been used, in order to connect the outputs of the four GTX. A different communication cable has been used mainly for 2 reasons:

- It is useful to have at least one port with SATA cable, which is easy to unplug during the codec operation to see how the link reacts.
- It may be useful to use different medium for separate lanes, in order to introduce a delay between them so that the multi lane synchronisation feature can actually be tested.

The design presented in section II occupies only a small fraction of the available programmable logic, as will be detailed in section IV. The rest of the logic may be used to implement several blocks or hardware accelerators, depending on the user needs. Also the processing system is heavily underused in the presented use case where it just performs I/O operations: the remaining computational power is available to the user for the design of customized applications.

In Figure 6 the assembled set-up is shown. It is possible to notice both the SATA and SMA cables. Please note that the *ZC706* is connected to an host PC, both with PCI-Express and Ethernet interfaces.

## IV. RESULTS

The proposed design has been synthesized on a *Xilinx ZYNQ-7 ZC706*. The results, shown in Table I, refer to a single SpaceFibre codec with 2 lanes and a variable number of VCs. Resources utilisation is presented by analysing the number of LUTs (Look Up Tables) and registers. The target clock

TABLE II
LAYERS RESOURCE UTILISATION, WITH 8 VCs AND 2 LANES

| Layer | LUT | util% LUT | Reg | util% Reg |
|---|---|---|---|---|
| Data Link | 12260 | 87.63% | 8964 | 93.43% |
| Multi Lane | 834 | 5.96% | 228 | 2.38% |
| Lane (x2) | 896 | 6.41% | 402 | 4.19% |

FPGA technology and may be used to build several application on top of them, using the remaining resources that the SoC offers to the designer. Future work will focus mainly on the development of a IP Core able to operate up to 16 lanes in parallel and will present synthesis results not only on commercial FPGA technology but also on space qualified devices.

### ACKNOWLEDGMENT

IngeniArs SpaceFibre technologies have been developed in the framework of the project SIMPLE (Spacefibre IMPLementation design & test Equipment). This project has received funding from the European Unions Horizon 2020 research and innovation programme under Grant Agreement No 757038.

### REFERENCES

[1] *Space engineering SpaceFibre Very high-speed serial link*, ECSS-E-ST-50-11C-DIR1, avaiable: http://ecss.nl/standard/ecss-e-st-50-11c-dir1/.
[2] S. Parkes, C. McClements, D. McLaren,a, A. F. Florit, A. G. Villafranca, *SpaceFibre: A multi-Gigabit/s interconnect for spacecraft onboard data handling*, IEEE Aerospace Conference Proceedings, 2015.
[3] RapidIO.org, *RapidIO Interconnect Specification Version 3.1.*, available: http://www.rapidio.org/wp-content/uploads/2014/10/RapidIO-3.1-Specification.pdf.
[4] C. P. Collier, J. Marshall, *Next Generation Space Interconnect Standard (NGSIS): A modular open standards approach for high performance interconnects for space*, IEEE Aerospace Conference, 2015
[5] *Space engineering Ground systems and operations*, ECSS-E-70C, avaiable: http://ecss.nl/standard/ecss-e-st-70c-ground-systems-and-operations/.
[6] A. G. Villafranca, S. Parkes, C. McClements, A. F. Florit, B. Yu, P. Scott, *A new generation of SpaceFibre test and development equipment: SpaceFibre, short paper*, 7th International SpaceWire Conference, 2016.
[7] *ZC706 Evaluation Board for the Zynq-7000 XC7Z045 All Programmable SoC User Guide*, avaiable: https://www.xilinx.com/support/documentation/boards_and_kits/zc706/ug954-zc706-eval-board-xc7z045-ap-soc.pdf.
[8] *FM-XM104 Connectivity Card User Guide*, avaiable: https://www.xilinx.com/support/documentation/boards_and_kits/ug536.pdf.
[9] A. F. Florit, A. G. Villafranca, S. Parkes, *SpaceFibre multi-lane*, 7th International SpaceWire Conference, 2016.

Fig. 6. Multi Lane implementation on a ZC706 Xilinx board

frequency is 62.5 MHz, resulting in a data rate of 2.5 Gbps per lane (5 Gbps total). The percentage of total LUTs and registers used is also present. In literature, the only implementation of a multi lane SpaceFibre codec is presented in [9], which shows results on space graded FPGAs (both Microsemi and Xilinx). Thus, the obtained results is not directly comparable with ours, as we decided to fit the system on commercial FPGAs, in order to provide information useful for future development of EGSE.

In Table II the impact of the layers in term of resources utilisation is analysed. The percentage of LUTs and registers refers to the SpFi codec total utilisation. It can be noted that the data link layer is by far the larger layer of the codec.

### V. CONCLUSIONS

A multi lane SpaceFibre codec implementation on a Xilinx ZC706 commercial board is presented in this paper. A multi lane SpaceFibre codec has been described as enabling technology for SpaceFibre compatible high speed EGSE. The chosen system architecture and the hardware set-up adopted have been presented. Finally, synthesis results have been shown for various VCs configuration. The obtained results about multi lane IP core shows resources utilisation on commercial

TABLE I
RESULT OF SYNTHESIS: SPFI CODEC RESOURCE UTILISATION

| VC | LUT | util% LUT | Reg | util% Reg |
|---|---|---|---|---|
| 1 | 6149 | 2.81% | 3521 | 1.01% |
| 2 | 7273 | 3.33% | 4806 | 1.10% |
| 4 | 8969 | 4.10% | 5942 | 1.36% |
| 8 | 13990 | 6.13% | 9594 | 2.31% |

# Design and implementation of a complete test equipment solution for SpaceWire links

Antonino Marino
*Dept. of Information Engineering*
*University of Pisa*
Pisa, Italy
antonino.marino@ing.unipi.it

Luca Dello Sterpaio
*Dept. of Information Engineering*
*University of Pisa*
Pisa, Italy
luca.dellosterpaio@ing.unipi.it

Luca Fanucci
*Dept. of Information Engineering*
*University of Pisa*
Pisa, Italy
luca.fanucci@unipi.it

*Abstract*— Spacecraft used in space mission have on-board a large number of peripherals such as instruments, mass-memory, downlink telemetry, sensors, actuators and processors. The State-Of-The-Art for the on-board spacecraft communication is the SpaceWire standard, which connects on-board devices directly, via point-to-point communication, or indirectly via router or switch.

This paper presents the architectural design and the implementation of a complete test equipment solution for SpaceWire link. This is designed in order to have in a single device the possibility to emulate both a single SpaceWire device and a portion of a SpaceWire network. In order to validate the proposed architecture, a demonstration prototype has been implemented.

*Keywords — SpaceWire, test equipment, high-speed serial link, EGSE (Electrical Ground Support Equipment), spacecraft communication*

## I. INTRODUCTION

Nowadays space missions for scientific purpose need to have a large number of peripherals, such as instruments, mass-memory, sensors, downlink telemetry, actuators and processors, on-board the spacecraft. The spacecraft's network consists of links, nodes and routers, and all the nodes have to communicate with each other, exchanging a large quantity of data.

The State-Of-The-Art for onboard spacecraft communication is the SpaceWire (SpW) standard [1]. SpW protocol is designed to connect together in a single network, promoting compatibility, all peripherals and on-board units used in the payload section of the spacecraft. A SpW network exploits either point-to-point (direct communication) links or routing switches (indirect communication link).

Typically, into a SpW network data transfers are managed by SpW routers to minimize the use of point-to-point links; to safeguard tolerance to link failures and path redundancy through the Group Adaptive Routing (GAR); reduce the relevant mass and volume budgets [2].

Many of actual test equipment for SpW units are all based on point-to-point communication. During the development of SpW units there is the need to emulate parts of SpW networks. For this reason the test equipment has to be able to: parse SpW packets intended for nodes of a SpW network, or to emulate part of the network in order to verify correct functionality of any SpW Device Under Test (DUT), including very particular corner cases.

This paper presents an architecture of an innovative test equipment for SpW standard. Proposed architecture offers a simple manner of extensively validate any SpaceWire-based system. Its features of most interest, presented SpW test equipment includes both SpW router interfaces in case of indirect communication link, and SpW codec for the point-to-point communication. It can also be used to: analyse SpW traffic; inject SpW packets; attest SpW standard conformance. Also, main uses cases for the SpW test equipment are presented. The presented SpW test equipment is designed upon the expertise of the IngeniArs S.r.l. (a University of Pisa spin-off company) on the SpW standard. In this work, the proprietary SpW interface and a SpW Router IP cores by IngeniArs S.r.l. has been used.

This paper is organized as follows:

- Section II presents an overview of the SpaceWire standard and the state-of-the-art study about SpaceWire test equipment field of applications.
- Section III describes the proposed test equipment solution for SpaceWire networks.
- Section IV presents the architectural implementation of SpaceWire router-based test equipment.
- Section V briefly reports the validation environment and some tests implemented on the system prototype.
- Finally, Section VI draws the conclusions.

## II. RELATED WORK

### SpaceWire Standard

SpW is the actual standard for on-board spacecraft communication, which connects together instruments, processing units, high data rate sensors, downlink telemetry and other on-board sub-system providing data-handling network. SpW provides high-speed serial link (from 2 to 400Mbps) bi-directional, full-duplex which connects together SpW enabled devices and sensors. The SpaceWire Standard is defined at several levels of protocol [1]:

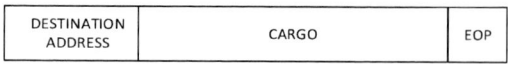

**Figure 1 - SpW packet**

- <u>Physical Layer</u>: it defines the cable and connectors format and EMC specifications.
- <u>Signal Level</u>: it covers signal noise margins, voltage levels and data rates.
- <u>Character Level</u>: it defines format of data and control characters.
- <u>Exchange Level</u>: it specifies the protocol initialization at connection, flow-control, fault-detection and link restart mechanisms.
- <u>Packet level</u>: it establishes how data are encapsulated in a packet and how messages are delivered from source to destination nodes.

The information across a SpW network are transferred in distinct packets. Figure 1 shows the format of a standard SpW packet.

The SpW packet is composed of a destination address, a cargo and a special control character used as End Of Packet (EOP) marker. The Destination Address is the first field of the packet to be sent. It can be done of zero or more data character depending of the type of communication link (point-to-point or through routing switches). In case of indirect communication, the destination address represents either the ID-code of destination node or the hops that the packet shall take to reach the final destination node. The cargo is the actual data meant to be transferred and, finally, EOP character is used to mark the end of the packet.

*SpaceWire Network*

SpW is a computer network and it makes use of links, nodes and routing switches for routing packets across the network. Figure 2 shows an example of SpW architecture. It presents different instruments and peripherals connected together through two SpW router. The different instruments on the network are connected each other with both the point-to-point link (such as the instrument 1 with the mass memory), and the connection link through the SpW routers, in order to communicate with all the remaining on-board instruments.

The SpW router connects together many nodes of a SpW network. Wormhole routing reduces both the required memory TX/RX buffers for SpW interfaces, and overall latency of the whole communication. When the SpW router receives a SpW packet, it immediately retrieves the output port to forward it. The SpW router implements both Path and Logical addressing schemes. Furthermore, the GAR is supported in order to guaranteed the fault tolerance to link failure [2][3][4].

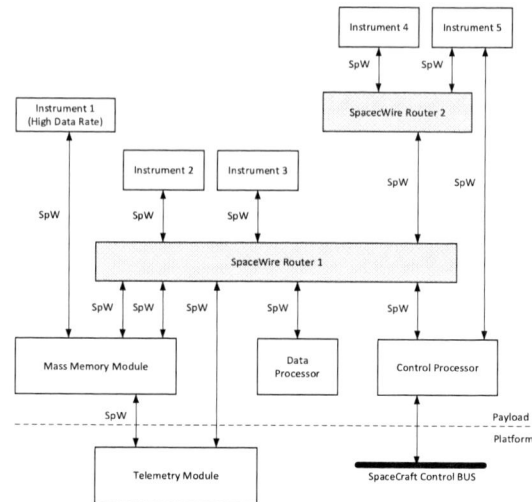

**Figure 2 - Example of SpaceWire architecture**

*State-Of-The-Art*

STAR-Dundee produced the SpW Router Mk2S [5], a routing device that provides a solution for exploring the SpW routing technology and help the development of a SpW equipment. The SpW router Mk2S accommodates eight SpW interfaces. Its main features are: time-code master, support for high-speed data transfer, various types of errors injection capabilities and support for low latency host-PC transmission of SpW packets. SpW can be used in two different operating modes:

- *interface mode* where the traffic received from the SpW links is automatically forwarded to host-PC with no routing required;
- *routing mode* where all the eight interfaces are capable to route SpW packets from one port to another one [5].

Despite SpaceWire devices in nowadays missions usually interacts through routers of a SpW network, currently no test equipment is available that includes SpaceWire Router functionalities and with routing-specific test and debugging features.

## III. SYSTEM DEFINITION

The SpW testing solution object of this paper, also referred to as "SpaceWire Router Analyser", comprises all the following features:

- Up to 28 SpW interfaces able to route packets between the SpW ports.

- Two SpW interfaces for point-to-point communication.

- Compliant with the SpW standard [1].

- Time-code Master.

*PRIME 2018, Prague, Czech Republic*                     *Session: Digital Circuits and Sub-Systems*

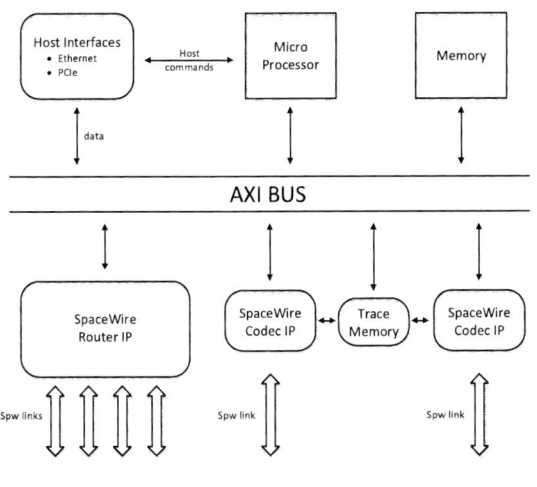

Figure 3 - SpaceWire Router Analyser implementation overview

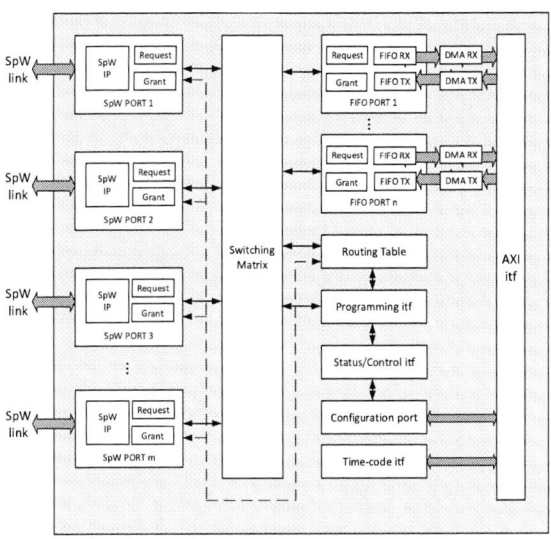

Figure 4 - Architecture of the SpaceWire Router IP

- Real-time Host-PC intercommunication through PCI Express (PCIe) interface.

- Unobtrusive monitoring of SpW links (Link Analyser mode).

- Electrical Ground Support Equipment (EGSE) capabilities for validation of SpW-based satellite sub-systems.

- Hardware accelerators for packet generation and consumption in order to saturate bandwidth of any SpW link and to allow stress testing of the SpW networks.

- Error injection capabilities.

- TX/RX trace memory to check the data streaming on the link.

## IV. IMPLEMENTATION

The proposed SpaceWire Router Analyser is implemented as a system based on Microprocessor. The use of Microprocessor guarantees the high-level flexibility of the system and the possibility to interface the SpaceWire Router Analyser to the host-PC. The Command Line Interface (CLI) allows to generate commands forwarded from the host interface to the Microprocessor. Furthermore, the core is entrusted with configuration and control of the SpW analyser system.

Figure 3 shows the high-level functional block diagram of the SpaceWire Router Analyser architecture. All the peripherals are connected to the same AXI bus. This configuration allows to achieve high level of flexibility and to move data at very high speed from one interface to another (either SpaceWire or Host-PC). AXI4 is the bus specification for high performance memory mapped systems and it allows achieving real-time data transfer inside the SpaceWire Router Analyser. Both Ethernet and PCIe host interfaces are available. The host-PC

interfaces enable the sending and receiving of high-speed data streams from/to host system to/from one, or multiple, SpW ports.

SpW Router and SpW CODEC IPs used in this work are provided from IngeniArs S.r.l. [6]. The IngeniArs Router IP core was extensively verified thanks to a complete and structured verification environment based on System Verilog/UVM.

Figure 4 shows the functional block diagram of the SpW Router. The VHDL model of the IP is parametric in terms of number of SpW interfaces (m) and FIFO ports (n).

The core of a SpW router is the switching matrix that is connected to m SpW interfaces, a configurable number of First-In-First-Out (FIFO) ports, one routing table, the programming interface module, the configuration port module, and the time-code interface block. The Router IP core was included into an AXI VHDL wrapper in order to enable the transmission/reception of data to/from the host-PC, through the Direct Memory Access (DMA) engine. The IngeniArs SpW CODEC IP core has been adopted already in successfully space projects, on ground segments and flight hardware. It is included into a VHDL AXI wrapper as depicted in Figure 5. The SpW CODEC is equipped with high-performance DMAs interfacing with the AXI interconnect. The hardware packet generator and consumer are implemented in order to avoid the interference with the other system peripherals. The SpW IP core is also able to inject on-demand errors every time the user set the related register from host-PC.

Finally, for each SpW interface a trigger can be programmed to receive the content of the associated trace memory. Triggers can occur on pre-defined SpW words and shows all words flown through the SpW interfaces within time window centred on the triggered word.

978-1-5386-5388-3/18 $31.00 © 2018 IEEE          179

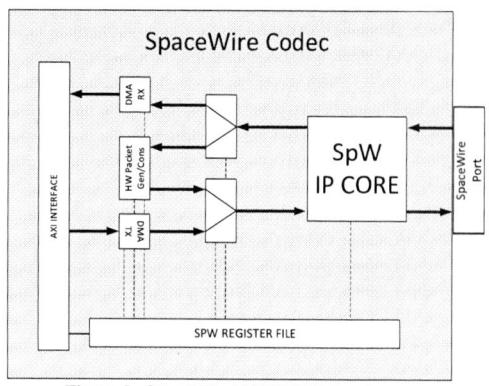

Figure 5 - SpaceWire CODEC IP structure

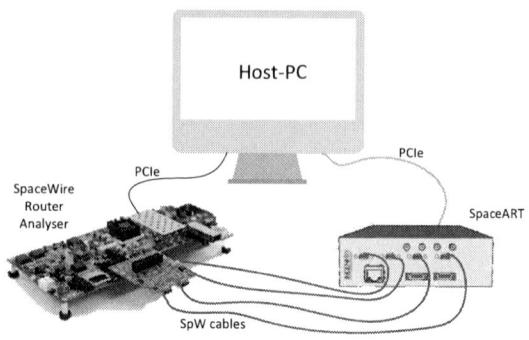

Figure 6 - SpaceWire Router Analyser demonstrator for testing

All the subsystems of the SpaceWire Router Analyser can be extensively configured at run time by the user, thanks to the possibility to access their internal register through a dedicated slave AXI interface.

## V. VALIDATION AND TEST

The SpaceWire Router Analyser has been tested and a prototype based on FPGA development board has been realized. The configuration of SpaceWire Routing Analyser used for the prototype is the following:

- two SpW interfaces related to the internal router;
- two FIFO ports;
- two SpW interfaces for the link-to-link communication that enable the capabilities of link monitoring and trace memory.

During the prototyping phase, the correct functionality of the SpaceWire Router Analyser was demonstrated through the interoperability with the SpaceWire Analyser Real-Time (SpaceART) equipment. Figure 6 shows an overview of the set-up used for the validation of the architecture. Both the SpaceWire Router Analyser and the SpaceART system are connected through PCIe interfaces to the host-PC. The four SpW interfaces of the SpaceWire Analyser Router are connected with the SpaceART equipment. The validation of the correct behaviour of all subsystems has been achieved through specific tests. Among them, the most remarkable are:

- Routing capabilities of the SpaceWire Router Analyser were intensively tested using the software file reader/writer capability of both SpaceART and SpaceWire Router Analyser.
- Hardware accelerator for packet generation/consumption were extensively tested for different packet sizes and transmission/reception bandwidths.
- Other EGSE capabilities such as: on demand error injection, link monitoring, trace memory and timecode master feature were all intensively tested.

## VI. CONCLUSION

An architecture of an innovative solution for test equipment for SpaceWire standard was presented in this paper. The described SpaceWire Router Analyser system includes in a single solution the capability to have SpW interface for both direct and indirect communication links. The test equipment with its embedded SpW router and other two SpW interfaces, can be productively used in a wide range of possible scenarios: it can be employed either as EGSE, or as a link analyser, or as emulator of a portion of SpW network, or simply as a SpW router.

To offer all the above-mentioned features, the SpaceWire Router Analyser embeds many powerful tools:

- a highly configurable SpW Router IP core;
- Integrated hardware packet generators/consumers;
- On-demand error injection capabilities.

Furthermore, the test equipment is able to interface itself with any host-PC in order to have the possibility to send data to each FIFO-port of the SpaceWire Router Analyser reading from file, and to log data received from the FIFO-ports into a file. Its wide set of features make the SpaceWire Router Analyser a powerful test equipment for the validation of any under-development SpaceWire-based device in many possible scenarios.

## REFERENCES

[1] European Cooperation for Space Standardization, "SpaceWire – Links, nodes, routers and networks," Issue 2.0, ESA Publications Division, Noordwyk (NL), July 2008, P/N ECSS E 50 12C.

[2] S. Saponara., L. Fanucci et al., "Radiation tolerant spacewire router for satellite on-board networking," IEEE Aerospace and Electronic Systems Magazine, vol 22, pp. 3-12, May 2007.

[3] S. M. Parkes and P. Armbruster, "SpaceWire: a spacecraft onboard network for real-time communications," in Real Time Conference, 14th IEEE-NPSS, pp. 6-10, Jun 2005.

[4] S. M. Parkes, C. McClements et al., "Spacewire router," in International Spacewire Seminar (ISWS 2003) vol. 4, Nov 2003.

[5] Star Dundee Ltd., https://www.star-dundee.com/

[6] IngeniArs S.r.l., https://www.ingeniars.com/

*PRIME 2018, Prague, Czech Republic*                    *Session: Radio Frequency Circuits and Systems II*

# A Novel Hybrid Polar-I/Q Modulation Method relaxing RF Phase Modulator Design Requirements

T. Buckel[*†], P. Preyler[*†], E. Hager[*†], T. Mayer[†], S. Tertinek[†], A. Springer[*] and R. Weigel[‡]

[*]University of Linz, Christian Doppler Laboratory for Digitally Assisted
RF Transceivers for Future Mobile Communications, Linz, Austria
[†]Intel Corporation, Linz, Austria
[‡]University of Erlangen-Nuremberg, Institute for Electronics Engineering, Erlangen, Germany
tobias.buckel@jku.at

*Abstract*—A novel modulation scheme enabling a seamless transition between digital-intensive quadrature and polar transmitter architectures by combining inphase / quadrature (I/Q) with phase modulation is presented. By incorporating phase modulation the digital-quadrature RF digital-to-analog converter (DAC) input codewords peak to average power ratio is reduced. Therefore, the quadrature RF-DAC can be operated in a lower back-off region resulting in higher average output power and drain efficiency. Simulation results for LTE uplink transmission including the novel modulation scheme in combination with a drain efficiency model for a shared-cell, switched-capacitor RF-DAC are shown. Compared to a digital-quadrature transmitter the average output power and efficiency can be increased, approaching the digital-polar limit while showing significantly lower tuning-range requirements on the phase modulation path.

*Index Terms*—Digital-polar transmitter, digital-quadrature transmitter, phase modulation, quadrature modulation, RF digital-to-analog converter (RF-DAC), RF digital power amplifier (RF-DPA), RF digital phase-locked loop (RF-DPLL).

## I. INTRODUCTION

Wireless transceivers for mobile handset applications are typically implemented in ultra-deep sub-micron CMOS technology. Therefore, digital-intensive transmitter (Tx) and receiver architectures have been developed to fully utilize the advantages of shrinking CMOS process structures. With the development of RF digital-to-analog converters (RF-DACs) and digital power amplifiers (PAs) directly operating at the channel frequency, digital-intensive polar and quadrature Tx architectures as shown in Fig. 1 became of focus [1], [2]. RF-DAC implementations with high dynamic range showing superior out-of-band spectral emission performance have been shown [3]. They enable the use of low complexity passive output matching networks based on integrated transformers avoiding the need for off-chip band-pass filters [4]. Whereas first implementations have been based on current-steering, latest implementations are based on switched-capacitor circuit topology [3], [5] further enhancing RF-DAC drain efficiency, linearity and scalability.

One major objective for the integration in mobile handsets is to achieve a sufficiently large Tx output power and high drain efficiency. This allows to get rid of a following internal or external PA stage or at least to relax its specifications. In this context it turns out that the digital-quadrature Tx

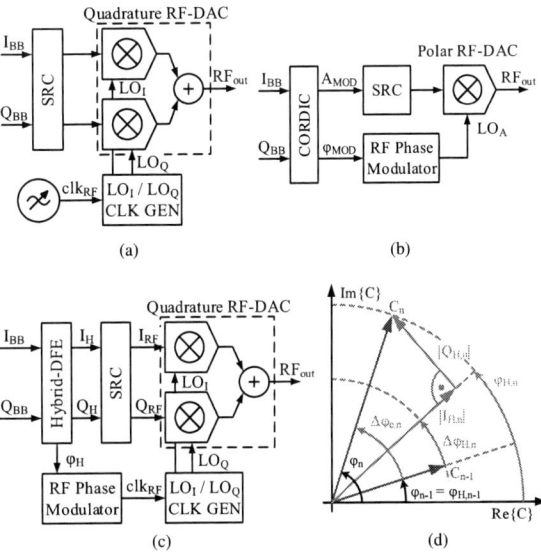

Fig. 1. Simplified digital-intensive Tx architectures for wireless handset applications. (a) digital-quadrature and (b) digital-polar Tx. (c) Proposed hybrid Tx combining I/Q and phase modulation. (d) Complex pointer diagram illustrating the proposed modulation methodology.

shows an inherent drawback since the RF signal is generated by orthogonal summing of up-converted inphase and quadrature (I/Q) components. As consequence, the RF-DAC has to operate in a higher back-off region due to the increased average and peak RF-DAC cell utilization. This results in a lower drain efficiency and average output power of the RF-DAC and therefore quadrature Tx. On the other hand, the digital-quadrature Tx usually shows a lower signal distortion since it does not suffer the bandwidth expansion due to the non-linear transformation from Cartesian to polar domain. This property becomes especially important with respect to the phase modulation path in combination with wide-band non-constant envelope modulation schemes [6]. For example, RF-DPLL phase modulators require a hole-punching of the complex modulation signal in order to avoid large excitation of the frequency modulation signal [7]. This results in a trade-

978-1-5386-5388-3/18 $31.00 © 2018 IEEE          181

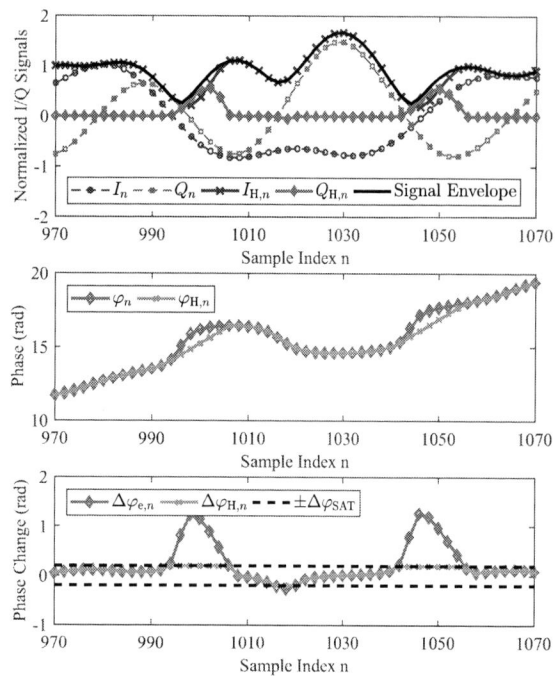

Fig. 2. Simulation results for processing a 20 MHz LTE uplink signal at $f_s = 307.2$ MHz and $\Delta\varphi_{\text{SAT}} = \pm 0.147$ rad corresponding to a frequency modulation saturating at 7.2 MHz.

off between increased signal distortion and relaxation of the phase modulator design requirements as well as a significantly more complex digital front-end implementation.

In this paper we show a hybrid modulation and RF direct up-conversion scheme combining I/Q and constrained phase modulation. The approach incorporates the superior RF-DAC drain efficiency and average output power of the digital polar Tx. By constraining the phase modulation while adjusting the I/Q components accordingly, tuning range requirements of the RF-DPLL based phase modulator can be relaxed significantly. The methodology is further explained in section II while the utilized shared-cell, switched-capacitor RF-DAC efficiency model for back-off operation is derived in section III. Simulation results characterizing the effect on RF-DAC drain efficiency, average output power and signal distortion are shown in section IV.

## II. MODULATION METHODOLOGY

A simplified block diagram of the digital-intensive Tx architecture implementing the proposed hybrid modulation method is shown in Fig. 1(c). The filtered and pre-processed complex baseband signal $C_n = I_n + jQ_n$ is split into hybrid inphase, qadrature $I_{\text{H},n}, Q_{\text{H},n}$ and constrained phase $\varphi_{\text{H},n}$ components. The constrained phase signal is applied to the phase modulator generating two phase modulated, orthogonal RF carriers $LO_\text{I}$ and $LO_\text{Q}$ for the quadrature RF-DAC. An

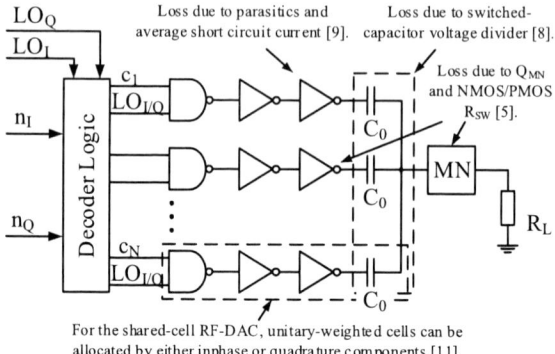

For the shared-cell RF-DAC, unitary-weighted cells can be allocated by either inphase or quadrature components [11].

Fig. 3. Exemplary switched-capacitor RF-DAC architecture with markup of the considered loss mechanisms.

efficient implementation of the phase modulator is given by means of a two-point modulated RF-DPLL with a frequency modulated digital-controlled oscillator (DCO) as shown in [10]. The hybrid DFE output signals $I_{\text{H},n}$ and $Q_{\text{H},n}$ are applied to a sample rate conversion (SRC) block. This is necessary to align the data to the phase modulated LOs. The SRC is also necessary to shift spectral replicas originating from the zero-order-hold operation of the RF-DAC to higher carrier offset frequencies in order to meet out-of-band spectral emission requirements.

The approach of combining a constrained phase with I/Q modulation is illustrated in Fig. 1(d) showing the transition between two complex baseband samples $C_{n-1}$ and $C_n$ of a complex modulation signal trajectory. The desired phase change $\Delta\varphi_{\text{e},n}$ between absolute phase $\varphi_n$ of $C_n$ and prior hybrid phase $\varphi_{\text{H},n-1}$ related to $C_{n-1}$ is given by

$$\Delta\varphi_{\text{e},n} = \varphi_n - \varphi_{\text{H},n-1} \tag{1}$$

and saturated to $\pm\Delta\varphi_{\text{SAT}}$. This results in the constrained phase change $\Delta\varphi_{\text{H},n}$ given by

$$\Delta\varphi_{\text{H},n} = \begin{cases} \pm\Delta\varphi_{\text{SAT}} & \text{if } \Delta\varphi_{\text{e},n} \gtrless \pm\Delta\varphi_{\text{SAT}} \\ \Delta\varphi_{\text{e},n} & \text{otherwise} \end{cases} \tag{2}$$

The constrained phase $\varphi_{\text{H},n}$ encodes the phase offset of the orthogonal hybrid I/Q components $I_{\text{H},n}, Q_{\text{H},n}$ forming the complex sample $C_n$. It turns out that $I_{\text{H},n}$ and $Q_{\text{H},n}$ can be calculated by a rotational transformation of the conventional I/Q components $I_n, Q_n$ by $\varphi_{\text{H},n}$, given by

$$\begin{pmatrix} I_{\text{H},n} \\ Q_{\text{H},n} \end{pmatrix} = \begin{pmatrix} \cos(-\varphi_{\text{H},n}) & -\sin(-\varphi_{\text{H},n}) \\ \sin(-\varphi_{\text{H},n}) & \cos(-\varphi_{\text{H},n}) \end{pmatrix} \begin{pmatrix} I_n \\ Q_n \end{pmatrix}. \tag{3}$$

The constrained phase $\varphi_{\text{H},n}$ results from the recursive equation

$$\varphi_{\text{H},n} = \Delta\varphi_{\text{H},n} + \varphi_{\text{H},n-1}. \tag{4}$$

From (2), the advantage of the proposed modulation method with respect to the phase modulation path becomes evident. The relation between phase and frequency is given by

$$f_{\text{MOD},n} = \frac{F_s}{2\pi}(\varphi_n - \varphi_{n-1}). \tag{5}$$

*PRIME 2018, Prague, Czech Republic*　　　　　*Session: Radio Frequency Circuits and Systems II*

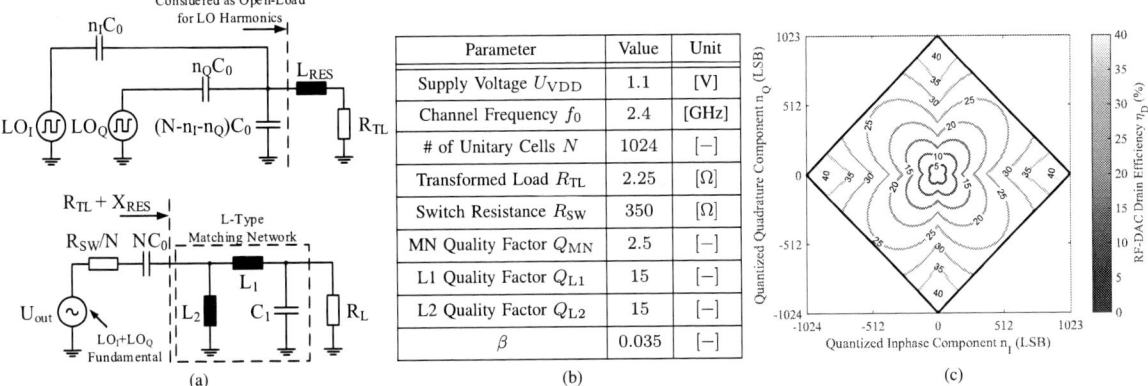

Fig. 4. (a) Equivalent circuit models for RF-DAC drain efficiency calculation with capacitive voltage divider related loss (top) and resistive as well as matching network insertion loss (bottom). (b) Exemplary RF-DAC parametrization (c) Contour plot showing the drain efficiency over quantized input codewords.

It can be seen that by saturating the phase change $\Delta\varphi_{\mathrm{H},n}$, a limitation of the frequency modulation $F_{\mathrm{MOD},n}$ and therefore required RF-DPLL tuning range results. According to (2) and (5) the frequency modulation is limited to

$$-F_{\mathrm{s}}\frac{\Delta\varphi_{\mathrm{SAT}}}{2\pi} \le f_{\mathrm{MOD},n} \le F_{\mathrm{s}}\frac{\Delta\varphi_{\mathrm{SAT}}}{2\pi} \qquad (6)$$

with $F_{\mathrm{s}}$ being the sampling frequency of the digital signal processing.

Exemplary signals to the approach illustrated in Fig. 1(d) are shown in Fig. 2. The simulation results are based on single-carrier frequency division multiple access utilized in LTE uplink transmission with 20 MHz channel bandwidth and 16-QAM symbol mapping at a sampling frequency of $F_{\mathrm{s}} = 307.2$ MHz. A saturation level of $\Delta\varphi_{\mathrm{SAT}} = \pm0.147$ rad has been chosen, corresponding to a frequency modulation saturating at $\pm7.2$ MHz. By comparing the conventional quadrature signal components $I_n$ and $Q_n$ with the hybrid signal components $I_{\mathrm{H},n}$ and $Q_{\mathrm{H},n}$ it can be seen that $I_{\mathrm{H},n}$ converges to the complex signal envelope for most of the time whereas $Q_{\mathrm{H},n}$ is kept small. This becomes evident by observing the constrained phase change $\Delta\varphi_{\mathrm{H},n}$ clipping at the specified saturation level. This results in the accumulated constrained phase $\varphi_{\mathrm{H},n}$ lagging the desired phase $\varphi_n$. Only for these instances, $Q_{\mathrm{H},n}$ is required to compensate for the increased phase error $\Delta\varphi_{e,n}$.

### III. RF-DAC Drain Efficiency Model

In the following, a code dependent capacitive RF-DAC drain efficiency model is derived in order to evaluate the effect of the hybrid modulation approach in a Tx context. In Section IV, this model is linked with the hybrid approach to derive RF-DAC drain efficiency, average output power and clipping related signal distortion for back-off operation. Fig. 3 depicts the considered switched-capacitor RF-DAC architecture based on shared, unitary-weighted power cells in combination with LO sign-switching and diamond-profile signal clipping as shown in [11]. Due to the cell sharing between the I/Q components

the total number of cells and therefore RF-DAC size can be reduced compared to implementations with dedicated I/Q cells. To calculate average output power and drain efficiency, several code dependent loss mechanisms as illustrated by the equivalent circuits shown in Fig. 4(a) have to be considered.

First, a dynamic loss in the CMOS inverter stage as described in [9] results from simultaneous switching of PMOS and NMOS transistors as well as due to parasitic input and output capacitances. The average power $P_{\mathrm{switch}}$ dissipated by the inverter calculates to

$$P_{\mathrm{switch}} = \beta\frac{(\frac{2}{\pi})^2 U_{\mathrm{VDD}}^2}{2R_{\mathrm{SW}}}(n_{\mathrm{I}} + n_{\mathrm{Q}}), \qquad (7)$$

with the technology parameter $\beta$ being independent from the CMOS inverter size [9].

Second, due to the switched capacitive voltage divider as shown in the top part of Fig. 4(a), charge is applied and removed at rising and falling $LO_{\mathrm{I}}$ and $LO_{\mathrm{Q}}$ clock edges. As shown in [8], the power loss $P_{\mathrm{cdiv}}$ for the quadrature RF-DAC calculates to

$$P_{\mathrm{cdiv}} = \big[n_{\mathrm{I}}(N - n_{\mathrm{I}}) + n_{\mathrm{Q}}(N - n_{\mathrm{Q}}) + 2n_{\mathrm{I}}n_{\mathrm{Q}}\big]\frac{U_{\mathrm{VDD}}^2 f_0}{N^2}. \quad (8)$$

Third, additional insertion loss results due to the effective inverter stage switch resistance given by $R_{\mathrm{SW}}/N$ and limited quality factors $Q_{\mathrm{MN}}, Q_{\mathrm{L1}}$ and $Q_{\mathrm{L2}}$ of the L-Type matching network [5]. Their contributions to the total RF-DAC drain efficiency are determined according to the equivalent circuit illustrated in the lower part of Fig. 4(a). The product of resistive loss and matching network related loss can be given as

$$\eta_{\mathrm{IL}} = \left(\frac{1 - Q_{\mathrm{MN}}/Q_{\mathrm{L1}}}{1 + Q_{\mathrm{MN}}/Q_{\mathrm{L2}}}\right)\left(\frac{R_{\mathrm{TL}}}{R_{\mathrm{TL}} + R_{\mathrm{SW}}/N}\right)^2. \quad (9)$$

Therefore, the code dependent RF-DAC output power $P_{\mathrm{out}}$ and drain efficiency $\eta_{\mathrm{D}}$ calculate to

$$P_{\mathrm{out}} = \eta_{\mathrm{IL}} \frac{(\frac{2}{\pi})^2 U_{\mathrm{VDD}}^2(n_{\mathrm{I}}^2 + 2n_{\mathrm{I}}n_{\mathrm{Q}} + n_{\mathrm{Q}}^2)}{2R_{\mathrm{TL}}N^2} \qquad (10)$$

978-1-5386-5388-3/18 $31.00 © 2018 IEEE

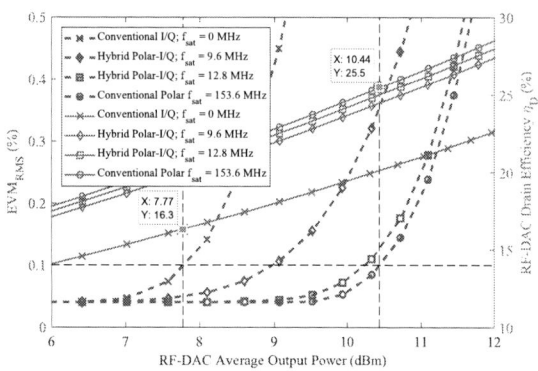

Fig. 5. $EVM_{RMS}$ and RF-DAC drain efficiency $\eta_D$ over average output power $P_{out}$ for different frequency saturation levels $f_{sat}$.

$$\eta_D = \frac{P_{out}}{P_{in}} = \frac{P_{out}}{P_{out}/\eta_{IL} + P_{switch} + P_{cdiv}} \quad . \quad (11)$$

## IV. SIMULATION RESULTS

For the RF-DAC parametrization given in Fig. 4(b) the code-dependent drain efficiency $\eta_D$ as shown in the heat-map plot of Fig. 4(c) results. By taking a closer look it becomes evident that a higher drain efficiency results for the hybrid approach since the quadrature component is kept small and the inphase is shifted towards a higher efficiency region. To emphasize the advantages of the hybrid modulation method compared to the conventional quadrature and polar approaches the resulting RF-DAC drain efficiency and in-band signal distortion over average output power $P_{out}$ is shown in Fig. 5 for different frequency saturation levels and 20 MHz LTE uplink transmission. The plot shows that given a certain distortion limit characterized by the signals error vector magnitude (EVM) of 0.1%, approximately 2.6 dB higher output power and 9% higher drain efficiency can be achieved compared to the conventional quadrature system. Furthermore, with the frequency saturation limit beeing less than 10% of the theoretical full scale range given by $\pm F_s/2$, only a fraction of the DCO fine-tuning range as would be required by a conventional polar system is necessary.

## V. CONCLUSION

A novel modulation method combining I/Q with constrained phase modulation and its impact on RF-DAC drain efficiency and average output power in back-off operation has been presented. Due to the constrained phase modulation a signifi-cant reduction of the RF-DPLL based phase modulator tuning range can be achieved compared to a digital-polar Tx. The hybrid I/Q signals are applied to a switched-capacitor RF-DAC with shared, unitary-weighted cells and diamond-shape signal clipping. Simulation results of the approach are presented for LTE uplink based on a minimalistic Tx model including a code-dependent RF-DAC drain efficiency model. The results show that compared to a conventional quadrature Tx, 2.6 dB additional average output power and 9% higher RF-DAC drain

efficiency can be achieved with less then 10% tuning range as would be required by a conventional polar system with RF-DPLL based two-point phase modulation.

## ACKNOWLEDGMENT

The authors wish to acknowledge Intel Corporation for supporting this work carried out at the Christian Doppler Laboratory for Digitally Assisted RF Transceivers for Future Mobile Communications. The financial support by the Aus-trian Federal Ministry of Science, Research and Economy and the National Foundation for Research, Technology and Development is gratefully acknowledged.

## REFERENCES

[1] R. B. Staszewski et al., "All-digital PLL and transmitter for mobile phones," *IEEE J. Solid-State Circuits*, vol. 40, no. 12, pp. 2469-2482, Dec. 2005.

[2] M. S. Alavi, A. Visweswaran, R. B. Staszewski, L. C. N. de Vreede, J. R. Long and A. Akhnoukh, "A 2-GHz digital I/Q modulator in 65-nm CMOS," in *Proc. IEEE Asian Solid-State Circuits Conf.*, 2011, pp. 277–280.

[3] M. Fulde et al., "A digital multimode polar transmitter supporting 40MHz LTE Carrier Aggregation in 28nm CMOS," in *Proc. IEEE Int. Solid-State Circuits Conf. Tech Dig.*, 2017, pp. 218-219.

[4] A. Passamani, D. Ponton, G. Knoblinger and A. Bevilacqua, "Analysis and design of a 1.1dB-IL third-order Matching Network for Switched-Capacitor PAs," in *Proc. IEEE Nordic Circuits and Systems Conf.*, 2015, pp. 1-4.

[5] Sang-Min Yoo, "A Switched-Capacitor RF Power Amplifier," *IEEE J. Solid-State Circuits*, vol. 46, no. 12, pp. 2977-2987, Dec. 2011.

[6] I. L. Syllaios, P. T. Balsara and R. B. Staszewski, "Recombination of Envelope and Phase Paths in Wideband Polar Transmitters," *IEEE Trans. Circuits Syst. I*, vol. 57, no. 8, pp. 1891-1904, Aug. 2010.

[7] J. Zhuang, K. Waheed and R. B. Staszewski, "A Technique to Reduce Phase/Frequency Modulation Bandwidth in a Polar RF Transmitter," *IEEE Trans. Circuits Syst. I*, vol. 57, no. 8, pp. 2196-2207, Aug. 2010.

[8] W. Yuan and J. S. Walling, "A Multiphase Switched Capacitor Power Amplifier," *IEEE J. Solid-State Circuits*, vol. 52, no. 5, pp. 1320-1330, May 2017.

[9] A. Passamani, D. Ponton, G. Knoblinger and A. Bevilacqua, "A linear model of efficiency for Switched-Capacitor RF Power-Amplifiers," in *Proc. 2014 10th Conference on Ph.D. Research in Microelectronics and Electronics (PRIME)*, 2014, pp. 1-4.

[10] T. Buckel et al., "A highly reconfigurable RF-DPLL phase modulator for polar transmitters in multi-band/multi-standard cellular RFICs," *2017 IEEE Radio Frequency Integrated Circuits Symposium (RFIC)*, Honolulu, HI, USA, 2017, pp. 104-107.

[11] Z. Deng et al., "9.5 A dual-band digital-WiFi 802.11a/b/g/n transmitter SoC with digital I/Q combining and diamond profile mapping for compact die area and improved efficiency in 40nm CMOS," in *Proc. IEEE Int. Solid-State Circuits Conf. Tech Dig.*, 2016, pp. 172-173.

*PRIME 2018, Prague, Czech Republic*

*Session: Radio Frequency Circuits and Systems II*

# A Sub-1V, 72 µW Stacked LNA-VCO for Wireless Sensor Network Applications

Ehsan Kargaran, Danilo Manstretta, and Rinaldo Castello

Microelectronics Laboratory, University of Pavia 27100, Italy

Ehsan.kargaran01@universitadipavia.it

*Abstract*— **Wireless sensor network (WSN) and Internet-of-Things (IoT) applications demand RF transceivers with extremely low power dissipation. An ultra-low power combined LNA-VCO is presented in this work with RF performance beyond the requirements of the intended application. The LNA current is lowered by factor of 12 in comparison with a standard common-gate LNA and it achieves impedance matching to 50 Ω thanks to current reuse combined with passive gain boosting. Stacking the VCO and the LNA voltage and power efficiency are improved even further. The power consumption of combined LNA-VCO is only 72 µW under a supply voltage of 0.9V. Simulation results in 40 nm CMOS technology at 2.4 GHz show NF and voltage gain of 3.1 dB and 23 dB, respectively, while the VCO operates at 4.8 GHz and has a phase noise of -105.8 dBc/Hz at 1 MHz offset frequency, corresponding to a FoM of 194.1 dBc/Hz.**

*Keywords*— *Ultra low power, gm-boosting, current reuse, LNA, VCO, WSN, IoT.*

## I. INTRODUCTION

The increasing demand for ubiquitous wireless and the rapid development of Internet-of-Things (IoT) and wireless sensor networks (WSNs) applications have generated a strong research interest toward ultra-low power (ULP) wireless transceivers. Some WSN applications including wireless medical telemetry, wireless body area network (WBAN) and Wearable-WSN (W-WSN) force strict limitations on the dissipated power of the RF transceiver to greatly extend battery lifetime or to harvesting energy from the environment, which requires low voltage designs. The adoption of a very low voltage supply adds even more challenges to the design of ULP RF transceivers. Preventing the use of stacked devices, it degrades amplifiers reverse isolation and maximum available gain, while the low overdrive voltage reduces the device cut-off frequency and hence the maximum achievable speed.

To minimize the power consumption of the receiver, numerous design techniques have been proposed. Recycling current in more than one block or reducing the supply voltage are frequently adopted. Fig.1 shows the possible configurations for sharing bias current between different blocks. In [1], shown in Fig.1(a), a Bluetooth Low-Energy (BLE) receiver front-end based on current reuse is reported. The same bias current is shared between the low-noise amplifier (LNA) and mixer, the voltage-controlled oscillator (VCO) and the baseband input stage. The combined structure consumes only 530 µA from a

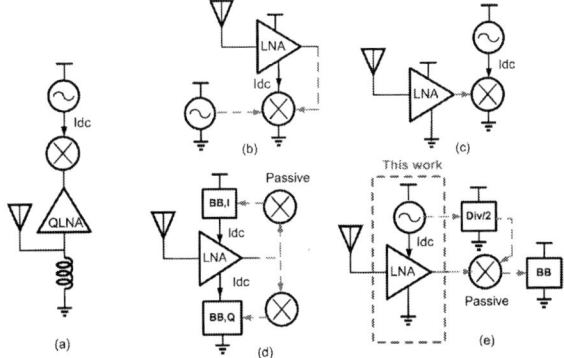

Fig.1. Current reuse receiver configurations: (a) QLMV [1], (b) Low noise converter [2] (c) Self-oscillating mixer [3], (d) Stacked LNA-BB [4], (e) Stacked LNA-VCO

0.8 V supply. When stacking several blocks performing different functions, the receiver operation is often plagued by isolation issues. Combining the LNA and the down-converter, the so-called low-noise converter is obtained, as shown in Fig.1(b) [2]. This configuration can achieve an adequate down-conversion gain but typically has a relatively poor noise figure (NF). Another current-reuse possibility is to combine oscillator and mixer [3] into a self-oscillating mixer (SOM), as shown in Fig.1(c). SOM typically have poor down-conversion gain and NF, which requires a low-noise pre-amplifier. In [4] a stacked LNA-baseband receiver for SoC coexistence is reported. As shown in Fig. 1. (d), the LNA and the first stage of both I and Q basebands are stacked under a 1.8 V supply and share the same current of 1.4 mA. Even though this approach leads to improved frontend linearity, it is not well-suited for true ULP operation.

Fig.1(e) shows the envisioned system configuration for ultra-low power dissipation. In a ULP front-end with moderate NF, LNA and VCO are typically the two blocks that dissipate the largest power. Stacking them such that they reuse the same current can significantly lower the overall power dissipation. The LNA is the first block of receiver, which not only has to preserve input power matching, but also needs to provide low noise amplification. It is quite difficult to drastically reduce its dissipation without compromising the performance. In WSN application, however, the receiver noise requirements are

978-1-5386-5388-3/18 $31.00 © 2018 IEEE

*Paper P43*

*PRIME 2018, Prague, Czech Republic*

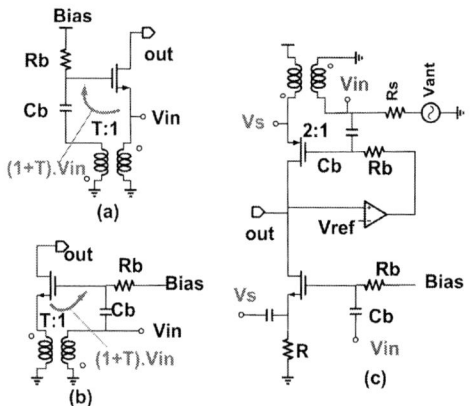

Fig.2. Passive $g_m$ boosting CG amplifiers (a) in [6], (b) in [7], (c) proposed current reuse passive $g_m$ boosting LNA

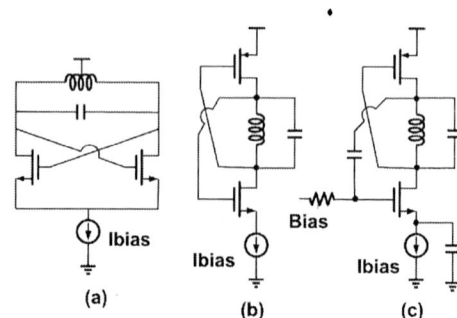

Fig.3. (a) Standard cross-coupled VCO, (b) current reuse VCO [9] (c) Proposed current reuse VCO

relaxed and the fundamental limitation on the LNA current dissipation comes from impedance matching [8]. This is due to the need to guarantee the proper input impedance matching and providing the reasonable gain to minimize the noise contribution of subsequent stages. On the other hand, considering the relaxed phase noise requirement of BT-LE (i.e. -102 dBc/Hz at 2.5 MHz offset frequency from carrier [5]), the main constrain on minimizing the VCO power dissipation comes from the Barkhausen criteria to start-up and sustain the oscillation. Therefore, merging and co-designing LNA-VCO would significantly lower the power consumption. This paper presents a stacked LNA-VCO achieving 3.1 dB NF at 2.4 GHz and a VCO FoM of -194.1 with only 80 μA.

## II. POWER EFFICIENT LNA DESIGN

Even though the Common Gate (CG) LNA is well-recognized for its intrinsic wideband nature, the impedance matching constraint on the device $g_m$ makes this topology power hungry. One of the effective methods to minimize power consumption in ULP LNAs is to boost the source impedance using a step-up impedance transformation. A transformer with a turns ratio 1:T can lower the power consumption by factor of $T^2$. In CG topologies, passive gain boosting, using a 1:T transformer across gate to source of the input device as shown in Fig.2(a), is another efficient approach to minimize the required device $g_m$ and to perform input power matching. This topology also improves the noise factor by 1+T [6]. An interesting approach was proposed in [7] to merge the above mentioned techniques, further lowering the required device $g_m$ to do input power matching, leading to very low power dissipation. As can be seen in Fig.2(b), the transformer primary is coupled to the LNA input and also the gate of input device, while the secondary is connected to the source of the same device. Considering an ideal transformer with k=1, source voltage can be increased by factor of T while the gate-source voltage is boosted by a factor of 1+T with respect to the input. Hence, the effective

transconductance, $Gm$, and input impedance of LNA are given by:

$$G_m = (1+T)g_m \tag{1}$$

$$Z_{in} = \frac{1}{g_m T(1+T)} \tag{2}$$

Considering a transformer with a turns ratio of 2 (T=2), to achieve a $Z_{in}$ of 50 Ω the device $g_m$ can be lowered to 1/(6$R_s$), which has two times better power efficiency compared with Fig.2(a) and is six time more power efficient in contrast to a simple a CG. Assuming lossless transformer, the noise factor of the LNA is:

$$F = 1 + \frac{T.\gamma}{1+T} \tag{3}$$

where $\gamma$ is the MOSFET noise parameter, for T=2, F=1+2/3$\gamma$. A current reuse scheme can be also applied to improve voltage efficiency, further lowering the power consumption. Stacking PMOS and NMOS devices, the signal is fed at both sources and gates via capacitors, doubling the effective $g_m$. In principle, this technique allows to halve the required bias current for the same input impedance and NF. Therefore, the required device $g_m$ for input impedance matching is lowered to 1/(12$R_s$). The proposed LNA, shown in Fig.2(c), burns only 80 μA and the noise factor of proposed LNA is:

$$F = 1 + \frac{\gamma T}{1+T} + \frac{T^2 R_s}{R_{loss}} + \frac{T^2 R_s}{R} \tag{4}$$

where the effect of transistor thermal noise is represented by the second term on the right-hand side, T is the transformer turns ratio (in this design T=2), losses of the transformer are modelled with a resistance $R_{loss}$ on the secondary, and R is the biasing resistor. In this design, the primary and secondary of transformer has Q of 9 and 14 respectively whilst occupying 0.065 mm$^2$[8].

## III. POWER EFFICIENT VCO DESIGN

As concluded in the previous section, the minimum required bias current to perform LNA input impedance matching while representing decent performance is only 80 μA. Let's us initially explore what is the minimum required bias current in VCO to sustain oscillation. In the conventional cross-coupled

978-1-5386-5388-3/18 $31.00 © 2018 IEEE       186

VCO shown in Fig.3(a), the cross-coupled pair provides a negative resistance, $-2/g_m$, to compensate for the LC tank losses. To sustain the oscillation, the minimum required device $g_m$ can be computed by,

$$g_m \geq \frac{2}{Q_L L \omega} \qquad (5)$$

where $Q_L$ is the quality factor of the inductor. Large inductance is preferable to minimize the required device $g_m$ and lower the power consumption. In general, the inductance cannot be chosen to be arbitrarily large since, for a given Q, the VCO phase noise will be eventually restricted by KT/C noise, where the C is the total tank capacitance. In fact, for a fixed oscillation frequency and constant Q, increasing the inductance leads to reduced tank capacitance and hence degraded VCO phase noise. However, in Bluetooth Low-Energy (BLE) standard, the phase noise requirement is very relaxed, -102 dBc/Hz at 2.5 MHz offset frequency from carrier operating at 2.4 GHz [5] (i.e. -96 dBc/Hz at twice the carrier frequency). Hence, the inductance can be chosen to be relatively large, in the favor of minimizing the power consumption. Targeting an oscillation frequency of 4.8 GHz, the 4.7 nH spiral inductor available from the design-kit (5 turns, winding W=4μm, S= 4μm) shows Q of 16, while the self-resonance frequency (SRF) is around 11 GHz and it occupies 0.026 mm². Thus, the minimum required device $g_m$ to sustain the oscillation is 0.88 mS and considering $g_m/I_D$ of 20 (i.e. biasing transistors in weak inversion), results in the minimum tail bias current of 88 μA. However, to ensure robust startup, 2 to 3 times higher bias current is required in practice. To improve the power efficiency, the stacked VCO topology originally proposed in [9], shown in Fig.3(b), can be adopted. Even though, for a given tank inductor, the required device $g_m$ is the same as in a standard cross-coupled topology, the total current is halved thanks to reusing the same current for both NMOS and PMOS devices. Targeting low voltage design and assuming equal threshold voltage ($V_{th}$) and overdrive ($V_{od}$) for both NMOS and PMOS, the minimum required supply voltage is $2V_{th}+3V_{od}$. Ac-coupling capacitors can be utilized to reduce the minimum supply voltage by one $V_{th}$. The schematic of the proposed VCO is shown in Fig.3(c). It is interesting to notice that, in contrast to the standard cross-coupled VCO, where in each half-cycle one device is on and the other one is off, in a CMOS current reuse VCO, both devices turn on and off in the same half-cycle.

## IV. STACKED LNA-VCO STRUCTURE AND SIMULATION RESULTS

The proposed stacked LNA-VCO is shown in Fig.4 and, as can be seen in Fig.3(c), the tail bias current of the current-reuse VCO is replaced by the LNA, shown in Fig.2(c). Therefore, the required supply voltage of the stacked structure is lower compared to the sum of the two separate supply voltages. The bias current for the LNA (80 μA) is almost 2 times higher than the minimum bias current of the VCO (44 μA), resulting in robust oscillation stat-up. Furthermore, a large capacitor, $C_F$, is placed between LNA and the VCO to ensure large isolation between the two blocks and prevent cross-talk effects. In fact, this large capacitor prevents LO re-radiation and acts as a signal

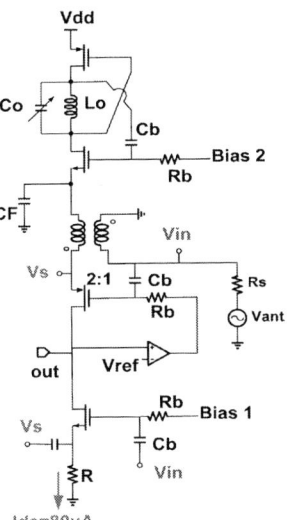

Fig.4. Proposed Stacked ULP LNA-VCO

ground for AC signals and noise coming from the LNA at twice the carrier frequency, avoiding their conversion into phase noise. On the other hand, the low-frequency LNA noise is converted into amplitude noise by the VCO and does not degrade its phase noise.

The proposed stacked LNA-VCO was designed in TSMC 40nm CMOS technology using Low-Vth devices. The dissipated power is just as low as 72 μW from 0.9 V supply voltage. Since the voltage at the common node of LNA and VCO is 0.47V, LNA dissipates only 38 μW while the power consumption of VCO is just 34 μW. Fig.5 shows the performance of the LNA while operating at 2.4 GHz. Return loss is good over a wide bandwidth: S11< -15 dB in the 2.3-2.7 GHz band. Down-conversion gain is 23 dB and NF is 3.1 dB. In the complete receiver the LNA will be loaded with low impedance associated with passive mixer followed by trans-impedance amplifier (TIA). Hence, the LNA linearity is simulated with low impedance load. Placing two tones at 2.4 GHz and 2.45 GHz, IIP3 of LNA is -9 dBm, and the LNA 1-dB compression point is - 15 dBm at 2.4 GHz. Moreover, the VCO operates at 4.8 GHz and at the offset frequency of 1 MHz, it has phase noise of -105.8 dBc/Hz and achieves FoM of 194.1. To further improve the phase noise, it is conceivable to employ the second harmonic filtering concept by placing transformer at the source nodes to resonate at twice the oscillation frequency as demonstrated in [10]. The differential peak swing of VCO is 190 mV. An ultra-low power inverter amplifier can be used after the VCO to enhance the swing for the following stages.

The overall performance of LNA and VCO are summarized and compared with that of state-of-the-art in Table I. In contrast with stacked LNA-VCO topology in [11], our proposed structure consumes 40% less power, has half the added noise (F-1), and 1.5 dB higher FoM for VCO. The proposed LNA is the least power hungry and exhibit better NF and higher gain in comparison with other presented work. Thanks to the extremely

978-1-5386-5388-3/18 $31.00 © 2018 IEEE

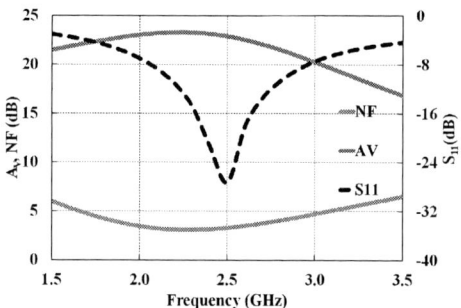

Fig.5. Simulation of LNA performance

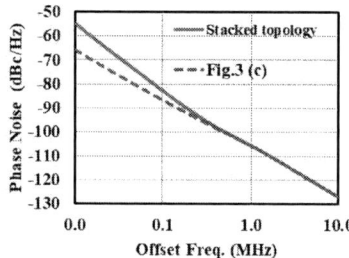

Fig.6. Simulated Phase noise of VCO, with 4.8 GHz carrier frequent

TABLE I: PERFORMANCE SUMMARY AND COMPARISON WITH STATE OF THE ART

| LNAs | Tech. (nm) | fo (GHz) | NF (dB) | Gain (dB) | IIP3 (dBm) | Pdc (µW) |
|---|---|---|---|---|---|---|
| [11] | 130 | 2.4 | 5.2 | 17.7 | NA | 84 |
| [12]* | 130 | 1 | 3.9 | 16.1 | -11.2 | 100 |
| [13]* | 130 | 2.4 | 5.3 | 13.1 | -12.2 | 60 |
| [16] | 90 | 0.-1.6 | 5.5 | 10.5 | -4.5 | 425 |
| **This work** | **40** | **2.4** | **3.1** | **23** | **-9** | **38** |

| VCOs | Tech. (nm) | fo (GHz) | PN (dBc/Hz) | Δf (MHz) | Pdc (µW) | FoM (dB) |
|---|---|---|---|---|---|---|
| [11] | 130 | 2.45 | -111 | 1 | 40.4 | 192.6 |
| [14]* | 65 | 2.53 | -126.1 | 3 | 280 | 190.2 |
| [15]* | 180 | 4.5 | -104 | 1 | 114 | 187 |
| [17] | 40 | 4.8 | -104 | 1 | 52 | 190.5 |
| **This work** | **40** | **4.8** | **-105.8** | **1** | **34** | **194.1** |

\* Measured results

low VCO power, our design achieves decent performance and the highest FoM.

## V. CONCLUSION

A design solution to combine LNA and VCO and obtain high performance and ultra-low overall power consumption is presented in this paper. Taking advantage of a combination of transformer-based passive $g_m$ boosting and current-reuse techniques results in a drastically reduced LNA bias current of only 80 µA. Moreover, the same bias current was recycled in the VCO to create a highly efficient current sharing configuration. The total power consumption is as low as 72 µW while achieving 3.1dB NF for LNA at 2.4 GHz, and FoM of 194.1 for the VCO. The proposed structure is well appropriate for WSN and IoT applications to greatly extend battery lifetime.

## REFERENCES

[1] A. Selvakumar, et al, " Sub-mW Current Re-Use Receiver Front-End for Wireless Sensor Network", *IEEE JSSC*, vol. 50, no. 12, Dec 2015.

[2] M. A. Abdelghany, R. K. Pokharel, H. Kanaya, and K. Yoshida, "Low-voltage low-power combined LNA-single gate mixer for 5GHz wireless systems," IEEE Radio Frequency Integrated Circuits Symposium, pp. 199-202, Baltimore, USA, June 2011.

[3] T.-P. Wang, C.-C. Chang, R.-C. Liu, et al., "A low-power oscillator mixer in 0.18-µm CMOS technology," IEEE Transactions on Microwave Theory and Techniques, vol. 54, no. 1, pp. 88-95, Jan. 2006.

[4] M. Ramella et al., "A SAW-Less 2.4GHz Receiver Front-End with 2.4mA Battery Current for SoC Coexistence", IEEE Journal of Solid-State Circuits, vol. 52, issue 9, pp. 2292-2305, September 2017.

[5] "Bluetooth Specification Version 4.1," Bluetooth 2013 [Online]. Available:http://www.bluetooth.com

[6] X. Li, S.Shekar, D.J. Allstot, "Gm boosted Common Gate LNA and differential Colpitts VCO/QVCO in 0.18um CMOS," ," IEEE Journal of Solid-State Circuits, vol. 40, , pp. 2609–2618, Dec. 2005.

[7] E. Kargaran, et al," A 30µW, 3.3dB NF CMOS LNA for Wearable WSN Application," in proceeding of IEEE International Symposium on Circuits and Systems (ISCAS), USA, 2017.

[8] E. Kargaran, et al.," Design and Analysis of 2.4 GHz 30µW CMOS LNAs for Wearable WSN Applications," IEEE Transactions on Circuits and Systems I: Regular Papers, vol.65, no.3, 2018.

[9] S. Yun, S. Shin, H. Choi, S. Lee, "A 1mW current-reuse CMOS differential LC-VCO with low phase noise," IEEE ISSCC, 2005.

[10] M. Garampazzi, P. Mendes, N. Codega, D. Manstretta and R. Castello, "A 195.6dBc/Hz peak FoM P-N class-B oscillator with transformer-based tail filtering," 40th European Solid State Circuits Conference (ESSCIRC), Venice Lido, 2014, pp. 331-334.

[11] T.Taris, et al., "A Low-Power 2.4-GHz Combined LNA-VCO Structure in 0.13-µm CMOS," in IEEE NEWCAS 2013.

[12] A. Shameli, P. Heydari," A Novel Ultra-Low Power (ULP) Low Noise Amplifier using Differential Inductor Feedback," IEEE ESSCIRC 2006.

[13] T. Taris, J.B. Begueret, Y. Deval," A 60µW LNA for 2.4 GHz Wireless Sensors Network Applications," IEEE RFIC symposium 2011.

[14] M. Taghivand, et al., "A Low Voltage Sub 300µW 2.5GHz Current Reuse VCO," in IEEE A-SSCC , 2012.

[15] K. Okada et al., "A 0.114-mW Dual-Conduction Class-C CMOS VCO with 0.2-V Power Supply", in proc. IEEE VLSI 2009.

[16] K. Allidina and M. El-Gamal, "A 1 V CMOS LNA for Low Power Ultra-Wideband Systems," in proceeding of IEEE International Conference on Electronics, Circuits and Systems (*ICECS*), Aug. 2008, pp. 165–168.

[17] E.Kargaran, D.Manstretta, and R.Castello, "A Sub-1V, 220 µW Receiver Frontend for Wearable Wireless Sensor Network Applications", in proceeding of IEEE International Symposium on Circuits and Systems (ISCAS), in Florence, Italy , 2018.

*PRIME 2018, Prague, Czech Republic*　　　　　　　*Session: Radio Frequency Circuits and Systems II*

# Low Power Locking Detector for Frequency Calibration of Multi-Frequency Injection Locked Oscillators

Abdessamad Boulmirat[1], Clément Jany[1]*, Alexandre Siligaris[1], José Luis Gonzalez Jimenez[1]

[1]Univ. Grenoble Alpes, CEA, LETI, *Stanford University

Grenoble 38000, France

abdessamad.boulmirat@cea.fr

*Abstract*—This paper presents a locking detector used as a first stage in the frequency calibration techniques for Multi-Frequency Injection Locked Oscillators (MFILO). It provides a high detection level and relaxes design constraints on the calibration circuit that follows. This technique will allow lower levels of power consumption with smaller occupied area for the frequency calibration circuits.

*Keywords*— *Locking detection, Frequency Calibration, Frequency Synthesizer, Injection Locked Oscillator (ILO), Frequency Multiplication, CMOS, Injection Locking, Phase Noise, mmW.*

## I. INTRODUCTION

With the unprecedented evolving of wireless communications, the amount of data rate needed for information exchange keeps on increasing. Many wireless standards are under discussion to satisfy data throughput requirements for the next generation of communication systems. IEEE 802.ay for instance, is targeting more than 100 Gb/s data rate using 60GHz band [1]. Fig. 1 shows some reported works on ultra–high speed transceivers enabling multi Gb/s wireless links during the last decade. The observed data traffic is 100 times higher every decade [8][9].

To meet all these requirements, the millimeter wave (mmW) band is very well placed with its 9 GHz unlicensed band between 57 and 66 GHz (In Europe). Using high order modulation (such as 64 Quadrature Amplitude Modulation: 64QAM) with channel bonding in these frequencies are expected to boost the data rate to more than 40 Gb/s. Four-channel bonding transceivers reported in [2] and [3] achieve 28.16 and 42.24 in 16QAM and 64QAM respectively. Another way to increase throughput is to perform multi band independent communications. This requires the use of multi-frequency synthesizer which generates several frequencies simultaneously. In both techniques, the front-end architecture is limited by several circuit impairments, especially phase noise of the Local Oscillator (LO).

To achieve the above results, a high quality millimeter wave frequency synthesizer is essential in the front-end architecture. A common solution is to use a Phase Locked Loop (PLL) in

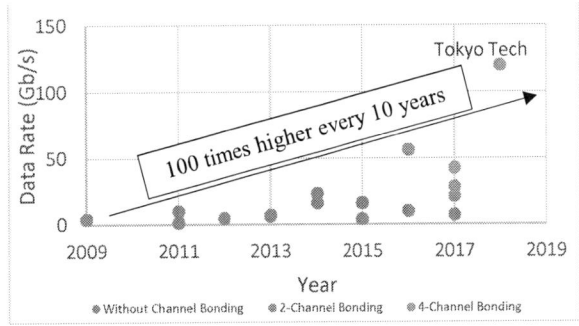

Fig. 1. Data traffic evolution of mmW CMOS transceivers.

desired frequency (more than 30GHz). However, when increasing the frequency, it is difficult to design a PLL with good spectral performances (more particularly phase noise)[5].

The frequency multiplication based on injection locking technique and using a low frequency reference PLL is widely used since it provides better spectral performances compared to a simple PLL [4]-[6]. The multiplication occurs using an Injection Locked Oscillator (ILO) acting as a high quality factor filter which selects the N$^{th}$ harmonic of the frequency reference while filtering the rest. Better phase noise performance is obtained using a low phase noise reference PLL with a high multiplication factor N [5]. Nevertheless, the ILO locking depends on the injection power of the desired harmonic, i.e. bigger injection power results in wider locking range [7], [8]. Process Voltage/Temperature (PVT) also deviates the free running frequency of the Voltage Controlled Oscillator (VCO), thus disabling the locking of the ILO on the desired harmonic. In both cases, the injection locking is not stable. As a consequence, a frequency calibration method is required to ensure the locking of this type of oscillators.

This paper presents the first stage (injection locking detector) of a proposed calibration architecture for Multi-Frequency Injection Locked Oscillators (MFILO). In section II, the state of the art is discussed before proposing theoretical analysis. Section III features the circuit schematic and simulations results of the proposed locking detector. Finally, a conclusion is drawn in last section.

978-1-5386-5388-3/18 $31.00 © 2018 IEEE

## II. SYSTEM ANALYSIS

### A. State of the art of frequency calibration methods

Many locking detection methods have been reported recently [10]-[12], [14]. Some conventional architectures for locking detectors are shown in Fig. 2. Three types of architectures can be differentiated: 1) envelope detector based architecture (fig. 2.(c), [11]), 2) replica-VCO based (Fig. 2.(a), [14]) and 3) phase-deviation detector based architectures (Fig. 2.(b) and 2(d), [10][12]). However, these methods present many practical problems such as high consumption level in [12] due to power hungry frequency dividers at high frequencies (Fig. 2.d). The detection level reported in [10], based on phase deviation detection between I/Q of the Quadrature ILO, is very low which increases post-detector circuit requirements (the use of high gain amplifiers) and hence increasing the power consumption and the complexity of the used circuit. Moreover, due to the low detection level of injection locking, the above reported frequency tracking techniques present destabilization risks of the locked state (confusion between two states). Besides, these three conventional techniques are not suitable for MFILOs.

### B. System analysis of the proposed detector

The MFILO conventional architecture is given in Fig. 3. This type of oscillator, proposed in [5], generates the output signal from a fixed sinusoidal reference $f_{ref}$ in three steps as shown in Fig. 3. First, the reference signal in the form of a square signal, i.e. $f_{ref}$-spaced multi-harmonic spectrum (purple signal in Fig. 3) periodically activates a mmW oscillator, operating near the desired output frequency $f_{out}$, resulting in a pulsed oscillations named PROT (Periodically Repeated Oscillation Train, blue signal in fig. 3). This last is injected into an Injection Locked Oscillator, with a free-running frequency close to desired $f_{out}$, performing a harmonic selection (occurs under certain conditions [6], [7]) of the N$^{th}$ harmonic of the frequency reference, i.e. $f_{out} = N * f_{ref}$. The use of the pulsed oscillator facilitate the injection locking on a high order harmonic $N * f_{ref}$ (N is around 30) by maximizing the power around the desired frequencies. This creates a high order programmable frequency multiplier [5].

The proposed injection locking detector is inspired from the above explanations. Indeed, the spectral form of the pulsed signal (four harmonic frequencies in a cardinal shape, blue signal fig. 3) is used as a reference to perform the detection. Fig. 4 shows the output power spectrum of the PROT and ILO signals in both oscillator states (injection locked and non-injection locked states). Hence, locking detection is realized by mixing the PROT signal with the ILO output signal. Fig. 5 shows the frequency calibration structure with the proposed locking detector.

From the signal patterns given in Fig. 4, two assumptions can be made to simplify analysis. First, sideband phase noise of the PROT harmonics is neglected, as a first approximation, compared to phase noise at the MFILO output. Also, only the harmonics N-2, N-1, N and N+1 are considered in the PROT signal. Hence, the two signals can be modeled as follow:

- ILO output signal:

---
This work is supported by CEA-LETI

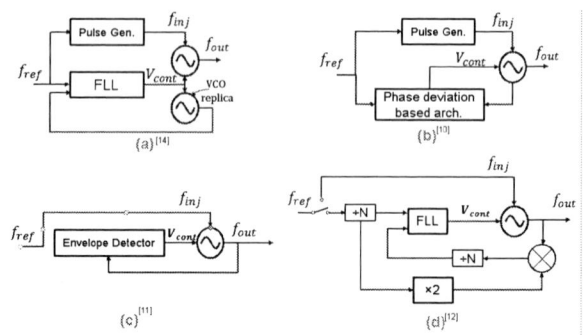

Fig. 2. ILO with different locking detection methods. a) replica-VCO method. b), d) phase deviation methods. c) envelope detector based architecture.

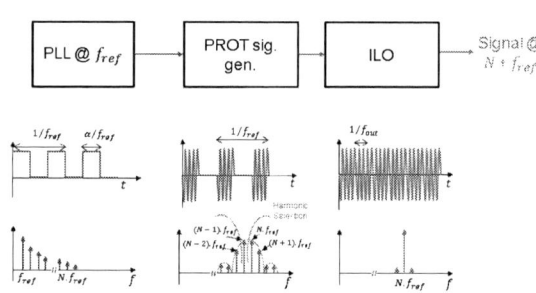

Fig. 3. Multi-Frequency Injection Locked Oscillator (MFILO) conventional architecture reported in [5].

Considering the above assumptions,

$$V_{ILO}(t) = (V_0 + \varepsilon(t))\,cos(\omega_{ILO}t + \varphi(t)) \qquad (1)$$

$V_0$ and $\omega_{ILO}$ are the amplitude and the instantaneous frequency of the ILO output signal respectively. $\varepsilon(t)$ and $\varphi(t)$ represents its amplitude and phase fluctuations respectively.

- PROT signal:

$$V_{PROT}(t) = \sum_{k=N-2}^{N+1} V_k\, cos(\omega_k t) \qquad (2)$$

$V_k$ and $\omega_k$ are, respectively, the amplitude and frequency of the k$^{th}$ harmonic of the pulsed signal (PROT signal, $\omega_k = k.\omega_{ref}$) where k = N-2, N-1, N, N+1.

The multiplication result of the two signals is given in (3).

$$V_{Det}(t) = \frac{1}{2}K_{MIX}\sum_{k=N-2}^{N+1} V_k\big(V_0 + \varepsilon(t)\big).cos(|\omega_{ILO} - \omega_k|t + \varphi(t)) \qquad (3)$$

$K_{MIX}$ represents the conversion gain of the mixer. Note that the resulted high frequency harmonics are filtered by a baseband filter with a cut-off frequency of $f_c = 2\ GHz$ (conf. Fig. 5). Thus, two states can be differentiated.

*PRIME 2018, Prague, Czech Republic*  *Session: Radio Frequency Circuits and Systems II*

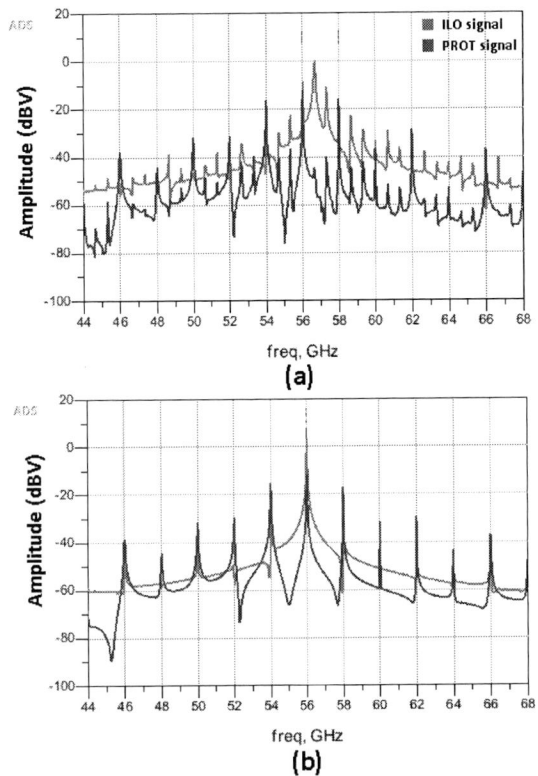

**(a)**

**(b)**

Fig. 4. Simulated output power spectrum of the MFILO output signal. Two different output states: (a) free running (or not lockced state) and , (b) locked.

- Injection locked state:

In this case, the ILO output frequency is equal to one of the selected harmonics: $\omega_{ILO} = k_0 . \omega_{ref} (k_0 =$ N-2, N-1, N or N+1). Moreover, for k ($k \neq k_0$): $|\omega_{ILO} - \omega_k| \geq 2\pi f_c$ (conf. Fig 4.b). Hence, the rest of the harmonics in (3) are highly attenuated by the baseband filter (cut-off frequency $f_c = 2 \ GHz$). Thus,

$$V_{Det}(t) \approx 0 \qquad (4)$$

In practice, the detector output signal in this case is equal to ILO sideband phase noise mixed with the one of PROT signal (in this case the detector acts as a phase noise measurement device [13]).

- Free running state:

In this case the ILO frequency is outside the locking range. This means: $\omega_{ILO} \neq k_0 . \omega_{ref}$ and $|\omega_{ILO} - \omega_{k_0;k_0+1}| < 2\pi f_c$. Hence, more than one Intermediate Frequency (IF) harmonic is amplified by the baseband active filter (actually when $k_0$ =N-1 or N, two IF harmonics appear at the baseband) and the resulted signal is given in (5).

$$V_{Det}(t) = \frac{1}{2} G_{BBF} K_{MIX} V_{k_0} (V_0 + \varepsilon(t)) . cos(|\omega_{ILO} - \omega_{k_0}|t + \varphi(t)) \qquad (5)$$

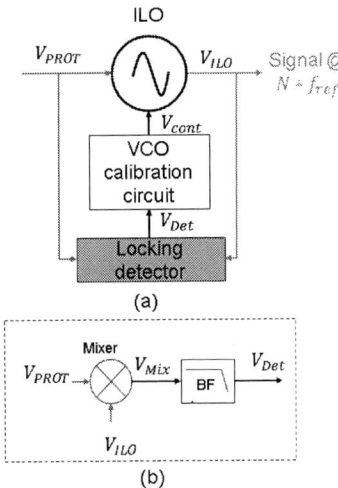

Fig. 5. Frequency calibration architecture (a) with the proposed locking detector (b).

$G_{BBF}$ is the baseband filter gain.
In this case, many spikes (or IF harmonics) are detected in the baseband at the output of the locking detector. Note that, in practice, in addition to these spikes, sideband phase noise of the PROT signal harmonics is also added. But, using the assumptions mentioned earlier, this sideband phase noise is neglected in both states of the ILO output signal. As a consequence, the detection level of the locking detector is determined by the ratio between $V_{Det}$ in (4) and (5) determines the detection level.

### III. CIRCUIT ARCHITECTURE

The detector architecture contains a mixer and an active baseband filter (conf. Fig. 5). The Gilbert Mixer (Fig. 6.(a)) is utilized to bring back the frequency difference to the baseband. The result is then filtered by an active baseband filter with a cut-off frequency of 2 $GHz$. The active filter deletes high frequency harmonics resulted from multiplication by the mixer. It also amplifies the mixer output signal to compensate conversion losses introduced by this latter.

The proposed locking detector circuit level is designed using 28nm CMOS simulation library. The simulation results are given in Fig. 6.(b). Detection result in Fig. 6.(b) is the output power spectrum of the proposed locking detector applied during both two states of the MFILO shown in Fig. 4.(b) and Fig. 4.(a). In the locked state, the output power of the locking detector is small and flat (red curve in Fig. 6.(b)). However, in the free running case (or not locked case of the ILO), many spikes are detected which were expected by the analysis in (5). Fig. 6.(b) also shows the comparison between the output power spectrum of the locking detector in the two ILO states.

The difference between the two injection states is 33 dB (Fig. 6.(c)) which demonstrates the high detection level of the proposed architecture. As a result, the probability of destabilizing the locked state is very low compared with the previous reported work. Furthermore, the design requirements of

978-1-5386-5388-3/18 $31.00 © 2018 IEEE          191

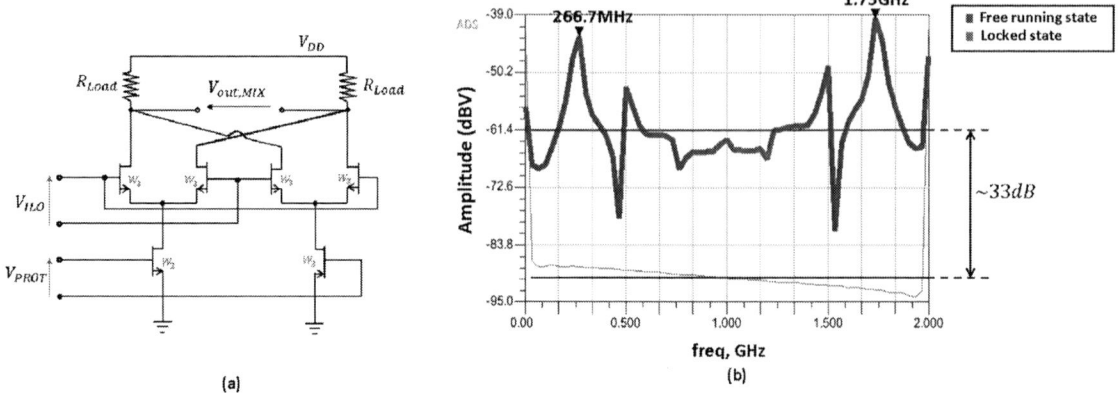

Fig. 6. Circuit schematic (a) and output power spectrum resulted from the simulation of the proposed locking detector. In (b), locked state (in red) and not locked state (in blue) of the MFILO are compared.

frequency calibration post-blocks are reduced which will further decrease power consumption and the occupied area.

## IV. CONCLUSION

This paper presents the first stage of a proposed frequency calibration architecture for Multi-Frequency Injection Locked Oscillators (MFILO). This block detects the frequency drifts of the VCO free running frequency with respect to the output frequency of the PROT signal. The proposed locking detector is suitable for multi-frequency ILOs. It represents a promising solution for frequency calibration techniques since it provides a high level of detection with optimized circuit architecture.

## ACKNOWLEDGMENT

This work is supported by the CEA-LETI, Grenoble, France.

## REFERENCES

[1] Y. Ghasempour, C. R. C. M. da Silva, C. Cordeiro, and E. W. Knightly, "IEEE 802.11ay: Next-Generation 60 GHz Communication for 100 Gb/s Wi-Fi," *IEEE Communications Magazine,* vol. 55, no. 12, pp. 186–192, Dec. 2017.

[2] J. Pang et al., "A 128-QAM 60GHz CMOS transceiver for IEEE802.11ay with calibration of LO feedthrough and I/Q imbalance," in 2017 *IEEE International Solid-State Circuits Conference (ISSCC),* 2017, pp. 424–425.

[3] R. Wu et al., "64-QAM 60-GHz CMOS Transceivers for IEEE 802.11ad/ay," *IEEE Journal of Solid-State Circuits,* vol. 52, no. 11, pp. 2871–2891, Nov. 2017.

[4] D. Shin and K. J. Koh, "An Injection Frequency-Locked Loop– Autonomous Injection Frequency Tracking Loop With Phase Noise Self-Calibration for Power-Efficient mm-Wave Signal Sources," *IEEE Journal of Solid-State Circuits,* vol. PP, no. 99, pp. 1–14, 2018.

[5] C. Jany, A. Siligaris, J. L. Gonzalez-Jimenez, P. Vincent, and P. Ferrari, "A Programmable Frequency Multiplier-by-29 Architecture for Millimeter Wave Applications," *IEEE Journal of Solid-State Circuits,* vol. 50, no. 7, pp. 1669–1679, Jul. 2015.

[6] B. Razavi, "A study of injection locking and pulling in oscillators," *IEEE Journal of Solid-State Circuits,* vol. 39, no. 9, pp. 1415–1424, Sep. 2004.

[7] R. Adler, "A Study of Locking Phenomena in Oscillators," *Proceedings of the IRE,* vol. 34, no. 6, pp. 351–357, Jun. 1946.

[8] (In press) K. K. Tokgoz et al., "A 120GHz 16QAM CMOS Millimeter-Wave Wireless Transceiver," *IEEE International Solid-State Circuits Conference (ISSCC),* 2018.

[9] A. Siligaris et al., "A 65nm CMOS fully integrated transceiver module for 60GHz wireless HD applications," in 2011 *IEEE International Solid-State Circuits Conference,* 2011, pp. 162–164.

[10] S. Yoo, S. Choi, J. Kim, H. Yoon, Y. Lee, and J. Choi, "A Low-Integrated-Phase-Noise 27-30-GHz Injection-Locked Frequency Multiplier With an Ultra-Low-Power Frequency-Tracking Loop for mm-Wave-Band 5G Transceivers," *IEEE Journal of Solid-State Circuits,* no. 99, pp. 1–14, 2017.

[11] D. Shin, S. Raman, and K. J. Koh, "2.8 A mixed-mode injection frequency-locked loop for self-calibration of injection locking range and phase noise in 0.13μm CMOS," in 2016 *IEEE International Solid-State Circuits Conference (ISSCC),* 2016, pp. 50–51.

[12] W. Deng, T. Siriburanon, A. Musa, K. Okada, and A. Matsuzawa, "A Sub-Harmonic Injection-Locked Quadrature Frequency Synthesizer With Frequency Calibration Scheme for Millimeter-Wave TDD Transceivers," *IEEE Journal of Solid-State Circuits,* vol. 48, no. 7, pp. 1710–1720, Jul. 2013.

[13] Schrer, D. "The art of phase noise measurement", *Proc. RF Microwave Measurement Symp. Exhib.,* pp 1-21, 1983

[14] A. Musa, W. Deng, T. Siriburanon, M. Miyahara, K. Okada, and A. Matsuzawa, "A Compact, Low-Power and Low-Jitter Dual-Loop Injection Locked PLL Using All-Digital PVT Calibration," *IEEE Journal of Solid-State Circuits,* vol. 49, no. 1, pp. 50–60, Jan. 2014.

# A novel true logarithmic amplifier in 0.25 μm GaN-on-SiC technology for radar applications

Alessandro Salvucci, Marco Vittori, Sergio Colangeli, Giorgio Polli, Ernesto Limiti
*Electronic Engineering Dept.*
*University of Roma Tor Vergata*
*Via del Politecnico 1, 00133, Rome, Italy*
*alessandro.salvucci@uniroma2.it*

*Abstract*—A new circuit topology for a true logarithmic amplifier (TLA) basic cell is presented. The basic cell is synthesized in quasi-distributed form as the cascade of two single-FET stages. Whereas the operating principle of the overall TLA is well-known (i.e., cascading several hard-limiting cells), the topology of the proposed basic cell is not common. The broadband characteristics and the extreme compactness of the proposed architecture make it particularly suitable for the realization of multi-stage TLAs. The proposed basic cell is then adopted to design, as a test vehicle, a six-stages TLA, using a 0.25 μm GaN-on-SiC HEMT technology provided by UMS foundry. The final MMIC exhibits a broadband behavior, in the range 1.2 GHz-2.2 GHz, with a global logarithmic error of ±1 dB over 60 dB of input dynamic range (IDR).

*Keywords—TLA, GaN, log error, logarithmic*

## I. INTRODUCTION

In modern radar systems, and more generally in EW receiver architectures, much attention is paid to increase performance, in terms of detection range and signal recognition, together with the demand for of improved features, such as determining the time and angle of arrival. The growing number of dual-use applications led to an increased use of the electromagnetic spectrum. Therefore, developing receiver systems that can handle signals of various nature and entity, especially in terms of amplitude, results to be critical. Because of the latter requirement, a high effort has been devoted to develop radar-oriented circuitry with wide power ranges, in particular capable of compressing enough the dynamic range of the received signal, in order to facilitate and improve the signal recognition and processing. Thus, to detect signals with amplitudes varying over decades and reaching very high values, amplifiers with a logarithmic input/output characteristic have been introduced in the Rx chains. Dynamic compression also allows to detect signals with high amplitude, thus avoiding saturation of the receiving system.

A particular version of logarithmic amplifier is the *true log amplifier* (TLA), defined "true" because of the ability to preserve the phase and frequency information of the received signal. Among the first TLA structures, in [1] an architecture called Twin-Logarithmic Amplifier Gain Stage is proposed, formed by the cascade of *N* identical amplification stages, where each stage is composed by the parallel connection of a unit gain amplification, (named *buffer*), and a limiting stage as shown in

Fig. 1. Various types of TLA have been proposed and realized over the years, using different technologies and operating on various frequency bandwidths, typically within the 0.5-18 GHz frequency span. Contributions reporting detailed analysis and further examples of block implementation for logarithmic amplifiers are [2] and [3]. One of the first examples of realization of a TLA with bipolar transistors is found in [4] and [5]. Subsequently, the development of new technology processes, together with the increase of the integration level for high-frequency circuits, led to further examples of TLA in integrated form such as in [6-8]. In open literature, further methods for the implementation of a logarithmic function are reported, such as the Successive-Detection Logarithmic Video-Amplifier (SDLVA) [9], which, however, does not exhibit the phase-preserving peculiarity of the TLA. In this work we propose a new circuit topology, with a distributed structure, for the synthesis of the basic cell. The proposed cell was finally used for the design of a six-stage TLA, as a test vehicle, designed for broadband operation in the range 1.2 GHz-2.2 GHz.

## II. THEORY

Starting from the basic configuration proposed in [1], the unit gain amplifier and the limiter amplifier work together, depending on the input signal level. For small input power levels the limiter operates in its linear region, but as input power increases, the cascade enters saturation, starting from the last basic cell to the first, and a fixed value of output voltage ($V_{OL}$) will result. In this way, the whole cascade has a piecewise linear $V_{out}$ vs. $V_{in}$ characteristic and defined by the following equation.

$$V_{out} = \left\{ N + \frac{1}{A} + log_{A+1}\left[ \frac{AV_{in}}{V_{OL}} \right] \right\} V_{OL} \qquad (1)$$

where $V_{in}$ and $V_{out}$ are the input and output voltages of the TLA, $N$ is the number of base cells adopted in the cascade, $A$ is the value of the voltage linear amplification of the basic cell, $V_{OL}$ is the value of the maximum output voltage for the basic cell. In this way, the whole cascade, schematically depicted in Fig. 2, has a piecewise linear characteristic, using a logarithmic scale for the input voltage (dBV) or, equivalently, for the input power expressed (dBm), as shown in Fig. 3. The $V_{out}/P_{in}$ behavior shown in the figure refers to a seven-cell TLA.

*Paper P74*

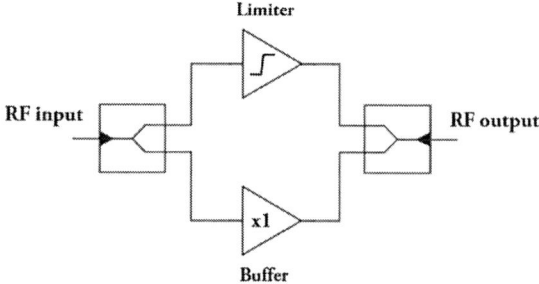

Fig. 1.   Dual gain stage basic cell for a TLA.

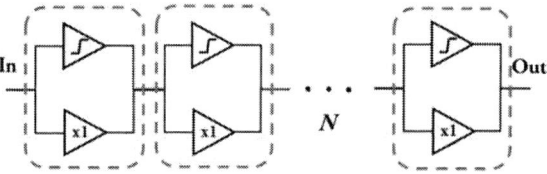

Fig. 2.   Block diagram of the multi-stage TLA.

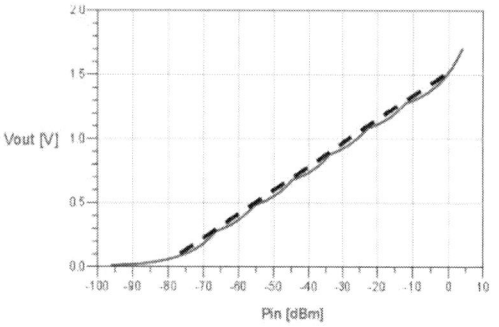

Fig. 3.   Example of a typical output logarithmic feature ($V_{out}$ / $P_{in}$) of an ideal seven-cell TLA (red curve), with a linear approximation overlapped (blue dashed line).

The logarithmic behavior of a TLA can also be expressed with other parameters of more practical utility for designers, considered as the figures of merit for a logarithmic amplifier, such as the *input dynamic range* (IDR) and *log error*. The IDR, defined in (2), represents the input dynamic range in which the logarithmic feature of the amplifier can be approximated with a straight line.

$$IDR_{dB} = 20 log(A + 1)^N \qquad (2)$$

The *log error*, instead, is an index that evaluates the difference between the actual characteristic of the amplifier (continuous curve in Fig. 3) and the best linear fit that approximates the same characteristic (blue curve in Fig. 3). Thus, considering the ideal, linear relationship between the output voltage of the amplifier and the input power

$$V_{out} = aP_{in} + b \qquad (3)$$

where $V_{out}$ is the output voltage, $P_{in}$ is the input power in dBm, *a* and *b* are the fitting parameters. The resulting log error is defined as:

$$log\ error = P - P_{in} \qquad (4)$$

in which *P* is the theoretical input power (obtained from the best linear fit of the input-output characteristic in semilog scale) and $P_{in}$ is the real value of the input power, both of them expressed in dBm. Based on the desired *log error*, the IDR value is automatically obtained. Clearly, it is desirable for a TLA to have a low log error over an input dynamic range as wide as possible.

III. DESIGN APPROACH

In this work a complete structure of a true logarithmic amplifier is proposed, based on a new basic cell topology. For the design of TLA, two operating regions must be considered, which are activated according to the level of the input signal, namely the input power $P_{in}$, namely a linear region and a limiting region. In the former one, the amplifier operates in class A, with a drain current varying linearly with the input signal, i.e. with no distortion. In the limiting region, the output voltage is truncated to a negative threshold by the device pinch-off. Ideally, the voltage should be hard-limited: in fact, self-bias components are generated, resulting in a less sharp cut. Cascading two inverting stages with the above characteristics will result in a truncation of the voltage in both its positive and negative half-waves. More in detail, the first FET cuts only a portion of the negative half-wave while leaving intact the positive one; after the sign inversion operated by the first stage, the original positive half-wave becomes negative and is cut by the second FET, biased in the same conditions as the first, i.e. near the pinch-off region. A graphic representation of the signal handling described is shown in the upper side of Fig. 4. Therefore, a necessary condition for the correct functioning is that the signal coming out from the first FET is a true replica of the input signal, apart for a time delay. To guarantee this condition, an instantaneous bandwidth as broad as possible must be pursued for each stage, therefore suggesting a distributed topology. However, since the basic cell has to be replicated several times to yield a TLA, the number of active devices should be minimized. The resulting trade-off, shown in Fig. 4, presents an almost distributed structure, i.e. with only one FET in each stage. The basic cell is completed, as typical distributed configurations proper, with two series inductors for each gate and drain lines, plus three series capacitances that perform the DC block function.

Based on the proposed basic cell features, a complete TLA was designed as a test vehicle of the new topology, with specifications listed in Table 1. The technological process adopted for the demonstrator is the *GH25-10* provided by United Monolithic Semiconductors (UMS): it is an AlGaN/GaN HEMT process on SiC substrate, featured by 250 nm gate length. This technology offers a 4 W/mm power density and targets the design of monolithic integrated robust low-noise and multi-stage high power high efficiency amplifiers up to 20 GHz.

The design process started taking into account the requested IDR: this is the main parameter and determines the number of basic cells to be adopted for the TLA. Typically, using a larger

978-1-5386-5388-3/18 $31.00 © 2018 IEEE

number of cascaded cells, the logarithmic characteristic of the whole amplifier is maintained for a wider input power range. In this case a cascade of six basic cells was used to obtain a TLA with an input dynamic range of 60 dBm.

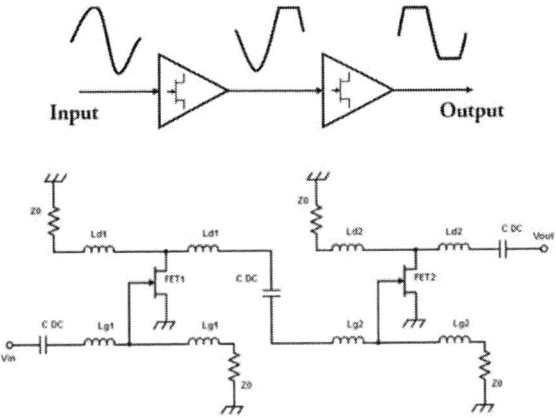

Fig. 4. Signal handling (up) and topology (down) of the distributed basic cell adopted in the TLA.

| Feature | Symbol | value | Unit |
|---|---|---|---|
| Frequency band | BW | 1.2-2.2 | GHz |
| Input Dynamic Range | IDR | 60 | dBm |
| Slope ($V_{out}/P_{in}$) | | $\approx 40$ | mV/dB |
| Log Error | | $\pm 2.5$ | dB |
| Power Consumption | $P_{diss}$ | <10 | W |

Table 1: TLA specifications.

## IV. RESULTS

The six cascaded cells of the designed demonstrator are appropriately arranged in the MMIC structure to optimize the RF and DC paths, and the overall occupied space. The layout of the MMIC, which has a size of 6x7 mm², is shown in Fig. 5, where the six cells are highlighted with dashed boxes. The designed logarithmic amplifier works in the 1.2-2.2 GHz band: this is much lower than the actual bandwidth of the unit cells, so that the signal is properly truncated and passed along otherwise unchanged. In the basic cell, two FETs with periphery of 2x125 μm and 6x70 μm were used for the first and second stage, respectively. The input RF port is on the left of the MMIC and placed at the same height with the RF output port on the right. The supply and bias voltages needed by the active devices are $V_{dd} = 10.4$ V and $V_{gg} = -3.19$ V, respectively. The layout was developed maintaining a symmetric structure, which was made possible by the even number of basic cells. Power supplies are provided, symmetrically, from two sides of the MMIC: on the upper side, the supply pads for the first three cells are placed and, similarly, on the lower side those for the last three cells. All cells share the same topology for the DC decoupling networks, which were designed with shunt C/RC structure, and include low-value resistors in the DC section, which value is optimized to provide the same bias voltage values to each stage.

The non-linear performance of the designed TLA was obtained through harmonic balance simulations. Based on the best linear fit of the input-output curve, the log error was then computed according to the expression (4). Typically, two types of log errors are defined for logarithmic amplifiers operating over significant bandwidths: a "local" log error, relative to the linear fit performed at the each input frequency, and a "global" log error, for which all the $V_{out}/P_{in}$ curves within the bandwidth are compared with a single linear characteristic. The performance presented below shows the obtained global log error, simulating frequencies in the whole design band (1.2 GHz-2.2 GHz), with steps of 0.2 GHz. Thus, in Fig. 6, the six $V_{out}/P_{in}$ transfer curves are shown, related to frequencies 1.2 GHz, 1.4 GHz, 1.6 GHz, 1.8 GHz, 2 GHz, 2.2 GHz, and the approximate linear characteristic (with a slope of ~ 50 mV/dB) assumed for the global log error calculation. Fig. 7 shows the log errors for the six simulated frequencies, calculated assuming the global linear fit of all transfer curves. The designed TLA exhibits a global log error contained within ±1 dB variation over an IDR of 60 dBm (-40 dBm through +20 dBm). Furthermore, the local log error values obtained over the same IDR at each simulated frequency are around ±0.6 dB. In Fig. 8 the small-signal parameters are shown, exhibiting a good level for input and output return loss, better than 10 dB and 26 dB, respectively. The small-signal gain obtained is around 40.5 dB with a flat behavior in the operating band (±0.8dB). The power consumption is calculated considering the worst case in which the circuit can operate, i.e. at maximum driving level (20 dBm), and results to be 9.85 W, within the specification limit.

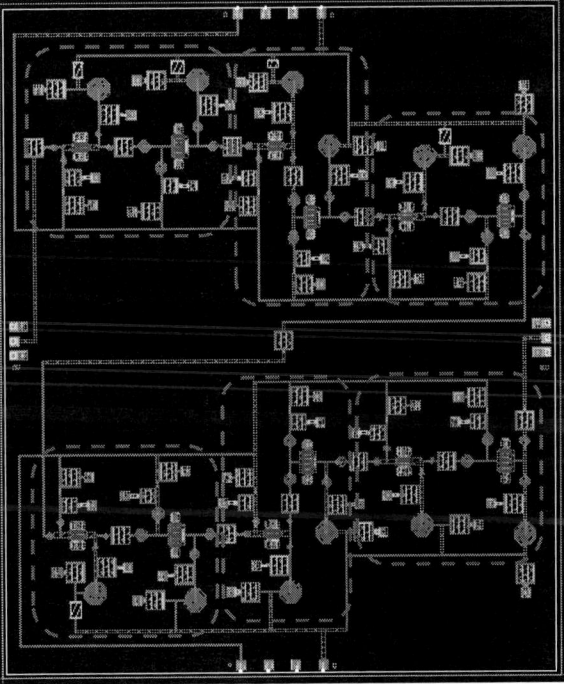

Fig. 5. Layout of the designed six-cell TLA. The MMIC has a size of 6x7 mm².

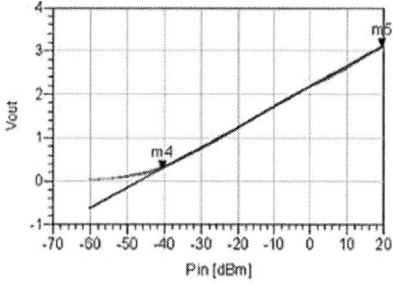

Fig. 6. Logarithmic behaviour obtained for the six-cell TLA in the 1.2 GHz - 2.2 GHz band. The graph shows six curves (red lines), related to the six simulated frequencies within the working band, with 0.2 GHz steps. Best global linear fit curve is traced (blue line) and taken as reference.

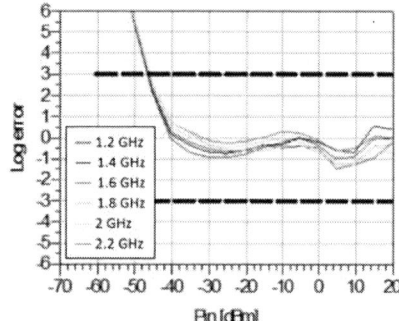

Fig. 7. Global log error calculated for signals in the working band, simulated with 0.2 GHz step, considering the best global linear fit on an IDR from -40 dBm to 20 dBm of $P_{in}$.

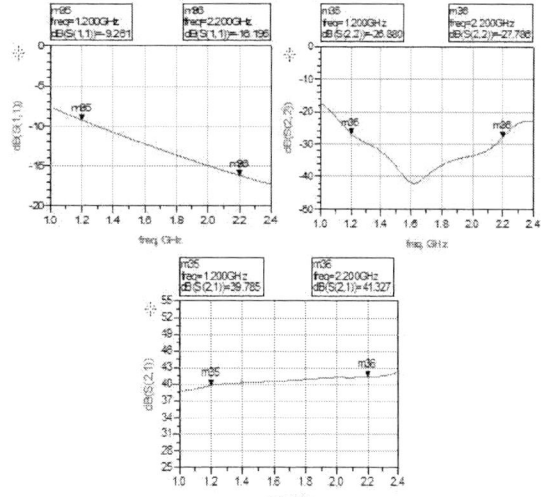

Fig. 8. Small signal parameters obtained for the six-cell TLA.

## V. CONCLUSION

In this paper a demonstrator for a true logarithmic amplifier (TLA) is presented, adopting a new basic cell topology. The proposed basic cell consists of the cascade of two quasi distributed stages, which perform the limiting operation typical of the logarithmic structures. Then, six basic cells where chained to yield a complete TLA. The design was carried out using a 0.25 µm GaN-on-SiC process provided by UMS foundry. The MMIC is designed to work over the 1.2-2.2 GHz band and exhibits a logarithmic characteristic ($V_{out}/P_{in}$) within an IDR of 60 dBm (-40 dBm through 20 dBm). Over the same IDR, a global log error within ±1 dB was obtained. A small-signal gain of 40.5 dB is achieved, together with input and output return losses better than 10 dB and 26 dB, respectively. The maximum power consumption expected is less than 10 W as required. The simulated performances make the designed TLA, with the adopted basic cell topology, particularly well-suited to receiver systems requiring a wide signal dynamic range.

## REFERENCES

[1] A. Woroncow and J. Croney, "A true i.f. logarithmic amplifier using twin-gain stages," in Radio and Electronic Engineer, vol. 32, no. 3, pp. 149-155, September 1966. doi: 10.1049/ree.1966.0068J.

[2] C. D. Holdenried, J. W. Haslett, J. G. McRory, R. D. Beards and A. J. Bergsma, "A DC-4-GHz true logarithmic amplifier: theory and implementation," in IEEE Journal of Solid-State Circuits, vol. 37, no. 10, pp. 1290-1299, Oct 2002. doi: 10.1109/JSSC.2002.803059

[3] G. Acciari, F. Giannini and E. Limiti, "Theory and performance of parabolic true logarithmic amplifier," in IEE Proceedings - Circuits, Devices and Systems, vol. 144, no. 4, pp. 223-228, Aug 1997. doi: 10.1049/ip-cds:19971290

[4] W. L. Barber and E. R. Brown, "A true logarithmic amplifier for radar IF applications," in IEEE Journal of Solid-State Circuits, vol. 15, no. 3, pp. 291-295, June 1980. doi: 10.1109/JSSC.1980.1051386

[5] A. K. Oki, M. E. Kim, G. M. Gorman and J. B. Camou, "High-performance GaAs heterojunction bipolar transistor monolithic logarithmic IF amplifiers," in IEEE Transactions on Microwave Theory and Techniques, vol. 36, no. 12, pp. 1958-1965, Dec. 1988.doi: 10.1109/22.17440

[6] M. A. Smith, "A 0.5 to 4 GHz true logarithmic amplifier utilizing monolithic GaAs MESFET technology," in IEEE Transactions on Microwave Theory and Techniques, vol. 36, no. 12, pp. 1986-1990, Dec 1988. doi: 10.1109/22.17443

[7] G. Acciari, F. Giannini and E. Limiti, "Novel decade-bandwidth microwave true logarithmic amplifier," in Electronics Letters, vol. 32, no. 5, pp. 464-, 1996. doi: 10.1049/el:19960309

[8] Y. J. Chuang, K. Cimino, M. Stuenkel, M. Feng, M. Le and R. Milano, "A Wideband InP DHBT True Logarithmic Amplifier," in IEEE Transactions on Microwave Theory and Techniques, vol. 54, no. 11, pp. 3843-3847, Nov. 2006. doi: 10.1109/TMTT.2006.883239

[9] L. Di Alessandro, M. Palomba, S. Colangeli and E. Limiti, "Robust GaN Successive-Detection Logarithmic Video-Amplifier for EW applications," 2015 Integrated Nonlinear Microwave and Millimetre-wave Circuits Workshop (INMMiC), Taormina, 2015, pp. 1-3.doi: 10.1109/INMMIC.2015.7330357

# Ka-/V-band self-biased LNAs in 70 nm GaAs/InGaAs Technology

Giorgio Polli, Marco Vittori, Walter Ciccognani, Sergio Colangeli, Ferdinando Costanzo, Alessandro Salvucci, Ernesto Limiti

*Electronic Engineering Dept.*
*University of Roma Tor Vergata*
*Via del Politecnico 1, 00133, Rome, Italy*
giorgio.polli@uniroma2.it, marco.vittori@uniroma2.it

**In this paper, two LNAs, designed to operate in Ka and V bands, and realized in a 70 nm GaAs/InGaAs technology, are presented. Both amplifiers have a 2-stage structure featured by source feedback and self-biasing networks to improve noise performance and to simplify the external circuitry, respectively. Total area occupation of the realized MMICs is 3x1.2 mm² and 3x1 mm². The Ka-band amplifier exhibits a noise figure lower than 1.5 dB over 27-31.5 GHz and a gain between 16 dB and 18 dB. The V-band LNA has a 1.7 dB noise figure in the 47-51 GHz band, with an associated gain between 14.5 dB and 15.5 dB.**

*Keywords — LNA; Ka-band; V-band;InGaAs/AlGaAs; self-bias*

## I. INTRODUCTION

In the last years, a growing number and complexity of information services have to be guaranteed to customers and users. Such huge number of services concerns several applications (telecom, earth observation, broadcasting and more) and requires an increase in data volume in a frequency spectrum already crowded. One possible solution is to move part of the system to higher frequency bands. This solution allows to increase also the amount of data per channel, for the same transmitted relative bandwidth [1], [2]. All the analog T/R circuitry has to be however redesigned and, moreover, the resulting chips have to guarantee competitive performance, also in terms of size, repeatability and cost, to find actual use in a competitive scenario.

As well known, for a radio communication system one of the main functionalities is low noise amplification. Even more so in the frequencies dealt with in this work (Ka and V-bands), where the atmospheric attenuation plays a key role in the link budget and the receiver's gain and noise performance are essential to provide a sufficiently covered distance. The low noise amplifier (LNA) de facto drives the noise figure of the whole receive system and, usually, is designed by exploiting the low noise figure (NF) of Gallium Arsenide (GaAs). This is true up to K/Ka-band applications [3], [4], then, for higher frequencies, parasitic effects take the lead, bringing to a performance degradation. In order to operate up to millimeter-wave frequencies, a process based on a high content of Indium in the active layer can be used, exploiting its properties in terms of linearity, noise performance and cut-off frequency [5], [6].

In this work, two self-biased LNAs, adopting a GaAs process with a high Indium content, are reported. These two amplifiers, operating in Ka- and V-band, have been also realized. The measured results are summarized and a comparison with the expected performance is provided.

## II. DESIGN ARCHITECTURE

The main figures of merit for LNAs are the noise figure and the small-signal available gain. Both have to be guaranteed while maintaining a high value of port matching for the whole operating bandwidth. In this work, two low noise amplifiers were designed to accomplish such figures of merit for two separate operating bands. The first one, named *Mercurius*, operates in the 26.5÷31.5 GHz band, the second one, named *Pluto*, in the 47÷52 GHz band. Both amplifiers are composed by two self-biased amplification stages: the first one has to minimize the noise while the second has to increase the gain; the LNA is ended by a gain flattening filter (GFF) that has the main purpose to shape the gain of the entire cascade. The general arrangement of the designed LNAs is shown in Fig. 1.

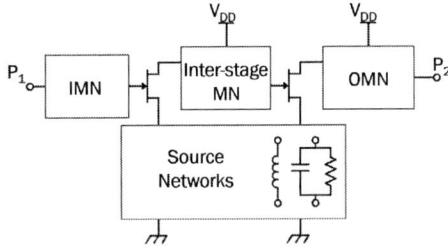

Fig. 1. Self-biased topology adopted for both LNAs.

Each stage was designed adopting the inductive degeneration methodology [7] to improve the noise performance. This method consists in inserting a series inductor between the source terminal and ground. Such inductor, properly sized, brings the optimum noise termination closer to the maximum gain load at the input port, helping to obtain simultaneously the input return loss and noise matching, while improving the stability of the amplifier. These results are obtained at the cost of a slight reduction of the maximum available gain and an increased complexity of the source networks.

Moreover, a self-biased topology [8]-[10] was adopted for both LNAs by inserting a parallel RC structure between the feedback inductor and ground for all the source networks. The resistors, in which drain currents flow, were sized to provide

978-1-5386-5388-3/18 $31.00 © 2018 IEEE

*Paper P23*

the selected positive DC voltages at the source terminals. Thus, fixing to zero the voltage at the gate terminal with a DC path to ground, any required negative gate-source voltage could be obtained for each active device. Meanwhile, shunt capacitors were adopted in order to provide a direct path to ground to the RF signal, reducing the detrimental effects of the resistors on gain and noise performance. Each LNA required the designing of several passive networks to implement the biasing, matching and stabilization functionalities. These tasks were mostly accomplished by using reactive elements, except for one small resistor connected to the drain of the second FET to increase the geometrical stability factor ($\mu$). Due to the high operating frequencies, all the passive networks were synthetized with the minimum number of elements to reduce the overall size, in order to limit their impact on the total noise figure and to obtain a simple layout with a minimum number of critical parts. A careful choice of lumped elements was required by the high operating frequencies of these LNAs. Therefore, microstrip line sections with optimized parameters were used to obtain low inductance values

These MMICs were designed to be subsequently inserted into more complex HMIC systems. Therefore, EM simulated bonding wires, and landing pads, were included into the LNAs' design, by appropriately shifting the input and output reference planes. More precisely, the contribution of the bonding wires was considered only at the output port for the Ka-band LNA and both at input and output ports for the V-band LNA. These conditions were fixed according to the specific applications the LNAs were designed for.

### A. Design choices and differences between the amplifiers

The MMICs were realized adopting the D007IH process provided by OMMIC foundry, whose Process Design Kit was complemented by specific active devices measurements performed at our microwave laboratory. The adopted process is based on a 70 nm technology with a GaInAs-InAlAs-GaInAs-InAlAs heterostructure. The high amount of Indium into the active layer, up to 70%, combined with the double mushroom gate topology adopted, guarantees the high cut-off frequencies and demanding noise performance required by this project.

The schematics of the designed LNAs are shown in Fig. 2, while the micro-photos of the realized MMICs are in Fig. 3 and Fig. 4, for Ka (hereinafter named Mercurius) and V-band (hereinafter Pluto) amplifiers, respectively. Chip area occupation is 3.0x1.2 mm² for the former and 3.0x1.0 mm² for the latter. Input and output are respectively on the left and on the right side, while DC voltage is provided via the pad on top. The input, inter-stage and output networks are highlighted in green along the RF path. The two blue boxes on the upper part of the MMICs contain the decoupling networks for drain terminals. Each one was realized by a first capacitor, sized to present the proper load at RF frequencies, and by subsequent ones, used to reduce undesirable effects of the off-chip bias networks at lower frequencies. The source networks (yellow boxes in the bottom part of the circuits) are composed by feedback inductors and RC structures. The inductive behavior was obtained through narrow micro-strip lines, where the line lengths resulted by a trade-off between the noise and the gain performance required by each LNA. The capacitance values

depended mainly by the frequencies of the RF signal. *Mercurius* LNA operates at lower frequencies and required a larger capacitor in order to by-pass the resistor. This capacitance value was realized by means of two parallel capacitors. Instead, for *Pluto* LNA, a grounded capacitor for each source network was sufficient to approach the required capacitance value. Resistor values depended directly on the operating condition for each FET and, in particular, on the drain bias current and the gate-source voltage required. About the Ka-band amplifier, both active devices have a gate periphery of 4x15 μm and operate at the same bias (i.e. $V_{ds} = 1$ V, $V_{gs} = -0.1$ V and $I_d = 14.3$ mA). According to the adopted self-bias approach, the value of the resistors in the source networks is 7 $\Omega$. Also the FETs adopted for the V-band amplifier have a gate periphery of 4x15 μm but each stage was designed to achieve a peculiar requirement. The first stage is designed to minimize NF values, while second stage was designed to fulfil specification of gain and output port match. Such conditions were obtained with $V_{ds} = 1$ V, $V_{gs} = -0.1$ V and $I_d = 14.3$ mA for the first stage and with $V_{ds} = 1$ V, $V_{gs} = -0.15$ V and $I_d = 10.3$ mA for the second. Therefore, the resistors are 7 $\Omega$ and 14.5 $\Omega$ for the first and second stage respectively. Finally, the gain flattening networks (red boxes contained in the output matching networks) are visible on the right side of the MMICs. This functionality was implemented through two different structures in the two LNAs. For the Ka-band LNA, a high-pass filter realized by a RC parallel structure was inserted along the RF-path. Instead, a pass-band network, composed by a grounded RLC chain, was used for the V-band LNA. Both filters have been designed to reduce the gain at lower frequencies, thus minimizing the resulting ripple in the operating band.

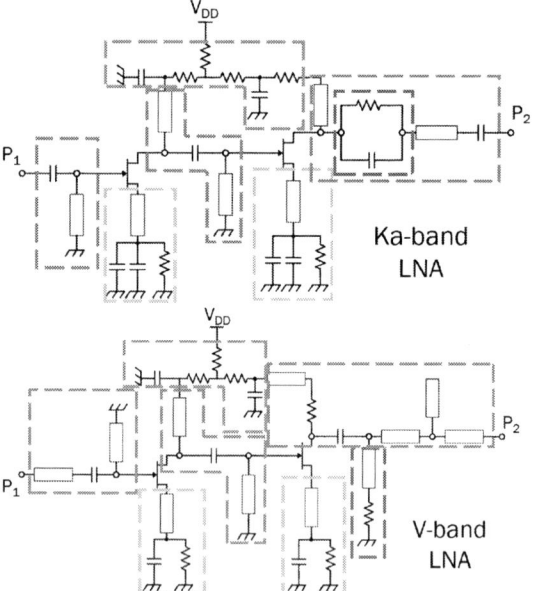

Fig. 2.    Schematics of the Ka-band (upper figure) and the V-band (lower figure) LNAs. Colored boxes highligth significative passive networks.

*PRIME 2018, Prague, Czech Republic*　　　　　　　　*Session: Radio Frequency Circuits and Systems II*

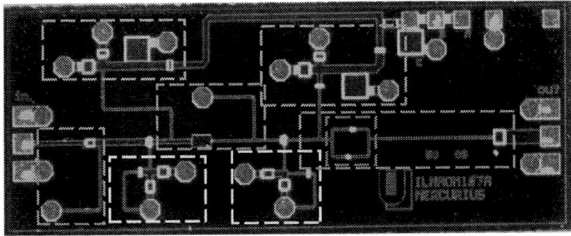

Fig. 3.　Micro-photo of the realized Ka-band LNA (*Mercurius*), colored boxes highlight significant blocks.

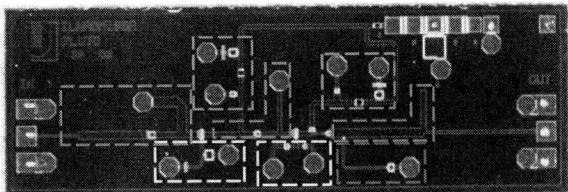

Fig. 4.　Micro-photo of the realized V-band LNA (*Pluto*), colored boxes highlight significant blocks.

## III. MEASUREMENTS

In the following, the comparisons between simulated and measured small-signal performance of the designed LNAs are summarized. The measurements were performed biasing the active devices at the operating points previously reported, applying an external DC voltage of 4 V on the DC pad of each LNA.

This value is a constraint fixed by the specific application the LNAs were designed for. Therefore, DC power consumption is around 115 mW for the Ka-band LNA and 99.2 mW for the V-band LNA. These values could be significantly lowered (by more than 70%) by reducing the size of resistors on the supply line and thus providing the drain voltages effectively required by the active devices bias.

Concerning *Mercurius*, the S-parameters comparison is shown Fig. 5 and Fig. 6. Dashed blue traces refer to simulations, while measurements are represented by the solid red lines. As can be noticed, simulations and measurements are in very good agreement. Minor deviations can be observed in the gain level, which suffered from a slight shift towards lower frequencies, remaining between 16.5 dB and 18 dB, with a gain variation lower than 0.5 dBpp/GHz over the whole operating bandwidth (26.5-31.5 GHz).

As to the noise figure, the measurements were carried out with an Agilent 346CK01 noise source connected to a receiver formed by the cascade of a Quinstar QLW-00504533-J0 pre-amplifier and an Agilent E4448A PSA. A 6 dB attenuator was inserted after the noise source to reduce the available noise power and, consequently, the resulting ENR. The resulting noise figures, for different biasing voltages, are reported in Fig. 7. As it can be noted, the measurement at the nominal bias conditions (the solid yellow trace) is close to the simulation (the dashed blue line), within the limits of the measurements uncertainty, and results below 1.6 dB.

Concerning *Pluto*, a comparison of linear simulations and measurements, is reported in Fig. 8 and Fig. 9.

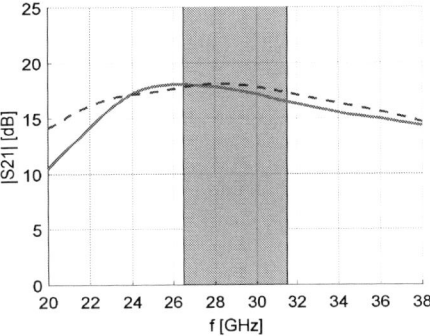

Fig. 5.　20-38 GHz measurated versus simulated gains ($|S_{21}|$ dB) for the Ka-band LNA. The solid red trace is the measurement, at the nominal bias-conditions. The dashed blue line is the simulation at the same reference planes.

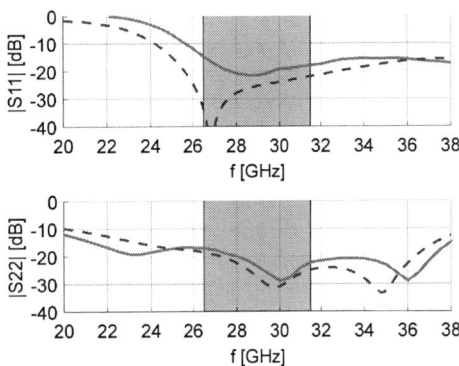

Fig. 6.　20-38 GHz measurated versus simulated return losses ($S_{11}$ on the top and $S_{22}$ on the bottom) of the Ka-band LNA. The solid red traces are the measurements, at the nominal bias-conditions. The dashed blue line are the simulations.

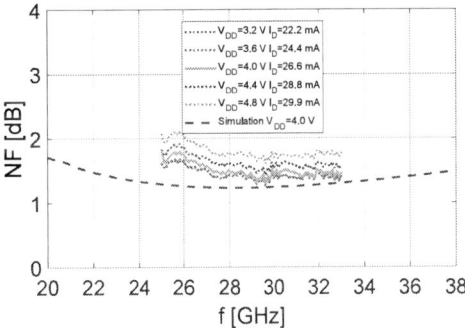

Fig. 7.　Measurements versus simulations at the SOLT probe tips reference-planes. |NF50| dB. The dotted traces are for the measurements, at different bias conditions, while the noise at nominal bias point is the yellow solid trace. The blue dashed line is the simulated noise figure.

The reduction of the in-band gain is a consequence of the frequency shifting phenomena of the matching performance.

978-1-5386-5388-3/18 $31.00 © 2018 IEEE

Unfortunately, the self-bias architecture precludes the option to tune the gate voltage in order to compensate this effect.
Noise measurements of *Pluto* LNA are superimposed with, or mostly below, the simulations, as shown in Fig. 10. The NF at the nominal $V_{DD}$ value, yellow trace, is always below 1.7 dB. The simulated 1dB compression point is 3 dBm and 2 dBm for the Ka-band and V-band LNAs, respectively.

## IV. CONCLUSION

The designs of two self-biased LNAs, operating at Ka and V-bands, have been presented. The realized chips, adopting a 70 nm GaAs process with high Indium content, occupy an area of 3x1.2 mm$^2$ and 3x1 mm$^2$. The amplifiers exhibit a noise figure lower than 1.6 dB and 1.7 dB in 27-31.5 GHz and 47-51 GHz bands respectively. Gain performance of 16.5 dB and 14.5 dB was achieved, for the Ka-band LNA and the V-band LNA respectively. The effects of bonding wires have been included into the design.

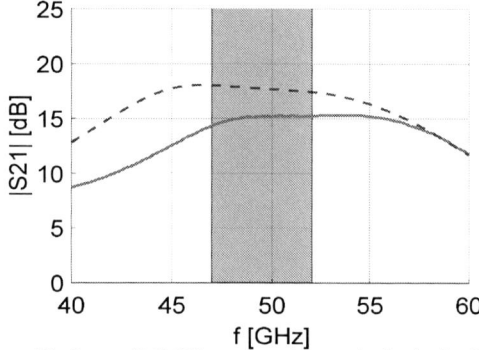

Fig. 8.    40-60 GHz measurated versus simulated gains ($|S_{21}|$ dB) of the V-band LNA. The red solid trace is the measurement, at the nominal bias-conditions. The blue dashed line is the simulation.

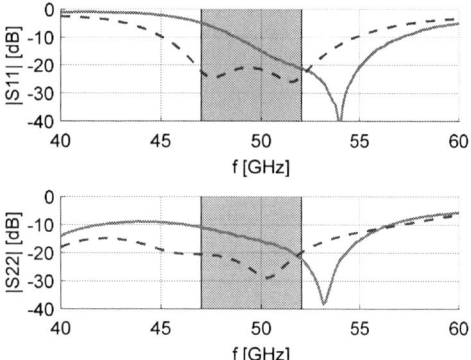

Fig. 9.    40-60 GHz measurated versus simulated return losses (S11 on the left and S22 on the right) of the Ka-band LNA. The red solid trace is the measurement, at the nominal bias-conditions. The blue dashed line is the simulation.

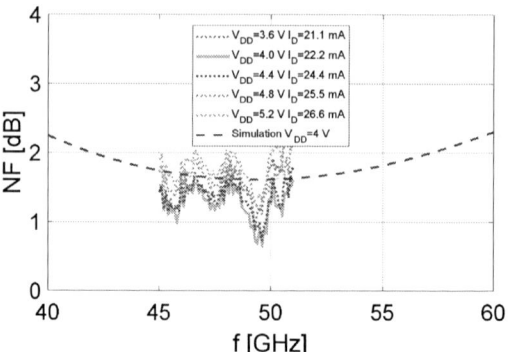

Fig. 10.    Measurements versus simulations at the SOLT probe tips reference-planes. |NF50| dB. The dotted traces are for the measurements at different bias conditions, while the noise at nominal bias point is the yellow solid trace.The blue dashed line is simulated noise figure.

## REFERENCES

[1]    K. Minot, M. Aust, R. Kasody, R. Katz, H. Wang, P. Rodgers, D. Smith, L. Shaw, K. Tan, N. Wang, S. Dow, and B. Allen, "A MMIC chip set for V-band crosslink communication systems," in *Proceedings of 1994 IEEE Microwave and Millimeter-Wave Monolithic Circuits Symposium*, May 1994, pp. 221–224.
[2]    A. D. Panagopoulos, P. D. M. Arapoglou, and P. G. Cottis, "Satellite communications at KU, KA, and V bands: Propagation impairments and mitigation techniques," *IEEE Communications Surveys Tutorials*, vol. 6, no. 3, pp. 2–14, Third 2004.
[3]    L. A. Samoska, M. Varonen, P. Kangaslahti, A. Fung, R. Gawande, M. Soria, R. Lai, and S. Sarkozy, "V-band MMIC LNAs and mixers for observing the early universe," in *2016 41st International Conference on Infrared, Millimeter, and Terahertz waves (IRMMW-THz)*, Sept 2016, pp. 1–2.
[4]    W. Ciccognani, S. Colangeli, E. Limiti, and L. Scucchia, "Millimeter wave low noise amplifier for satellite and radio-astronomy applications," in *2012 IEEE First AESS European Conference on Satellite Telecommunications (ESTEL)*, Oct 2012, pp. 1–4.
[5]    W. Ciccognani, E. Limiti, P. E. Longhi, and M. Renvoise, "MMIC LNAs for radioastronomy applications using advanced industrial 70 nm metamorphic technology," *IEEE Journal of Solid-State Circuits*, vol. 45, no. 10, pp. 2008–2015, Oct 2010.
[6]    M. V. Aust, T. W. Huang, M. Dufault, H. Wang, D. C. W. Lo, R. Lai, M. Biedenbender, and C. C. Yang, "Ultra low noise Q-band monolithic amplifiers using InP- and GaAs-based 0.1 µm HEMT technologies," in *IEEE 1996 Microwave and Millimeter-Wave Monolithic Circuits Symposium. Digest of Papers*, June 1996, pp. 89–92.
[7]    Y. Wang, C. C. Chiong, J. K. Nai, and H. Wang, "A high gain broadband lna in gaas 0.15-µm phemt process using inductive feedback gain compensation for radio astronomy applications," in *2015 IEEE International Symposium on Radio-Frequency Integration Technology (RFIT)*, Aug 2015, pp. 79–81.
[8]    Y. T. Chou, C. C. Chiong, and H. Wang, "A Q-band LNA with 55.7% bandwidth for radio astronomy applications in 0.15- µm GaAs pHEMT process," in *2016 IEEE International Symposium on Radio-Frequency Integration Technology (RFIT)*, Aug 2016, pp. 1–3.
[9]    Q. Wang and Y. Guo, "Ka-band self-biased monolithic GaAs pHEMT low noise amplifier," in *2011 IEEE International Conference on Microwave Technology Computational Electromagnetics*, May 2011, pp. 261–263.
[10]    M. Vittori, G. Polli, W. Ciccognani, S. Colangeli, A. Salvucci, and E. Limiti, "Q-band self-biased MMIC LNAs using a 70 nm InGaAs/AlGaAs process," in *2017 IEEE Asia Pacific Microwave Conference (APMC)*, Nov 2017, pp. 630–633.

PRIME 2018, Prague, Czech Republic

Session: Radio Frequency Circuits and Systems II

# Single MMIC receivers for C-band T/R module in 0.25 μm GaN technology

Alessandro Salvucci, Giorgio Polli, Aurora De Padova, Sergio Colangeli, Ferdinando Costanzo, Walter Ciccognani, Ernesto Limiti

*Electronic Engineering Dept.*
*University of Roma Tor Vergata*
*Via del Politecnico 1, 00133, Rome, Italy*
*alessandro.salvucci@uniroma2.it, giorgio.polli@uniroma2.it*

*Abstract*— In this contribution two different versions of MMIC LNAs integrating the limiting function are presented. The chips are designed with 0.25 μm gate length GaN on SiC technology as provided by Leonardo foundry, and arrange the receiving circuitry of a T/R module operating in C-band, specific for AESA systems. The final performance show the differences of the two circuits, which were designed with different methodologies. For the first version the constant mismatch circles method was applied, while for the second version, the typical design method for LNAs was adopted. Both circuits use the same switch (absorptive SPST), that exhibits an isolation level better than 30 dB and insertion loss lower than 0.6 dB. The first version shows a wider operation bandwidth, with a noise figure of 2.2 dB, a gain of 35.5 dB, and excellent levels of return loss (22 dB for input and 23 dB for output). The second version exhibits a noise figure of 2.1 dB, a gain of 35.5 dB, with return losses of 22 dB and 20 dB for input and output respectively.

*Keywords—TRM, AESA, LNA, GaN, C-Band.*

## I. INTRODUCTION

In recent years, high-frequency radar systems have drawn increasing attention for new multifunctional architecture, especially for the aerospace scope, such as the AESA (Active Electronically Scanned Array) [1]. These radar systems allow a fully electronic scanning, driving many integrated T/R modules which include, besides the switch circuitry, a low noise amplification function (LNA) for the Rx mode, and a high power amplifier (HPA) for the Tx mode [2]. Challenges involved in the actual implementation features the miniaturization of the module, in order to mount easily the complete AESA system on various avionic platforms, and the generation of ever-increasing power, to enlarge the area covered by the radar system. For these reasons a strong push towards the implementation of each function in monolithic technology and integrated in a single package, has been introduced [3]. In this context, an essential role is also played by the adopted technology. GaAs-based T/R structures [4], even if they exhibit good performance in terms of noise or power, are now outdated and replaced by new technologies based on Gallium Nitride (GaN). The inherently high power density, together with a wideband operation and the high breakdown voltage makes GaN more suitable for the realization of more compact and efficient transmit/receive modules [5][6].

In this contribution, the design of two versions of the Rx circuitry of a T/R module is presented, and in particular the low noise amplification and switching functionalities, integrated into a monolithic microwave integrated circuit (MMIC) and developed in GaN-on-SiC technology. The two versions were designed with two different methodologies, aiming at the same specifications.

The adopted technology process is the *SGN25*, and provides 0.25 μm GaN-HEMTs on a 100 μm Silicon Carbide substrate (SiC), as provided by Leonardo foundry. This process uses an AlGaN/GaN-based heterojunction and exhibits a good behavior for both high-power and low-noise applications. The typical transconductance for the active devices is 300 mS/mm, while cut-off frequency ($f_T$) is 28 GHz, since a decreasing of parasitic effects was attained: gate-source capacitance ($C_{GS}$) of 2.7 pF/mm is achieved with $V_{gs}$ of 0 V). Pinch-off and break down voltages are -3.3 V and 125 V respectively.

## II. DESIGN & GENERAL ARRANGMENT

In the literature, various types and architectures for T/R modules have been proposed and realized. Typically the scheme of a T/R module is composed by a circulator that directs the signal coming from the transmission chain towards the antenna port, while sending the received signal to the low-noise section. The Tx chain is typically composed by a driver and a high-power amplifier, while the Rx functionality is implemented by a low-noise amplifier following a limiter, realised by means of an absorptive switch, used as protection of the receiving circuitry from high level impinging signals, as depicted in Fig. 1. Design requirements for the receiving section are summarized in Table 1.

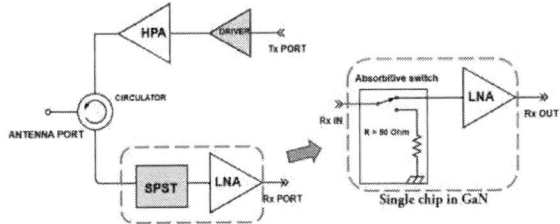

Fig. 1. Block diagram of the T/R module, with detail of the receiver chip in GaN.

978-1-5386-5388-3/18 $31.00 © 2018 IEEE    201

| Feature | Symbol | value | Unit |
|---------|--------|-------|------|
| **Bandwidth** | BW | 250 | MHz |
| **Noise Figure** | NF | 2,2 | dB |
| **Rx Gain** | $G_{RX}$ | 34,5 | dB |
| **Return loss** | RL | 20 | dB |

Table 1: Receiver mode specifications.

## A. SPST

The same absorptive single pole single-throw (SPST) switch is used in both MMIC versions. As visible from the topology in Fig. 2, the switching function is performed by a set of 5 cold-biased FETs, all driven by the same control voltage $V_c$. The absorptive chain is composed by two parallel FETs, in series configuration, *sw1* and *sw2* of 8x200 μm periphery, that drive the signal towards a 50 Ω load (which is composed by two shunt resistors of 100 Ω to withstand high currents), when $V_C$= 0 V. For the Rx path, a chain composed by a λ/4 line and three shunt FETs (*sw3*, *sw4*, *sw5*) with a 5x175 μm periphery has been placed along the RF path to guarantee the high isolation required (> 30 dB). The ladder structure, composed by the FETs, line 1 and line 2, was sized to show almost a short-circuit in Tx-mode, while maintaining an insertion loss lower than 0.6 dB, in Rx-mode ($V_C$= -50 V). The quarter-wave line section is inserted to transform the load shown by the ladder structure into an almost open-circuit. In this way, the load at the input port, during Tx-mode, is almost entirely due to the absorptive chain resistors.

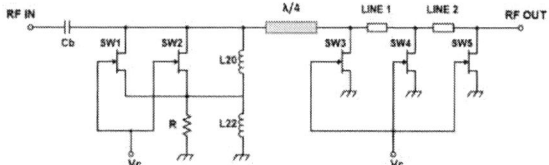

Fig. 2.   Simplified schematic of the SPST functionality.

## B. LNA

The schematic topology of both LNAs is shown in Fig. 3. In order to achieve the required gain level, a cascade of three active stages was used, on which the inductive degeneration feedback method was applied. Low-noise amplification functionalities of the two MMICs were designed with two different methods.

More specifically, in the first version a completely deterministic design method for LNAs was applied, the constant mismatch circles method, specifically tailored for two-stage LNAs [7]. This method allows to provide the minimum noise achievable by the adopted FETs, and a simultaneous perfect input and output match working on inductive source feedback values and synthesis of passive networks. Once the source inductances have been selected, several passive networks can be synthesized in order to share the same inter-stage mismatch level, showing the optimal source load for the minimum noise measurement of each

active device. This choice also allows a simultaneously matched chain, which leads to have perfectly matched input and output sections (i.e. $IM_1=OM_2=0$). We refer to [8] for the analytical details of the applied method. Finally, the third stage was designed to obtain matching between its input section and the output section of the two-stage cascade, being optimized for gain.

Instead, for the second MMIC, the typical design method for LNAs was adopted, using the degenerative source feedback and synthesizing the optimum loads of the active devices. Finally, a tradeoff between gain and port match performance of the whole cascade was carried out.

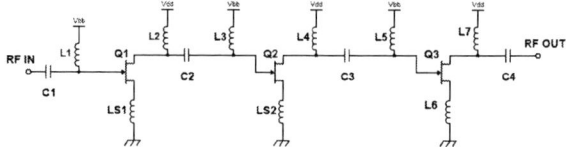

Fig. 3.   Simplified schematic of the LNA functionality.

All passive networks were initially designed with the minimum number of discrete elements, although further series elements were then inserted in the final optimization phase to improve the circuit stability. Specifically, further resonant networks were inserted along the RF path of the second version to improve the stability of each stage, while the first version did not require these structures. Both MMICs use the same supply voltages. Control voltage $V_c$ of the SPST is -50 V for *on* condition (Rx mode) and 0 V for *off* condition (Tx mode). The LNA function adopts drain and gate supply voltages of 10 V and -3 V respectively, for all active devices. The two LNA versions for the second stage periphery: a 4x50 μm FET was used in the first version, (the same as the first stage, as due to the adopted design method), whereas a 4x75 μm FET used for the second version. Instead, both versions share the same gate peripheries for first and third stage (i.e. 4x50 μm and 4x25 μm). Thus, regarding power consumption, the two LNAs differ by 50 mW, (250 mW and 300 mW respectively for first and second version), as summarized in Table 2.

| | Version 1 | | | Version 2 | | |
|---|---|---|---|---|---|---|
| | *1st stg* | *2nd stg* | *3rd stg* | *1st stg* | *2nd stg* | *3rd stg* |
| **Periphery** | 4x50 μm | 4x50 μm | 4x25 μm | 4x50 μm | 4x75 μm | 4x25 μm |
| **$V_{dd}$** | 10 V | | | 10 V | | |
| **$I_d$** | 10 mA | 10 mA | 5 mA | 10 mA | 15 mA | 5 mA |
| **$P_{DC}$** | 100 mW | 100 mW | 50 mW | 100 mW | 150 mW | 50 mW |
| **$P_{DCtot}$** | 250 mW | | | 300 mW | | |

Table 2: LNAs features.

## III. LAYOUT & FINAL RESULTS

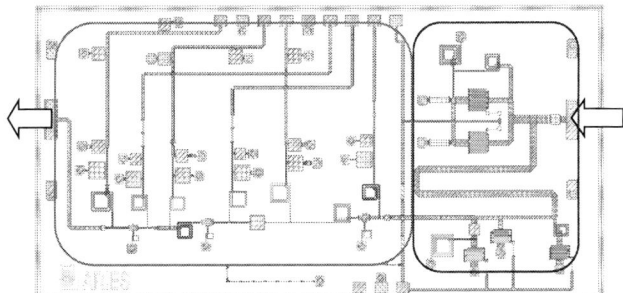

Fig. 4. Layout of receiver chip in GaN version 1 "*Aries*". LNA in red, switch in blue.

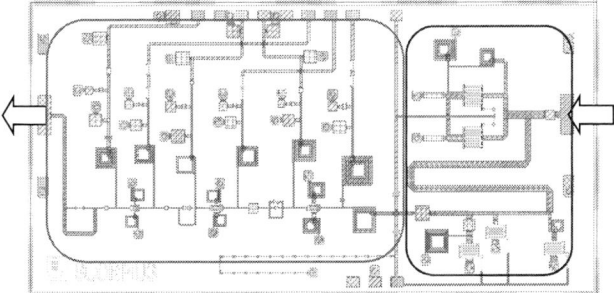

Fig. 5. Layout of receiver chip in GaN version 2 "*Scorpius*". LNA in red, switch in blue.

The final layouts of both versions are shown in Fig. 4 and Fig. 5. Starting from the RF input port, on the right, the switching circuitry is placed before the low-noise amplification block. Power supply and control voltage pads are located in the upper side of the chip. Both MMICs occupy the same area, e.g. 6.5x3.3 mm2. In the following graphs, the final performances are presented, comparing the main parameters of the two chips. The circuit design took into account bonding wires placed at both input and output port of the MMICs, and here considered in the reported simulations. Moreover, the SPST for both versions exhibits an insertion loss of 0.6 dB, that directly affects the chip noise figure, and an isolation of 33.5 dB in Tx mode. The noise figure of both versions is shown in Fig. 6, where, inside the C-band (5.28 GHz-5.53 GHz), NFs lower than 2.2 dB and 2.1 dB are achieved. Gain performance, in Fig. 7, is very similar for both circuits, exhibiting an average value of 35.5 dB, with a very flat behavior in the operating band (±0.33 dB and ±0.5 dB). Input and output port return losses of the two circuits are compared in Fig. 8 and Fig. 9 respectively. Thanks to the peculiarities of the applied design method, since very high matching levels are imposed at single design frequency (i.e. middle of the operating band), the first version exhibits a more broadband behavior with respect to the second version. For both versions, input port match is lower than -22 dB within the operating band. At output port, a $|S_{22}|$ lower than -23 dB is achieved by the first version and lower than -20 dB by the second version, within the operating band. Considering noise and gain behavior, the first version maintains better performance even outside the operating band,

as compared to the second one. In conclusion, in Table 3, the obtained performance of the two MMICs are compared with similar works reported in literature. The two receiver designs are actually being manufactured by Leonardo foundry. Measured performance will be presented and compared to simulations as soon as realization is completed.

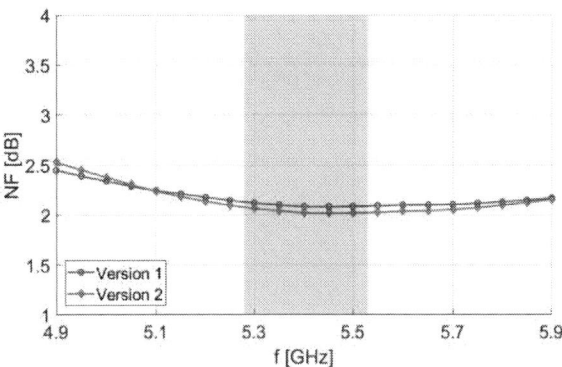

Fig. 6. MMICs noise figure comparison.

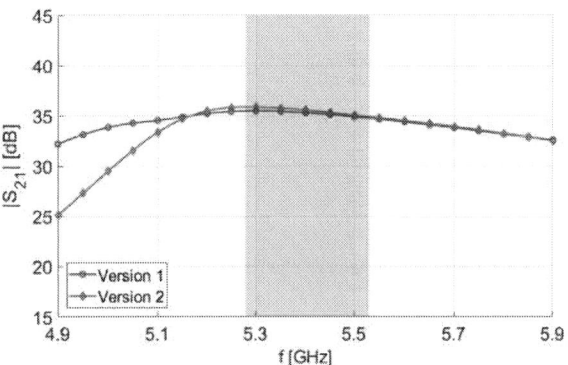

Fig. 7. MMICs gain comparison.

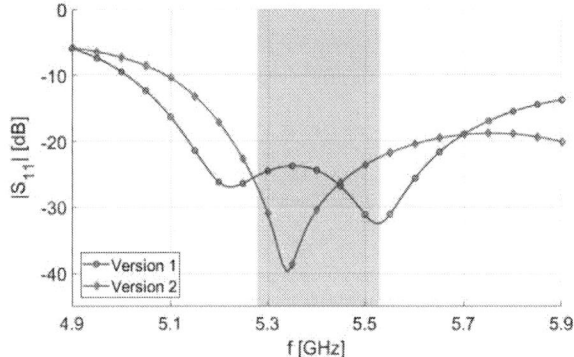

Fig. 8. MMICs input port match comparison.

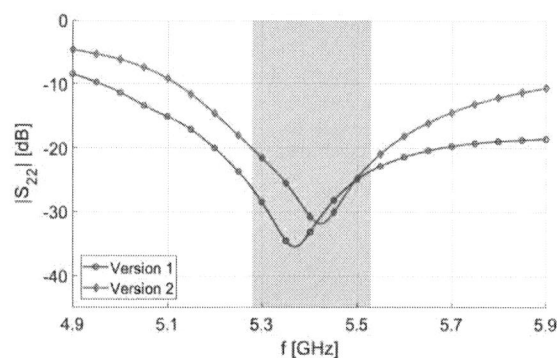

Fig. 9. MMICs output port match comparison.

| | BW (GHz) | NF (dB) | Gain (dB) | $|S_{11}|,|S_{22}|$ (dB) |
|---|---|---|---|---|
| **MMIC #1[a]** | 5.28–5.53 | 2.1 | 35.5; 3-stages | -22, -23 |
| **MMIC #2[a]** | 5.28–5.53 | 2.1 | 35.5; 3-stages | -21, -20 |
| [9] | 6 | 1 | 23.3; 3-stages | -15, -23 |
| [10] | 5 – 6 | 5.5 | 12; Stacked 2-stages | -8, |
| [11] | 3 – 7 | 2.3 | 20 | -5, -10 |
| [12] | 6 | 2.7 | 10.8 | -5, -10 |
| [13] | 6 | 1.6 | 10.9; 1-stage | -5.5@6GHz, -18@7GHz |
| [14][a] | 5–6.7 | 1.2 | 10; 1 stage | -7, -6 |

[a] Considering switch effort.

Table 3: LNAs performance comparison.

## IV. CONCLUSION

Two different versions of MMICs receiving chip for a T/R module operating in the C-band have been presented. The two test vehicles have been designed with Leonardo 0.25 μm GaN-on-SiC HEMT technology and developed with two different design methodologies. For the first version a completely deterministic method has been used, that allows to have the minimum noise measurement together with an excellent port matching. In second version, the typical method has been used, synthesizing the optimum noise loads for the adopted FETs. Both versions fulfil the specifications, exhibiting notable NF and gain performance on the overall bandwidth, respectively of ~2.1 dB and ~33.5 dB. The first version exhibits better performances in terms of port matching on a wider band. Return losses are better than 22 dB for input and 20 dB for output within the operating band (5.28 GHz-5.53 GHz). The high level of integration and the good performance obtained make the two designed MMICs suitable for the Rx chain of AESA systems.

## ACKNOWLEDGMENT

This research work was supported by European Space Agency (ESA contract no. 4000119200 "High Power GaN C-Band Transmit/Receive Module Demonstrator").

## REFERENCES

[1] A. Caronti et al., "The future of Italian ground and naval active electronically scanned arrays (AESA) radars," 2016 IEEE International Symposium on Phased Array Systems and Technology (PAST), Waltham, MA, 2016, pp. 1-7. doi: 10.1109/ARRAY.2016.7832540

[2] P. Schuh et al., "T/R-module technologies today and future trends," The 40th European Microwave Conference, Paris, 2010, pp. 1540-1543. doi: 10.23919/EUMC.2010.5616412

[3] T. Boles, D. J. Carlson and C. Weigand, "MMIC based phased array radar T/R modules," 2011 IEEE International Conference on Microwaves, Communications, Antennas and Electronic Systems (COMCAS 2011), Tel Aviv, 2011, pp. 1-4.doi: 10.1109/COMCAS.2011.6105784

[4] C. S. Cheng, C. C. Wei, H. C. Chiu, Y. C. Chiang, J. S. Fu and C. S. Wu, "A Ka-Band Monolithic CPW-Mode T/R Modules Using 0.15 μm Gate-Length GaAs pHEMT Technology," 2008 Global Symposium on Millimeter Waves, Nanjing, 2008, pp. 87-90. doi: 10.1109/GSMM.2008.4534565

[5] A. Barigelli et al., "Development of GaN based MMIC for next generation X-Band space SAR T/R module," 2012 7th European Microwave Integrated Circuit Conference, Amsterdam, 2012, pp. 369-372.

[6] R. Rieger, A. Klaaßen, P. Schuh and M. Oppermann, "GaN based wideband T/R module for multi-function applications," 2015 European Microwave Conference (EuMC), Paris, 2015, pp. 514-517.doi: 10.1109/EuMC.2015.7345813

[7] S. Colangeli, W. Ciccognani, A. Salvucci and E. Limiti, "Deterministic design of simultaneously matched, two-stage low-noise amplifiers," 2017 IEEE Asia Pacific Microwave Conference (APMC), Kuala Lumpar, 2017, pp. 558-561. doi: 10.1109/APMC.2017.8251506I.S

[8] W. Ciccognani, P. E. Longhi, S. Colangeli and E. Limiti, "Constant Mismatch Circles and Application to Low-Noise Microwave Amplifier Design," in IEEE Transactions on Microwave Theory and Techniques, vol. 61, no. 12, pp. 4154-4167, Dec. 2013. doi: 10.1109/TMTT.2013.2288696

[9] E. M. Suijker et al., "Robust AlGaN/GaN Low Noise Amplifier MMICs for C-, Ku- and Ka-Band Space Applications," 2009 Annual IEEE Compound Semiconductor Integrated Circuit Symposium, Greensboro, NC, 2009, pp. 1-4. doi: 10.1109/csics.2009.5315640

[10] C. Andrei, O. Bengtsson, R. Doerner, S. A. Chevtchenko and M. Rudolph, "Robust stacked GaN-based low-noise amplifier MMIC for receiver applications," 2015 IEEE MTT-S International Microwave Symposium, Phoenix, AZ, 2015, pp. 1-4. doi: 10.1109/MWSYM.2015.7166766

[11] M. Rudolph, R. Behtash, K. Hirche, J. Wurfi, W. Heinrich and G. Trankle, "A Highly Survivable 3-7 GHz GaN Low-Noise Amplifier," 2006 IEEE MTT-S International Microwave Symposium Digest, San Francisco, CA, 2006, pp. 1899-1902. doi: 10.1109/MWSYM.2006.249786

[12] Zhiqun Cheng, Yong Cai, Jie Liu, Yugang Zhou, Kei May Lau and K. I. Chen, "Monolithic integrated C-band low noise amplifier using AlGaN/graded-AlGaN/GaN HEMTs," 2005 Asia-Pacific Microwave Conference Proceedings, 2005, pp. 4 pp.-. doi: 10.1109/APMC.2005.1606470

[13] Hongtao Xu, C. Sanabria, A. Chini, S. Keller, U. K. Mishra and R. A. York, "A C-band high-dynamic range GaN HEMT low-noise amplifier," in IEEE Microwave and Wireless Components Letters, vol. 14, no. 6, pp. 262-264, June 2004. doi: 10.1109/LMWC.2004.828020

[14] A. Mattamana, W. Gouty, W. Khalil, P. Watson and V. J. Patel, "Multi-Octave and Frequency-Agile LNAs Covering S-C Band Using 0.25 μm GaN Technology," 2016 IEEE Compound Semiconductor Integrated Circuit Symposium (CSICS), Austin, TX, 2016, pp. 1-5. doi: 10.1109/CSICS.2016.7751057

# Decreasing the Actuation Voltage in Electrowetting on Dielectric With Thin and Micro-Structured Dielectric

Semih Türk,
Erik Verheyen,
and Reinhard Viga
Electronic Components and Circuits
University Duisburg Essen
Duisburg, Germany
Email: semih.tuerk@uni-due.de

Sonja Allani,
Andreas Jupe,
and Holger Vogt
CMOS Microsystem Technologies
Fraunhofer Institute for
Microelectronic Circuits and Systems
Duisburg, Germany
Email: holger.vogt@ims.fraunhofer.de

*Abstract*—**This work presents the analysis of the minimum actuation voltage $V_{min}$ for droplet actuation with electrowetting on dielectric (EWOD). First, the fundamentals of electrowetting are described. In the second chapter, the impact on the actuation voltage in EWOD is shown by a dielectric deposited with atomic layer deposition (ALD) and micro-structured surface. In the last part, results of a simulation with COMSOL Multiphysics® are presented to verify the hypothesis and a short discussion about the results is given.**

## I. INTRODUCTION

In several digital microfluidic application, electrowetting is used to modify the surface tension of a fluid and to control the contact angle between a solid surface and a vapor-liquid interface. This opens up new areas of application like an optical microlense [1], a novel flexible display [2] or a parallel cell sorter with higher sorting speed [3]. By utilizing electrowetting the surface stress of a sessile droplet is changed by an electrical field between two electrodes, which are enclosed by dielectric materials [4]. This results in a change of the contact angle and can be described by the Lippmann-Young equation [5]:

$$\cos(\theta) = \cos(\theta_0) + \frac{\epsilon \cdot V^2}{2 \cdot \gamma \cdot d} \tag{1}$$

Here, $\theta$ is the resulting contact angle, $\theta_0$ is the Young-contact angle defined by the Young-equation [5], $\epsilon$ is the permittivity of the dielectric, $V$ is the applied actuation voltage, $\gamma$ is the surface tension of the liquid droplet and $d$ is the thickness of the dielectric.

A closed EWOD system without applied voltage is shown in figure 1a and the resulting contact angle is $\theta_0$. By switching the electrical field to the next electrodes like shown in figure 1b, the force $F_{EWOD}$ acts on the droplet caused by the Maxwell stress tensor and the resulting contact angle is $\theta$ [5]. If $F_{EWOD}$ is bigger than the force of the resistivity due to the contact angle hysteresis [6], the droplet will move to the next electrodes.

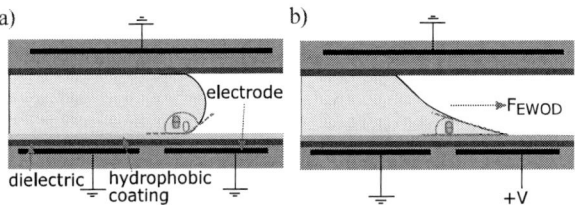

Figure 1. Closed EWOD system with a water droplet. In a) no electric voltage is applied and the ground potential is both on the top and the bottom electrode. In b) electric voltage is applied. The potential difference between the first and the second electrode leads the droplet moving to the right.

Since the contact angle saturates at $V = 40 - 50\,\text{V}$, the actuation voltage must be decreased for an even wider range of the contact angle's modification. The reason for the saturation effect has not yet been researched [5].

In this work, at first the impact on the minimum actuation voltage $V_{min}$ for EWOD is analyzed analytically by decreasing the thickness of the dielectric with ALD deposition and increasing the Young-contact angle by micro-structuring of the surface. Next the Young-contact angle of a water droplet on a deposited polymer is measured and is continued to be used in a finite element method (FEM) based simulation with COMSOL Multiphysics® to verify the hypothesis.

## II. MINIMUM ACTUATION VOLTAGE $V_{min}$ FOR EWOD

In the following the impact of the dielectric's thickness and micro-structured surface on the actuation voltage is shown.

### A. State of the art

In literature a wide range of work exists about decreasing the actuation voltage in EWOD by e.g. varying the type and size of the liquid droplet, the atmosphere in which the droplet is moving and the dielectric layer. Both Fair et al.

[7] and Polack et al. [8] used for insulation of the electrodes an approximately 800 nm coating of parylene C and showed in their experiments that an approximately 900 nL droplet of 0.1 M KCL solution in air moved at $V = 40 - 50$ V. However, Cho et al. [4] used 400 nL water droplet in air atmosphere on 100 nm silicon dioxide and could decrease the voltage to $V = 18 - 25$ V. Moreover Samad et al. [9] moved water in air on 1.8 $\mu m$ polyvinylidene difluoride dielectric with $V = 14.8\ V$ and they could decrease the actuation voltage by optimize the electrode shape and using multi-layer dielectric coating to $V = 12.5$ V [10]. In all works Teflon™ is used as a hydrophobic coating over the dielectric layer.

In this work the focus will be on the thickness of the dielectric and micro-structuring of the surface for a further reduction of the actuation voltage. Furthermore a water droplet in air atmosphere is chosen because it requires higher actuation voltages than in silicone oil [8]. So the results of this work can easily be transferred to water droplet actuation in silicone oil.

### B. Dielectric's thickness

A minimum actuation voltage $V_{min}$ is needed due to a hysteresis of the contact angle. If $V_{min}$ and the hysteresis angle $\alpha$ are negligible small, than $V_{min}$ is approximately [5]:

$$V_{min} \approx 2\sqrt{\frac{\gamma \alpha \sin(\theta_0)}{C}} \qquad (2)$$

here, $C$ is the capacitance of the dielectric and $\alpha$ is the hysteresis angle.

Since the capacitance is anti-proportional to the dielectric's thickness, the minimum actuation voltage $V_{min}$ is proportional to the dielectric thickness and anti-proportional to the Young-contact angle $\theta_0$. A method to deposit a few atomic layers of dielectric is ALD.

### C. Micro-structured surface

One method to increase the contact angle of an even hydrophobic layer is micro-structuring of the surface. He et al. [11] showed that layers with contact angles of $140°$ and higher have been fabricated, which are called super-hydrophobic. There are two models describing the contact angle on a micro-structured surface. The Wenzel model defines a droplet filling the graves between the nanopillars of the structure [11]. This results in decreasing a hydrophilic contact angle and increasing a hydrophobic contact angle, as shown in figure 2a.

In this work the focus is on the Cassie-Baxter model. It describes a droplet sitting above the nanopillars of the structure with the length $a$, the spacing $b$ and the height $H$ in figure 2b. Here, the contact angle is

$$\cos\left(\theta_{0,C}\right) = -1 + f \cdot \left(1 + \cos\left(\theta_0\right)\right) \qquad (3)$$

where $f = \frac{a^2}{(a+b)^2}$ is the ratio of the contact area of the droplet and the surface to the total horizontal area of one nanopillar. So a droplet sitting on the nanopillars always increases the contact angle. He et al. [11] used in their work

nanopillars in $\mu m$-dimension and analyzed the corresponding contact angle. The result is that the calculated theoretical values of the contact angle fits best to the experimental values at a ratio of $\frac{a}{b} \approx 1$.

The combination of decreasing the thickness of the dielectric by ALD and increasing the contact angle by micro-structuring of the surface leads to an minimum actuation voltage, as shown in figure 3.

### D. Determining the Young-contact angle

The Young-contact angle on a smooth hydrophobic coating is determined in initial experiments in the laboratories of the Fraunhofer IMS in Duisburg. For this, a silicon wafer is used and an octafluorocyclobutane ($C_4F_8$) plasma gets ignited in a chemical vapour deposition (CVD). In figure 4a the energy dispersive X-ray spectroscopy (EDX) measurement result is shown and the deposited polymer contains twice as much fluorine (F) atoms than carbon (C) atoms, which indicates that the deposited layer is a polytetrafluoroethylene (PTFE). In addition, the contact angle measurement result in figure 4b shows that a water droplet on the deposited PTFE coating has a contact angle of $\theta_0 = 109°$, which matches to the contact angle of water on PTFE coating from other works [12].

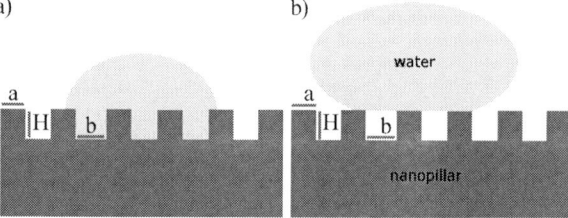

Figure 2. In a) the droplet fills the graves and in b) the droplet sits above the nanopillars of the structure. H is the height, a is the length and b is the spacing between the nanopillars.

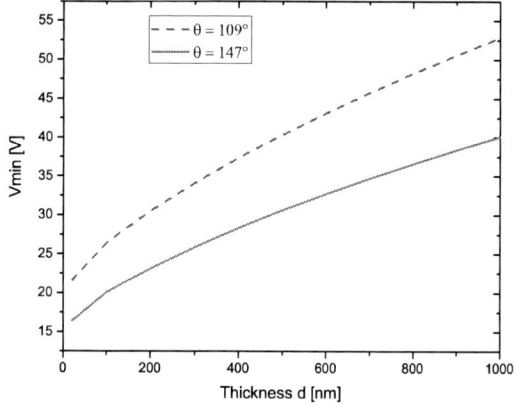

Figure 3. Calculated values of the minimum actuation voltage $V_{min}$ as a function of the thickness of $Al_2O_3$ dielectric for two different contact angles.

## III. SIMULATION OF A DROPLET TRANSPORT

To investigate the impact of decreasing the dielectric's thickness and increasing the Young-contact angle on the minimum actuation voltage $V_{min}$ for EWOD, 2-D simulations for a water droplet transport in air atmosphere are done with COMSOL Mutliphysics® 5.3a, as shown in figure 5. COMSOL Multiphysics® is a finite element method (FEM) based simulation tool. FEM is a numerical method to solve physical problems with differential equations. An object or domain is subdivided into finite elements which consist of simplified geometric shapes like triangular and quadrilateral. The physics of these geometries is easy to calculate using initial functions [13]. The physical behavior of the entire object is dependent on continuity conditions which must fulfill the initial functions. To highlight the difference between very thin dielectric with micro-structured surface and thick dielectric with smooth surface, two cases were defined. In case 1 the droplet movement is done on a micro-structured $Al_2O_3$ dielectric layer with a thickness of $20\,nm$ and in case 2 on a smooth $Al_2O_3$ dielectric layer with a thickness of $1000\,nm$. The results of both cases are compared.

### A. Model definition and boundary conditions

To simulate a water droplet movement in air with EWOD, the velocity and pressure field for incompressible fluids and a two-phase flow is solved by the equation for continuity and Navier-Stokes [14].

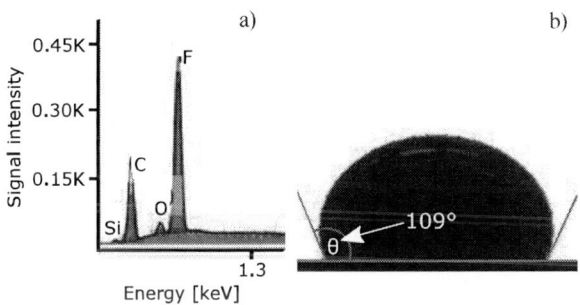

Figure 4. Measurement result of a) the EDX of the deposited polymer and b) the contact angle between a water droplet and the deposited polymer.

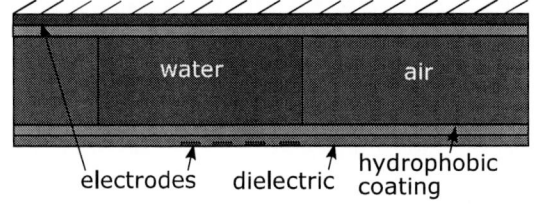

Figure 5. Overview of the geometry used in the simulation. At the beginning the water droplet is defined as an rectangle, which deforms after the simulation starts.

For this, the Two-Phase-Flow, Phase Field interface has been selected. Besides, on the bottom and top surface, the wetted wall condition and navier-slip condition are selected. Also the contact angle $\theta_0$ is defined on the bottom and top surface. For case 2 the previously determined contact angle $\theta_0 = 109°$ from figure 4b is continued to be used. For the micro-structured surface in case 1, the contact angle $\theta_{0,C} \approx 147°$ is determined analytically by the Cassie-Baxter equation (3) for nanopillars with equal length and spacing of $10\,\mu m$. Additionally on the activated electrode the contact angle $\theta$ is measured with the Lippmann-Young equation (1). This separates the electrostatic and the fluid flow. The electrical potential of each electrode is time dependent to get an potential difference between adjacent electrodes. To keep the simulation effort as low as possible, the electrodes are represented as a flat area both for the micro-structured surface and the smooth surface. On the vertical boundaries an inlet and an outlet is defined to compensate hydrostatic pressure and suppress backflow. All used parameters for the simulation are shown in table I. Further details of the used physics can be found in [15].

### B. Results and discussion

Both simulations start at the same position, like indicated in the mass fraction in figure 6a. According to the phase field method, the blue phase represents the water droplet and the red phase represents air. After the simulation has started, the change of contact angle results in droplet moving. The different contact angles of booth simulations can be seen at the curvature of the interface from air to water.

As described in figure 3, the calculated theoretical value for the droplet actuation voltage is for $\theta_0 = 109°$ approximately $50\,V$ and for $\theta_0 = 147°$ approximately $16\,V$. Nevertheless, in the simulation the actuation voltages are fewer. For $\theta_0 = 109°$ only $V_{min} = 20\,V$ was required and for $\theta_0 = 147°$ only $V_{min} = 3\,V$ was required. The mass fraction of both cases is shown in figure 6b and 6c.

In figure 7 the distance of the droplet actuation is shown as a function of time. In case 1, the droplet has traveled a distance of $2500\,\mu m$ in approximately $48\,ms$. In case 2, approximately $120\,ms$ are needed for the same distance. So in case 1 the droplet is more than twice as fast as in case 2. Also the course

Table I. Parameter Used for the Simulation With Comsol Multiphysics

| Parameter name | value (case 1) | value (case 2) |
|---|---|---|
| actuation voltage [V] | $V(1 - n \cdot t_{step})$ | $V(1 - n \cdot t_{step})$ |
| $t_{step}$ [ms] | 0.5 | 0.5 |
| $\epsilon_{r,Al_2O_3}$ | 9 | 9 |
| $\epsilon_{r,PTFE}$ | 2 | 2 |
| microfluidic channel, height [$\mu m$] | 100 | 100 |
| microfluidic channel, length [$\mu m$] | 3000 | 3000 |
| water droplet, height [$\mu m$] | 100 | 100 |
| water droplet, length [$\mu m$] | 200 | 200 |
| surface tension [$mN/m$] | 72 | 72 |
| electrode, length [$\mu m$] | 40 | 40 |
| contact angle [°] | 147 | 109 |
| dielectric thickness [$nm$] | 20 | 1000 |

of the lines are showy. In booth cases, the velocity is higher at the beginning and decreases to a nearly constant velocity.

## IV. CONCLUSION AND OUTLOOK

The presented analysis shows that the actuation voltage for EWOD can be decreased by both thin deposition of dielectric structures with ALD and increasing the contact angle $\theta_0$ by micro-structuring of the surface. First, a theoretical analysis has been done and second the Young-contact angle $\theta_0$ is determined on a deposited polymer. Micro-structuring the surface increases the contact angle from $109°$ to $147°$. The analytically analyzed minimum actuation voltage is in combination of both micro-structured surface and $20\ nm\ Al_2O_3$ dielectric $V_{min} = 16\ V$. However, simulation results performed with COMSOL Multiphysics® show that the droplet actuation is possible with only $V_{min} = 3\ V$. This is the lowest actuation voltage for EWOD in literature, considering that it is a simulation result. This divergence can be caused by neglecting the contact angle hysteresis in the simulation. Hu et al. [16] used the dynamic contact angle instead of the static contact angle. This considers the change of the contact angle during the impact on the surface, which will be taken into account in

further works. Moreover, the speed of the droplet movement could be increased. In the simulation the droplet movement on micro-structured $20\ nm\ Al_2O_3$ dielectric with $V_{min} = 3\ V$ is twice as fast as on smooth $1000\ nm\ Al_2O_3$ dielectric with $V_{min} = 20\ V$. Nevertheless, further physical characteristics of the dielectric should be considered in further works, e.g. the dielectric breakdown or the dielectric constant. Higher dielectric constants could decrease the voltage even further.

## ACKNOWLEDGMENT

This research work is sponsored by the German Research Foundation (DFG) under contract No. VO 802/4-1 and 638556.

## REFERENCES

[1] B. Berge, "Liquid lens technology: principle of electrowetting based lenses and applications to imaging," *18th IEEE International Conference on Micro Electro Mechanical Systems*, pp. 227-230, Florida, USA, 2005.

[2] D. Y. Kim and A. J. Steckl, "Electrowetting on Paper for Electronic Paper Display," in *ACS Applied Materials & Interfaces*, vol.2, pp. 3318-3323, 2005.

[3] S. Kahnert, "Entwicklung einer Mikrochip-navigierten Zellsortieranlage," *Ph.D. dissertation*, electronic components and circuits, University Duisburg-Essen, 2016.

[4] S. K. Cho, H. Moon and C. J. Kim, "Creating, transporting, cutting, and merging liquid droplets by electrowetting-based actuation for digital microfluidic circuits," in *Journal of Microelectromechanical Systems*, vol. 12, no. 1, pp. 70-80, 2003.

[5] J. Berthier, *Micro-Drops and Digital-Microfluidics*, 2nd ed. Oxford, England: Elsevier, 2013.

[6] H. Tavana and A. Neumann, "On the question of rate-dependence of contact angles," in *Colloids and Surfaces A: Physicochemical and Engineering Aspects*, vol.282, pp. 256-262, 2006.

[7] R. B. Fair, M. G. Pollack, R. Woo, V. K. Pamula, R. A. Hong, T. Zhang and J. T. Venkatraman, "A micro-watt metal-insulator-solution-transport (MIST) device for scalable digital bio-microfluidic systems," *International Electron Devices Meeting, Technical Digest (Cat. No.01CH37224)*, pp. 16.4.1-16.4.4, Washington, DC, USA, 2001.

[8] M. G. Pollack, A. D. Shenderov and R. B Fair, "Electrowetting-based actuation of droplets for integrated microfluidics," in *Lab Chip, The Royal Society of Chemistry*, vol. 2, pp. 96-101, 2002.

[9] M. F. Samad, A. Z. Kouzani, M. M. Rahman, K. Magniez and A. Kaynak, "Design and fabrication of an electrode for low-actuation-voltage electrowetting-on-dielectric devices," in *Procedia technology*, vol. 20, pp. 20-25, 2015.

[10] M. F. Samad, A. Z. Kouzani, M. F. Hossain, M. I. Mohammed and M. N. H. Zainal, "Reducing electrowetting-on-dielectric actuation voltage using a novel electrode shape and a multi-layer dielectric coating," in *Microsystem technologies*, vol. 23, pp. 3005-3013, 2017.

[11] B. He, N. A. Patankar, and J. Lee, "Multiple Equilibrium Droplet Shapes and Design Criterion for Rough Hydrophobic Surfaces," in *Langmuir*, vol. 19, Art. 12, pp. 4999-5003, 2003.

[12] M. Kalin and M. Polajnar, "The wetting of steel, DLC coatings, ceramics and polymers with oils and water: The importance and correlations of surface energy, surface tension, contact angle and spreading," in *Applied Surface Science*, vol. 293, pp. 97-108, 2014.

[13] H. J. Bungartz and M. Griebel, "Sparse grids," in *Acta Numerica*, vol. 13, pp. 1-123, Cambridge, England: Cambridge University Press, 2004.

[14] J. Hu, R. Jia, X. Huang, X. Xiong and K. T. Wan, "Numerical Simulation of the Dynamics of Water Droplet Impingement on a Wax Surface," in *ASEE Northeast Section Conference*, Boston, USA, 2015.

[15] R. W. Pryor, "Multiphysics Modeling Using COMSOL®: A First Principles Approach," *Ph.D. dissertation*, Institute of nano- and medical electronic, TU Hamburg-Harburg, Hamburg, Germany, 2009.

[16] J. Hu, X. Xiong, H. Xiao and K. T. Wan "Effects of Contact Angle on the Dynamics of Water Droplet Impingement," *COMSOL Conference*, Boston, USA, 2015.

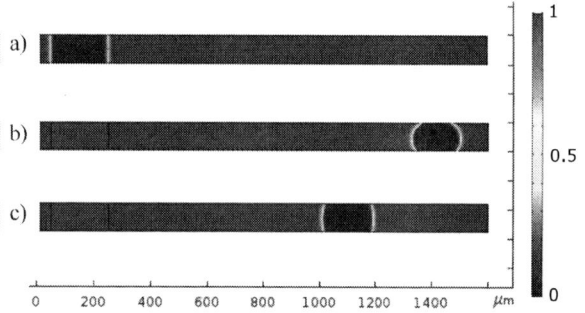

Figure 6. Mass of fraction a) at $t = 0\ ms$, b) after $t = 50\ ms$ in case 1 and c) after $t = 50\ ms$ in case 2.

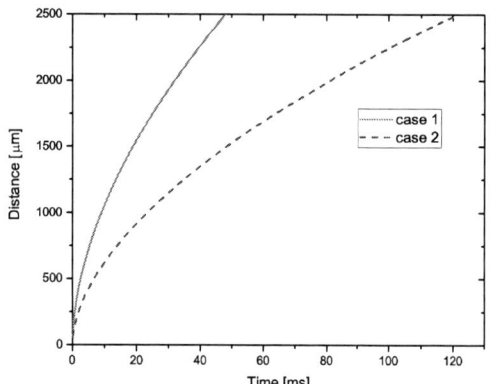

Figure 7. Distance of the droplet movement as a function of time for case 1 and case 2.

*PRIME 2018, Prague, Czech Republic*

*Session: Sensing and Biomedical Circuits I*

# Exploiting nonlinearities to improve the linear region in an electrostatic MEMS demodulator

Jeremy Scerri, Barnaby Portelli, Ivan Grech, Edward Gatt, Owen Casha

Microelectronics & Nanoelectronics Department,
Faculty of ICT, University of Malta,
Msida, Malta.
jeremyscerri@gmail.com

*Abstract*—**This paper presents a technique whereby the overall nonlinear behavior of an electrostatically actuated and sensed MEMS is linearised for most of its usable range. The nonlinear characteristics are first analysed theoretically. This analysis reveals that the nonlinearity can be 'neutralised' by replacing the spring with a nonlinear - cubic stiffness - spring. Finding a feasible solution requires finding a compromise between a large number of geometric dimensions and constraints; this was achieved by making extensive use of MATLAB's optimization toolbox. The device having optimal dimensions was manufactured using the SOIMUMPs process and lab measurements confirmed that the overall nonlinearity was practically eliminated for actuation voltages of 4 volts and upwards.**

*Keywords— nonlinear MEMS, cubic stiffness, hardening spring, electrostatic nonlinearity*

## I. INTRODUCTION

With MEMS and nanoscale structures, nonlinear behavior becomes inevitable. Generally, nonlinear behavior is considered detrimental as it reduces the operating range of the device. However, nonlinearities can also be harnessed such that the device performs some required nonlinear function. In literature, one can find numerous studies involving improved modelling of nonlinear MEMS devices [1], studies that make use of these nonlinearities [2, 3] and studies that focus on reducing nonlinearities to improve the linear region [4].

The next section details how process constraints affect the design decisions for the electrostatic demodulator topology. The subsequent two sections describe the mathematical model, the static nonlinear behavior and how redesigning of the springs reduce the overall nonlinearity. The last section presents a comparison between actual measurements and theoretical predictions.

## II. DESIGN OPTIONS

### A. Electrostatic Mixing – Actuation and Sensing

The demodulator was designed within the SOIMUMPs process constraints. The SOIMUMPs process provides choice between two silicon layer heights, 10 μm & 25 μm. This layer is insulated from the substrate by an oxide layer. The 25 μm option was chosen such that all the capacitances designed have the largest vertical plates possible.

Mixing or frequency shifting is achieved by applying the local oscillator (LO) voltage, $v_{LO}$, and the radio frequency (RF) voltage, $v_{RF}$, to the plates of a capacitor ($C_{LR}$). The force generated is proportional to the voltage squared, which force is used to create a displacement with the required frequency components [5]. On the sensing side, a differential setup is adopted such that parasitics are minimized, hence, two capacitors are required - $C_{S1}$ and $C_{S2}$ - such that when the gap on one increases, the gap on the other decreases.

The SOIMUMPs process does not offer the possibility to design two plates which are electrically isolated in the horizontal direction while being mechanically connected. Due to this, the RF plate is used for both actuation and sensing - Fig. 1.

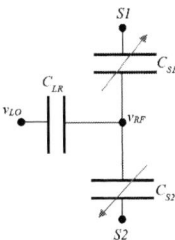

Fig. 1. Actuation and Differential Sense Capacitors.

### B. Topological Considerations

For practically realistic intermediate frequency (IF), the inertia of the mechanical structure has to be kept as low as possible, the spring constant has to be maximised, all the while keeping sufficient capacitive area and displacement for both actuation and sensing. This requires that actuation and sensing gaps make use of the smallest available gap of 2 μm in SOIMUMPs which means that a smaller gap for the stoppers has to be used resulting in design rule violation. In the SOIMUMPs process, the 2 μm gap is usually used for the stopper and a 2.25 μm to 2.5 μm gap is used for actuation or sensing capacitors [6]. In this work, this problem is circumvented by adopting a rotational setup. With this setup, the actuation and sensing gaps can be kept at the minimum of 2 μm while the stoppers' gap is also at the minimum of 2 μm however the stoppers are positioned at a larger radial distance. This amplifies the movement such that the stoppers close the gap first. To make use of the smallest gap allowed of 2 μm, the

978-1-5386-5388-3/18 $31.00 © 2018 IEEE

process specification requires that the gaps are in the orthogonal direction. This implies that under rotation, the capacitive gap does not remain parallel. For this effect to be negligible (< 0.01% of nominal capacitance), the radius of the combs has to be greater than 600 μm [7] for a finger length of 100 μm (the maximum length allowed for a 2 μm structure). Fig. 2 shows the adopted layout.

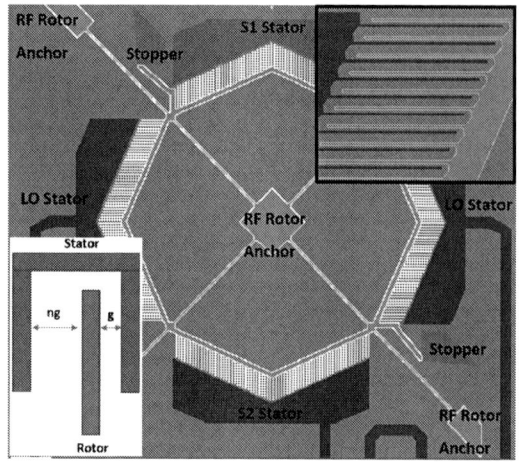

Fig. 2. The stators and rotor with insets showing finger spacing

## III. MATHEMATICAL MODELLING

For actuation, the structure makes use of two combs for the LO signal and these interact with the RF combs on the rotor. The number of fingers on each stator comb is $N$. Since the two LO stator combs are on opposite sides, the finger gaps are in such a way that one would be pulling and the other would be pushing the rotor to achieve rotation. As shown in Fig. 2 inset, there are two gaps that control the generated force - the force produced by the larger gap should ideally be negligible. The larger gap is a multiple of the smaller gap 'g' where the constraints on the multiplier 'n' are discussed next.

### A. Displacement Differential Sensing

The sensing stator combs, S1 and S2, have $N$ fingers each. There are two rotor-to-stator finger gaps (Fig. 2 inset) and these can be considered as two capacitors in parallel. The total capacitance in the S1 stator is given in (1):

$$C_{S1}(x) = \varepsilon AN \left[ \frac{1}{ng + x} + \frac{1}{g - x} \right] + C_f \quad (1)$$

where, $\varepsilon$ is the absolute permittivity of air, $A$ is the finger overlap area, and $x$ is the varying displacement and $C_f$ the fringe capacitance. To achieve differential sense for the S2 stator, ideally $C_{S2}(x) = C_{S1}(-x)$. The function $C_{S2}(x)$ has a minimum at $x = g(n-1)/2$. It was required to have both $C_{S1}(x)$ & $C_{S2}(x)$ monotonic for the whole range of motion ($0 < x < g/3$) and this gave a lower limit on $n$ of 5/3.

### B. Actuation

The same setup as in Fig. 2 inset diagram was used for actuation. Hence the net force per finger $\Delta F_f$ can be described with (2):

$$\Delta F_f = \varepsilon A \frac{\Delta V^2}{2} \left[ \frac{1}{(g - x)^2} - \frac{1}{(ng + x)^2} \right] \quad (2)$$

where $\Delta V = v_{RF} - v_{LO}$. Since the fingers are not at the same distance from the centre of the rotor, each finger gives a different torque contribution. The total torque, $T$, of the rotor can be expressed as a sum as in (3):

$$T = 4 \sum_{i=1}^{N/2} \Delta F_f \, d_i = 4\Delta F_f \sum_{i=1}^{N/2} a + b_i + c \quad (3)$$

where, $c = 50$ μm (half the finger length) and $a = r - l\cos\alpha$, are both constants (Fig. 3) and for an octagon $\alpha = 67.5°$. Only $b_i$ is a function of the finger position, $i$, $b_i = l \cos\alpha \, (N-2i)/N$. Equation (3) sums up to (4):

$$T = 2N\Delta F_f D \quad (4)$$

where $D = (a + c + l/2 \cos \alpha)$.

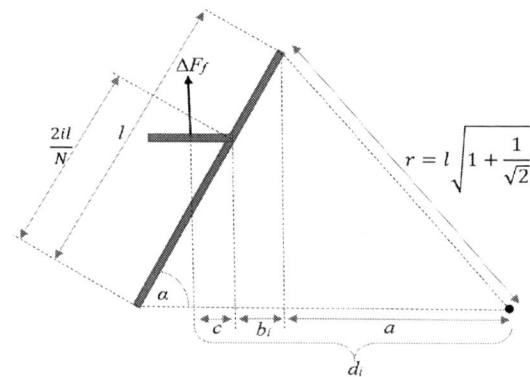

Fig. 3. One side of the octagon showing the $i$th finger.

Under static conditions, this torque is acting against the four clamped-clamped beam springs. Each beam has length $2r$. The force generated at mid-span of each beam is $T/4r$. Equation (5) gives the deflection, $x$, at mid-span:

$$x = Tr^2 / 96EI \quad (5)$$

where $E$ is the Young's modulus and $I = tw^3/12$, the second moment of area of the spring rectangular section, $w$ the spring width and beam depth, $t$, is fixed by the SOI thickness at 25 μm. From (5), the rotational deflection $\theta$ of the rotor would be:

$$\theta = T / (4kr^2) \quad (6)$$

where $k = 192EI / (2r)^3$ is the transverse stiffness of each beam. A closed form expression for the inertia of the rotor, $J$, as a function of rotor dimensions was obtained using the parallel axis theorem and a summing technique similar to the one used in (3). It is not reproduced here due to space constraints. From this an estimate of the rotor resonant frequency, $f_r$, can be obtained as given by (7):

$$f_r = (1/2\pi)\sqrt{(4kr^2 / J)} \quad (7)$$

## C. Using MATLAB's Nonlinear Constrained Optimization

Having this set of equations describing actuation, sensing and resonance made it possible to optimize for some required objective while at the same time respecting process and geometric constraints. This involved creating two scripts: one as a main script using MATLAB's *multistart* function, the second containing the description of the constraints i.e. equations (1) to (7) and also the parameters that constrain the design. MATLAB's *multistart* is a function that starts a simulation a number of times from different points in the search space and gives the optimal solutions reached from each starting point. The use of this function is instrumental in finding several sets of dimensions satisfying the constraints.

Three parameters are used to constrain the problem, $f_r$, $x_{max}$ (pull-in) @ $\Delta V = 10$ v and the change in capacitance at each sensor comb $\Delta C_S$. The solutions are then further studied for sensitivity and those solutions which are found to be highly sensitive to small changes (order of magnitude of the process tolerance) are discarded. The dimensions that were eventually selected and confirmed to be within 5% accuracy with finite element analysis (FEA) are listed in Table I. These gave $f_r = 25$ kHz, $x_{max} = g/3$ @ $\Delta V = 12$ v, and $\Delta C_{s1} = 200$ fF (closing gap) with $C_s = 1$ pF for no displacement as in Fig. 4.

TABLE I.     GEOMETRIC PARAMETERS ADOPTED IN SI UNITS

| $n$ | $g$ | $l$ | $r$ | $w$ | N | $A$ |
|---|---|---|---|---|---|---|
| 3 | $2\times10^{-6}$ | $458\times10^{-6}$ | $598\times10^{-6}$ | $10\times10^{-6}$ | 71 | $2.25\times10^{-9}$ |

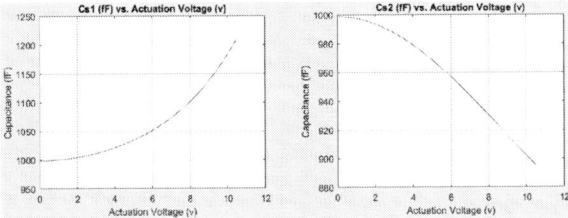

Fig. 4. The resulting Sensor capacitance for different Actuation voltages

## IV.    NONLINEARITIES IN SENSING AND ACTUATION

As is evident in Fig. 4, the capacitance for both sensing combs is nonlinear with voltage, especially on the closing gap side ($C_{S1}$). On the widening gap sensor comb ($C_{S2}$), for $\Delta V > 4$ V, the relationship is approximately linear. This was investigated further.

## A. Nonlinear Contributions

The equations that govern actuation to sensing are all nonlinear except (6) which is a linear relationship between torque and rotation. Fig. 5 breaks down the two composite functions $C_{S1}(x(\Delta F_f (\Delta V)))$ & $C_{S2}(x(\Delta F_f (\Delta V)))$. The reason for having a higher nonlinear behaviour for $C_{S1}$ can be understood by observing that $\Delta F_f (\Delta V)$ has increasing positive gradient just like $C_{S1}(x)$ while $C_{S2}(x)$ has decreasing negative gradient for $x > 0$.

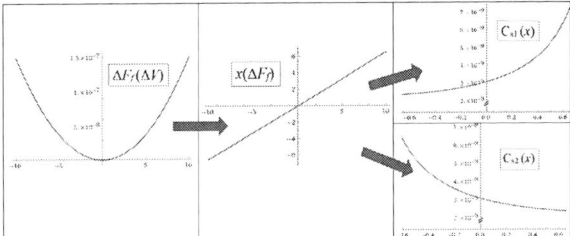

Fig. 5: Breakdown of the composite functions

## B. Reducing the overall nonlinearity

The spring stiffness function, $x(\Delta F_f)$, was modified such that it acts to neutralise the steep gradient in the closing gap ($C_{S1}(x)$). This was achieved by introducing a hardening spring i.e. a cubic stiffness term together with the linear term for the total torque provided by the four springs as in (8).

$$T = k_l \theta + k_c \theta^3 \qquad (8)$$

where $k_l$ and $k_c$ are the linear and cubic stiffness coefficients respectively. For this to be achieved, redesign of the springs was required. MATLAB optimization was employed again, this time to find ratios of $k_l / k_c$ that tone down the steep gradient of the closing gap while keeping adequate overall

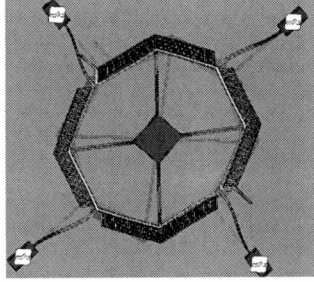

Fig. 6: Alternative spring layout provides both transverse and axial stiffness

sensitivity. The objective is to minimise the second derivative of $C_{S1}(x)$ and $C_{S2}(x)$ and to maximise the first derivative of $\Delta C_{S1}$ and $\Delta C_{S2}$. The ratios for $k_l / k_c$ have to be physically realizable and hence an alternative spring layout that provides cubic stiffness is employed as shown in Fig. 6. In this new layout, each cantilever spring of length $q$ offers transverse stiffness $k_t = 12EI/q^3$ however this layout offers also axial stiffness as the spring is prone to elongation apart from bending. Hence, a component of axial stiffness $k_a = E(tw)/q$ contributes towards rotor rotation and the resulting total torque is as in (9);

$$T = 4 (k_t x + k_a \Delta q \gamma) r \qquad (9)$$

where $\Delta q$ is the spring elongation and $\gamma$ is the angle subtended by the spring as shown in Fig. 7.

Fig. 7: Cantilever spring geometry and parameters

Replacing linear with rotational displacement and $\Delta q$ with an approximation [8] of $r\theta$ $(r\theta/q)^2/2$ results in (10),

$$T = 4k_l r^2\,\theta + (2k_a r^4/q^2)\,\theta^3 \qquad (10)$$

Optimal values obtained from MATLAB for $k_l$ and $k_c$ were fitted to (10) and these fixed actual spring dimensions that gave a better overall linearity for $C_{S1}(\Delta V)$ with only a slight degradation on $C_{S2}(\Delta V)$. In [8], an H shaped spring fixture was used to control the linear and cubic stiffness ratio, however, this was not needed here. Only dimensions for the spring were changed; dimensions related to the octagonal rotor were held the same. For the cantilever spring, $w = 9\ \mu m$ and $q = 458\ \mu m$ were selected. Fig. 8 shows the improvement. This was also confirmed with FEA.

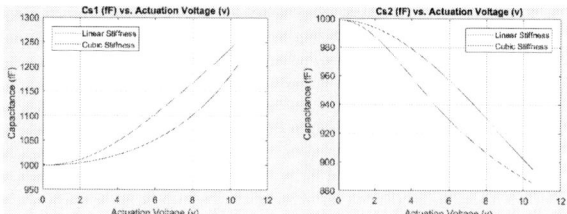

Fig. 8: Improved linearity by introducing a nonlinear spring

## V. Measurements and Experimental Results

Figure 9 shows the manufactured device. On visual inspection, it was noticed that some critical parameters had considerable inaccuracies albeit within process tolerances.

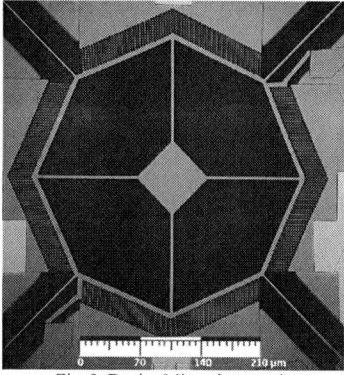

Fig. 9: Device Microphotograph

The actual comb fingers were at $1.5\ \mu m$ width rather than $2\ \mu m$. This changed the gap, $g$, to $2.5\ \mu m$, and the resulting $n$ was 2.6 (still $> 5/3$). Spring width $w$ was also found smaller at $8.4\ \mu m$. FEA simulations were carried out with these new parameters and sensing comb capacitances were measured for different actuation voltages. Fig. 10 shows the actual measurements when compared to the cubic and linear stiffness models.

Capacitance measurements were performed with an LCR meter (Agilent E4980A) at $2\ MHz$ with 256-point averaging. Attention was given to reduce parasitics as much as possible and measurements were repeatable to within $1\ fF$. The measured results are closer to the cubic stiffness model and hence more linear. To take advantage of the achieved linear

range, it is planned that the RF signal is DC shifted to 6 V such that the device input-output relationship is effectively linear.

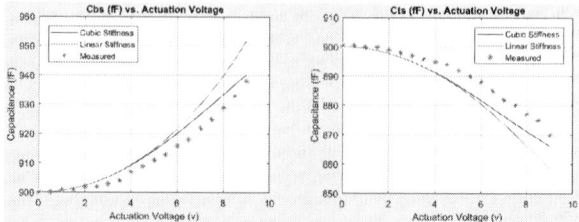

Fig. 10: Actual measurements vs. Linear and Cubic Stiffness models

With the actual dimensions, FEA simulation resulted in a resonant frequency of 20.4 kHz while experimental results yielded 20.7 kHz which is well within 5% of error. Pull-in occurred at a higher voltage of 13.78 V (due to a larger $g$).

## Conclusions

The results show that it is in fact possible to neutralise nonlinear behavior using another nonlinearity. Although linearity was not achieved for the whole range of operation, the technique still provided a large enough linear range for practical application. The process of finding optimal parameters that satisfy linearity and sensitivity was sped up using MATLAB's optimization toolbox. With this technique, FEA, which is much more computationally intensive, was only used at the very last step for confirmation. Moreover, although the actual demodulator dimensions differed from the original design dimensions, the cubic stiffness still managed to provide a large enough linear region. This is attributed to the fact that the optimal set of dimensions selected where tested for sensitivity to dimensional tolerances and the solution that was sent for manufacturing was one which was optimal for a 'broad' range of dimensions.

Currently, the device is undergoing tests to characterize the dynamics and its efficacy as a demodulator.

## References

[1] A. M. Elshurafa, et al., "Nonlinear Dynamics of Spring Softening and Hardening in Folded-MEMS Comb Drive Resonators," in *Journal of Microelectromechanical Systems*, vol. 20, no. 4, 2011.

[2] J. Scerri, I. Grech, E. Gatt and O. Casha, "Reduced-order model for MEMS PZT vibrational energy harvester exhibiting buckling bistability," in *Electronics Letters*, vol. 51, no. 5, pp. 409-411, 3 5 2015.

[3] Rhoads JF, Shaw SW, Turner KL., "Nonlinear Dynamics and Its Applications in Micro- and Nanoresonators,", ASME 2008.

[4] E. Tatar, T. Mukherjee and G. K. Fedder, "Nonlinearity tuning and its effects on the performance of a MEMS gyroscope," *18th International Conference on Solid-State Sensors, Actuators and Microsystems*, 2015.

[5] J. Scerri, I. Grech, E. Gatt and O. Casha, "Suppression of spurious products in an electrostatic RF MEMS downconverter having differential drive and sense," MELECON 2016.

[6] Rosana A Dias and Luis A Rocha, "Improving capacitance/damping ratio in a capacitive MEMS transducer," IOP Publishing, Journal of Micromechanics and Microengineering, Volume 24, Number 1, 2014.

[7] Patla Biju, "Small Angle Approximation for Non-parallel Plate Capacitors with Applications in Experimental Gravitation" arXiv:1208.2984, 2013.

[8] S. Boisseau, G. Despesse, B. Ahmed Seddik, "Adjustable Nonlinear Springs to Improve Efficiency of Vibration Energy Harvesters", arXiv:1207.4559,2015

# Multi-Channel Electrotactile Stimulation System for Touch Substitution: A Case Study

Hoda Fares[1,3], Lucia Seminara[1], Hussein Chible[3], Strahinja Dosen[2], Maurizio Valle[1]

[1]Cosmic Lab, DITEN, University of Genoa, Genova, Italy
[2]Center for Sensory-Motor Interaction, Aalborg University, Denmark
[3]MECRL Lab, EDST, Lebanese University, Beirut, Lebanon

*Abstract* — Reconstructing the sense of touch in prosthetics is a long-standing research challenge. To this aim, the prosthesis can be supplied with sensory arrays to measure the tactile interaction with the environment. In addition, a reliable feedback system is required to code and transmit the measured somatosensory information to the residual limb. This paper presents a multichannel electrotactile stimulation interface. Two coding schemes (mixed and uniform coding) were tested to assess the ability of the subject to localize the stimulation (identify the active pad). The outcome measures were position recognition and frequency discrimination. Our preliminary results show high accuracies in discriminating different frequency levels, i.e., 80% for low-level frequencies and 87% for high-level frequencies. In addition, the mixed coding has substantially improved the spatial localization. These are important insights regarding the development of multichannel sensing and stimulation systems for feedback in prosthetics.

*Keywords: Electrotactile stimulation, frequency coding, stimulation electrodes, restoration of sense of touch, somatosensory feedback, tactile perception.*

## I. INTRODUCTION

The contemporary upper limb prostheses have made a substantial progress in reproducing the dexterity of the human hands. There are several commercial systems that offer individual finger control, and they are operated intuitively, by translating electrical muscle activity into prosthesis commands using traditional two-channel control or pattern classification [1]. However, none of the commercial systems, apart from one recent example, restore somatosensory feedback to the user. Therefore, the amputees do not "feel" their prosthetic limbs [2]. Sensory feedback can be restored using sensory substitution. The prosthesis is equipped with tactile sensors, from which the data are acquired, suitably coded and communicated to the user by activating spared tactile sensory structures using either invasive (e.g. direct nerve or brain stimulation) or non-invasive interfaces (e.g. electrotactile and mechanical stimulation) [3].

Electrotactile stimulation evokes tactile sensation by delivering the electrical current pulses to the skin through surface electrodes [4]. Electrotactile feedback information can be transmitted by modulating the quality and intensity of the elicited sensations i.e. by changing the stimulation parameters (pulse width, amplitude, and frequency coding) and/or location of the stimulation (spatial coding). Typically, a single stimulation unit is used to convey prosthesis variable, which is often the grasping force. In some cases, multichannel interfaces have been used to encode up to two discrete variables such as grasping force and aperture [5].

Distributed electrotactile stimulation interfaces that comprise matrices of stimulation unites placed over skin surface are believed to be able to provide high-bandwidth information to the user. The quality of the tactile sensation depends on the stimulation parameters, which can elicit vibration, tickling, tingling, buzz, pressures, etc. [6]. The major features of the stimulation that could be modulated to transmit electrotactile feedback information include i) electrical parameters i.e. current amplitude (1-20mA), pulse waveform (monophasic/biphasic), frequency (1Hz-5KHz), pulse width, duration; ii) electrode properties i.e. small or large sized, conducting material; and iii) skin characteristics i.e. location, thickness. Stimulators can be either current or voltage regulated; the last decreases skin burns that could appear from high current intensity stimulation.

Since there are no moving mechanical parts, the electrotactile systems consume less power, have low weight, produce less noise, and respond faster compared to other tactile feedback systems (e.g., vibration motors, force applicators). On the other hand, if not setup properly the electrical stimulation can lead to uncomfortable sensations. Moreover, the stimulation can interfere with the recording of electrical signals (e.g. EMG) for prosthesis control [7].

We aim to restore the sense of touch for prosthetic hand using electrotactile stimulation. However, contrary to the conventional approach where the feedback communicates one or two global prosthesis variable, our goal is to mimic the human sense of touch. Providing high-resolution tactile information comparable to that of the human sense of touch requires adequate artificial sensing systems, which integrate a high-density network of sensing units. To realize this goal, we propose the approach illustrated in Figure 1. The system would comprise three main compartments: 1) a high-density sensory array attached to hand palm and fingertips to acquire and measure distributed tactile data, 2) a socket enclosed embedded electronic system to decode and transmit structured information, and 3) Electrotactile stimulation electrodes to convey the information to the residual limb of the amputee [8],[9].

In this perspective, this paper presents an assessment of the feasibility of using multi-channel electrotactile stimulation interface to convey information to the prosthetic user. We investigate how the modulation of stimulation parameters e.g. frequency coding can improve the perception of the elicited tactile sensation and increase the identification rate of the spatial recognition of the delivered stimulus. Two psychometric experiments were conducted: 1) position recognition test and 2)

978-1-5386-5388-3/18 $31.00 © 2018 IEEE 213

frequency discrimination test. Our preliminary results show high accuracies in discriminating different frequency levels, 80% for low-level frequency (10 Hz) and 87% for high-level frequency (400 Hz).

## II. ELECTROTACTILE STIMULATION SYSTEM

### A. System Description

A programmable computer-controlled stimulator (IntFES Ver 2 MAXSENS, Tecnalia Serbia) is employed. The stimulator is battery powered; it generates current-controlled pulses with pulse intensity in the range of 0-5mA with 0.1mA step, frequency from 1 to 400 Hz with 1Hz step, and pulse width from 50 to 1000µs with 10 µs step. It produces charge-balanced biphasic continuous electrostimulation pulses in any combination of electrodes simultaneously or individually in each electrode. The stimulator has been tested by an oscilloscope and the waveforms with different amplitudes are shown in figure 3.

The host PC adjusts the stimulation parameters by sending text commands to the stimulator via Bluetooth. The stimulation pulses were delivered to the pre-moistened volar side of subject's forearm through a flexible rectangular electrode matrix with total area of $11 \times 5$ cm2. The stimulation electrode matrix comprised $6 \times 4 = 24$ oval fields with longitudinal radius of 5 mm and transversal radius of 3 mm each. The fields were aligned in four vertical lines and six horizontal rows; the center-to-center distance between two adjacent fields was 20 mm in the longitudinal and 14 mm in the transversal directions. Each pad was coated with circular conductive hydrogel elements of 5 mm radius (AG730, Axelgaard, DK) to improve the electrical contact between the fields and the skin and to assure the delivery of comfortable electrotactile stimulation (see figure 4). An insulation coating covers the top of the whole electrode matrix, excluding the fields. The conductive fields acted as cathodes whereas a single self-adhesive electrode (ValuTrode Foam) placed on the dorsal side of the forearm acted as the common anode. The ValuTrode electrode was made of glycerin, water and poly (acrylate) co-polymer.

## III. METHODS AND EXPERIMENTAL SETUP

### A. Subjects and experimental setup

Six healthy subjects (2 females and 4 males, $28 \pm 8$ years) participated in two experimental sessions performed on two separate days after providing informed consent under a protocol approved by the the Region Liguria Ethical Committee (approval ID 172REG2016, approval date September 13, 2016). Each session lasted for approximately 1 hour and comprised three phases.

### B. Subject preparation

At the beginning of each session (phase zero), the subject was comfortably seated on a chair in front of a table in a quiet environment to avoid distraction. With the forearm of the non-dominant hand on the table surface and the volar side oriented upwards, the electrode matrix was mounted on the volar side of the moistened forearm distally from the elbow at one third of the forearm length. The matrix lines (columns) were aligned with the four fingers of the subject hand. The experimenter waited for 5 min for the skin–electrode to stabilize after wrapping it with a

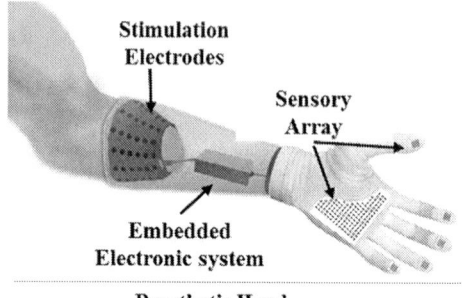

Figure 1. Application scenario for restoring the sense of touch in prosthetic hand using electrotactile stimulation

medical bandage. Then, the experimental procedure was explained to the subject and the subject received the stimulation for 5 min at a self-selected comfortable intensity to familiarize him/her with electrotactile stimulation. Whenever a prickly sensation, referred sensation or muscle contraction was reported, the electrode location was adjusted by moving it few millimeters. The electrode matrix location remained fixed during the three phases for all main experimental sessions, because the electrode location affects the perceptual thresholds and the qualitative aspects of the electrotactile percepts.

### C. Testing Methodology

Each experimental session was divided into three phases:

1) **First Phase**: After the warming up phase, the subject was asked to define a clearly perceived stimulation intensity for each activated pad while avoiding discomfort and pain. In order to determine the clear sensation, the stimulation intensity was increased systematically, starting from zero, in 0.1mA steps, until the subject reported the clear and comfortable sensation. The pads were activated line by line from left to right (i.e. L1, L2, L3, L4 in Fig. 4. respectively) and in each line the pads were activated from the lowest to the highest number while skipping adjacent pads (i.e. 1-3-6-4-2-5) to avoid adaptation. During this, the subjects looked in a real size sketch of the electrode matrix (see Figure 2) to associate the stimulation to the electrode location. Only one electrode was activated at a time.

2) **Second Phase:** The training phase consisted of two stages. In the first stage, the subject was trained to perceive the applied

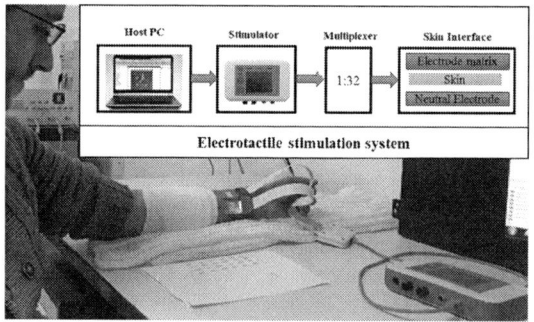

Figure 2. Experimental setup and block diagram of Electrotactile stimulation system.

stimulation and build a tactile mental map between this sensation and the position of the activated pad in the electrode matrix. To this aim, the experimenter announced to the subject the pad that will be activated (line and number) and then started the stimulation. In total, 24 stimulation trials were presented to the subject and the pads were activated from top to bottom sequentially (6 stimulations over each line). The stimulation for each pad lasted 2 sec and was followed by 2 sec pause before next stimulation. In the second cycle, a reinforced learning was performed. The pad was randomly selected and activated. The subject was asked to recognize which pad was active by reporting the line and number, and then he/she received a verbal feedback about the correct answer.

**3) Third Phase**: after a 5 min break, the test phase was started. The stimulation was delivered in a pseudorandom order, and each pad was presented 2 times (i.e. 24×2 =48 stimulation trails). The participant's task was to provide an estimate of the position of the stimulated pad by reporting the line and pad number. The subjects received no verbal feedback on performance. The subjects were allowed to look at both the forearm and the sketch of the electrode matrix. When the subjects could not decide about the pad location, this was reported as a missed sensation. Each trail consisted of 2-second continuous stimulation with fixed intensity (determined in the first phase of each session and specific for each pad) and the pulse width was set to 200 µsec.

*D. Outcome Meaures:*

Electrotactile stimulation systems can be used as an alternative data channel to provide the user with information normally communicated through a different sense to attain efficient functional recovery of somatosensory feedback. To evaluate the feasibility of the proposed multi-channel stimulation system in delivering information to the user, we analyzed the outcome measures computed from the experiments conducted: 1) position recognition and 2) frequency discrimination.

**1) Position Recognition:** The position recognition outcome investigates how well the subject identifies the position of the activated pad (i.e. stimulation position). To evaluate the variation of position recognition, we used the Correct Identification Rate (CIR) which is defined as the ratio of correctly identified trails to the total applied trails:

$$CIR = \frac{Number\ of\ correctly\ identified\ trails}{Number\ of\ total\ trials} \times 100\% \qquad (1)$$

CIR was calculated for perfectly identified active pads (i.e. right guess, $CIR_r$), any pad adjacent to the activated pad (i.e. first neighbor, $CIR_{1n}$), any pad other than first two mentioned (i.e. second neighbor, $CIR_{2n}$) and missed guess (i.e. not predicted, $CIR_m$). The rates were calculated for both coding themes (i.e. uniform frequency coding and mixed frequency coding).

**2) Frequency discrimination**: This outcome evaluates the accuracy of discriminating two stimulation frequency levels, low-level frequency and high-level frequency respectively 10 Hz and 400 Hz. The recognition was considered successful if the subject correctly recognized the applied frequency (i.e. true guess). When the subjects were mistaken and wrongly recognized the applied frequency i.e. predict high frequency as low or vice versa, it was reported as false guess. In certain trails the subjects could not give a guess about the stimulation frequency so it was recorded as missed. It was limited to the second experimental session only.

## IV. DISCUSSION AND RESULTS

As the sensation changes with parameter modulation, it is significant to study the recognition of stimulation, also to what extent and in what way does the sensation change. Two coding schemes (mixed and uniform coding) were tested to assess the ability of the subject to localize the stimulation (identify the active pad).The outcome measures are position recognition and frequency discrimination. These coding themes were tested over two experimental sessions, which may be used for transferring tactile information from prosthesis to upper-limb amputees in future. Session1: one stimulation frequency 50 Hz - uniform frequency coding was applied over the whole electrode matrix. Session2: Two alternating stimulation frequencies 10 Hz and 400 Hz (Mixed frequency Coding) successively applied from the left most side to the right most side of the matrix (i.e. the lines $L_1$, $L_2$, $L_3$, $L_4$ stimulated with 10, 400, 10 and 400 Hz, respectively).

The experimental results of the position recognition and frequency discrimination are shown in Figures 5 and 6 respectively. Figure 5 reports the results for the position recognition outcome for the uniform frequency coding and mixed frequency coding. The average correction identification rate for perfectly identified active pads $CIR_r$ were 21% and

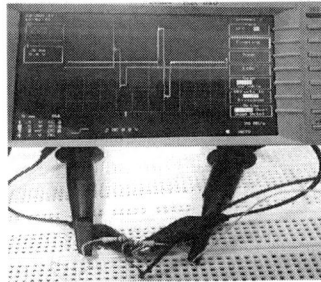

Figure 3. Different waveforms of two active channels with different amplitudes measured using 500Ω resistance.

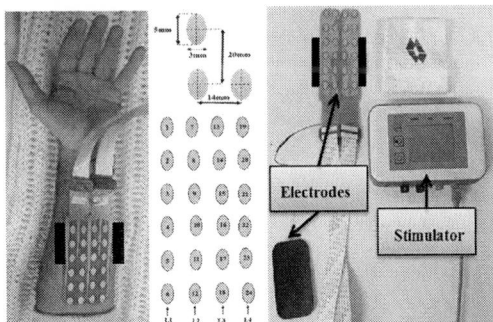

Figure 4 . Rectangular electrode matrix used for skin stimulation attached to the forearm of the left hand.

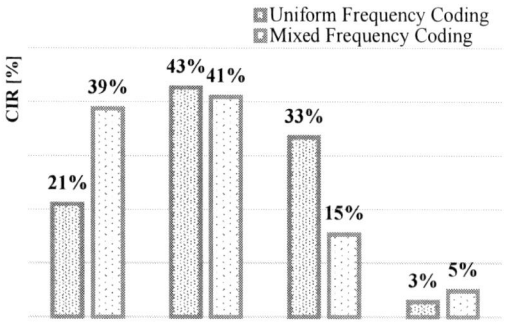

Figure 5. CIR, for uniform (50Hz) and mixed (10Hz - 400Hz) frequency coding.

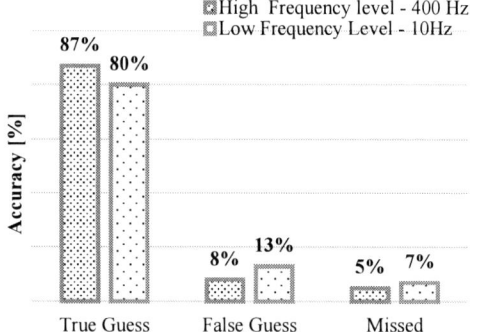

Figure 6. Accuracy for frequency level discrimination of 10 Hz and 400 Hz.

39%. For first neighbors (i.e. identifying a pad adjacent to the activated pad – one position error) $CIR_{1n}$ rates were 43% and 41%, while for second neighbors (i.e. identifying any pad other than the activated pad and its $1^{st}$ neighbor- two position error) the $CIR_{2n}$ rates were 33% and 15%. The missed guessed stimulations for both coding schemes were 3% and 5 % as reported in Table 1. Figure 6 depicts the overall accuracies for frequency discrimination. The percentages of successfully recognized frequencies were 80% for low-level frequency (10 Hz) and 87% for high- level frequency (400 Hz). 8% of the stimulations were predicted as low frequency instead of high (i.e. substitute 400 Hz with 10 Hz) and 13% were mistaken for high frequency instead of low (i.e. substitute 10 Hz with 400 Hz). The reported missed stimulations, where the subject was confused and could not provide an answer, were distributed as 5% and 7 % for high frequency level and low frequency level

TABLE 1.  CIR FOR DIFFERENT CODING METHODS

| Coding Methods | Correction Identification Rate (CIR) | | | |
|---|---|---|---|---|
| | $CIR_r$ | $CIR_{1n}$ | $CIR_{2n}$ | $CIR_m$ |
| Uniform frequency (50 Hz) | 21% | 43% | 31% | 3% |
| Mixed frequency (10 Hz - 400 Hz) | 39% | 41% | 15% | 5% |

respectively. These results demonstrate that mixed frequency coding can be used to improve the perception of electrotactile feedback delivered through multi-channel interface. In addition, it shows how certain coding schemes (e.g. mixed frequency coding) can help in better spatial recognition of the elicited tactile sensation. The proposed high-resolution interface can be used as a research platform to explore different possible scenarios for high-bandwidth sensory transmission to the prosthesis user. Moreover, the high accuracies in discriminating different frequency levels may be used to code different touch modalities for example low frequency stimulus for light touch and high- frequency level for strong touch.

## V.  CONCLUSION AND FUTURE WORK

In this paper, we investigated the feasibility of the proposed multi-channel electrotactile stimulation interface to deliver the spatial information (activated pad). The results have shown that frequency coding could improve the spatial recognition of the stimulus (e.g. the $CIR_r$ increased from 21% with uniform frequency coding to 39% with coding mixed frequency). In addition, high accuracies were recorded for discriminating different frequency levels (80% for low-level frequency (10 Hz) and 87% for high-level frequency (400 Hz). As future work, understanding how to modulate the stimulus by applying different coding techniques (i.e. intensity, waveforms) may lead to develop predictable and reliable presentation of tactile sensation to the user. Also, it may improve the information communication through electrotactile stimulation.

## ACKNOWLEDGMENT

We would like to thank Tecnalia Serbia ltd, Belgrade, Serbia for the MaxSens stimulator.

## REFERENCES

[1]  https://www.coaptengineering.com

[2]  K. Li, Y. Fang, Y. Zhou and H. Liu, "Non-Invasive Stimulation-Based Tactile Sensation for Upper-Extremity Prosthesis: A Review," in IEEE Sensors Journal, vol. 17, no. 9, pp. 2625-2635, May1, 1 2017.

[3]  C. Antfolk, M. D'Alonzo, B. Rosén, G. Lundborg, F. Sebelius, and C. Cipriani, "Sensory feedback in upper limb prosthetics.," Expert Rev. Med. Devices, vol. 10, no. 1, pp. 45–54, Jan. 2013.

[4]  K. A. Kaczmarek, "Electrotactile adaptation on the abdomen: preliminary results," in IEEE Transactions on Rehabilitation Engineering, vol. 8, no. 4, pp. 499-505, Dec 2000

[5]  H. J. Witteveen, H. S. Rietman, and P. H. Veltink, "Vibrotactile grasping force and hand aperture feedback for myoelectric forearm prosthesis users.," Prosthet. Orthot. Int., vol. 39, no. 3, pp. 204–12, Jun. 2015

[6]  L. Seminara; M. Franceschi; L. Pinna; A. Ibrahim; M. Valle; S. Dosen; D. Farina,"Restoring the Sense of Touch in Hand Prosthetics", 2017 IEEE International Symposium on Circuits and Systems (ISCAS), Baltimore, 2017

[7]  K. A. Kaczmarek, J. G. Webster, P. Bach-y-Rita and W. J. Tompkins, "Electrotactile and vibrotactile displays for sensory substitution systems," in IEEE Transactions on Biomedical Engineering, vol. 38, no. 1, pp. 1-16, Jan. 1991.

[8]  H. Fares et al., "Distributed Sensing and Stimulation Systems for Sense of Touch Restoration in Prosthetics," 2017 New Generation of CAS (NGCAS), Genova, 2017, pp. 177-180.doi: 10.1109/NGCAS.2017.54

[9]  M. Franceschi, L. Seminara, S. Dosen, M. Strbac, M. Valle, and D. Farina, "A system for electrotactile feedback using electronic skin and flexible matrix electrodes: Experimental evaluation," IEEE Trans. Haptics, pp. 1–1, 2016

PRIME 2018, Prague, Czech Republic

Session: Sensing and Biomedical Circuits I

# Modeling of a capacitive sensor dedicated to drug injection

Sylvain JOLY, Albrecht LEPPLE-WIENHUES
Innovation group
Valtronic Technologies SA
Lausanne, Switzerland
sjoly@valtronic.ch

Catherine DEHOLLAIN
RFIC Research group
Ecole Polytechnique Fédérale de Lausanne
Lausanne, Switzerland

*Abstract*— to follow up patients who inject themselves insulin with injection pen is a quite complex task. Due to the repetition of injections, and the numerous influent parameter the patient can easily made an injection mistake and it is difficult for health-takers to follow these injections and can detect errors. In order to compensate this problem, a device which measures the volume of drug by using electric fields has been developed. A theoretical model, simulations and experiments have been conducted to validate this capacitive measurement technique. The experiments confirm Theory and simulations results. A sensitivity around 5fF for 10uL have been found and a good repeatability between three caps have been achieved which is equal to an error of 0.44fF for 10uL of insulin and correspond to an error of ±4.12%.

*Keywords—Capacitive; Injection pen; electrodes; volume measurement; Simulation*

## I. INTRODUCTION

Presently, the "Internet Of Things" (IOT) is rapidly expanding due to the evolution of miniaturization and the improvement of low power consumption circuitry. More and more devices can send and share information and feed the "Big Data". This "Data mining" is a powerful tool in the domain of medical care. With the evolution of telemedicine, more and more medical device are becoming connected. This is a real advantage because patients who are far from medical centers have an easier access to medical care and the medical staff has a better follow up of the patients. Following this tendency, the pharmaceutical industry has developed injection pens as shown in Figure 1. This allows patients who follow therapies requiring multiple repeated injections to perform their own injections without the health-takers. Treatments against several diseases need this kind of injector pen. In this research work, our focus will be on Diabetes therapy.

Figure 1 : Injection pen [9]

In 2014, 422 Million people suffered from the diabetes in the world [1]. This is 108 million people more than 1980,

indicating this pathology concern more and more people. The people who have diabetes have reduced insulin production (Diabetes type 1) or insulin resistance (Diabetes type 2). Insulin is a hormone which allows the body to metabolized sugar. This disease can have dramatic consequences. In 2012, 1.5 Million people died directly from diabetes [2]. In order to limit the effects of this disease, synthetic insulins are produced by large pharmaceutical companies. Thanks to the injection pens, patients can live a quite normal life [3]. But they need to pay attention about their injections because it is not an easy task. The insulin injection depends on many parameters (e.g. last meal, age, activity after injection). It is quite complex to record a real follow up of when was the last injection and how many units have been injected, and it is an even more complex problem for care-givers. For example, parents can wish to know when and how much drugs their children have auto-injected during the school day. A simple solution up to this day is to write the injection dose and time into a document or a smartphone application. But people can easily forget to fill this kind of document or make a mistake due to an injection problem like bubbles or leakages.

The objective of this research work is to help with the development of a system called "Smart cap" which can measure and log automatically the quantity of drug inside an injection pen. This will have two main advantages. Firstly, the measurement will be absolute and not relative, limiting the errors due to injection problems, leakage or bubbles. Secondly, the automatic log will provide a better and error-free follow up of the injections for the patient, doctors, and families.

The following constraints must be respected:
- no contact with the drugs
- low power consumption
- dimension of the measuring device close to the original protective cap

Different types of detection technologies have been investigated including
- Using photo-emitters and photo-receivers to find the position of the plunger and determine the

978-1-5386-5388-3/18 $31.00 © 2018 IEEE      217

remaining volume. But this technique needs a lot of photo-emitters/receivers in order to discriminate a submillimeter displacement, and it would be vulnerable to dirt or a modification of the pen transparency.

- Using sound wave propagation to assess the filling level of the pen. But this technique would be severely dependent on the mechanical tolerances of the pen and it would be complicated to characterize the volume of drugs versus the response of the acoustic wave. Furthermore the acoustic coupling of the drug container would be quite cumbersome.

The selected solution is to use an electrostatic measurement. This choice was guided by the high electric permittivity of water (~80) versus air (~1).

Using capacitance variation to measure a level of water is quite an old concept, and it is well established for large water tanks [4] or for humidity detection [5]. This technique is even used to see the different states of water [6]. The particularity of the sensor in the drug measurement device is that the electrode plates are half-cylinder shaped. However, due to the pen geometry, a model of the system has been established with the help of [7]. The objective is to be able to measure a variation of 1 International Unit (IU) of insulin solution which corresponds to 10µl with 34.7µg of insulin inside. Subsequently, this device needs to resolve at least half a dose (5µl) to be able to discriminate an injection of 1 IU.

The device is separated in three parts which are:

- **Electrodes** : This part is presented in the next section, it is the key point of this device and the challenge is the optimization of the sensitivity
- **Capacitance measurement**: This part is the second important part of this device. The measurement depends of the capacitance precision and accuracy on the measurement system. The precision and accuracy need to be better than the required sensitivity in order to resolve a dose step.
- **Control and Communication**: This part allows the user to interact with the device and to follow the drug level inside the pen or manage the battery charge level

## II. ELECTRODES MODELISATION

A first model was present in [8] but has some drawbacks like the difficulty to modify a dimension due to the separation of the different zones. This new approach of modeling allows easiest modification of layers.

The pen inside the cap is divided in different longitudinal parts, in this case three parts have been considered. For each part, a multiple layer of capacitor is considered. This capacitor network can be modified easily.

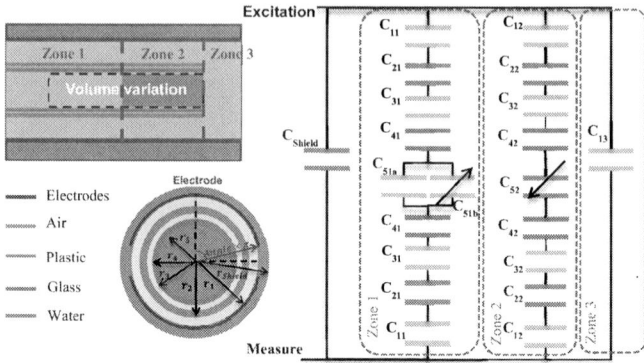

Figure 2 : Model of the pen inside the cap

So with simple electrical equations and capacity formulas, an approximation of this model can be found.

| Zone | Empty Pen [pF] | Full Pen [pF] |
|------|----------------|---------------|
| 1 | 26.37 | 13.2 |
| 2 | 0 | 14 |
| 3 | 6.61 | |
| All | 32.98 | 33.81 |

TABLE I : MATHEMATICAL MODEL RESULTS

According to this model, the sensitivity of the capacity by dose or volume of drug can be approach using (1).

$$Sensitivity = \frac{Pen_{Full} - Pen_{Empty}}{Nb_{Dose}} \quad (1)$$

Thanks to this formula, it is possible to estimate the variation of sensitivity depending on different parameters (geometry variations, matters...). The width of the electrodes has been tested and the results are presented in Figure 3.

PRIME 2018, Prague, Czech Republic

Session: Sensing and Biomedical Circuits I

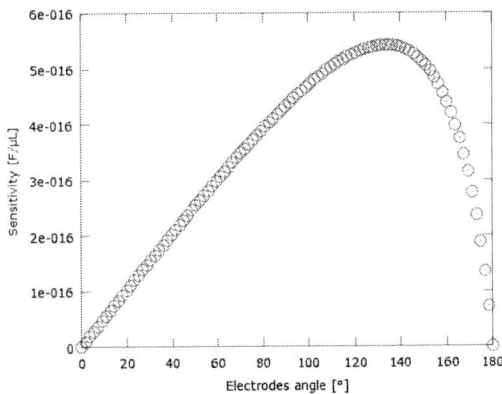

Figure 3 : Electrodes width effect on the sensitivity

The results obtained with this mathematical model are presented in Table 1. These results show that a standard dose of insulin; which is 10µL of the solution, correspond to a variation of capacitance equal to 5.4fF

## III. SIMULATION

To confirm mathematical results, some simulations have been realized with the software FEMM®. The results are presented in Figure 4, voltage potential iso-line are represented. The electric field is concentrate on two points located between excitation electrodes and shield. Values obtain in Table 2 come from the (2) which link the stored energy to the capacitance.

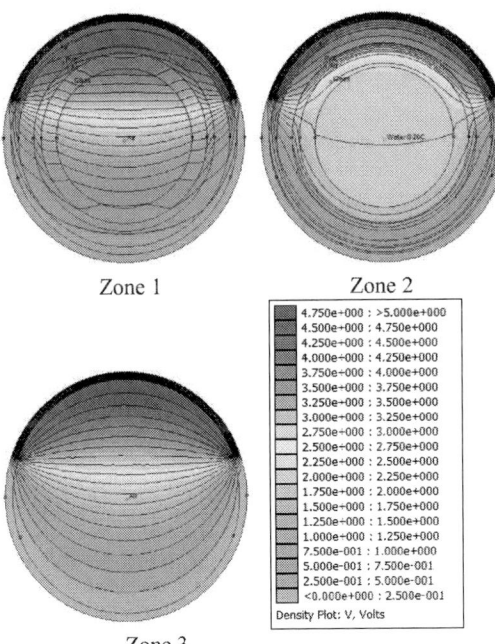

Figure 4 : Simulation results

Using (1), the sensitivity given by simulation is 0.46fF/µL. For 10 µL of solution, corresponding 1IU insulin, the capacitance change by 4.6fF. This is a little bit less than the mathematical model, but this difference could be due to some model approximations.

| Zone | Energy | | Capacitance | |
|---|---|---|---|---|
| | Empty Pen [pJ] | Full Pen [pJ] | Empty Pen [pF] | Full Pen [pF] |
| 1 | 365 | 182 | 29.16 | 14.58 |
| 2 | 0 | 191 | 0 | 15.27 |
| 3 | 88.69 | | 7.1 | |
| All | 453 | 462 | 36.26 | 36.94 |

TABLE 2 : SIMULATION RESULTS

$$J = \frac{1}{2}CV^2 \rightarrow C = \frac{2J}{V^2} \text{ with :} \quad \begin{array}{l} J : \text{Stored Energy} \\ V : \text{Voltage} \\ C : \text{Capacitance} \end{array} \quad (2)$$

## IV. EXPERIMENTS

### A. Demonstrator

Three demonstrators have been assembled as it is shown in Figure 5. They are based on a commercial Capacitance to Digital converter chip which provides sensitivity 4 aF in the best case. This chip is controlled by a micro-controller and data are communicated to a computer via Bluetooth Low Energy. The electrodes are composed by a 0.1mm thickness sheet of copper cut by laser and mechanical parts are 3D printed.

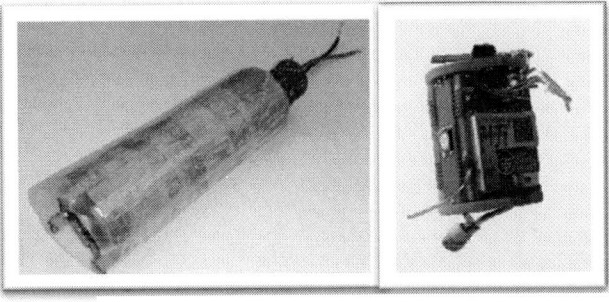

Figure 5 : First demonstrator

### B. Experimental protocol

To tests these caps, this injection protocol have been followed:

• Priming the pens in order to remove any bubbles

CAP 1 and 2:

• 3 x 160 µl
• 3 x 200 µl
• 3 x 60 µl

CAP 3:

• 3 x 20 µl
• 3 x 40 µl
• 3 x 60 µl

978-1-5386-5388-3/18 $31.00 © 2018 IEEE      219

The climatic conditions have been considered constant and the pens have been weighed after each injection. 20 points have been taken by injection steps.

### C. Results

The results of Figure 6 show that the capacitance variation is proportional to the volume of drugs inside the pen with a mean correlation factor greater than 99.9%. The repeatability between these three caps is quite good, the slope is varying about 0.0442fF/µL, and that correspond to an error of around ±12IU for a full pens injections which have 300 IU. It corresponds to a global error of ±4.12%. The mean slope is 0.536fF/µL. So, for 10uL variation, the capacitance will change by 5, 36 fF.

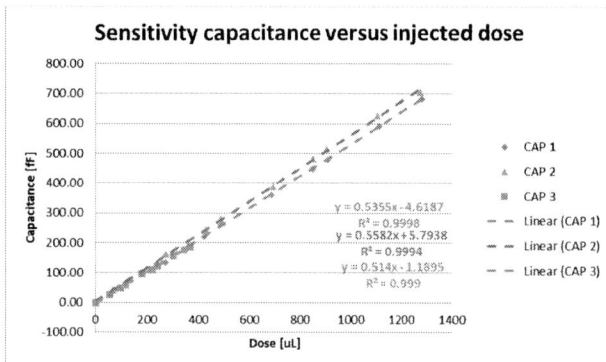

Figure 6 : Experiments results

## V. DISCUSSION

The theoretical model and simulations are in the same range of value and validate each other their results. Effectively for the theoretical model the sensitivity is 5.4fF/Dose and for simulation it is 4.8fF/Dose. Experiments give the value of 5.36fF/dose which is close to the theoretical analysis. The difference between values can come from the fact that some approximation have been done in theoretical analysis and simulation and also imperfections of the devices for experiments. To be able to discriminate one dose the resolution needed is, at least, half dose, so less than 2.68fF.

## VI. CONCLUSION

To conclude this work, this model is more flexible than the previous one. The results given by the mathematical model, was confirmed by simulation and experimentation and it could be used to do various calculation.

These tools will be used to change some parameters in order to find the best configuration of the cap (Radius of the cap, length of the electrodes…) with the objective to improve the sensitivity. A study on the impact of various disturbances (electrodes misalignment, ampoule not centered…) will be conducted. Finally, a solution to create model with multiple electrodes will be studied.

## VII. ACKNOLEDGEMENT

This study will not be possible without the financing of Valtronic Technologies and the support of the CTI grant n°25386.1 PFLS-LS. Special thanks to Matteo Simoncini, Younouss Faye, Antoine Spahr and Jean Baptiste Orhan for their help in this project. Thanks to the developer of FEMM (www.femm.info).

## VIII. BIBLIOGRAPHY

[1] World Health Organization, Global report on Diabetes, Geneva: World Health Organization, 2016.

[2] World Health Organization, Definition, diagnosis and classification of Diabetes mellitus and its complications, Geneva: World Health Organization, 1999.

[3] B. J. Anderson and M. J. Redondo, "What can we learn from patient-reported outcomes of insulin pen device," *Journal of diabetes science and technology,* vol. 5, no. 6, pp. 1563-1571, 2011.

[4] K. Liozou, E. Koutroulis, D. Zalikas and G. Liontas, "A low-cost capacitive sensor for water level monitoring in large-scale storage tanks," in *Industrial Technology (ICIT),* Seville, 2015.

[5] C. K. Huang and Y. W. Liu, "A capacitance sensor for the water content of desiccant wheels," in *International conference on Sensing Technologie (ICST),* Auckland, 2015.

[6] H. C. Cho, X. Zhi, B. Wang, C. H. Ahn and J. S. Go, "Development of a capacitive ice sensor to measure ice growth in a real time," in *Solid-State Sensors, Actuators and Microsystems,* Anchorage, 2015.

[7] L. K. Baxter, Capacitive Sensors Design and Applications, New York: IEEE, 1997.

[8] S. Joly, A. Lepple-Wienhues and C. Dehollain, "Capacitance measurement applied to the medical injection pen," in *PRIME,* Giardini Naxos, 2017.

[9] Lilly USA, "Humalog insulin lispro Injection," 08 2017. [Online]. Available: https://www.humalog.com/index.aspx. [Accessed 02 2018].

# LOCK-IN BASED DIFFERENTIAL FRONT-END FOR RAMAN SPECTROSCOPY APPLICATIONS

A. Ragni, G. Sciortino, M. Sampietro, G. Ferrari
*Dipartimento di Elettronica, Informazione e Bioingegneria*
*Politecnico di Milano*
Milano, Italy
andrea.ragni@polimi.it

F. Crisafi, V. Kumar, D. Polli
*Dipartimento di Fisica*
*Politecnico di Milano*
Milano, Italy
dario.polli@polimi.it

*Abstract*—**The intrinsic sensitivity limit of Stimulated Raman Spectroscopy (SRS) is given by the shot noise of the optical stimulation. However, it is seldom reached due to the electronic noise of the front-end amplifier and the intensity fluctuations of the laser source. Here, we present and test a low-noise pseudo-differential amplifier, for Raman spectroscopy applications, able to compensate the common-mode fluctuations given by the laser and to reach a sensitivity better than 10 ppm thanks to the lock-in technique.**

*Index Terms*—**Raman Spectroscopy, lock-in, differential, transimpedance, amplifier, photodiode, laser, noise**

## I. INTRODUCTION

Raman spectroscopy is a label-free technique for non-invasive and non-destructive imaging, with growing applications in biomedicine and in materials science for the identification and analysis of molecules and organic compounds. It is based on the Raman scattering effect of molecules that was discovered by Indian scientist C.V. Raman in the early 1930s. In stimulated Raman spectroscopy (SRS), samples under test interact with two synchronized pulsed lasers, called Pump and Stokes, which have different wavelengths. The Pump excites the sample molecules to a virtual state, which then relax to the ground state emitting photons with lower energy at Stokes wavelength. The presence of this latter enhances the relaxation transition because when the Pump-Stokes frequency difference matches a vibrational frequency of the molecule, all the molecules in the focal volume are resonantly excited [1]. This provides signal enhancement by many orders of magnitude with respect to spontaneous Raman scattering where only Pump laser is used [2]. Consequently, acquisition speed is significantly improved opening new possibilities to the video-rate imaging [3].

Different vibrational modes of molecules can be investigated with a wavelength scan of Stokes or Pump laser beam. Since each molecule has a specific Raman signature, it is possible to analyse the sample composition by measuring the vibrational behaviour on a wide energy range (i.e. the Raman Spectrum), typically with a Raman wavenumber spanning from $100 \ cm^{-1}$ to $3500 \ cm^{-1}$. The SRS signal over the Stokes beam, whose average power is typically greater by a factor of $10^4$ or more [4], is finally acquired with a photodiode and amplified with a transimpedance amplifier to retrieve the sample Raman spec-

trum. The pump, having no information, is instead optically filtered.

In this work we address the problem of designing a valid front-end for Raman spectroscopy applications combining high sensitivity, to detect the weak Raman signal, and high dynamic range to correctly manage the Stokes average power. The latter is limited to less than 1 mW for organic samples to avoid the damage of the molecules during the measurement. A commercial $1.55\mu m$ femtosecond Er:fiber 40MHz-oscillator followed by two erbium-doped fiber amplifiers (EDFAs) is used to generate two coherent pulsed trains beams:

- the Pump with $\lambda_p = 770nm$ and average power of $\sim$1mW
- the Stokes with a $\lambda_s$ tunable from $950nm$ to $1050nm$ and average power of $\sim$100$\mu$W.

This latter, being in the near-infrared range, is detected with a silicon photodiode providing a sufficient responsivity ($\sim$0.5A/W) in the whole wavelength range.

For a better understanding of the physical quantities involved in the experiment, a laser noise measurement was performed. In fact, intensity fluctuations of the Stokes beam are often the main contribution to the noise in SRS [5, p. 474]. Figure 1 shows the Relative Intensity Noise (RIN) of the two beams, the Pump, used to excite the molecules and the Stokes, containing the Raman information. The frequency dependence behavior of the noise, decreasing with $f$, suggests need for modulation in the MHz range of the Raman signal. In the setup used in the experiment, this is done by modulating the Pump source before the sample under test, with an acousto-optic modulator (AOM). A lock-in based front-end (LIA) is then used for demodulation and amplification. A tuned amplifier (TAMP) used in [6] is also valid solution but only when the modulation frequency $f_m$ is fixed.

In the MHz range, the Stokes has $RIN \approx -115dBHz^{-1}$ which results in a fluctuation of the optical power with a spectral density of

$$S_{Stokes}(f) \approx (177.8 \frac{pW}{\sqrt{Hz}})^2 \qquad (1)$$

with an average power of $\sim$100$\mu$W. The corresponding shot noise, only depending on the Stokes average power [8] and

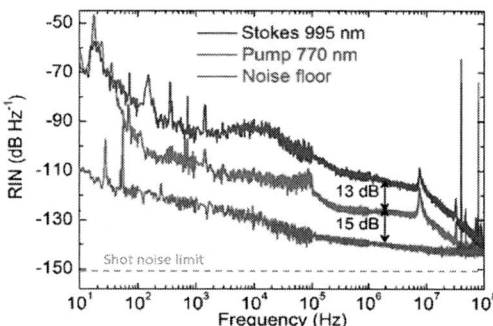

Fig. 1. Measured Relative Intensity Noise (RIN) of Stokes (Blue) and Pump (Red) with 1.6mW Pump power [7]. The Raman signal adds to the Stokes which gives the main contribution to the noise if no compensation technique is adopted (Stokes $RIN \approx -115 dBHz^{-1}$ in the MHz range).

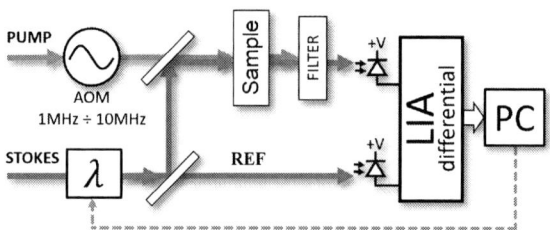

Fig. 2. Differential acquisition scheme for single frequency stimulated Raman spectroscopy over a sample. The PC can control the Stokes wavelength over the interval $950nm \div 1050nm$ in order to investigate the whole Raman spectrum.

setting the intrinsic limit of sensitivity, is:

$$S_{shot}(f) \approx (6.3 \frac{pW}{\sqrt{Hz}})^2 \qquad (2)$$

This comparison confirms that the Stokes fluctuations are the main contribution to the noise, moreover they depend on working conditions (i.e. wavelength, temperature, humidity) resulting in an unpredictable SNR on the final measurement. To compensate the Stokes fluctuations we have adopted a differential architecture where the Stokes noise becomes a common mode contribution. The difference of the signals and the lock-in demodulation are done in the analog domain to overcome the resolution limit of digital lock-in amplifiers [9]. Moreover, an analog approach is better suited to have a compact system operating in parallel on different wavelengths, as required by video-rate imaging applications [3].

## II. ANALYSIS OF THE PROPOSED SOLUTION

Figure 2 shows the differential setup adopted in this work. The Pump is modulated with an AOM at $f_m = 1MHz \div 10MHz$ and it is blocked after the sample with an optical filter. The Stokes is split in two parallel laser beams of $\sim 40\mu W$ each: the first one crosses the sample under investigation and it will result in a Raman gain while the second, having the same power fluctuations, is used as a reference. Both get acquired by two identical photodiodes (TeMd5020X01 - Vishay) producing

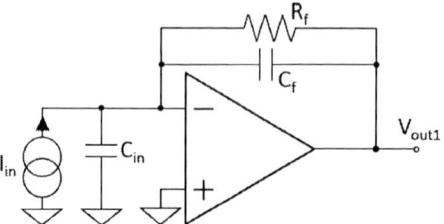

Fig. 3. Photodiode connected to a standard TIA configuration. $C_{in}$ is the sum of all capacitances connected to the virtual ground node of the amplifier (major contribution is given by the photodiode capacitance.

two current signals $I_p$ and $I_{pRef}$ mainly composed by the following terms:

$$I_p = I_{DC} + i_{train} + i_{Snoise} + \sqrt{i_{shot}^2} + i_{Raman} \qquad (3)$$

$$I_{pRef} = I_{DC} + i_{train} + i_{Snoise} + \sqrt{i_{shot,ref}^2} \qquad (4)$$

where $I_{DC}$ is the current proportional to the average Stokes power through the photodiode responsivity (R=$\sim$0.5 A/W at 1000nm), $i_{train}$ consists in the AC 40MHz pulsed-train response and $i_{Snoise}$ represents the common mode fluctuations (equal for both $I_p$ and $I_{pRef}$). Then, $i_{shot}$ is the shot noise which is uncorrelated between the two beams and finally $i_{Raman}$ represents the weak Raman signal ($\sim 10^{-5}$ respect to $I_{DC}$) which is only in the branch where the sample is present.

The difference between $I_p$ and $I_{pRef}$ (i.e. the Raman signal with the intrinsic noise given by the shot) is amplified with a differential transimpedance amplifier while the Stokes fluctuations, being a common mode contribution, are in principle cancel out. The $I_{DC}$ signal is important for normalization, so its value must be acquired before the differential amplifier.

If a photodiode is connected to a standard TIA, see Fig. 3, the output voltage in the ideal case is a single-pole transfer function given by the following expression:

$$\frac{V_{out1}(s)}{I_{in}(s)} = -\frac{R_f}{1 + sR_f C_f} \qquad (5)$$

The feedback capacitor $C_f$ sets the bandwidth of the TIA and its value should, in principle, be minimized. Considering the stage stability, however, there is a low limit due to the input node capacitance $C_{in}$ that includes all the stray capacitances related to the virtual ground node of the amplifier. The main contribution is given by the photodiode capacitance which is about $10pF$ with a reverse voltage $V_R = \sim 15V$. Moreover, because of parasitic capacitances are in the order of few hundreds of femtofarad for discrete components technology, it is not possible to control lower value of $C_f$. It follows that, in practice, the feedback resistor $R_f$ is setting both the DC gain and the bandwidth of the TIA, resulting in a trade-off. Its value should be maximized in order to reduce the equivalent input current noise (proportional to $1/R_f$), but this way the bandwidth it is also reduced. Furthermore, since the output voltage $V_{out1}$ is proportional to the short and powerful

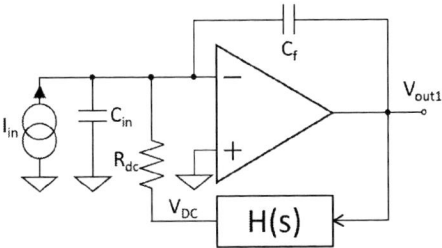

Fig. 4. The advanced TIA [10] used in this work connected to the photodiode modelled with a current source $I_{in}$. The DC feedback network (with $R_{DC} = 47k\Omega$) manages the DC current while the Raman signal is amplified with $C_f = 0.5pF$. $C_{in} = 20pF$ and it includes the photodiode capacitance $C_{pd} \approx 10pF$.

Stokes pulses, $R_f$ value is strongly limited by the operational amplifier output dynamic range to prevent clipping.

These limitations are overcome by connecting the silicon photodiode to an integrator stage with an additional feedback network needed to manage the DC bias current. Figure 4 shows the schematic of the advanced TIA topology [10] of the readout path. The feedback network consists in an amplifier $H(s)$, designed to have high gain in DC and high attenuation in the signal bandwidth, in series to the resistance $R_{DC}$. While the AC term $i_{ac}$, including the pulsed train current and the Raman signal, results amplified by the integrator at the output node:

$$V_{out1}(s) = -\frac{1}{sC_f} \cdot i_{ac}(s) \qquad (6)$$

the DC current $I_{DC}$ of the photodiode (related to the Stokes average power) flows through the resistor $R_{DC}$ resulting in the voltage:

$$V_{DC}(s) = -R_{DC} \cdot I_{DC}(s) \qquad (7)$$

Having the two terms amplified on different nodes strongly relax the output dynamics. Moreover, the amplifier in $H(s)$ can have a greater power supply voltage because it operates at low frequency while the operational amplifier in the integrator stage can only be selected for speed and noise performance. This translates in a greater $R_{DC}$ that will benefit the input noise. This stage has no more the gain-bandwidth trade-off seen in the previous example because, referring to the integrator, the bandwidth is proportional to the gain-bandwidth product irrespective of the $R_{DC}$ value. With the operational amplifier LTC6268-10 the bandwidth is ~100MHz. Finally, the information about the DC current, useful for the Raman signal normalization, is already available on the $V_{DC}$ node.

The input referred current noise of the front-end is:

$$\overline{i}_{neq}^2 \approx \overline{i}_n^2 + \frac{\overline{e}_{H(s)}^2}{R_{DC}^2} + \overline{e}_n^2\omega^2(C_f + C_{in})^2 + \frac{\overline{e}_n^2}{R_{DC}^2} + \frac{4kT}{R_{DC}} \quad (8)$$

where $\overline{i}_n^2$ and $\overline{e}_n^2$ are the operational amplifier current and voltage equivalent noise, respectively, and $\overline{e}_{H(s)}^2$ is the $H(s)$ network equivalent noise. By using a FET operational amplifier and $R_{DC} > 1k\Omega$, at the signal frequency of 1 MHz the

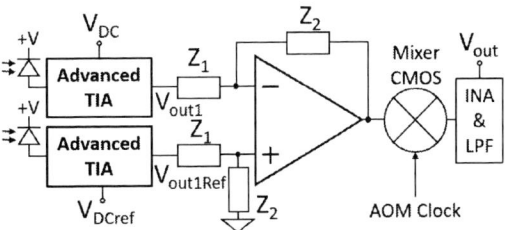

Fig. 5. Diagram of the balanced acquisition system developed in this work. The differential stage and INA have a gain $G_{diff} = 40$ and $G = 10$ respectively while the low-pass filter has a cut-off frequency $f_c \approx 480Hz$.

noise of the first stage is mainly due to the input capacitance $C_{IN}$ and the thermal noise of $R_{DC}$:

$$\overline{i}_{neq}^2 \approx \overline{e}_n^2\omega^2(C_f + C_{in})^2 + \frac{4kT}{R_{DC}} \qquad (9)$$

In the frequency range $1MHz \div 10MHz$ and considering an input capacitance of $\approx 15$ pF and $R_{DC} = 20k\Omega$, the input-referred current noise $i_{neq}^2$ goes from $(850\frac{fA}{\sqrt{Hz}})^2$ to $(5\frac{pA}{\sqrt{Hz}})^2$. These values are comparable to the shot noise of the optical signal (2 divided for the squared responsivity, ~0.5A/W), thus the designed front-end satisfies the noise specification.

The balanced acquisition, required for the laser noise cancellation, is obtained by connecting a couple of the advanced TIAs mentioned above, to a second stage in differential amplifier configuration with a gain $G_{diff} = 40$ as shown in Fig. 5. In principle, only the Raman signal and the uncorrelated noise sources (as shot and thermal noise) get amplified. With a custom CMOS mixer driven by the AOM clock, the signal is then demodulated to baseband and finally amplified and filtered with an instrumentation amplifier (INA with $G = 10$) followed by a low-pass filter (LPF with $\tau \approx 330\mu s$).

## III. EXPERIMENTAL VALIDATION AND DISCUSSION

To characterize and experimentally evaluate the proposed differential LIA, a single-frequency Raman spectroscopy experiment has been carried out on a Methanol sample (MeOH) in liquid phase with the acquisition scheme of Fig. 2. The modulation frequency was chosen to $f_m = 1MHz$ to meet the best working condition for the amplifier. In fact, it results to an input equivalent power noise equal to $S_{el} \approx (3.4\frac{pW}{\sqrt{Hz}})^2$ (considering the photodiode responsivity R=0.5A/W) which is lower than Stokes shot noise $S_{shot} \approx (8\frac{pW}{\sqrt{Hz}})^2$ with $P_s \approx 40\mu W$. Note that the input equivalent power noise includes a factor 2 given by the sum of the uncorrelated noise in the signal and reference path, as well another factor 2 given by the lock-in technique. Referring to Fig. 5, the output voltages $V_{out}$ with the Raman information and $V_{DC}$ for the normalization were sampled with a data acquisition system (NI USB-6259 by National Instruments) and sent to a PC. A LabVIEW interface was used to control the laser and perform the Stokes wavelength scan in the interval 950nm to 1050nm.

The normalized Raman spectrum of Methanol, plotted in Fig. 6, was correctly acquired by the proposed front-end

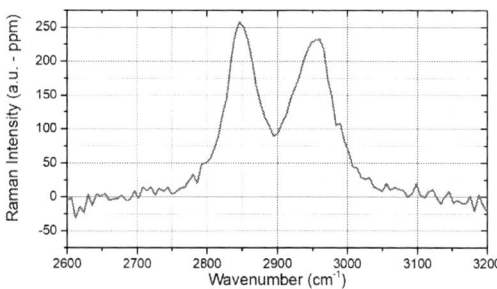

Fig. 6. Normalized Raman spectrum ($\Delta I_s/I_s$) of Methanol (MeOH) obtained with the proposed amplifier operating with $\tau = 330\mu s$ and with $\sim 40\mu W$ Stokes power.

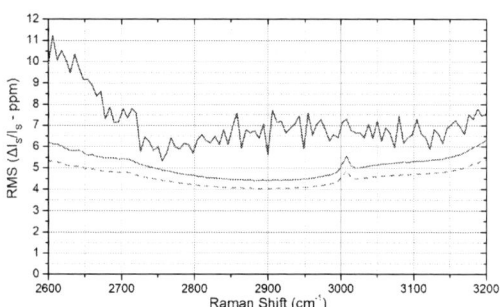

Fig. 7. Normalized RMS noise spectra: measured (blue), theoretical noise (red) which includes shot noise (dotted red) and electronics noise.

Fig. 8. Experimental setup for single frequency SRS with the proposed differential lock-in amplifier (red circled)

veloping a custom integrated circuit able to readout four differential channels including the analog front-end and the lock-in demodulation.

## ACKNOWLEDGMENT

We acknowledge support by the European Research Council Consolidator Grant VIBRA (ERC-2014-CoG No. 648615) - Very fast Imaging by Broadband coherent RAman - headed by Dario Polli in the Department of Physics of Politecnico di Milano.

## REFERENCES

[1] F. Crisafi, V. Kumar, T. Scopigno, M. Marangoni, G. Cerullo, and D. Polli, "In-line balanced detection stimulated raman scattering microscopy," *Scientific Reports*, vol. 7, no. 1, p. 10745, 2017.
[2] V. Kumar, M. Casella, E. Molotokaite, D. Gatti, P. Kukura, C. Manzoni, D. Polli, M. Marangoni, and G. Cerullo, "Balanced-detection raman-induced kerr-effect spectroscopy," *Physical Review A*, vol. 86, no. 5, p. 053810, 2012.
[3] D. Polli, "http://www.vibra.polimi.it/,"
[4] C. W. Freudiger, W. Min, B. G. Saar, S. Lu, G. R. Holtom, C. He, J. C. Tsai, J. X. Kang, and X. S. Xie, "Label-free biomedical imaging with high sensitivity by stimulated raman scattering microscopy," *Science*, vol. 322, no. 5909, pp. 1857–1861, 2008.
[5] W. Demtröder, *Laser spectroscopy: basic concepts and instrumentation.* Springer Science & Business Media, 2013.
[6] C.-S. Liao, M. N. Slipchenko, P. Wang, J. Li, S.-Y. Lee, R. A. Oglesbee, and J.-X. Cheng, "Microsecond scale vibrational spectroscopic imaging by multiplex stimulated raman scattering microscopy," *Light: Science & Applications*, vol. 4, no. 3, p. e265, 2015.
[7] N. Coluccelli, V. Kumar, M. Cassinerio, G. Galzerano, M. Marangoni, and G. Cerullo, "Er/tm: fiber laser system for coherent raman microscopy," *Optics letters*, vol. 39, no. 11, pp. 3090–3093, 2014.
[8] C. W. Freudiger, W. Yang, G. R. Holtom, N. Peyghambarian, X. S. Xie, and K. Q. Kieu, "Stimulated raman scattering microscopy with a robust fibre laser source," *Nature photonics*, vol. 8, no. 2, p. 153, 2014.
[9] M. Carminati, G. Gervasoni, M. Sampietro, and G. Ferrari, "Note: Differential configurations for the mitigation of slow fluctuations limiting the resolution of digital lock-in amplifiers," *Review of Scientific Instruments*, vol. 87, p. 026102, feb 2016.
[10] G. Ferrari and M. Sampietro, "Wide bandwidth transimpedance amplifier for extremely high sensitivity continuous measurements," *Review of Scientific Instruments*, vol. 78, no. 9, p. 094703, 2007.
[11] J. Réhault, F. Crisafi, V. Kumar, G. Ciardi, M. Marangoni, G. Cerullo, and D. Polli, "Broadband stimulated raman scattering with fourier-transform detection," *Optics express*, vol. 23, no. 19, pp. 25235–25246, 2015.
[12] Y. Yu, Y. Wang, K. Lin, N. Hu, X. Zhou, and S. Liu, "Complete raman spectral assignment of methanol in the c–h stretching region," *The Journal of Physical Chemistry A*, vol. 117, no. 21, pp. 4377–4384, 2013.

with a filter time constant $\tau = 330\mu s$. The spectrum exhibits two peaks typical of C-H stretching region, respectively $\sim 2850 cm^{-1}$ and $\sim 2950 cm^{-1}$ [11], [12].

The experimental data also indicates that the measured noise is almost equal to the shot noise, as shown in Fig. 7, leading to a sensitivity lower than 10ppm. This demonstrates the ability of the proposed front-end to strongly compensate Stokes intensity fluctuations. As further confirmation of this experimental results, a subsequent measurement has indicated a $CMRR \approx 56$ at 1MHz, enough to make noise of Eq. 1 negligible with respect to the intrinsic limit given by the shot noise. A photo of the experimental setup is shown in Fig. 8.

## IV. CONCLUSIONS

In this paper, a novel lock-in based differential amplifier, for Raman Spectroscopy applications, is presented and tested. The spectrum of liquid phase Methanol is correctly acquired, with a time constant $\tau \approx 330\mu s$, and normalized over the average intensity. The measurement is practically shot noise limited thanks to the low-noise advanced TIA - $S_{el} \approx (3.4\frac{pW}{\sqrt{Hz}})^2$ at $f_m = 1MHz$ - and to the balanced acquisition with $CMRR \approx 56$. A sensitivity lower than 10ppm is reached, thus experimentally validating the proposed topology. The front-end can be used with common cathode photodiodes arrays making ideal for broadband Raman Spectroscopy where a multi-channel architecture is needed. Indeed, we are de-

# Electric Vehicle Battery Management System Using Power Line Communication Technique

Arash Pake Talie
*Infineon Technologies*
Graz, Austria
Arash.PakeTalei@infineon.com

Wolfgang A. Pribyl
*Institute of Electronics*
*Graz University of Technology*
Graz, Austria
Wolfgang.Pribyl@TUGraz.at

Guenter Hofer
*Infineon Technologies*
Graz, Austria
Guenter.Hofer@infineon.com

*Abstract*—**this paper presents the analysis and test of a power line communication system targeting the communication between each battery cell and the battery management system located in an electric vehicle. The battery cell type which is used for analysis and for the lab measurement is 18650 and the objective is to use a stack of batteries as a channel to transmit and receive digital data with high data rate using FSK modulation with 10MHZ carrier frequency.**

*Keywords—power line communication, PLC, electric vehicle, battery management system, BMS*

## I. Introduction

The state-of-the-art technology for energy storage systems in electric vehicles is the lithium-ion-battery. These single cells show a slightly different electrochemical behavior, depending on production quality, exposed temperatures etc. A battery management system (BMS) is responsible for monitoring and controlling the functionality of each single cell continuously to keep the whole battery in optimal condition all time long. To achieve a sufficiently high measurement quality of the parameters of each individual battery cell in the entire stack, electronic hardware in proximity to the battery cells is needed. This local hardware is responsible for sensing and conditioning of physical parameters such as temperature, state of charge, state of health etc., and sending them to the main BMS for further data processing. Normally, the transmission and receiving of data is being done by a twisted pair wire between each cell and the main BMS unit. Since the number of cells on an electric car can be quite high [1] (around 400), using twisted pair (or any types of dedicated wire) will increase the complexity and cost of the system. Power line communication (PLC) can be used as a communication technique by using the dedicated high voltage power line as the communication channel between the BMS and each cell. By using PLC methods, smart data transmission techniques must be used, specifically for lithium-ion batteries with the capability of efficiently controlling and monitoring each cell up to 400 cells. To have an easier understanding of the whole system, an overall block diagram is depicted in Fig 1 that explains the connection of each block.

This paper is organized as follows. In section II of this paper the HF modeling of 18650 [2] cell is explained which is followed by section III with a proposed coupling/decoupling circuit and an overall circuit simulation. In section IV a demo board is tested in the lab for the proof of concept.

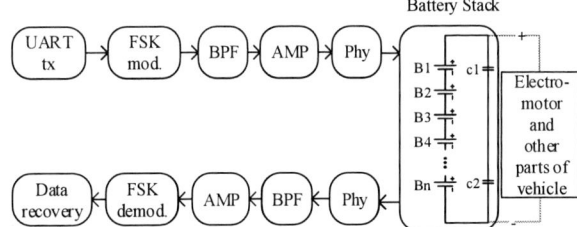

Fig 1 - overall block diagram

## II. High frequency model of the 18650 battery cell

Before sending any type of carrier signal through any communication channel, a good understanding of electrical characteristics of that channel is essential. In this case 18650 battery cell and the cables in between are of our main interest. Cables can be easily modeled with the existing formula since they are not so complex. However we cannot say that for the battery cell.

By using E5071C ENA vector network analyzer [3], the impedance of an 18650 battery cell is measured from 300 kHz to 30 MHZ and the result is shown on Fig 2. The top graph is the magnitude in ohm and the bottom graph displays the phase of the impedance in degrees.

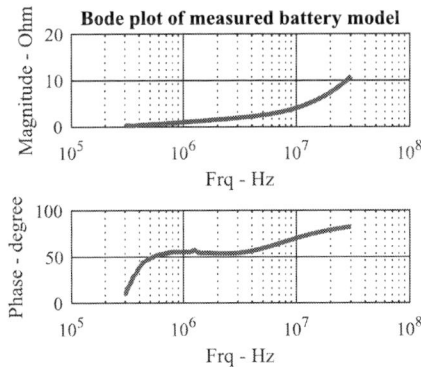

Fig 2 - impedance of 18650 battery cell from 300 kHz to 30 MHZ

978-1-5386-5388-3/18 $31.00 © 2018 IEEE

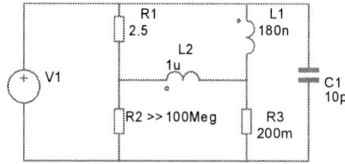

Fig 3 - equivalent circuit of the model

In the beginning when the frequency is low, the battery cell behaves as almost pure resistive circuit with a very low resistance around 0.28 ohm. However, when the frequency increases, the phase and the impedance increase which tells us that the battery cell is behaving more like an inductance, yet the phase stops increasing from 1MHz up to 3MHz notifying us that a capacitor is compensating the phase. This condition changes after 3MHz and the phase rises again because of a second inductor that is affecting the circuit.

Taking these evidences into account we can estimate a model as Fig 3 where V1 is the excitation voltage and the rest are the elements creating the high frequency behavior shown in Fig 2. To calculate the model of this circuit, the matrix of impedance is entered as follow.

$$\begin{pmatrix} R1 + R2 & -R1 & -R2 & 0 \\ -R1 & R1 + XL2 + XL1 & -XL2 & -XL1 \\ -R2 & -XL2 & R2 + XL2 + R3 & -R3 \\ 0 & -XL1 & -R3 & XL1 + R3 + XC1 \end{pmatrix} \quad (1)$$

R2 does not have an influence on the overall impedance. The geometric symmetry of the circuit is the sole purpose of this resistor.

By substituting the values from Table 1 into ( 1) , an overlay of impedance measurement of modeled circuit versus the real measurement can be plotted in Fig 4.

| Element | Value |
|---------|-------|
| R1 | 2.5Ω |
| R2 | 100 MΩ |
| R3 | 0.2 Ω |
| L1 | 180nH |
| L2 | 80nH |
| C1 | 10pF |

Table 1 - values that are substituted in the matrix form

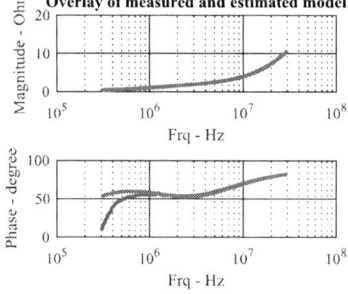

Fig 4 - Bode plot comparison of measured model and estimated model of 18650 battery cell

## III. COUPLING / DECOUPLING CIRCUIT

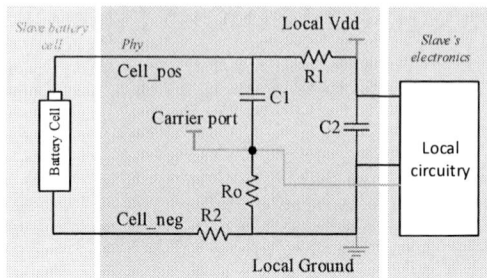

Fig 5 - connection between physical layer, battery cell and slave's circuitry

By looking at the magnitude on Fig 4 one can see that the two curves match perfectly together, however the phase graph shows a different behavior for frequencies lower than 1MHZ. since 10MHZ is the frequency of interest, we can ignore this mismatch and move forward.

Sending and receiving data to and from each battery cell (slave) requires each slaves' local electronics to be independent from each other while creating a stable supply. This stable supply is very important as it has to power up the local circuitry while exhibiting as low ripple as possible. These nodes are denoted with L_vdd (local vdd) and L_gnd (local ground).

The circuit that is shown on Fig 5 is responsible for creating the local supply and separating the carrier from the DC supply voltage. Stacking the system in Fig 5, results in the whole system which is shown on Fig 6 with the matching box colors. When the master sends the modulated signal using carrier generator and couples it to the power line (positive and negative poles of the battery), this signal is added to the DC voltage of battery cell (blue box) and enters the physical layer (green box).

Please note that the GND node is not connected to the chassis and is only a local ground.

The two wires Cell_pos and Cell_neg carry the DC voltage of the local battery cell as well as the carrier signal which is send from the master. R1 and C2 form a low-pass filter that filters out the carrier signal and creates the stable DC voltage for the local circuitry while C1 and Ro form a high-pass filter removing the DC voltage from carrier signal which then is fed to the demodulator. Table 2 contains the values of passive components.

| Component | Value |
|-----------|-------|
| R1 | 10 |
| R2 | 10 |
| Ro | 100k |
| C1 | 100nF |
| C2 | 470nF |

Table 2 - values of the passive components from Fig 5

Fig 6 displays how the battery cells are connected in series and power up the whole vehicle while connected to each local physical layer as well as to one master that monitors all of them. In this figure the thick red and blue wires, represent the high DC voltage which is the result of n batteries connected in series.

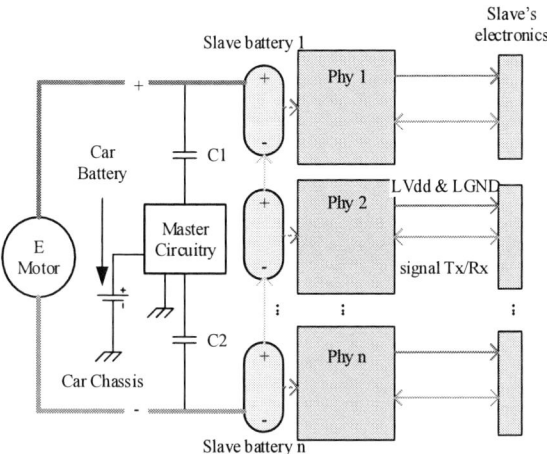

Fig 6 - overall PLC network

Thin solid red lines are local supply and green lines are wires that carry Tx/Rx modulated signal and red dashed lines are carrying dc supply together with the modulated signal to each slave. C1 and C2 are DC isolating capacitors as well as a communication link between the master and slaves. The master is being powered from the battery that is meant to power up different parts of the car (lights, windshield wipers etc.).

Slave's electronics (pink rectangle) is simulated in the SIMetrix tool [4] with the schematic that is shown in Fig 7. In Fig 7, a square wave is being created by PLL block which then enters the buffer section that is feeding the band-pass filter (BPF), in order to convert the square wave to a more sine shaped wave. The reason for this filtering stage is that the transmission channel shows resonances at high frequency components and as a result, injecting a square wave would lead to a much stronger amplification of high frequencies compared to lower frequencies (e.g. the carrier signal). This would make it difficult to extract the main carrier at the receiver side.

The filtered signal then enters a push-pull stage that feeds the physical layer (through the green arrow). This carrier signal then passes through the stack of batteries and closes its way from the power line and the DC isolator capacitor.

## IV. DEMO BOARD

The demo board has been implemented following the block diagram on Fig 1. Digital data is being generated by BASYS3 FPGA board [5] and it is in UART format.

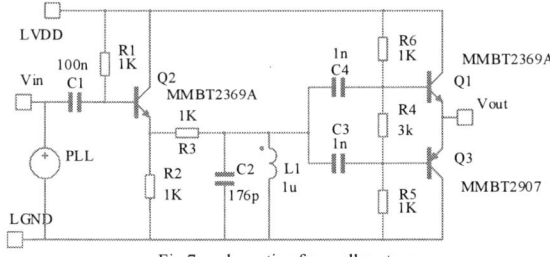

Fig 7 - schematic of overall system

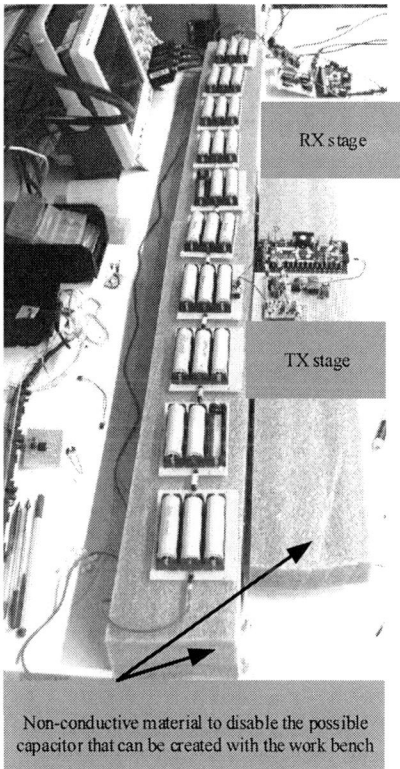

Fig 8 - TX stage

The baud rate and 16bit data are controllable on the fly, by the push buttons and switches that are implemented on the FPGA board. This UART data enters the FSK modulator that is made of CMOS 4046 PLL family with a square wave output signal and frequency of 10MHz which feeds the buffer stage and the band-pass filter (BPF) respectively.

The output of the filter then is amplified by a push-pull amplifier (AMP). This amplifier feeds the physical layer which is depicted in Fig 5 and completes the transmitter stage.

Fig 9 – overall look on the demonstration setup

*Paper P6*

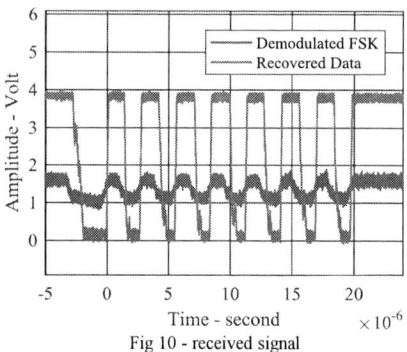

Fig 10 - received signal

On the receiver side, after being decoupled from the battery cell, the signal enters the band-pass filter and is amplified by an amplifier which then feeds a logic detection unit made up from a fast comparator. This signal then enters to the receiver side FPGA and is decoded to be displayed on the available LEDs. The overall setup is shown in Fig 9.

To put the system into test, a 16bit data word (b1010101010101010) is transferred. One start bit and one stop bit is needed which adds up to 18 bits in total that need to be transmitted. The received signal on Fig 10 is showing the demodulated analog signal (blue) and the digital data extracted from the demodulated signal (red).

Looking at Fig 11 which displays the same waveform that is extracted from the oscilloscope, each symbol is about 1.45us long. Accordingly the symbol rate in this test is about 687ks/s.

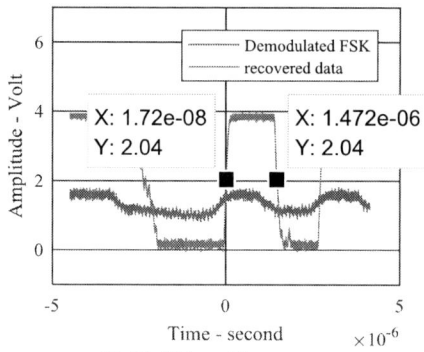

Fig 11 - bit/sec of the test setup

## V. CONCLUSION

In this paper a PLC system targeting BMS has been simulated and tested in the lab with good correlation of the results.

We started by modeling 18650 type battery. The result was used to simulate the whole PLC system using SIMetrix simulation tool. The same circuit then was implemented in the lab and proved to be working well by sending 16bit data through the battery stack.

For the next steps, a deep noise analysis, specifically for this system, must be done to help us to predict the bit rate error. This can be followed by testing different types of modulation (BPSK, QPSK etc.) and studying the advantages and disadvantages of each.

## REFERENCES

[1] I. Ouannes, P. Nickel, and K. Dostert, "Cell-wise monitoring of Lithium-ion batteries for automotive traction applications by using power line communication: Battery modeling and channel characterization," *IEEE ISPLC 2014 - 18th IEEE Int. Symp. Power Line Commun. Its Appl.*, pp. 24–29, 2014.

[2] "SAMSUNG- ICR18650-26F." .

[3] "E5071C ENA Vector Network Analyzer." [Online]. Available: https://www.keysight.com/en/pdx-x202270-pn-E5071C/ena-vector-network-analyzer?cc=US&lc=eng.

[4] "SIMetrix." [Online]. Available: www.simetrix.co.uk/products/.

[5] "Basys 3 Artix-7 FPGA Trainer Board." [Online]. Available: https://store.digilentinc.com/basys-3-artix-7-fpga-trainer-board-recommended-for-introductory-users/.

# Evaluation of Frontend Readout Circuits for High Performance Automotive MEMS Accelerometers

Alice Lanniel*[†], Thomas Alpert*, Tobias Boeser*, Maurits Ortmanns[†]

*Robert Bosch GmbH, Automotive Electronics, Reutlingen, Germany
Email: alice.lanniel, thomas.alpert, tobias.boeser@de.bosch.com
[†]Institute of Microelectronics, University of Ulm, Ulm, Germany
Email: maurits.ortmanns@uni-ulm.de

*Abstract*—This paper reviews and evaluates different capacitive sensor readout frontend circuits for high performance MEMS applications. Fully-differential, pseudo-differential, single-ended and open-loop architectures are considered. Furthermore, the primary design parameters and the tradeoffs are presented. The focus is placed on area, linearity, EMC robustness, noise considerations and sensitivity. This work analyses existent readout architectures and derives the best for automotive specifications.

## I. INTRODUCTION

MEMS accelerometers are micro-electromechanical devices that measure static and dynamic accelerations. The demand for high performance devices has increased tremendously in the recent years thanks to their low cost, small size and low power dissipation. MEMS accelerometers can be found in various applications. For example, they are used in cars for airbag application or ESC (electronic stability control) and in general for navigation, space gravity instruments and consumer applications as for example smartphones or headsets for virtual reality. Depending on the application, the design varies due to distinct requirements. Whereas in consumer electronics the emphasis is put on size, cost and power consumption, the critical automotive requirements are safety, robustness, long term supply, reliability, quality and functioning in harsh environment. Several readout interfaces can be found in literature, each of them optimized for specific parameters. This paper focuses on four readout circuit architectures and automotive key performances as linearity or EMC robustness are evaluated and compared. The readout architectures are introduced in section II and simulation results are presented in section III.

## II. ARCHITECTURE OVERVIEW

Accelerometer readout interfaces can be classified into two main groups, which are force-feedback and open-loop. Force-feedback is realised by applying a feedback force, generally an electrostatic force, to oppose displacement of the proof mass from its nominal position, whereas in open-loop circuits the suspended proof mass is allowed to move freely [1]. Force-feedback improves several characteristics of a sensor. It allows for higher dynamic range, higher linearity, higher sensitivity and wider signal bandwidth [2]. Open-loop architectures are smaller, have a lower power consumption and are easier to implement as force-feedback circuits. A compromise between these parameters is obtained with the charge-balanced architecture. This third interface architecture is a combination of open-loop and force-feedback. It employs a feedback to adjust the voltage applied on the sensor, though this is not used to prevent movements of the proof mass. A more detailed description of the charge-balanced concept can be found in section II.B. Throughout this paper, the charge-balanced architecture is preferred to the force-feedback one since it has a similar linearity while consuming less power [3].

Open-loop and charge-balanced systems can both be designed as fully-differential, pseudo-differential and single-ended configurations. The fully-differential configuration has two sensors and therefore two cores throughout the whole front-end. The pseudo-differential configuration has one sensor and one dummy placed on chip. The dummy is a copy of the sensor but it does not measure any acceleration so it stays constant. Nevertheless, the dummy sensor enables to have a fully-differential readout circuit. The single-ended configuration has one sensor and therefore one core all along. The four frontend architectures presented in this paper consist of three stages: The sensor, which measures an acceleration, the C/V (charge to voltage) stage converting this signal into a voltage and finally a sigma delta modulator, which digitizes the output signal of the C/V stage into a bitstream. The bitstream is then processed by a digital backend, which is not addressed here.

### A. Sensor

The type of sensing considered in this paper is the capacitive sensing. It has the main advantages of low temperature coefficient and high sensitivity [4]. The challenge in the design of capacitive sensors is the sensitivity to parasitics. The sensor in Fig. 1 is composed of a moving proof mass, which is suspended by a spring over a substrate and located between two fixed electrodes [1]. A differential pair of capacitors is formed between the proof mass and the electrodes. When an acceleration appears, the proof mass moves, which makes the capacitances value change. The arrows in the two capacitors yielding in opposite direction indicate their mutual inverse sensitivity to an external acceleration. Thanks to the differential capacitor scheme, the output is proportional to the displacement of the proof mass. However, this is only true for small displacements, in which case $\Delta x^2$ at the denominator can be ignored, as shown below:

978-1-5386-5388-3/18 $31.00 © 2018 IEEE

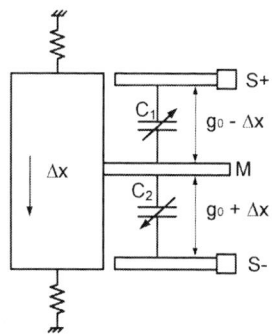

Fig. 1. Principle of capacitive acceleration sensing with a different pair of capacitances [1]

$$\Delta C(\Delta x) = \frac{\epsilon * A}{g_0 - \Delta x} - \frac{\epsilon * A}{g_0 + \Delta x} = \frac{\epsilon * A * 2 * \Delta x}{g_0^2 - \Delta x^2}$$

$$\propto \frac{\epsilon * A * 2 * m * a_{ext}}{g_0^2 * k} \qquad (1)$$

$$with \quad \Delta x = \frac{m * a_{ext}}{k}$$

where $\epsilon$ is the permittivity, A is the area of the plates, $g_0$ is the initial gap between the electrodes and the sensor capacitors, $\Delta x$ is the deflection of the proof mass due to an acceleration, m is the mass, k is the stiffness of the spring and $a_{ext}$ is the external acceleration. The sensor is in fact composed of several of these electrodes and therefore contains several capacitance pairs to increase the sensitivity. The capacitive change due to an acceleration has then to be detected by a readout circuit.

### B. C/V Stage

The C/V stage detects the sensor output charge and converts it to a voltage. In this paper, two C/V stage concepts are considered: The $\Delta C/C$ (Fig. 2a) [1] and the $\Delta C$ concept (Fig. 2b) [5].

For the $\Delta C/C$ concept, voltage pulses of equal magnitude and opposite polarity are applied on the end of the capacitive bridge and the voltage at the common electrode is observed. By replacing the capacitance value by the expression of a parallel plate capacitor, as described in (1), it illustrates that the output voltage, without the parasitic capacitance $C_p$, is linearly proportional to the displacement of the proof mass even for large displacement $\Delta x$ as below:

$$V_o = V_s * \frac{C_1 - C_2}{C_1 + C_2} \propto \frac{\Delta x}{g_0} \qquad (2)$$

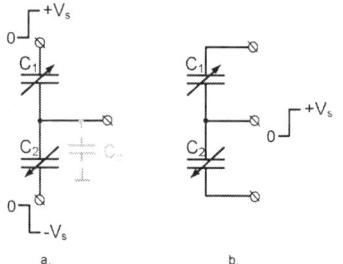

For the $\Delta C$ concept in Fig. 2b, the common electrode is driven and the output signal is taken on the fixed electrodes. By charge conservation, the output charge is:

$$Q_0 \propto V_s * (C_1 - C_2) \qquad (3)$$

Equation (3) shows that a circuit is still needed for the $\Delta C$ concept to convert the output charge into a voltage whereas for the $\Delta C/C$ concept the output is already a voltage. However, the case presented in (2) is an ideal case. In reality, parasitic capacitances $C_p$ in Fig. 2a create distortion. These parasitics are an order of magnitude higher than the sensing capacitances and the issue is that they take away charge from the capacitive divider so that the net electrostatic force on the proof mass is not zero as it was the case without parasitics [1]. Moreover, the output voltage is not linearly proportional to the displacement of the proof mass anymore.

[1] presented the self-balancing bridge principle to solve this nonlinearity issue. It uses a negative feedback to regulate the voltage driving the ends of the capacitive bridge so that in a steady state no charges flow away from the capacitive divider, Fig. 3. This C/V stage is used for the charge-balanced configuration, especially for the $\Delta C/C$ concept.

For the open-loop configuration, the $\Delta C$ concept, the charge obtained from the sensor has to be converted into a voltage. In Fig. 4 a circuit is shown. The charge from the sensor is integrated on $C_i$ so that the output voltage becomes:

$$V_o = -\frac{C_1 - C_2}{C_i} * V_s * (1 - \frac{C_0}{C_0 + C_i + C_p}) \qquad (4)$$

where $C_0$ is the nominal capacitance.

The offset caused by the parasitic $C_p$ can be removed with an input common mode feedback [5].

### C. Sigma Delta Modulator

The ADC used to digitize the output of the C/V stage is a sigma delta converter. The sigma delta modulation is attractive for providing a digital output while achieving high linearity and large bandwidth. Moreover, it can be implemented in high density CMOS technologies [6][7]. The sigma delta modulator used for the evaluation in this paper is a standard second order CIFF single bit modulator. In the following, this work focuses on the differences resulting from the four different C/V stages configurations, and therefore the readout circuits presented here all use the same sigma delta modulator.

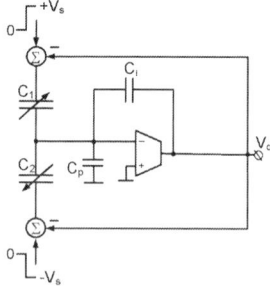

Fig. 2. a) $\Delta C/C$ concept, b) $\Delta C$ concept

Fig. 3. Self balancing bridge [1]

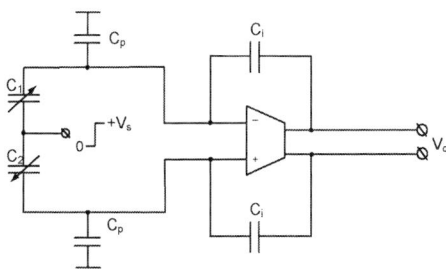

Fig. 4. Open-loop C/V stage [5]

III. SIMULATIONS

The circuits used for the simulation correspond to Fig. 3 and Fig. 4 with the parasitic capacitance $C_p$ in the following three different configurations: Fully-differential, pseudo-differential and single-ended. The simulations presented in this paper were done with MATLAB/Simulink and Cadence ADE on system level. All the components in the circuit simulation are written in Verilog-A. MATLAB/Simulink enables the behavioral-level modeling. On the other side, the circuit simulation allows a more detailed transient analysis at each node of the circuit. The presented results are divided into two parts for each analysed parameter. In the first part, the charge-balanced and the open-loop architectures are compared in a fully-differential configuration. In the second part, the fully-differential, pseudo-differential and single-ended configurations are compared in a charge-balanced architecture. The reason for this is that the different results between charge-balanced and open-loop architectures are independent of their configuration as fully-differential, pseudo-differential or single-ended and the reverse is true as well.

A. Estimated Front-end and MEMS Area

Table I and II show an evaluation of the expected area for the different architectures. The charge-balanced fully-differential configuration is considered as the reference and the sizes of the other architectures are compared to it. The open-loop configuration has a smaller C/V stage than the charge-balanced configuration since it only has an integrator and no summing amplifier (Fig. 3 and 4). The pseudo-differential architecture has a slightly higher ASIC area than the fully-differential configuration since the dummy sensor is added on it. However, the dummy size is negligible to the overall size

TABLE I
AREA COMPARISON FOR CHARGE-BALANCED AND OPEN-LOOP

|  | Charge-balanced | Open-loop |
|---|---|---|
| MEMS | Reference | Same area |
| C/V stage | Reference | 1/3 of area |

TABLE II
AREA COMPARISON FOR FULLY-, PSEUDO-DIFFERENTIAL AND SINGLE-ENDED

|  | Fully-differential | Pseudo-differential | Single-ended |
|---|---|---|---|
| MEMS | Reference | Area halved | Area halved |
| C/V stage | Reference | Same area | Area halved |

and therefore the ASIC size of both configurations can be considered as similar.

B. Achievable Linearity

Fig. 5 shows the output voltage for different input accelerations for the charge-balanced and open-loop configurations. The x-axis is the input acceleration normalized to g = 9.8m/s².

The charge-balanced configuration, in this case implemented as $\Delta C/C$ as in Fig. 2a, is highly linear until 60g and saturates at 70g, whereas the open-loop configuration, using a $\Delta C$ concept as in Fig. 2b, gets nonlinear from 25g and saturates at 50g. The clipping occurs due to the chosen circuit parameters and could be adapted. However, the nonlinearity of the $\Delta C$ concept is difficult to improve and the achieved linearity is as expected much worse than for the charge-balanced concept. Indeed, the $\Delta C/C$ circuit intrinsically has a higher linearity since it is proportional to the displacement of the proof mass, (2), whereas the $\Delta C$ concept is equivalent to (1), which is non-linear for a high displacement $\Delta x$.

The fully-differential, pseudo-differential and single-ended implementations have a similar linearity. Nevertheless, the integral non-linearity (INL) is simulated for the three configurations and Fig. 6 shows that the pseudo-differential architecture has a lower linearity than the fully-differential or single-ended configuration. The non-linearity does not occur because of the internal MEMS modeling since this behavior is also observed with an ideal differential pair of capacitors with the parasitic capacitance $C_p$ as depicted in Fig. 2a. By determining the transfer function of the C/V stage with sensor and evaluating from there the linearity, it appears that this behavior is due to the parasitic capacitance $C_p$ and is only observed for the pseudo-differential configuration.

C. Electro-Magnetic Interference

Capacitive MEMS readout circuits are very sensitive to electromagnetic interferences picked up on the wires between

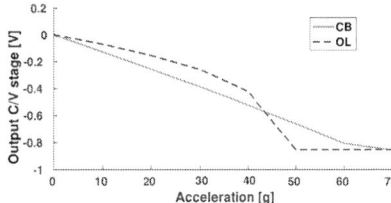

Fig. 5. Linearity charge-balanced and open-loop

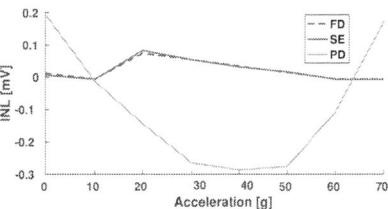

Fig. 6. INL for fully-differential, pseudo-differential and single-ended

the sensor and the ASIC. This is because these bondwires are connected to high impedance capacitive nodes [1].

A corresponding simulation is performed by applying a voltage of different frequencies on the wire between the sensor and the C/V stage through a capacitance of 18fF. An interference of 2kHz was applied and the output of the sigma delta modulator for the different configurations was observed. Regarding EMC, no difference could be noted between a charge-balanced and an open-loop architecture. Fig. 7 shows the output spectrum for the fully-differential and the pseudo-differential configurations. The interference does not appear on the fully-differential configuration since it is rejected thanks to the symmetry. However, the interference can be observed at 2kHz for the pseudo-differential architecture. The single-ended architecture has the same behavior, regarding EMC, as the pseudo-differential configuration. However, for pseudo-differential configurations various methods such as chopping can be used to reduce this interference.

### D. Noise Performance

The noise performance was evaluated by calculating each noise source of the readout circuit and referring it to the input. The main noise source comes from the first integrator of the C/V stage, which is also influenced by the parasitic capacitance of the wire connecting the sensor and the C/V stage. The noise of the summing amplifier is reduced thanks to the gain of the integrator preceeding it. The sigma delta modulators noise is dominated by the noise of the first integrator. The pseudo-differential architecture has the worst noise performance. Indeed, it has twice the number of capacitance as a single-ended configuration but still the same output voltage and therefore about $\sqrt{2}$ more noise. The fully-differential configuration has also twice the number of capacitance compared to the single-ended one, but it has twice the output voltage.

### E. Summary

Table III and IV summarize the performance for the four configurations regarding area, linearity, EMC robustness, noise and sensitivity where ++ corresponds to the best performance, - - to the worst and / to equivalency. The sensitivity corresponds to the minimum detectable capacitance which is determined by thermal noise floor. Therefore, it has the same results in the table III and IV as the noise. Thereby, the sensitivity is a function of the parasitic capacitances.

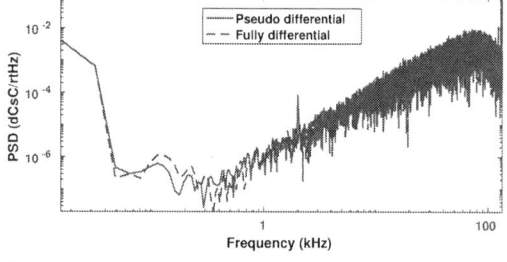

Fig. 7. Output spectrum for EMC perturbation for pseudo-differential and fully-differential

TABLE III
SUMMARY FOR CHARGE-BALANCED AND OPEN-LOOP

|  | Charge-balanced | Open-loop |
|---|---|---|
| Size | - - | ++ |
| Linearity | ++ | - - |
| EMC robustness | / | / |
| Noise | - | + |
| Sensitivity | - | + |

TABLE IV
SUMMARY FOR FULLY-DIFFERENTIAL, PSEUDO-DIFFERENTIAL AND SINGLE-ENDED

|  | Fully-differential | Pseudo-differential | Single-ended |
|---|---|---|---|
| Size | - - | - | + |
| Linearity | ++ | - | ++ |
| EMC robustness | ++ | + | - |
| Noise | ++ | - | + |
| Sensitivity | ++ | - | + |

## IV. CONCLUSION

In this work, an overview of different architectures for accelerometer readout circuits was presented. Advantages and disadvantages were compared for each of the four configurations. Especially, this paper has shown that the pseudo-differential configuration has per concept a lower linearity than the fully-differential architecture. A fully-differential charge-balanced architecture could be used in automotive applications, where linearity and EMC robustness are essential requirements. On the other side, an open-loop single-ended concept might be the better choice for an application where the size, power consumption and noise are the major concern. Each parameter could be improved by adapting several circuit properties as for example the capacitance values or the sampling frequency. Though, improvement on one parameter is usually at the cost of another. Moreover, the simulations presented here were performed on ideal circuits and for more detailed analysis the different circuits should be simulated at transistor level as well. However, these simulations give an indication of the concept performance for each architecture which enables to choose the most appropriate one for a specific application.

## REFERENCES

[1] V. P. Petkov, G. K. Balachandran, and J. Beintner, "A fully differential charge-balanced accelerometer for electronic stability control," *IEEE Journal of Solid-State Circuits*, vol. 49, no. 1, pp. 262–270, January 2014.

[2] C. C. Jonathan Soena, Alina Vodab, "Controller design for a closed-loop micromachined accelerometer," *Control Engineering Practice*, vol. 15, pp. 57–68, March 2007.

[3] S. Amini and D. A. Johns, "A flexible charge-balanced ratiometric open-loop readout system for capacitive inertial sensors," *IEEE Transactions on Circuits and Systems II: Express Briefs*, vol. 62, no. 4, pp. 317–321, April 2015.

[4] B. E. Boser, "Electronics for micromachined inertial sensors," in *Transducers 97. 1997 Internation Conference on Solid-State Sensors and Actuators*, June 1997.

[5] B. E. B. Mark Lemkin, "A three-axis micromachined accelerometer with a cmos position-sense interface and digital offset-trim electronics," *IEEE Journal of Solid state circuits*, vol. 1, pp. 428–478, 1999.

[6] C. W. W. Q. Liu Yuntao, Liu Xiaowei, "Design and noise analysis of a sigma-delta capacitive micromachined accelerometer," *Journal of Semiconductors*, vol. 31, no. 5, May 2010.

[7] P. G. Weijie Yun, R.T. Howe, "Surface micromachined, digitally force-balanced accelerometer with integrated cmos detection circuitry," in *Solid state sensor and actuator workshop*, vol. 5, August 1992.

# Real Time Defect Detection of Wheel Bearing by Means of a Wirelessly Connected Microphone

Erica Raviola, Franco Fiori
*Electronics and Telecom. Dpt.*
*Politecnico di Torino*
Torino, Italy
{erica.raviola, franco.fiori}@polito.it

*Abstract*—In this work, an electronic system aiming to automatically monitor the state of health of wheel bearings, is proposed. The focus is on designing a low cost, small size and wirelessly interfaced module. Acoustic emissions are exploited to detect defects by means of a low cost micro electro-mechanical system (MEMS) microphone. A microcontroller was used to evaluate the frequency spectrum and to interface the system through a wireless data link. The designed module successfully achieved the proposed goals. Finally, a novel measurement process is presented to evaluate the system performance under realistic conditions.

*Index Terms*—Fault detection, Wheel bearings, BLE, MEMS microphone

## I. INTRODUCTION

Rotating mechanical components are widely present in vehicles, and they are often used in conjunction with roller bearings, which unavoidably affect the system reliability. For this reason, a number of different techniques have been presented in the last years to detect unhealthy bearings.

An accelerometer is typically used to sense the vibrations generated by the mechanical component to be monitored, and the corresponding signal is processed to determine the state of health of the bearing. Recently, methods based on the processing of acoustic emissions (AE) have grown in popularity to address the limitations of vibration analysis, since AE results in better early damage defects and low speed detection [1]. Consumer microphones, in substitution of more sophisticated and expensive ultrasound probes, were identified as suitable candidates to perform bearing monitoring, as described in [2]-[3]. Furthermore, a micro electro-mechanical system (MEMS) microphone was successfully exploited to monitor an automation system by detecting the acoustic signal [6]. Numerous investigations have been conducted to determine the signal processing technique providing the most reliable results. In most works, the acoustic signal has been acquired by means of a data acquisition board, and the signal processed with Matlab. References [4], [5] used a field programmable gate array (FPGA) and a digital signal processor, respectively, to accelerate computations. However, to the authors knowledge, cost and size of the monitoring system have not been yet addressed as designed criteria.

This work aims to design an electronic system that checks automatically the state of health of a mechanical component, and faces the challenges of low cost and small size hardware.

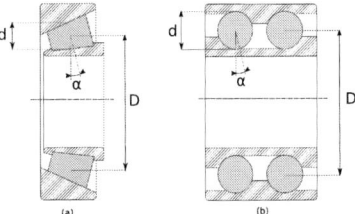

Fig. 1. Tapered roller bearing (a) and double row ball bearing (b) section.

For these reasons, a MEMS microphone and a microcontroller, instead of a more sophisticated sensor and a PC, were used. Furthermore, a wireless data link was introduced for the first time in a bearing monitoring system, thus eliminating the cost related to data wiring.

The paper is organized as follows: in Section II, roller bearing defects and the corresponding defect fundamental frequencies are introduced; in Section III, an optimized version of the Fast Fourier Transform algorithm is exploited to evaluate the frequency spectrum and then, in Section IV, the designed system and a first prototype is described. Moreover, some preliminary experimental results are presented in Section V and finally, in Section VI, concluding remarks are drawn.

## II. BEARING FAULT DETECTION

Wheel bearings are necessary components in the vehicle steering system, reducing friction between wheel hubs and constant-velocity joints, and remaining possible for the wheels to steer. Double-row ball and tapered roller bearings are preferred for this task, because of their capability to support heavy radial and axial loads [8]. A damaged bearing typically shows scratches, and when the rolling elements impact on them, pulsed noise is delivered. Knowing the bearing geometry and the shaft angular velocity, the defect fundamental frequencies can be evaluated. Depending on the scratch locations, a fundamental frequency is associated to the specific defect: ball pass inner frequency ($BPFI$), ball pass outer frequency ($BPFO$) and ball spin frequency ($BSF$).

Assuming no slip, a general expression for $BPFI$ and $BPFO$ can be derived [7] i.e.,

$$BPFI = \frac{Z f_r}{2}\left(1 + \frac{d}{D}cos\alpha\right) \qquad (1)$$

978-1-5386-5388-3/18 $31.00 © 2018 IEEE          233

$$BPFO = \frac{Zf_r}{2}\left(1 - \frac{d}{D}cos\alpha\right) \qquad (2)$$

where $Z$ is the number of rolling elements, $f_r$ the inner angular frequency expressed in $Hz$, $d$ is the rolling element diameter, $D$ is the pitch diameter and $\alpha$ the contact angle. Bearing sections are shown in Fig. 1, where $d$ and $D$ dimensions are highlighted. As far as automotive applications are concerned, all the parameters in Eqn.(1), (2) are assumed to be known a priori.

## III. FFT ALGORITHM

Different algorithms have been proposed to efficiently compute the Discrete Fourier Transform (DFT). They are all based on the Cooley-Tukey radix-2 Fast Fourier Transform (FFT), which exploits a divide-et-impera approach to gather the required operations in stages and butterflies. The most popular solutions are the radix-4 FFT [10], split radix FFT [10] and quick Fourier Transform [11]. By comparing these algorithms in terms of performance, [10] noticed that the radix-2 is the one requiring the least amount of memory to store data. Since all the computations will be performed by a microcontroller, whose internal memory is limited, the radix-2 approach has been preferred for this task. Moreover, since the acquired samples are real, DFT symmetry properties can be exploited to process a N-length real value sequence by means of a N/2 FFT, plus some pre-FFT and post-FFT stage. In this way, the Cooley-Tukey radix-2 kernel is surrounded by a pre-FFT and post-FFT stage, reducing the number of operations, thus time and power [9]. Given a real valued sequence $g[n]$, with $n = 0, 1, ..., 2N-1$, a complex value sequence is derived as $x[n] = g[2n] + jg[2n+1]$ where $n = 0, 1, ..., N-1$. Then, exploiting linearity and symmetry properties, the spectrum of the starting sequence $g[n]$ can be expressed as

$$G[k] = A[k]X[k] + B[k]X^*[N-k] \qquad k = 0, 1, ...N-1, \quad (3)$$

where

$$A(k) = \frac{1}{2}\left(1 - jW_{2N}^k\right) \quad B(k) = \frac{1}{2}\left(1 + jW_{2N}^k\right). \quad (4)$$

The $G[k]$ spectrum can be obtained by $X[k]$ spectrum at the cost of some extra additions and multiplications by $A[k], B[k]$ coefficients. In such a case, the number of complex operations is strongly reduced, as reported in Table I. It is worth noticing that the percentage reduction reaches almost 40% for the best case ($N = 65536$).

Since the bearing fault detection is based on the evaluation of the spectrum magnitudes, an important aspect to be considered is the spectral leakage phenomenon, which may leads to an underestimation of the spectral magnitudes, resulting in an false positive. To minimize this effect, a windowing function was exploited, in particular the top flat, instead of Hanning or Blackman windows, was chosen because it provides the best peak accuracy. All the $n$ time samples, with $n = 0, 1, ...N-1$, have to be multiplied by the corresponding value of the top

TABLE I
NUMBER OF COMPLEX OPERATIONS REQUIRED BY COOLEY-TUKEY RADIX 2 FFT COMPARED TO REAL DATA FFT.

| Samples $N$ | Classic FFT | Real valued FFT | | Saving(%) |
|---|---|---|---|---|
| | kernel | kernel | post FFT stage | |
| 256 | 20486 | 8965 | 5632 | 28,7 |
| 1024 | 102408 | 46087 | 22528 | 33,0 |
| 4096 | 491530 | 225289 | 90112 | 35,8 |
| 16384 | 2293772 | 1064971 | 360448 | 37,9 |
| 65536 | 10485774 | 4915213 | 1441792 | 39,4 |

flat function, which expression in reported in Eqn. (5), before applying the FFT algorithm [12].

$$w(n) = \sum_{i=0}^{4} a_i cos\left(\frac{2\pi n}{N}i\right) \qquad (5)$$

where $a_0 = 0.21, a_1 = -0.42$, $a_2 = 0.28$, $a_3 = -0.08$ and $a_4 = 0.007$.

## IV. SYSTEM ARCHITECTURE

Aiming to design a bearing monitoring system, different hardware and software modules are required. The block level view of the designed system is that shown in Fig. 2, and each sub-part will be discussed in details hereafter. The MEMS microphone is responsible for sensing the acoustic noise generated by the bearing, then an amplifier adapts the signal to the analog to digital converter (ADC) input range. After that, the frequency spectrum evaluation and the fault detection are performed by the developed firmware, which also manages the bluetooth low energy (BLE) interface.

### A. Data acquisition

The MEMS microphone used in this work is characterized by bandwidth 100 Hz - 10 kHz, and it is directly mounted on the bottom side of the PCB board. The signal conditioning circuit consists of an audio amplifier, having a transfer function characterized by a pole at 10 kHz, a zero at 5 Hz and bandwidth gain equals to 200. The sampling frequency of the ADC can be chosen in the range $100\,Hz - 50\,kHz$, and the use of an external SPI volatile memory allows to collect up to 32768 time samples.

### B. Fast Fourier Transform algorithm

The algorithm used to evaluate the signal spectrum was described in Section III, i.e. the Cooley-Tukey radix-2 optimized for real data values. Twiddle factors, as well as post stage coefficients, are computed on board every time to reduce the amount of required memory. Fractional point representation has been chosen in order to better exploit microcontroller performance, with no need for a floating point unit.

### C. Fault detection

Once the amplitude spectrum is obtained, it has to be compared against some thresholds. When a defect arises, the corresponding fundamental frequency amplitude will increase, thus by monitoring the range of frequencies corresponding to

*PRIME 2018, Prague, Czech Republic*                    *Session: Automotive Circuits and Systems*

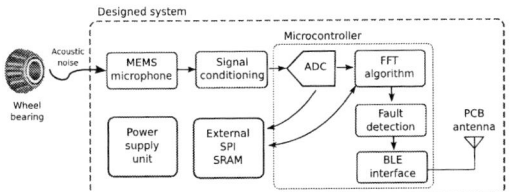

Fig. 2. Block diagram representing the system architecture.

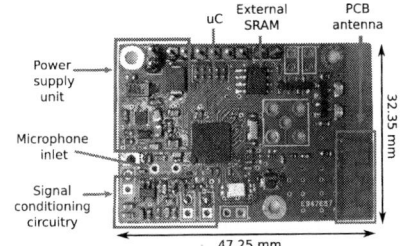

Fig. 3. System prototype.

the defect fundamental frequencies, faults can be detected. Depending on the bearing type and current vehicle speed, the central unit should provide the range of frequencies to be monitored, together with the amplitude thresholds.

*D. Bluetooth Low Energy Interface*

Between the different standards operating at 2.4 GHz, BLE has become a popular solution in IoT world. With a nominal range equals to 10 m, it allows a point-to-point wireless connection between the developed system and a central unit. Defining a proper generic attribute layer (GATT), it is possible to customize the interface exhibited by the module according to the target application needs. In this way, the central unit may change in real time the configuration parameters of the module, as well as be notified on the spectrum amplitudes greater than the thresholds.

*E. Prototype*

A PCB board, with all the components necessary for the system to work by itself, was realized. A photo of the prototype is presented in Fig.3. In operating condition, only power supply wires are required, since configuration parameters and bearing faults are transmitted over BLE. The prototype can be supplied directly by a 12 V battery.

## V. PRELIMINARY RESULTS

As far as bearing monitoring systems are concerned, measurements are usually carried out connecting an healthy and a damaged bearing to an electric motor. However, in this scenario the major acoustic noise sources are the bearing and the motor themselves, which are not sufficiently representative of the real sound near the wheels of a moving vehicle. In addition to that, it would be useful to modify the detected amplitudes, for testing purposes. A different approach is thus

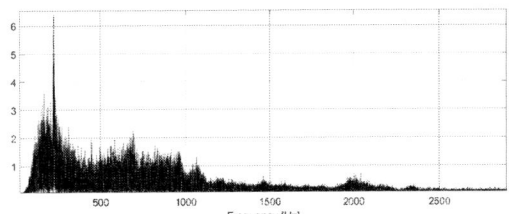

Fig. 4. Spectrum of the noise recorded. The vehicle was moving at 80 $km/h$.

TABLE II
CONFIGURATION OF THE SYSTEM DURING MEASUREMENT PROCESS.

| ADC sampling frequency | $6\,kHz$ |
|---|---|
| Number of acquired samples | 16384 |
| Windowing function | Top flat |
| Monitored range 1 | $[110.5, 112.5]\,Hz$, threshold $10\,mV$ |
| Monitored range 2 | $[196.6, 198.6]\,Hz$, threshold $10\,mV$ |

proposed to evaluate the system performance. Firstly, it is required to collect a representative noise of a moving vehicle. By means of a smartphone patched to the vehicle frame just upper the front left wheel, the acoustic noise along a straight was recorded, and the corresponding spectrum is shown in Fig. 4. As it can be noticed, it is pretty populated in the 100 Hz - 1 kHz frequency range. Secondly, an inner and an outer raceway defect were simulated by adding two tones to the noisy spectrum, at the corresponding defect fundamental frequencies. A tapered roller bearing was taken as reference NTN. It is characterized by 20 cylindrical elements, $d = 9.75\,mm$, $D = 63.5\,mm$ and $\alpha = 32\,$deg. The fundamental frequencies can be evaluated recalling Eqn. (1), (2), obtaining $BPFO = 111,5\,Hz$ and $BPFI = 197.6\,Hz$. During the recording phase, the speed of the vehicle was set to 80 km/h, allowing the defect fundamental frequencies to be in the microphone bandwidth. Regarding the added tone amplitudes, they were not kept constant, but different values were tested. The average of the amplitude spectrum, called $A_{aver}$, in the range 100 Hz - 1 kHz of the noise signal was evaluated, since this is the most populated region. Then the amplitude of the two tones is computed according to

$$A_{tones} = A_{aver}\left(1 + 10^{\frac{A_{dB}}{20}}\right) \qquad (6)$$

where $A_{dB}$ was set to $40, 35, 30, 25, 20, 15, 10$ in the different tests. In this way, a new audio file is synthesized, containing both the vehicle noise and the peaks related to the bearing defects. Using a speaker, this new signal is reproduced few centimeters closed to the microphone. The designed system was configured as reported in Table II, where all parameters were setted using the BLE interface. The spectra, evaluated by the system, are reported in Fig. 5, with the area corresponding to the two tones zoomed. The purple and the green markers correspond to the BPFO and the BPFI defect, respectively. Referring to the monitored ranges configured via BLE, the system successfully recognizes the defects only if $A_{dB} = [+40, +30]\,dB$ for BPFI and $A_{dB} = [+40, +20]\,dB$

978-1-5386-5388-3/18 $31.00 © 2018 IEEE          235

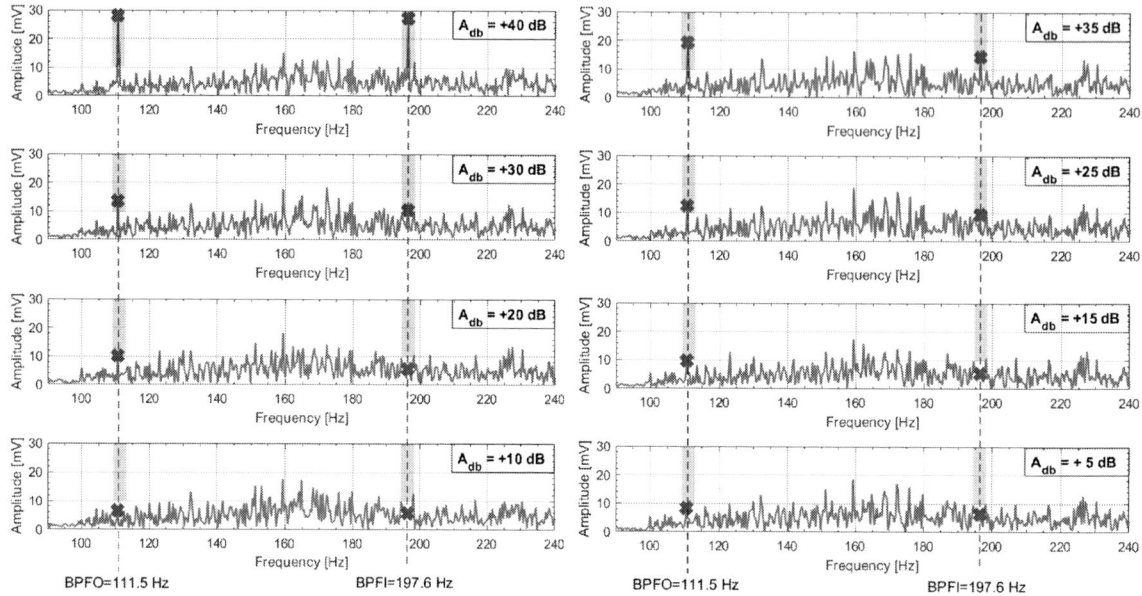

Fig. 5. System evaluated spectra at different tones amplitude. The purple marker deals with the BPFO defect frequency, the green one to the BPFI. The two tones were superimposed to the noise recorded, with the vehicle moving at $80\ km/h$. The monitored ranges, defined in Table II, are reported as red boxes. When the magnitude spectrum is greater then a certain threshold in a given frequency range, the system recognizes the bearing as a defected one and notifies the failure via BLE.

for BPFO. Further development should be aimed at enhancing the defect detection capability.

## VI. CONCLUSION

In this work, a system that automatically monitor a wheel bearing is proposed. It is based on the sensing of acoustic noise, generated by the bearing, by means of a MEMS microphone. The acquired signal is digitally converted and its frequency spectrum evaluated using a real data valued FFT algorithm. Comparing the obtained spectrum against some thresholds for a given sets of frequencies, unhealthy bearings can be recognized. Finally, a BLE interface allowed to transfer data, implementing on the fly configuration and notification in case of defect detection. The challenges were successfully achieved, since the designed system is characterized by low-cost hardware, small dimensions and a wireless interface. The preliminary results, based on a real moving vehicle acoustic noise, shows the capability of the system to detect the faulty frequencies, provided that they rise sufficiently with respect to the background noise.

## REFERENCES

[1] A. b. Ming, W. Zhang, Z. y. Qin and F. l. Chu, "Dual-Impulse Response Model for the Acoustic Emission Produced by a Spall and the Size Evaluation in Rolling Element Bearings", *IEEE Transactions on Industrial Electronics*, vol. 62, no. 10, pp. 6606-6615, Oct. 2015.

[2] J. Grebenik, Y. Zhang, C. Bingham and S. Srivastava, "Roller element bearing acoustic fault detection using smartphone and consumer microphones comparing with vibration techniques", *2016 17th International Conference on Mechatronics - Mechatronika (ME)*, Prague, 2016, pp. 1-7.

[3] P. Rzeszucinski, M. Orman, C. T. Pinto, A. Tkaczyk and M. Sulowicz, "A signal processing approach to bearing fault detection with the use of a mobile phone", *2015 IEEE 10th International Symposium on Diagnostics for Electrical Machines, Power Electronics and Drives (SDEMPED)*, Guarda, 2015, pp. 310-315

[4] M. Kang, J. Kim, I. K. Jeong, J. M. Kim and M. Pecht, "A Massively Parallel Approach to Real-Time Bearing Fault Detection Using Sub-Band Analysis on an FPGA-Based Multicore System", *IEEE Transactions on Industrial Electronics*, vol. 63, no. 10, pp. 6325-6335, Oct. 2016.

[5] W. Liu, T. Zhao, D. He and D. Liu, "DSP based module for processing vibration signals of rotation machinery", *2017 IEEE International Instrumentation and Measurement Technology Conference (I2MTC)*, Turin, 2017, pp. 1-6.

[6] C. M. Wang, Y. T. Lin and C. H. Lin, "Monitoring of micro-stamping procedures utilizing MEMS microphones under acoustic vibration transduction", *2017 IEEE 12th International Conference on Nano/Micro Engineered and Molecular Systems (NEMS)*, Los Angeles, CA, 2017, pp. 106-109.

[7] R. B. Randall, J. Antoni, "Rolling element bearing diagnostics A tutorial", *Mechanical Systems and Signal Processing*, Volume 25, Issue 2, pp 485-520, 2011.

[8] D. Knowles, "Automotive suspensions and steering system", 4th edition.

[9] H. Sorensen, D. Jones, M. Heideman and C. Burrus, "Real-valued fast Fourier transform algorithms", *IEEE Transactions on Acoustics, Speech, and Signal Processing*, vol. 35, no. 6, pp. 849-863, 1987.

[10] Manish Soni, Padma Kunthe, "A General Comparison Of FFT Algorithms", *9th National Conference*, Pioneer Journal.

[11] Haitao Guo, G.A. Sitton, C.S. Burrus, "The quick Fourier transform: an FFT based on symmetries", *IEEE Transactions on Signal Processing*, vol. 46, no. 2, Feb. 1998.

[12] F.J. Harris, "On the use of windows for harmonic analysis with the discrete Fourier transform", *Proceedings of the IEEE*, vol. 66, issue 1, Jan. 1978.

[13] NTN Bearing Corporation, Tapered Roller Set - Metric Series, product code 30207.

*PRIME 2018, Prague, Czech Republic*                                    *Session: Automotive Circuits and Systems*

# Multi object detection in direct Time-of-Flight measurements with SPADs

Jan F. Haase[1], Maik Beer[1], Jennifer Ruskowski[1], Holger Vogt[12]

[1]Fraunhofer Institute for Microelectronic Circuits and Systems (IMS), Finkenstr. 61, 47057 Duisburg, Germany
[2]University of Duisburg-Essen, Department of Electronic Components and Circuits, Bismarkstr. 81, 47057 Duisburg, Germany
E-Mail: Jan.Haase@ims.fraunhofer.de

*Abstract*—**We present several contributions of our test system to detect multiple targets with the direct time-of-flight technique. With a precise time-to-digital-converter it is possible to capture the time-of-flight of a short light pulse reflected by a target with high temporal resolution. Based on this technique we can relate the single events to its resulting distance and separate the different objects.**

*Keywords—light detection and ranging (LIDAR); single-photon avalanche diode (SPAD); time-of-flight (TOF); range imaging*

## I. INTRODUCTION

In robotics, automation engineering or autonomous driving it is indispensable to detect the distance of an object in a reliable mode with three dimensional imaging. For higher resolution in imaging the acquisition is done by light-detection-and-ranging (LiDAR) [1]. For LIDAR a light source with a wavelength in the near infrared range (e.g. 905 nm) is used typically in the automotive sector, as it is invisible for the human eyes [2]. It is also possible to use other wavelengths like in the visible range if a visible light beam does not affect the application.

A short laser pulse is emitted and the light reflected by the target is measured. A sensor with highly sensitive single-photon-avalanche-diodes (SPADs) can detect the reflected light beam and hence the distance to the target can be calculated from the time-of-flight (TOF). In CMOS processes SPAD-based sensors can be fabricated cost efficiently with on-chip electronics [3]. In contrast to other photo diodes SPADs are highly sensitive with time resolution in the tens of picosecond range [4], therefore they can be used to detect single photon events.

The method to record the first photon is called direct TOF (see Section II) [5]. With a highly accurate stopwatch the TOF is recorded directly. In the indirect TOF technique the light of the reflected target is integrated in several time windows of equal width to calculate the TOF from the ratio of the intensity [5]. It is not possible to separate single objects, because measuring the mean intensity causes a loss of information. On the other hand, the electronics and the analysis are simpler. We will investigate the characteristic of multi object echoes in the direct TOF. The direct TOF is chosen because the timestamp of each individual event is available for signal processing and object separation which is not possible in the indirect TOF.

The test system consists of a CMOS integrated sensor based on the direct TOF method and a laser source (see Section III).

The first measurement includes a separation of two objects arranged one behind the other. In the second case the first object is replaced by an acrylic glass plate. The results are presented in Section IV. With these measurements we will show that the separation of multiple targets in one pixel is possible in the direct TOF method.

## II. DIRECT TIME-OF-FLIGHT TECHNIQUE

In the direct TOF measurement the time between the emission and the reception of a short light pulse reflected by an object is measured using a time-to-digital-converter (TDC). Typically, the TDC starts with the emission of the light pulse and stops with the detection of the first photon [5]. One problem of the direct TOF method is that also a photon from the background illumination can stop the TDC which results in a false measurement. To suppress false measurements a certain number of single measurements are collected in a histogram (e.g. Fig. 3). This gives a distribution of the measured distances and the possibility to get the true distance of the target out of the focus of the histogram.

The probability to detect the first photon in a direct TOF measurement is given by the probability density function (PDF) for the first photon $P_1$ as

$$P_1(t) = \lambda(t) \cdot \left(1 - \int_0^t P_1(\tau)\, d\tau\right). \tag{1}$$

$\lambda$ represents the photon detection rate at the sensor [6]. The solution of Eq. (1) with an assumed constant photon detection rate $\lambda$ results in an exponential decay of the PDF according to

$$P_1(t) = \lambda \cdot e^{-\lambda \cdot t}. \tag{2}$$

The integral over Eq. (2) of a defined time interval results in the probability to detect a photon in this period of time. Therefore the possibility to detect the first photon at a later time will be decreased.

If the light pulse is reflected by more than one target and detected in the same pixel of the sensor, it is possible to separate the targets with the direct TOF method. This is presented in Fig. 1 schematically. Within the timestamp of the first photon of several measurements, there is a probability to get the first photon from the first object as well as from the object behind.

978-1-5386-5388-3/18 $31.00 © 2018 IEEE                    237

*Paper P32*                                                    *PRIME 2018, Prague, Czech Republic*

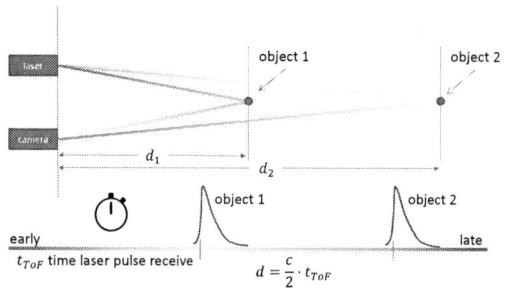

Fig. 1. Schematic of a scene with two objects detected in one pixel of a SPAD-based-sensor with the direct TOF measurement.

Another separation of targets is possible, if the first object is partially transparent, for example a glass plate. In this case the photons are reflected and transmitted by the transparent object. The transmitted photons can be reflected by a target behind the glass plate and detected by the sensor.

Eq. (1) shows, that the signal of the farther object is damped in comparison to a single target signal, since photons reflected by the closer object have a higher probability to stop the time measurement. Fig. 2 shows the PDF calculated with Eq. (1) of a two object situation. It is presumed, that the event rate of the reflected light is the same for both objects independent of their distances. From theory, the peaks can be separated if the distance between the objects is greater than the pulse width.

### III. TEST SYSTEM

For our test setup we used the chip presented in [7]. It is a sensor designed for the direct TOF measurement with CMOS integrated SPADs and chip-on electronics. The sensor includes two lines with 192 pixels each, implemented in a flash LiDAR camera demonstrator. Some important specifications are shown in Tab. 1.

The CMOS chip and the laser modules are placed together in an aluminum housing controlled by a Field Programmable Gate Array (FPGA). The data is transferred by USB and the analysis and evaluation of the data is done on a Computer.

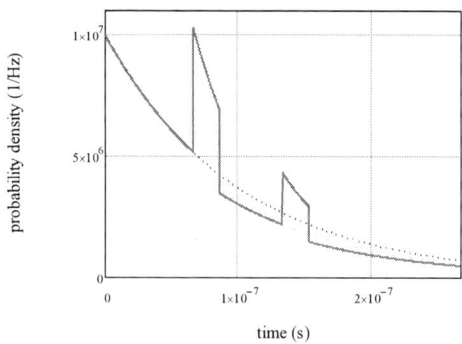

Fig. 2. The figure shows the probability density function of the first photon in the direct TOF technique. The peaks represent two targets in 10 m (66.7 ns) and 20 m (133.4 ns) distance, respectively. The rates of the reflected light and the background illumination are 10 MHz each and the pulse width is 20 ns.

TABLE I.        CHIP AND TEST SETUP SPECIFICATIONS

| Parameter | Value | Unit |
|---|---|---|
| Sensor resolution | 2 x 192 | pixel |
| Pixel pitch | 40.56 | µm |
| Clock frequency | 200 | MHz |
| TDC resolution | 312.5 / 46.8 | ps / mm |
| Pixel row FOV | 0.8 x 34.6 | ° |
| Vertical distance rows | 4.7 | ° |
| Laser peak power | 75 | W |
| Laser pulse width | ~20 | ns |
| Laser wavelength | 905 | nm |
| Number of Lasers | 2 | |
| Repetition rate | 10 | kHz |
| Frame rate[a] | 25 | Hz |

[a] 400 measurements each frame

The clock frequency of 200 MHz delayed in 16 steps gives a timing resolution of 312.5 ps corresponding to a distance of 46.8 mm. Two laser modules expose each line of the system at the same time. Hence, the complete scene can be measured in flash mode simultaneously without moving parts. With the used optical lens a field of view (FOV) of 0.8° x 34.6° for each line with 192 pixels is achieved. Therefore, the horizontal resolution of one pixel is about 0.18°. The lines are arranged with a gap of 4.7°.

### IV. MEASUREMENTS

For the first measurement two objects were one behind the another. In this setup the pixel's FOV covers the edge of the first object and the second object behind simultaneously. Depending on the angle of view the covered parts of the first and second object vary.

In the second setup a target was placed behind an acrylic glass plate. Since the glass is partially transparent, it is possible to see the target through it. Different angles of the acrylic glass plate and the reflected light were investigated. The results of the measurement are presented in Fig. 3 and 4, respectively.

All measurements are done indoor with a low backlight illumination.

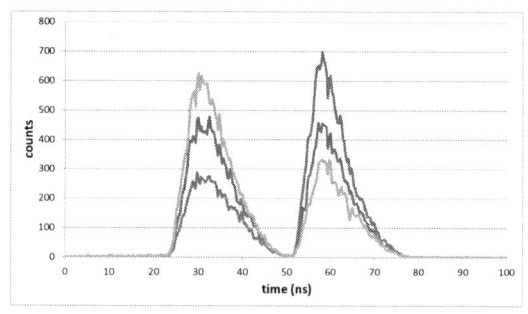

Fig. 3. Setup 1 of two objects in one pixel. The pixel sees the edge of an object, so that the object behind is also seen in the same pixel. A measurement with a little part of the first object in the pixel is represented in the blue line, in the green line, the part in the pixel of the first object is much higher. The red line is a representation of both objects in about the same part in the pixel.

978-1-5386-5388-3/18 $31.00 © 2018 IEEE          238

*PRIME 2018, Prague, Czech Republic*                                    *Session: Automotive Circuits and Systems*

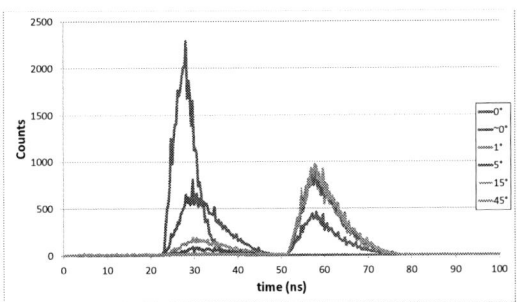

Fig. 4. Setup 2 of two objects captured by one pixel. The first object is a transparent acrylic glass plate. It was rotated and measurements were taken. The second object was not changed.

Fig. 3 shows the histogram recorded over 40,000 single measurements with two distinct peaks of the two objects. The beginning of the peaks corresponds to the position of the object at about 3.5 m and 7.7 m, respectively. The total number of counts is approximately the same at all three conditions in Fig. 3. Only the distribution of the counts corresponding to the first and the second object differs.

In the second measurement, an acrylic glass was rotated to investigate the reflected light beam depending on different angles. In Fig. 4 the results from the total reflection at an incident angle of 0° to a lower reflection at an angle of 45° is shown.

Since the shiny surface of the glass behaves like a mirror for a certain part of the incident light at an angle of 0° a lot of the light beam was reflected directly back to the sensor causing a high peak in the histogram. Because the high pulse intensity causes the time measurement to stop at almost every measurement, there is no visible peak of the target behind. When the glass is twisted just a little bit, the peak of the glass is reduced and the peak of the target behind becomes visible. At higher rotation angles, the peak of the target behind becomes visible comparable to a non-existing glass in front and the first peak of the glass disappears.

## V. Outlook

The next steps are to write a secure algorithm to identify the different peaks in the histogram. The algorithm has to evaluate the height and the shape of the peak to decide, if the measurement is valid and if it is possible to separate the peaks in case they are quite close together. To allow a better separation of the peaks at short distances, it is important to have a short laser pulse with only a few nanoseconds in width.

## VI. Summary

We have presented measurements to demonstrate the ability of the direct TOF measurement technique to separate multiple targets in different distances detected by the same pixel. We investigated two situations: In the first case each of the two objects was partially covered by the pixel's FOV and in the second case the object was located behind a transparent glass plate. The measurements were carried out using our test system with the direct TOF method. The characteristics were investigated in one pixel and the measurement results have been presented. In both situations the peaks of the different objects could be separated and the distances of the objects could be determined.

## References

[1] E. Ackerman, "Lidar that will make self-driving cars affordable [News]," *IEEE Spectrum*, vol. 53, no. 10, pp. 14–14, Oct. 2016.

[2] D. Stoppa, L. Pancheri, M. Scandiuzzo, L. Gonzo, G.-F. Dalla Betta, and A. Simoni, "A CMOS 3-D Imager Based on Single Photon Avalanche Diode," *IEEE Transactions on Circuits and Systems I: Regular Papers*, vol. 54, no. 1, pp. 4–12, Jan. 2007.

[3] D. Bronzi, F. Villa, S. Bellisai, B. Markovic, S. Tisa, A. Tosi, F. Zappa, S. Weyers, D. Durini, W. Brockherde, and U. Paschen, "Low-noise and large-area CMOS SPADs with timing response free from slow tails," in *Solid-State Device Research Conference (ESSDERC), 2012 Proceedings of the European, 2012, pp. 230–233.*

[4] P. Seitz and A. J. P. Theuwissen, Eds., *Single-photon imaging.* Heidelberg ; New York: Springer, 2011.

[5] F. Remondino and D. Stoppa, *Tof range-imaging cameras.* New York: Springer, 2013.

[6] M. Beer, B. J. Hosticka, and R. Kokozinski, "SPAD-based 3D sensors for high ambient illumination," in Conference on Ph.D. *Research in Microelectronics and Electronics (PRIME)*, 2016, pp. 1–4.

[7] M. Beer, O. M. Schrey, J. F. Haase, J. Ruskowski, W. Brockherde, et al., „SPAD-based flash LiDAR sensor with high ambient light rejection for automotive applications", *Proc. SPIE Vol. 10540*, Jan. 2018

# Super-capacitors for implantable medical devices with wireless power transmission

Pablo Mendoza-Ponce[1,2], Bibin John[1], Dietmar Schroeder[1] and Wolfgang H. Krautschneider[1]

[1]Institute of Nano- and Medical Electronics, Hamburg University of Technology, Hamburg, Germany
[2]Electronics Engineering Department, Instituto Tecnológico de Costa Rica, Cartago, Costa Rica

*Abstract*—**Patients using implantable devices with wireless power transmission (such as inductive telemetry) as the energy source are required to carry external hardware and antennas on their body continuously during the monitoring period. This discomfort can be reduced by integrating tiny super-capacitors as an energy source for the device. The external wireless telemetry unit used to activate and interrogate the implantable device charges the super capacitors integrated on it. When the wireless telemetry unit is removed, the super capacitor powers the implantable device. This paper shows the benefits of integrating super capacitors onto an existing pressure sensing medical implant that utilises an inductive telemetry link. Test results show that the implant can be powered for a full day using an 88 mF super capacitor. This super capacitor can be fully charged using the inductive telemetry link at a wireless distance of 20 cm in 81 seconds.**

*Index Terms*—**wireless power transmission, super-capacitor, inductive coupling, implantable device**

## I. INTRODUCTION

An active medical implant containing various electronic components can be powered using three available methods: battery, energy scavenging or wireless power transmission. Battery and wireless power links are commonly used as energy sources for implantable devices since the energy scavenging technique is limited by its energy density [1].

Although the size and lifetime of batteries limit their usefulness for implantable devices, battery powered devices work as fully autonomous systems without the need for external hardware on the patient's body in contrast to devices using wireless power transmission. This advantage of independent device operation can be implemented on a wirelessly powered implantable device by using super-capacitors as an energy storage element.

Environmental friendliness and long lifetime ($\approx 10^6$ charge cycles) make super capacitors ideal for medical implantable devices. Although super capacitors lack the energy density of batteries, they can be charged at a faster rate [2]. Super capacitors of small size (2.5 mm x 3 mm x 0.9 mm) [3] with high power density are particularly suitable for miniaturised medical implant devices.

Remote wireless power transmission has been used for pressure monitoring devices such as heart failure implants, aneurysm and glaucoma monitoring implants as well as measuring strain in metallic hip implants [4]. Inductive coupling is most often used (rather than capacitive coupling or acoustic powering) as the wireless power transmission technique.

Implantable devices with inductive coupling as the power transmission technique require carrying electronic hardware and antennas on the patient's body in order to provide energy to the implant. In an inductive telemetry link a loop antenna, typically wound around the body of the patient at the location of the device, is used for powering the implant [5]. This wireless setup, together with the need for continuous monitoring, causes discomfort and limits freedom of movement due to the powering circuitry and antenna that have to be carried constantly (including while sleeping). Additionally, this setup cannot be used during some activities, such as air travel and sports, resulting in interruption to the patient monitoring.

Another issue with inductive links is that the efficiency of the power transmission falls drastically when the antenna on the implant and the external power transmitter antenna are oriented at different angles (factor of $cos^2\alpha$ where $\alpha$ is the angle between antennas [6]). This issue limits the usage of inductive coupling for real-time neural recordings with laboratory animals like mice since the mice move freely inside the cage [7].

A super capacitor powering the implantable device for longer periods (in hours) can overcome these limitations and improve patient comfort. Whenever the external inductive telemetry unit is within the range of the implantable device, the super-capacitor is recharged and the measured sensor data is transferred wirelessly. Once the telemetry unit is removed, the super capacitor powers the implantable device. Thus the patient only needs to use the wireless telemetry unit for a few minutes per day instead of wearing it on the body constantly. This approach can also be used with laboratory animals so that their movement is not limited to a small cage containing a wireless power transmission unit.

This paper investigates the benefits of integrating super-capacitors into an existing implantable device. The autonomous operating time of the new design for an implantable device (i.e. when running using the energy stored in the super-capacitor) was measured for a set of super-capacitors of various capacitance values and for multiple sampling times. The charging time of the super-capacitors was also measured using a wireless power link setup.

## II. IMPLANTABLE DEVICE

Figure 1 shows the block diagram of the implantable system prototype. The pressure sensing, power harvesting and data transmission blocks are based on the implantable device

978-1-5386-5388-3/18 $31.00 © 2018 IEEE

presented in [8]. The processing core of the designed device is able to run using the energy from the power harvesting unit. Furthermore, the unit has been optimised for ultra low power consumption so that it can run for hours using the energy stored in the super-capacitor when the wireless power is not available.

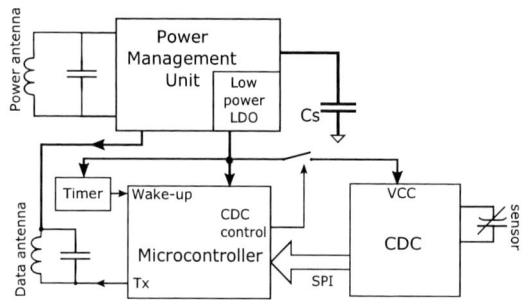

Fig. 1. Circuit diagram of the implant device

The implant uses inductive coupling for both wireless power reception and data transfer. The power antenna coil tuned to the frequency of the external RF sender receives the wireless energy. The power acquisition unit translates the AC output of the receiver antenna into a regulated DC power supply. This DC power is used for charging the super-capacitor $C_s$ [3] and enabling the wireless data transmission.

When the inductive telemetry link is active, the super-capacitor $C_s$ is charged to its maximum rated voltage of 3.3 V. Upon removal of the wireless power link, the super-capacitor $C_s$ powers the implantable device through an ultra-low power voltage regulator. The low power regulator provides a stable voltage of 1.8 V. The core of the implant consists of a micro-controller, a low power timer, a capacitive to digital converter (CDC) and the pressure sensor.

The device operates in either of two phases (active or idle) when powered by the super-capacitor. The system remains primarily in the idle phase due to the low sampling rates used by the proposed implantable device. This fact means that the average power consumption of the device is mainly defined by the power consumption of the idle mode. Because of this, taking the approach of minimising the energy use during this phase leads to a maximisation of the operating time of the device on a single full charge of $C_s$.

During the idle phase the CDC and the micro-controller are placed in a shutdown mode leaving only the timer active to trigger the sampling. As previously mentioned, the power consumption of this mode has to be minimised as much as possible. For this reason, an external low power timer is utilised. The power consumption of the external timer averages 90 nW, while using a timer integrated in the micro-controller results in a power consumption in the order of several μW. The current consumption of the implantable device during the idle phase is around 670 nA.

A trigger signal generated by the timer activates the system. For the experiment the timer was programmed to signal at various intervals between 1 s to 30 s. The active phase consists of the micro-controller turning on the CDC and triggering the sensor, acquiring the digital value available at the CDC output via a serial interface, saving the data to memory, turning off the CDC and going back to the idle phase. The system is in active phase for $7ms$ and during this phase the device consumes an average current of 600 μA.

The power receiver inductor antenna in the implant device has dimensions of 10 mm x 2.2 mm x 1.2 mm (complying with the geometrical constraints of the implantable device in [8]) with an inductance of 25.2 μH and is tuned to the wireless power frequency of the RF power sender. The power antenna is custom made by winding 60 turns of enamelled copper over two ferrite cores of 1 mm in diameter. The fabricated device using a single super-capacitor is shown in figure 2.

Fig. 2. Test device using a single super-capacitor

## III. WIRELESS-POWER SUPER-CAPACITOR CHARGING

The implant can be located in various positions inside the body based on the desired application. Some devices are utilised near the surface of the body while others are implanted deep inside. Because of this, each application requires the device to have a different wireless range. For a glaucoma implant like SENSIMED Triggerfish ® the range is less than 5 cm while for active aneurysm [9] and hip implant devices [10] it is 10 cm.

The commercially developed passive pressure transducers such as CardioMEMS EndoSure™and ImPressure™for the treatment of aneurysm have a wireless range of up to 20 cm [11] [12]. The implantable device with super-capacitors was tested for a wireless power range of 20 cm to meet this maximum depth.

Since inductive coupling was chosen as the wireless power transmission method, a setup consisting of a power amplifier and a loop antenna (with a diameter of 40 cm in order to achieve the wireless range of 20 cm ) was used as external hardware. The loops consist of three turns of enamelled copper wire. The loop antenna is connected to the power amplifier using an impedance matching network. The L-section capacitor-capacitor impedance matching network is made using high Q (quality factor) capacitors in order to match the impedance of the loop antenna to the power amplifier for maximum power transfer. A waveform generator supplies the power amplifier with a signal of the desired RF frequency.

*PRIME 2018, Prague, Czech Republic*                    *Session: Sensing and Biomedical Circuits II*

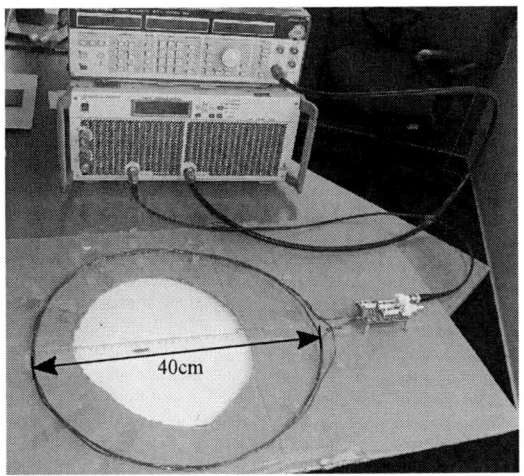

Fig. 3. Setup for wirelessly charging the super-capacitors in the implantable device.

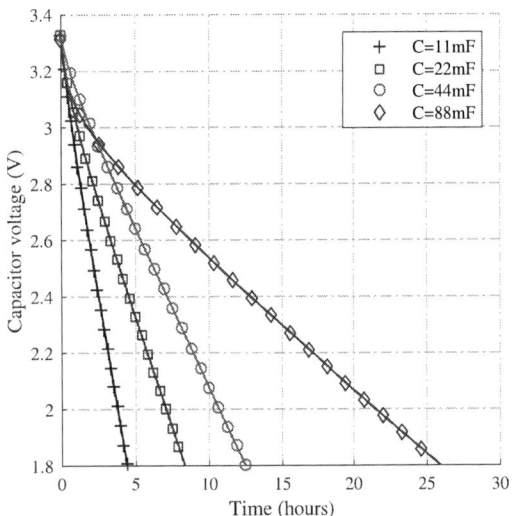

Fig. 4. Super-capacitor voltage over time for different capacitance values (measurement interval is 30 s).

The test setup for measuring the wireless charging time of the super-capacitors is shown in figure 3. Wireless power transmission at a frequency of 2 MHz has been used for this test. The voltage at the super-capacitors was monitored using a voltmeter. Before testing, the capacitors were fully discharged, in this way the charging time is the amount of time required for the capacitor voltage to move from 0 V to 3.3 V. The measured charging time for different capacitance values (made using super-capacitors of 11 mF connected in parallel) and using various values for the power of the external RF transmitter are shown in table I.

working whenever the supply voltage falls below 1.8 V. The data indicates that it could be possible to power the implant device for a little over a day with a capacitance of 88 mF. For these results it is important to note that in order to track the capacitor voltage, a voltmeter with a voltage logging feature was connected across the $C_s$. This voltmeter drains an average current of 300 nA.

TABLE I
WIRELESS CHARGING TIME: RF POWER VERSUS CAPACITANCE

| RF power | 11 mF | 22 mF | 44 mF | 88 mF |
|---|---|---|---|---|
| 4 W | 16 sec | 41 sec | 88 sec | 220 sec |
| 10 W | 8 sec | 20 sec | 48 sec | 108 sec |
| 15 W | 5 sec | 13 sec | 37 sec | 81 sec |

## IV. CAPACITOR DISCHARGING OVER TIME

Once the capacitor array is charged and the data stored on the device has been collected, the wireless telemetry unit is removed. At this point the implant device starts working by using the charge stored on the super-capacitor. Since the active and idle phase currents are already fixed by the hardware on the implant device, the only factors that determine the time required to deplete the capacitor charge are the implant measurement interval and the total capacitance used to store the energy.

Figure 4 shows the time required to deplete the super-capacitors for various capacitance values. For this test the implant device was configured to fetch one pressure sensor measurement every 30 seconds. The capacitance $C_s$ is increased from 11 mF to 88 mF . The implant device stops

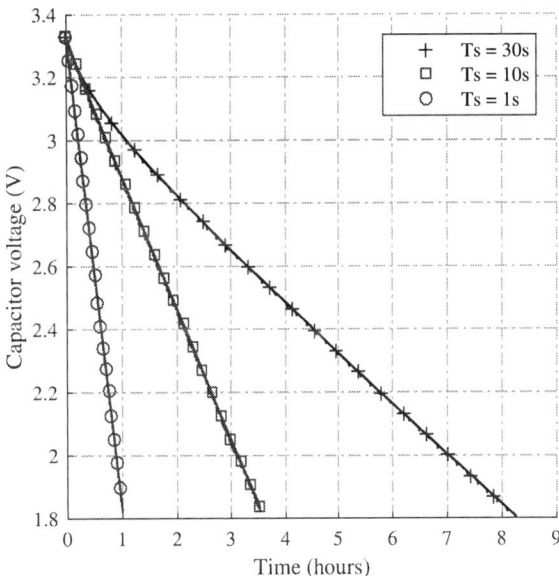

Fig. 5. Super-capacitor voltage over time for different sensor measurement intervals ($T_s$) for capacitance of 22 mF

978-1-5386-5388-3/18 $31.00 © 2018 IEEE      243

In the second test set, the implantable device was configured for three different measurement intervals: $1\,\mathrm{s}$, $10\,\mathrm{s}$ and $30\,\mathrm{s}$. For this test the capacitance $C_s$ was set to $22\,\mathrm{mF}$. The test results are shown in figure 5. The test results show that larger measurement intervals extend the autonomous operating time of the implant.

It is important to note that the run-times shown in Figures 4 and 5 are lower than what can be expected when using the ideal capacitor formula: $\Delta t = C(\Delta V/I)$. The first factor to consider in evaluating energy loss in the super-capacitor is its self-discharge. Figure 6 presents the self-discharge curve for a $11\,\mathrm{mF}$ capacitor used on the system. The curve was obtained by first charging an individual super-capacitor to $3.3\,\mathrm{V}$ and then disconnecting it from the power source. During the test the super-capacitor was not connected to any other devices except for short times when its voltage was measured. The curve presents the first 70 hours after disconnecting the super-capacitor from the power source. As shown, the capacitor voltage has dropped around $315\,\mathrm{mV}$ in a day. Also significant is that the rate of discharge due to self-discharging currents in the capacitor is not constant and is reduced as time passes and the capacitor voltage drops.

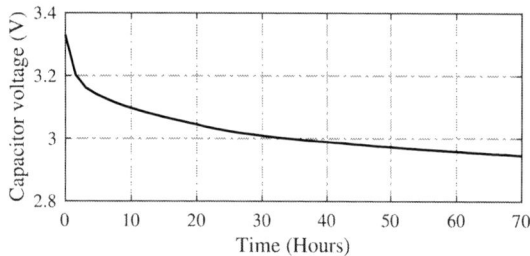

Fig. 6. Self-discharging curve for the $11\,\mathrm{mF}$ capacitor (initial voltage $3.275\,\mathrm{V}$)

The second factor to take into account are the nonlinearities present in super-capacitors [13] [14]. The effective capacitance of the system is dependant on the capacitor voltage and also on the discharging frequency [15]. The capacitance dependency can be noticed on the self-discharging curve in figure 6 and also on the discharge curves in figures 5 and 4. The discharging current has a major influence on voltage drop across capacitor immediately after disconnecting the power source from the implant (abrupt change on the capacitor current).

The test results show that an implantable device can be powered by means of super-capacitors. The design has to account for the trade-off between the capacitance value, the measurement intervals and the operating time required by the device. Also, the designer has to note that the use of small super-capacitors requires ultra-low power design, where the power consumed in idle states plays a major role.

## V. CONCLUSIONS

Integrating tiny super capacitors into an existing implantable device permits fast charging and long operating times (in

hours) as well as improves patient freedom of movement by removing the discomfort of being constantly connected to the external wireless telemetry unit. Test results show that the implant can run for up to one day using a super-capacitor which can be charged wirelessly at a range of 20 cm in 81 seconds. The fast charging, small size and large energy density make these components ideal for future miniaturised implantable devices.

## REFERENCES

[1] M. A. Hannan, S. Mutashar, S. A. Samad, and A. Hussain, "Energy harvesting for the implantable biomedical devices: issues and challenges," *Biomedical engineering online*, vol. 13, no. 1, p. 79, 2014.

[2] A. Fahad, T. Soyata, T. Wang, G. Sharma, W. Heinzelman, and K. Shen, "Solarcap: super capacitor buffering of solar energy for self-sustainable field systems," in *SOC Conference (SOCC), 2012 IEEE International*. IEEE, 2012, pp. 236–241.

[3] Seiko Instruments Inc., "Chip Capacitors CPH3225A," http://www.sii.co.jp/en/me/datasheets/chip-capacitor/cph3225a/, accessed: 2016-11-30.

[4] J. A. Potkay, "Long term, implantable blood pressure monitoring systems," *Biomedical Microdevices*, vol. 10, no. 3, pp. 379–392, 2008.

[5] B. John, *Smart system for invasive measurement of biomedical parameters*, ser. Wissenschaftliche Beiträge zur Medizinelektronik, Bd. 7, W. Krautschneider, Ed. Logos Verlag Berlin GmbH, 09 2017.

[6] D. C. Yates, A. S. Holmes, and A. J. Burdett, "Optimal transmission frequency for ultralow-power short-range radio links," *IEEE Transactions on Circuits and Systems I: Regular Papers*, vol. 51, no. 7, pp. 1405–1413, 2004.

[7] L. Melo-Thomas, K.-A. Engelhardt, U. Thomas, D. Hoehl, S. Thomas, M. Wöhr, B. Werner, F. Bremmer, and R. K. Schwarting, "A wireless, bidirectional interface for in vivo recording and stimulation of neural activity in freely behaving rats," *Journal of visualized experiments: JoVE*, no. 129, 2017.

[8] B. John, R. Ranjan, C. Spink, D. Schroeder, A. Koops, G. Adam, and W. H. Krautschneider, "Wireless Blood Pressure Measurement Implant Electronics for integration in a Stent Graft," in *Biomedical Engineering (BioMed 2016)*, 02 2016.

[9] F. Springer, R. Schlierf, J.-G. Pfeffer, A. H. Mahnken, U. Schnakenberg, and T. Schmitz-Rode, "Detecting endoleaks after endovascular aaa repair with a minimally invasive, implantable, telemetric pressure sensor: an in vitro study," *European Radiology*, vol. 17, no. 10, pp. 2589–2597, 2007.

[10] C. Moss, W. Sass, and J. Mueller, "Receiver for inductively coupled communication with an implantable sensor," in *Proceedings of International Symposium on Applied Sciences in Biomedical and Communication Technologies*. ACM, 2011, p. 31.

[11] M. G. Allen, "Micromachined endovascularly-implantable wireless aneurysm pressure sensors: from concept to clinic," in *Proceedings of International Conference on Solid-State Sensors, Actuators and Microsystems*, vol. 1, June 2005, pp. 275–278.

[12] S. H. Ellozy, A. Carroccio, R. A. Lookstein, M. E. Minor, C. M. Sheahan, J. Juta, A. Cha, R. Valenzuela, M. D. Addis, and T. S. Jacobs, "First experience in human beings with a permanently implantable intrasac pressure transducer for monitoring endovascular repair of abdominal aortic aneurysms," *Journal of vascular surgery*, vol. 40, no. 3, pp. 405–412, 2004.

[13] P. Kurzweil, M. Chwistek, and R. Galley, "Electrochemical and spectroscopic studies on rated capacitance and aging mechanisms of supercapacitors," *Procedings 2nd European Symposium on Super Capacitors and Applications*, 2006.

[14] T. J. Freeborn, B. Maundy, and A. S. Elwakil, "Measurement of supercapacitor fractional-order model parameters from voltage-excited step response," *IEEE Journal on Emerging and Selected Topics in Circuits and Systems*, vol. 3, no. 3, pp. 367–376, 2013.

[15] A. S. Elwakil, A. Allagui, T. Freeborn, and B. Maundy, "Further experimental evidence of the fractional-order energy equation in supercapacitors," *AEU - International Journal of Electronics and Communications*, vol. 78, pp. 209 – 212, 2017. [Online]. Available: http://www.sciencedirect.com/science/article/pii/S1434841117302236

# Design and Modelling of a Super-Regenerative Receiver for Medical Implant Devices

Naci Pekcokguler
Electrical and Electronics Engineering,
Bogazici University, TR-34342
Istanbul, Turkey
Email: naci.pekcokguler@boun.edu.tr

Gunhan Dundar
Electrical and Electronics Engineering,
Bogazici University, TR-34342
Istanbul, Turkey
Email: dundar@boun.edu.tr

Catherine Dehollain
RFIC Group
Ecole Polytechnique Fédérale de Lausanne
CH-1015 Lausanne, Switzerland
Email: catherine.dehollain@epfl.ch

*Abstract*—**Medical implant devices have been widely used in recent years. The Super-Regenerative Receiver has been on preferred architecture due to its power advantage over other architectures. We present a detailed analysis of the circuits and their equivalent models to be used in system level design of a Super-Regenerative Receiver in this paper. Designs were carried out in UMC 180nm process with a center frequency of 416MHz for MedRadio band. The study is concluded with the simulation results of the circuits and the equivalent models.**

## I. INTRODUCTION

Medical implant devices are state-of-art equipment that are used in medicine for diagnostic and therapeutic purposes. Since these devices work with a battery or are powered remotely, they must have ultra-low power consumption to prevent any surgical operation to maintain the functionality over their lifetimes. Moreover, these devices must have a wireless link for the comfort of patient. Because the most power-hungry part in these devices is generally the RF front-end, the main focus is on finding a low power receiver architecture to extend the battery life. The first regulation on medical implant communication was made by the U.S. Federal Communications Commission (FCC) in 1999 [1] and regulations were finalized under the name of Medical Device Radiocommunications Service (MedRadio) [2].

Most receivers employ continuously running local oscillators which dissipate most of the power. On the other hand, the Super-Regenerative Receiver (SRR) architecture, which was first proposed in 1922 by E. H. Armstrong [3], offers a solution to this problem by periodically turning on and off, in other words by quenching, the oscillator. At the start-up, the oscillator provides very large gain which regenerates the received weak signal.

General block diagram of an SRR is depicted in Fig. 1 for On-Off Keying (OOK) modulation which was chosen because of the need for simplicity in the receiver. Low noise amplifier (LNA) output is directly connected to the Super-Regenerative Oscillator (SRO) rather than a mixer in the SRR architecture. Quenching is adjusted such that the SRO does not start oscillating unless an RF signal having the same frequency with the oscillator is present at the antenna. Envelope of the regenerated signal is obtained and converted into digital data with the help of an amplifier, a comparator, and a D

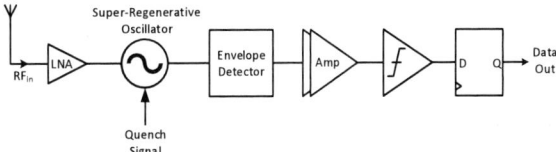

Fig. 1. General block diagram of the Super-Regenerative Receivers.

Flip-Flop (FF). This work focuses on the RF part of the SRR, which contains LNA, SRO and envelope detector. In the following section, analytic solutions of these blocks as well as the equivalent models are presented. Simulation results of the circuits and models are shown in Section III. Finally, conclusions are drawn in Section IV.

## II. CIRCUIT MODELING AND DESIGN

### A. Low Noise Amplifier

LNA is the most critical block in a receiver in terms of noise performance. Additionally, it should provide enough reverse isolation to prevent any SRO leakage to the antenna, which will interfere with the environment and affect other wireless systems.

The circuit schematic for an inductively degenerated differential cascode LNA is shown in Fig. 2a. This architecture was chosen since it is possible to obtain impedance matching and noise matching simultaneously. Thanks to the fully differential topology, half circuit can be used in the analysis (see Fig. 2b). Small signal analysis can be done with model shown in Fig. 2c. Indices of the components are ignored in small signal analysis for the sake of simplicity. Input impedance, $Z_{in}$ can be determined by finding $V_{in}$ in terms of $I_{in}$ as [4],

$$V_{in} = j\omega L_g I_{in} + \frac{I_{in}}{j\omega C_{gs}} + \left(I_{in} + I_{in}\frac{g_m}{j\omega C_{gs}}\right)j\omega L_s \quad (1)$$

$$C_{gs} = C_{add1,2} + C_{gs,M1,M2} \quad (2)$$

$$Z_{in} = \frac{V_{in}}{I_{in}} = j\omega\left(L_g + L_s\right) + \frac{1}{j\omega C_{gs}} + \frac{g_m L_s}{C_{gs}} \quad (3)$$

The following conditions must be met to have impedance matching between the antenna and LNA.

$$j\omega_0\left(L_g + L_s\right) + \frac{1}{j\omega_0 C_{gs}} = 0 \quad , \quad \frac{R_s}{2} = \frac{g_m L_s}{C_{gs}} \quad (4)$$

978-1-5386-5388-3/18 $31.00 © 2018 IEEE

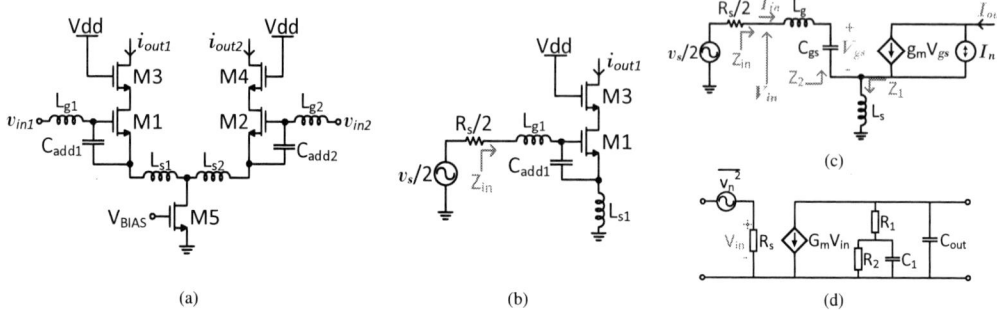

Fig. 2. Schematics of (a) the fully differential LNA, (b) the half circuit, (c) the small signal model and (d) the proposed equivalent model.

where $\omega_0 = 1/\sqrt{(L_g + L_s)C_{gs}}$.

Since the cascode transistor has almost no effect on current gain and contributes almost no noise, input-output relationship and noise transfer characteristics can be found as,

$$G = \left|\frac{I_{out}}{V_s}\right| = \frac{1}{2\omega_0\left(L_s + \frac{C_{gs}R_s}{2g_m}\right)} = \frac{\omega_T}{\omega_0} \cdot \frac{1}{2R_s} \quad (5)$$

$$|I_{n,out}|_{M1} = |I_n|\frac{R_sC_{gs}}{2g_mL_s + R_sC_{gs}} = \frac{|I_n|}{2} \quad (6)$$

where $\omega_T \approx g_m/C_{gs}$ and $2g_mL_s/C_{gs} = R_s$ from (4). Noise contribution of M1 can be expressed as,

$$\overline{I^2_{n,out}}|_{M1} = \frac{\overline{I^2_n}}{4} = kT\gamma g_m \quad (7)$$

From above results, noise factor can be calculated as,

$$F = 1 + \frac{\gamma R_s}{2g_m}\omega_0^2C_{gs}^2 = 1 + \frac{2\gamma}{g_mR_s}\frac{1}{Q^2_{in}} \quad (8)$$

where $Q_{in} = 2/\omega_0 R_s C_{gs}$, which is the quality factor of the input matching network. Optimum source impedance $R_{s,opt}$ for noise matching can be found by taking partial derivative of the noise factor and equating to zero. The result is,

$$\frac{\partial F}{\partial R_s} = 0 \implies R_{s,opt} = \frac{2g_mL_s}{C_{gs}} \quad (9)$$

which is the same impedance as the one required for impedance matching.

Output impedance seen from the drain of M1 can be found by ignoring $i_n$, taking $r_o$ into account, equating $v_s$ to zero, and connecting a test source to the output in the circuit shown in Fig. 2c. If the test source has voltage of $V_x$ and current of $I_x$, output impedance, $Z_{out1}$, can be found as,

$$Z_{out1} = \frac{V_x}{I_x} = \frac{\overbrace{r_{o1}(Z_1 + Z_2)}^{\text{part 1}} + \overbrace{Z_1Z_2}^{\text{part 2}} + \overbrace{g_mr_{o1}Z_1\frac{1}{j\omega C_{gs}}}^{\text{part 3}}}{Z_1 + Z_2} \quad (10)$$

Part 2 in the above equation is very small compared to others and can be ignored. Part 1 is simply equal to $r_{o1}$. So, $Z_{out1}$ simplifies to,

$$Z_{out1} = r_{o1} + r_{o1}\frac{2g_mL_s}{R_sC_{gs}} \xrightarrow{@\,\omega_0} Z_{out1} = 2r_{o1} \quad (11)$$

Output impedance of the LNA can be found with the effect of cascode transistor as,

$$Z_{out} = r_{o3} + (1 + g_{m3}r_{o3})\left(2r_{o1} \parallel \frac{1}{j\omega_0C_{par}}\right) \quad (12)$$

where, $C_{par}$ is the total capacitance at the source node of M3.

Equivalent model of the LNA is depicted in Fig. 2d and the component values are calculated as follows,

$$R_s = 50\Omega \quad , \quad G_m = 2G = -j\frac{\omega_T}{\omega_0}\frac{1}{R_s} \quad (13)$$

$$\overline{v_n^2} = \frac{2\overline{i^2_{n,out}}|_{M1}}{|G_m|^2} = 2kT\gamma g_mR_s^2\left(\frac{\omega_0}{\omega_T}\right)^2 \quad (14)$$

$$R_1 = 2r_{o2} \quad , \quad R_2 = 4(1 + g_{m3}r_{o3})r_{o1} \quad (15)$$

$$C_1 = \frac{C_{par}}{2(1 + g_{m3}r_{o3})} \quad , \quad C_{out} = \frac{C_{db2} + C_{gd2}}{2} \quad (16)$$

### B. Super-Regenerative Oscillator

The most crucial block in the proposed receiver is the SRO. Unlike traditional local oscillators used in almost all of the receivers, SRO is not allowed to reach steady-state oscillation. It is periodically turned on and off and high gain of the oscillator at the start-up is used to regenerate the received signal. Operation of the oscillator is illustrated in Fig. 3 under square wave quench and will be derived later in this section.

For the SRO, traditional cross-coupled oscillator topology was chosen because of the decent phase noise performance and suitability for quenching. Schematic of the oscillator is shown in Fig. 4a. In small-signal operation, the conductance seen from these transistors by the LC tank is equal to $-g_m/2$. Thus, the equivalent model for the circuit is obtained with a negative conductance as shown in Fig. 4b. Component values shown in the model are obtained from the circuit as $G_0 = 1/Q^2R_s$, $L = L_p$, $C = Cp$ where, $Q$ is the quality factor of the inductor at $\omega_0$ and $R_s$ is the loss of inductor. The governing differential equation is obtained from the model as [5],

$$\ddot{v}(t) + \frac{G(t)}{C}\dot{v}(t) + \left(\frac{1}{LC} + \frac{\dot{G}(t)}{C}\right)v(t) = \frac{i(t)}{C} \quad (17)$$

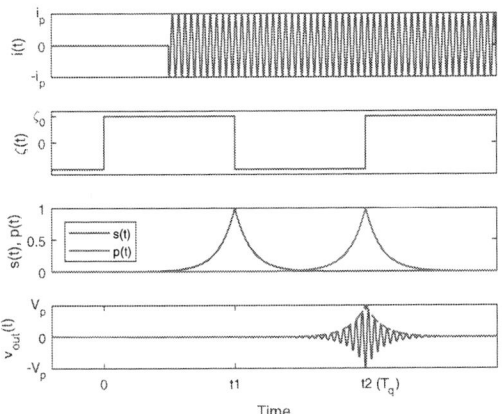

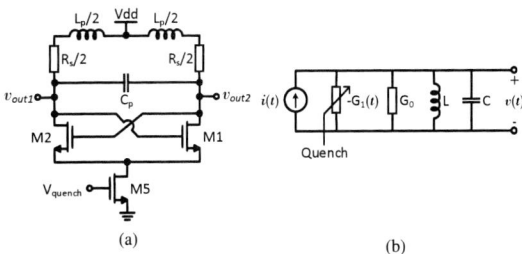

Fig. 4. (a) Schematic and (b) the equivalent model of the SRO.

In a properly designed SRO, the homogeneous solution should be too small compared to the forced response. Envelope of the SRO output is determined by $p(t)$ and an output is generated only if there exists an RF signal during the sensitivity period $s(t)$. Gain, output envelope and sensitivity are determined by the amplitude and the shape of damping factor. Solutions obtained above are plotted via MATLAB and shown in Fig. 3.

### C. Envelope Detector

The simplest and lowest power solution to demodulate the OOK data is the envelope detector. It simply rectifies the RF signal and filters the high frequency component, thus generating the envelope. Active pseudo-differential envelope detector is the best choice for the given technology. MOSFETs are operated in weak inversion which has the highest non-linear characteristics and is favorable to obtain envelope of the signal. Schematic of the envelope detector is shown in Fig. 5.

Drain current formula for the MOSFET is given as [6],

$$i_{D1} = \underbrace{I_0 e^{\frac{V_G - V_{th}}{nV_T}} \left( e^{-\frac{V_S}{V_T}} - e^{-\frac{V_D}{V_T}} \right)}_{I_Q} e^{\frac{v_{inp}}{nV_T}} = I_Q e^{\frac{v_{inp}}{nV_T}} \quad (25)$$

where $V_T = k\frac{T}{q}$, $I_0 = 2n\beta V_T^2$, $\beta = \mu_n C_{ox} \left( \frac{W}{L} \right)_1$, and $n$ is the slope correction factor and for M1. For small $v_{inp}$, we can express the drain current of M1 as,

$$i_{D1} \approx I_Q + \left( \frac{\partial I_d}{\partial V_g} \right) v_g + \left( \frac{\partial^2 I_d}{\partial V_g^2} \right) \frac{v_g^2}{2} \quad (26)$$

$$= I_Q + \frac{I_Q}{nV_T} v_g + \frac{I_Q}{2(nV_T)^2} \frac{v_g^2}{2} \quad (27)$$

With above results, total output current can now be calculated for $v_{inp} = -v_{inm} = v_{id}/2$ as,

$$i_{out} = i_{D9} + i_{D10} = 2I_Q + \frac{I_Q}{(nV_T)^2} \frac{v_{id}^2}{4} \quad (28)$$

If an RF signal of $v_{id} = A\sin\omega t$ is applied to the circuit,

$$i_{out} = 2I_Q + \frac{I_Q}{4(nV_T)^2} \frac{A^2}{2} \left( 1 - \cos 2\omega t \right) \quad (29)$$

Applying the low-pass characteristics of the envelope detector which stem from the output resistance and capacitor $C$,

$$i_s = \frac{I_Q A^2}{8(nV_T)^2} \quad (30)$$

Fig. 3. RF input signal $i(t)$, damping factor $\zeta(t)$, sensitivity curve $s(t)$, output envelope $p(t)$, and SRO output.

To gain more insight on the operation of the SRO, above equation is converted to,

$$\ddot{v}(t) + 2\zeta(t)\omega_0 \dot{v}(t) + \left( \omega_0^2 + 2\omega_0 \dot{\zeta}(t) \right) v(t) = \frac{\dot{i}(t)}{C} \quad (18)$$

where, $\zeta(t) = G(t)/2C\omega_0$, $G(t) = G_0 - G_1(t)$, and $\omega_0 = 1/\sqrt{LC}$. (18) is a second order time-varying coefficient ODE and homogeneous solution can be obtained by the method of change of variable. Following assumptions were made to solve the homogeneous equation, which must already be guaranteed for the proper operation of the SRO.

$$\zeta^2(t) \ll 1 \quad , \quad \left| \dot{\zeta}(t) \right| \ll \omega_0 \quad (19)$$

First assumption requires under-damped response which is already the case in an oscillator. Second assumption corresponds to a slow-varying damping factor. Thus, homogeneous solution is obtained as,

$$v_h(t) = 2Re \left[ V_1 e^{j\omega_0 t} \right] \cdot e^{-\omega_0 \int_0^t \zeta(\tau)d\tau} \quad (20)$$

which is a damped cosine whose damping is determined by the integral of the overall damping factor. $V_1$ is a constant to be determined from the boundary conditions. The forced response is found by the method of variation of parameters as,

$$v_p(t) = \frac{-1}{\omega_0 C} K_s p(t) \int_0^t \dot{i}(\lambda)s(\lambda)\sin\omega_0(\lambda - t)d\lambda \quad (21)$$

where,

$$s(t) = e^{\omega_0 \int_{t_1}^t \zeta(\tau)d\tau} \quad , \quad p(t) = e^{-\omega_0 \int_{t_2}^t \zeta(\tau)d\tau} \quad (22)$$

$$K_s = e^{-\omega_0 \int_{t_1}^{t_2} \zeta(\tau)d\tau} \quad (23)$$

For an input signal of $i(t) = A\cos(\omega_0 t + \phi)$, the output becomes,

$$v_p(t) \approx \frac{A}{2C} K_s p(t)\cos(\omega_0 t + \phi) \int_0^t s(\lambda)d\lambda \quad (24)$$

*Paper P90*  PRIME 2018, Prague, Czech Republic

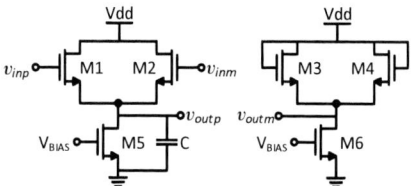

Fig. 5. Schematic of the pseudo-differential active envelope detector.

Output voltage can be calculated by multiplying output current with output resistance as,

$$R_{out} = \frac{1}{g_{m1}} \parallel \frac{1}{g_{m2}} = \frac{1}{2g_m} = \frac{nV_T}{2I_Q} \implies v_{out} = \frac{A^2}{16nV_T} \tag{31}$$

### III. SIMULATION RESULTS

Circuits and their equivalent models were designed using Cadence Virtuoso in UMC $0.18\mu m$ process. Operation frequency is $416MHz$ and a quench frequency of $10MHz$ was used, which corresponds to $10Mbps$ data rate. Targeted specs for LNA were $1.2dB$ NF, minimum power consumption, and maximum gain. Simulation results of both circuit and model are shown in Fig. 6. Model fits to the circuit behavior in the vicinity of the operation frequency which is due to the fact that the calculations were made at $\omega_0$. Output impedance was simulated with the LC tank of the SRO which will be the load of the LNA in the completed system. At operation frequency, errors in gain, output impedance and noise factor are 1.7%, 1%, 0.1%, respectively. Total power consumption of the LNA is $2.7mW$ including biasing.

Simulation results of the whole receiver is shown in Fig. 7 for both the model and the circuits. Transient noise bandwidth was set to $10GHz$. Damping factors were obtained from the calculated parameters and the difference between the model and the circuit stems from this fact. Small spikes seen at the envelope detector output in the circuit results arise from the common mode voltage fluctuations because of the quenching but they are small enough to be ignored. (30) suggests $12.9mV$ amplitude at output and simulation yields $12mV$ which corresponds to 7% error. SRO dissipates an average power of $1.62mW$ and envelope detector dissipates $2.7\mu W$. Thus, total power consumption appears to be $4.323mW$. Time elapsed to perform four quench cycle long transient simulation is $221s$ for the circuit, whereas it takes only $1.33s$ for the model, which corresponds to 166 times improvement.

### IV. CONCLUSION

A low-power super-regenerative receiver was presented in this paper for medical implant devices. Additionally, the crucial blocks were modeled and their analysis was shown. This models help the designer to optimize the whole system for a given condition and to determine required specs for each block. Furthermore, simulation time to obtain the whole system response was shortened thanks to the simplicity of the models. Finally, models were validated with simulation results

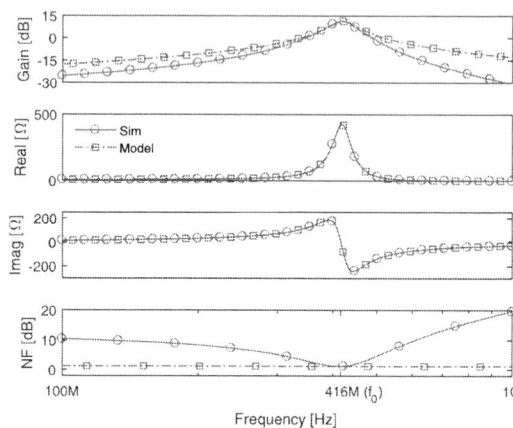

Fig. 6. Simulation results of gain, real and imaginary part of the output impedance, and NF of LNA and the proposed model.

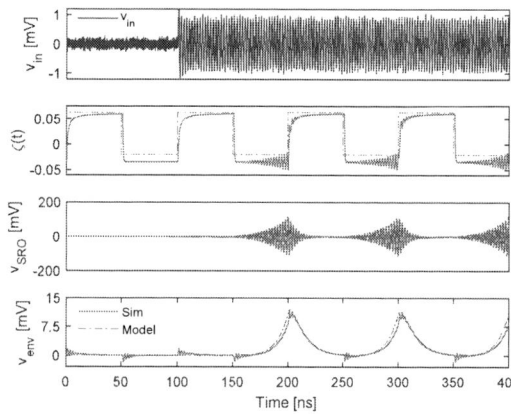

Fig. 7. Simulation results of input voltage, damping factor, SRO and envelope detector outputs of the whole receiver and the model.

of the actual circuits. Further optimization of the design, both at circuit level and at system level is ongoing.

### ACKNOWLEDGMENT

The authors would like to thank Engin Afacan and Ismail Kara for their invaluable contributions and motivations.

### REFERENCES

[1] F. Rules, "Regulations. mics band plan," 2003.
[2] ——, "Regulations 47 cfr part 95," *Subparts E (95.601-95.673) and I (95.1201-95.1219) Personal Radio Services*, 2002.
[3] E. H. Armstrong, "Some recent developments of regenerative circuits," *Proceedings of the Institute of Radio Engineers*, vol. 10, no. 4, pp. 244–260, 1922.
[4] B. Razavi and R. Behzad, *RF microelectronics*. Prentice Hall New Jersey, 1998, vol. 2.
[5] F. X. Moncunill-Geniz, P. Pala-Schonwalder, and O. Mas-Casals, "A generic approach to the theory of superregenerative reception," *IEEE Transactions on Circuits and Systems I: regular papers*, vol. 52, no. 1, pp. 54–70, 2005.
[6] E. A. Vittoz, "Weak inversion for ultra low-power and very low-voltage circuits," in *Solid-State Circuits Conference, 2009. A-SSCC 2009. IEEE Asian*. IEEE, 2009, pp. 129–132.

*PRIME 2018, Prague, Czech Republic*

*Session: Sensing and Biomedical Circuits II*

# A Fully Fail-Safe Capacitive-Based Charge Metering Method for Active Charge Balancing in Deep Brain Stimulation

Reza Ranjandish[1], Omid Shoaei[2], and Alexandre Schmid[3]

[1,3]Microelectronic Systems Laboratory, Swiss Federal Institute of Technology (EPFL), Switzerland

[2]Nano-Electronic Center of Excellence, School of Electrical and Computer Engineering, University of Tehran, Tehran, Iran

*Abstract*—**Related to safety issues, charge balancing is a major concern in neural and functional electrical stimulation. This paper presents a capacitive-based charge metering method as a low-power and precise charge balancing method used in Deep Brain Stimulation (DBS). In contrast to the previously presented capacitive-based charge metering methods, the proposed method does not need any precise and high-speed comparator for net-zero charge detection. It is proven that this method is insensitive to the delay and the offset of its components. Consequently, using ultra-low power components in the charge balancer is feasible. Furthermore, the proposed method properly supports any stimulation mode and waveform. The proposed approach along with voltage and current mode pulse generators was validated in a 0.9% saline solution by using a DBS lead.**

*Keywords*—*Electrical stimulation; current mode stimulation; voltage mode stimulation; charge balancing; implantable medical devices.*

## I. INTRODUCTION

Deep Brain Stimulation (DBS) is an application of electrical stimulation to suppress movement disorders such as Parkinson disease. DBS Implantable Pulse Generators (IPGs) apply voltage or current stimulus through electrodes to different part of Basal Ganglia motor pathways based on the type of neurological disorders (e.g., Parkinson's disease). In DBS, the stimulation amplitude may reach to 25.5 mA in current mode and 10.5 V in voltage mode [1]. Mostly because of better control over the injected charge, constant current stimulation (CCS) is a dominant method over constant voltage stimulation (CVS). However, CVS is much more power efficient thanks to the selection of the supply voltage more closely to the stimulation voltage.

A biphasic stimulation is needed to prevent tissue damage and any long-term effects such as pH shift and erosion of the electrodes [2]. Biphasic stimulation consists of a cathodic phase followed by an anodic phase. During the cathodic phase, the cell membrane is depolarized. Then, the anodic phase neutralizes the charge which has been injected in the cathodic phase. For exerting a safe stimulation, the voltage across the electrode must be constraint within a specific window. In addition, to block any direct current passing through the tissue, a large off-chip capacitor, namely blocking capacitor, is placed in series with the stimulation electrode. This capacitor blocks any DC current to flow in case of semiconductor failure through the tissue which makes the stimulator fail-safe [3]. However, in multichannel stimulators such as retinal or cochlear implants a large silicon area cannot be allocated to the

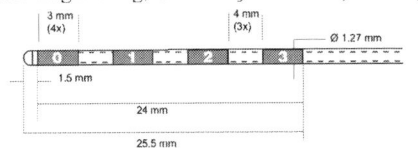

Fig. 1. Schematic drawing of Medtronic's DBS lead model 3391 [4]. A DBS lead contains four electrodes (lead contacts).

large blocking capacitors. Since our stimulation target is Deep Brain Stimulation (DBS), the limited number of the electrodes allows us to use blocking capacitors to ensure safety of the stimulator. In addition, employing capacitive-based charge metering is feasible in the applications with limited number of channel such as DBS. A typical DBS electrode shape and its sizes are shown in Fig. 1. This commercial electrode has only four channels.

## II. ELECTRODE-ELECTROLYTE INTERFACE MODEL

A comprehensive electrical model of the electrode-electrolyte interface is shown in Fig. 2a in which $R_s$ is the spreading resistance which is a linear resistance, $R_{ct}$ is the charge transfer resistance which is a non-linear resistance and $CPE$ is the constant phase element that models the capacitive behavior of the interface. In order to simplify the comprehensive model, a linearized model is used as shown in Fig. 2b. The linearized model simplifies the calculations without significant effect on the accuracy of the results.

## III. LITERATURE REVIEW

Active charge balancing systems have been developed in recent years. Different active charge balancing methods are

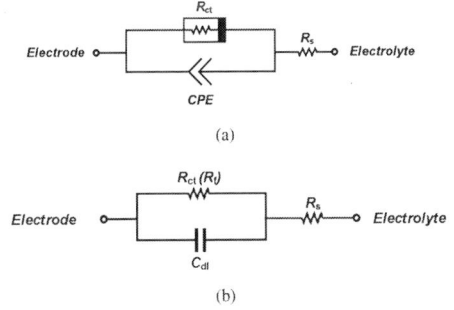

Fig. 2. Electrical model between electrode and electrolyte: (a) comprehensive model and (b) linearized model.

978-1-5386-5388-3/18 $31.00 © 2018 IEEE

*Paper P46*                                    *PRIME 2018, Prague, Czech Republic*

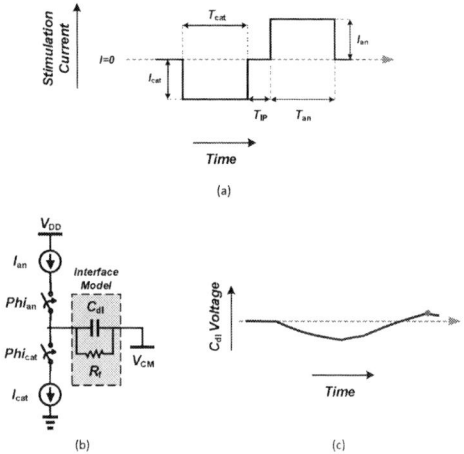

Fig. 3. Effect of a leaky electrode interface on well-matched anodic and cathodic phases.

introduced in literature. One of the active charge balancing methods is charge metering in which the charge balancing is done by measuring the charge injected during the cathodic and anodic phases. In this method, a sensor senses the amount of charge that passes through the tissue in the anodic and cathodic phases. Then, the charge is integrated and when the amount of charge in the anodic phase is equal to the cathodic phase, the charge balancer terminates the anodic phase. Using this method, any type of stimulation waveform can be employed, since the charge integration is not affected by the stimulation waveform. In [5], a resistor is employed to sense the current. Then, the current is integrated to calculate the amount of charge that is injected to the tissue. When the net-charge that passes through the tissue is zero, the anodic phase is terminated. In this design, the comparator and integrator should be fast enough to reduce the amount of error charge due to the delay of the charge balancer. The main drawback of this design is the lack of electrode shortening after the anodic phase. As we prove in the following, electrode shortening is very important in charge metering methods and without electrode shortening these methods are completely unsafe. In [6] a capacitive-based charge metering method is proposed. Since the capacitor inherently acts as an integrator, there is no need for an integrator circuitry. A Field-Programmable-Analog-Array (FPAA) is used for the physical implementation of the method. For acceptable accuracy an error amplifier is used to amplify the voltage on the sensing capacitor. The system is tested by placing a resistance at the output of the stimulator and it is not tested using a leaky tissue model (Fig. 2b). Likewise [5], this method also lacks of an electrode discharging phase which is very crucial for the safety of the stimulator. Hence, the method cannot be consider as a safe method.

## IV. IMPORTANCE OF THE ELECTRODE SHORTENING PHASE FOR THE CHARGE METERING METHODS

In order to show the importance of the electrode shortening phase in the charge metering methods, we use the linearized model of the tissue interface. We calculate the remaining voltage on the tissue after a biphasic waveform. For this sake,

we eliminate $R_s$ from tissue model, since it has no effect on the calculation of the voltage at the electrode-electrolyte interface. We assume that a current waveform as shown in Fig. 3a is injected to the tissue model as depicted in Fig. 3b which generates a voltage waveform on $C_{dl}$ as illustrated in Fig. 3c. We want to calculate the voltage on $C_{dl}$ exactly after anodic phase (the red dot in Fig. 3c). Using the superposition theorem, the remaining voltage after finishing the anodic phase can be calculated as follows:

$$V_{rem} = I_{cat} \times R_f \times (1 - e^{-\frac{T_{cat}}{C_{dl}R_f}}) \times e^{-\frac{T_{IP}+T_{an}}{C_{dl}R_f}} \\ - I_{an} \times R_f \times (1 - e^{-\frac{T_{an}}{C_{dl}R_f}}) \quad (1)$$

In an ideal case we can consider that $R_f \times C_{dl}$ is much larger than $T_{cat}$ and $T_{an}$. Consequently, by using Taylor series, Eq. 1 can be rewritten as:

$$V_{rem} = \frac{I_{cat} \times T_{cat}}{C_{dl}} - \frac{I_{an} \times T_{an}}{C_{dl}} \quad (2)$$

Consequently, with an ideal charge metering circuitry, a remaining voltage equal to zero is achievable if $R_f \times C_{dl}$ is much larger than $T_{cat}$ and $T_{an}$. However, in practice neither the charge balancer is ideal nor $R_f \times C_{dl}$ is much larger than $T_{cat}$ and $T_{an}$. Therefore, a remaining voltage on the tissue is always present which may accumulate on the tissue and after few stimulation cycle it may exceed the safety limits. Hence, this remaining charge should be depleted using an electrode shorting phase. Consequently, electrode shorting is

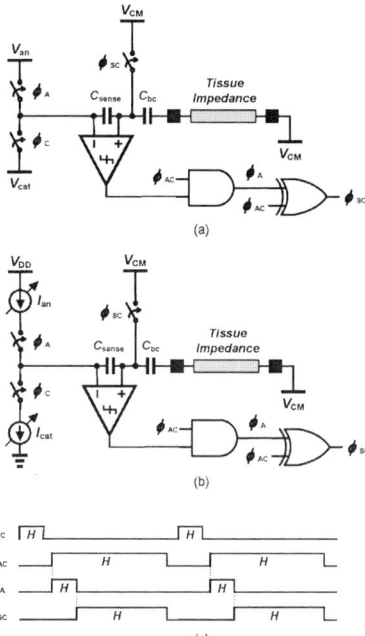

Fig. 4. Schematic of the proposed charge balancing working in (a) voltage mode stimulation and (b) current mode stimulation.

978-1-5386-5388-3/18 $31.00 © 2018 IEEE          250

*PRIME 2018, Prague, Czech Republic*                                      *Session: Sensing and Biomedical Circuits II*

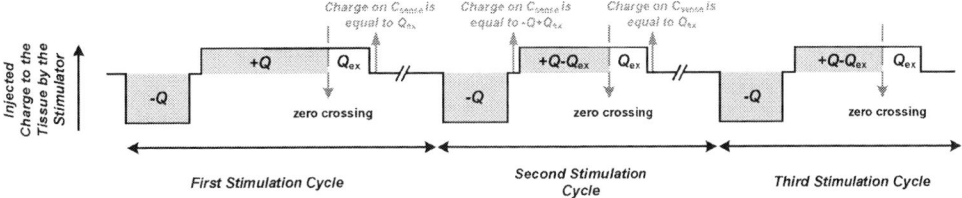

Fig. 5.   Stimulation charge transfer digram in a charge balanced current mode stimulation by the proposed charge balancer

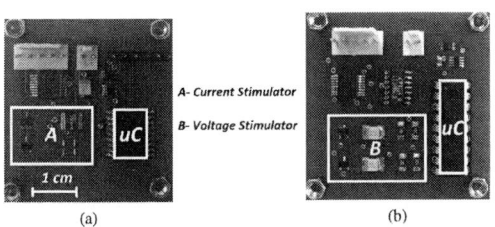

Fig. 6.   Implemented two-sided PCBs for testing the proposed charge balancer. (a) Top side of the current mode and, (b) top side of the voltage mode stimulators.

a mandatory part of a charge metering method and without electrode shorting, the charge balancer is completely unsafe. It should be highlighted again that the previous published capacitive-based charge metering techniques proposed in [5] and [6] are not safe since they don't have electrode shorting phase to deplete the remained charge due to the charge leakage at the electrode interface.

## V.   PROPOSED CHARGE BALANCER

The proposed charge balancer circuitry is depicted in Fig. 4 supporting both voltage mode and current mode stimulation. The system has two capacitors in series; one of them is used as a sensor ($C_{sense}$) and another one is the blocking capacitor ($C_{bc}$) to protect the tissue in case of failure. The system operates in three phases. At first, the cathodic phase starts with the LOW $\phi_{AC}$, LOW $\phi_A$ (initial condition) and HIGH $\phi_C$. Consequently, the cathodic current passes through the tissue. So the sensing capacitor ($C_{sense}$) which is placed in series with the tissue starts to integrate the current. Hence, the voltage of the sensing capacitor increases and the comparator output goes HIGH. With LOW $\phi_C$, the cathodic phase ends. After a brief inter-pulse delay, for a HIGH $\phi_{AC}$, the AND gate output becomes HIGH because the comparator output remained HIGH from the previous cathodic phase. Therefore, $\phi_A$ goes HIGH and the anodic phase starts. Subsequently, a reverse current will flow. This current depletes the charge accumulated on the sensing capacitor in the previous cathodic phase. Consequently, the voltage of the sensing capacitor decreases during this phase until the comparator detects a zero crossing of this voltage. Then, the comparator output is changed to LOW and as a result, the AND gate output becomes LOW, which ends the anodic phase ($\phi_A$=LOW). Therefore, this system controls the anodic phase duration according to the arbitrary anodic amplitude. In addition, the method insensitive to determined non-idealities in the comparator and

### TABLE I.   OPERATING VALUES FOR THE STIMULATORS

| Programmable parameter | Operating range and resolution |
|---|---|
| Amplitude (Voltage mode) | 0-10.5 V with 0.1 V res. |
| Amplitude (Current mode) | 0-25.5 mA with 0.1 mA res. |
| Rate | 50-250 Hz with 10 Hz res. |
| Pulse width | 64-448 $\mu$s with 32 $\mu$s res. |

other components, such as delay and input offset. The reason is that the occurrence of any one of the mentioned non-idealities translates into more or less delivered charge ($\pm Q_{ex}$) in one cycle, which is automatically compensated in the next cycles by reducing or increasing the pulse duration of the anodic phase. To illustrate this concept, we assume that in the current mode stimulation there is an extra charge after the completion of the first anodic phase on the sensing capacitor due to some non-idealities ($-Q + Q + Q_{ex} = Q_{ex}$), as depicted in Fig. 5. Since the cathodic phase duration is fixed (and so its injected charge, $-Q$), the capacitor will be charged to $-Q + Q_{ex}$ after the cathodic phase in the second stimulation cycle. Now, in the second anodic phase, a charge equal to $-(-Q + Q_{ex})$ must be discharged from the sensing capacitor for happening a zero crossing at the voltage, but due to the existence of the mentioned non-idealities, an extra charge equal to $Q_{ex}$ is depleted from the sensing capacitor, as well. Therefore, the amount of the charge remained on the sensing capacitor would still be $Q_{ex}$ at the end of the second stimulation cycle. Thus, after the first cycle of stimulation, the mentioned non-idealities will be compensated by the system and the anodic phase's charge will be equal to the cathodic one. In fact, $Q_{ex}$ is a sample of the non-idealities. Hence, for all cycles of stimulation any non-ideality would lead to a non-zero charge on the sensing capacitor (like $Q_{ex}$ in the above example). The accuracy of this technique is enough to allow us not to use any error amplifier before comparator, in contrast to [6].

In order to deplete the remaining charge on the tissue caused after the first stimulation cycle as well as the remaining charge due to the leaky nature of the electrode interface (as proved with Eq. (1)), an electrode shorting phase is generated after termination of the anodic phase. Electrode shorting phase is produced by an XOR gate with $\phi_{AC}$ and $\phi_A$ as inputs. With HIGH $\phi_{SC}$ (after the end of $\phi_A$) the electrode is shortened and this unfavorable charge is depleted during this phase. The extra charge on the blocking capacitor is also depleted by the auxiliary switch during $\phi_{SC}$. This switch does not affect the performance of the $C_{BC}$ as a blocking capacitor.

### VI.   EXPERIMENTAL RESULTS

The proposed charge balancer is implemented making use of some selected off-the-shelf components to implement a

978-1-5386-5388-3/18 $31.00 © 2018 IEEE

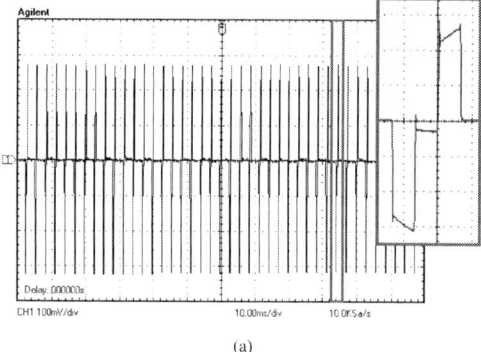

(a)

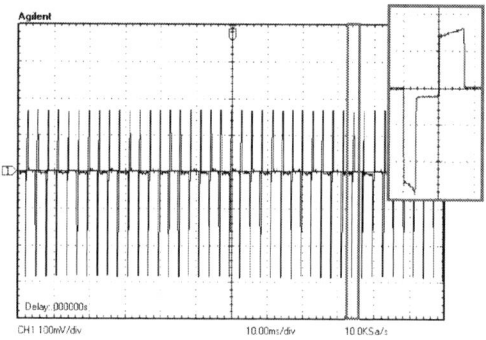

(b)

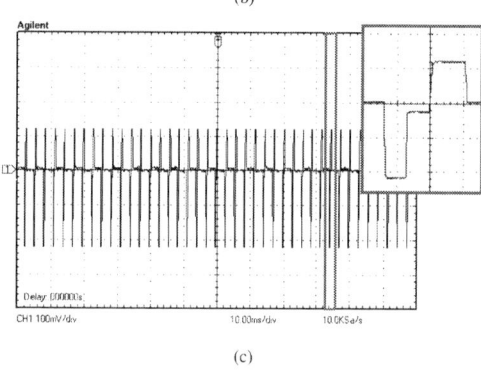

(c)

Fig. 7. Experimental results for stimulation frequency of 250 Hz. (a) Electrode voltage for a symmetric current mode stimulation (1 mA for both cathodic and anodic currents). (b) Electrode voltage for an asymmetric current mode stimulation (1 mA for cathodic current and 0.5 mA for anodic one). (c) Electrode voltage for an asymmetric voltage mode stimulation (220 mV for cathodic voltage and 120 mV for anodic one

low-budgeted power consumption system. A MAX9019 is used for the comparator which has ultra-low quiescent supply current (0.85 $\mu$A) and 4 mV internal hysteresis. All switches are implemented with MAX4622 (high voltage compatible) with 0.5 $\mu$A quiescent current per switch. A PIC16f1827 is used for controlling both stimulators and also producing PWM pulses that is used for generating cathodic and anodic phases. In order to save power, microcontroller is set to sleep mode. The microcontroller uses 31.25 KHz internal oscillator. Consequently, PWM section has a 32 $\mu$s resolution. The

specifications of the implemented stimulators are presented in Table I. Fig. 6 shows the top side of the printed circuit boards (PCBs) of the two stimulators operating in the current mode (Fig. 6(a)) and the voltage mode (Fig. 6(b)) along with the proposed charge balancer. The output is delivered to a DBS electrode (Medtronics Model 3391) which is dipped into a saline solution (0.9% NaCl). The remaining voltage was measured using an instrumentation amplifier and a 24-bit ADC. Measurements shows an average remaining voltage of 6.6 mV with a standard deviation of 1.46 mV for 1200 samples taken from a symmetric current stimulation with an amplitude of 1 mA. The output voltages for different stimulations between two adjacent electrodes on the DBS lead are measured by a DSO3202A Agilent oscilloscope and shown in Fig. 7. Fig. 7(a) shows electrode voltage for a symmetric current mode stimulation. Fig. 7(b) shows the electrode voltage for a current mode stimulation in which the amplitude of the cathodic phase is twice the amplitude of the anodic phase. The electrode voltage in voltage mode stimulation is illustrated in Fig. 7(c).

## VII. CONCLUSION

A charge balancing technique is proposed for an Implantable Pulse Generator (IPG) that is used for Deep Brain Stimulation (DBS). The proposed charge balancer can produce an anodic pulse to balance the injected charge accurately and consumes a less power for both current and voltage stimulation in comparison with the previously published charge balancers. It itself can determine the anodic pulse width according to the amount of the charge injected in the cathodic phase as well as the amplitude of the stimulation in the anodic phase. At the same time, this system is not sensitive to the non-idealities in the charge balancing circuit. We also show that charge metering methods need electrode shorting phase no matter how precise they are. The proposed method is validated using discrete-component implementation. The accuracy of the charge balancer is validated by testing it in saline solution.

## VIII. ACKNOWLEDGMENT

This work has been supported by the Swiss NSF under grant number 200021-157090.

## REFERENCES

[1] "Activa RC IPG, Multi-program rechargable neurostimulator implant manual," Medtronic, Minneapolis, USA.

[2] R. Ranjandish and O. Shoaei, "A simple and precise charge balancing method for voltage mode stimulation," in *Biomedical Circuits and Systems Conference (BioCAS), 2014 IEEE*. IEEE, 2014, pp. 376–379.

[3] R. Ranjandish and A. Schmid, "High frequency self-oscillating current switching for a fully integrated fail-safe stimulator output stage," in *Ph. D. Research in Microelectronics and Electronics (PRIME), 2016 12th Conference on*. IEEE, 2016, pp. 1–4.

[4] "DBS lead model 3391, Lead kit for Deep Brain Stimulation implant manual," Medtronic, Minneapolis, USA.

[5] X. Fang, J. Wills, J. Granacki, J. LaCoss, A. Arakelian, and J. Weiland, "Novel charge-metering stimulus amplifier for biomimetic implantable prosthesis," in *Circuits and Systems, 2007. ISCAS 2007. IEEE International Symposium on*. IEEE, 2007, pp. 569–572.

[6] F. Kolbl, R. Guillaume, J. Hasler, S. Joucla, B. Yvert, S. Renaud, and N. Lewis, "A closed-loop charge balancing fpaa circuit with sub-nano-amp dc error for electrical stimulation," in *Biomedical Circuits and Systems Conference (BioCAS), 2014 IEEE*. IEEE, 2014, pp. 616–619.

# Current Controlled CMOS Stimulator with Programmable Pulse Pattern for a Retina Implant

Pascal Raffelberg*, Roman Burkard*, Reinhard Viga*,
Wilfried Mokwa†, Peter Walter‡, Anton Grabmaier*§ and Rainer Kokozinski*§

*University Duisburg Essen, Department of Electronic Components and Circuits, Duisburg, Germany
†RWTH Aachen University, Institute of Materials in Electrical Engineering 1, Aachen, Germany
‡RWTH Aachen University, Department of Ophthalmology Aachen University Hospital, Aachen, Germany
§Fraunhofer Institute for Microelectronic Circuits and Systems, Duisburg, Germany

*Abstract*—In this work the constant current stimulator of a new epiretinal implant is presented. It consists of a digital waveform generator device, which permits to modify the pulse pattern via a programming interface, a digital–to–current converter, which translates the digital waveform into current pulses with adjustable amplitude, and an output driver, which combines the function of an electrode multiplexer and a high voltage current source for driving large resistive loads. For each of those subcircuits the simulated performance and its designed layout is presented.

## I. INTRODUCTION

As shown in [1] a new wide angle retinal implant approach has been presented, which allows for electrical stimulation of neuronal cells for patients who suffer and have gone blind from retinopathia pigmentosa. This implant consists of a flexible foil containing several independent ASICs (Application Specific Integrated Circuit). Each IC (Integrated Circuit) includes an integrated image sensor, a signal processing unit and a constant current stimulator device which electrically evokes neuronal action potentials.

The electrode–electrolyte interface, which has to be considered for this stimulation, is described by the Randles model in figure 1 [2]. The values of the electrical components vary in a wide range and depend on the used electrodes, the electrode–tissue contact and the tissue itself. In this project a double layer capacitance of $C_{dl} < 100\,\mathrm{nF}$, a series resistance of $R_s = 20\,\mathrm{k\Omega}$ and a charge transfer resistance of $R_{ct} > 1\,\mathrm{M\Omega}$ can be assumed in optimal conditions for the experiments of stimulating tissue [3], but to account for the variation of the electrode–electrolyte interface the stimulator has to be able to drive higher resistive loads [4][5].

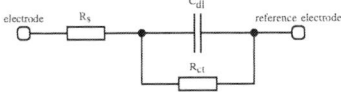

Figure 1. The Randles model for an equivalent circuit of the electrode–electrolyte interface. This model describes the electrical behavior of the electrodes when stimulating the tissue.

The stimulation waveform consists of charge balanced biphasic current pulses with omission of an inter phase delay between the two pulses as shown in figure 2. In [6] it is shown

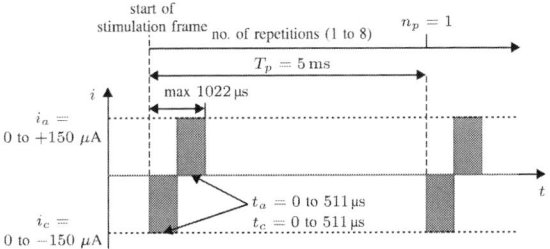

Figure 2. Timing diagram of the selected stimulation waveform. The waveform consists of charge balanced biphasic current pulses with a duration of $t_c$ for the cathodic and $t_a$ for the anodic pulse. $i_c$ and $i_a$ are the independent cathodic and anodic phase currents. The pattern is repeated up to eight times with a repetition frequency of 200 Hz.

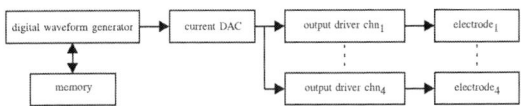

Figure 3. Design of the stimulator approach. The digital waveform is converted into an analog current and then driven by an individual output driver for each electrode.

that this waveform shape provides a good probability to evoke an action potential with a low risk of damaging the tissue or the electrode. The stimulation waveform is described by a set of parameters to allow stimulation with a variety of different current values and waveform shapes. The stimulator shall be capable of generating current pulses of at least $100\,\mu\mathrm{A}$ with a duration of $200\,\mu\mathrm{s}$ [5], and the parameters were selected to match these conditions. The parameterized pattern is stored in a memory inside the IC.

In this approach, a digital waveform generator creates the waveform from the memorys content and controls a digital–to–current converter, which generates the output current. This current is then led to a specific electrode by its corresponding output driver, as shown in figure 3. The output driver is designed using 16 V type transistors in a 350 nm CMOS–process to drive high output currents through the series resistance of the Randles model. To reduce the size and power consumption of the ASIC, the rest of the circuit is designed using 3.3 V type transistors with a 3.3 V power supply.

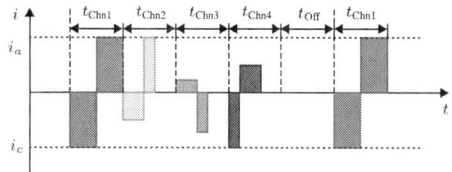

Figure 4. Timing diagram of a series of stimulation waveforms of four different channels created with one stimulator using time-division multiplexing. The stimulation waveform of each channel is independent in stimulation duration, intensity and polarity. After 5 ms the first repetition starts.

## II. DIGITAL WAVEFORM GENERATOR

To reconstruct the stimulation pattern from the data stored in the memory a digital waveform generator is implemented. This circuit basically consists of a digital counter and two comparators to trigger the change of the output current via the digital–to–current converter.

Within the stimulation pattern the length and the intensity of both, the cathodic and the anodic pulse can be varied. The duration of each of the pulses $t_c$ and $t_a$ can be adjusted from $0\,\mu s$ to $511\,\mu s$ and the current values $i_c$ and $i_a$ can be chosen from $0\,\mu A$ to $150\,\mu A$. Furthermore the polarity $p_p$ (anodic/cathodic or cathodic/anodic) can be choosen and the pulses can be repeated. A full stimulation sequence consists of up to eight pulses, controlled by the parameter $n_p$ with a repetition frequency $f_p$ of 200 Hz. The full stimulation pattern is repeated with a period of 40 ms, so that the pulse repetition count within this frame can be used to modify the ratio of time when stimulation is enabled or disabled. All parameters are listed in table I and shown in figure 2.

Table I
LIST OF ADJUSTABLE STIMULATION PARAMETERS AND THEIR RANGE

| Parameter | Symbol | Range |
|---|---|---|
| cathodic / anodic phase duration | $t_c/t_a$ | $0\,\mu s - 511\,\mu s$ |
| cathodic / anodic phase amplitude | $i_c/i_a$ | $0\,\mu A - \pm150\,\mu A$ |
| pulse polarity | $p_p$ | AC / CA |
| no. of pulses | $n_p$ | 1 - 8 |

Using this set of parameters, the waveform shape can be modified to be monophasic, charge balanced biphasic, charge imbalanced biphasic or charge balanced biphasic with slow reversal. Only the charge balanced biphasic shape with fast reversal can not be formed by these parameters, but as shown in [6], this shape is very inefficacious in evoking action potential.

Since the repetition frequency is set to 200 Hz, the stimulation signal has a pause of approximatly 4 ms after each pulse pair. This time is used to stimulate three more electrodes with independent waveforms, utilizing TDM (time-division multiplexing) as shown in figure 4. Therefore only one waveform generator device has to be implemented to drive up to four electrodes.

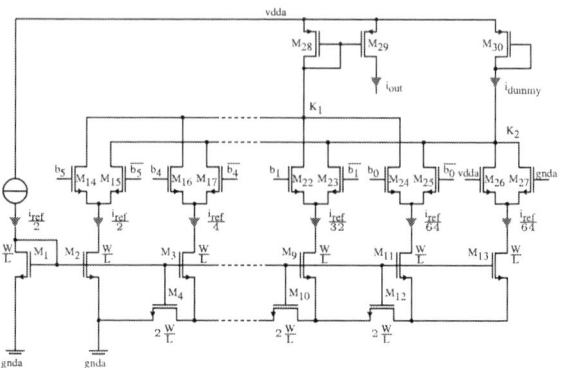

Figure 5. Design of the W–2W current steering DAC. The reference current is mirrored and binary divided by the Transistors $M_1$ to $M_{13}$ building the reference currents $\frac{i_{ref}}{2}$ to $\frac{i_{ref}}{64}$. Those currents can be switched between node $K_1$, where they get mirrored to the output, and node $K_2$, where they get sourced by a dummy load.

## III. W–2W CURRENT STEERING DAC

Figure 5 shows the schematic of the 6–bit W–2W current steering DAC (Digital-to-Analog Converter), which is used to generate an output current of $0\,\mu A$ to approximately $25\,\mu A$ in steps of approximately $390\,n A$. This current is amplified by a factor of 6 in the high voltage output driver stage which offers a total output current of up to $150\,\mu A$ with a stepsize of $i_{LSB} = 2.34\,\mu A$. To reduce the power consumption and size of the DAC circuit, it is supplied by a $v_{dda} = 3.3\,V$ power source and build up using 3.3 V type transistors.

The circuit uses a reference current of $\frac{i_{ref}}{2} = 12.5\,\mu A$ which is mirrored by the transistors $M_1$ and $M_2$ as the highest single output current. Furthermore five identical divider stages are added where each stage divides the current of the previous stage by 2. Each stage consists of two MOSFETs (e.g. $M_3$ and $M_4$) with one of them having the same $\frac{W}{L}$ ratio as $M_1$ (e.g. $M_3$) and the other one having twice the ratio (e.g. $M_4$) to compensate the resistive load of the previous stage.

With this cascade the current values from $\frac{i_{ref}}{4}$ to $\frac{i_{ref}}{64}$ are provided, but since all those current values sum up to just $\frac{31}{64}i_{ref}$, a single last stage made of $M_{13}$ has to be added with the current $\frac{i_{ref}}{64}$, so all current values sum up to $\frac{i_{ref}}{2}$. This last current is not to be used for the converters output as shown in [7], and it should not be added to the output current.

Because all currents have to flow in order to create the correct fractions, each of the control bits decides if the responding current is switched to node $K_1$ or node $K_2$. The current at node $K_1$ gets mirrored by the PMOS–mirror circuit of $M_{28}$ and $M_{29}$ as the output current $i_{out}$, while the unselected current at node $K_2$ flows through the dummy load $M_{30}$. The current of the last stage ($M_{13}$) is always connected to node $K_2$.

Figure 6 shows the nonlinearity of the provided circuit. It is shown that the DAC offers a monotone output signal with a maximum DNL (differential nonlinearity) of 0.22 LSB

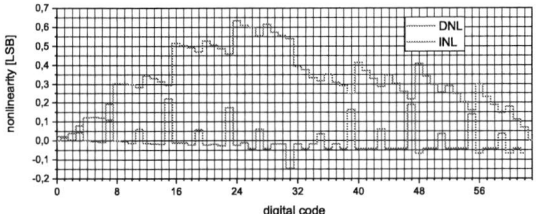

Figure 6. Simulation result of the nonlinearity of the output current of the DAC. The maximum DNL is 0.22 LSB while the maximum INL is 0.63 LSB. The DAC offers a monotone output signal.

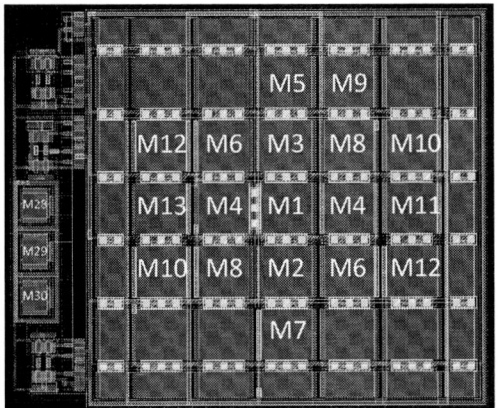

Figure 7. Layout of the W–2W current steering DAC. The transistors of the W–2W ladder are placed in a $5 \times 5$ transistor matrix with point symmetry in respect to the center transistor $M_1$. The matrix is framed with dummy transistors. On the left-hand side of the layout the MOS-switches and PMOS current mirrors are placed. This layout has a size of $95\,\mu m \times 77\,\mu m$.

and a maximum INL (integral nonlinearity) of 0.63 LSB. The simulation includes the extracted parasitics from the layout and uses the output driver stage as load for the DAC.

Figure 7 shows the layout of the current steering DAC. The matrix arrangement includes all the transistors used by the W–2W ladder, which are placed point symmetrically around the center transistor $M_1$. Transistors which have a high impact on the output current, like $M_2$ to $M_4$, are placed next to the center, while the transistors of the lower bits are placed in the outer ring. The unused transistors in the corners and at the borders are dummy transistors to achieve good matching between the transistors. The switches and the output current mirrors are placed on the left-hand side of the matrix.

## IV. HIGH VOLTAGE OUTPUT DRIVER

Figure 8 shows the high voltage output driver stage. To drive higher resistive loads the output voltage can be increased to $v_{ddhv} = 16\,V$. To reduce the size and power consumption of this circuit, a combination of 3.3 V type transistors, drawn in black, and 16 V type transistors drawn in red, is used. Since the current pattern from figure 2 requires only a sourcing or a sinking current at a time, both modes are driven by the same

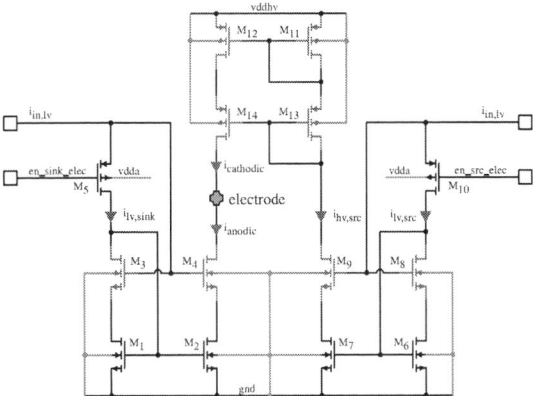

Figure 8. Design of the high voltage output driver with 3.3 V type transistors drawn in black, 5 V type transistors drawn in blue and 16 V type transistors drawn in red. The driver uses two wide swing cascode current mirrors to create a sinking current in the 16 V domain and a highside current mirror to create a sourcing current.

input current $i_{in,lv}$ which is provided by the current output $i_{out}$ of the discussed current DAC.

When in current sinking mode, the output current from the digital–to–current converter flows through the enable transistor on the left hand side of the circuit and is mirrored with a ratio of 1:6 by the NMOS wide swing cascode current mirror circuit made of $M_1$ to $M_5$. The cascode of $M_3$ and $M_4$ consists of 16 V type transistors and protects $M_2$ from high drain–source voltages. The maximum output sinking current is $i_{anodic,max} = -138\,\mu A$ which is less than determined due to the mirror mismatch, but it is still enough to fit the application. When the current sink is disabled, the remaining current is $i_{anodic,disabled} = -112\,pA$.

To reduce the mismatch between the current values of $i_{anodic}$ and $i_{cathodic}$, the same NMOS wide swing cascode current mirror topology with the same output current of the DAC is used in current sourcing mode. A PMOS-cascode current mirror is used to provide the positive output current $i_{cathodic}$. It consists of small 5 V type PMOS transistors as the current mirror circuit, which allow high voltages between their n-well and bulk contact. Additional 16 V type PMOS transistors are used building the cascode circuit to prevent $M_{12}$ from high drain–source voltages. The maximum output sourcing current is $i_{cathodic,max} = 137\,\mu A$.

The complete current mismatch between the anodic and the cathodic current pulse of the DAC and the output driver is shown in figure 9. This current is less than one LSB and leads to a charge injection of less than 2 pC per microsecond of stimulation time, which gives a maximum of 1 nC charge injection for the maximum stimulation time. To further reduce the current mismatch it can be measured at the electrode output during the calibration phase. The charge injection can then be compensated by adjusting the duration parameters $t_a$ or $t_c$ and achieve charge neutrality at the end of the stimulation phase.

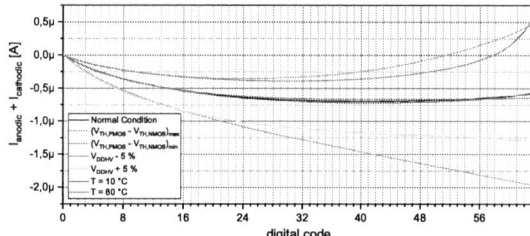

Figure 9. Simulation result of the overall mismatch between the anodic and the cathodic current at the electrode output with different PVT conditions.

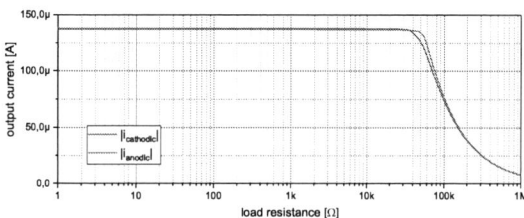

Figure 10. Maximum output current of the combined current DAC and the high voltage output driver over the load resistance with a reference electrode potential of 8 V. The driver is capable of driving the specified output current of 100 μA with resistive loads of up to 70 kΩ.

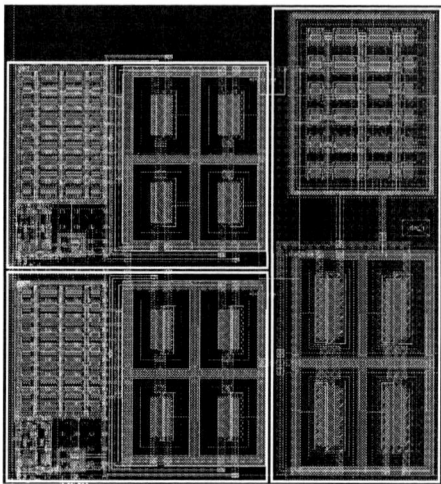

Figure 11. Layout of the high voltage output driver. This layout has a size of approximately $105\,\mu m \times 110\,\mu m$. On the left–hand side, both wide swing cascode current mirrors are placed. On the right hand side the highside PMOS current mirror is placed.

Using this method a total charge injection of 2 pC could be achieved because the pulse duration is adjustable in steps of $t_{LSB} = 1\,\mu s$.

Figure 10 shows the maximum output current over the load resistance. For high resistive loads the output current is limited by the supply voltage of 16 V. The maximum output current of $\pm 137\,\mu A$ can be driven into loads of approximately 33 kΩ. An output current of $\pm 100\,\mu A$, as specified in the introduction, can be driven into loads of up to 70 kΩ.

Figure 11 shows the layout of the output driver circuit, in which the twelve 16 V type transistors require most of the area. Therefore the design has been optimized to use a 1:1 ratio for the high voltage transistors, while the current gain of 1:6 is provided by the smaller low voltage current mirror circuit on the left-hand side.

## V. CONCLUSION

With the here presented design of an integrated constant current pulse generator circuit, it is possible to stimulate retinal tissue with a variety of different signal shapes, as monophasic, charge balanced biphasic, charge imbalanced biphasic or charge balanced biphasic with slow reversal. The stimulator is capable of stimulating up to four different electrodes with independent current pulses of different duration, current values, repetition counts and with selectable pulse phase order. The total charge injection can be minimized by independently adjusting each stimulation phase.

This circuit requires only a small area because it is designed using low voltage transistors. The output driver shifts the output signal into the 16 V domain, to allow the output current

to drive high resistive loads, as they appear when the electrode–tissue contact gets worse during the stimulation. In this case the driver circuit can keep the output current constant until the series resistance of the Randles model exceeds approximately 70 kΩ.

## ACKNOWLEDGMENT

This research project is sponsored by the German Research Foundation (DFG) under contract No. 623545.

## REFERENCES

[1] Raffelberg, P., Waschkowski, F., Viga, R., Mokwa, W., Walter, P. and Kokozinski, R., "Design of a CMOS image sensor and stimulation IC for a wide-angle retina implant," 2017 13th Conference on Ph.D. Research in Microelectronics and Electronics (PRIME), Giardini Naxos, 2017, pp. 309-312. doi: 10.1109/PRIME.2017.7974169

[2] Meza-Cuevas, M. A., "Stimulation of Neurons by Electrical Means" ISBN 978-3-8325-4152-1 (2015).

[3] Erbslöh, A., Viga, R., Walter, P., Kokozinski, R. and Grabmaier, A., "Implementation of a Charge-Controlled Stimulation Method in a Monolithic Integrated CMOS-Chip for Excitation of Retinal Neuron Cells," 2017 Austrochip Workshop on Microelectronics (Austrochip), Linz, 2017, pp. 47-52. doi: 10.1109/Austrochip.2017.9

[4] Weiland, J. D., Humayun, M. S., "Visual Prosthesis," in Proceedings of the IEEE, vol. 96, no. 7, pp. 1076-1084, July 2008. doi: 10.1109/JPROC.2008.922589

[5] Mahadevappa, M., Weiland, J. D., Yanai, D., Fine, I., Greenberg, R. J., Humayun, M. S., "Perceptual thresholds and electrode impedance in three retinal prosthesis subjects," in IEEE Transactions on Neural Systems and Rehabilitation Engineering, vol. 13, no. 2, pp. 201-206, June 2005. doi: 10.1109/TNSRE.2005.848687

[6] Merrill, D. R., Bikson, M., Jefferys, J. G. R., "Electrical stimulation of excitable tissue: design of efficacious and safe protocols", Journal of Neuroscience Methods 141 (2005) 171–198.

[7] Gupta, S., Saxena, V., Campbell, K. A., Baker, R. J., "W-2W Current Steering DAC for Programming Phase Change Memory", 2009 IEEE Workshop on Microelectronics and Electron Devices, Boise, ID, 2009, pp. 1-4. doi: 10.1109/WMED.2009.4816148

# A Laser Diode-Based Wireless Optogenetic Headstage

Alireza Mesri
*Department of Electronics, Information and Bioengineering (DEIB)*
Politecnico di Milano
Milan, Italy
alireza.mesri@polimi.it

André B. Cunha
*Department of Physics*
*University of Oslo*
Oslo, Norway
andre.cunha@fys.uio.no

Ørjan G. Martinsen
[1]*Department of Physics*
[2]*Department of Clinical and Biomedical Engineering*
[1]*University of Oslo*, [2]Oslo University Hospital
Oslo, Norway
ogm@fys.uio.no

Marco Sampietro
*Department of Electronics, Information and Bioengineering (DEIB)*
Politecnico di Milano
Milan, Italy
marco.sampietro@polimi.it

Giorgio Ferrari
*Department of Electronics, Information and Bioengineering (DEIB)*
Politecnico di Milano
Milan, Italy
giorgio.ferrari@polimi.it

*Abstract*— A power efficient, battery powered optogenetic headstage for doing in-vivo experiments with freely moving genetically modified animals is presented. The proposed system is designed with commercial off-the-shelf components, and is based on a Bluetooth Low Energy (BLE) System-on-Chip (SoC) with an integrated antenna and a programmable ARM Cortex-M3 microprocessor core able to control the circuit. The optical signal is generated using a compact laser diode (LD) suitable for a wearable headstage. LD produces light in a highly concentrated way considerably improving the LD-optical fiber coupling efficiency. The proposed optogenetic system is shown to provide 120 mW/mm$^2$ at the fiber tip with a current consumption of 60mA, considerably lower than LED-based systems. The system is remotely controlled by a smartphone app where the user can define optical stimulations patterns settings (optical power, frequency, duty cycle, etc.). It is also powerful enough to be ready to house additional optogenetics functionalities, like electrochemical sensing of the cell response, without significant modifications, thus being the basis of an integrated optogenetic platform.

*Keywords—optogenetics, laser diode, optical stimulation, headstage, channelrhodopsin*

## I. Introduction

Optogenetics, first introduced in 2005 [1], is a ground breaking biomedical technique, which combines optics and genetics to control the activity of cells at in vitro and in vivo conditions. It is based on introducing opsins, light sensitive proteins, into neurons through genetic engineering approaches. Optogenetics has emerged as a tool that enables the activation or inhibition of specific cell populations in the brain by the use of different wavelengths of light. For excitatory Channelrhodopsins, and inhibitory Halorhodopsins, blue and yellow wavelengths are needed, respectively[2,3].

The stimulation of deep brain areas is necessary for treatment of some neurodegenerative disorders, like Parkinson

The research is under Training4CRM project which is funded by the European Union Horizon 2020 Programme (H2020-MSCA-ITN-2016) under the Marie Skłodowska-Curie Initial Training Network and Grant Agreement No.722779. Project website : www.training4crm.eu

Disease (PD), and Epilepsy (EPI). Electrical deep brain stimulation (DBS) affects all cells in the stimulated area in an indiscriminative way and the treatment loses efficacy over time. Also, confounding thermal effects, mood alteration, motor and sensory control issues or depression are some of side effects derived from the application of that method [4]. In contrast, optogenetic allow us to stimulate genetically modified cells in a more precise and selective way, and can address those challenges [1,2,4,5]. In the framework of the European Union project "Training4CRM", we are developing a platform to exploit the potential of optogenetics for the treatment of PD and EPI [6].

Channelrhodopsin-2 (ChR2), the main optogenetic protein, is introduced to modulate neuronal activities [1]. For the activation of ChR2, blue light in the spectral range from 440 to 490nm is needed, operating on the millisecond timescale with 1mW/mm$^2$ optical power. Higher power source should anyhow be used due to the strong light absorption by the brain tissue. As an example, for 473nm wavelength, the intensity of light inside the brain tissue will drop to about 1% after 1mm of travelling [4,5].

Different light-delivery methods have been suggested and are used for the manipulation of neuronal activity. LEDs have become progressively interesting for optogenetic experimentation in freely moving animals. Low cost, small dimension, reliability, and stability are some of their advantages. But, one serious disadvantage of LEDs is poor LED-fiber coupling[2]. Different methods are proposed to improve LED fiber coupling [7-9]. In [7], authors proposed silicon housing technique to improve LED-fiber coupling, achieving 1.71 mW/mm$^2$ with a driving current of 30 mA. In [8], they have suggested a micro-lens coupled LED, which improves light intensity by 99%. All these techniques, need specific fabrication approaches. In [9], the LED is coupled to an implanted 200 μm diameter optical fiber. For a 150 mA LED current, 70 mW/mm$^2$ is achieved.

A laser light source can deliver light with low divergence improving the coupling with the optical fiber, and is able to activate genetically modified neurons within millimeters of the fiber. However, the laser-based systems reported in the literature use a bulky laser source with a long fiber from the lab bench to the animal. Therefore, it tethers the animals, and restricts their natural behavior[1,5].

Here, we report a new power-efficient wireless optogenetic headstage based on a laser diode (LD) to provide the optical power for stimulation, which significantly reduces the required current for a given optical power in comparison with previously reported works. With 60 mA stimulation current, it can provide 120 mW/mm² at the fiber tip. It also features a Bluetooth Low Energy (BLE) which allows the adjustment of the optical power stimulation through any a BLE enabled client (smartphone, tablet, PC, smartwatches).

The paper is organised as follows. The proposed optogenetic system is presented in Section II. Section III presents measurement results, which validate the performance of the platform. Comparisons with other methods from the literature are reported in Section IV. Finally, section V summarizes the features of the proposed optogenetic system.

## II. System Description

### A. Hardware

Fig. 1 shows the optogenetic system structure. We have developed the optogenetic circuit in two boards. PL 450B, an OSRAM blue LD, is used as a light source for the optogenetic system. 3.8 mm package size, 450 nm emission wavelength, and 80 mW power are the main features of the LD. At 25°C, with 24 mA driving current, it can provide an optical power of about 11 mW [10].

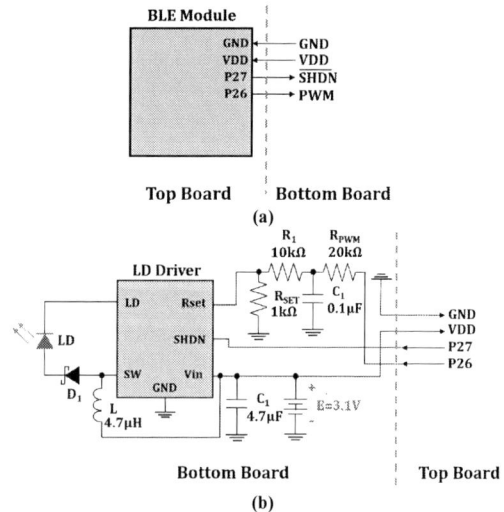

Fig. 1. Schematic of the proposed optogenetic headstage: (a) Top board, and (b) Bottom board.

The headstage consists of two interconnected boards. The top board (Fig. 1a) is dedicated for the BLE System-on-Chip (SoC). The board will be connected through 4 pins to the bottom board, which has the LD, the LD driver IC, and two batteries in series to provide 3.1 V supply voltage. Two pins (VDD, GND) provide supply voltage from batteries in the bottom board for the BLE SoC. The other two pins provide the stimulation pattern, and shutdown (SHDN) signal for the LD driver IC. Fig. 1b shows the schematic for the LD driver stage. SHDN pin allow us to turn on the LD driver IC during stimulation. $R_{SET}$ defines the maximum current at the LD pin.

Anaren's BLE SoC A20737 was selected to control the circuit. It is based on an ARM Cortex-M3 microprocessor core and features an integrated antenna facilitating the PCB design [11]. This programmable SoC includes a non-changeable kernel which accepts instructions written on an extended application programming interface (API) of the C programming language. This kernel provides an event-driven architecture on top of which Anaren has built an abstraction layer API which allows loop based programming (Atmosphere). This API takes care of the entire BLE stack in both client (smartphone) and server (optogenetic implant) and simplifies numerous tasks. It works in 2.4 GHz industrial, scientific and medical (ISM) frequency band, and its package size is $11 \times 13 \times 2.5$ mm³. The weight is around 0.5 grams and can work from 1.7 to 3.6 V power supply voltages. Its average current consumption during sleep and no sleep modes is 1.5mA, and 4.5 mA, respectively. Pin 26, which provides the PWM signal with the programmable frequency and duty cycle is used to generate the stimulation pattern. Pin27, is used to turn on the LD driver IC during stimulation.

We have selected LT1932 from Linear Technology as the LD driver. It is a constant frequency step-up DC/DC converter which operates as a constant-current source. The value of $R_{SET}$ defines the LD current between 5 mA and 40 mA. There are different methods to change the current value of the LD pin. We have applied the PWM signal to the $R_{SET}$ pin in order to change the current of the LD pin. In this case, increasing the duty cycle of the PWM signal will reduce the LD current and thus its brightness [12].

The power supply is provided by two Zinc/monovalent Silver Oxide batteries SR927W, from Renata company. They have a nominal voltage of 1.55 V, a capacity of 55 mAh and a weight as low as 0.77 g. A solution with a single battery and a DC/DC converter to increase the voltage was discarded due to the excessive increase of the current absorption during the optical stimulation.

### B. Software

The proposed firmware is built on the loop structure Anaren provides. It is designed as a finite state machine using a control counter which also functions as the main timer for the system (see Fig. 2).

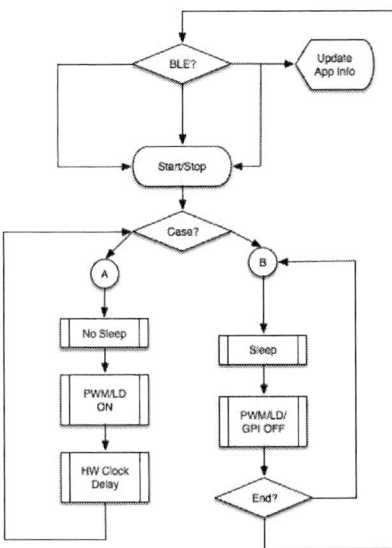

Fig. 2. Flowchart for system firmware.

The structure of the finite state machine is based on a nested if/else structure which being a basic C structure, translates itself into less machine code which results in less computational power while being more reliable as proven by other tested alternatives. To ensure PWM precision without compromising power efficiency, the SoC is set on sleep mode only during dead periods. BLE notifications are only broadcasted while connected through BLE. To ensure reliability, no stimulation periods are a 0% duty cycle PWM rather than disabling it. The app is developed through Anaren software tools which generate a Cordova source portable to both Android and iOS.

### III. MEASUMENT RESULTS

Fig. 3 shows a preliminary prototype of the proposed optogenetic headstage. Dimension of the headstage is $25 \times 20 \times 17$ mm$^3$, and its weight is 6.8 grams where 1.5 grams is due to 2 batteries. We have designed the prototype using the BLE-Anaren A20737-KD Circuits which has access to all pins, 2 buttons and a LED for programming and testing purposes.

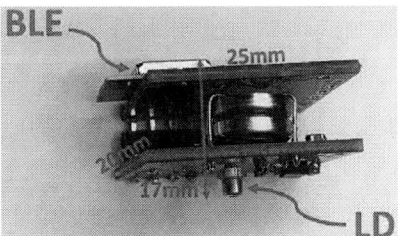

Fig. 3. Proposed optogenetic headstage

To investigate the performance of the headstage, we have emulated a standard optogenetic experiment by applying 150 pulses at 10Hz every 6 minutes, repeated for 40 times (4 hours). Duty cycle of 10 Hz pulses is 5% generating stimulation pulses of 5ms. During the stimulation time, it is possible to change the optical stimulation power by changing the duty cycle in the app. The PWM signal frequency, and the duty cycle were selected to be 5 kHz, and 5%, respectively.

The optical power at the tip of a 200µm diameter optical fiber (Thorlabs-CFMXB20) was measured using a Centronic Series 5T photodiode coupled to DHPCA-100 variable gain transimpedance amplifier (TIA). The photodiode was calibrated using the digital optical meter PM110D by Thorlabs. For 450nm as a wavelength of LD, the photo diode responsivity is about 0.25A/W. The output optical power at the fiber tip is equal to :

$$Pout_{opt} = \alpha Vout/(0.25 \times 100) \qquad (1)$$

Where Vout is the output voltage of the TIA, and 100 $\Omega$ is its gain. Also, $\alpha$ is the optical power level dependent parameter, and is obtained by calibration . It should be mentioned that the coupling efficiency of the LD-fiber is about 45%. Due to LD's package which makes coupling of the cannula to light source simpler, getting higher than 35% is easily achievable, while we achieved about 15% for 0402 package LEDs, which also generate much lower optical power.

Fig. 4 shows the total current consumption of the headstage and the measured TIA output voltage of a single optical pulse. During stimulation, the current consumption is about 65mA, where 60mA is given by the LD driver circuit and the LD. The output voltage of the TIA is about 80mV. By considering the photodiode sensitivity, the optical power at the fiber tip is 3.77mW(120mW/mm$^2$).

### IV. COMPARISON

Table I provides a comparison between the proposed optogenetic stimulation headstage and other recent works. Programmability, current consumption, and optical power at the fiber tip are the main features of the stimulation stage.

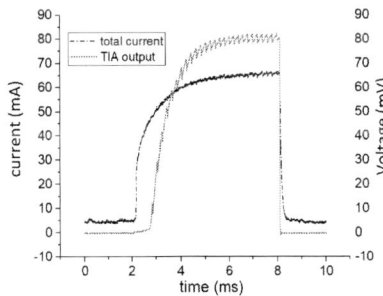

Fig. 4. Total current of the headstage (left axis) and TIA output voltage (right axis)

TABLE I.     COMPARISON TABLE

| | **This work** | **[9]** | **[13]** |
|---|---|---|---|
| Light source | LD | LED | LED |
| Wavelength | 450 nm | 465 nm | 455 nm |
| Optical fiber diameter | 200 μm | 200 μm | 200 μm |
| Stimulation current | 60mA | 150 mA | 700 mA |
| Optical power at the fiber tip | 120mW/mm$^2$ (3.77mW) | 70 mW/mm$^2$ (2.2 mW) | 76.4mW/mm$^2$ (2.4 mW) |
| Headstage size | 25×20×17mm$^3$ (With battery, w/o signal recording circuits) | 17×18×10mm$^3$ (w/o battery) | 12.5 × 12.5 × 5.5 mm$^3$ (w/o battery, antenna, LED and, ferrule holder) |
| Weight   Total | 6.8g | 4.9g | 5.6g |
| Weight   w/o battery | 5.3g | 2.8g | 1.9g |
| Battery | 1.55V,55mAh, × 2 | 3.7V, 100mAh, | --, 100mAh, |

All systems are based on the coupling of the blue light source to an optical fiber which delivers the light into the area of interest in the brain. Also, they are using the battery to power up the system, and they can be mounted directly on the head of the animal. Based on the provided data in the table, while our circuit is using much less current in comparison with other systems, it provides the highest optical power at the fiber tip. Light emission of the LD in a highly concentrated way is the main reason for such considerable improvement. It should be mentioned that in the reported headstage in [9], which performs both optical stimulation and electrophysiological recording (with 32 recording channels) simultaneously, about 48% of total power consumption is related to stimulation stage. Size of the proposed headstage is bigger than previous works. But it is based on the BLE board provided by KD Circuits, which has been designed to consider the usage of all pins and functions of the BLE SoC. Therefore, by designing new board as well as ordering a lower thickness for the FR4 layer of the PCB, not only we can reduce the size of the boards, but also the weight of the system.

## V. CONCLUSIONS AND FUTURE WORKS

In this paper, a new power efficient battery powered optogenetic system is proposed based on only commercial off-the-shelf components. Thanks to the LD as a light source, significant improvements in terms of optical power and current consumption are achieved. It can provide 120mW/mm$^2$ at the fiber tip while it consumes a current of 60 mA. Also, due to the selected structure and the BLE SoC, the users can control the system and the optical power of the stimulation pattern through the app. Finally, by using a PWM signal to adjust the optical power for stimulation, the circuit complexity is reduced.

The next step is to do in-vitro and in-vivo experiments with modified version of the headstage. Moreover, the functionalities of the headstage will be enlarged including the possibility to perform electrochemical measurements to sense the neurotransmitter releases stimulated by the optical signal. The BLE SoC has been already selected powerful enough to manage cyclic voltammetry, and amperometry. For this purpose, we will use the ADC of the BLE SoC to measure and send data to a platform(smart device, cloud, PC). Also, a SPI bus will be used to program an external DAC in order to produce the required signals for doing electrochemical measurements. We will use a carbon based optoelectrical waveguide to deliver both light to the brain and doing electrochemical measurements which will correspond to the released dopamine.

## ACKNOWLEDGMENT

The authors would like to thanks Prof. Merab Kokaia, and Dr. Fredrik Berglind at Epilepsy Center of Lund University for their useful discussion about the optogenetic systems and experiments, and Prof. Jenny Emnéus for leading the Training4CRM project.

## REFERENCES

[1] N. Li, and P. Miao, "Let there be light: A tutorial on optogenetics," in *IEEE Pulse*, vol. 5, no. 4, pp. 55-59, July-Aug. 2014.

[2] M. R. Warden, J. A. Cardin, and K. Deisseroth, "Optical neural interfaces," Annu. Rev. Biomed. Eng., vol. 16, pp. 103-29, Jul. 2014.

[3] H. Zhao, "Recent progress of development of optogenetic implantable neural probes," Int. J. Mol. Sci., vol. 18, no. 8, pp. 1-21, Aug. 2017.

[4] R. Pashaie, P. Anikeeva, J. H. Lee, R. Prakash, O. Yizhar, et al., "Optogenetic brain interfaces," in *IEEE Reviews in Biomedical Engineering*, vol. 7, pp. 3-30, 2014.

[5] M Aravanis, L. P.Wang, F. Zhang, L. A. Meltzer, M. ZMogri, et al., "An optical neural interface: *in vivo* control of rodent motor cortex with integrated fiberoptic and optogenetic technology,"J. Neural Eng., vol 4, no.3, pp. S143–S156, Sep. 2007.

[6] Training4CRM project. Available online : http://www.training4crm.eu/

[7] M. Schwaerzle, P. Elmlinger, O. Paul and P. Ruther, "Miniaturized tool for optogenetics based on an LED and an optical fiber interfaced by a silicon housing," *2014 36th Annual International Conference of the IEEE Engineering in Medicine and Biology Society*, Chicago, IL, pp. 5252-5255, 2014.

[8] X. Bi, T. Xie, B. Fan, W. Khan, Y. Guo and W. Li, "A flexible, micro-lens-coupled LED stimulator for optical neuromodulation," in *IEEE Transactions on Biomedical Circuits and Systems*, vol. 10, no. 5, pp. 972-978, Oct. 2016.

[9] G. Gagnon-Turcotte, Y. LeChasseur, C. Bories, Y. Messaddeq, Y. De Koninck and B. Gosselin, "A wireless headstage for combined optogenetics and multichannel electrophysiological recording," in *IEEE Transactions on Biomedical Circuits and Systems*, vol. 11, no. 1, pp. 1-14, Feb. 2017.

[10] OSRAM PL 450B laser diode. Available online : https://www.osram.com/os/ecat/TO38%20PL%20450B/com/en/class_pim_web_catalog_103489/global/prd_pim_device_2220052/

[11] Anaren Integrated Radio(AIR). Available online : https://app.atmosphereiot.com/wiki/A20737_Module

[12] Linear Technology. LT1932 Constant-Current DC/DC LED Driver in ThinSOT. Available online : http://cds.linear.com/docs/en/datasheet/1932f.pdf

[13] Multichannel systems, w2100-hs4-opto. Available online : https://www.multichannelsystems.com/products/w2100-hs4-opto

PRIME 2018, Prague, Czech Republic

Session: Sensing and Biomedical Circuits II

# An Active Charge Balancing Method Based on Chopped Anodic Phase

Reza Ranjandish[1], and Alexandre Schmid[2]

[1,2]Microelectronic Systems Laboratory, Swiss Federal Institute of Technology (EPFL), Switzerland

*Abstract*—A new method for safe electrical neural stimulation is proposed. A negative feedback is used to automatically produce short anodic pulses according to the remaining voltage. The period of the short anodic pulses varies according to the remaining voltage. Thanks to the structure of the static comparator, the system produces pulses with high duty-cycle at the beginning of the anodic phase and low duty-cycle at the end of the anodic phase. Hence, the system can accurately balance the charge. In contrast with previous methods based on the chopped anodic phase, this method employs a new structure that reduces the area of charge balancer for integration and increases the accuracy for a safe electrical stimulation. Lowering the power consumption of the charge balancing circuit is another outcome of the proposed method. An integrated circuit implementing the proposed method is designed using a 0.18 $\mu$m technology. The simulation results prove the accuracy and proper performance of the proposed method in both current and voltage mode stimulation.

*Keywords—Electrical stimulation; current mode stimulation; voltage mode stimulation; charge balancing; implantable medical devices.*

## I. INTRODUCTION

Implantable medical devices are one new category of systems proposed to treat some medicamental refractory diseases or disabilities. These devices include DBS for Parkinson disease, retinal implants to restor vision to the blind, pacemakers to control cardiac irregularities, cochlear implants for profound deafness treatment and functional electrical stimulation (FES) to restore functionality to paralyzed organs. The aim of implantable electrical tissue stimulators is not only limited to help people with neural or muscular disabilities but also to treat the patients with refractory depression. Initiating an action potential (AP) upon the transfer of charge into the tissue is the way these stimulators operate. The injection of charge can be carried out either by a current or voltage stimulator . Because of the controlled amount of charge, current-mode stimulators have becomes the most popular stimulators in recent years [1]. However, making use of voltage mode stimulators leads to a more power efficient stimulation due to the selection of a supply voltage close to the stimulation voltage [2]. Regarding biological safety issues, the injected net charge in an electrical stimulation must be equal to zero. Any nonconformities leads to a pH shift or tissue damage [3]. in addition, a pH shift causes electrode dissolution due to the electrolysis and consequently leads to releasing toxic substances into the biological environment [4]. Therefore, stimulators must ensure that the remaining net charge on the electrode after each stimulation period is equal to zero or is kept within a safe window (e.g. $\pm 100$ mV), making use of a charge balancer.

This paper introduces a new method for precise charge balancing with minimum number of components. The orga-

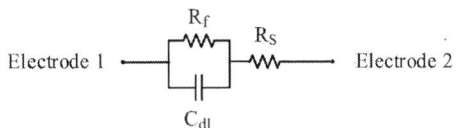

Fig. 1. Simplified model of the two-electrode system.

nization of this paper is as follows: A short description of the electrode modeling is introduced in Section II. Different types of common stimulus waveforms are described in Section III. In Section IV, the charge balancing methods based on chopped anodic phase are explained, briefly. The proposed charge balancing method is introduced in Section V. In Section VI, simulation results are presented to show the validity of the proposed charge balancer.

## II. ELECTRODE-ELECTROLYTE INTERFACE

In a metal, charge is transfered by electrons, however, in an electrolyte ions are responsible to transfer the charge. Hence, an interface between the electrode and electrolyte should associate these two types of charge carriers, in case considering passing a current from the electrode through the tissue (electrolyte). Hence, types of reaction occur at the electrode-electrolyte interface to insure that the charge transfer happens between the electrode and electrolyte: Faradaic and non-Faradaic reactions. Non-Faradaic or capacitive reactions can be modeled as an ideal capacitor, namely the double-layer capacitor ($C_{dl}$). Faradaic charge transfer reactions are considered as reduction-oxidation (redox) reactions that support charge exchange between the electrode and electrolyte in addition to the non-Faradaic charge transfer. Faradaic reactions are due to electrochemical reactions that create chemical species which may either remain to the vicinity of the electrode surface or go to the solution, far away from the electrode surface [3].

As shown in Fig. 1, in a monopolar stimulation using a double voltage topology and a large return electrode, a tissue model consists of a $C_{dl}$ as a representative of non-Faradaic reactions, $R_f$ as a representative of Faradaic reactions and $R_s$ as spreading resistance. In most electrical stimulation applications, $R_f$ is negligible and a series model of $R_s$ and $C_{dl}$ is an appropriate model. In order to test the proposed system, a tissue model consists of an $R_S = 10\ k\Omega$ and a $C_{dl} = 100\ nF$ is used in this paper.

978-1-5386-5388-3/18 $31.00 © 2018 IEEE        261

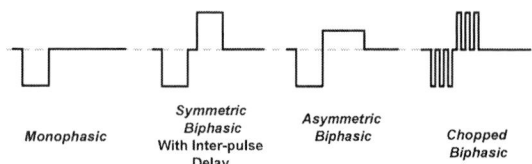

Fig. 2. Common stimulus waveforms used in electrical stimulation.

## III. STIMULUS WAVEFORMS

In principle, in a biphasic cathodic-first stimulation, cathodic charges are responsible for AP initiation and anodic charges are responsible to neutralize the injected charge from the interface due to the safety issues. However, it is shown that the waveform and even the anodic pulse-shape can affect the efficacy of the AP initiation. Fig. 2 shows prevalent waveforms used in electrical stimulation. Monophasic stimulation is the most efficient waveform for AP initiation, however, due to safety issues it cannot be used in chronic stimulations and commercial implantable devices (IMD). The symmetric biphasic waveform (with or without inter-pulse delay) is the most prevalent waveform among the stimulation waveforms, thanks to the inherent charge balanced nature of this waveform. The asymmetric biphasic waveform has better efficiency in AP initiation [3]. Using chopped anodic and/or cathodic phases is a way to increase the efficiency of the stimulation [5]. This type of stimulation waveform is used in the design of efficient stimulators or in charge balancers [6]. Other types of waveforms are also used in the electrical stimulation like imbalanced biphasic, rising exponential biphasic, triphasic, etc. which may have different effects on the action potential initiation and/or stimulation safety.

## IV. CHARGE BALANCING METHODS BASED ON CHOPPED ANODIC PHASE

For safety reasons, the tissue interface voltage should be kept in a safe voltage window after each stimulation cycle by using a charge balancing method. Charge balancing can be performed as an active or passive technique. In active charge balancing, active circuits are used to insure that the voltage on the tissue interface is controlled to its initial value after cathodic phase. In a passive charge balancer no active components is used and charge cancellation is carried out by shorting the electrode.

Different circuits are proposed to balance the charge. One method which was first introduced in [7], is charge balancing based on chopped anodic phase. After each cathodic phase, the anodic current is exerted as short pulses. During the period in which no current flows through the tissue, the balancer checks the voltage remaining on the tissue and stops generating anodic short pulses when the remaining voltage becomes positive (after anodic phase the remaining voltage is negative). There are major differences between this method and the pulse insertion method proposed in [6]. Pulse insertion is a current-mode charge balancer since it employs symmetric current pulses before the injection of the short pulses. In voltage-mode stimulation, this latter technique cannot be used since a symmetric voltage stimulation waveform may result in a

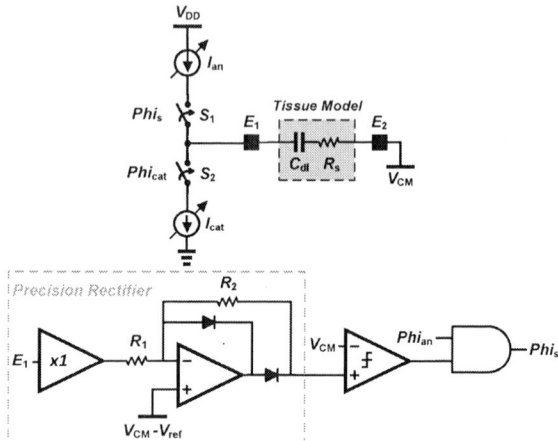

Fig. 3. Charge balancing technique proposed in [8].

large remaining voltage, as proved in [7]. In addition, the stability problems and successful charge balancing condition are described in [8] which show that the pulse insertion technique is a sensitive method to the tissue impedance and stimulation parameters. Furthermore, the polarity of the short pulses is not predictable and it depends on the polarity of the mismatch. Hence, the short pulses may affect the stimulation. Another difference is the number of comparators used in this technique. Since a window comparator is used in the pulse insertion technique, two comparators are used and the charge balancer forces the remaining voltage to stay in the safe window defined by the window comparators (which can affect the stability of the charge balancer [8]). However, in the charge balancing technique based on chopped anodic phase, a single comparator is used to detect the change in the polarity of the remaining voltage. Therefore, the area and power that is consumed by the charge balancing circuit is less than the pulse insertion technique. In addition, no instability condition is introduced by using the charge balancing technique based on chopped anodic phase.

The method proposed in this paper is based on the technique introduced in [8], introducing major modification to reduce the area and power consumption and increase the accuracy. The schematic implementation of the method proposed in [8] is shown in Fig. 3 in which a negative feedback to the switch $S_1$ is an essential element supporting the method. In order to control the pulse-width of $Phi_s$, a precision rectifier is used. The input of the precision rectifier is the buffered voltage on the tissue. The precision rectifier does not allow the input voltage of the comparator to exceed an specific value defined by the values of the resistors $R_1$ and $R_2$ and the reference voltage of $V_{ref}$. The rectification is done to control the on-time of the signal $Phi_s$ and produce a PFM signal. Although this technique provide faster charge balancing than [7], it introduces some disadvantages. First, the amplification path should be much faster than the comparator. Failing to satisfy this constraint affect the on-time and off-time of the anodic pulses and increase the complexity of the system. This means that the amplification path is a power hungry part in the system.

*PRIME 2018, Prague, Czech Republic*                    *Session: Sensing and Biomedical Circuits II*

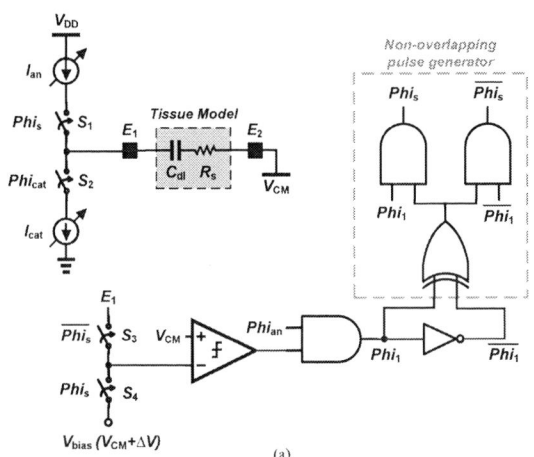

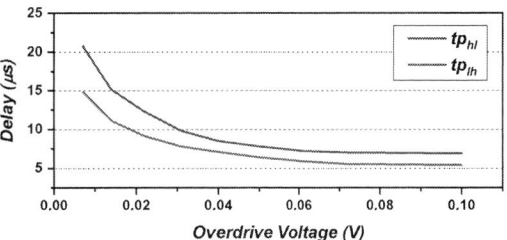

Fig. 5.   Propagation delay curves versus input overdrive voltage.

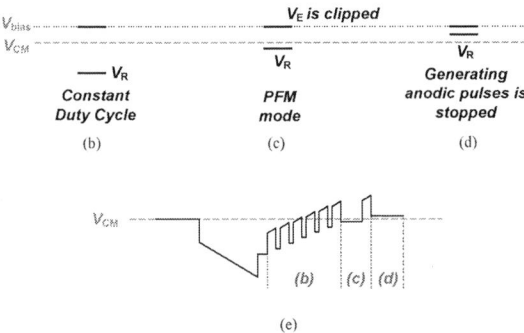

Fig. 4.   Proposed charge balancer: (a) schematic of its architecture, (b-d) its different working regions and (e) a sample waveform.

Secondly, using a precision rectifier that has an acceptable offset requires large silicon area which is not indicated in multichannel stimulators. Thirdly, the precision rectifier affects the accuracy of the charge balancing. In order to demonstrate this statement, we should calculate the transfer function of the precision rectifier. While the output of the precision rectifier is larger than $V_{CM} - V_{ref}$ the relation between the input ($V_{in}$) which is the buffered voltage of the electrode and output ($V_{out}$) of the precision rectifier is equal to:

$$V_{out} = (V_{CM} - V_{ref}) - \frac{R_2}{R_1} \times (V_{in} - (V_{CM} - V_{ref})) \quad (1)$$

Otherwise, the output is clipped at $V_{CM} - V_{ref}$. These equations show that the generation of anodic pulses stops prior to fully discharging the electrode interface by solving the Eq. (1) by assuming $V_{out} = V_{CM}$. However, the values of $R_1$, $R_2$ and $V_{ref}$ can be selected in a way that the termination of the anodic pulses occurs when the remaining voltage is within a safe window (ex. $\pm 100$ mV).

In the new technique proposed in this paper, the precision

rectifier and the buffer are replaced by a switching scheme. The switching scheme controls the pulse-width of the chopped anodic pulses. Removing precision rectifier and the voltage buffer reduces the power consumption as well as the area of the charge balancing circuit. Furthermore, it increases the precision of charge balancing.

## V.   PROPOSED CHARGE BALANCING SYSTEM

Using a voltage clipper increases the power consumption and the area of the charge balancing circuit. Therefore, in this paper, a new method based on the switching of the input of the comparator is presented. The schematic of the proposed method is depicted in Fig. 4. In order to describe how the charge balancer performs, it is assumed that the stimulator is a current mode stimulator. The proposed charge balancer can also operate in voltage mode stimulation. First, the voltage on the tissue remaining after the cathodic phase is negative. At the inter-pulse delay, $Phi_{an}$ is LOW, and consequently $Phi_1$ and $Phi_s$ are LOW. Therefore, switch $S_3$ is closed and the input of the comparator is connected to the tissue. At this point no anodic current passes through the tissue ($S_1$ is open). At start of the anodic phase ($Phi_{an}$ is HIGH), since $Phi_s$ is HIGH, both switches $S_1$ and $S_4$ are closed and anodic current passes through the tissue. At this point, since the negative input voltage of the comparator is higher than the positive input, the comparator output changes to LOW. Nevertheless, this change is not fast due to the delay of the comparator. Biasing $V_{bias}(V_{CM} + \Delta V)$ close to the $V_{CM}$ increases the pulse width of the anodic short pulses Since the delay of the comparator increases with lower differential overdrive. Biasing this node to a specific value controls the comparator fixing the on-time of $S_1$; hence, anodic short pulses have a fixed pulse width. The output of the comparator becomes LOW after the specific amount of delay which is defined by $V_{bias}$ has elapsed. Consequently, both switches $S_1$ and $S_4$ are opened and the comparator measures the remaining voltage. As shown in Fig. 4(b), for large negative remaining voltages (at the beginning of the anodic phase), the input overdrive is large and the delay of the comparator is small. Hence, the duty cycle is large, due to the fixed pulse width. By injecting anodic short pulses, the remaining voltage increases and the input overdrive decreases; therefore, the delay of the comparator increases which leads to generating pulses with smaller duty cycle than the pulses generated at the beginning of the anodic phase. This type of response of the charge balancing circuit introduces pulse frequency modulation (PFM). The generation of the anodic short pulses stops whenever the remaining voltage crosses

978-1-5386-5388-3/18 $31.00 © 2018 IEEE          263

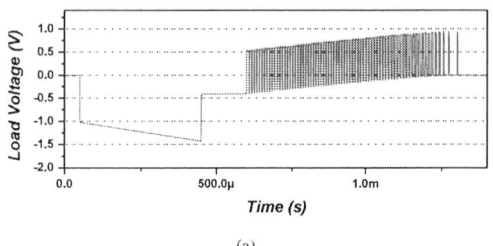

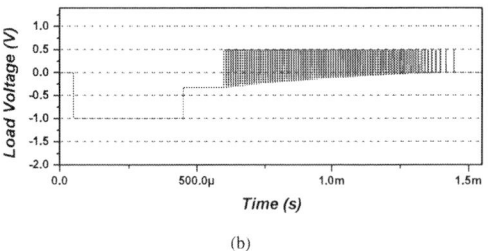

Fig. 6. Load voltage across the electrical model due to the electrical stimulation in (a) current mode stimulation and (b) voltage mode stimulation.

$V_{CM}$ (Fig. 4(d)). Fig. 4(e) shows a conceptual example output of the proposed charge balancer. In order to insure that $S_3$ and $S_4$ are not simultaneously closed, in any case, a non-overlapping pulse generator is used as presented in Fig. 4(a).

At this point, the proposed technique can be compared with the technique presented in [8]. Removing precision rectifier and using two switches instead, reduces the area of the charge balancing circuit. Furthermore, there is no need for high-speed and precise amplifier in the proposed method; hence, power consumption is dramatically reduced in the proposed charge balancer. The proposed method stops generating anodic pulses when the remaining voltage crosses $V_{CM}$. Consequently, the proposed method has better accuracy in comparison to [8].

## VI. SIMULATION RESULTS

The proposed method is designed using a 0.18 $\mu$m technology using I/O transistors (3.3 V). The charge balancer is tested with both current mode and voltage mode stimulator. The performance of the charge balancer is validated using a tissue model consisting of a $R_s$ =10 k$\Omega$ in series with a $C_{dl}$ =100 nF. The comparator propagation delay curve for both rising ($tp_{lh}$) and falling ($tp_{hl}$) edges versus the input overdrive is presented in Fig. 5. The load voltage across the tissue model in case of a current mode stimulation with a cathodic current of $I_{cat}$=100 $\mu$A and cathodic pulse width of 400 $\mu$s is shown in Fig. 6(a). The anodic current, $I_{an}$, has an amplitude that is 10% smaller than the cathodic current to simulate the mismatch between anodic and cathodic currents. At the end of the anodic phase, as described above, the remaining voltage is close to $V_{CM}$ which increases the off-time of the anodic switch ($S_1$ in Fig. 4(a)). Consequently, the duty cycle at the end of the anodic phase is reduced. Fig. 6(b) shows the load voltage across the tissue model in case of a voltage mode stimulation with a cathodic voltage of $V_{cat}$=-1 V and cathodic pulse width of 400 $\mu$s. The anodic

TABLE I. COMPARISON OF THE CHARGE BALANCING METHODS BASED ON CHOPPED ANODIC PHASE.

| Method | This Work | [8] | [7] |
|---|---|---|---|
| Technology | 0.18 $\mu$m Simulation | Discrete Implementation | Discrete Implementation |
| Modulation | PFM | PWM/PFM | No Modulation |
| Rectification Method | Switching Method | Buffering & Precision Rectifier | No Rectification Method |

voltage, has an amplitude equal to 0.5 V. Simulation results validates the performance of the charge balancer with both current and voltage mode stimulation. The charge balancing circuit (excluding the current DAC) only consumes 1.85 $\mu$A which results in a 6.1 $\mu$W power consumption.

## VII. CONCLUSION

A new active charge balancing method for electrical stimulation is introduced. A negative feedback loop is used in order to generate short anodic pulses until the remaining voltage exceeds $V_{CM}$. Short anodic pulses are generated without using a pulse generator. The proposed circuit shows an inherent PFM behavior in which the remaining voltage is modulated into the duty cycle of the short anodic pulses. The switching method eliminates the need of a prescision rectifier which is area and power consuming. Employing this method, reduces the power and area consumption by the charge balancer and represents better accuracy than the original method. The performance of the charge balancer is validated using a 0.18 $\mu$m technology. Table I compares different charge balancing methods based on chopped anodic phase.

## VIII. ACKNOWLEDGMENT

This work has been supported by the Swiss NSF under grant number 200021-157090.

## REFERENCES

[1] J. Vidal and M. Ghovanloo, "Towards a switched-capacitor based stimulator for efficient deep-brain stimulation," in *Engineering in Medicine and Biology Society (EMBC), 2010 Annual International Conference of the IEEE*. IEEE, 2010, pp. 2927–2930.

[2] M. Ghovanloo and K. Najafi, "A compact large voltage-compliance high output-impedance programmable current source for implantable microstimulators," *Biomedical Engineering, IEEE Transactions on*, vol. 52, no. 1, pp. 97–105, 2005.

[3] D. R. Merrill, M. Bikson, and J. G. Jefferys, "Electrical stimulation of excitable tissue: design of efficacious and safe protocols," *Journal of neuroscience methods*, vol. 141, no. 2, pp. 171–198, 2005.

[4] J. Xu, R. K. Shepherd, R. E. Millard, and G. M. Clark, "Chronic electrical stimulation of the auditory nerve at high stimulus rates: a physiological and histopathological study," *Hearing research*, vol. 105, no. 1, pp. 1–29, 1997.

[5] R. K. Shepherd and E. Javel, "Electrical stimulation of the auditory nerve: Ii. effect of stimulus waveshape on single fibre response properties," *Hearing research*, vol. 130, no. 1, pp. 171–188, 1999.

[6] M. Ortmanns, A. Rocke, M. Gehrke, and H.-J. Tiedtke, "A 232-channel epiretinal stimulator asic," *IEEE Journal of Solid-State Circuits*, vol. 42, no. 12, pp. 2946–2959, 2007.

[7] R. Ranjandish and O. Shoaei, "A simple and precise charge balancing method for voltage mode stimulation," in *Biomedical Circuits and Systems Conference (BioCAS), 2014 IEEE*. IEEE, 2014, pp. 376–379.

[8] R. Ranjandish and A. Schmid, "An active charge balancing method based on self-oscillation of the anodic current," in *Biomedical Circuits and Systems Conference (BioCAS), 2016 IEEE*. IEEE, 2016, pp. 496–499.

# Fabrication of Full-3D Printed Electronics RF Passive Components and Circuits

A. Salas-Barenys, N. Vidal, J. Sieiro, J.M. López-Villegas
Department of Electronics
Universitat de Barcelona
Barcelona, Spain

B. Medina-Rodríguez and F.M Ramos
FAE - Francisco Albero Electrónica S.A.U.
Hospitalet de Llobregat, Spain

*Abstract*—**This paper presents a process for full-3D circuit and RF passive component fabrication based on two main steps: additive manufacturing of the plastic or ceramic substrate, through a stereolitographic 3D printer, and a copper electroless plating metallization process. The metallization results are discussed in terms of resistivity, comparing them to the State-of-the-Art on 3D printed electronics. The capabilities and accuracy of the process have been demonstrated and discussed through the fabrication of conical inductors.**

*Keywords*—**3D printing; Electroless plating; Conical inductors.**

## I. Introduction

The principal motivation of this project comes from the necessity of controlling the manufacturing process while fabricating prototypes. In the radiofrequency (RF) electronics field, as in others, this issue is crucial for proper prototyping, because the performance of the final products is very sensible to changes in dimensions and symmetries. Additive Manufacturing (AM) offers the possibility of local prototyping, eliminating the dependence on external providers that may hijack project schedules with an opaque manufacturing process. Otherwise, the design loop *"design → prototype → test → redesign"* flows better with significantly shortened idle times and a manufacturing process completely transparent to the designer.

A part from that, the large amount of benefits of AM is discussed in the literature [1], [2], [3]. For that reason, there already exist several works on the topic of 3D printed electronics in the academic field [4], [5] and [6]. Moreover, there can be found some RF applications in 3D antenna fabrication [7], [8], [9] and [10]. Also, in the commercial State-of-the-Art (SoA), some companies already offer 3D electronic printing equipment [11], [12].

All these works use different AM technologies and metallization strategies. Nevertheless, among all this SoA, there are still some weak points:

This work was supported in part by the Spanish Secretary of State for Research, Development and Innovation Under Projects TEC2013-40430-R and TEC2017-83524-R.

- Achieving **high conductivity** on the metallized area, as much closed to bulk copper or silver as possible, maintaining an **affordable price** .
- Even solving the previous issue, it is hard to find a method that also permits to print in **full-3D geometries**, taking profit of all the capabilities that AM is offering. Most of them are limited to 2D, 2.5D or only some 3D surfaces.

Therefore, the aim of this work is to develop an electronics manufacturing process, adaptable to any 3D geometry, and fabricate some demonstrators in order to show a possible RF application.

## II. Fabrication process

In 3D design, there is a wide range of available CAD tools. For this work, the EMPro modeller has been used due to its parametric modelling capabilities; it also permits to make electromagnetic simulations, what is indispensable for RF applications. The output files must be exported in .stl format to open it with the NAUTA slicer software, through which the object is properly adapted for the XFAB 3D printer, from DWS Systems, that offer their own catalogue of self-made printable materials.

The process, previously explained in [13], is based on the electroless plating concept, that uses a reduction-oxidation reaction where (in this case) copper is deposited over a catalyst surface, which must be electrically conductive [14]. Then, the prototype fabrication requires these 4 basic steps:

1) During the design, the area to be metallized must be sank about 300 microns.
2) Cover the whole sample with a metallic layer using a (nickel) spray.
3) Polish the sample to remove the metal from the area that has not to be metallized. The trenches made in the first step prevent the conductive layer to be harmed where it should not.
4) Deposit copper over the metallic seed using the electroless plating process, that is dependent on the solution temperature and the time the sample is submerged.

## III. Metallization results

There are different issues concerning the quality of the metallization stage. Probably, the most important in this case is to characterize the achieved resistivity, that has been determined measuring the sheet resistance and the copper layer thickness. Since the electroless reaction depends on time and temperature [14], the study of the achievable conductivity has been made taking different values of these variables.

The sheet resistance is defined as

$$R_s = \rho t \tag{1}$$

where $r$ is the resistivity and $t$, the metal layer thickness. The Van der Pauw method[15] allows to calculate the sheet resistance of any sample by taking a few measures from four probes and using this equation below

$$R_s = \frac{\pi}{\ln 2} R_{MN,OP} \tag{2}$$

as long as the sample shape is symmetric and the $M$, $N$, $O$ and $P$ probes are symmetrically placed. $R_{MN,OP}$ is the resistance measured between the point $M$ and $N$ while applying a current between $O$ and $P$. Errors related to probe misplacement or sample fabrication deviations can be eliminated by using shapes with smooth edges[16] as in figure 1.

Fig. 1. One of the Van der Pauw samples used in the experiments, already metallized.

The measures have been carried out by the B1500 Semiconductor Device Parameteer Analyze from Keysight, with four probes, and the values have been calculated using the EasyEXPERT Application Test for Van Der Pauw measurement installed in the same equipment. The thickness of the copper layer has been measured taking images of the sample cross-section, using the ultra-high resolution TM3000 TableTop Scanning Electron Microscope (SEM). The obtained results were quite disparate, presenting values from from 30 to $120 \mu m$, and without a clear correlation with time and temperature. Nevertheless, from the expression 1, the resistivity has been extracted and depicted in figure 2 in terms of the solution temperature and the time the sample is submerged.

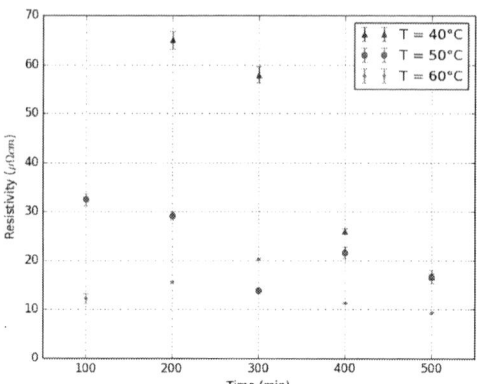

Fig. 2. Resistivity achieved depending on the time the samples were inside the electroless solution and temperature it was at.

It can be seen that the temperature is crucial at the beginning, while in the last measures the difference is drastically reduced. In fact, the samples at $60°C$ do not observe much resistivity variance along time. Looking at the $60°C$ points or at the $50°C$ ones from the 300-minute step ahead, it can be stated that a resistivity about $9-20 m\Omega cm$ has been achieved. These are very good results, among the best within the found SoA, as it is shown in table I.

| Company/author | Resistivity ($\mu\Omega cm$) | Full 3D | Price (€) |
|---|---|---|---|
| A.J. Lopes [4] | 4-80 | Full 3D | - |
| E. Saleh [5] | 48 | 2.5D | - |
| G. Saada [6] | 27.16 | Full-3D | - |
| Optomec [12] | 5 | Full-3D | 400,000 |
| Nano Dimension [12] | Not availaible | 2D | 250,000 |
| Voxel8 [12] | 50 | 2.5D | 7,300 |
| This work | 9-20 | Full-3D | 10,050 |

TABLE I

COMPARISON BETWEEN THE DIFFERENT METHODOLOGIES FOUND IN THE SOA AND THE RESULTS OBTAINED IN THIS WORK. THE PRICE COLUMNS IS RELATED TO THE COMMERCIAL MACHINES, THUS IT IS NOT CONSIDERED FOR ACADEMIC WORKS.

The discrepancies with bulk copper resistivity, which is $1.68\mu\Omega cm$, are provoked by the micro-gaps within copper layers. Electroless plating proceeds by the coalescence of the metal particles, forming granular layers. The hole density is not very high in this case, as it can be seen in figure 3a. That is because the picture has been taken from one of the best samples, where the copper has been deposited during 500 minutes at $50°C$. The biggest holes can reach diameters around $50\mu m$ in this case (figure 3b).

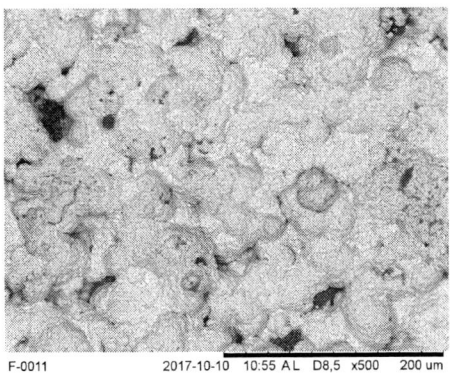

(a) Zoom = x500.

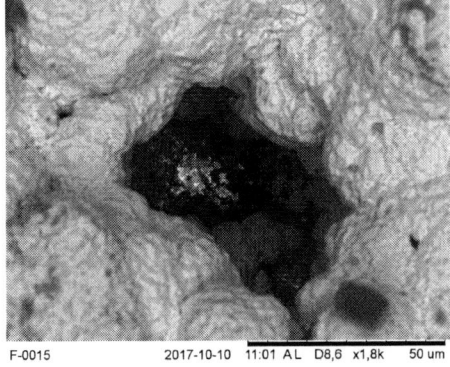

(b) Zoom = x1800.

Fig. 3. SEM picture of the surface of a Van der Pauw sample. Time and tempereature of the electroless bath equal to 500 minutes and $50°C$.

The cost of the process in table I has been calculated only from the equipment, since the consumable (printable materials, nickel spray, electroless solution,...) cost was negligible. In addition, it must be taken in account that most of the price comes from the 3D printer and a small fume hood extractor, which are general-purpose equipment that might already be present in any research facilities, decreasing the actual cost.

IV. DEMONSTRATORS

The feasibility of this process to fabricate any kind of full-3D circuits, with either through hole and SMD component soldering, was already demonstrated in [13].

Conical inductors present an improvement in bandwidth for high inductances, comparing to standard surface-mount ones[17]. A possible application for the process detailed in this work is the fabrication of conical inductors, as explained in [18].

The design has been performed using the EMPro environment from Keysight Technologies Inc, as well as the simulation under the Finite Difference Time Domain method. The material for the substrate is Therma 289, offered by

the same company of the 3D printer, that presents a real relative permittivity equal to $\epsilon_r' = 2.9$, while the loss tangent is $\tan \delta = 0.01$ as it was found in [13]. In figure 4, one of the fabricated inductors is presented. Their main design parameters are the angle $\alpha$, the spire number $N$, the pitch between turns $p$ and the metal width $w$.

Fig. 4. Example of conical Inductor mounted on the sample holder for experimental characterization. In this case the geometry of the inductor correspond to $\alpha = 60°, N = 3, p = 5mm$ and $w = 2.5mm$.

The measurements have been taken using a E5071C 4 port Vector Network Analyzer from Keysight Technologies Inc. The comparison between the simulation and the experimental results for the equivalent inductance and the quality factor of the sample is depicted in figure 5. There, a very good agreement between both data sets is demonstrated.

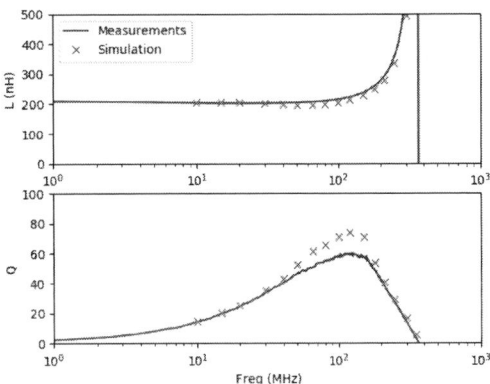

Fig. 5. Equivalent Inductance and Quality Factor of a Conical Inductor with the following geometrical parameters: $\alpha = 60°$, $N = 5$, $p = 3mm$ and $w = 1.5mm$. Continuous lines correspond to experimental measuremets and symbols to simulation results.

The small discrepancies between simulations and measurements for the quality factor, are probably due to the maximum precision about determining the thickness of the metal layer, which is not homogeneous.

V. CONCLUSION

In this work, a full-3D fabrication process for passive components and circuits is presented. The metallization stage has been characterized, where resistivities around $9-20\mu\Omega$ are achieved, which are very good results comparing to the SoA. Finally, an application on RF broadband conical inductors demonstrate the capabilities of this process.

REFERENCES

[1] M. Attaran, "The rise of 3-d printing: the advantages of additive manufacturing over traditional manufacturing," *Business Horizons*, vol. 60, no. 5, pp. 677–688, 2017, ISSN: 00076813. DOI: 10.1016/j.bushor.2017.05.011. [Online]. Available: http://dx.doi.org/10.1016/j.bushor.2017.05.011.

[2] P. C. Priarone and G. Ingarao, "Towards criteria for sustainable process selection: on the modelling of pure subtractive versus additive/subtractive integrated manufacturing approaches," *Journal of Cleaner Production*, vol. 144, pp. 57–68, 2017, ISSN: 09596526. DOI: 10.1016/j.jclepro.2016.12.165. [Online]. Available: //dx.doi.org/10.1016/j.jclepro.2016.12.165.

[3] J. Watson and K. Taminger, "A decision-support model for selecting additive manufacturing versus subtractive manufacturing based on energy consumption," *Journal of Cleaner Production*, 2015, ISSN: 09596526. DOI: 10.1016/j.jclepro.2015.12.009. [Online]. Available: http://linkinghub.elsevier.com/retrieve/pii/S0959652615018247.

[4] A. J. Lopes, I. H. Lee, E. Macdonald, R. Quintana, and R. Wicker, "Laser curing of silver-based conductive inks for in situ 3d structural electronics fabrication in stereolithography," *Journal of Materials Processing Technology*, vol. 214, no. 9, pp. 1935–1945, 2014, ISSN: 09240136. DOI: 10.1016/j.jmatprotec.2014.04.009. [Online]. Available: http://dx.doi.org/10.1016/j.jmatprotec.2014.04.009.

[5] E. Saleh, F. Zhang, Y. He, J. Vaithilingam, J. L. Fernandez, R. Wildman, I. Ashcroft, R. Hague, P. Dickens, and C. Tuck, "3d inkjet printing of electronics using uv conversion," *Advanced Materials Technologies*, vol. 2, no. 10, p. 1700134, 2017, ISSN: 2365709X. DOI: 10.1002/admt.201700134. [Online]. Available: http://doi.wiley.com/10.1002/admt.201700134.

[6] G. Saada, M. Layani, A. Chernevousky, and S. Magdassi, "Hydroprinting conductive patterns onto 3d structures," *Advanced Materials Technologies*, vol. 2, no. 5, p. 1600289, 2017, ISSN: 2365709X. DOI: 10.1002/admt.201600289. [Online]. Available: http://doi.wiley.com/10.1002/admt.201600289.

[7] M. Mirzaee, S. Noghanian, L. Wiest, and I. Chang, "Developing flexible 3d printed antenna using conductive abs materials," *IEEE Antennas and Propagation Society, AP-S International Symposium (Digest)*, vol. 2015-Octob, pp. 1308–1309, 2015, ISSN: 15223965. DOI: 10.1109/APS.2015.7305043.

[8] J. J. Adams, E. B. Duoss, T. F. Malkowski, M. J. Motala, B. Y. Ahn, R. G. Nuzzo, J. T. Bernhard, and J. A. Lewis, "Conformal printing of electrically small antennas on three-dimensional surfaces," *Advanced Materials*, vol. 23, no. 11, pp. 1335–1340, 2011, ISSN: 09359648. DOI: 10.1002/adma.201003734.

[9] H. Tsang, J. Barton, R. Rumpf, and C. Garcia, "Effects of extreme surface roughness on 3d printed horn antenna," *Electronics Letters*, vol. 49, no. 12, pp. 734–736, 2013, ISSN: 0013-5194. DOI: 10.1049/el.2013.1528. [Online]. Available: http://digital-library.theiet.org/content/journals/10.1049/el.2013.1528.

[10] M. Liang, J. Wu, and X. Yu, "3d printing technology for rf and thz antennas - ieee xplore document," *Ieee*, no. 2, pp. 536–537, 2016. [Online]. Available: http://ieeexplore.ieee.org/document/7821190/.

[11] Neotech AMT, *Neotech amt*. [Online]. Available: http://www.neotech-amt.com/.

[12] J. Kerns, "Who's who in 3d printing electronics," pp. 1–4, 2017. [Online]. Available: http://www.machinedesign.com/3d-printing/who-s-who-3d-printing-electronics.

[13] A. Salas-Barenys, N. Vidal, J. Sieiro, J. López-Villegas, B. Medina-Rodríguez, and F. Ramos, "Full-3d printed electronics process using stereolitography and electroless plating," *Submitted for publication*, 2018.

[14] M. Paunovic, "Electroless deposition of copper," *Modern Electroplating*, vol. 1, pp. 433–446, 2010. DOI: 10.1002/9780470602638.

[15] L. J. van der Pauw, *A method of measuring the resistivity and hall coefficient on lamellae of arbitrary shape*, 1958. DOI: 537.723.1:53.081.7+538.632:083.9. arXiv: arXiv:1011.1669v3. [Online]. Available: http://www.mendeley.com/research/method-measuring-resistivity-hall-coefficient-lamellae-arbitrary-shape/.

[16] D. W. Koon, M. Hemanová, and J. Náhlík, "Electrical conductance sensitivity functions for square and circular cloverleaf van der pauw geometries," *Measurement Science and Technology*, vol. 26, no. 11, 2015, ISSN: 13616501. DOI: 10.1088/0957-0233/26/11/115004.

[17] T. A. Winslow, "Conical inductors for broadband applications," *IEEE Microwave Magazine*, vol. 6, no. 1, pp. 68–72, 2005, ISSN: 15273342. DOI: 10.1109/MMW.2005.1418000.

[18] J. López-Villegas, N. Vidal, J. Sieiro, A. Salas-Barenys, B. Medina, and F. Ramos, "Study of 3d printed conical inductors for broadband rf applications," *Submitted for publication*,

PRIME 2018, Prague, Czech Republic

Session: Emerging and Non-CMOS Technologies

# Modeling and Simulation of Novel GaN-based Light Emitting Transistor for Display Applications

Sang Myung Lee
*Electrical and Electronic Engineering*
*Yonsei University*
Seoul, Republic of Korea
leesm508@yonsei.ac.kr

Ilgu Yun
*Electrical and Electronic Engineering*
*Yonsei University*
Seoul, Republic of Korea
iyun@yonsei.ac.kr

*Abstract*—**For the research of next-generation displays, technology of shrink device size is the most attractive and important technology. It is possible to manufacture high-performance display products by using high integrated devices such as mobile application. However, there is a certain limitation to the downsizing technology. Therefore, new device synthesis techniques are becoming important. In this paper, we propose a device design that combines inorganic material based light-emitting diode (LED) and thin-film transistor (TFT). By integrating the LED and TFT devices into one region, it is possible to highly integrate the devices, which can greatly reduce the size of the entire device. To investigate a possibility of device implementation, technology computer-aided design (TCAD) simulation is used. After that, an optical and electrical characteristic of the device are analyzed. Finally, the light-emitting transistor (LET) is proposed.**

*Keywords*—*inorganic LET, TCAD simulation and modeling, device prototype implementation*

## I. INTRODUCTION

For a next generation display, development of micro- or nano-scaled light-emitting diode (LED) is important [1]. A micro size LED can be used for augmented reality (AR) or virtual reality (VR) applications. Also, a thin-film transistor (TFT), which operate as a switch and drive device in display, is core device. For a device size shrinking technology, many researchers studied the various methodology such as nanowire LED [2, 3]. However, shrinking technology of device size also limited due to the material characteristics or difficulty of the fabrications. Therefore, device synthesis technology become important issue. A organic light-emitting transistor (OLET), which is synthesis technology of the organic based LED and TFT, is one of the attractive device [4]. However, organic material based LED have a reliability issues such as burn-in effect and fabrication complexity. Due to the burn-in effect, brightness is limited when using an organic based device for display [5]. In addition, the overall price increases due to the complexity of the process. In this paper, we suggest inorganic material based light-emitting transistor instead of organic material. A inorganic material based LED have not burn-in effect. Therefore, when applying to a display, there is not limitation on the brightness. Also, inorganic material such as gallium nitride (GaN) based LED is very classical. Therefore, many other GaN based devices have already been developed and many studies on device fabrication and physical properties have been made such as high electron mobility transistor

(HEMT) device [6, 7]. As a result, it is very easy to adapted in industry. For the research, 2D based technology computer-aided design (TCAD) tool is used for a device simulation. By using the TCAD, it is possible to analyze the characteristics of the device before the actual device manufacturing process, thereby greatly reducing time and cost is possible through device optimization. In this simulation, a structure of the inorganic based light-emitting transistor (LET) is proposed. After that, an optical and electrical characteristics of the inorganic based LET are analyzed. Using the simulation based study, we can see the feasibility of implementing inorganic based LET.

## II. MODELING SCHEME

### A. Light-emitting transistor structure scheme

In this research, Atlas TCAD tool, which developed by Silvaco International Incorporation, is used for a 2D device simulation. Initially, a classical GaN based LED is designed. The single-quantum wells (SQW) layer of the LED is then utilized as a channel of the TFT. Finally, the anode of the LED and the source of the TFT are connected. A designed LET schematics are shown in Fig. 1.

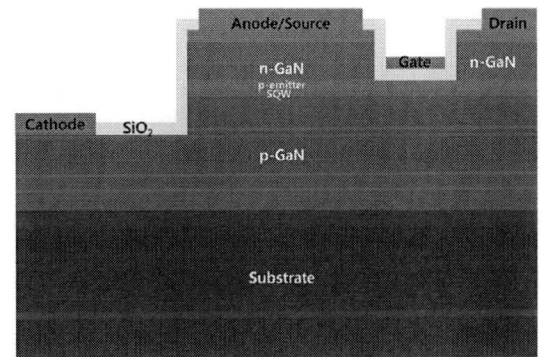

Fig. 1 Schematic structures of LET.

A simulated GaN based LED is blue light type. Therefore, a sapphire is used as a substrate layer. A n-type GaN and p-type GaN are used as a n-type and p-type contact. Here, the aluminum gallium nitride (AlGaN) is used as a p-emitter, and the indium gallium nitride (InGaN) is used as a SQW. In this research, LET device is designed to share the SQW layer with

978-1-5386-5388-3/18 $31.00 © 2018 IEEE        269

TABLE I. MODEL PARAMETERS IN DEVICE SIMULATION

| Epi-layer | Material | Type | Thickness (nm) | Doping concentration (cm$^{-3}$) |
|---|---|---|---|---|
| p-contact | GaN | p | 500 | $2\times10^{17}$ |
| p-emitter | Al$_{0.2}$GaN | p | 100 | $2\times10^{17}$ |
| SQW | In$_{0.2}$GaN | - | 3 | - |
| n-contact | GaN | n | 2000 | $2\times10^{17}$ |
| n-buffer | GaN | - | 30 | - |
| substrate | Sapphire | - | 6000 | - |

TFT areas, and this SQW layer also operates as a channel in the TFT. For a insulator layer, silicon dioxide (SiO$_2$) is used and the indium tin oxide (ITO) is used as electrodes in this simulation.

### B. Model parameters

In the LET device simulation, the total SQW length is 2.5 µm and the channel length of TFT area is 1 µm. The SiO$_2$ layer thickness is 100 nm, and the total device length is 3.5 µm. The doping concentration of p-contact, p-emitter, and n-contact is $2\times10^{17}$ cm$^{-3}$. The other epi-layers are not doped. To design a blue LED, materials of the SQW layer are selected as In$_{0.2}$GaN with 3 nm thickness. The specific parameters are shown in Table 1.

For a device simulation, various simulation models are used [8]. For the drift-diffusion simulation, the k·p model is used. Also, the incomplete ionization model is used for accounts for the dopant freeze-out. For the band to band recombination, the optical model is used. The Lorentz distribution is used for a diffusive reflection. In addition, the spontaneous model is used for include the radiative recombination rates derived from Chuang model into the drift diffusion.

### C. Operating bias conditions

Usually, the electrode at the anode side is positively biased and that of the cathode side is negatively biased. However, the increased number of electrode and bias conditions add complexity to device circuit design and have a negative effect on commercialization. Therefore, we tried to reduce the number of the electrode and bias conditions as much as possible. In order to reduced electrode number, a cathode and gate electrode are swept from 5 V to -5 V at the same time.

TABLE II. OPERATING BIAS CONDITIONS IN SIMULATION

| Electrode | Operating section | | |
|---|---|---|---|
| | *LED only* | *TFT only* | *LED+TFT (LET)* |
| Cathode | -5 V | 0 V | 5 ~ -5 V |
| Anode/Source | 0 V | 0 V | 0 V |
| Gate | 0 V | 5 ~ -5 V | 5 ~ -5 V |
| Drain | 0 V | 2.1 V | 2.1 V |

In order to confirm the effect of the each area, the simulation was performed by dividing each area. Initially, only LED region was simulated and then only TFT region was simulated. Finally, the entire device was simulated.

To operate LED region only, the anode, gate and drain voltages are set to 0 V, and the cathode voltage is set to – 5V. In addition, the cathode and anode voltages are set to 0 V to test the TFT region only and the drain voltage is set to 2.1V, and gate voltage is swept from 5 V to – 5 V. Finally, the entire LET was simulated with combined the cathode and gate voltages. The measurement conditions are summarized in Table II.

### III. RESULT AND DISCUSSION

#### A. LED characteristics

The simulated device is operated with different bias conditions to confirm LED area characteristics. To analyze a optical characteristics of the LED, the power spectral density per wavelength is used. The simulated power spectral density results are shown in Fig. 2.

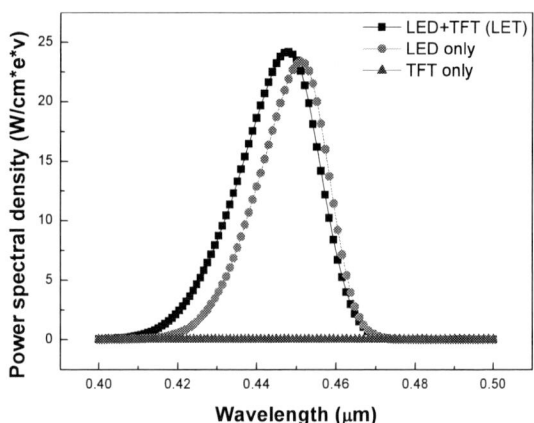

Fig. 2 Simulated power spectral density results with wavelength

When only LED area is activated, the maximum power spectral density results are shown at the wavelength of 450 nm. However, when only the TFT region is activated, the drain current is not enough to drive the LED. Therefore, the power spectral density results are failed. Finally, when the both LED and TFT are activated, the power spectral density shows similar tendency to the 'LED only' case. In this case, the maximum power spectral density results are shown at the wavelength of 448 nm. In this research, the blue type LED is simulated. Therefore, it is appropriate for blue light characteristics with the wavelength of 430 ~ 480 nm.

### B. TFT characteristics

In order to confirm the TFT characteristics in the LET device, only the TFT region is simulated separately, and the result is compared with the entire LET device simulated result. To analyze the TFT characteristics of the LET, the transfer curve is used. The simulated transfer curve results are shown in Fig. 3.

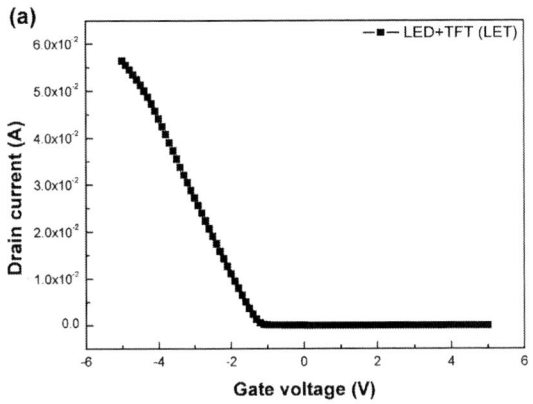

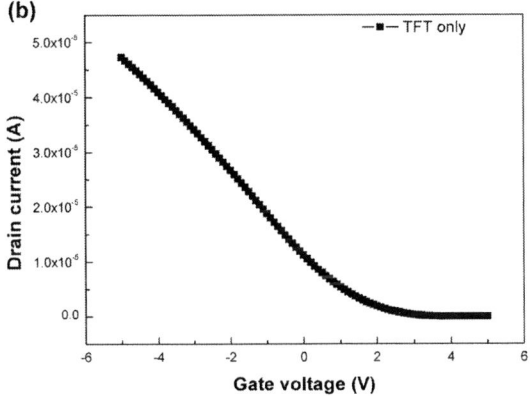

Fig. 3 Simulated transfer curve results with (a) LET, and (b) TFT only case.

In the LET operation case, the on current and the threshold voltage is significantly increased compare to the only TFT operation case. According to the LET operation, the bias is applied to the cathode and the gate at the same time. As a result, the number of carriers of the channel, which is SQW, is increased, and the total current is also increased. Also, the threshold voltage is related with insulator capacitance as (1),

$$V_{th} = V_{gs} - \frac{I_{ds}}{G_m} - \frac{V_{ds}}{2} \qquad (1)$$

where $V_{th}$ is the threshold voltage, $I_{ds}$ is the drain current, $V_{ds}$ is the drain voltage, and $G_m$ is the transconductance. Also, $G_m$ can be calculated with (2)

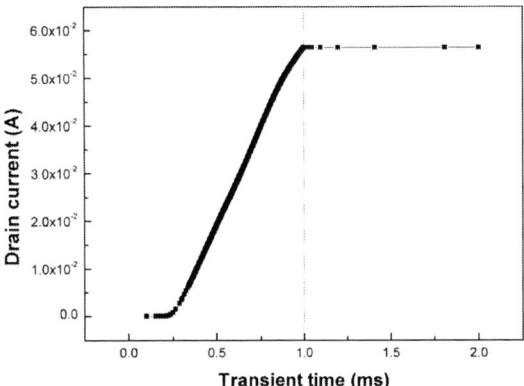

Fig. 4 Simulated transient response results of LET device.

$$G_m = \mu_{lin} \times C_{ox} \frac{W}{L} V_{ds} \qquad (2)$$

where $\mu_{lin}$ is the mobility on a linear region, $C_{ox}$ is the capacitance of the gate insulator, $W$ is the channel width and $L$ is the channel length.

Therefore, the threshold voltage also increased as the capacitance increases. The subthreshold swing value of the TFT only case was three times larger than LET case. In the result, the characteristics of the TFT are improved except for the threshold voltage when operating LET.

### C. Transient time result

In order to apply the device to the circuit level, the transient time is also one of the important parameters. The simulated transient response graph is shown in Fig. 4.

Fig. 4 shows the transient response of the drain current. The drain current was 56.41 mA, when the gate voltage and cathode voltage reached -5V. The time to reach the same drain current was 1 ms. A output rise time is the time from the value of 10% of the final value to 90%. A result of the output time is 0.567 ms in this simulation.

Based on the optical and electrical simulation results, it show that the proposed LET device can replace LED devices and TFT devices for display application and the higher performance display can be implemented by using the proposed LET device.

## IV. Conclusion

In this paper, the feasibility of the inorganic based light-emitting transistor device implementation was investigated. For this study, TCAD modeling and simulation scheme was used and the SQW was implemented in the LED area using quantum mechanics. In addition, the device size was reduced through the combination of LED and TFT by using SQW as a TFT channel simultaneously. For verification of the device operation, both LED and TFT regions were activated separately, and confirmed that each device works well. The gate of the TFT and the cathode of the LED were driven by applying the voltage at the same time. Compare with the case of operating LED only, both case showed blue light, and it can be investigated using the simulated power spectral density result. However, it is not enough to activated the LED with drain current only. Compared with the case of operating TFT only, the drain current was significantly increased. Based on the result, a feasibility of the device size reduction with proposed inorganic based LET was confirmed. However, a 2D based simulation result had a certain limitation especially on the device structures. Therefore, the structure design, electrical and optical characteristics of LET device using 3D simulation should be studied further and they remain as a future work.

## Acknowledgment

This work was supported by Institute of BioMed-IT, Energy-IT and Smart-IT Technology (BEST), a Brain Korea 21 plus program, Yonsei University. Also, the authors would like to thank J.H. Kim of LG Display for supporting with concept idea.

## References

[1] J. Yoon, S.M. Lee, D. kang, M.A. Meitl, C.A. Bower, and J.A. Rogers, "Heterogeneously integrated optoelectronic devices enabled by micro - transfer printing," Adv. Opt. Mater. Germany, vol. 3(10), pp. 1313-1335, September 2015.

[2] S.D. Hersee, M. Fairchild, A.K. Rishinaramangalam, M.S. Ferdous, L. Zhang, P.M. Varangis, B.S. Swartzentruber, and A.A. Talin, "GaN nanowire light emitting diodes based on templated and scalable nanowire growth process," Electron. Lett. vol. 45(1), pp. 75-76, January 2009.

[3] M.S. Gudiksen, L.J. Lauhon, J. Wang, D.C. Smith, and C.M. Lieber, "Growth of nanowire superlattice structures for nanoscale photonics and electronics," Nature, vol. 415(6872) , pp. 617, February 2002.

[4] A. Hepp, H. Heil, W. Weise, M. Ahles, R. Schmechel, and H. von Seggern, "Light-emitting field-effect transistor based on a tetracene thin film," Phys. Rev. Lett. New York, vol. 91(15), pp. 157406, October 2003.

[5] A. Buckley, Organic light-emitting diodes (OLEDs): Materials, devices and applications, Elsevier, 2013.

[6] T. Fujii, Y. Gao, R. Sharma, E. Hu, S. DenBaars, and S. Nakamura, "Increase in the extraction efficiency of GaN-based light-emitting diodes via surface roughening," Appl. Phys. Lett. New York, vol. 84(6), pp. 855-857, February 2004.

[7] U.K. Mishra, P. Parikh, and Y.-F. Wu, "AlGaN/GaN HEMTs-an overview of device operation and applications," Proc. IEEE. vol. 90(6), pp. 1022-1031, November 2002.

[8] Silvaco Int., ATLAS User's Manual, Santa Clara, CA, ver. 5, 2011..

# Improving Deep Learning with a customizable GPU-like FPGA-based accelerator

Mirko Gagliardi, Edoardo Fusella, Alessandro Cilardo

Department of Electrical Engineering and Information Technologies, University of Naples Federico II

via Claudio 21, 80125 Napoli, Italy, Email: mirko.gagliardi@unina.it

*Abstract*—An ever increasing number of challenging applications are being approached using Deep Learning, obtaining impressive results in a variety of different domains. However, state-of-the-art accuracy requires deep neural networks with a larger number of layers and a huge number of different filters with millions of weights. GPU- and FPGA-based architectures have been proposed as a possible solution for facing this enormous demand of computing resources. In this paper, we investigate the adoption of different architectural features, i.e. SIMD paradigm, multithreading, and non-coherent on-chip memory for Deep Learning oriented FPGA-based accelerator designs. Experimental results on a Xilinx Virtex-7 FPGA show that the SIMD paradigm and multithreading can lead to an improvement in the execution time up to $5\times$ and $3.5\times$, respectively. A further enhancement up to $1.75\times$ can be obtained using a non-coherent on-chip memory.

## I. INTRODUCTION AND RELATED WORK

The emerging wave of the Big Data [1] is paving the way for the widespread adoption of Deep Learning techniques in diverse application domains including image recognition [2], sound processing [3], medical systems [4], gaming [5], and others. However, despite the huge potential of Deep Learning, most of these algorithms rely on a large number of performance-hungry convolutions limiting the usability of these techniques. In addition, Deep Neural Networks (DNNs) require a training phase that is a very compute intensive task. For instance, training a popular architecture like, e.g. GoogLeNet [6], can easily take several days on a standard GPU. Because of these requirements, the applicability of Deep Learning is becoming increasingly performance- or power-constrained.

Not surprisingly, the industry and academia are continuously introducing new architectures dictating the evolution of Deep Learning techniques. First-generation solutions consist of large-scale distributed systems comprised of tens of thousands of CPU cores [7]. However, the growing demand for high-parallel energy-efficient architectures has led to an increasing interest in GPUs and FPGAs [8], [9], [10]. For example, many entries in the annual ImageNet Large Scale Visual Recognition Challenge (ILSVRC) [11] use GPUs and FPGAs to implement DNNs. FPGAs are an attractive alternative and provide an intermediate point between Application-Specific Integrated Circuits (ASIC) and standard GPUs, enabling higher efficiency even compared to high-end GPUs [12]. In addition, the higher flexibility allows their use in different compute problems. There is thus a tradeoff between power-hungry high-performance GPUs and energy-efficient application-specific solutions.

In this paper we investigate the adoption of different architectural features, i.e. SIMD paradigm, multithreading, and non-coherent on-chip memory for Deep Learning oriented FPGA-based accelerator designs. We designed and implemented a customizable GPU-like SIMD architecture as a solution to support architecture-level exploration for Deep Learning oriented systems. Architectural customization plays a key role, as it enables unprecedented levels of resource-efficiency compared to GPUs. The accelerator was synthesized into a Xilinx Virtex-7 2000T XC7V2000T FPGA chip. Experimental results show that such a customization leads to significant improvements over non-customized architectures.

## II. GPU-LIKE ARCHITECTURE

This work relies on an experimental platform, called nu+, providing a parameterizable GPU-like architecture inspired by modern GPUs, yet exposing full customization capabilities for architectural exploration. The heart of the platform is a RISC in-order core oriented to highly data-parallel kernels with a lightweight control infrastructure, shown in Figure 1. Most of its resources are dedicated to computation-intensive operations on massive datasets. Such accelerator blends together a hardware multithreading paradigm with a vector processor model. Each hardware thread has private internal resources such as PC, register file, and control registers along with a private memory stack, although all threads share the same compute units, L1 cache and an on-chip non-coherent memory. The thread control unit implements an interleaved multithreading scheduling in a fine-grain way. An internal round robin arbiter issues instructions for different threads in a fair mode after every cycle with a low architectural impact. Execution datapaths and register files are designed to exploit data-level parallelism. The architecture implements an instruction set containing instructions that operate on arrays of data. Computational units are organized in hardware vector lanes, with each scalar operator being instantiated $N$ times. Dually, each thread is equipped with a vectorial register file, where each register can store up to $N$ scalar data in order to satisfy the execution pipeline data throughput. Such a data parallelism allows each thread to perform SIMD operations on $N$ independent data simultaneously.

The proposed GPU-like architecture has an $n$-way set-associative write-back L1 cache strictly coupled with a light

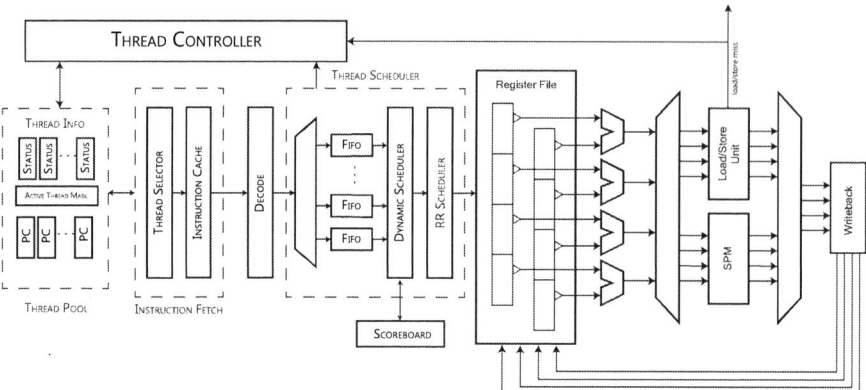

Fig. 1. Simplified overview of the developed GPU-like accelerator. This figure highlights both data and thread level parallelism. Beside memory configuration parameters, this architecture has different customization levels, such as the number of hardware lanes and the number of hardware threads implemented.

cache controller which implements a simple valid/invalid coherence mechanism. Such cache controller handles misses and memory transactions and it also provides both request serialization and merging mechanisms in order to correctly manage concurrent requests from different threads. The cache line width matches the internal hardware lanes capability, thus a read memory request loads $N$ scalar data from main memory and stores them into a vectorial register at once, minimizing requests and exploiting the internal parallelism of the GPU-like accelerator.

As in general purpose platforms, the performance of custom accelerators is also critically dependent on data movement and memory accesses. In that respect, hardware coherence mechanisms introduce both architectural and data management overheads, which are not always necessary in some applications. Many modern parallel architectures utilize fast non-coherent on-chip memories, called *scratchpad memories* (SPMs). Since NVIDIA Fermi family, GPUs are equipped with this kind of memories, that are intensively used to facilitate communication across threads and to store partial outputs or temporary data that are not requested to be synchronized back into main memory.

The nu+ core supports such kind of high-throughput non-coherent scratchpad memory, which is divided in a parameterized number of banks based on a user-configurable mapping function in order to support multiple memory accesses. The scratchpad memory is organized in multiple independently-accessible memory banks providing a high data access parallelism. Therefore, if all memory accesses request data mapped to different banks, they can be handled in parallel. However, when multiple requests are made for data within the same bank, conflicts occur and a resolution logic handles and serializes each request resulting in a significantly performance loss. In fact, whenever an $n$-way conflict is detected, such a serialization logic notifies to the GPU-like accelerator control logic that the memory is not able to receive any further request, then it splits the conflicting requests into $n$ conflict-free sub-requests issued serially in the next $n$ cycles.

The implemented architecture comes with a toolchain based on the LLVM project and includes a custom version of the Clang frontend and a native nu+ backend. The Clang frontend allows users to compile traditional C/C++ source code in a fast way and with a low memory usage. On the other hand, the toolchain is deeply customized for exploiting the core internal data parallelism and reaching the maximum throughput. The compiler has a complete vision of the SIMD nature of the datapath. It supports custom vector types, thus standard arithmetic and bitwise operators are available for both scalar and vector operations. Furthermore, the custom version of Clang supports ad-hoc builtin functions that are required to fully exploit target specific features, such as thread synchronization and special SIMD operations.

Combining a non-coherent on-chip memory approach, high data parallelism, and a fine-grain thread control, this GPU-like accelerator provides a significant speed-up in compute-intensive and data demanding workloads.

## III. CONVOLUTION ALGORITHM

Deep Learning is a class of machine learning algorithms using convolutional neural networks (CNN) which are inspired by the behavior of optic nerves. Deep Learning gives state-of-the-art accuracy for many computer vision tasks, such as image classification and image search engine in data centers.

CNN employs a feedforward process for recognition and a backward path for training. Consequently a typical CNN is composed of multiple computation layers, and the output $y$ is the sum of multiple different convolutions between the input $x$ and the filter $k$:

$$y[n] = x[n] \cdot k[n] = \sum_k x[n] \cdot k[n-k]$$

Our work focuses on the exploration of different architectural features in a custom GPU-like accelerator targeted at convolution operations. In fact, these account for over 90%

of the processing in CNNs for both inference/testing and training [13].

The pseudo code of a convolution with a $K \times K$ filter with no stride, bi-dimensional input and output matrices, respectively of $N \times N$ and $M \times M$ where $M = N - K$, can be written as in the following listing:

```
for(row = 0; row < M; row++)
  for(col = 0; col < M; col++)
    for(krow = 0; krow < K; krow++)
      for(kcol = 0; kcol < K; kcol++)
        y[row][col] += k[krow][kcol] *
                       x[row + krow][col + kcol];
```

This scalar single-thread version of the convolution algorithm has been adapted to our target architecture exploiting both thread and data level parallelism. Output matrix row calculations are equally spanned across all threads, i.e., for each thread, the outer loop starts with the thread ID (*thid*) and increments by the number of threads (*thnumb*). On the other hand, both input and output matrices are organized in target specific vector types, becoming vectors of vectors. Such an organization results in a distribution of the $M$ column partial results on the $M$ hardware lanes; each thread calculates $M$ partial results every cycle. The vectorization makes the second cycle unnecessary. The inner cycle, however, scrolls the input matrix in the scalar version. This can be replaced by a vectorial shift operation supported by the target architecture on the input row, which shifts each scalar element inside the hardware vector by $n$ positions. The resulting pseudo code of the algorithm optimized for a GPU-like accelerator, can be written as in the following listing:

```
for(row = thid; row < M; row += thnumb)
  for(krow = 0; krow < K; krow++)
    for(kcol = 0; kcol < K; kcol++) {
      y[row] += k[krow][kcol] *
                x[row + krow];
      x[row + krow] = x[row+krow] << 1;
    }
```

## IV. EVALUATION

We carried out our experiments on a proFPGA MB-4M FPGA board by ProDesign, equipped with one Xilinx Virtex-7 2000T XC7V2000T FPGA. The GPU-like accelerator has been developed in SystemVerilog hardware description language (HDL) and synthesized using the Vivado design suite provided by Xilinx. The design has been validated with the Verilator RTL simulator tool [14] and with an in-house event-driven cycle-accurate emulator. Finally, the convolution algorithm was written in C and compiled using our toolchain.

The convolutions were performed on $16 \times 16$, $32 \times 32$, and $64 \times 64$ input images with filter kernels of size between $3 \times 3$ and $7 \times 7$. We first carried out a set of experiments to assess speedup over a naive scalar single-thread implementation and estimate the performance boost of the SIMD paradigm. Figure 2 depicts the results. By sweeping the size of the input image from $16 \times 16$ to $64 \times 64$, we observe a great increase in

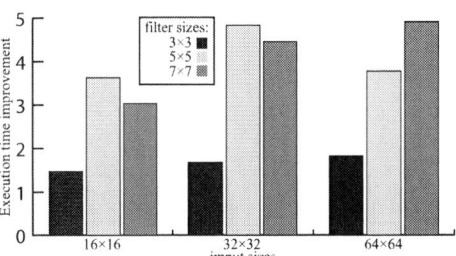

Fig. 2. Speedup over naive scalar single-thread implementation on $16 \times 16$, $32 \times 32$, and $64 \times 64$ input images with $3 \times 3$, $5 \times 5$, and $7 \times 7$ filter kernels.

the effective acceleration (up to 5×). This is due to the higher number of multiply-add operations that need to be performed. Unsurprisingly, small input images with filter kernels of size $7 \times 7$ have a reduced speedup due to the unbalanced sizes of the images and the filter causing a suboptimal use of the hardware lanes.

Then, in a second set of experiments we evaluated the benefits of using multithreading. Figure 3 shows the speedup over a single-thread implementation. Two threads ensure a speedup between 1.3× and 2×, while a higher number of threads leads to better performance (up to 3.5×). However, this trend is not constant since in case of small images and/or small filters, a higher number of threads may be useless. This is because hardware multithreading involves some overhead for handling the different stacks (one for each thread) and for thread scheduling and synchronization. For instance, in case of a $16 \times 16$ input image and filters with a size of $3 \times 3$ and $7 \times 7$, the optimum number of threads is respectively two and six. This is because convolutions performed on a $16 \times 16$ input image with filter kernels of size $7 \times 7$ require 2.6 more arithmetic operations than in case of $3 \times 3$ filters.

Finally, in the last set of experiments, we evaluated the benefits of using the scratchpad memory. The results in terms of speedup over an accelerator with a standard memory subsystem are summarized in Figure 4. In case of smaller filters, we observe better results since there is a lower need to swap data between the scratchpad and the main memory. In general, the achieved speedup scales up with the number of threads up to 1.75×. This is due to the higher efficiency of the scratchpad memory, which does not implement the coherence functionalities of traditional cache memories, as well as the lower number of cache misses when using the scratchpad memory (up to 30%).

## V. CONCLUSIONS

In this work, we investigated various architectural features for the definition of an innovative GPU-like accelerator targeted at Deep Learning. In particular, we evaluated how those features impact the performance of convolutions, the dominant operation in Deep Learning applications. Thread level parallelism and SIMD operation achieve a great increase in the acceleration efficiency reaching respectively a speed-up of 3.5× and 5× over a scalar single-thread implementation of

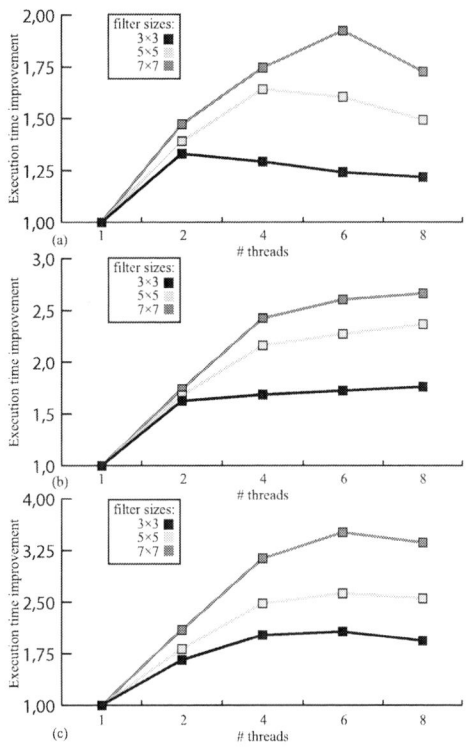

Fig. 3. Speedup over single-thread implementation when varying the number of threads on $16 \times 16$, $32 \times 32$, and $64 \times 64$ input images with $3 \times 3$, $5 \times 5$, and $7 \times 7$ filter kernels.

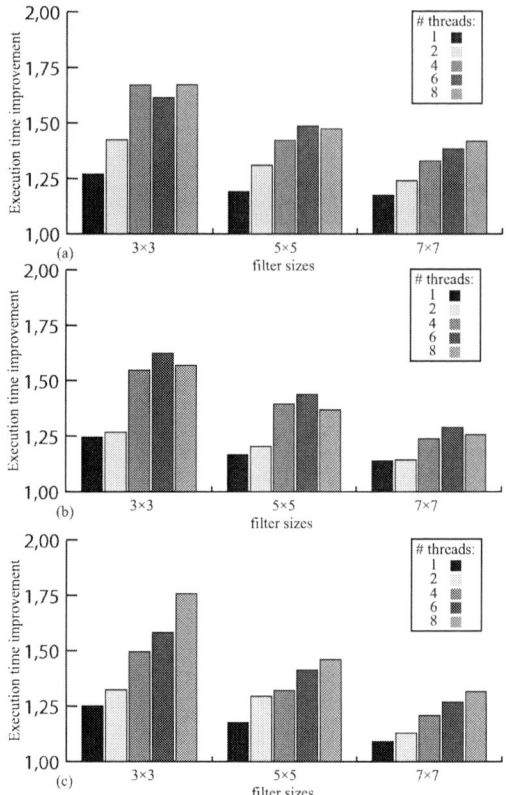

Fig. 4. The speedup achieved using scratchpad memory when varying the number of threads on $16 \times 16$, $32 \times 32$, and $64 \times 64$ input images with $3 \times 3$, $5 \times 5$, and $7 \times 7$ filter kernels.

the convolution algorithm. The use of non-coherent on-chip memory can further enhance performance to some extent, due to the increased locality and lower number of cache misses. In conclusions, the findings of our work may be particularly impactful for driving the long-term evolution of current accelerator architectures towards improved specialization and workload-specific customizability.

## REFERENCES

[1] K. Kambatla et al., "Trends in big data analytics," Journal of Parallel and Distributed Computing, vol. 74, no. 7, pp. 2561–2573, 2014.

[2] K. He et al., "Deep residual learning for image recognition," in Proceedings of the IEEE conference on computer vision and pattern recognition, 2016, pp. 770–778.

[3] J. Salamon and J. P. Bello, "Deep convolutional neural networks and data augmentation for environmental sound classification," IEEE Signal Processing Letters, vol. 24, no. 3, pp. 279–283, 2017.

[4] A. Esteva et al., "Dermatologist-level classification of skin cancer with deep neural networks," Nature, vol. 542, no. 7639, p. 115, 2017.

[5] D. Silver et al., "Mastering the game of go with deep neural networks and tree search," nature, vol. 529, no. 7587, pp. 484–489, 2016.

[6] C. Szegedy et al., "Going deeper with convolutions." Cvpr, 2015.

[7] J. Dean et al., "Large scale distributed deep networks," in Advances in neural information processing systems, 2012, pp. 1223–1231.

[8] Y. Chen et al., "Dadiannao: A machine-learning supercomputer," in Proceedings of the 47th Annual IEEE/ACM International Symposium on Microarchitecture. IEEE Computer Society, 2014, pp. 609–622.

[9] C. Farabet et al., "Neuflow: A runtime reconfigurable dataflow processor for vision," in Computer Vision and Pattern Recognition Workshops (CVPRW), 2011 IEEE Computer Society Conference on. IEEE, 2011, pp. 109–116.

[10] C. Zhang et al., "Optimizing fpga-based accelerator design for deep convolutional neural networks," in Proceedings of the 2015 ACM/SIGDA International Symposium on Field-Programmable Gate Arrays. ACM, 2015, pp. 161–170.

[11] O. Russakovsky et al., "Imagenet large scale visual recognition challenge," International Journal of Computer Vision, vol. 115, no. 3, pp. 211–252, 2015.

[12] C. Murphy and Y. Fu, "Xilinx all programmable devices: A superior platform for compute-intensive systems," Xilinx White Paper, 2017.

[13] Y.-H. Chen et al., "Eyeriss: An energy-efficient reconfigurable accelerator for deep convolutional neural networks," IEEE Journal of Solid-State Circuits, vol. 52, no. 1, pp. 127–138, 2017.

[14] W. Snyder, P. Wasson, and D. Galbi, "Verilator-convert verilog code to c++/systemc," 2012.

*PRIME 2018, Prague, Czech Republic*

*Session: Emerging and Non-CMOS Technologies*

# Analysis and Verification of Identical-Order Mixed-Matrix Fractional-Order Capacitor Networks

Aslihan Kartci[1], Agamyrat Agambayev[2], Norbert Herencsar[1], Khaled N. Salama[2]

[1]*Faculty of Electrical Engineering and Communication, Brno University of Technology, Brno, Czech Republic*
[2]*King Abdullah University of Science and Technology, Thuwal, Saudi Arabia*
kartci@feec.vutbr.cz, agamyrat.agambayev@kaust.edu.sa, herencsn@feec.vutbr.cz, khaled.salama@kaust.edu.sa

*Abstract*—In the open literature while capacitors are introduced with −90 degrees phase angle, here we described our fabricated polymer composite, mixed matrix, as a fractional-order capacitor (FoC). The effect on phase and pseudo-capacitance using a detailed numerical and experimental study of circuit network connections of three identical-order FoCs is shown. The used devices have excellent feature such as constant phase angle in the frequency range 200 KHz – 20 MHz.

*Keywords*—*constant phase element, fractional-order capacitor, FoC, fractional-order circuit, identical-order connection, stability measurement*

## I. INTRODUCTION

The lossy nature of the dielectric material in capacitors and their electrical conductivity does not allow us to treat them as ideal capacitors since their impedance show a complex frequency-dependent behavior. Due to this fact a fractional-order capacitor (FoC), also called as constant phase element, possess both a real and imaginary impedance part

$$Z(s) = 1 \Big/ \omega^\alpha C_\alpha \left[ \cos\left(\frac{\alpha\pi}{2}\right) + j\sin\left(\frac{\alpha\pi}{2}\right) \right]$$ while its phase is

frequency independent that differs from the series connected resistor and capacitor. However, an ideal capacitor has only an imaginary part [1], as demonstrated in Fig. 1. This is particularly important, if the proposed application requires a configuration using capacitors, where errors accumulate the metrics of the individual components. Thus, applications of FoCs are geared towards engineering in many application areas. Particularly, the research was focused on improvement of peaking in passband of filters [2], increase of oscillation frequency [3], electronic tunability of resonators [4], [5], modelling fractional-order impendance of neural systems [6] or employing a real effective capacitances by using a model consisting a FoC in energy storage devices [7]. Recently, a ferroelectric polymer [8] and carbon nanotube-polymer composite [9] based FOCs were introduced. Their easy integration on printed circuit board and constant phase angle are two main advantages of these fabricated FoCs.

The circuit network connections of FoCs and the order of each element have a crucial role in calculating the equivelant impedance in supercapacitor bank design [7]. Therefore, this paper aims to study the equivalent magnitude, phase, and pseudo-capacitance

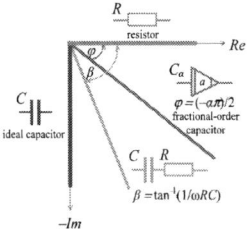

Fig. 1. Nyquist diagram of fractional-order capacitor.

of three identical-order series- and parallel-connected fabricated polymer composite FoCs in the fractional domain.The phase is found to be constant with ±4 degrees phase angle deviation in the measured frequency range of interest 200 KHz – 20 MHz. Moreover, similarly to given single FoCs, the pseudo-capacitance in the parallel connection was higher than those of series-connected FoCs, which agrees well with the related integer-order theory.

## II. FABRICATED FoCS

The behavior of three samples of fabricated ferroelectric polymer (TP, P) and carbon nanotube-polymer (CP) based FoCs of different orders (respectively $\alpha = \{0.69, 0.92, 0.62\}$) was confirmed using an Impedance Analyzer. The magnitude, phase, and pseudo-capacitance at center frequency $f_c = 2$ MHz of all the devices are $\{2.24, 6.43, 1.51\}$ k$\Omega$, $\{-61.86, -82.59, -55.45\}$ degrees, and $\{5.89$ n, $47.52$ p, $28.1$ n$\}$ Farad·sec$^{\alpha-1}$, respectively. An illustration of fabricated PCB compatible fabricated FoC and its cross- sectional SEM image are shown in Fig. 2. Each device contains nine FoCs with identical order and pseudo-capacitance.

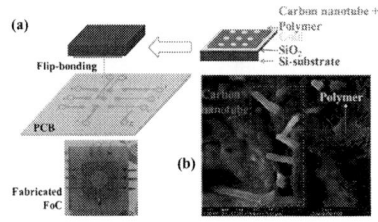

Fig. 2. (a) PCB compatible FoC from carbon nanotube-polymer, (b) its cross-sectional SEM image.

The research described in this paper was financed by the National Sustainability Program under grant LO1401 and by the Czech Science Foundation under grant no. 16-06175S. For the research, an infrastructure of the SIX Center was used.

978-1-5386-5388-3/18 $31.00 © 2018 IEEE

TABLE I. IMPEDANCE, MAGNITUDE AND PHASE DESCRIPTIONS OF THREE SERIES-CONNECTED FOCS.

| $Z_{Ceq,s}(s)$ | $Z_{\alpha_1} + Z_{\alpha_2} + Z_{\alpha_3} = \dfrac{s^{\alpha_1+\alpha_2}\dfrac{1}{C_{\alpha_3}} + s^{\alpha_1+\alpha_3}\dfrac{1}{C_{\alpha_2}} + s^{\alpha_2+\alpha_3}\dfrac{1}{C_{\alpha_1}}}{s^{\alpha_1+\alpha_2+\alpha_3}}$ |
|---|---|
| $\|Z_{Ceq,s}(s)\|$ | $\dfrac{\sqrt{\omega^{2(\alpha_1+\alpha_2)}\dfrac{1}{C_{\alpha_3}^2} + \omega^{2(\alpha_1+\alpha_3)}\dfrac{1}{C_{\alpha_2}^2} + \omega^{2(\alpha_2+\alpha_3)}\dfrac{1}{C_{\alpha_1}^2} + 2\left\{\dfrac{\omega^{2\alpha_1+\alpha_2+\alpha_3}}{C_{\alpha_2}C_{\alpha_3}}\cos\left[(\alpha_2-\alpha_3)\dfrac{\pi}{2}\right] + \dfrac{\omega^{\alpha_1+2\alpha_2+\alpha_3}}{C_{\alpha_1}C_{\alpha_3}}\cos\left[(\alpha_1-\alpha_3)\dfrac{\pi}{2}\right] + \dfrac{\omega^{\alpha_1+\alpha_2+2\alpha_3}}{C_{\alpha_1}C_{\alpha_2}}\cos\left[(\alpha_1-\alpha_2)\dfrac{\pi}{2}\right]\right\}}}{\omega^{\alpha_1+\alpha_2+\alpha_3}}$ |
| $\angle Z_{Ceq,s}(s)$ | $\tan^{-1}\left[\dfrac{\dfrac{\omega^{\alpha_1+\alpha_2}}{C_{\alpha_3}}\sin\left[(\alpha_1+\alpha_2)\dfrac{\pi}{2}\right] + \dfrac{\omega^{\alpha_1+\alpha_3}}{C_{\alpha_2}}\sin\left[(\alpha_1+\alpha_3)\dfrac{\pi}{2}\right] + \dfrac{\omega^{\alpha_2+\alpha_3}}{C_{\alpha_1}}\sin\left[(\alpha_2+\alpha_3)\dfrac{\pi}{2}\right]}{\dfrac{\omega^{\alpha_1+\alpha_2}}{C_{\alpha_3}}\cos(\alpha_1+\alpha_2) + \dfrac{\omega^{\alpha_1+\alpha_3}}{C_{\alpha_2}}\cos\left[(\alpha_1+\alpha_3)\dfrac{\pi}{2}\right] + \dfrac{\omega^{\alpha_2+\alpha_3}}{C_{\alpha_1}}\cos\left[(\alpha_2+\alpha_3)\dfrac{\pi}{2}\right]}\right] - \left[(\alpha_1+\alpha_2+\alpha_3)\dfrac{\pi}{2}\right]$ |

## III. CASE STUDIES AND THEIR EXPERIMENTAL VERIFICATION

This section introduces to the literature specific case studies based on general approach previously presented by the authors [10]. In this paper for the first time three identical-order solid-state passive FoCs are connected in series, parallel, and their special inter-connection. During experiments via impedance analyzer, the devices introduced in Section II were used with applied 500 mV AC voltage in the frequency range 200 KHz –20 MHz. A sinsoidal input of frequency of 1 MHz was applied, while the common node of FoCs was grounded ($V_g = 0$ V).

### A. Connection of FoCs in Series

Considering three series-connected FoCs, as shown in Figs. 3, and assuming each with orders $\alpha_1$, $\alpha_2$, $\alpha_3$, and impedances $Z_{\alpha_1}(s) = 1/s^{\alpha_1}C_{\alpha_1}$, $Z_{\alpha_2}(s) = 1/s^{\alpha_2}C_{\alpha_2}$, and $Z_{\alpha_3}(s) = 1/s^{\alpha_3}C_{\alpha_3}$, the equivalent impedance, magnitude, and phase responses can be calculated from the equations in Table I. It is clear that the order of each element has a significant effect on the impedance of each FoC and their equivalent capacitance cannot be considered. Therefore, to verify the theoretical analysis, the behavior of the three identical FoCs has been verified experimentally. The phase, magnitude, and pseudo-capacitance responses of the equivalent impedances $Z_{Ceq,s}(s)$ are shown in Fig. 4. For reference, both pseudo-capacitance and phase for an individual FoC have been plotted inside same figures. Compared to a single device, we note that the phase response of three identical order FoCs connected in series remains same. However, its magnitude of the equivalent impedance is tripled while pseudo-capacitance is one-third as shown in the inset of Fig. 4(a), and Fig. 4(b).

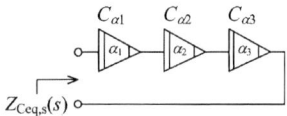

Fig. 3. Series-conneted three FoCs.

TABLE II. COMPARISON OF IDENTICAL-ORDER SERIES-CONNECTED FOCS: MEASURED AND CALCULATED RESULTS.

| No. | #1 | #2 |
|---|---|---|
| Connection of Orders | 0.69 + 0.69 + 0.69 | 0.92 + 0.92 + 0.92 |
| Equivalent Impedance @ $f_c$ [kΩ] | 6.58 (6.65) | 10.54 (22.09) |
| Phase [°] | −60.26 (−62.13) | −83.99 (−81.64) |
| Relative Phase Error [%] | −3.00 | 2.88 |
| Equivalent Order $\alpha$ [−] | 0.67 (0.69) | 0.93 (0.91) |
| Pseudo-Capacitance [Farad·sec$^{\alpha-1}$] | 2.68 (1.89) n | 22.48 (16.46) p |

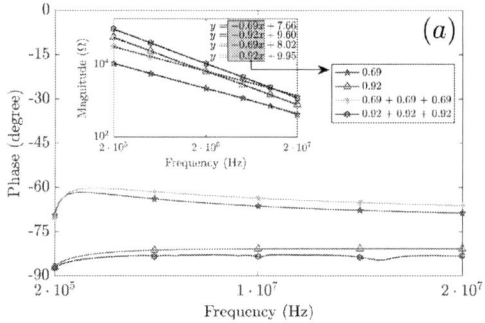

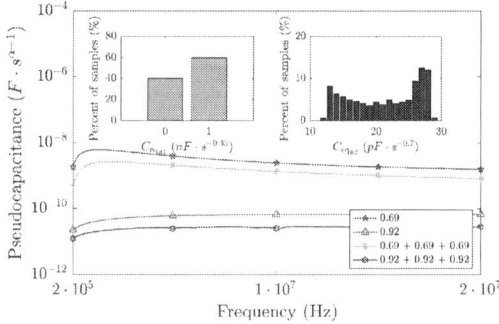

Fig. 4. Experimental verification of three series-connected FoCs: (a) phase, magnitude, (b) pseudo-capacitance responses.

TABLE III. Impedance, Magnitude and Phase Descriptions Of Three Parallel-Connected FoCs.

| $Z_{Ceq,p}(s)$ | $\dfrac{1}{s^{\alpha_1}C_{\alpha_1}+s^{\alpha_2}C_{\alpha_2}+s^{\alpha_3}C_{\alpha_3}}$ |
|---|---|
| $\lvert Z_{Ceq,p}(s)\rvert$ | $\dfrac{1}{\sqrt{\omega^{2\alpha_1}C_{\alpha_1}{}^2+\omega^{2\alpha_2}C_{\alpha_2}{}^2+\omega^{2\alpha_3}C_{\alpha_3}{}^2+2\left[\omega^{\alpha_1+\alpha_2}C_{\alpha_1}C_{\alpha_2}\cos\left(\alpha_1-\alpha_2\right)\dfrac{\pi}{2}+\omega^{\alpha_1+\alpha_3}C_{\alpha_1}C_{\alpha_3}\cos\left(\alpha_1-\alpha_3\right)\dfrac{\pi}{2}+\omega^{\alpha_2+\alpha_3}C_{\alpha_2}C_{\alpha_3}\cos\left(\alpha_2-\alpha_3\right)\dfrac{\pi}{2}\right]}}$ |
| $\angle Z_{Ceq,p}(s)$ | $-\tan^{-1}\left[\dfrac{\omega^{\alpha_1}C_{\alpha_1}\sin\left(\alpha_1\dfrac{\pi}{2}\right)+\omega^{\alpha_2}C_{\alpha_2}\sin\left(\alpha_2\dfrac{\pi}{2}\right)+\omega^{\alpha_3}C_{\alpha_3}\sin\left(\alpha_3\dfrac{\pi}{2}\right)}{\omega^{\alpha_1}C_{\alpha_1}\cos\left(\alpha_1\dfrac{\pi}{2}\right)+\omega^{\alpha_2}C_{\alpha_2}\cos\left(\alpha_2\dfrac{\pi}{2}\right)+\omega^{\alpha_3}C_{\alpha_3}\cos\left(\alpha_3\dfrac{\pi}{2}\right)}\right]$ |

The comparison of measured values @ $f_c$ = 2 MHz and expected results, i.e, calculated via MATLAB are evaluated in Table II. For cases #1 → #2 given in Table II, equivalent magnitudes vary in ranges (31.79 → 1.36) kΩ and (95.20 → 1.19) kΩ, respectively. The table also includes calculated relative phase error and corresponding pseudo-capacitance of each connection. The magnitude and pseudo-capacitance responses are plotted in the logarithmic scale meanwhile the phase is in linear scale. Moreover, to estimate the equivalent order $\alpha$ (or phase), the measured magnitude data are fitted to the function $\log\lvert Z\rvert = \alpha\log f + \log(2\pi)^{\alpha}C_{\alpha}$ using the linear least squares (LLS) method. The obtained equivalent equations from fitting the magnitude is equal to measurement samples that are provided inside Fig. 4(a). As a result, the orders of single devices TP, P2, i.e. 0.69, 0.92 with corresponding phases −61.86, −82.59, [degrees] are evidently respond to their equivalent orders from series connections that are found to be 0.67, 0.93 (corresponding to Table II cases #1 → #2 with phases −60.26, −83.99).

### B. Connection of FoCs in Parallel

In case of parallel connection of three identical-order FoCs as in Fig. 5, the equivalent impedance, magnitude and phase responses can be expressed as in Table III. In order to demonstrate the behavior of an equivalent FoC with impedance $Z_{Ceq,p}$, the phase, magnitude and pseud-capacitance responses was experimentally verified. The obtained measurement results are shown in Fig. 6 while the comparison of measured values @ $f_c$ = 2 MHz and calculated results via MATLAB are listed in Tables IV, respectively. Inspecting the obtained results, it is evident that the phase remains identical to initial single FoCs phase and the only change is in the magnitude response, which reflects the pseudo-capacitance $C_{\alpha}$. Obviously, the magnitude is the one-third of individual FoC. The equivalent orders, which are obtained using the LLS fitting and given in Fig. 6(a) as an inset, are found to be 0.69 and 0.92 as equal to related FoC order. The calculated relative phase errors for these cases are −1.63% and −0.08%. It is worth to note that the accuracy of above theoretical analyzes are verified and showed a flexibility and degree of freedom to work with any order of FoCs with a random connection.

TABLE IV. Comparison Of Identical Order Parallel Connected FoCs: Measurement and Calculated Results.

| No. | #3 | #4 |
|---|---|---|
| Connection of Orders | 0.69 ‖ 0.69 ‖ 0.69 | 0.92 ‖ 0.92 ‖ 0.92 |
| Equivalent Impedance @ $f_c$ [kΩ] | 0.700 (0.737) | 2.30 (2.34) |
| Phase [°] | −61.12 (−62.13) | −82.14 (−82.20) |
| Relative Phase Error [%] | −1.63 | −0.08 |
| Equivalent Order $\alpha$ [−] | 0.68 (0.69) | 0.91 (0.91) |
| Pseudo-Capacitance [Farad·sec$^{\alpha-1}$] | 21.55 (17.04) n | 144.45 (140.23) p |

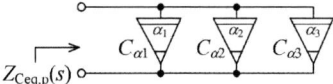

Fig. 5. Parallel-conneted three FoCs.

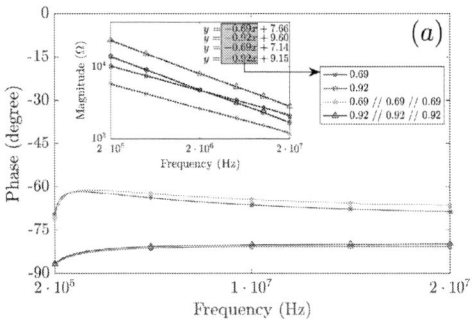

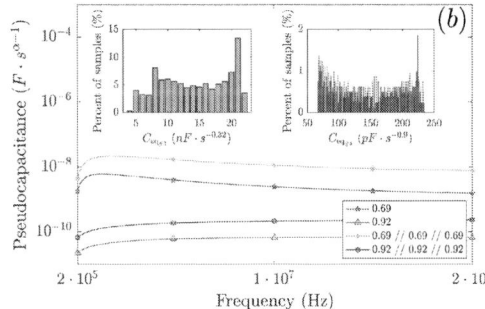

Fig. 6. Experimental verification of three parallel-connected FoCs: (a) phase, magnitude, (b) pseudo-capacitance responses.

TABLE V. COMPARISON OF INTER-CONNECTED FoCs: MEASURED AND CALCULATED RESULTS.

| No. | #5 |
|---|---|
| Connection of Orders | $(0.69 \parallel 0.69) + (0.92 \parallel 0.92) + (0.62 \parallel 0.62)$ |
| Equivalent Impedance @ $f_c$ [kΩ] | 7.23 (4.96) |
| Phase [°] | −76.09 (−73.94) |
| Relative Phase Error [%] | 2.90 |
| Equivalent Order $\alpha$ [−] | 0.85 (0.82) |
| Pseudo-Capacitance [Farad·sec$^{\alpha-1}$] | 137.72 (196.33) p |

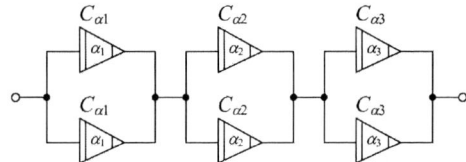

Fig. 7. Inter-connection (series-parallel) of FoCs.

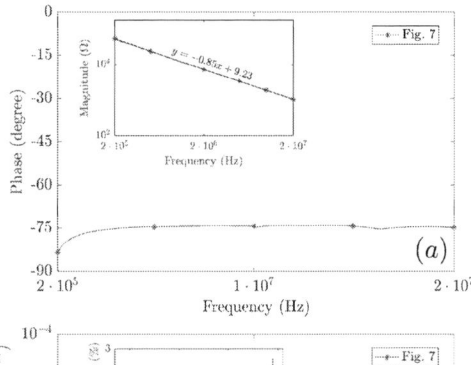

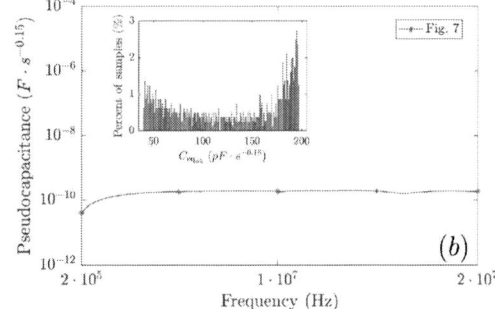

Fig. 8. Experimental verification of inter-connected FoCs: (a) phase, magnitude, (b) pseudo-capacitance responses.

## C. Inter-connection of FoCs

Considering the network shown in Fig. 7, its equivalent impedance is given as:

$$Z_{eq\#5}(s) = \frac{2s^{\alpha_1+\alpha_2}C_{\alpha_1}C_{\alpha_2} + 2s^{\alpha_1+\alpha_3}C_{\alpha_1}C_{\alpha_3} + 2s^{\alpha_2+\alpha_3}C_{\alpha_2}C_{\alpha_3}}{s^{\alpha_1+\alpha_2+\alpha_3}C_{\alpha_1}C_{\alpha_2}C_{\alpha_3}}. \quad (1)$$

The measured phase, magnitude, and pseudo-capacitance responses of the inter-connected FoCs are shown in Fig. 8 and the detailed comparison of results @ 2 MHz are given in Table V. The calculated relative phase error and equivalent order of interconnection are +2.90% and 0.85, respectively. It is clear that small deviation in phase effects on pseudo-capacitance lot due to the power of order. By simplifying Eq. (1) with assuming $\alpha_1 = \alpha_2 = \alpha_3$, then the equivalent impedance would be $2C_\alpha/3$.

## IV. CONCLUSION

The stability measurement of mixed-matrix identical-order FoC connections and the effect of frequency on the capacitive performance of the devices were studied. This is due to the change of the order (dispersion parameter) $\alpha$ from the high frequency (more resistive) to a low-frequency region (more capacitive)which results in the proper use of FoCs, either individually or more importantly in FoCs connections. It was also shown that $C_\alpha$ match well with classical circuit theory formulae (equivalent capacitance of series- or parallel-connected FoCs, identical order phase, magnitude, and pseudo-capacitance) expected by application designers. As future work, the experimental results of series- and parallel-connected arbitrary- and identical-order FoCs will be correlated with dielectric properties in different microstructures. Therefore, the FoC characteristics with connection of different dielectric materials will be the key point. Moreover, these fabricated FoCs will be modelled with different structures and optimized using different algorithms as the next step. This study is the very first step to achieve understanding of natural behavior of FoCs and move forward to computational part.

## REFERENCES

[1] S. Westerlund and L. Ekstam, "Capacitor theory," IEEE Transactions on Dielectrics and Electrical Insulator, vol. 1, pp. 826–839, 1994.

[2] A. Adhikary, S. Sen, and K. Biswas, "Practical realization of tunable fractional order parallel resonator and fractional order filters," IEEE Trans. on Circuits and Systems I; vol. 63, pp. 1142–1151, 2016.

[3] A. Kartci, N. Herencsar, J. Koton, L. Brancik, K. Vrba, G. Tsirimokou, C. Psychalinos, "Fractional-Order Oscillator Design Using Unity-Gain Voltage Buffers and OTAs," In Proc. 60th IEEE Int. Midwest Symp. Circuits Syst. (MWSCAS), Boston, MA, USA, pp. 555–558, 2017.

[4] G. Tsirimokou, C. Psychalinos, A. S. Elwakil, and K. N. Salama, "Electronically tunable fully integrated fractional-order resonator," IEEE Trans. on Circuits and Systems II, vol. 65, pp.166–170, 2018.

[5] L. Kadlčik and P. Horský, "A CMOS Follower-Type Voltage Regulator With a Distributed-Element Fractional-Order Control," IEEE Trans. on Circuits and Systems I, in print, pp. 1–11, 2018.

[6] C. M. Ionescu, "Emerging tools in engineering: fractional order ladder impedance models for respiratory and neural systems," IEEE Journal on Emerging and Selected Topics in Circuits and Systems,vol. 3, pp. 425–31, 2013.

[7] A. Allagui, A. S. Elwakil, B. J. Maundy, and T. J. Freeborn, "Spectral capacitance of series and parallel combinations of supercapacitors," ChemElectroChem, vol. 3, no. 9, pp. 1429–1436, 2016.

[8] A. Agambayev, S. Patole, M. Farhat, A. Elwakil, H. Bagci, and K. N. Salama, "Ferroelectric fractional-order capacitors," ChemElectroChem, vol. 4, pp. 2807–2813, 2017.

[9] A. Agambayev, K. H. Rajab, A. H. Hassan, M. Farhat, H. Bagci, and K. N. Salama, "Towards fractional-order capacitors with broad tunable constant phase angles: Multi-walled carbon nanotube-polymer composite as a case study," J. of Physics D: App. Physics, vol. 51, pp. 1–6, 2018.

[10] A. Kartci, A. Agambayev, N. Herencsar, and K.N. Salama, "Series-, parallel-, and inter-connection of solid-state arbitrary fractional-order capacitors: theoretical study and experimental verification," IEEE Access, vol. 6, pp. 10933–10943, 2018.

*PRIME 2018, Prague, Czech Republic*

# Author Index

Agambayev, Agamyrat, 45, 277
Ahamed, Raju, 145
Al-Attar, Talal, 37
Alam, Naushad, 57
Allani, Sonja, 205
Almansouri, Abdullah, 37
Alpert, Thomas, 229
Alturki, Abdullah, 37
Arif, Syed Waqas, 165
Aymerich, Joan, 25

Bagci, Hakan, 45
Baldanzi, Luca, 77
Baltus, Peter G. M., 129
Barile, Gianluca, 9
Barquinha, Pedro, 33
Bau, Plinio, 105
Beer, Maik, 237
Belay, Yilkal Andualem, 73
Bellizia, Davide, 157
Berkol, Gönenç, 129
Berroth, Manfred, 13, 149
Bieg, Robert, 149
Bigalke, Steve, 89
Boera, Filippo, 109
Boeser, Tobias, 229
Bonizzoni, Edoardo, 109
Boulmirat, Abdessamad, 189
Bruschi, Paolo, 1
Buccolini, Luca, 117
Buckel, Tobias, 181
Bulusu, Anand, 57, 101
Burkard, Roman, 253

Cabrini, Alessandro, 73
Cantatore, Eugenio, 129
Caratelli, Alessandro, 49
Carnevale, Berardino, 77
Casha, Owen, 121, 133, 209
Casper, Thorben, 89
Castello, Rinaldo, 185
Catania, Alessandro, 1

Cavarra, Andrea, 153
Celik, Umut, 141
Cenkeramaddi, Linga Reddy, 97
Ceresa, Davide, 49
Chible, Hussein, 213
Ciccognani, Walter, 197, 201
Cilardo, Alessandro, 273
Colangeli, Sergio, 193, 197, 201
Colin, Davy, 105
Conti, Massimo, 117
Correia, Ana, 33
Coskun, Adem, 165
Costanzo, Ferdinando, 137, 197, 201
Cougo, Bernardo, 105
Cousineau, Marc, 105
Crisafi, Francesco, 221
Crocetti, Luca, 77
Cunha, André B., 257

Dündar, Günhan, 245
Da Silva, Mathieu, 85
Davalle, Daniele, 173
De Caro, Davide, 169
De Cataldo, Giacinto, 133
De Padova, Aurora, 201
Dehollain, Catherine, 217, 245
Dei, Michele, 25
Del Cesta, Simone, 1
Dello Sterpaio, Luca, 177
Di Meo, Gennaro, 169
Di Natale, Giorgio, 85
Dinelli, Gianmarco, 173
Dosen, Strahinja, 213
Dupuis, Sophie, 85

El Adel, El Mostafa, 53
Errico, Vito, 113
Esposito, Darjn, 169

Fanucci, Luca, 41, 77, 173, 177
Fares, Hoda, 213
Fariborzi, Hossein, 37
Ferrari, Giorgio, 221, 257

*Author Index*                                                 *PRIME 2018, Prague, Czech Republic*

Ferri, Giuseppe, 9, 113
Fiori, Franco, 233
Flandre, Denis, 61
Fleischhacker, Christian, 5
Flottes, Marie-Lise, 85
Fusella, Edoardo, 273

Gagliardi, Mirko, 273
Galea, Francarl, 121
Garbuglia, Federico, 117
Gatt, Edward, 121, 133, 209
Gauci, Jordan Lee, 133
Giofrè, Rocco, 137
Giovinazzo, Cecilia, 69
Goes, João, 33
Gonzalez Jimenez, José Luis, 189
Grözing, Markus, 13, 149
Grabmaier, Anton, 253
Granja, Rodrigo, 29
Graton, Guillaume, 53
Grech, Ivan, 121, 133, 209
Guilherme, Jorge, 29

Haase, Jan Frederik, 237
Hager, Ehrentraud, 181
Halonen, Kari, 145
Harpe, Pieter J. A., 129
Hassan, Ali. H, 45
Heinen, Stefan, 125
Herencsar, Norbert, 45, 277
Hoang, Trong-Thuc, 161
Hofer, Guenter, 225
Horta, Nuno, 29

Idsøe, Henning, 97
Indra Kumar, Chaudhry, 101
Inoue, Katsumi, 161

Jany, Clément, 189
John, Bibin, 241
Joly, Sylvain, 217
Jung, Seong-Ook, 65
Jupe, Andreas, 205

Kale, Izzet, 165
Kampus, Vahur, 5
Kargaran, Ehsan, 185
Kartci, Aslihan, 45, 277
Kelz, Sebastian, 13
Kilic, Mustafa, 21
Kim, Suk-Min, 65
Kiss, Jozef, 5
Kloukinas, Kostas, 49

Knaller, Daniel, 5
Kokozinski, Rainer, 81, 253
Korabi, Taki Eddine, 53
Korak, Markus, 5
Krautschneider, Wolfgang, 241
Kulkarni, Pranav, 97
Kumar, Vikas, 221

Lanniel, Alice, 229
Leblebici, Yusuf, 21, 49, 69
Lee, Sang-Myung, 269
Leoni, Alfiero, 9, 113
Lepple-Wienhues, Albrecht, 217
Lienig, Jens, 89
Limiti, Ernesto, 137, 193, 197, 201
Lopez Villegas, Jose Maria, 265

Müller, Kai-Uwe, 81
Maloberti, Franco, 109
Manstretta, Danilo, 185
Marino, Antonino, 177
Marques, João, 33
Martinsen, Ørjan G., 257
Mayer, Thomas, 181
Medina, Beatriz, 265
Meinerzhagen, Bernd, 17
Mendoza Ponce, Pablo, 241
Meoni, Gabriele, 41
Mesri, Alireza, 257
Meyer, Alexander, 125
Micallef, Joseph, 121, 133
Mokwa, Wilfried, 253

Najmussadat, Md, 145
Nannipieri, Pietro, 173
Napoli, Ettore, 169
Nawaz, Kashif, 61
Nguyen, Hong-Thu, 161
Nguyen, Xuan-Thuan, 161
Nocera, Claudio, 153

Oh, Tae-Woo, 65
Ortmanns, Maurits, 229
Ouladsine, Mustapha, 53

Pake, Arash, 225
Pallotti, Antonio, 113
Palmisano, Giuseppe, 153
Palumbo, Gaetano, 157
Papotto, Giuseppe, 153
Parveg, Dristy, 145
Pekçokgüler, Naci, 245
Petra, Nicola, 169

Pham, Cong-Kha, 161
Pilato, Luca, 41
Pinaton, Jacques, 53
Piotto, Massimo, 1
Polli, Dario, 221
Polli, Giorgio, 137, 193, 197, 201
Portelli, Barnaby, 209
Preyler, Peter, 181
Pribyl, Wolfgang A., 225

Qunaj, Valdrin, 141

Raffelberg, Pascal, 253
Ragni, Andrea, 221
Ragonese, Egidio, 153
Ramos, Francisco, 265
Rang, Toomas, 5
Ranjandish, Reza, 249, 261
Raviola, Erica, 233
Reynaert, Patrick, 141
Ria, Andrea, 1
Ricci, Mariachiara, 113
Richardeau, Frederic, 105
Rouger, Nicolas, 105
Rouzeyre, Bruno, 85
Ruskowski, Jennifer, 237

Saalfeld, Tobias, 125
Safari, Leila, 9
Saggio, Giovanni, 113
Salama, Khaled Nabil, 37, 45, 277
Salas Barenys, Arnau, 265
Salimath, Arunkumar, 109
Salvucci, Alessandro, 137, 193, 197, 201
Sampietro, Marco, 221, 257
Sandrini, Jury, 69
Santos, Mauro, 29
Sarti, Luca, 77
Scarfi, Simone, 49
Scerri, Jeremy, 209
Schöps, Sebastian, 89
Schmid, Alexandre, 249, 261
Schmidt, Martin, 149
Schrey, Patrick, 93
Schroeder, Dietmar, 241

Schulte Bocholt, Eva, 125
Sciortino, Giuseppe, 221
Scotti, Giuseppe, 157
Seminara, Lucia, 213
Serra-Graells, Francisco, 25
Shahrabi, Elmira, 69
Sharma, Arvind, 57
Shoaei, Omid, 249
Sieiro, Javier, 265
Siligaris, Alexandre, 189
Song, Byungkyu, 65
Soumya, J., 97
Springer, Andreas, 181
Standaert, François-Xavier, 61
Stanitzki, Alexander, 81
Stornelli, Vincenzo, 9, 113
Strollo, Antonio, 169

Türk, Semih, 205
Terés, Lluís, 25
Tertinek, Stefan, 181
Torelli, Guido, 73
Trifiletti, Alessandro, 157

Ulrich, Robin, 81
Unterhorst, Matteo, 117
Uran, Arda, 21

Valea, Emanuele, 85
Valle, Maurizio, 213
Van Brandt, Léopold, 61
Varonen, Mikko, 145
Veda Bhanu, P., 97
Veigel, Thomas, 13
Verheyen, Erik, 205
Vidal, Neus, 265
Viga, Reinhard, 205, 253
Vittori, Marco, 193, 197
Vogt, Holger, 205, 237

Walter, Peter, 253
Weigel, Robert, 181
Wenger, Yannick, 17
Wunderlich, Ralf, 125

Yun, Ilgu, 269

**IEEE**
445 Hoes Lane
Piscataway, NJ 08854-4141

ISBN 978-1-5386-5388-3